Bears of the World

Ecology, Conservation and Management

Bears have fascinated people since ancient times. The relationship between bears and humans dates back thousands of years, during which time we have also competed with bears for shelter and food. In modern times, bears have come under pressure through encroachment on their habitats, climate change, and illegal trade in their body parts, including the Asian bear bile market. The IUCN lists six bears as Vulnerable or Endangered, and even the Least Concern species, such as the brown bear, are at risk of extirpation in certain countries. The poaching and international trade of these most threatened populations are prohibited, but still ongoing. Covering all bear species worldwide, this beautifully illustrated volume brings together the contributions of 200 international bear experts on the ecology, conservation status, and management of the Ursidae family. It reveals the fascinating long history of interactions between humans and bears and the threats affecting these charismatic species.

Vincenzo Penteriani is a researcher at the Spanish National Research Council (CSIC), Senior Editor of *Animal Conservation*, Associate Editor of *Ursus*, and is leading the Cantabrian Brown Bear Research Group at the Research Unit of Biodiversity (UMIB), Spain. His current research focuses on the ecology and behavior of brown bears in human-modified landscapes, as well as on large carnivore attacks on humans.

Mario Melletti is an independent researcher and member of the African Buffalo Initiative Group, and the Wild Pig Specialist Group, which are part of the International Union for the Conservation of Nature (IUCN SSC). He has contributed to several projects and surveys on both mammals and birds in Africa and Europe. He has edited two books with Cambridge University Press: *Ecology, Evolution and Behaviour of Wild Cattle* (2014) and *Ecology, Conservation and Management of Wild Pigs and Peccaries* (2017).

Bears of the World

Ecology, Conservation and Management

Edited by

Vincenzo Penteriani

Spanish National Research Council (CSIC)

Mario Melletti

WPSG (Wild Pig Specialist Group), IUCN SSC

CAMBRIDGE
UNIVERSITY PRESS

CAMBRIDGE
UNIVERSITY PRESS

Shaftesbury Road, Cambridge CB2 8EA, United Kingdom

One Liberty Plaza, 20th Floor, New York, NY 10006, USA

477 Williamstown Road, Port Melbourne, VIC 3207, Australia

314–321, 3rd Floor, Plot 3, Splendor Forum, Jasola District Centre, New Delhi – 110025, India

103 Penang Road, #05–06/07, Visioncrest Commercial, Singapore 238467

Cambridge University Press is part of Cambridge University Press & Assessment, a department of the University of Cambridge.

We share the University's mission to contribute to society through the pursuit of education, learning and research at the highest international levels of excellence.

www.cambridge.org
Information on this title: www.cambridge.org/9781108483520
DOI: 10.1017/9781108692571

First published 2021 (version 3, April 2024)

Printed in Great Britain by CPI Group (UK) Ltd, Croydon CR0 4YY, April 2024

A catalogue record for this publication is available from the British Library

ISBN 978-1-108-48352-0 Hardback

To those Apennine brown bears who, in the eighties, showed a young boy how to live following the rhythm of the seasons; to my father Stefano, my daughter Giulia and María del Mar, who have always understood my long absences from bear den exit to the beginning of hibernation.
Vincenzo Penteriani

To my mother Carla, for the amazing woman she is, whose support, advice, and encouragement throughout my life helped me to pursue my dreams!

To my family, Marco, Lucia, Luca, Chiara, Jacopo, Diletta, Alessandro, Kuma, India, Rhum, Kelly, and Kira for your help and patience during my absences.
Mario Melletti

Contents

Contributors

Fakhar-i-Abbas
Bioresource Research Centre

Isaac Goldstein Aizman
Wildlife Conservation Society (WCS), New York, USA

Naim Akhtar
Nature Activity Centre for Environment Education, Ahmedabad, India

Hüseyin Ambarlı
Department of Wildlife Ecology and Management, Faculty of Forestry, Düzce University, Düzce, Turkey

Jon Martin Arnemo
Department of Forestry and Wildlife Management, Inland Norway University of Applied Sciences, Elverum, Norway and Department of Wildlife, Fish, and Environmental Studies, Swedish University of Agricultural Sciences, Umeå, Sweden

Todd C. Atwood
US Geological Survey, Alaska Science Center, Anchorage, AK, USA

Harendra Singh Bargali
The Corbett Foundation, Uttarakhand, India

Leonid Baskin
Institute of Ecology and Evolution, Russian Academy of Sciences, Moscow, Russia

Carlos Bautista
Institute of Nature Conservation, Polish Academy of Sciences, Krakow, Poland

Jon P. Beckmann
Wildlife Conservation Society, North America Program and University of Nevada, Reno, USA

Stanislav E. Belikov
All-Russian Research Institute for Environment Protection, Sarov, Russia

Eva Bellemain
SPYGEN, Savoie Technolac, Le Bourget du Lac, France

Giorgio Bertorelle
Department of Life Sciences and Biotechnology, University of Ferrara, Ferrara, Italy

Luigi Boitani
Department of Biology and Biotechnologies "Charles Darwin", Rome, Italy

Giulia Bombieri
Research Unit of Biodiversity, Oviedo University – Campus Mieres, Mieres, Spain

Museo delle Scienze, Sezione Zoologia dei Vertebrati, Trento, Italy

Mark S. Boyce
Department of Biological Sciences, University of Alberta, Edmonton, Canada

Stewart W. Breck
USDA-National Wildlife Research Center, Department of Fish, Wildlife, & Conservation Biology, Colorado State University, CO, USA

Jan Schmidt-Burbach
The World Society for the Protection of Animals, London, UK

Özgün Emre Can
Wildlife Conservation Research Unit (WildCRU), Department of Zoology, University of Oxford, UK

Armando Castellanos
Andean Bear Foundation, La Isla, Quito, Ecuador

TSG (Tapir Specialist Group) IUCN SSC, Gland, Switzerland

Wenbo Chen
Graduate School of Media and Governance, Keio University, Japan

Nature & Science Consulting Co., Kanagawa, Japan

Anwaruddin Choudhury
The Rhino Foundation for Nature in NE India, Guwahati, India

Paolo Ciucci
Department of Biology and Biotechnologies "Charles Darwin", Università di Roma "La Sapienza", Rome, Italy

Melanie Clapham
Applied Conservation Science Laboratory, Department of Geography, University of Victoria, Victoria, Canada

Joseph D. Clark
Southern Appalachian Field Branch, Northern Rocky Mountain Science Center, US Geological Survey, Knoxville, TN, USA

Cecily C. Costello
Research Wildlife Biologist, Montana Fish, Wildlife & Parks, Helena, USA

Pamela L. Coy
Minnesota Department of Natural Resources, Minnesota, USA

Brian Crudge
Free The Bears, Luang Prabang, Lao PDR

Department of Natural Resources and Environmental Health, University of South-Eastern Norway, Telemark, Norway

Neil D'Cruze
PanEco, Yayasan Ekosistem Lestari, Sumatran Orangutan Conservation Programme, Pro Natura Foundation, Medan, Indonesia

Pjotr I. Danilov
Laboratory of Zoology, Institute of Biology, Karelian Research Centre of Russian Academy of Sciences, Petrozavodsk, Russia

María del Mar Delgado
Research Unit of Biodiversity, Oviedo University – Campus Mieres, Mieres, Spain

Miguel Delibes
Department of Conservation Biology, Estación Biológica de Doñana, CSIC (Spanish Council of Scientific Research), Seville, Spain

Andrew E. Derocher
Department of Biological Sciences, University of Alberta, Edmonton, Canada

Nishith Dharaiya
Department of Life Sciences, Hemchandracharya North Gujarat University, Patan, India

Rune Dietz
Faculty of Science and Technology, Department of Bioscience, Arctic Research Center, Aarhus University, Aarhus, Denmark

Xiang Ding
Key Laboratory of Southwest China Wildlife Resources Conservation, College of Life Sciences, China West Normal University, Nanchong, China

Mark A. Ditmer
Department of Fisheries, Wildlife and Conservation Biology, University of Minnesota, MN, USA

Tomoko Doko
Graduate School of Media and Governance, Keio University, Fujisawa, Japan

Nature & Science Consulting Co., Ltd., Yokohama, Japan

David C. Douglas
US Geological Survey, Alaska Science Center, Juneau, AK, USA

George M. Durner
US Geological Survey, Alaska Science Center, Anchorage, AK, USA

Trishna Dutta
Wildlife Sciences, Faculty of Forest Sciences and Forest Ecology, University of Goettingen, Goettingen, Germany

Zachary Dvornicky-Raymond
Global Health Program, Smithsonian Conservation Biology Institute, Front Royal, VA, USA

Marcus Elfström
EnviroPlanning AB, Gothenburg, Sweden

Norwegian University of Life Sciences, Faculty of Environmental Sciences and Natural Resource Management, Ås, Norway

Santiago Espinosa
Facultad de Ciencias UASLP, San Luis Potosí, México

Alina L. Evans
Department of Forestry and Wildlife Management, Inland Norway University of Applied Sciences, Elverum, Norway

José María Fedriani
Centro de Investigaciones sobre Desertificación (CIDE-CSIC), Moncada (Valencia), Spain

Centre for Applied Ecology "Prof. Baeta Neves"/InBIO, Institute of Agronomy, University of Lisbon, Lisbon, Portugal

Néstor Fernández
German Centre for Integrative Biodiversity Research (iDiv), Halle-Jena-Leipzig, Germany

Alberto Fernández-Gil
Department of Conservation Biology, Estación Biológica de Doñana, CSIC (Spanish Council of Scientific Research), Seville, Spain

Eva Filipczykova
School of Health, Medical and Applied Sciences, Gladstone, Australia

Czech University of Life Sciences Prague, Faculty of Environmental Sciences, Prague, Czech Republic

Cheryl Frederick
School of Biodiversity Conservation, Unity College, Unity, ME, USA

Gabriella M. Fredriksson
Institute for Biodiversity and Ecosystem Dynamics/Zoological Museum, University of Amsterdam, Amsterdam, The Netherlands

Hiromichi Fukui
International Digital Earth Applied Science Research Center, Chubu Institute for Advanced Studies, Chubu University, Kasugai, Japan

Pedro José Garrote
Centre for Applied Ecology "Prof. Baeta Neves"/InBIO, Institute of Agronomy, University of Lisbon, Lisbon, Portugal

David L. Garshelis
Minnesota Department of Natural Resources, MN, USA

Michael L. Gibeau
Mountain National Parks, Parks Canada, Lake Louise, Canada

Enrique González-Bernardo
Research Unit of Biodiversity, Oviedo University – Campus Mieres, Mieres, Spain

Tabitha Graves
US Geological Survey, Northern Rocky Mountain Science Center, West Glacier, MT, USA

Claudio Groff
Servizio Foreste e Faune, Settore Grandi Carnivori Via G. B. Trener, Trento, Italy

Kerry A. Gunther
Bear Management Office, Yellowstone National Park, WY, USA

Mark A. Haroldson
US Geological Survey, Northern Rocky Mountain Science Center, Interagency Grizzly Bear Study Team, MT, USA

Zhang Hemin
China Conservation and Research Center for the Giant Panda, Chengdu Dujiangyan, China

Miguel de Gabriel Hernando
Department of Biodiversity and Environmental Management, Faculty of Biological and Environmental Sciences, Universidad de León, León, Spain

Anne G. Hertel
Senckenberg Biodiversity and Climate Research Centre, Frankfurt (Main), Germany

Yiling Hou
Key Laboratory of Southwest China Wildlife Resources Conservation (Ministry of Education), College of Life Sciences, China West Normal University, Nanchong, China

Qiongyu Huang
Smithsonian Conservation Biology Institute, Front Royal, VA, USA

Djuro Huber
Institute of Nature Conservation of Polish Academy of Sciences, Adama Mickiewicza, Krakow, Poland

Vanessa Hull
Department of Wildlife Ecology and Conservation, University of Florida, Gainesville, FL, USA

Paul A. Iaizzo
Department of Surgery, University of Minnesota, Saint Paul, MN, USA

David Jackson
Andean Bear Foundation, La Isla, Quito, Ecuador

Kajal K. Jadav
School of Wildlife Forensic and Health, Nanaji Deshmukh Veterinary Science, Jabalpur, India

Yonten Jamtsho
Wangchuck Centennial National Park, Department of Forest and Park Services, Ministry of Agriculture and Forest, Bhutan

Ashish Kumar Jangid
Tiger Cell, Wildlife Institute of India, Dehradun, India

Axel Janke
Department of Evolutionary Biology, Evolutionary Biology Centre, Uppsala University, Uppsala, Sweden

Bjørn M. Jenssen
Department of Biology, Norwegian University of Science and Technology, Trondheim, Norway

Klemen Jerina
Department of Forestry and Renewable Forest Resources, Biotechnical Faculty, University of Ljubljana, Ljubljana, Slovenia

Alexandros A. Karamanlidis
ARCTUROS – Civil Society for the Protection and Management of Wildlife and the Natural Environment, Aetos, Greece

Katherine C. Kendall
US Geological Survey, Reston, VA, USA

Jonas Kindberg
Norwegian Institute for Nature Research, Trondheim, Norway and Department of Wildlife, Fish, and Environmental Studies, Swedish University of Agricultural Sciences, Umea, Sweden

Andrew C. Kitchener
Department of Natural Sciences, National Museums Scotland, Edinburgh, UK

Shinsuke Koike
Institute of Agriculture, Tokyo University of Agriculture and Technology, Tokyo, Japan

Ilpo Kojola
Natural Resources Institute Finland (Luke), Rovaniemi, Finland

Lydia Kolter
IUCN SSC Bear Specialist Group, Cologne, Germany

Alexander Kopatz
Norwegian Institute for Nature Research, Trondheim, Norway

Chinatsu Kozakai
Central Region Agricultural Research Center, National Agriculture and Food Research Organization (NARO), Ibaraki, Japan

Miha Krofel
Department of Forestry and Renewable Forest Resources, Biotechnical Faculty, University of Ljubljana, Ljubljana, Slovenia

Josip Kusak
Department of Biology, University of Zagreb, Zagreb, Croatia

Verena E. Kutschera
Science for Life Laboratory, Department of Biochemistry and Biophysics, Stockholm University, Solna, Sweden

Andrés Laguna
Direction of Research Department, Big Mammals Conservation Company, Ecuador

Babu Ram Lamichhane
NTNC – Biodiversity Conservation Center, Sauraha, Nepal

Evolutionary Ecology Group, Faculty of Sciences, University of Antwerp, Antwerp, Belgium

Institute of Cultural Anthropology and Development Sociology, Leiden University, Leiden, The Netherlands

Timothy G. Laske
Department of Surgery, University of Minnesota, Saint Paul, MN, USA

Martin Leclerc
Applied Conservation Science Laboratory, Department of Geography, University of Victoria, Victoria, Canada

Bruce D. Leopold
Wildlife Ecology, Michigan State University, East Lansing, MI, USA

Robert J. Letcher
Environment and Climate Change Canada, Wildlife and Landscape Science Directorate, Ecotoxicology and Wildlife Health Division, National Wildlife Research Centre, Carleton University, Carleton, Ottawa, Canada

Dihua Li
Urban and Regional Ecology and Planning, Landscape Architecture, Landscape Sociology, College of Architecture and Landscape, Peking University, Beijing, China

Renqiang Li
Key Laboratory of Ecosystem Network Observation and Modeling, Institute of Geographic Sciences and Natural Resources Research, Chinese Academy of Sciences, Beijing, China

Anne E. Loosen
Department of Biological Sciences, University of Alberta, Edmonton, Canada

Frank T. van Manen
US Geological Survey, Northern Rocky Mountain Science Center, Interagency Grizzly Bear Study Team, MT, USA

Tsutomu Mano
Nature Conservation Division, Institute of Environmental Sciences, Environmental and Geological Research Organization, Japan

Prakash Mardaraj
IUCN/Sloth Bear & Human–Bear Conflict Expert Team, Amity Institute of Forestry & Wildlife, Noida, India

María C. Mateo-Sánchez
ETSI Montes, Forestal y del Medio Natural, Technical University of Madrid, Ciudad Universitaria, Madrid, Spain

Bruce McLellan
Emeritus Scientist, British Columbia Forest Service, D'Arcy, British Columbia, Canada

William M. McShea
Conservation Ecology Center, Smithsonian Conservation Biology Institute, National Zoological Park, VA, USA

Mario Melletti
Wild Pig Specialist Group (WPSG), IUCN SSC, Gland, Switzerland

Yorgos Mertzanis
"CALLISTO" NGO, Wildlife and Nature Conservation Society Mitropoleos, Thessaloniki, Greece

Randi Meyerson
Detroit Zoological Society, Detroit, MI, USA

Sterling D. Miller
Alaska Department of Fish and Game and National Wildlife Federation, Lolo, MT, USA

Cyril Milleret
Faculty of Environmental Sciences and Natural Resource Management, Norwegian University of Life Sciences, Ås, Norway

Paola Morini
Sirente Velino Natural Regional Park – Viale XXIV Maggio snc, Rocca di Mezzo (AQ), Italy

Karen V. Noyce
Minnesota Department of Natural Resources, MN, USA

Martyn Obbard
Wildlife Research and Monitoring Section, Ontario Ministry of Natural Resources and Forestry, DNA Building, Trent University, Peterborough, Canada

Kirsty Officer
Free the Bears, Phnom Penh, Cambodia

Andrés Ordiz
Faculty of Environmental Sciences and Natural Resource Management, Norwegian University of Life Sciences, Ås, Norway

Megan A. Owen
Recovery Ecology, Institute for Conservation Research, San Diego Zoo Global, Escondido, CA, USA

Dave Paetkau
Wildlife Genetics International, Nelson, British Columbia, Canada

Santiago Palazón
Fauna and Flora Service, Government of Catalonia, Barcelona, Spain

Himanshu Shekhar Palei
Aranya Foundation, Bhubaneswar, India

Jamshid Parchizadeh
Global Wildlife Conservation Center, State University of New York College of Environmental Science and Forestry, Syracuse, NY, USA

Paolo Pedrini
Muse – Museo delle Scienze, Trento, Italy

Michael R. Pelton
Institute of Agriculture, University of Tennessee, Knoxville, TN, USA

Vincenzo Penteriani
Spanish National Research Council (CSIC), Research Unit of Biodiversity (UMIB), Mieres, Spain

James P. Phillips
Oregon State University, Corvallis, OR, USA

Karine E. Pigeon
Geomatics and Landscape Ecology Laboratory (GLEL), Carleton University, Ottawa, Canada

Nicholas W. Pilfold
Applied Animal Ecology, Institute for Conservation Research, San Diego Zoo Global, Escondido, CA, USA

Francesco Pinchera
CISDAM – Via Liberata 1, Rosello (CH), Italy

Michael F. Proctor
Birchdale Ecological, Kaslo, Canada

Pierre-Yves Quenette
Office Français de la Biodiversité, Direction de la Recherche et de l'Appui Scientifique, Villeneuve de Rivière, France

Shaenandhoa Garcia Rangel
United Nations Environment World Conservation Monitoring Program, Cambridge, UK

Shyamala Ratnayeke
Department of Biological Sciences, Sunway University, No. 5 Jalan University, Selangor, Malaysia

Eloy Revilla
Pyrenean Institute of Ecology, Zaragoza, Spain

Karyn D. Rode
US Geological Survey, Alaska Science Center, Anchorage, AK, USA

Manuel Ruiz-García
Laboratorio de Genética de Poblaciones Molecular-Biología Evolutiva, Departamento de Biología, Facultad de Ciencias, Pontificia Universidad Javeriana, Bogotá DC, Colombia

Luca Francesco Russo
Dipartimento Bioscienze e Territorio, Contrada Fonte Lappone, Pesche, Italy

Murray Rutherford
School of Resource and Environmental Management, Simon Frazer University, Burnaby, Canada

Veronica Sahlén
Faculty of Environmental Sciences and Natural Resource Management, Norwegian University of Life Sciences, Ås, Norway

Ryo Sakurai
College of Policy Science, Ritsumeikan University, Osaka, Japan

Valentina Salomashkina
A.N. Severtsov Institute of Ecology and Evolution, Russian Academy of Sciences, Moscow, Russia

National Research Center for Hematology, Ministry of Healthcare of the Russian Federation, Moscow, Russia

Håkan Sand
Swedish University of Agricultural Sciences, Uppsala, Sweden

Jenny Santiestevan
Global Health Program, Smithsonian Conservation Biology Institute, Front Royal, VA, USA

Yoshikazu Sato
Rakuno Gakuen University, Ebetsu, Hokkaido, Japan

Lorraine Scotson
IUCN Saola Working Group, Vinh University, Vinh, Vietnam

Nuria Selva
Institute of Nature Conservation, Polish Academy of Sciences, Krakow, Poland

Agnieszka Sergiel
Institute of Nature Conservation, Polish Academy of Sciences, Krakow, Poland

Christopher Servheen
Dept. of Ecosystem and Conservation Sciences, W.A. Franke College of Forestry and Conservation, University of Montana Missoula, MT, USA

Ivan Seryodkin
Laboratory of Animals Ecology and Conservation, Pacific Geographical Institute FEB RAS, Vladivostok, Russia

Ravi Kumar Sharma
Tiger Cell, Wildlife Institute of India, Dehradun, India

Sandeep Sharma
Department of Conservation Biology, Johann-Friedrich-Blumenbach Institute of Zoology, University of Goettingen, Goettingen, Germany

Thomas Sharp
Conservation and Research/Wildlife SOS – USA, Salt Lake City, UT, USA

Scott Silver
National Audubon Society, Constitution Marsh Audubon Center and Sanctuary, New York, USA

Therese Ramberg Sivertsen
Norwegian Institute for Nature Research, Oslo, Norway

A.K. Skidmore
Department of Natural Resources, ITC, Faculty of Geo-Information Science and Earth Observation of the University of Twente, Enschede, The Netherlands

Michaela Skuban
Carpathian Wildlife Society, Zvolen, Slovakia

Tom Smith
Department of Plant and Wildlife Sciences, Brigham Young University, Provo, UT, USA

Melissa Songer
Smithsonian Conservation Biology Institute, Front Royal, VA, USA

Christian Sonne
Arctic Research Center, Department of Bioscience, Faculty of Science and Technology, Aarhus University, Aarhus, Denmark

Thomas Spady
California State University San Marcos, San Marcos, CA, USA

Robert Steinmetz
WWF-Thailand, Bangkok, Thailand

Gordon B. Stenhouse
Foothills Research Institute, Hinton, Alberta, Canada

Sam M.J.G. Steyaert
Faculty of Biosciences and Aquaculture, Nord University, Steinkjer, Norway

Ole-Gunnar Støen
Norwegian Institute for Nature Research, Oslo, Norway

Faculty of Environmental Sciences and Natural Resource Management, Norwegian University of Life Sciences, Ås, Norway

Ashok Subedi
National Trust for Nature Conservation, Annapurna Conservation Area Project, Pokhara, Nepal

Ronald R. Swaisgood
Recovery Ecology, Institute for Conservation Research, San Diego Zoo Global, Escondido, CA, USA

Jon E. Swenson
Faculty of Environmental Sciences and Natural Resource Management, Norwegian University of Life Sciences, Ås, Norway

Aimee Tallian
Norwegian Institute for Nature Research, Trondheim, Norway

Hidetoshi B. Tamate
Department of Biology, Yamagata University, Yamagata, Japan

Gregory W. Thiemann
Faculty of Environmental Studies, York University, Toronto, Canada

Konstantin F. Tirronen
Laboratory of Zoology, Institute of Biology, Karelian Research Centre of Russian Academy of Sciences, Petrozavodsk, Russia

A.G. Toxopeus
Department of Natural Resources, ITC, Faculty of Geo-Information Science and Earth Observation of the University of Twente, Enschede, The Netherlands

Andrew N. Tri
Minnesota Department of Natural Resources, Grand Rapids, MN, USA

Morten Tryland
UiT The Arctic University of Norway, Arctic and Marine Biology, Arctic Infection Biology, Tromsø, Norway

Toshio Tsubota
Laboratory of Wildlife Biology and Medicine, Department of Environmental Veterinary Sciences, Graduate School of Veterinary Medicine, Hokkaido University, Sapporo, Japan

Hifumi Tsuruga
Hokkaido Research Organization, Japan

Research Institute of Energy, Environment and Geology, Industrial Technology and Environment Research Department, Sapporo, Hokkaido, Japan

Reina Uno
Institute for Advanced Biosciences, Keio University, Tsuruoka, Japan

Marc Valitutto
Smithsonian Conservation Biology Institute's Global Health Program, Front Royal, VA, USA

Ximena Velez-Liendo
Wildlife Conservation Research Unit, Recanati-Kaplan Centre, Oxford, UK

Chester Zoo, Cedar House, Chester, UK

Dag Vongraven
Norwegian Polar Institute, Tromsø, Norway

Petter Wabakken
Department of Forestry and Wildlife Management, Inland Norway University of Applied Sciences, Elverum, Norway

Sangay Wangchuk
Ugyen Wangchuck Institute for Conservation and Environment Research, Bhutan

Fuwen Wei
Institute of Zoology, Chinese Academy of Sciences (CAS), Key Laboratory of Animal Ecology and Conservation Biology, Beijing, China

Lars Werdelin
Department of Palaeobiology, Swedish Museum of Natural History, Stockholm, Sweden

David Wildt
Center for Species Survival, Smithsonian Conservation Biology Institute, Front Royal, VA, USA

Seth M. Wilson
W.A. Franke College of Forestry & Conservation – University of Montana, Missoula, MT, USA

Northern Rockies Conservation Cooperative, Jackson, WY, USA

Wai-Ming Wong
Panthera, New York, USA

Ming Xu
Department of Ecology, Evolution, and Natural Resources, Grant F. Walton Center for Remote Sensing & Spatial Analysis, School of Environmental and Biological Sciences, Rutgers University, New Brunswick, NJ, USA

Masami Yamanaka
Shari Town Museum, Japan

Koji Yamazaki
Forest Ecology Laboratory, Department of Forest Science,
Faculty of Regional Environmental Science,
Tokyo University of Agriculture,
Tokyo, Japan

Alejandra Zarzo-Arias
Research Unit of Biodiversity, Oviedo University – Campus Mieres, Mieres, Spain

Andreas Zedrosser
Department of Natural Resources and Environmental Health
University of South-Eastern Norway, Bø, Norway

Jindong Zhang
Key Laboratory of Southwest China Wildlife Resources Conservation, China West Normal University,
Ministry of Education, Nanchong, China

Zejun Zhang
China West Normal University, Nanchong, China

Illustrators

Giulia Bombieri
Research Unit of Biodiversity, Oviedo University – Campus Mieres, Mieres, Spain
Museo delle Scienze, Sezione Zoologia dei Vertebrati,
Trento, Italy

Enrico Chiarelli
Rome, Italy

Alberto Mattia Perronace
Rome, Italy

Francesco Rinaldi
Sansepolcro, Italy

Luciano Toma
Dipartimento di Malattie Infettive, Reparto di Malattie Trasmesse da Vettori, Istituto Superiore di Sanità, Rome, Italy

Distribution maps

Daniele Baisero
Key Biodiversity Area Secretariat, Wildlife Conservation Society, Cambridge, UK

Foreword

Bears of the World brings together for the first time much of what is known of the evolution, ecology, behavior, and conservation status of the eight existing species of bears in a single volume. Its 29 chapters are written mostly by collaborating groups of scientists from multiple institutions and countries and provide biologists and managers concerned with bears with easy access to balanced, up-to-date accounts of different aspects of the biology, ecology, and conservation of bears and a single go-to source of information. The first five chapters deal with general comparative topics – systematics, mating strategies, reproduction, comparative anatomy, and ecology of bears – while the next nine provide focused reviews of the existing knowledge of the biology, ecology, and conservation status of different bear species: giant pandas, Andean bears, sun bears, sloth bears, Asiatic and North American black bears, brown bears (including grizzlies and Kodiak bears), and polar bears. Red pandas, once regarded as allied with bears, are not included because recent genetic analyses place them outside the Ursidae in a separate family more closely allied with the Procyonids, and show that their resemblance to pandas is the result of convergent evolution. The remaining 15 chapters explore interactions between bears and humans and issues relating to their conservation and management.

Six of the eight contemporary bear species (giant pandas, sun bears, sloth bears, Asiatic black bears, brown bears, and polar bears) are found in Eurasia; three in North America (North American black bears, brown bears, and polar bears); and one (Andean bears) in South America. Recent genetic phylogenies suggest that bears diverged from the ancestors of the Ursidae between 23 and 35 million years ago, while the giant panda and its extinct relatives diverged from other bears between 12 and 20 million years ago. Andean bears diverged from the ancestors of the other species between 7 and 13 million years ago, while the remaining species (sun bears, sloth bears, North American and Asiatic black bears, and polar bears) appear to have diverged from each other within the last 5 million years.

There are striking ecological differences between bear species that have important implications for their behavior, evolution, and conservation. Habitats range from tropical forests (in sun bears) to montane woodlands and grasslands (Andean bears), northern forests, woodlands and grasslands (black and brown bears), and the marine margins and sea ice of the Arctic (polar bears). The average size of annual ranges runs from less than 10 km^2 in some populations of sun bears to hundreds of square kilometers in brown bears and tens of thousands in polar bears. Diets are exclusively vegetarian in giant pandas (99% of whose diet is constituted by bamboo), consist of varying combinations of fruit, roots, foliage, insects, and vertebrates in sun bears, sloth bears, black bears, and brown bears, while polar bears are specialized hunters living principally on three seal species and also scavenging the carcasses of cetaceans and walruses. The smaller species are partly arboreal while the larger ones are principally terrestrial, and polar bears are accomplished swimmers. While black bears, brown bears, and Asiatic black bears (only in northern latitudes) hibernate in dens in winter (although the duration of denning varies with latitude), Andean bears, sun bears, sloth bears, giant pandas, and polar bears are active throughout the year.

As in many other groups of mammals, the available information varies widely between species. Of all the bear species, North American black bears are the most widespread and most easily accessible to scientists, habituating relatively easily to observers, and long-term, individual-based studies of black bears provide detailed records of the lives of large numbers of recognizable individuals that make it possible to explore relationships between variation in ecology, development, and behavior. They demonstrate the benefits of long-term, individual-based studies and help to shed light on ecological processes that probably occur in many of the other species for which less complete information is available. Because of their larger ranges and lower densities, studies of brown bears and polar bears face greater logistical problems, but here, too, long-term studies are starting to provide extensive information on the lives of individuals and on the effects of differences in feeding behavior and hunting tactics on development, survival, and reproduction. In contrast, much less is known of the lives of Andean bears, sun bears, and sloth bears, whose habitat and activity patterns make detailed studies on individuals more difficult – and further research on their ecology and behavior is badly needed.

The lifestyles and behavior of all eight bears show substantial similarities. Adults usually forage alone or (in the case of females) with dependent young and occupy overlapping home ranges, which are typically larger for males than females. During breeding seasons, males range widely in search of receptive females. The prolonged dependence of juveniles on their mothers reduces the availability of receptive females, skewing the ratio of adults ready to breed towards males and generating intense competition between males for mating opportunities and strong selection for increased body size in males. Implantation is delayed, litter size is small (1–3 cubs, exceptionally 4), and the young are guarded and provisioned by mothers until they are ready to disperse. Male infanticide has been observed in several species and can represent a relatively common cause of cub mortality. Offspring adopt separate ranges from their mothers as they develop and, as in many other mammals, sons typically move further away than daughters, although the situation is reversed in giant pandas.

Because of their fabled strength and ferocity and their upright stature when threatened, bears have had a prominent position in human religion, mythology, and iconography for over 2000 years. All species apart from giant pandas have been known to attack humans and they are widely feared, but attacks are uncommon, are very seldom predatory, and mostly occur when individuals are threatened or cornered or when mothers with dependent young are surprised. Experts recommend that the safest response for humans threatened by bears is to lie face down with hands behind your head and legs apart, so that you cannot be flipped over, and wait until the bear leaves. In contrast, cases of predation on domestic animals and exploration of human waste are much more common, especially in black and brown bears, stimulating hunting to control local problems.

Bears are still hunted for sport in many countries, including North America, where black bears are a common target species in many US states. Throughout much of human history, they have also played a central role in popular entertainment. Since before Roman times, captive bears have been used in staged fights with dogs throughout much of Eurasia, and although bear baiting is now illegal in most countries, it still persists in a few. In addition, captive bears have commonly been taught to dance and have been displayed at fairs and games, and bear parts and products feature prominently in Chinese traditional medicine. Bears continue to be hunted and captured for both purposes.

Conservation biologists often (and rightly) complain of the lack of connection between fundamental studies of the ecology and behavior of mammals and the needs of conservation biologists and managers for guidance on the optimal form of conservation or management strategies. They can have no such complaints about this book, for a large part of its second half is devoted to analyzing the direct and indirect effects of human impact on bear populations and assessing different ways of reducing human impact on existing bear populations. Chapters describe how overhunting, poaching, and reductions in their natural habitat have reduced the historical ranges of all bear species and how increasing human populations and changes in land use now threaten the persistence of many populations. In addition, human-induced changes in climate jeopardize the future of both pandas and polar bears. In polar bears, the impact of rising winter temperatures on the extent of sea ice affects their ability to hunt seals and has already led to substantial reductions in their numbers. Predictions suggest that, if current changes in winter temperatures continue, many polar bear populations are unlikely to persist to the end of this century. Climate change also threatens the remaining populations of giant pandas, as the distribution of over a third of the bamboo species that they rely on is likely to be affected by rising temperatures, and a substantial part of the protected areas set aside for pandas may no longer be suitable habitat by the end of the century. As a result, there is now an urgent need to identify and protect suitable habitat to replace them.

At a time when the rate of publication is rising rapidly and scientists are relying more and more on electronic sources, large, taxonomically focused books might initially appear superfluous. However, the increasing volume of available information and the diversity of its origins and reliability make the need for expert syntheses of the available information more pressing. By bringing together authoritative, multi-authored reviews of the information available for different species, integrative books like *Bears of the World* make it possible to identify contrasts and similarities in the biology and ecology of different species and to compare the experience of different strategies for conserving and managing remaining populations, providing a clear view of the woods as well as the trees.

Tim Clutton-Brock
University of Cambridge

Acknowledgments

For their continuous effort, patience and dedication, as well as their beautiful drawings, we are extremely grateful to our illustrators Luciano Toma, Francesco Rinaldi, Enrico Chiarelli, Alberto Mattia Perronace, and Giulia Bombieri. We are indebted to Daniele Baisero for his great dedication and patience, and also for making the distribution maps of the eight bear species. Dr. Craig Hilton-Taylor, Head of the Red List Unit of IUCN, granted us permission to use the IUCN data for the distribution maps, a huge thanks to him!

We would like to warmly thank the Cambridge staff for their support, in particular Dominic Lewis, Commissioning Editor, for his advice during the different stages of the book, Aleksandra Serocka, Senior Editorial Assistant, and Jenny van der Meijden, Senior Content Manager, for their continuous help and useful advice throughout the entire preparation of this volume. Working with them has been as smooth as one could hope for, and their help has been invaluable throughout the entire process. We also thank the Press Syndicate at Cambridge University Press for approval of our title.

We are also grateful to Yorgos Mertzanis and two anonymous reviewers who helped us to improve our proposal during the submission of the book plan to Cambridge University Press.

Last, but not least, we are always in debt to our families for their continuous support.

Introduction

Vincenzo Penteriani and Mario Melletti

I consider the mountainside a special place, a place with power, as I do certain other valleys and basins ... where grizzlies still roam. I return to these places year after year, to keep track of the bears and to log my life. The bears provided a calendar for me when I got back from Vietnam, ... I had trouble with a world whose idea of vitality was anything other than the naked authenticity of living or dying. The world paled, as did all that my life had been before, and I found myself estranged from my own time. Wild places and grizzly bears solved this problem.
Doug Peacock, 1990, Grizzly Years

Bears have fascinated people since ancient times. The relationship between bears and humans dates back tens of thousands of years, during which time we have also competed with bears for shelter and food. Our strong link with bears is also attested to by the Neanderthal burial of "Le Regourdou," in France, where the skeleton of a Neanderthal in a fetal position was found under a funeral slab surrounded by the bones of a brown bear, probably sacrificed for the burial. Bears were also represented in rock paintings in caves inhabited by our ancestors in Europe. The bears depicted by our ancestors were cave bears, which roamed Eurasia until about 24,000 years ago when they became extinct during the Last Glacial Maximum. Recently, gene flow between extinct cave bears and brown bears has been discovered, providing direct evidence for ancestral hybridization between the two species which resulted in the modern *Ursus arctos* that we all know (Chapter 1).

In human culture, bears also represent an important figure in Native American mythologies. For example, the bear is a symbol of power and strength. In fact, warriors of some tribes wore necklaces of bear claws. Bears also play a major role in several religious ceremonies in many North American tribes, which used to have a bear dance as part of their tribal traditions, and they represent an important clan animal in some native cultures, for example in tribes such as the Cherokees, Creeks, Hurons, and Navajos. Furthermore, bears are often found carved on totem poles of several tribes of north-western North America.

Bears have also influenced the culture of many tribes in Asia. In fact, they are important animals for some tribes in Siberia, and the people of the Hokkaido and Ryukyu islands in Japan. For example, the Ainu people in Japan consider the bear as the "Spirit of the Mountains." In Russia, at a Fat'yanovo cultural site dated to around 1500 BC, necklaces made with bear teeth were found, and other Neolithic findings have been discovered as far north as Lake Ilmen, in the Russian oblast of Nóvgorod. Several bear claws with bronze mounting, dated between the ninth and eleventh centuries, were also discovered among a Finno-Ugrian group located along the River Tsna (a river in the Tambov and Ryazan oblasts of Russia), whereas another Finno-Ugrian group in the Urals venerated the bear as a symbol of heroism.

The images of bears in popular culture have helped them to become an icon that most people know and love. The most famous example is the teddy bear, which has been one of the most popular stuffed animals since the early 1900s and continues to be a favorite of children. Developed almost simultaneously by toymakers Morris Michtom in the US and Margarethe Steiff in Germany, and then named after US President Theodore "Teddy" Roosevelt, the teddy bear is an iconic children's toy celebrated in stories, songs, and films. More recently, Baloo from *The Jungle Book*, Winnie the Pooh, Yogi and Bubu, and Masha and the Bear tell us that the strong link between people and bears, which started more than 80,000 years ago, continues today.

At the present time eight bear species are recognized, from the very popular polar bear, giant panda, brown bear, and American black bear to the lesser-known Andean bear, sun bear, sloth bear, and Asiatic black bear. In Chapter 1, the authors follow a different taxonomy for sun bear (*Helarctos malayanus*) and Asiatic black bear (*Ursus thibetanus*), ascribing them to the genus *Melursus*. However, we recognize that further genetic and morphometric studies are still required in order to fully understand the taxonomy of these two bear species. Therefore, and pending further investigation on *Melursus* taxonomy, we will still refer in this book to the genus *Helarctos* for sun bear and *Ursus* for Asiatic black bear, acknowledging that they might deserve a different genus in future. The conservation issue facing some of these species is a big conservation challenge today. Human activities, population encroachment, and poaching in bear landscapes continue to represent serious threats for some bear species or populations. As an example, in more recent times, human–bear conflicts have been exacerbated by the increasing number of people sharing the same landscapes with bears. Such coexistence has engendered an increase in conflicts, such as damage to livestock, crops, and apiaries, as well as the fear of bear attacks (Chapters 15–17, 20, 28). Moreover, in some regions people continue to keep bears in terrible captive conditions to extract bile and other body parts, mainly in Asia where the trade in these kinds of products still flourishes (Chapter 26). For instance, in China and Vietnam, thousands of bears are kept in tiny cages and bred for their bile. Furthermore, an emerging threat to bear conservation is related to the effect of global warning. This is the case, for example, for polar bears, which are now more endangered than even a few decades ago due to the dramatic reduction of sea ice. In fact, polar bears depend on the ice shelf for feeding, breeding, and movement, and they can only persist where the temporal and spatial availability of sea ice provides adequate access to their marine mammal prey (Chapter 21). The panda, another iconic bear species, is also threatened by climate change. The panda is an extremely specialized species that relies on bamboo for 99% of its diet, occurs in a very restricted range, and has one of the lowest reproduction rates among bears (Chapters 6 and 21). The challenge that bear conservation and management represents around the world makes this book extremely important and timely because it provides informative and complete accounts of everything you want to know on bears to a broad audience. However, even if our general knowledge of bear ecology and behavior has significantly increased in the past decades, we still have a lot to learn about this group of species, and in particular tropical bears (Chapters 4, 7–10, 22, 24–28).

The idea to edit a major volume on the ecology and conservation of bears started in 2017 from the need to provide a comprehensive book on all the bear species that inhabit our planet, as well as a useful tool for both the general public and people more directly involved in the fields of animal ecology, behavior, and conservation such as researchers, wildlife managers, conservationists, stakeholders (such as farmers and hunters), and students. In our minds, this book would also been able to demonstrate why the study of human–bear interactions, and stakeholder perception and involvement, are crucial for bear conservation and management, with detailed examples and case studies from all the continents inhabited by bears. Such a book would also highlight the urgency of conservation actions that need to be put into practice in the different regions of the world, mainly for lesser-known bear species in developing countries. Additionally, and despite the long-term interest in bears, as well as the many groups working on bears for decades and the large literature on these species around the world, a comprehensive and very detailed book on all the bear species in the world had never been published before. To make this idea a reality, we started by contacting more than 250 people among the best bear biologists in the world. Many of them replied enthusiastically to our invitation, saying that such a book was very much welcomed and long overdue. Of course, this positive feedback gave us further motivation to move forward with this project, in which 200 authors ultimately participated, many of them having spent their entire lives studying bears. The authors come from 33 countries spread across five continents, and work at very diverse institutions, such as research centers, universities, IUCN SSC Bear Specialist Group and Polar Bear Specialist Group and IBA, and non-governmental organizations (NGOs). To give you an idea of the heterogeneity of the book's contributors, people involved in this project are from (in alphabetical order): Bhutan, Bolivia, Canada, China, Croatia, Denmark, Ecuador, Finland, France, Germany, Greece, Hungary, India, Iran, Italy, Japan, Malaysia, Mexico, Nepal, the Netherlands, Norway, Pakistan, Poland, Romania, Russia, Slovenia, Spain, Sri Lanka, Sweden, Turkey, UK, USA, and Venezuela.

Together with the most important information on the ecology and behavior of bear species, the volume also includes specific chapters on taxonomy, phylogeny and genetics, population status and trends, as well as conservation status, management, and climate changes. The book is composed of 28 chapters subdivided into four sections: Part I – Systematics, Ecology, and Behavior (Chapters 1–5); Part II – Species Accounts (Chapters 6–14); Part III – Human–Bear Coexistence (Chapters 15–19); and Part IV – Conservation and Management (Chapters 20–28).

We hope that you will enjoy this book at least as much as we have enjoyed its long preparation and our close collaboration with chapter contributors, and that the huge effort made by all the authors will be appreciated by the public and scientific community. We will consider that our work has achieved its goal if it is rewarded by an increase in the understanding of bears and their effective conservation. However, the future of bear species will depend on our capacity to find pragmatic solutions that should represent a trade-off between human growth and the needs of bears and their habitats. What bears, among the most charismatic creatures on the planet, do for people has an inestimable value for our most intimate life, and their loss will create a void impossible to fill. Being in a bear country captivates our minds and, at the same time, offers a lesson in humility by giving us the feeling that something more powerful than us is out there.

Chapter 1

Systematics, Evolution, and Genetics of Bears

Andrew C. Kitchener, Eva Bellemain, Xiang Ding, Alexander Kopatz, Verena E. Kutschera, Valentina Salomashkina, Manuel Ruiz-García, Tabitha Graves, Yiling Hou, Lars Werdelin, and Axel Janke

Introduction

Bears, Family Ursidae, are among the largest of the Carnivora, but the Ursidae is among the least speciose carnivoran family with only eight extant species. Owing to the bears' widespread distribution in Eurasia, North America, and north-western South America, they occupy many habitats from the northern polar ice cap to tropical rainforests. Despite having few species, relationships between the Ursidae and other Carnivora and between ursid species (and indeed what constitutes an ursid) have been controversial. New molecular genetic techniques in recent years have allowed relationships to be explored in new ways, leading to some clarification, but debates continue owing to incomplete lineage sorting and ancient hybridization.

Bears are threatened by humans; conflict with human land uses and exploitation for fur and other body parts, coupled with habitat loss and extirpation through direct hunting, have led to fragmentation of populations of all species. Molecular genetics are key to understanding current and historical relationships between isolated populations, including species' colonization during glacial–interglacial cycles, to determine viability of local populations, needs for habitat corridors, and other aspects of population management, especially where bears are harvested for sport, etc. As natural habitats shrink, some bear species will inevitably require high levels of management, perhaps combining captive and wild populations following the IUCN's One Plan Approach. In this chapter we review the systematics of the Ursidae and its relationships with other Carnivora, the molecular phylogenetics of extant ursid species, the phylogeography of and morphological variation within each species, and the use of molecular genetics to monitor bear populations for management and conservation.

Systematics of the Ursidae

Wozencraft (1989) summarized the systematic history of the Carnivora, hence only a brief overview is given here. The classification of the Order Carnivora based on the morphology of the basicranium began with Turner (1848), was further developed by Flower (1869), and has since been refined, e.g. by Hunt (1974), who characterized the development and evolution of auditory bullae. The Carnivora is characterized by morphological characters, including carnassial teeth, i.e. the fourth upper premolar (P4) and first lower molar (m1), which form two shearing blades for slicing through body tissues, and the presence of three elements in the auditory bulla: the caudal and rostral entotympanic and ectotympanic bones (Hunt 1998; Wozencraft 1989).

Based on basicranial characteristics, the Carnivora comprises two suborders: the Feliformia (felids, viverrids, hyaenas, and related families) and the Caniformia (canids, ursids, pinnipeds, and musteloids). The Caniformia comprises two Infraorders, the Cynoidea (or Canoidea), including the Canidae, and the Arctoidea, including the remaining caniforms. Rose (2006) characterized the Arctoidea as having a suprameatal fossa (a hollow in the dorso-lateral wall of the middle-ear cavity) and loss of the third upper molar (M3), and that most arctoids have a single-chambered auditory bulla comprising mostly the ectotympanic bone. Some relationships within the Arctoidea remain controversial, although a consensus is growing. The main contentious areas include the relationship between the giant panda, *Ailuropoda melanoleuca*, the red panda, *Ailurus fulgens*, and the Ursidae (see below), whether the superfamily Pinnipedia is monophyletic, and the relationships between pinnipeds, the Ursidae, and Superfamily Musteloidea. Traditionally, the systematics of the Carnivora were based on morphological studies of extant and fossil taxa (e.g. Wozencraft 1989), but new molecular techniques have clarified many relationships, especially where homoplasy or convergence gave ambiguous or incorrect affinities between taxa. However, some relationships remain uncertain, owing to incomplete lineage sorting and ancient hybridization. The problematic systematics of the Arctoidea are discussed below.

The Superfamily Pinnipedia includes seals (Family Phocidae), walrus (Family Odobenidae), and sea lions and fur seals (Family Otariidae). Morphological studies since the nineteenth century suggested a sister relationship between the Ursidae and Otariidae/Odobenidae based, for example, on the structure of the auditory bulla, and between the Mustelidae and Phocidae, based, for example, on the absence of an alisphenoid canal, thereby supporting a diphyletic origin of the Pinnipedia. However, recent morphological studies demonstrated a monophyletic Pinnipedia (e.g. Berta et al. 2015, but see Koretsky et al. 2016).

Table 1.1 Classification of the extant Ursidae (Tedford 1976; Bryant 1996; Wagner 2010).

Classification
Class Mammalia Linnaeus, 1758; 1
Order Carnivora Bowdich, 1821; 33
Suborder Caniformia Kretzoi, 1943; 16, 194
Infraorder Arctoidea Flower, 1869; 15
Superfamily Ursoidea Batsch, 1788; 110
Family Ursidae Batsch, 1788; 110
Subfamily Ursinae Batsch, 1788; 110
Genus *Ursus* Linnaeus, 1758; 47
Ursus arctos Linnaeus, 1758; 47
Ursus maritimus Phipps, 1774; 185
Ursus americanus Pallas, 1780; 5
Genus *Melursus* Meyer, 1793; 155
Melursus ursinus (Shaw, 1790; pls. 58–59)
Melursus malayanus (Raffles, 1821; 254)
Melursus thibetanus (G. Cuvier, 1823; 325)
Subfamily Arctotheriinae F. Ameghino, 1902; 236
Genus *Tremarctos* Gervais, 1855; 20
Tremarctos ornatus (F. Cuvier, 1825)
Subfamily Ailuropodinae Grevé, 1894; 217
Genus *Ailuropoda* Milne-Edwards, 1870; 342
Ailuropoda melanoleuca (David, 1869; 12–13)

There is still no consensus on the relationship between the Ursidae and other Arctoidea. Several studies showed a sister relationship between the Ursidae and the Pinnipedia, but excluded the Musteloidea (i.e. Mustelidae, Procyonidae, Ailuridae, and Mephitidae) (Luan et al. 2013), while in most others the Ursidae was basal with the Pinnipedia and the Musteloidea grouped together (Eizirik et al. 2010; Nyakatura & Bininda-Emonds 2012; Doronina et al. 2015) (Figure 1.1), or grouped the Ursidae with the Musteloidea (Luan et al. 2013).

The Ursidae comprises eight extant species, which many authors consider to form two subfamilies (Wozencraft 1989; Goswami 2010; Wagner 2010): the Ailuropodinae includes the basal and highly derived giant panda, *Ailuropoda melanoleuca*, which is adapted to feeding on bamboo; and the Ursinae, which comprises two tribes, the Tremarctini, including the Andean bear, *Tremarctos ornatus*, and the Ursini, comprising the remaining six ursid species. Hunt (1998) divided the Ursidae into four subfamilies: the Ursinae containing the extant bears; the Ailuropodinae containing the giant panda; and two extinct subfamilies, the Amphicynodontinae and the Hemicyoninae. He further divided the Ursinae into three tribes – Ursini, Tremarctini, and the ancestral Ursavini. Most recent authors place the Andean bear in the subfamily Tremarctinae (e.g. Soibelzon et al. 2005; Mitchell et al. 2016), but Wagner (2010) suggested that the Tremarctinae Merriam and Stock, 1925 is pre-dated by the Arctotheriinae Ameghino, 1903, but it actually dates to Ameghino (1902) (Table 1.1).

Currently, the extant Ursinae includes three genera: *Helarctos*, for the sun bear, *H. malayanus*; *Melursus*, for the sloth bear, *M. ursinus*; and *Ursus*, comprising the brown bear, *U. arctos*, polar bear, *U. maritimus*, Asian black bear, *U. thibetanus*, and American black bear, *U. americanus*. Previously, three genera were recognized for Asian bears, including *Melursus* (sloth bear), *Selenarctos* (Asian black bear), and *Helarctos* (sun bear). A recent molecular phylogenetic study by Kumar et al. (2017) showed that the sloth and sun bears are sister species in a clade with the Asian black bear, which is basal. This suggests that either all Ursinae should be in the genus *Ursus* as proposed previously (e.g. Wozencraft 1989; Hunt 1998) or that the three Asian bears may be in a separate genus, *Melursus*. Recently, Kitchener et al. (2017) employed Hennig's Rule, which discriminates genera if divergence times are *c.*5 million years or more. Using Hennig's Rule divides the Ursinae into *Melursus* for Asian bears and *Ursus* for the rest (Table 1.2).

The Ursidae has a primitive dental formula (I3/3 C1/1 PM4/4 M2/3) and a Type A auditory bulla (Hunt 1974), comprising mostly the ectotympanic but also rostral and two

Table 1.2 Classification of the Ursidae, including extinct taxa.

Family Ursidae Batsch, 1788
Subfamily Hemicyoninae Frick, 1926
Tribe Cephalogalini Bonis, 2013
Adelpharctos Bonis, 1971
Cyonarctos Bonis, 2013
Phoberogale Ginsburg & Morales, 1995
Filholictis Bonis, 2013
Cephalogale Jourdan, 1862
Tribe Phoberocyonini Ginsburg & Morales, 1995
Plithocyon Ginsburg, 1955
Phoberocyon Ginsburg, 1955
Tribe Hemicyonini Frick, 1926
Zaragocyon Ginsburg & Morales, 1995
Dinocyon Jourdan, 1861
Hemicyon Lartet, 1851
Subfamily Ursavinae Hendey, 1980
Ballusia Ginsburg & Morales, 1998
Ursavus Schlosser, 1899
Subfamily Agriotheriinae Kretzoi, 1929
Agriotherium Wagner, 1837
Subfamily Ailuropodinae Grevé, 1894
Tribe Indarctini Abella et al., 2012
Miomaci Bonis et al., 2017
Indarctos Pilgrim, 1913
Tribe Ailuropodini Grevé, 1894
Kretzoiarctos Abella et al., 2012
Agriarctos Kretzoi, 1942
Ailurarctos Qi et al., 1989
Ailuropoda Milne-Edwards, 1870
Subfamily Arctotheriinae Ameghino, 1903
Plionarctos Frick, 1926
Arctodus Leidy, 1854
Arctotherium Burmeister, 1879
Tremarctos Gervais, 1855
Subfamily Ursinae Batsch, 1788
Ursus Linnaeus, 1758
Melursus Meyer, 1793

caudal entotympanic bones (Hunt 1998). The internal carotid artery loops inside the petrosal venous sinus, covered by the basioccipital, and may act as a countercurrent heat exchanger to cool the brain (Hunt 1998) (Figure 1.2).

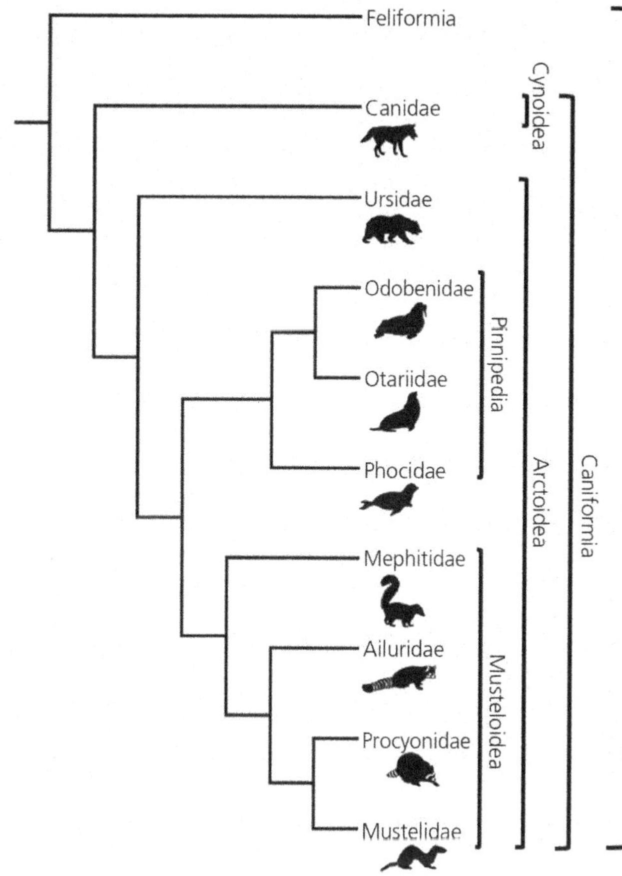

Figure 1.1 Phylogenetic tree showing a version of the relationships among the Arctoidea (Koepfli et al. 2017).

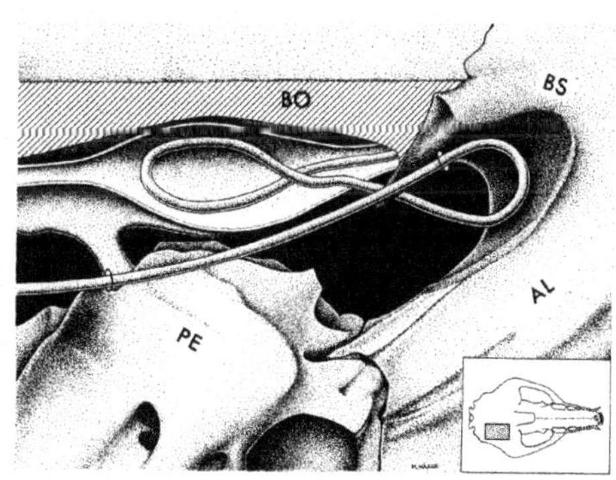

Figure 1.2 Dissected auditory region of a bear, showing the internal carotid artery looped within the inferior petrosal venous sinus on the lateral edge of the basioccipital bone. AL, alisphenoid; BO, basioccipital; BS, basisphenoid; PE, petrosal (Hunt 1977).

The Ursinae and Arctotheriinae share several morphological characters; a reduced lacrimal forming a vestigial bony rim around the naso-lacrimal foramen; a flat auditory bulla, which is much smaller than the mastoid, squamosal, and basioccipital processes; reduced premolars, whereby P4 has lost most of its shearing function; the upper and lower molars are quadrate (with four main cusps: paracone, metacone, protocone, and a large metaconule, which migrated to the postero-internal border to form the molars' oblong shape), and the rear border of M2 is greatly enlarged and elongated, forming a distinctive heel and making it the longest tooth in the maxilla (Wozencraft 1989; Hunt 1998).

The enamel on the surface of the talonid and trigonid basins of m2 of arctotheriines is tuberculated and forms a "z" pattern on the talonid (Hunt 1998). The skull is shorter than those of the Ursinae and there is a pre-masseteric fossa in the mandible in *Tremarctos* and *Arctodus*, but not *Plionarctos*. The diploid ($2n$) chromosome number is 52 biarmed autosomes in *Tremarctos* compared with $2n = 74$ mostly acrocentric chromosomes of the Ursinae, and $2n = 42$ mostly biarmed autosomes in the giant panda (Chorn & Hoffmann 1978).

The giant panda's classification was controversial since its description by David (1869), who regarded it as a bear, but Milne-Edwards (1870) classified it as a procyonid. Later authors proposed placing it in a family (Ailuridae) with the red panda, or in a monotypic family (Ailuropodidae). The controversy continued as different morphological features were presented for and against the two main hypotheses until molecular studies demonstrated unequivocally that the giant panda is basal to the Ursinae and Arctotheriinae. It is usually classified in the Ailuropodinae, although some still place it in its own family (although not recently, e.g. Thenius 1989), owing to its highly derived morphology and deep divergence time of *c.*12.5 million years ago (Ma) (Kutschera et al. 2014).

A key morphological character of the Ailuropodinae is a greatly enlarged radial sesamoid in the wrist, the so-called panda's thumb, which grasps stems, such as those of bamboo. In most other extant ursids the radial sesamoid is not enlarged, but recently the Andean bear's moderately enlarged radial sesamoid was described (Salesa et al. 2006). Therefore, enlarged radial sesamoids are either plesiomorphic (i.e. an ancestral character shared by two or more taxa), or they evolved independently in primarily herbivorous ursids. Other morphological characteristics of ailuropodines include the highly domed skull for attachment of massive jaw muscles, and enlarged molars and premolars for crushing tough bamboo stems and shoots.

Fossil History of the Ursidae

There is no agreement regarding quite what a bear is, because opinion differs as to which basal taxa should be included in the Ursidae. For any mammal group the deeper one descends into the fossil record, the more similar taxa become, until the origin of the group is reached. Today's bears differ greatly from those of the late Eocene and Oligocene (*c.*35–23 million years ago). Consequently, opinion differs as to which basal taxa should be included in the Ursidae (Table 1.2). A group of genera, including *Amphicynodon*, *Amphicticeps*, *Parictis*, *Kolponomos*, *Allocyon*, and *Pachycynodon*, distributed in Europe, Asia, and North America, constitute the amphicynodonts. This is either a paraphyletic stem lineage within the Ursidae or a monophyletic family Amphicynodontidae that may or may not be sister taxon to the Ursidae. The latter view point has gained credence in recent years, but without consensus. Here, we consider the amphicynodonts to constitute a distinct family and exclude them from further discussion.

The earliest Ursidae are from the earliest Oligocene. The family is subdivided into a number of subfamilies: Hemicyoninae, Ursavinae, Agriotheriinae, Ailuropodinae, Arctotheriinae, and Ursinae. The Hemicyoninae is the basal subfamily and encompasses about 10 genera in three tribes. Among these is the genus *Cephalogale*, often called "the earliest bear." This genus has recently been revised (de Bonis 2013), which showed that traditional "*Cephalogale*" is an assemblage of variably related, primitive Ursidae, within the tribe Cephalogalini, which now comprises five genera – *Cephalogale*, *Filholictis*, *Cyonarctos*, *Adelpharctos*, and *Phoberogale* – of jackal- to wolf-sized, rather canid-like animals. Cephalogalini is known from the earliest Oligocene (*c.*34 Ma) to the early Miocene (*c.*19 Ma). Their geographical range extended from Western Europe to the Midwestern United States, with *Cephalogale* and *Phoberogale* occurring on both continents. De Bonis (2013, p. 810) notes that no antecedents of Cephalogalini are known from the Eocene and surmises that they (and the Ursidae) may have had an Asian origin, where Cephalogalini are poorly known.

A second tribe of Hemicyoninae is the Phoberocyonini, revised by Ginsburg and Morales (1998). This tribe comprises *Phoberocyon* (not to be confused with *Phoberogale*) and *Plithocyon*, two genera with wide geographical ranges, both known from France to Florida. *Phoberocyon* is the older genus, with a stratigraphic range encompassing the late Oligocene to late middle Miocene (*c.*29–14 Ma), whereas *Plithocyon* is restricted to the middle Miocene (*c.*16–11.5 Ma). These two genera include larger, more derived species than the Cephalogalini, with the largest, e.g. *Phoberocyon aurelianensis*, within the size range of ursids today. Although phylogenetically close, *Phoberocyon* evolved toward hypercarnivory, with taller cusps on the lower carnassial and a narrow, trenchant talonid, whereas *Plithocyon* became more hypocarnivorous, with a low, wide carnassial. The reasons for this ecological divergence are debatable and complicated by the presence of both genera on multiple continents, each with different competitors. Nevertheless, the Ursidae would never be as hypercarnivorous again.

The Hemicyonini comprises two genera, *Hemicyon* and *Dinocyon*. *Hemicyon* includes generalized, medium- to large-sized hemicyonines of early and middle Miocene age with a broad geographical range from western Europe (where it is relatively common) to the midwestern United States, although Hunt (1998) suggests that the latter (*Hemicyon barbouri*) represents a distinct genus. *Hemicyon* is also recorded from Kenya by a single upper carnassial from the early Miocene (Schmidt-Kittler 1987). This tooth is the only record of a bear from Africa prior to *Agriotherium* in the latest Miocene (see below), despite decades of searching at productive localities, raising many presently unanswerable questions.

Dinocyon is much less common than *Hemicyon*. It is exclusively known from western Europe, in sediments dating from the middle Miocene to the earliest late Miocene (*c*.16–11 Ma). *Dinocyon* includes some of the largest bears of all time and, at least in tooth dimensions, rivaled giants such as *Agriotherium africanum* and *Arctotherium angustidens* (see below).

Closer to crown-group Ursidae, relationships between subfamilies are a little better understood (Abella et al. 2012), although the first, Ursavinae, is clearly a paraphyletic stem lineage. It comprises the genera *Ballusia* and *Ursavus*. The former is exclusively European from the early Miocene, whereas the latter comprises several species from the early to late Miocene of Eurasia and the middle Miocene of North America. *Ursavus* is traditionally considered ancestral to more derived bears. They are all small to medium-sized animals with generalized omnivore characteristics.

Following the Ursavinae is the subfamily Agriotheriinae, which includes only the genus *Agriotherium*. *A. africanum* is the first unequivocal bear to be described from sub-Saharan Africa, by Hendey (1980). Until that time, bears were only thought to have reached North Africa in the late Pleistocene, but never to have crossed the Sahara. *Agriotherium africanum* was considerably larger and more carnivorous than modern *Ursus*.

Crown-group Ursidae (the common ancestor of all living bears and all of its descendants) includes three subfamilies. The Arctotheriinae encompasses the South American ursid radiation, with *Tremarctos ornatus* as its surviving representative. The arctotheriines have antecedents in North America, where the ancestral taxon is *Plionarctos* of the late Miocene to Pliocene. *Tremarctos* also originated in North America, where *T. floridanus* overlapped in time with *Plionarctos*. However, most characteristic of the arctotheriine radiation are the short-faced bears, *Arctodus* (Pliocene of North America) and *Arctotherium* (Pleistocene of South America). Among the latter, *Arctotherium angustidens* may have been the largest bear (and carnivoran) of all time, with an estimated body mass of up to 1500 kg (Soibelzon & Schubert 2011). Curiously, the South American short-faced bear was more closely related to the living Andean bear (which is not short-faced) than it was to the North American short-faced bear. This is a striking example of convergence in size and morphology to ecological circumstance.

The Ursinae includes the extant genera *Melursus* and *Ursus*. The antecedents of *Melursus* are poorly known except for the Asian black bear. However, the genus *Ursus* is extremely well known in the fossil record, particularly the cave bear, *U. spelaeus*, which is studied from many thousands of specimens across large parts of western Eurasia, where it hibernated (and died) in caves. Numerous other species of *Ursus* have been described, including fossil relatives of the American black bear.

The final subfamily, the Ailuropodinae, includes the giant panda and relatives. The palaeontology of this subfamily has progressed greatly recently (Abella et al. 2012; de Bonis et al. 2017). The giant panda, once thought to be very isolated phylogenetically among the Ursidae, is now known to be among a large diversity of genera and species in the Ailuropodinae. Two tribes are recognized, the Indarctini and the Ailuropodini. The former includes the early genus *Miomaci* of the late Miocene in Europe and *Indarctos*, a genus known from the late Miocene of Eurasia and North America. The Ailuropodini comprises several genera, the oldest and most primitive of which are the European *Kretzoiarctos* and *Agriarctos* of late middle and early late Miocene age. Together with the Chinese *Ailurarctos*, these genera form a clade that is sister to *Ailuropoda*, culminating in the extant giant panda. This lineage shows a strong trend toward increased size and specialization to herbivory.

Molecular Phylogenies

The relationships among today's bear species have been unclear or contradictory depending on which methods were used. In particular, the relationships between the six ursine species remained unresolved until recently, because early molecular studies contradicted each other, mostly because of insufficient data (Talbot & Shields 1996; Yu et al. 2004). However, even larger molecular data sets that included nuclear genes did not confidently resolve the bears' phylogeny (Pages et al. 2008; Kutschera et al. 2014), resulting in a "forest of gene trees" with contradicting phylogenies. This suggested that bear evolution was either shaped by incomplete lineage sorting or that ancient hybridizations complicated phylogenetic reconstruction (Kutschera et al. 2014).

The first analyses of whole-genome data from all ursine and arctotheriine species have recently resolved relationships among bears (Figure 1.3; Kumar et al. 2017). Previously, the position of the American and Asian black bears with respect to their relatives was much debated (Pages et al. 2008; Kutschera et al. 2014). In the bifurcating tree (Figure 1.3), the American black bear is the sister species to polar and brown bears, while

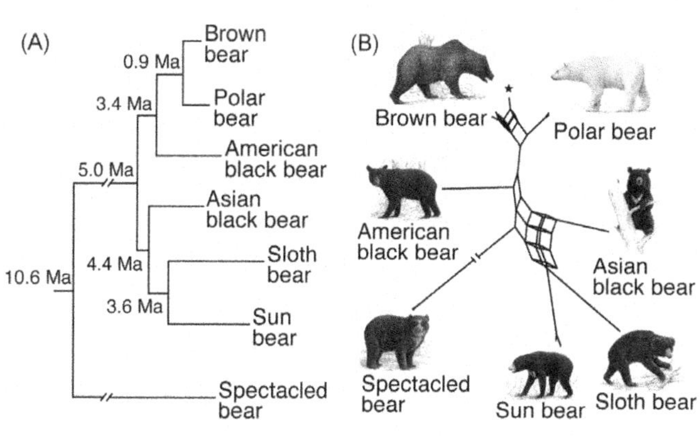

Figure 1.3 Phylogenomic analyses of 18,621 maximum likelihood trees estimated from non-overlapping 100-kb genome fragments from whole genome data of seven bear species (Kumar et al. 2017). (A) Schematic of coalescent species tree with 100% bootstrap support of all branches. Divergence time estimates in million years ago (Ma) from Kumar et al. (2017) were mapped onto the tree that was obtained from an analysis of 5.2 million bp coding sequences in MCMC tree (PAML) with multiple calibration points. (B) Split network analysis at a 7% threshold level. Ma, million years ago; ABC, brown bear from the Admiralty, Baranof, and Chichagof (ABC) Islands. Paintings by Jon Baldur Hlidberg (www.fauna.is). Modified from Kumar et al. (2017), figure 1.3, licensed under CC BY 4.0.

sun and sloth bears are grouped with the Asian black bear. The Arctotheriinae forms the sister group to all ursine bears. The giant panda was consistently placed as sister group to the Ursinae/Arctotheriinae in previous phylogenetic analyses (e.g. Kutschera et al. 2014).

However, analyses of bear genomes have shown that gene flow among species typifies the bears' evolutionary history (Kumar et al. 2017). Although harder to interpret, the bears' evolutionary radiation is actually a network, where signals from deviating phylogenetic trees become evident, with the American black bear placed between Asian bears and polar/brown bears (Figure 1.3B; Kumar et al. 2017). A brown bear from the Admiralty, Baranof, and Chichagof (ABC) Islands, off the western Canadian coast, is positioned between polar and brown bears, consistent with gene flow between ABC brown bears and polar bears (Miller et al. 2012; Cahill et al. 2015).

This network represents best the genome-wide evolutionary signal among extant ursids (Bapteste et al. 2013), because frequent hybridization among bears means their genomes are mosaics of evolutionary histories. A similar mosaic can result from incomplete lineage sorting, where ancestral gene lineages are randomly sorted along the species tree, and where phylogenetic signals may also differ from the species tree (Pamilo & Nei 1988). Only if all phylogenetic information in the evolutionary network is forced into one bifurcating tree is the more conventional view of bear evolution recovered (Figure 1.3A). While this tree may serve many purposes, such as taxonomy, conservation genetics, or molecular clock analyses, it provides an incomplete understanding of the bears' evolutionary history and it may also explain why some morphological characters are difficult to place on a simple, bifurcating tree.

Genetic differences among the bears can be converted into time, assuming that substitutions accumulate relatively constantly between species, as in a molecular clock (Zuckerkandl & Pauling 1962). Molecular divergence times among bears have been studied extensively (e.g. Kutschera et al. 2014; Kumar et al. 2017), with particular interest in the divergence of polar and brown bears (e.g. Hailer et al. 2012; Miller et al. 2012). Genome-wide estimates suggest the deepest split in the Ursinae (*Ursus* versus *Melursus*) was 5.0 (4.5–6.0) million years ago (Ma), and the divergence of polar and brown bears was 0.9 (0.6–1.1) Ma (Kumar et al. 2017).

Phylogenetic reconstruction of bears exemplifies the complications that gene flow creates for understanding evolutionary relationships. One of the first molecular studies on bear species used mitochondrial DNA (mtDNA) to study the relationship between polar and brown bears (Cronin et al. 1991). This analysis suggested that polar bears originated from brown bears from the ABC Islands, questioning their validity as a distinct species and leaving *Ursus arctos* paraphyletic. However, some 20 years later, a comprehensive analysis of nuclear gene loci demonstrated that polar bears are a separate, ancient lineage that diverged from the common ancestor of polar and brown bears about 600,000 years ago (ya) (Hailer et al. 2012). Therefore, the mtDNA of ABC Islands brown bears probably transferred to polar bears by hybridization, making it the first example of gene flow among bear species. For this transfer of mtDNA, female brown bears and male polar bears must have hybridized some 150,000 ya, coinciding with the penultimate glacial period (*c.*185,000–135,000 ya; Marine Isotope Stage 6; Margari et al. 2010). The brown bear mitochondrial haplotype has been identified in hundreds of polar bears studied for this locus so far (e.g. Cronin et al. 1991; Hailer et al. 2012). Under this scenario, the brown bear mitochondrial haplotype spread to all polar bears in a comparatively short time, also implying that brown bear mtDNA was selectively advantageous for polar bear survival, or the population bottleneck was very severe. An alternative explanation for high genetic similarity of polar and brown bear mtDNA is that gene flow occurred from polar bears into some brown bear populations (Cahill et al. 2013, 2018; Liu et al. 2014; Hassanin 2015). Indeed, gene flow between polar and brown bears was repeatedly detected in analyses of entire genomes (e.g. Miller et al. 2012; Cahill et al. 2015; Kumar et al. 2017). Up to 8.8% of the brown bear genome is of polar bear ancestry, but polar bear genomes harbor no significant genetic introgression from brown bears (Cahill et al. 2015). Cahill et al. (2018) suggested that brown bears carried some polar bear alleles away from where admixture occurred through male-dominated dispersal. It is unclear if this gene-flow asymmetry reflects continuing hybridization between the two species, is a remnant from the speciation event, or results from strong selection pressure on the highly specialized polar bear.

Genome data from all bear species show that gene flow was much more common than assumed for mammalian genera (Kumar et al. 2017). Whether hybridization is current or ancestral among bears, different parts of their genomes have different evolutionary histories. This complicates our understanding of bear evolution. For example, some parts of the American black bear's genome support a close relationship with the polar and brown bears, while other parts place the American black bear closest to the Asian black bear (Figure 1.3B). Phylogenetic reconstruction among the Asian bears is even more complicated and consistent with natural hybridization between Asian black bears and sun bears (Galbreath et al. 2008). Technical advances in analyses of ancient DNA allow studies of entire genomes of extinct species. Recently, a study found gene flow from extinct cave bears, *Ursus spelaeus*, into brown bears (Barlow et al. 2018), providing direct evidence for ancestral hybridization. Such interspecific gene flow might have been advantageous for adapting to changing environments.

Phylogeography and Intraspecific Taxonomy

Although relationships between species are now better understood, relationships between populations within species and the taxonomic status of those populations are often unresolved in the Ursidae despite descriptions of many different subspecies and even species. For example, 232 recent and 39 fossil species and subspecies of today's *Ursus arctos* have been described (see Erdbrink 1953; Hall 1981), which Kurtén and Anderson (1980) considered "a waste of taxonomic effort, which, as far as we know, is unparalleled." Phylogeographical studies allow better understanding of how populations are related to each other, where refugia existed during glaciations, and how species have recolonized continental ranges. However, lack of genetic and morphological data and their frequent apparent incongruence, along with high intraspecific variability, have led to poor resolution of intraspecific taxonomy and relationships in some species. For example, polar and Andean bears are considered monotypic, despite geographical variation in size and some genetic structuring of populations, although ESUs (evolutionary significant units) or MUs (management units) are recognized owing to genetically discrete populations, which occupy different habitats or experience different climatic conditions. In contrast, 8–12 extant brown bear subspecies are described; some are historically and widely recognized, but recent ones are not fully accepted, and for some, their geographical ranges are poorly known. Below, we briefly review the phylogeography and current status of subspecies within each bear species.

Ailuropoda melanoleuca

The giant panda was widely regarded as monotypic until Wan et al. (2003) described *A. m. qinlingensis*, from the Qinling Mountains, Shaanxi Province, based on genetic differentiation from *A. m. melanoleuca* in Sichuan. This proposed new subspecies appears to have a browner pelage and smaller skull with larger molars (Wan et al. 2005). However, the distribution gap between the two putative subspecies is caused by human impacts, so genetic differences may represent genetic drift in now separate populations or clinal variation in a previously continuous range.

The giant panda's phylogeography shows a mismatch between mitochondrial and nuclear genetic studies. Early studies of short mtDNA sequences showed no haplogroups, even using longer sequence lengths up to 680 base pairs (bp) (Zhang et al. 2007; Hu et al. 2010a), which distinguish brown bear subclades. Recent studies, using complete mitochondrial genomes, revealed at least three extant and two extinct clades in giant pandas (Barlow et al. 2019). However, although mitochondrial clades are well defined and calibrated, they are not restricted to any population and show no geographical subdivision. In contrast, nuclear DNA reveals quite finely resolved geographical patterns with the Qinling Mountains, home to the most divergent population (Zhao et al. 2013), and Xiaoxiangling as a probable refugium (Chen et al. 2013). This discordance in genetic markers probably results from female-biased dispersal (Hu et al. 2010b). Also, considering drastic range decreases and population fragmentation, it is probable that many other genetic lineages are extinct, which could have highlighted the species' history and phylogeography.

Tremarctos ornatus

Although subspecies were proposed previously (García-Rangel 2012), today the Andean bear is regarded as monotypic. However, genetic studies show that Andean bears comprise two ESUs: the Northern Andean clade (NAC; Venezuela, Colombia, Ecuador, and north-central Peru) and the Southern Andean clade (SAC; southern Peru and the northern and central Bolivian Andes), which split around 500,000 ya (Ruiz-García 2013; Ruiz-García et al. 2020a). Additionally, in Bolivia, some Andean bears from Santa Cruz Department were more related to the NAC than the SAC. These results and the slightly higher-level genetic diversity in the SAC show that the latter was the ancestral Andean bear population, contradicting palaeontological finds, which support the most northern Andean bear population as ancestral. Haplotypes vary in how they spread through the Colombian Andean Cordilleras. Gene-flow estimates were relatively high; therefore, geographical barriers have not prevented dispersal of Andean bears there. However, significant genetic heterogeneity was found in northern Colombian populations, with significant spatial autocorrelation for the Andean bear in Colombia (Ruiz-García et al. 2020b). In contrast, no genetic differentiation was found between different Cordillera Provinces, and northern–southern regions in Ecuador (Ruiz-García et al. 2020c).

The first molecular genetic studies of Andean bears used nuclear markers (microsatellites; Ruiz-García 2003, 2007; Ruiz-García et al. 2005) and estimated low to medium genetic diversities. However, mitochondrial markers yielded high levels of genetic diversity. This paradox is probably explained by "ascertainment bias" (Ellegren et al. 1995), whereby the species for which microsatellites were developed shows higher genetic diversity than that of the non-target species. In this case, the microsatellites used for Andean bears were originally designed for American black bears (Kumar et al. 2017).

Melursus malayanus

Usually two subspecies of sun bears are recognized: *M. m. malayanus* on the Asian mainland and Sumatra, and *M. m. euryspilus* from Borneo (Corbet & Hill 1992). Meijaard (2004) analyzed skull morphometrics of sun bears and confirmed that the Bornean population, which has a smaller skull and relatively longer upper tooth row, is distinct from Sumatran and mainland populations.

Very little is known about sun bear genetics. Meijaard (2004) cited an unpublished mtDNA study by L. Waits on sun bears from Sumatra and Borneo that yielded five clades with no geographical structure, suggesting some gene flow between islands during the Last Glacial Maximum (LGM). From published data, two mtDNA lineages are reported from Borneo (divergence 5.6 ± 1.9 base pairs (bp)), but no geographical separation (Onuma et al. 2006). Yu et al. (2007) and Krause et al. (2008) obtained mtDNA sequences from three putative continental Asian sun bears that could represent another subspecies, and which yielded two additional haplotypes that diverged from the two Bornean lineages by 14.39 ± 3.39 bp and 16.26 ± 3.57 bp, respectively. The differences between these mtDNA clades suggest a robust structure exists, dividing continental and island bears, but this is not yet fully resolved.

Melursus ursinus

Usually two subspecies of sloth bear are recognized: *M. u. ursinus* from peninsular India and *M. u. inornatus*, which has a smaller skull, from Sri Lanka (Corbet & Hill 1992). However, sample sizes for cranial lengths were very small. There may be clinal variation in skull size with latitude from larger northern animals to smaller southern animals, but this has not yet been studied.

The sloth bear's phylogeography has not been studied. However, in GenBank there are seven sequences of mtDNA fragments, including Zhang and Ryder 1993 (captive); Yu et al. 2007 (no locality); Talbot and Shields 1996 (captive); Mohan et al. unpublished (India); and Krause et al. 2008 (no locality). Not all these sequences could be aligned, although a comparison reveals at least three highly divergent haplogroups, lacking any geographical information.

Melursus thibetanus

Asian black bears are widely distributed in southern and eastern Asia; several subspecies are usually recognized, including *M. t. thibetanus* from Nepal to SE Asia, *M. t. japonicus* from Japan, *M. t. formosanus* from Taiwan, *M. t. gedrosianus* from Pakistan, *M. t. ussuricus* from the Russian Far East, Korea and northern China, *M. t. mupinensis* from southern and central China, and *M. t. laniger* from Kashmir and Punjab (Ellerman & Morrison-Scott 1953).

The Asian black bear's phylogeographical structure, based on mtDNA, is robust, but not yet fully resolved. It includes several haplogroups roughly congruent with seven putative subspecies (Hwang et al. 2008; Kadariya et al. 2018). The Japanese subspecies forms a distinct basal mitochondrial clade (Wu et al. 2015; Kadariya et al. 2018), which probably colonized Japan in the Middle Pleistocene, and is further substructured among the Honshu bears (Ohnishi et al. 2009; Kim et al. 2011). The mainland Asian clade probably represents a latitudinal cline, with increasing adaptation to more tropical climates from north to south, including the Taiwanese population, but the exact phylogenetic order of the matrilines is uncertain (Ohnishi et al. 2009; Kim et al. 2011; Wu et al. 2015; Kadariya et al. 2018). Recently, Kadariya et al. (2018) showed that the Himalayan population is a distinct basal lineage to mainland Asian populations. This suggests that three subspecies are recognizable: *M. t. thibetanus* from mainland Asia, *M. t. laniger* from the Himalayas, and *M. t. japonicus* from Japan. This species' phylogeographical structure is consistent with female philopatry and it seems very likely that it was influenced by the Pleistocene and Holocene glaciations throughout its range.

Although the Ussuri black bear (known as *M. t. ussuricus*) is separated by more than 1500 km from other populations, it is not genetically distinct. The Ussuri population arose very recently, separating *c.*20,000 years ago (peak LGM), and adapting to colder climates. The distribution gap, cf. the tiger, *Panthera tigris*, arose even later, owing to forest loss since the early Neolithic, *c.*9000 years ago, and the exploitation of bears for traditional medicines by early civilizations in northern China since at least the Shang dynasty *c.*3500 years ago (Barnes 1999; Ren 2000). Owing to its ecological distinctiveness and geographical isolation, the Ussuri population should be considered as a separate MU.

Ursus americanus

Owing to its widespread distribution, coupled with polychromatism (Rounds 1987), many subspecies of *U. americanus* (16 listed by Hall 1981) were recognized, with *U. a. americanus* occupying most of the species range.

For American black bears phylogeographical structures are described for both mitochondrial and nuclear DNA (Wooding & Ward 1997; Pelletier et al 2011; Puckett et al. 2015). Analysis of mtDNA revealed two lineages; the continental lineage is subdivided into eastern and western subclades, while the coastal lineage has a much simpler structure along the Pacific coast of North America. Nuclear SNP (single nucleotide polymorphism) data also showed quite clear geographical structure, including three main continental nuclear clusters (Alaska, Eastern, and Western), which were divided into nine subdivisions (Puckett et al. 2015). However, nuclear and mitochondrial clusters mostly did not coincide. The current phylogeographical structure is believed to have originated before the LGM [31–67 kya (thousand years ago) for nuclear clusters and 169 kya to 1.07 Ma for the mitochondrial ones] in several refugia in northern and southern North America, including Beringia, the north-west Pacific, the Southwest, and the Southeast (Puckett et al. 2015).

Neither mitochondrial nor nuclear structures match subspecies delineation very well (Puckett et al. 2015). Indeed, it is difficult to discern subspecies owing to complex admixture following expansion from these refugia, although Puckett et al. (2015) suggested that the nine subclusters could be associated with subspecies, which still seems too many for a highly mobile species showing extensive admixture. Perhaps the three mtDNA clusters, which appear mostly geographically discrete, would be better recognized as subspecies, including *U. a. americanus* (clade A-east) in the east, *U. a. emmonsii* in the west (clade B) and *U. a. cinnamomum* (clade A-west) from the central range.

Ursus maritimus

The polar bear is a monotypic species, although subspecies were previously proposed. Wilson (1976) found clinal variation in skull morphometrics across North America, with the possible exception of a distinct population in South Alaska, which was proposed as a subspecies. Currently, 19 MUs are recognized (Durner et al. 2018).

For the polar bear, mtDNA reveals no clear genetic structure, although particular locations may differentiate in haplotype frequencies (Campagna et al. 2013; Kutschera et al. 2016). This lack of structure is partially attributed to long-distance movements of individuals and partially to the relatively recent origin/divergence of the species' mtDNA lineages (see above). In contrast, microsatellite and SNP data show four to six geographical clusters, which are consistent between studies (Kutschera et al. 2016; Malenfant et al. 2016) and show differentiation between adjacent populations (Crompton et al. 2008; Campagna et al. 2013; Viengkone 2016). Putative populations are often identified as MUs, although microsatellite clusters have broader geographical ranges and may include several putative MUs. A comprehensive reconstruction of the species' history, cf. brown bear and American black bear, is prevented by current unavailability of sufficient genetic markers with enough phylogenetic signal.

Ursus arctos

The brown bear is the most widespread ursid species and, given the wide variety of habitats it occupies, it is unsurprising that its morphology is so variable. As a result, many subspecies have been recognized, but the principal extant ones are: Eurasian brown bear, *U. a. arctos*; Syrian bear, *U. a. syriacus*; Isabelline bear, *U. a. isabellinus*; blue bear, *U. a. pruinosus*; Kamchatkan bear, *U. a. beringianus/piscator*; Siberian brown bear, *U. a. collaris*; Japanese brown bear, *U. a. lasiotus*; Kodiak bear, *U. a. middendorffii*; and grizzly bear, *U. a. horribilis* (Ellerman & Morrison-Scott 1953). However, brown bears show much individual, seasonal, age, and sexual variation, and thus it can be challenging to ascertain subspecies taxonomy based on morphology alone.

Although there have been many phylogeographical studies on brown bears, very few used nuclear DNA, so phylogeographical patterns often reflect female natal philopatry. The phylogeographical structure of the brown bear is well resolved. Several mitochondrial haplogroups are recognized, some of which are widely distributed and sympatric, whereas others are restricted to particular geographical regions. The picture is further complicated (see above), because polar bears share a mitochondrial DNA haplogroup with ABC Islands brown bears and an extinct Irish population (Edwards et al. 2011). Some studies found that current phylogeographical patterns do not match distributions of mtDNA haplotypes from fossil specimens (e.g. Paetkau et al. 1998; Barnes et al. 2002; Hofreiter et al. 2004; Valdiosera et al. 2007; Ersmark et al. 2019), such that today's patterns result from female philopatry, population expansion and fragmentation, and genetic drift. MtDNA reveals at least six major clades grouped into two main branches. The first branch includes Western European (clade 1) and the Iranian clade from south-west Asia and the ABC Islands (clade 2; including the polar bear) (Davison et al. 2011; Ashrafzadeh et al. 2016; Salomashkina et al. 2017). The second branch includes bears from southern North America and Hokkaido (clades 4 and 3d), Tibet (clade 5), and superclade 3, which occurs throughout most of the Eurasian and Northern American species range and is divided into two (3a and 3b) to four (3a, 3b, and two Caucasian) subclades (Murtskhvaladze et al. 2010; Çilingir et al. 2016; Salomashkina et al. 2017) (Figure 1.4). There is also a recently discovered Himalayan clade that appears to form a third main branch (Lan et al. 2017). Particular traits of the brown bear's phylogeography allow reconstruction of its complex colonization history, which was heavily influenced by glacial cycles, involving multiple cycles of range contractions followed by lineage sorting and subsequent range expansions (Davison et al. 2011; García-Vázquez et al. 2017; Anijalg et al. 2018). Therefore, care should be taken in making taxonomic decisions based on contemporary genetic data alone without a deeper time perspective. MtDNA reflects matrilinear history, but it is not directly connected with environmental adaptation, and thus in some cases animals inhabiting the same area may morphologically evolve into a homogeneous population while still carrying very divergent matrilines through different colonizations and subsequent population introgression. For example, before data on their mtDNA were available, the genetically distinct ABC bears were not a recognized subspecies, while clearly morphologically distinct bears from Kamchatka and Kodiak were recognized as subspecies, but are just branches inside the 3a subclade. The Syrian bear comprises three major mtDNA clades (and seven subclades), and the Siberian brown bear seems to be a mixture of two. The recognition of local populations in the Apennines (*U. a. marsicanus*) and the Pyrenees (*U. a. pyrenaicus*) cannot be supported based on current genetic evidence (Loy et al. 2008; Colangelo et al. 2012).

Paternal phylogeography is a new genetic approach that could be applied to this species, which is based on SNPs and short tandem repeats (STRs) located in the non-recombining fragment of the Y-chromosome (Bidon et al. 2014; Hirata et al. 2017). However, there are currently insufficient data for a comprehensive coverage of the brown bear's widespread range to understand its geographical structure.

Population Genetic Assessment and Management of Bear Populations for Conservation

Conservation genetics based on non-invasive genetic sampling has revolutionized how individuals, groups, and populations of bears are characterized and monitored (Waits et al. 1999; Schwartz et al. 2007). The majority of population genetic analyses in bears used microsatellites or STRs. However, markers based on genomic approaches, such as SNPs and genome sequencing of individuals, have been increasingly applied. Substantial improvements in methodology have led to the extensive use of non-invasive genetic sampling to

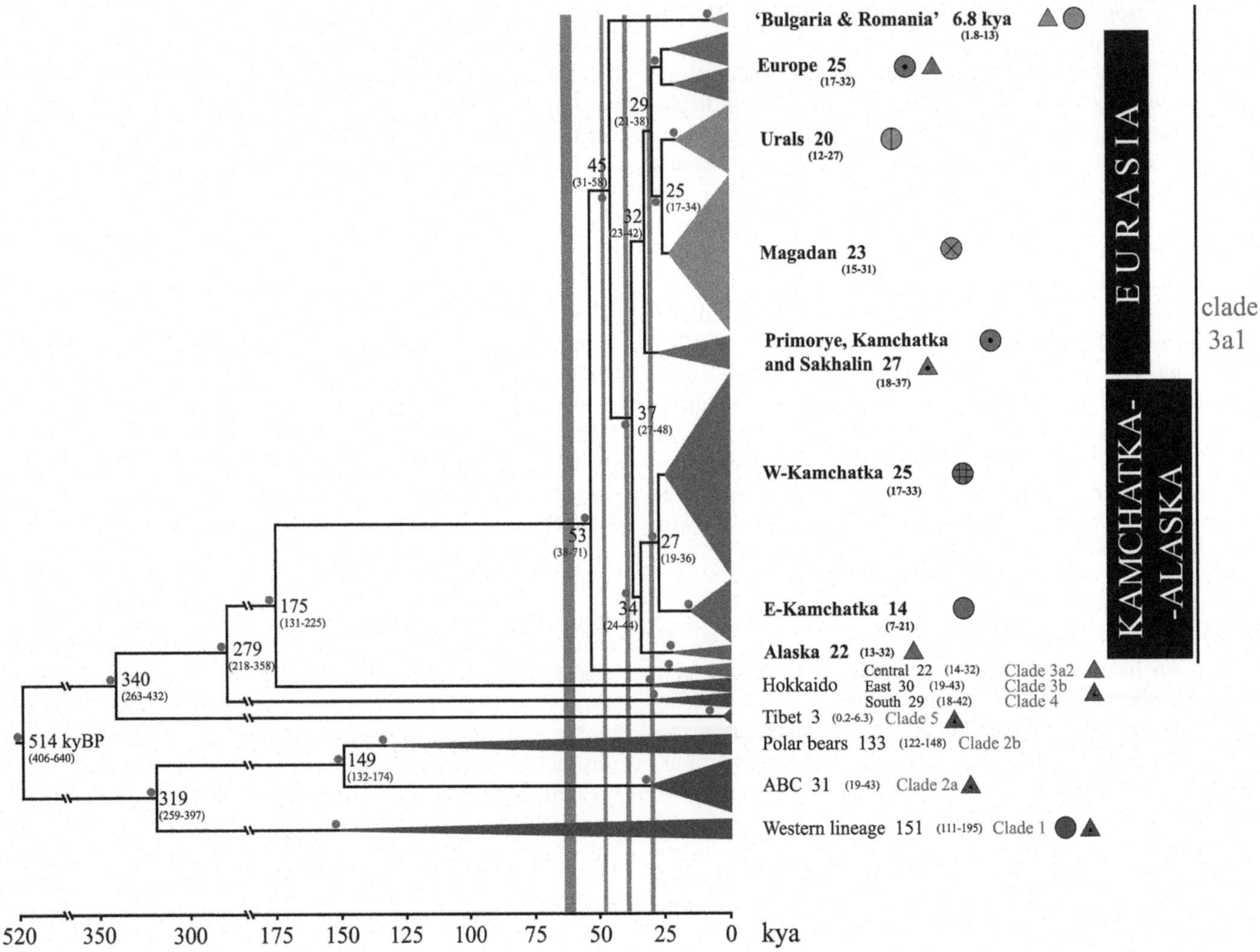

Figure 1.4 Bayesian phylogenetic tree of brown bears based on mitochondrial genomes (Anijalg et al. 2018). Numbers on nodes and after clades represent posterior mean estimates of the ages of different lineages, with numbers in parentheses representing 95% credibility intervals. Vertical gray lines represent long glacial periods. The recently described Himalayan brown bear clade is not shown (Lan et al. 2017), but is basal to clade 1, Western lineage.

identify and determine the numbers of bears in areas, countries, or even transborder regions (Bischof et al. 2016). Non-invasive samples are used for genetic assessments of populations, including estimation of genetic diversity and substructure, inbreeding, effective population size and thus population viability, and conservation status. Further, regular sampling allows development of databases containing many individual genotypes and also comprehensive ecological monitoring data of the same individuals over several years [e.g. Norway (see Figure 1.5) and the USA; see Chapter 13].

DNA-Based Monitoring of Bears

Today, many techniques are used in DNA research and monitoring of bears. Different genetic markers and assessments may show different results owing to the resolution of genetic data. Wildlife researchers and managers must define a specific genetic sampling and monitoring scheme, and choose the genetic assessment method which fits their needs and delivers the information needed to achieve management or research goals. Early studies of populations were based on high-quality DNA samples from blood or tissues collected from captured or dead bears. From such samples many genetic markers can be used to identify individuals, ascertain sex and assess genetic diversity, structure, kinship, and gene flow among individuals and populations, or to obtain long fragments of mtDNA to study phylogeny.

When studying large, small or endangered bear populations, or when budgets are limited, non-invasive sampling (feces, hairs) provides opportunities to obtain genetic data without disturbing or even observing individuals and may be less expensive. The collected samples are usually of low-quality, somewhat degraded DNA, which constrains the number of markers that can be studied and the length of the studied DNA fragment, and therefore options for population genetic analyses (see Table 1.3). However, when samples are successfully genotyped, individual identification allows estimation of home ranges, population size through capture–mark–recapture

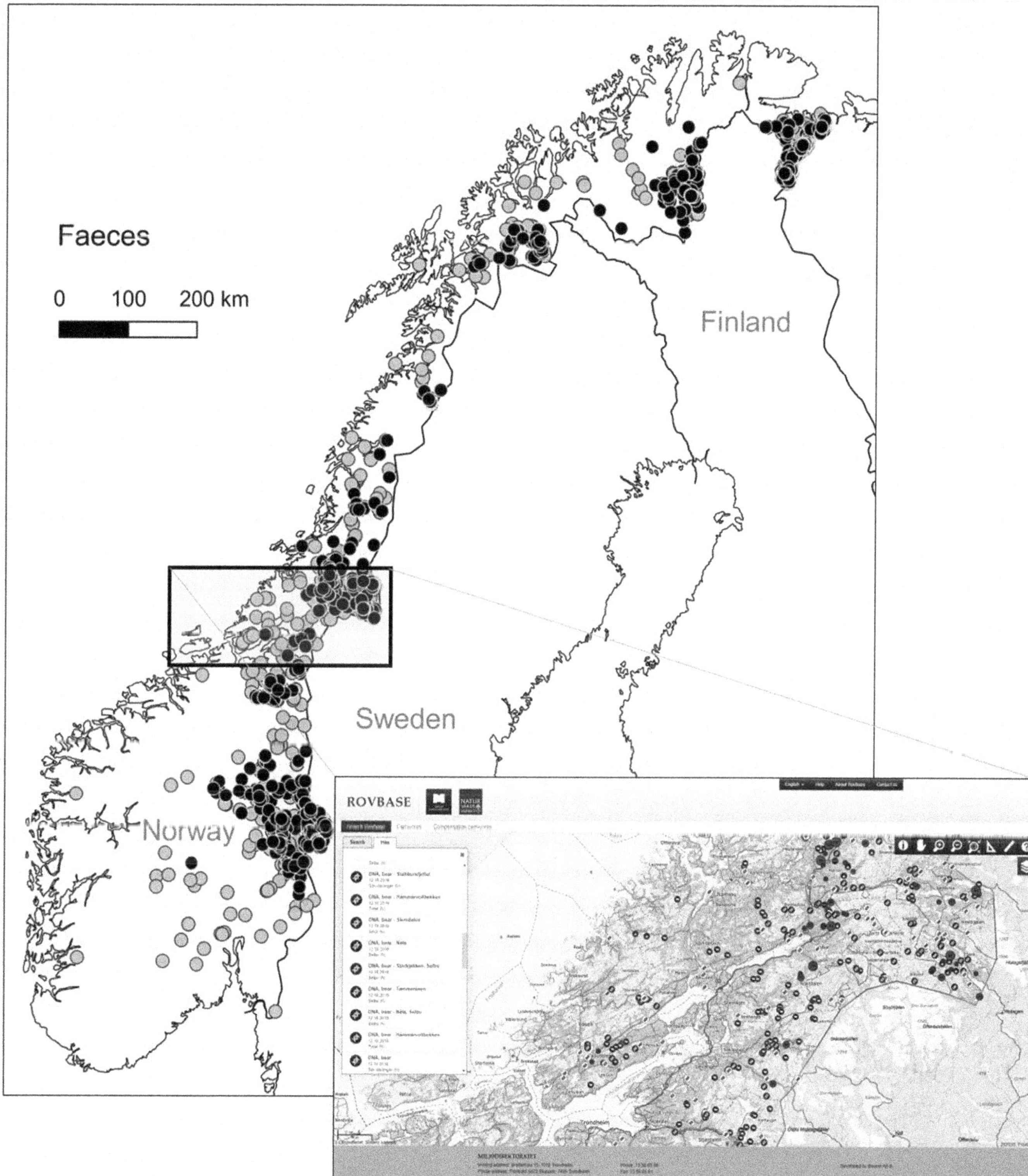

Figure 1.5 The Norwegian Large Predator Monitoring Program operated by Rovdata. Rovdata is responsible for the analyses and reports the results and monitoring data on brown bears in the country. Roughly 1000 non-invasive samples (mainly feces and hairs) are collected annually to assess the number of bears to guide the management by estimating the annual reproductive output (Bischof & Swenson 2012). Results of individual identifications, and other data, are accessible to the public at www.rovbase.no. From 2016 to 2019, for example, 2867 feces samples were collected and analyzed: successfully genotyped samples (n = 1770; black), negative samples (n = 1097; gray); figure by Kopatz et al. (in preparation, modified).

Table 1.3 Applications for DNA-based studies of bears depending on the type of sample obtained. x, commonly observed in the literature; (x), observed in recent studies or potential application.

		Tissue samples	Fecal samples/hairs	Ancient DNA samples	eDNA samples/ (mixed genetic material)
Species (mitochondrial DNA)	Species identification	x	x	x	x
	Lineage identification	x	x	x	
	Interspecific phylogeny	x		(x)	
Individual (nuclear DNA)	Genotyping	x	x		(x)
	Sex identification	x	x		(x)
	Genetic diversity	x	x		
	Abundance estimates	x	x		
	Structure/gene flow	x	(x)		
	Kinship analyses	x	(x)		

methods, or population trends of identified groups or subpopulations based on genetic parameters (e.g. Kendall et al. 2009). Importantly for isolated and endangered species and populations, such genotyping data can also be used to identify MUs or ESUs.

In scenarios where collection of tissues, feces or hairs is impossible, species or even individuals can be identified from environmental DNA (eDNA) samples, i.e. biological material collected from substrates such as soil, water, or snow that bears have encountered. Several recent studies applying eDNA technologies in various conditions and environments highlight the potential for this technology. For example, Wheat et al. (2016) collected saliva swabs from salmon carcasses and showed that this is a more cost-effective, rapid method to identify individual bears than the use of scats. Von Duyke et al. (2017) simulated fresh tracks by pressing the foot of a subsistence-harvested polar bear into snow. The genotype from the snow sample, based on next-generation sequencing (De Barba et al. 2017), matched the genotype obtained from this harvested bear. Further technical developments might allow researchers to obtain bear DNA from tracks in the field, from different substrates (such as soil around the den), or indirectly through invertebrate- or parasite-derived DNA (iDNA; Abrams et al. 2019). The possibility of identifying individuals with these new approaches could revolutionize the monitoring and management of bear populations.

Population Genetic and Genomic Assessments

Genetic diversity provides a metric of genetic health, an evaluation of historical reductions in population size, and insights into the potential for long-term viability through adaptive capacity. Genetic diversity varies considerably across bear populations and is affected by the number and variability of loci along with sample size. For example, in brown bears, heterozygosity ranges from 0.27 to 0.81 (Skrbinšek et al. 2012). As expected, low diversity occurs in small and isolated populations (e.g. brown bears in the Gobi Desert, Cantabrian Mountains, and Kodiak Island; Andean bears in Ecuador). Theoretical concepts and methodology to estimate effective population sizes in natural populations have been evaluated (Skrbinšek et al. 2012; Kamath et al. 2015; Kopatz et al. 2017) to determine whether they would be practical to assess and track the number of individuals effectively contributing to populations (Pierson et al. 2018). A study on a population of grizzly bears in Greater Yellowstone Ecosystem, USA demonstrated that numbers of bears and effective population size have increased, despite relatively low and stable genetic diversity (Kamath et al. 2015). While limited evidence for deleterious inbreeding effects exists in natural populations (but see Dunbar et al., 1996), decreased litter sizes and albinism have been demonstrated in captive brown bears (Laikre et al. 1996). Similarly, male Scandinavian brown bears with high interrelatedness had lower yearly reproductive success (Zedrosser et al. 2007), and, in the Apennine mountains, alleles predicted to be deleterious were fixed in brown bears, including several associated with health problems in humans (Benazzo et al. 2017). Low diversity has also been demonstrated in the captive-bred giant panda population (Shen et al. 2009) and genetic bottlenecks have been illustrated for American black bears (Murphy et al. 2017). However, genetic diversity can increase rapidly when populations and genes mix (e.g. Mikle et al. 2016).

The importance of isolation has led many studies to assess population structure, often in combination with assignment tests and coalescent approaches, to evaluate connectivity between subpopulations. For example, Dutta et al. (2015) identified migrants between populations of Indian sloth bears.

Many studies use landscape–genetic methods to evaluate resistance of various landscape features, with the aim of managing landscapes to maintain or enhance connectivity between separated subpopulations. Draheim et al. (2018) found that American black bears showed a stronger signal of isolation by resistance over time as forest fragmentation increased. Parentage approaches have been used to evaluate movement (e.g. Graves et al. 2014), as well as to understand the impact of interactions of behavior and relatedness on fitness, through mechanisms such as multiple paternity, sexually selected infanticide, and site fidelity (e.g. Norman et al. 2017). Analyses of population structure are also used in forensics to identify the source of illegally killed individuals or bear body parts (Frosch et al. 2014). With the development of genomics, new approaches to diversity, inbreeding, and other questions are becoming more common, as well as the ability to evaluate the role of selection and adaptive capacity in population survival and evolution. Genetic information enables studies on relatedness and kinship. For example, Mikle et al. (2016) used a pedigree approach to document mechanisms behind changes in diversity, where initial low diversity in a population resulted from a few reproductively dominant individuals, but increases in migration between subpopulations over time led directly to increased diversity. Norman et al. (2017) developed a method to identify fine-scale population genetic structure, which is sensitive to spatial and temporal changes in the landscape.

Conclusions

Despite the few extant ursid species, the evolution of bears has been relatively well studied. This is partly because of their large size, so they are common in the fossil record, but also because they compete for space and resources with people, and have been and continue to be targets for direct exploitation, including entertainment, sport hunting, and to provide fur and other tissues for human use. In recent years genetics have provided a better understanding of relationships among bear species, which are complicated owing to ancient hybridizations, and with other carnivorans, although some uncertainties remain. Genetics have also improved our knowledge of variation within species, some of which, such as the brown bear, have very extensive geographical ranges, so that species' histories and colonizations can be resconstructed in deeper time by comparison with the fossil record.

Genetics are vital for the future successful conservation and management of bear populations, especially with the advent of new techniques in recent years. Whereas the population genetics of the European brown and grizzly bears, as well as the American black bear to some extent, are quite well studied in terms of population genetics, there are few genetic assessments for other bear species, e.g. sun bears, sloth bears, and Asian black bears. Increased human expansion and a changing climate will cause further fragmentation, stress populations, and endanger bear species across the world, which will challenge current conservation, management, and research. Therefore, the genetic assessment of bear populations and their status and connectivity will increase in importance. Attempts to rewild certain areas may include the release of bears, which should incorporate genetic monitoring programs. Bears will continue to play an important role in genetic research because they (especially brown bears) are ecologically among the most intensively studied species, which enables comprehensive studies on the interplay of wildlife ecology and genomics. Specifically, we expect future genetic studies to investigate further adaptation and natural selection of bears, inbreeding, and evaluate circumstances for which effective population size might be appropriate for tracking abundance of bears. As new tools that use genetic information for conservation questions emerge, and sampling, genotyping, and sequencing techniques improve and become cheaper, genetics will become even more essential for bear research, conservation, and management.

References

Abella, J., Alba, D. M., Robles, J. M., et al. (2012). *Kretzoiarctos* gen. nov., the oldest member of the giant panda clade. *PLoS ONE* 7: e48985.

Abrams, J. F., Hoerig, L., Brozovic, R., et al. (2019). Shifting up a gear with iDNA: from mammal detection events to standardized surveys. *Journal of Applied Ecology* 56: 1637–1648. https://doi.org/10.1111/1365-2664.13411

Ameghino, F. (1902). L'age des formations sédimentaires de Patagonie. *Anales de la Sociedad Científica Argentina* 54: 220–249.

Ameghino, F. (1903). *L'age des formations sédimentaires de Patagonie.* Buenos Aires: Imprimerie Coni Frères.

Anijalg, P., Ho, S. Y. W., Davison, J., et al. (2018). Large-scale migrations of brown bears in Eurasia and to North America during the Late Pleistocene. *Journal of Biogeography* 45: 394–405. https://doi.org/10.1111/jbi.13126

Ashrafzadeh, M. R., Kaboli, M. & Naghavi, M. R. (2016). Mitochondrial DNA analysis of Iranian brown bears (*Ursus arctos*) reveals new phylogeographic lineage. *Mammalian Biology-Zeitschrift für Säugetierkunde* 81(1): 1–9.

Bapteste, E., van Iersel, L., Janke, A., et al. (2013). Networks: expanding evolutionary thinking. *Trends in Genetics* 29: 439–441.

Barlow, A., Cahill, J. A., Hartmann, S., et al. (2018). Partial genomic survival of cave bears in living brown bears. *Nature Ecology & Evolution* 2: 1563–1570.

Barlow, A., Sheng, G. L., Lai, X. L., Hofreiter, M. & Paijmans, J. L. (2019). Once lost, twice found: combined analysis of ancient giant panda sequences characterises extinct clade. *Journal of Biogeography* 46: 251–253.

Barnes, G. L. (1999). *The rise of civilization in East Asia: The archaeology of China, Korea and Japan.* London: Thames & Hudson.

Barnes, I., Matheus, P., Shapiro, B., Jensen, D. & Cooper, A. (2002): Dynamics of Pleistocene population extinctions in Beringian brown bears. *Science* 295: 2267–2270.

Batsch, A. J. G. C. (1788). *Versuch einer Anleitung, zur Kenntniß und Geschichte der Thiere und Mineralien, für akademische Vorlesungen entworfen, und mit den nöthigsten Abbildungen versehen.* Jena: Akademische Buchhandlung.

Benazzo, A., Trucchi, E., Cahill, J. A., et al. (2017). Survival and divergence in a small group: the extraordinary genomic history of the endangered Apennine brown bear stragglers. *Proceedings of the National Academy of Sciences* 114: E9589–E9597.

Berta, A., Sumich, J. L. & Kovacs, K. M. (2015). *Marine mammals: Evolutionary biology.* London: Academic Press.

Bidon T., Janke A., Fain S. R., et al. (2014). Brown and polar bear Y chromosomes reveal extensive male-biased gene flow within brother lineages. *Molecular Biology and Evolution* 31: 1353–1363. https://doi.org/10.1093/molbev/msu109

Bischof, R. & Swenson, J. E. (2012). Linking noninvasive genetic sampling and traditional monitoring to aid management of a trans-border carnivore population. *Ecological Applications* 22(1): 361–373.

Bischof, R., Brøseth, H. & Gimenez, O. (2016). Wildlife in a politically divided world: insularism inflates estimates of brown bear abundance. *Conservation Letters* 9: 122–130.

de Bonis, L. (2013). Ursidae (Mammalia, Carnivora) from the Late Oligocene of the “Phosphorites du Quercy” (France) and a reappraisal of the genus *Cephalogale* Geoffroy, 1862. *Geodiversitas* 35: 787–814.

de Bonis, L., Abella, J., Merceron, G. & Begun, D. R. (2017). A new late Miocene ailuropodine (giant panda) from Rudabánya (north-central Hungary). *Geobios* 50: 413–421.

Bowdich, T. E. (1821). *An analysis of the natural classifications of Mammalia, for the use of students and travellers.* Paris: J. Smith.

Bryant, H. N. (1996). Explicitness, stability, and universality in the phylogenetic definition and usage of taxon names: a case study of the phylogenetic taxonomy of the Carnivora (Mammalia). *Systematic Biology* 45(2): 174–189.

Cahill, J. A., Green, R. E., Fulton, T. L., et al. (2013). Genomic evidence for island population conversion resolves conflicting theories of polar bear evolution. *PLoS Genetics* 9(3): e1003345.

Cahill, J. A., Stirling, I., Kistler, L., et al. (2015). Genomic evidence of geographically widespread effect of gene flow from polar bears into brown bears. *Molecular Ecology* 24: 1205–1217.

Cahill, J. A., Heintzman, P. D., Harris, K., et al. (2018). Genomic evidence of widespread admixture from polar bears into brown bears during the Last Ice Age. *Molecular Biology and Evolution* 35: 1120–1129.

Campagna L., Van Coeverden de Groot, P. J., Saunders, B. L., et al. (2013). Extensive sampling of polar bears (*Ursus maritimus*) in the Northwest Passage (Canadian Arctic Archipelago) reveals population differentiation across multiple spatial and temporal scales. *Ecology and Evolution* 3: 3152–3165.

Chen, Y.-Y., Zhu, Y., Wan, Q.-H., et al. (2013). Patterns of adaptive and neutral diversity identify the Xiaoxiangling Mountains as a refuge for the giant panda. *PLoS ONE* 8(7): e70229.

Chorn, J. & Hoffmann, R. S. (1978). *Ailuropoda melanoleuca. Mammalian Species* 110: 1–6.

Çilingir, F. G., Akın Pekşen, Ç., Ambarlı, H., Beerli, P. & Bilgin, C. C. (2016). Exceptional maternal lineage diversity in brown bears (*Ursus arctos*) from Turkey. *Zoological Journal of the Linnean Society* 176: 463–477.

Colangelo, P., Loy, A., Huber, D., et al. (2012). Cranial distinctiveness in the Apennine brown bear: genetic drift effect or ecophenotypic adaptation? *Biological Journal of the Linnean Society* 107: 15–26.

Corbet, G. B. & Hill, J. E. (1992). *The mammals of the Indomalayan region: A systematic review.* London: Natural History Museum Publications and Oxford: Oxford University Press.

Crompton, A. E., Obbard, M. E., Petersen, S. D. & Wilson, P. J. (2008). Population genetic structure in polar bears (*Ursus maritimus*) from Hudson Bay, Canada: implications of future climate change. *Biological Conservation* 141: 2528–2539.

Cronin, M. A., Amstrup, S. C., Garner, G. W. & Vyse, E. R. (1991). Interspecific and intraspecific mitochondrial DNA variation in North American bears (*Ursus*). *Canadian Journal of Zoology* 69: 2985–2992.

Cuvier, F. (1825). Ours des Cordilères du Chili. In: Geoffroy-Saint-Hilaire, É. & Cuvier, F. (Eds.) *Histoire naturelle des mammifères, avec des figures originales, coloriées, dessinées d'après des animaux vivans*, pt. 3, vol. 5(5), 2 pp. Paris: A. Belin.

Cuvier, G. (1823). Des ossemens d'ours. In *Recherches sur les ossemens fossiles, où l'on rétablit les caractères de plusieurs animaux dont les révolutions du globe ont détruit les espèces*, vol. 4(3), pp. 311–380. Paris: G. Dufour et E. d'Ocagne.

David, A. (1869). Voyage en Chine, troisième partie. *Nouvelles archives du Muséum d'histoire naturelle de Paris* 5: 3–13.

Davison, J., Ho, S. Y., Bray, S. C., et al. (2011). Late-Quaternary biogeographic scenarios for the brown bear (*Ursus arctos*), a wild mammal model species. *Quaternary Science Reviews* 30: 418–430.

De Barba, M., Miquel, C., Lobréaux, S., et al. (2017). High-throughput microsatellite genotyping in ecology: Improved accuracy, efficiency, standardization and success with low-quantity and degraded DNA. *Molecular Ecology Resources* 17: 492–507.

Doronina, L., Churakov, G., Shi, J., et al. (2015). Exploring massive incomplete lineage sorting in arctoids (Laurasiatheria, Carnivora). *Molecular Biology and Evolution* 32: 3194–3204.

Draheim, H. M., Moore, J. A., Fortin, M. J. & Scribner, K. T. (2018). Beyond the snapshot: landscape genetic analysis of time series data reveal responses of American black bears to landscape change. *Evolutionary Applications* 11: 1219–1230.

Dunbar, M. R., Cunningham, M. W., Wooding, J. B. & Roth, R. P. (1996). Cryptorchidism and delayed testicular descent in Florida black bears. *Journal of Wildlife Diseases* 32: 661–664.

Durner, G. M., Laidre, K. L. & York, G. S. (Eds.) (2018). *Polar bears: Proceedings of the 18th Working Meeting of the IUCN/SSC Polar Bear Specialist Group, 7–11 June 2016, Anchorage, Alaska.* Gland, Switzerland and Cambridge, UK: IUCN.

Dutta, T., Sharma, S., Maldonado, J. E., Panwar, H. S. & Seidensticker, J. (2015). Genetic variation, structure, and gene flow in a sloth bear (*Melursus ursinus*) meta-population in the Satpura-Maikal landscape of Central India. *PLoS ONE* 10(5): e0123384.

Edwards C. J., Suchard, M. A, Lemey, P., et al. (2011). Ancient hybridization and an Irish origin for the modern polar bear matriline. *Current Biology* 21(15): 1–8.

Eizirik, E., Murphy, W. J., Koepfli, K. P., et al. (2010). Pattern and timing of diversification of the mammalian order Carnivora inferred from multiple nuclear gene sequences. *Molecular Phylogenetics and Evolution* 56: 49–63.

Ellegren, H., Primmer, C. R. & Sheldon, B. C. (1995). Microsatellite evolution: directionality or bias? *Nature Genetics* 11: 360–362.

Ellerman, J. R. & Morrison-Scott, T. C. S. (1951). *Checklist of Palaearctic and Indian mammals 1758–1946*. London: Trustees of the British Museum.

Erdbrink, D. P. (1953): *A review of fossil and recent bears of the Old World with remarks on their phylogeny based on their dentition*, 2 vols. Deventer, the Netherlands: Jan de Lange.

Ersmark, E., Baryshnikov, G., Higham, T., et al. (2019). Genetic turnovers and northern survival during the last glacial maximum in European brown bears. *Ecology and Evolution* 9: 5981–5905. https://doi.org/10.1002/ece3.5172

Flower, W. H. (1869). On the value of the characters of the base of the cranium in the classification of the order Carnivora, and on the systematic position of *Bassaris* and other disputed forms. *Proceedings of the Zoological Society of London* 37(1): 4–37.

Frosch, C., Dutsov, A., Zlatanova, D., et al. (2014). Noninvasive genetic assessment of brown bear population structure in Bulgarian mountain regions. *Mammalian Biology* 79: 268–276.

Galbreath, G. J., Hunt, M., Clements, T. & Waits L. P. (2008). An apparent hybrid wild bear from Cambodia. *Ursus* 19: 85–86.

García-Rangel, S. (2012). Andean bear *Tremarctos ornatus* natural history and conservation. *Mammal Review* 42: 85–119.

García-Vázquez, A., Pinto Llona, A. C. & Grandal-d'Anglade, A. (2017). Post-glacial colonization of Western Europe brown bears from a cryptic Atlantic refugium out of the Iberian Peninsula. *Historical Biology* XX: 1–13.

Gervais, P. (1855). *Histoire naturelle des mammifères avec l'indication de leurs moeurs et de leurs rapports avec les arts, le commerce et l'agriculture*. Paris: L. Curmer.

Ginsburg, L. & Morales, J. (1998). Les Hemicyoninae (Ursidae, Carnivora, Mammalia) et les formes apparentées du Miocène inférieur et moyen d'Europe occidentale. *Annales de Paléontologie* 84: 71–123.

Goswami, A. (2010). Introduction to Carnivora. In Goswami, A. & Friscia, A. (Eds.), *Carnivoran evolution. New views on phylogeny, form and function*, pp. 1–24. Cambridge: Cambridge University Press.

Graves, T., Chandler, R. B., Royle, J. A., Beier, P. & Kendall, K. C. (2014). Estimating landscape resistance to dispersal. *Landscape Ecology* 29: 1201–1211.

Greve, C. (1894). Die geographische Verbreitung der jetzt lebenden Raubthiere. *Kaiserlich Leopoldinisch-Carolinische Deutsche Akademie der Naturforscher. Nova Acta* 63(1): 7–280.

Hailer, F., Kutschera, V. E., Hallström, B. M., et al. (2012). Nuclear genomic sequences reveal that polar bears are an old and distinct bear lineage. *Science* 336(6079): 344–347.

Hall, E. R. (1981): *The mammals of North America*, Vol. II (2nd edn.). New York, NY: John Wiley.

Hassanin, A. (2015). The role of Pleistocene glaciations in shaping the evolution of polar and brown bears. Evidence from a critical review of mitochondrial and nuclear genome analyses. *Comptes rendus biologies* 338: 494–501.

Hendey, Q. B. (1980). *Agriotherium* (Mammalia, Ursidae) from Langebaanweg, South Africa, and relationships of the genus. *Annals of the South African Museum* 81: 1–109.

Hirata, D., Mano, T., Abramov, A. V., et al. (2017). Paternal phylogeographic structure of the brown bear (*Ursus arctos*) in northeastern Asia and the effect of male-mediated gene flow to insular populations. *Zoological Letters* 3 (1): 21.

Hofreiter, M., Serre, D., Rohland, N., et al. (2004). Lack of phylogeography in European mammals before the last glaciation. *Proceedings of the National Academy of Sciences* 101: 12,963–12,968.

Hu, Y., Qi, D., Wang, H. & Wei, F. (2010a). Genetic evidence of recent population contraction in the southernmost population of giant pandas. *Genetica* 138: 1297–1306.

Hu, Y., Zhan, X., Qi, D. & Wei, F. (2010b). Spatial genetic structure and dispersal of giant pandas on a mountain-range scale. *Conservation Genetics* 11: 2145–2155.

Hunt Jr., R. M. (1974). The auditory bulla in Carnivora: an anatomical basis for reappraisal of carnivore evolution. *Journal of Morphology* 143(1): 21–75.

Hunt Jr., R. M. (1977). Basicranial anatomy of *Cynelos* Jourdan (Mammalia: Carnivora), an Aquitanian amphicyonid from the Allier Basin, France. *Journal of Paleontology* 51(4): 826–843.

Hunt Jr., R. M. (1998). Ursidae. In Janis, C. M., Scott, K. M. & Jacobs, L. L. (Eds.), *Evolution of Tertiary mammals of North America*, pp. 174–189. Cambridge: Cambridge University Press.

Hwang, D.-S., Ki, J.-S., Jeong, D.-H., et al. (2008). A comprehensive analysis of three Asiatic black bear mitochondrial genomes (subspecies *ussuricus*, *formosanus* and *mupinensis*), with emphasis on the complete mtDNA sequence of *Ursus thibetanus ussuricus* (Ursidae). *Mitochondrial DNA* 19: 418–429.

Kadariya, R., Shimozuru, M., Maldonado, J. E., Moustafa, M. A. M., Sashika, M. & Tsubota, T. (2018). High genetic diversity and distinct ancient lineage of Asiatic black bears revealed by non-invasive surveys in the Annapurna Conservation Area, Nepal. *PLoS ONE*, 13(12): e0207662.

Kamath, P. L., Haroldson, M. A., Luikart, G., et al. (2015). Multiple estimates of effective population size for monitoring a long-lived vertebrate: an application to Yellowstone grizzly bears. *Molecular Ecology* 24: 5507–5521.

Kendall, K. C., Stetz, J. B., Boulanger, J., et al. (2009). Demography and genetic structure of a recovering grizzly bear population. *The Journal of Wildlife Management* 73(1): 3–16.

Kim, Y. K., Hong, Y. J., Min, M. S., et al. (2011). Genetic status of Asiatic black bear (*Ursus thibetanus*) reintroduced into South Korea based on mitochondrial DNA and microsatellite loci analysis. *Journal of Heredity* 102: 165–174.

Kitchener, A. C., Breitenmoser-Würsten, Ch., Eizirik, E., et al. (2017). A revised taxonomy of the Felidae. The final report of the Cat Classification Task Force of the IUCN/SSC Cat Specialist Group. *Cat News Special Issue* 11: 1–80.

Koepfli, K.-P., Dragoo, J. W. & Wang, X. (2017). The evolutionary history and molecular systematics of the Musteloidea. In: Macdonald, D. W., Newman, C. & Harrington, L. A. (Eds.), *Biology and conservation of the musteloids*, pp. 92–128. Oxford: Oxford University Press.

Kopatz, A., Eiken, H. G., Schregel, J., et al. (2017). Genetic substructure and admixture as important factors in linkage disequilibrium-based estimation of effective number of breeders in recovering wildlife populations. *Ecology and Evolution* 7: 10,721–10,732.

Koretsky, I. A., Barnes, L. G. & Rahmat, S. J. (2016). Re-evaluation of morphological characters questions current views of pinniped origins. *Vestnik Zoologii* 50: 327–354.

Krause, J., Unger, T., Nocon, A., et al. (2008). Mitochondrial genomes reveal an explosive radiation of extinct and extant bears near the Miocene–Pliocene boundary. *BMC Evolutionary Biology* 8(1): 220.

Kretzoi, M. (1943). *Kochitis centennii* n. g. n. sp. az egeresi felső oligocénből [in Hungarian]/*Kochitis centenii* n.g. n.sp., ein altertümlicher Creodonte aus dem Oberoligozän Siebenbürgens [in German]. *Földtani Közlöny* **73**: 10–17 [Hungarian] / 190–195 [German].

Kumar, V., Lammers, F., Bidon, T., et al. (2017). The evolutionary history of bears is characterized by gene flow across species. *Scientific Reports* 7: 46487.

Kurtén, B. & Anderson, E. (1980). *Pleistocene mammals of North America.* New York, NY: Columbia University Press.

Kutschera, V. E., Bidon, T., Hailer, F., et al. (2014). Bears in a forest of gene trees: phylogenetic inference is complicated by incomplete lineage sorting and gene flow. *Molecular Biology and Evolution* 31: 2004–2017.

Kutschera, V. E., Frosch, C., Janke, A., et al. (2016). High genetic variability of vagrant polar bears illustrates importance of population connectivity in fragmented sea ice habitats. *Animal Conservation* 19: 337–349.

Laikre, L., Andrén, R., Larsson, H. O. & Ryman, N. (1996). Inbreeding depression in brown bear *Ursus arctos. Biological Conservation* 76: 69–72.

Lan, T., Gill, S., Bellemain, E., et al. (2017). Evolutionary history of enigmatic bears in the Tibetan Plateau–Himalaya region and the identity of the yeti. *Proceedings of the Royal Society B: Biological Sciences* 284 (1868): 20171804.

Linnaeus, C. V. (1758). *Systema Naturae per regna tria naturae. Secundum classes, ordines, genera, species, cum characteribus, differentiis, synonymis, locis*, 10th edition, Vol. 1. Stockholm: Lars Salvius.

Liu, S., Lorenzen, E. D., Fumagalli, M., et al. (2014). Population genomics reveal recent speciation and rapid evolutionary adaptation in polar bears. *Cell* 157: 785–794.

Loy, A., Genov, P., Galfo, M., Jacobone, M. G. & Vigna Taglianti, A. (2008). Cranial morphometrics of the Apennine brown bear (*Ursus arctos marsicanus*) and preliminary notes on the relationships with other southern European populations. *Italian Journal of Zoology* 75: 67–75.

Luan, P. T., Ryder, O. A., Davis, H., Zhang, Y. P. & Yu, L. (2013). Incorporating indels as phylogenetic characters: impact for interfamilial relationships within Arctoidea (Mammalia: Carnivora). *Molecular Phylogenetics and Evolution* 66: 748–756.

Malenfant, R. M., Davis, C. S., Cullingham, C. I. & Coltman, D. W. (2016). Circumpolar genetic structure and recent gene flow of polar bears: a reanalysis. *PLoS ONE* 11(3): e0148967.

Margari, V., Skinner, L. C., Tzedakis, P. C., et al. (2010). The nature of millennial-scale climate variability during the past two glacial periods. *Nature Geoscience* 3: 127–131.

Meijaard, E. (2004): Craniometric differences among Malayan sun bears (*Ursus malayanus*); evolutionary and taxonomic implications. *The Raffles Bulletin of Zoology* 52: 665–672.

Merriam J. C. & Stock C. (1925). Relationships and structure of the shortfaced bear, *Arctotherium*, from the Pleistocene of California. *Contributions to Palaeontology, Carnegie Institution of Washington* 347: 1–35.

Meyer, F. A. (1793). *Systematisch-summarische Uebersicht der neuesten zoologischen Entdeckungen in Neuholland und Afrika.* Leipzig: Dykischen Buchhandlung.

Mikle, N., Graves, T. A., Kovach, R., Kendall, K. C. & Macleod, A. C. (2016). Demographic mechanisms underpinning genetic assimilation of remnant groups of a large carnivore. *Proceedings of the Royal Society of London B* 283(1839): 20161467.

Miller, W., Schuster, S. C., Welch, A. J., et al. (2012). Polar and brown bear genomes reveal ancient admixture and demographic footprints of past plimate change. *Proceedings of the National Academy of Sciences* 109: E2382–E2390.

Milne-Edwards, A. (1870). Note sur quelques mammifères du Thibet oriental. *Annales des Sciences Naturelles, Zoologie et Paléontologie*, 5th series, 13, article 10.

Mitchell, K. J., Bray, S. C., Bover, P., et al. (2016). Ancient mitochondrial DNA reveals convergent evolution of giant short-faced bears (Tremarctinae) in North and South America. *Biology Letters* 12(4): 20160062.

Murphy, S. M., Augustine, B. C., Ulrey, W. A., et al. (2017). Consequences of severe habitat fragmentation on density, genetics, and spatial capture-recapture analysis of a small bear population. *PLoS ONE* 12(7): e0181849.

Murtskhvaladze M., Gavashelishvili A. & Tarkhnishvili D. (2010). Geographic and genetic boundaries of brown bear (*Ursus arctos*) population in the Caucasus. *Molecular Ecology* 19: 1829–1841.

Norman, A. J., Stronen, A. V., Fuglstad, G. A., et al. (2017). Landscape relatedness: detecting contemporary fine-scale spatial structure in wild populations. *Landscape Ecology* 32: 181–194.

Nyakatura, K. & Bininda-Emonds, O. R. (2012). Updating the evolutionary history of Carnivora (Mammalia): a new species-level supertree complete with divergence time estimates. *BMC Biology* 10(1): 12.

Ohnishi, N., Uno, R., Ishibashi, Y., Tamate, H. B. & Oi, T. (2009). The influence of climatic oscillations during the Quaternary Era on the genetic structure of Asian black bears in Japan. *Heredity* 102: 579.

Onuma, M., Suzuki, M. & Ohtaishi, N. (2006). Possible conservation units of the sun bear (*Helarctos malayanus*) in Sarawak based on variation of mtDNA control region. *Japanese Journal of Veterinary Research* 54: 135–139.

Paetkau, D., Shields, G. F. & Strobeck, C. (1998). Gene flow between insular, coastal and interior populations of brown bears in Alaska. *Molecular Ecology* 7: 1283–1292.

Pages, M., Calvignac, S., Klein, C., et al. (2008). Combined analysis of fourteen nuclear genes refines the Ursidae phylogeny. *Molecular Phylogenetics and Evolution* 47: 73–83.

Pallas, P. S. (1780). *Spicilegia zoologica: quibus novae imprimis et obscurae animalium species iconibus, descriptionibus atque commentariis illustrantur*, fascicle 14, pp. 3–24. Berlin: Joachim Paul.

Pamilo, P. & Nei, M. (1988). Relationships between gene trees and species trees. *Molecular Biology and Evolution* 5: 568–583.

Pelletier, A., Obbard, M. E., White, B. N., Doyle, C. & Kyle, C. J. (2011). Small-scale genetic structure of American black bears illustrates potential postglacial recolonization routes. *Journal of Mammalogy* 92: 629–644.

Phipps, C. J. (1774). *A voyage towards the North Pole undertaken by His Majesty's command.* London: J. Nourse.

Pierson, J. C., Graves, T. A., Banks, S. C., Kendall, K. C. & Lindenmayer, D. B. (2018). Relationship between effective and demographic population size in continuously distributed populations. *Evolutionary Applications* 11: 1162–1175.

Puckett, E. E., Etter, P. D., Johnson, E. A. & Eggert, L. S. (2015). Phylogeographic analyses of American black bears (*Ursus americanus*) suggest four glacial refugia and complex patterns of postglacial admixture. *Molecular Biology and Evolution* 32: 2338–2350.

Raffles, T. S. (1821). Descriptive catalogue of a zoological collection, made on account of the Honourable East India Company, in the island of Sumatra and its vicinity, under the direction of Sir Thomas Stamford Raffles, Lieutenant-Governor of Fort Marlborough; with additional notices illustrative of the natural history of those countries. *Transactions of the Linnean Society of London* 13(1): 239–274.

Ren, G. (2000). Decline of the mid- to late Holocene forests in China: climatic change or human impact? *Journal of Quaternary Science* 15: 273–281.

Rose, K. D. (2006). *The beginning of the age of mammals.* Baltimore, MD: Johns Hopkins University Press.

Rounds, R. C. (1987). Distribution and analysis of colourmorphs of the black bear (*Ursus americanus*). *Journal of Biogeography* 14: 521–538.

Ruiz-García, M. (2007). Genética de Poblaciones: Teoría y aplicación a la conservación de mamíferos neotropicales (Oso andino y delfín rosado). *Boletín de la Real Sociedad Española de Historia Natural* 102: 99–126.

Ruiz-García, M. (2013). The genetic demography history and phylogeography of the Andean bear (*Tremarctos ornatus*) by means of microsatellites and mtDNA markers. In: Ruiz-García, M. & Shostell, J.M. (Eds.), *Molecular population genetics, evolutionary biology and conservation of Neotropical carnivores* (pp. 129–158). New York, NY: Nova Science Publishers.

Ruiz-García, M., Orozco-terWengel, P., Payán, E. & Castellanos, A. (2003). Genética de Poblaciones molecular aplicada al estudio de dos grandes carnívoros (*Tremarctos ornatus* – Oso andino, *Panthera onca* – jaguar): lecciones de conservación. *Boletín de la Real Sociedad Española de Historia Natural* 98: 135–158.

Ruiz-García, M., Orozco-terWengel, P., Castellanos, A. & Arias, L. (2005). Microsatellite analysis of the spectacled bear (*Tremarctos ornatus*) across its range distribution. *Genes and Genetics Systems* 80: 57–69.

Ruiz-García, M., Arias, J. Y., Castellanos, A., Kolter, L. & Shostell, J. M. (2020a). Molecular evolution (mitochondrial and nuclear microsatellites markers) in the Andean bear (*Tremarctos ornatus*; Ursidae, Carnivora): How many ESUs are there? In: Ortega, J. & Maldonado J. E. (Eds.), *Mammalian Conservation Genetics* (pp. 165–194). Berlin: Springer-Verlag.

Ruiz-García, M., Arias, J. Y., Restrepo, H., Cáceres-Martínez, C. & Shostell, J. M. (2020b). The genetic structure of the Andean bear (*Tremarctos ornatus*; Ursidae, Carnivora) in Colombia by means of mitochondrial and microsatellite markers. *Journal of Mammalogy* (in press).

Ruiz-García, M., Castellanos, A., Arias, J.Y. & Shostell, J.M. (2020c). Genetics of the Andean bear (*Tremarctos ornatus*; Ursidae, Carnivora) in Ecuador: When the Andean Cordilleras are not an obstacle. *Mitochondrial DNA Part A* 31: 194–212.

Salesa, M. J., Siliceo, G., Antón, M., et al. (2006). Anatomy of the "false thumb" of *Tremarctos ornatus* (Carnivora, Ursidae, Tremarctinae): phylogenetic and functional implications. *Estudios Geológicos* 62: 389–394.

Salomashkina, V. V., Kholodova, M. V., Semenov, U. A., Muradov, A. S. & Malkhasyan, A. (2017). Genetic variability of brown bear (*Ursus arctos* L., 1758). *Russian Journal of Genetics* 53: 108–117.

Schmidt-Kittler, N. (1987). The Carnivora (Fissipedia) from the lower Miocene of East Africa. *Palaeontographica. Abteilung A, Paläozoologie, Stratigraphie* 197: 85–126.

Schwartz, M. K., Luikart, G. & Waples, R. S. (2007). Genetic monitoring as a promising tool for conservation and management. *Trends in Ecology & Evolution* 22: 25–33.

Shaw, G. (1790). *The naturalist's miscellany, or coloured figures of natural objects,* vol. 2. London: Nodder & Co.

Shen, F., Zhang, Z., He, W. E. I., et al. (2009). Microsatellite variability reveals the necessity for genetic input from wild giant pandas (*Ailuropoda melanoleuca*) into the captive population. *Molecular Ecology* 18: 1061–1070.

Skrbinšek, T., Jelenčič, M., Waits, L., et al. (2012). Monitoring the effective population size of a brown bear (*Ursus arctos*) population using new single-sample approaches. *Molecular Ecology* 21: 862–875.

Soibelzon, L. H. & Schubert, B. W. (2011). The largest known bear, *Arctotherium angustidens*, from the Early Pleistocene Pampean Region of Argentina: with a discussion of size and diet trends in bears. *Journal of Paleontology* 85: 69–75.

Soibelzon, L. H., Tonni, E. P. & Bond, M. (2005). The fossil record of South American short-faced bears (Ursidae, Tremarctinae). *Journal of South American Earth Sciences* 20: 105–113.

Talbot, S. L. & Shields, G. F. (1996). A phylogeny of the bears (Ursidae) inferred from complete sequences of three mitochondrial genes. *Molecular Phylogenetics and Evolution* 5: 567–575.

Tedford, R. H. (1976). Relationships of pinnipeds to other carnivores (Mammalia). *Systematic Zoology* 25: 363–374.

Thenius, E. (1989). Molekulare und adaptive Evolution, Kladistik und Stammesgeschichte1: Ergänzungen zu einer Arbeitshypothese. *Journal of Zoological Systematics and Evolutionary Research* 27: 94–105.

Turner, H. N. (1848). Observations relating to some of the foramina at the base of the skull in Mammalia, and on the classification of the order Carnivora. *Proceedings of the Zoological Society of London* 16: 63–88.

Valdiosera, C. E., García, N., Anderung, C., et al. (2007). Staying out in the cold: glacial refugia and mitochondrial DNA phylogeography in ancient European brown bears. *Molecular Ecology* 16: 5140–5148.

Viengkone, M., Derocher, A. E., Richardson, E. S., et al. (2016). Assessing polar bear (*Ursus maritimus*) population structure in the Hudson Bay region using SNPs. *Ecology and Evolution* 6: 8474–8484.

Von Duyke, A., Bellemain, E., Dejean, T., et al. (2017). The future is now; science in a spoonful of snow. Using eDNA to monitor a polar bear (*Ursus maritimus*) population in the Chukchi Sea. *22nd Biennial Conference on the Biology of Marine Mammals.*

Wagner, J. (2010). Pliocene to early Middle Pleistocene ursine bears in Europe: a taxonomic overview. *Journal of the National Museum (Prague), Natural History Series* 179(20): 197–215.

Waits, L., Paetkau, D. & Strobeck, C. (1999). Genetics of the bears of the world. In: Servheen, C., Herrero, S. & Peyton, B. (Eds.) *Bears. Status survey and conservation action plan*, pp. 25–32. Bern and Cambridge: IUCN.

Wan, Q. H., Fang, S. G., Wu, H. & Fujihara, T. (2003). Genetic differentiation and subspecies development of the giant panda as revealed by DNA fingerprinting. *Electrophoresis* 24: 1353–1359.

Wan, Q.-H., Wu, H. & Fang, S.-G. (2005). A new subspecies of giant panda (*Ailuropoda melanoleuca*) from Shaanxi, China. *Journal of Mammalogy* 86: 397–402.

Wheat, R. E., Allen, J. M., Miller, S. D., Wilmers, C. C. & Levi, T. (2016). Environmental DNA from residual saliva for efficient noninvasive genetic monitoring of brown bears (*Ursus arctos*). *PLoS ONE* 11(11): e0165259.

Wilson, D. E. (1976). Cranial variation in polar bears. In Pelton, M. R., Lentfer, J. W. & Folk, G. E. (Eds.) *Bears: Their biology and management. A selection of papers from the third international conference on bears*, pp. 447–453. Morges, Switzerland: IUCN.

Wooding, S. & Ward, R. (1997). Phylogeography and Pleistocene evolution in the North American black bear. *Molecular Biology and Evolution* 14: 1096–1105.

Wozencraft, W. C. (1989). The phylogeny of the recent Carnivora. In: Gittleman, J. L. (Ed.) *Carnivore behaviour, ecology and evolution* (pp. 495–535). London: Chapman and Hall.

Wu, J., Kohno, N., Mano, S., et al. (2015). Phylogeographic and demographic analysis of the Asian black bear (*Ursus thibetanus*) based on mitochondrial DNA. *PLoS ONE* 10(9): e0136398.

Yu, L., Li, Q.-W., Ryder, O. A. & Zhang, Y.-P. (2004). Phylogeny of the bears (Ursidae) based on nuclear and mitochondrial genes. *Molecular Phylogenetics and Evolution* 32: 480–494.

Yu, L., Li, Y. W., Ryder, O. A. & Zhang, Y. P. (2007). Analysis of complete mitochondrial genome sequences increases phylogenetic resolution of bears (Ursidae), a mammalian family that experienced rapid speciation. *BMC Evolutionary Biology* 7(1): 198.

Zedrosser, A., Støen, O. G., Sæbø, S. & Swenson, J. E. (2007). Should I stay or should I go? Natal dispersal in the brown bear. *Animal Behaviour* 74: 369–376.

Zhang, B., Li, M., Zhang, Z., et al. (2007). Genetic viability and population history of the giant panda, putting an end to the "evolutionary dead end"? *Molecular Biology and Evolution* 24: 1801–1810.

Zhang, Y. P. & Ryder, O. A. (1993). Mitochondrial DNA sequence evolution in the Arctoidea. *Proceedings of the National Academy of Sciences of the United States of America* 90: 9557–9561.

Zhao, S., Zheng, P., Dong, S., et al. (2013). Whole-genome sequencing of giant pandas provides insights into demographic history and local adaptation. *Nature Genetics* 45(1): 67.

Zuckerkandl, E. & Pauling L. (1962). Molecular disease, evolution and genetic heterogeneity. In: *Horizons in biochemistry* (pp. 189–225). New York, NY: Academic Press.

Chapter 2

Mating Strategies

Sam M.J.G. Steyaert, Andreas Zedrosser, Ronald R. Swaisgood, Eva Filipczykova, Brian Crudge, Trishna Dutta, Sandeep Sharma, Shyamala Ratnayeke, Shinsuke Koike, Martin Leclerc, Andrew E. Derocher, Melanie Clapham, Thomas Spady, Bruce McLellan, Andrés Ordiz, Alberto Fernández-Gil, Miguel Delibes, and Jon E. Swenson

Introduction

The mating system and mating strategies of a species refer to the behavioral strategies used to obtain reproductive partners and ensure reproductive success (Emlen & Oring 1977). Common determining factors of mating systems and strategies are the manner of mate acquisition, the number of mates obtained by an individual, as well as the absence or presence and duration of parental care (Emlen & Oring 1977). In mammals, the energetic investments in gametes and rearing offspring are typically larger for females than for males (Trivers 1972; Andersson 1994). Mate selection is thus a much more important decision for females than for the rather indiscriminate males (Trivers 1972). This dichotomy results in sexual selection, which in turn is determined by male–male competition for access to females, as well as female mate choice (Andersson 1994). Because receptive females generally are considered as the limiting resource in reproduction, males face intrasexual competition for mates (Trivers 1972). In a multitude of mammalian species, including bears, this has resulted in pronounced sexual size dimorphism and polygamous mating systems (Andersson 1994). Despite common characteristics (e.g. sexual size dimorphism, polygamy), variation in mating systems and strategies occurs among bear populations and species.

Mating Systems of Bears of the World

Giant Panda

The giant panda's mating system is best classified as scramble competition polygyny. Males occupy home ranges that are not mutually exclusive, overlapping with other male and female ranges. Average home range size is ~7 km^2, and male ranges are about 1.5 times larger than female ranges (Connor et al. 2016). Females enter estrus asynchronously during the well-defined mating season from March to May (Schaller et al. 1985; Nie et al. 2012b; Pan 2014). When a female enters estrus, she displays increased movement and scent marking with urine and anogenital gland secretions, which serve to attract multiple males (Schaller et al. 1985). Up to five or six males congregate around the estrous female and compete vigorously for access to the female, and injuries are common. Males appear to gain access to females through a combination of their ability to locate females (scramble competition) and direct physical competition with other males, with the largest, strongest males securing mating access (contest competition). Once relative competitive ability has been established among males through indirect assessment and escalated fighting where necessary, aggression among males subsides, but all males appear to remain with the aggregation until after the female has mated (Nie et al. 2012b). Females typically mate within their home range, although they occasionally move long distances outside of their home range during the mating season (Zhang et al. 2014). When a female enters the fertile period and is accompanied by several males, she typically climbs a tree and observes the competing males on the ground beneath her. Although difficult to determine, these observations may be used in mate selection.

Consistent with this mating strategy that emphasizes male competitive ability, pandas show body size sexual dimorphism, with males weighing ~15% more than females. Given the importance of male competitive ability in determining reproduction outcomes, it is somewhat surprising that size dimorphism is not greater. Larger male body size may be constrained by the energetic requirements of participating in scramble and direct competition in a species that is highly energetically challenged (Nie et al. 2015). Pandas are obligate bamboo specialists, and bamboo comprises >98% of their diet (see also Chapter 6). Although the panda gut contains microbial symbionts that help break down cellulose and extract more nutritional value from bamboo, bamboo remains a very low-energy food source (Zhu et al. 2011). Thus, it is plausible that the panda's mating strategy is constrained energetically. Indeed, wild male pandas do not appear able to maintain energetically expensive elevated androgens during the mating season, as observed in most seasonally breeding mammals, and instead show elevated androgen only when challenged by the presence of an estrous female (Nie et al. 2012a). Females typically mate with only one male during a given mating season (Schaller et al. 1985; Nie et al. 2012b; Pan 2014), male pandas participate in several mating aggregations each season, and the most competitive males may secure copulations with multiple females.

The interbirth interval for females is 1–3 years and the female fertile period lasts 2–3 days, producing a highly skewed operational sex ratio, the driving force behind male–male competition. Although signs of intrasexual selection are easily observed, there is anecdotal evidence of female choice in the wild (Pan 2014). In captivity, studies have shown that both male and female mate choice occurs, and the opportunity to choose one's own mate has substantial consequences for reproductive success in the form of copulation and birth rates (Martin-Wintle et al. 2015; see also Chapter 23). Personality – repeatable individual behavioral differences across time and situations – is one of the factors used in mate choice in pandas (Martin-Wintle et al. 2017).

Reproductive maturity is reached in males and females at approximately 4.5 years of age. First reproduction for females typically occurs at age 6, and females can produce 7–8 cubs during their lifetime. The age of first reproduction in males is not known, but is presumed to be 6–7 years, when they gain sufficient size and competitive ability. The physiological parameters of the female reproductive cycle have been well documented, including temporal patterns of estrogens, progestagens, corticoids, and other hormones that characterize the estrous cycle, ovulation, pregnancy, and birth (Chapter 6). Pandas are spontaneous ovulators and, like other ursids, experience a variable period of embryonic diapause lasting from 50 to 108 days (Spady et al. 2007).

Female litter size is 1–2 cubs (rarely 3), with twins occurring about half of the time (Zhu et al. 2001). Pandas also appear to engage in obligate litter reduction, abandoning one cub when there are twins, although there are rare reported cases of females rearing twins in the wild and captivity (Schaller et al. 1985). Pandas give birth to highly vulnerable offspring weighing about 1/900 the weight of the mother, making them the most altricial of all eutherian mammals (Gittleman 1994). This relatively small litter size and birth weight compared to other ursids is believed to be the consequence of energy limitation stemming from the panda's low-energy bamboo diet.

Giant panda females gained an unfavorable reputation as poor mothers from early challenges in captive breeding programs, but it now seems clear that panda maternal care is highly specialized and competent (Zhu et al. 2001; Pan 2014). Following birth, the female remains with the cub for two or more weeks before leaving briefly to drink, urinate, and defecate. Unlike many bear species, panda mothers actively hold their single cub on the ventrum, and remain attentive to infant vocalizations and postural adjustments. As the cub ages and gains independence, the mother begins to leave the cub behind for several hours while she forages. Older cubs typically spend this time safely sequestered in a tree. Cubs remain dependent on the mother for 1.5 years, occasionally 2.5 years. Unlike most other mammals, pandas exhibit female-biased dispersal, with subadult females dispersing out of their natal area (Zhang et al. 2014).

Andean Bear

Andean bears are sexually size dimorphic (Garcia-Rangel 2012). Adult males measure 1.5–2 m and weigh 100–175 kg (Peyton 1980; Garshelis 2009; see Chapter 7), and females reach about 2/3 of the adult male size (Garshelis 2009; Hunter & Barrett 2019). Females typically reach sexual maturity between 4 and 7 years old, and males usually mature when they are about 5 years old (Rosenthal 1999; Garshelis 2009; Garcia-Rangel 2012).

Reproductive seasonality in Andean bears has mainly been studied in captivity (Spady et al. 2007). Females are polyestrous and reproductive cycles take at least 9 months, albeit three cycles sometimes follow each other within 24 months (Rosenthal 1999; Enciso & Guimarães 2013). The further away from the Equator, the stronger Andean bears exhibit seasonality in the timing of the breeding season (Appleton et al. 2018). Generally, mating in wild Andean bears occurs throughout the year, but peaks during the fruiting season (March–October; Garcia-Rangel 2012). However, Appleton et al. (2018) showed that Andean bears in the dry forests of Peru regularly mate in December and January. In zoos, mating occurs from February to September, depending on the latitude and photoperiod (Spady et al. 2007). Appleton et al. (2018) suggested that Andean bears display phenotypic plasticity in the timing of reproduction, possibly related to climate and food availability. The breeding season of Andean bears is relatively short. Appleton et al. (2018) reported that the breeding season in the wild generally lasts for about 6 weeks. Bears in captivity typically mate during a period of one to two weeks (Garcia-Rangel 2012). Captive females are in estrus for 3–14 days, which can be recognized by a swollen vulva. On average 2–8 copulations occur per day, each taking about 10–45 minutes. During courtship, couples play nonaggressive games accompanied by various rumbles. Andean bears communicate their reproductive and dominance status in the wild through scent marking (Filipczyková et al. 2017; Kleiner et al. 2018), although more evidence is needed about the functional significance of their marking behavior.

Female Andean bears have delayed implantation. Gestation time is short and difficult to estimate, due to embryonic diapause (Garcia-Rangel 2012). In captivity, gestation takes from 160 to 210 days, but sometimes as long as 250 days (Rosenthal & Haggerty 1989). It appears that parturition, as with mating, occurs seasonally (Garcia-Rangel 2012; Appleton et al. 2018), with the majority of captive births occurring during autumn or winter. In the tropics, most births occur during the autumn, whereas most births outside the tropics occur during winter. Seasonality in births of wild Andean bears has been recorded in Peru (August–October) and in Ecuador (April), and might be related to the climate and plant phenology (Appleton et al. 2018; Castellanos et al. 2018). Females usually give birth for the first time when they are about 5 years old. As for other ursids and mammals in general, litter size depends on the body condition of the female and on food availability (Peyton 1980). The litter size ranges from one to four, but two cubs is most common (Rosenthal & Haggerty 1989). Birth occurs in a den and newborn cubs weigh approximately 275–380 g (Garcia-Rangel 2012). The mother and her dependent cubs stay in or in the immediate vicinity of the maternal den for

Figure 2.1 A female Andean bear with her cub, high up in a tree in Cali Zoo, Colombia. The visible nipples of the mother suggest that she is still nursing her cub, or had recently stopped nursing at that time of photographing. (Photo by E. Filipczykova.)

3–4 months (Castellanos et al. 2018). It has been reported that after this time, wild females with cubs change dens several times, possibly because of food availability and safety reasons. Mothers lactate for up to approximately 12 months, possibly even up to 14 months (Garcia-Rangel 2012; Castellanos et al. 2018; Figure 2.1; see Chapter 7). Cubs stay with their mothers on average for about two years, and females communicate with their offspring through a series of sounds. After weaning, captive females reportedly enter estrus after about 4–6 weeks. Annual survival is low (59–64%) for newborn captive Andean bear cubs (Garcia-Rangel 2012).

Andean bears can coexist in multigenerational groups. In Northern Ecuador, for example, Castellanos and Vasquez (2017) reported that a single subadult female from a previous litter joined her mother, which had cubs-of-the-year. The subadult female interacted with both her mother and the cubs, and accompanied the cubs when the mother left to search for food. The subadult female stayed with the group until the mother separated from them. Later, the subadult female stayed within her mother's home range and produced offspring of her own. Such observations suggest that Andean bear populations may have complex and kin-related social–spatial structures.

Sun Bear

Sun bears are unique among ursids in several aspects of reproduction. Little is known about the sun bear's reproductive behavior and social organization in the wild, but some insight on reproductive biology has been gained, predominantly from studies involving captive bears.

There is limited sexual dimorphism in the body size of sun bears. In Borneo, male and female body mass averages about 40 and 25 kg, respectively. Healthy adults in captivity at a rescue facility in Cambodia (Free the Bears, https://freethebears.org/) typically weigh about 70 kg for males and 60 kg for females. The weights of these bears revealed that adult females varied little seasonally, whereas males exhibited a pronounced fluctuation, with a decline (*ca.* 12%) during the hot dry season (March–May; Crudge et al. 2019).

Sun bears do not have a distinct breeding season (Schwarzenberger et al. 2004; Spady et al. 2007; Frederick et al. 2012; see also Chapter 8). Although they are capable of breeding throughout the year, more cubs are born during the wet season compared to the dry season in some parts of their range (Schwarzenberger et al. 2004). In a captive population, births occur during all months of the year, but least frequently during spring and most frequently during autumn and winter (Frederick et al. 2012). The absence of clear reproductive seasonality may be an adaptation for living in habitats with unpredictable food resources, like the populations that inhabit aseasonal evergreen rainforests in Peninsular Malaysia, Sumatra, and Borneo. These populations experience synchronized fruit masting events, followed by intermast periods with little fruiting. These large-scale fluctuations occur at intervals of 2–10 years. During extended periods of low fruit availability, loss of body mass and starvation occurs (Wong et al. 2005; Fredriksson & Wich 2006; Fredriksson et al. 2007). The unpredictability of fruit availability in such habitats may favor opportunistic, rather than seasonal, reproductive strategies (Frederick et al. 2012).

With the exception of family groups, sun bears are solitary and associate only for mating (Ngoprasert et al. 2012). In captivity, sun bears have been shown to be spontaneous ovulators, but the presence of males had an influence on hormone metabolite concentrations and cytological profiles in females. The presence of males also coincided with an increased likelihood that females would cycle (Frederick et al. 2013). In a captive population, the age range for females giving birth was 4–27.8 years (mean = 10.5) and the reproductive age range for males was 4.2–26.6 years (mean = 11.0) (Frederick et al. 2013). Nonpregnant sun bears are polyestrous throughout the year, with up to three estrous cycles per year (Frederick et al. 2010). Like other bear species, sun bears routinely exhibit pseudopregnancy (Schwarzenberger et al. 2004; Frederick et al. 2010). Sun bears are capable of initiating estrus following neonatal mortality (Frederick et al. 2012). Gestation in sun bears is approximately 95–107 days, which is less variable than the other ursids (Schwarzenberger et al. 2004). Studies suggest that sun bears do not exhibit delayed implantation (see "Reproductive physiology"; Schwarzenberger et al. 2004; Frederick et al. 2012). Litters typically consist of one cub, or occasionally twins. In the wild, cubs will likely remain with their mother until fully grown at about 2 years of age (Schwarzenberger et al. 2004).

Sloth Bear

Sloth bears exhibit size dimorphism with adult males weighing 80–150 kg and females weighing 60–100 kg (Prater 1965; Garshelis et al. 1999; Yoganand 2005). Home ranges of adult and subadult sloth bears overlap within and between sexes (Joshi et al. 1999; Yoganand 2005; Ratnayeke et al. 2007), in accordance with their polygamous mating system (see also Chapter 9).

Some of the best evidence of mating behavior in the wild comes from Chitwan National Park, Nepal, where behavioral observations of sloth bears were made from the backs of elephants (Laurie & Seidensticker 1977; Joshi et al. 1999). Males may make long excursions out of home range boundaries (>8 km) to mate. Females in estrus are accompanied by several males and may mate multiple times with certain males. Although dominance relationships among males may determine which male has priority of access to a receptive female, fights and loud vocalizations are a common feature of mating behavior in sloth bears (Laurie & Seidensticker 1977; Phillips 1984; Joshi et al. 1999). Joshi et al. (1999) report some evidence of female mate choice in such mating aggregations, including that of a female moving 2 km out of her home range to mate with a large male.

Females may breed first when 4 years old, but do not necessarily produce cubs following their first breeding season. Pregnant females use dry caves among boulders and rock outcrops (Phillips 1984), or create dens by excavating holes in the banks of dry streambeds (Joshi et al. 1999). Females remain in dens for almost 8 weeks, emerging periodically to feed in the ninth to tenth week, and bring cubs out of the den shortly after. A litter size of two cubs is most common (Laurie & Seidensticker 1977; Phillips 1984; Joshi 1996). Although individual cubs have been observed, it is suspected that these are a result of early mortality of the second cub (Joshi 1996). Cubs stay with their mothers for 1.5–2.5 years, becoming independent just before the breeding season (Joshi et al. 1999). Thus, females breed at either 2- or 3-year intervals. The overall survival rate of cubs up to 2 years in Nepal was 69% (Joshi et al. 1999).

Sloth bears exhibit mating seasonality. The mating season in sloth bears from mainland India spans May–July, and females are reported to give birth from November to January (Laurie & Seidensticker 1977; Gopal 1991; Joshi et al. 1999; Spady et al. 2007), usually after a period of delayed implantation (Puschmann et al. 1977). In Central India, young cubs have been observed on the back of females during February and March (Figure 2.2), which suggests that birth occurs in winter (Sharma and Dutta, personal observation). Phillips (1984) indicated that there was no conspicuous breeding season in Sri Lanka, although Ratnayeke (unpublished data from remote cameras and hunter surveys) concluded that most births in Sri Lanka matched the pattern reported for bears on the Indian mainland.

An unusual characteristic of sloth bears is that females habitually carry the cubs (Norris 1969; Laurie & Seidensticker 1977; Joshi et al. 1999). Cubs ride on their mothers' backs almost fulltime for the first 3 months (Figure 2.2), descending periodically to play and feed thereafter, but continue to be carried at least part of the time until they are 6–9 months old (Joshi et al. 1999). Cubs ride cross-wise or heads-forward, usually on the rump and lower back of the female. This behavior, which is shared by other myrmecophagous mammals, may have evolved independently in sloth bears to guard against other large sympatric predators, such as tigers and leopards (Joshi et al. 1999). Tigers and leopards kill sloth bears occasionally (Kurt & Jayasuriya 1968; Joshi et al. 1999), and dholes and jackals could be a threat to cubs (Yoganand et al. 2013). Cubs can climb trees to feed on honey and fruits, but have never been observed to use trees to escape from predators (see also Chapter 9).

Figure 2.2 A sloth bear cub riding on its mother's back in Satpura National Park, India. (Photo by J. E. Swenson.)

In Chitwan National Park, Nepal, adult males and lone females are nocturnal, whereas females with cubs and subadults of both sexes are mostly diurnal, a behavior proposed to safeguard young from nocturnal predators and infanticidal conspecifics (Joshi et al. 1996). However, Yoganand (2005) found that activity patterns for females with dependent cubs did not differ from those of adult males and solitary adult females in the dry deciduous forest of Panna National Park, probably due to lower sloth bear density and the lack of concentrated rich sources of food.

Asiatic Black Bear

The Asiatic black bear is a size-dimorphic and polygamous species (Yamamoto et al. 1998), with adult body mass typically ranging between 100 and 200 kg for males and 50 and 125 kg for females. Growth patterns differ by sex, with males growing until about 6–8 years of age and females until about 3–5 years (Nakamura et al. 2011).

Males and females generally reach sexual maturity around 2–3 and 4 years old, respectively (Komatsu et al. 1994; Katayama et al. 1996), but females in nutritionally good condition can reproduce earlier. Although there are no data available, the age of first reproduction in the wild is probably delayed for males, because of strong male–male competition for females. Anecdotal reports suggest that females are reproductively active until their early twenties (Katayama et al. 1996; Nakamura et al. 2011; see also Chapter 23).

In males, spermatogenesis starts during hibernation in March or April, and peaks in May and June to correspond with female receptivity (Komatsu et al. 1994). Although detailed observations of mating behavior in the wild have not been reported, mating behavior and aggregations of adult bears have been observed in June and July. The mating season lasts from mid-June to early August for bears in captivity (Yamamoto et al. 1998). During the mating season, sexually mature males tend to be more active than females (Kozakai et al. 2013), and they search for receptive females and aggregate at locations where females are likely to be found (Oi et al. 2008). Males sometimes commit infanticide, which can induce mating opportunities with the victimized mother (Yamazaki 2017). The frequency of infanticide, its influence on cub mortality, and the survival rate of cubs-of-the-year is unknown (see also Chapter 10). Many males captured during this period carry wounds or injuries, indicating contest competition among males for females and/or sexual conflict.

The length of the estrous period varies widely among females, ranging from 12 to 35 days, and estrous and anestrous periods tend to be repeated in an irregular pattern. Hence, multiple paternity litters can occur (Yamamoto et al. 2012). Females are believed to be induced ovulaters. Male reproductive behavior coincides with estrus. Younger males (4–5 years old) mount females less often that older males, and among adult males (>5 years old), the larger ones mate most frequently (Yamamoto et al. 1998). Females have delayed implantation, which occurs in late November–early December. Cubs are born from late January to early February. Newborn cubs weigh about 300 g. They grow fast, and accumulate around 2–3 kg of body mass over the approximately 90 days between birth and the end of hibernation (i.e. around 20–33 g/day).

Based on observations of corpora lutea, corpora albicantia, and placental scars in killed bears, as well as the number of cubs observed in the wild, the mean number of ovulations per reproductive cycle is estimated to be 1.9, the mean number of implantations 2.0, and mean litter size 1.9 (Katayama et al. 1996). In contrast, the mean litter size in captive animals is reported to be 1.4 (Iibuchi et al. 2009). Although no observations of the interlitter interval in wild individuals currently exist, it is thought to be 2–3 years.

Mothers and cubs-of-the-year remain in the vicinity of their place of hibernation until mid-May, and continue to remain together in the following months. The exact lactation period is unknown; however, lactation has been observed in adult females captured as late as October, meaning that it is possible that cubs are not weaned until autumn. Cubs-of-the-year are thought to overwinter with their mothers during the next winter and become independent during the breeding season of the following year, but the precise timing of the separation is unknown (see also Chapter 10).

American Black Bear

The mating activities of American black bears (*Ursus americanus*) start in May, peak in June, and extend until July (Jonkel & Cowan 1971). Males usually are 10–20% longer and 20–70% heavier than females (Larivière 2001; Kovach & Powell 2003). During a single mating season, males mate with multiple females and females mate with multiple males (Jonkel & Cowan 1971). Therefore, the American black bear is considered a seasonal polygamous breeder (Kovach & Powell 2003; see also Chapter 11).

Male testicular activity starts during the late denning period in February or March. The peak of testicular activity occurs at the onset of the mating season and decreases slowly until October, after which the testicular activity ceases (Garshelis & Hellgren 1994). Males reach sexual maturity when around 3 years old, but they rarely successfully reproduce at that age because of competition with older and heavier males. Females' age at first reproduction varies from 2 to 8 years, but most females produce their first litter at 5 or 6 years of age (Jonkel & Cowan 1971). Females are polyestrous with interestrous intervals of about 10 days, and multiple paternity litters are common (Himelright et al. 2014; Ombrello et al. 2016). Induced ovulation has been suggested (Boone et al. 1998, 2004).

During receptive periods, females are usually followed by multiple males, but larger and heavier males tend to guard females during the peak of their estrus (Kovach & Powell 2003). Smaller and lighter males gain access to females mostly outside of the females' receptive periods, or can obtain mating opportunities when multiple females experience estrus simultaneously (Kovach & Powell 2003; Costello et al. 2009). Consequently, paternity is often skewed toward a few dominant males in an area (Moore et al. 2015). Females have delayed implantation, which occurs in late November or early December (Tsubota et al. 1998). Once implanted, blastocyst(s) develop(s) for 60–70 days and parturition occurs in January or early February (Larivière 2001). Females produce litters of 1–4 cubs that weigh about 400 g each, and older and heavier females usually produce larger litters (Kolenosky 1990; Larivière 2001).

Parental care is provided by the female only, which has lactational anestrus during the first mating season following cub birth (Larivière 2001). Lactation lasts for about 7 months, but cubs stay with their mother their entire first year. Cub survival varies from 25% to 50% and most cubs die from natural causes, such as disease, den flooding, infanticide, or predation by brown bears, wolves, coyotes, and bobcats (LeCount 1987; Kolenosky 1990; Larivière 2001). Family breakup occurs before or during the mating season, when offspring are approximately 16 months old (Jonkel & Cowan 1971). Depending on food availability and female body condition, litter intervals range from 1 to 4 years, but females usually breed every second year, as they can reproduce during the mating season following family breakup (Kolenosky 1990; Larivière 2001).

Brown Bear

The brown bear has a polygamous mating system in the broadest sense of the term, as both sexes can associate and mate with several partners during the mating season. Mating strategies, however, can vary across populations (Steyaert et al. 2012). It is a size-dimorphic species, in which adult males can be over twice the size of adult females. Such dimorphism is characteristic for species with strong sexual selection (Steyaert et al. 2012). Brown bears are seasonal breeders with a mating season that lasts from spring to early summer. The exact timing of the breeding season, however, varies across populations, as it correlates with photoperiod (Spady et al. 2007), and reproductive behavior during the autumn has occasionally been observed (Vaisefeld & Chestin 1993; see also Chapters 12 and 13).

Brown bears, particularly males, have large home ranges that can extend over several 1000 km^2 and overlap the ranges of many other males and females (McLoughlin et al. 2000). During the mating season, both sexes roam-to-mate, but especially males tend to increase their movements for doing so (Dahle & Swenson 2003; Krofel et al. 2010). Reproductive associations are usually comprised of one male and one female, but associations with several males and one female, or vice versa, are also common (Stenhouse et al. 2005). When traveling in associations, males monitor the females' estrous status, which is probably signaled by pheromones (Craighead et al. 1995). Associations can last for just a few hours or up to several weeks. Single males sometimes sequester individual females at remote sites (e.g. mountain ridges) and over prolonged periods of time, which may be a strategy to assure paternity for the sequestering male (Hamer & Herrero 1990; Edwards & Derocher 2015). Multiannual mating areas as small as just few hectares may develop in small or low-density populations (e.g. Cantabrian Mountains, Spain; see also Chapter 19). Such areas can facilitate individuals to find each other and reproduce, and counteract the challenge of finding mates at low population densities (Fernández-Gil et al. 2006; Fernández-Gil 2013). In populations where bears aggregate around clustered food resources during the mating season (e.g. garbage dumps), social hierarchies can develop, where the most dominant male can monopolize reproductive females and prevent subdominant males from mating (Craighead et al. 1995).

Strong competition for females exists among males. Prior to copulation, both scramble and contest competition occur, in which the latter can inflict severe injuries or cause death, especially in younger or smaller individuals (Craighead et al. 1995). Body size and condition, age, and aggression typically determine the dominance status of males (Zedrosser et al. 2007). Competition after copulation may occur as sperm competition within the reproductive tract of the female brown bears, although no conclusive evidence for this exists (Bellemain et al. 2006b). Competition can even extend after females give birth, as sexually selected infanticide (SSI) (Bellemain et al. 2006a; Davoli et al. 2018), a reproductive strategy in which males kill dependent conspecific offspring for obtaining mating opportunities (Hrdy 1979). Females control partner choice and mating to some extent and sometimes initiate mating. In populations with SSI, they face a dilemma between selecting for the best-quality males or mating with all nearby males to reduce the risk for infanticide (Bellemain et al. 2006b), as promiscuity may confuse paternity and is considered as a counterstrategy to infanticide (Ebensperger 1998). Larger, older, and more heterozygous males that live in the vicinity of a female typically sire her offspring (Bellemain et al. 2006b; Zedrosser et al. 2007), and many males do not reproduce during a given mating season (Steyaert et al. 2012).

Reproductive parameters show large variation among populations. Those parameters are typically related to habitat quality, which varies tremendously across the brown bears' geographical distribution, from rich temperate ecosystems with access to abundant spawning salmon to harsh desert ecosystems (Steyaert et al. 2012; see also Chapters 12 and 13). Males reach sexual maturity around the age of 5–6 years, but occasionally reproduce at a younger age (Zedrosser et al. 2007). Testicular activity has a circannual rhythm, with testes size and function being greatest during the mating season (Spady et al. 2007). Female age of primiparity generally varies between about 5 and 10 years of age, although some females give birth at 3 years of age (Steyaert et al. 2012). Their prime reproductive age is between 9 and 20 years of age, and post reproductive survival is limited (Schwartz et al. 2003). Females are polyestrous and multiple paternity litters are common (Bellemain et al. 2006b). Females have lactational anestrus, and can (re)enter estrus only a few days after litter loss during the mating season (Steyaert et al. 2014). Induced ovulation has been suggested. Delayed implantation occurs in November–December and parturition occurs after 6–8 weeks of active gestation, around January–February (Spady et al. 2007; Friebe et al. 2014). Females usually give birth to 1–3 altricial cubs, but litter sizes up to 6 have been reported. Mean interlitter intervals typically range between 2.4 and 5.7 years and maternal care lasts for about 1.4–3.5 years in most populations (Figure 2.3), but can be as long as 4.5 years (Steyaert et al. 2012).

Figure 2.3 A female brown bear with her yearling at the beginning of the autumn. (Photo by V. Penteriani.)

Polar Bear

The mating season of the polar bear may vary geographically, but ranges between March and June (Smith & Aars 2015). The mating system is best described as serial female defense polygyny (Ramsay & Stirling 1986). A form of a polygynous mating system is supported by marked sexual dimorphism, with males typically 2–3 times heavier than females (Derocher et al. 2010). Due to the prolonged mother–offspring bond of about 2.5 years, fewer females are available to breed in any given year than there are males (2–3 males per female) (Molnár et al. 2008) (see also Chapter 14). The male-biased operational sex ratio results in intense intrasexual conflicts for access to estrous females and these are noted in tooth breakage (mainly canines), cuts, and scars (Ramsay & Stirling 1986; Derocher et al. 2010). Injuries can be severe and can include, for example, broken bones and loss of vision (Derocher et al. 2010).

Polar bears live at low densities and have large overlapping home ranges (usually >100,000 km^2; Mauritzen et al. 2001). This low density, in addition to low predictability of mates and dynamic sea ice conditions, results in males performing long-distance searches for available females (Stirling et al. 2016). Males track females over long distances, presumably to assess their breeding status based on scents in the tracks (Owen et al. 2015). Once a male locates an estrous female, he may try to herd her toward a remote area, or keep her in a small area to reduce the likelihood of detection by other males (Wiig et al. 1992; Stirling et al. 2016). Most observations of mating occur on the sea ice (Figure 2.4), but some pairs move onto land. Males in mated pairs tend to be slightly older than unpaired males (Derocher et al. 2010), but the importance of age may vary by population (Zeyl et al. 2009). Estrous females are usually accompanied by a single male, but observations of 2–5 males near an estrous female are common. Successful access to a female is thought to occur largely through contest competition, although some pairings may result from scramble competition (Derocher et al. 2010). Female mate choice is poorly understood in polar bears, but the wide range in body mass of males with females (1:1 to 3:1) suggests that it may be limited. Male–female associations normally last about 2 weeks, but can range up to 18 days (Stirling et al. 2016).

Figure 2.4 A mating pair of polar bears during spring in the Beaufort Sea, Canada. (Photo by A. E. Derocher.)

Males can produce sperm at 2 years of age (Rosing-Asvid et al. 2002), but rarely sire offspring until the age of 6 years, after which a steep increase in mating success can occur. Spermatogenesis starts in February and testes size is largest during the breeding season. Male mating success peaks at around 14 years of age, after which a steep decline occurs (Richardson 2014). Lifetime mating success in male polar bears is heavily skewed, with only few males siring many offspring (up to at least 9 successful matings; Richardson 2014). Females typically mate first when 4–6 years old and produce their first litter the following year (Ramsay & Stirling 1988). An increase in reproduction occurs until about 14 years of age, followed by a slow decline, and they can reproduce until their late 20s (Derocher & Stirling 1994). Polar bears are thought to be induced ovulators, and undergo delayed implantation during autumn. Females may switch partners while in estrus, which can result in multiple paternity litters (Zeyl et al. 2009). Inbreeding in polar bears appears to be low (Zeyl et al. 2009).

Parturition occurs between November and January and the cubs are altricial (Amstrup & DeMaster 2003). Litter size depends on maternal age and body condition and varies from one to four, with singletons or twins being most common (Ramsay & Stirling 1988). Adoption sometimes occurs (Derocher & Wiig 1999b). Estimates of cub survival range from below 20% to over 95% between years (Bromaghin et al. 2015) and are influenced mainly by environmental conditions and maternal factors (Derocher & Stirling 1996). Although the typical reproductive interval for females is three years, intervals of one or two years are common if a litter is lost (Ramsay & Stirling 1988). Infanticide occurs in polar bears and appears associated with nutritional gain, as it usually occurs outside the mating season (Derocher & Wiig 1999a; Stone & Derocher 2007). Therefore, there is limited evidence for sexually selected infanticide as a reproductive tactic. As with other ursids, there is no paternal investment in rearing of young.

Chemical Communication

Chemical communication involves sending and receiving signals and cues via organic chemicals. Chemicals exit the body through glandular secretions and excretions and are either passively released or actively directed toward environmental objects, forming a scent mark (Adams 1980). Scent marking is particularly prevalent in mammals and functions broadly in territory defense, group cohesion (for social species), recognition of individuals and kin, and assessment of mates or competitors (Muller-Schwarze 2006).

Bears have evolved a variety of physical traits and behavioral strategies to communicate efficiently using organic compounds. Glandular secretions from modified sweat and

Figure 2.5 Scent marking in bears is biased toward males. Here, however, an Alaskan female brown bear with her three cubs-of-the-year investigated and scent marked a rubbing tree. After the mother rubbed the tree, the cubs imitated her behavior, suggesting that female bears may teach their cubs to scent-mark. (Photo by E. Filipczykova.)

sebaceous glands appear to be the primary sources of scent used for marking, along with urine (Sergiel et al. 2017). Bears are selective in where, how, and when they place scent marks within their environment (Clapham et al. 2014). For those bear species studied, all appear to increase marking behavior around the mating season. Observations indicate that adult males scent mark more than other age–sex classes (Tattoni et al. 2015; Figure 2.5), although this varies in giant pandas. Due to the general lack of territoriality in bears and their solitary living, chemical communication appears to be focused primarily on sexual behavior. Giant pandas have been the focus of most of the research conducted on identifying the chemical signals and cues contained within bear odor, which include: sex, age, estrous state, kinship, and individuality (Swaisgood et al. 2002; Liu et al. 2008). In addition, compounds coding for sex have been identified in brown bears (Rosell et al. 2011) and behavioral discrimination of sex and estrous state has been shown in polar bears (Owen et al. 2015).

The role of chemical signaling in the mating system of bears appears to function to facilitate reproductive advertisement and male intrasexual competition. Female bears seem to rely on chemical cues to advertise reproductive condition, and males likely select and locate mates based on this knowledge (Clapham et al. 2012; Owen et al. 2015). Behavioral studies suggest that male bears use scent marking to signal competitive ability to other males for access to females, which may involve year-round signaling of dominance (Nie et al. 2012a). By signaling competitive ability, male bears may be able to reduce the risk of injury from physical aggression by influencing competitors into subordination, as seen in other mammals. Population density, environmental factors, and reproductive biology all likely play a role in the varying strategies that bears use to communicate chemically (Owen et al. 2015; Lamb et al. 2017). In evolutionary terms, chemical signaling ensures that the females' energetic investment in gamete production is not wasted and reduces the risk to males by reducing the necessity for physical contests.

Large gaps in our knowledge of the function of chemical signaling and its role in the mating systems of bears remain, particularly for those species inhabiting tropical regions. The apparent similarities in chemical signaling and marking behavior among several bears suggest that the chemical signaling in sexual behavior is an ancestral trait of the Ursidae.

Reproductive Physiology

Bears possess a number of physiologic adaptations of the reproductive system that allow them to increase reproductive success and maximize genetic diversity. The ovary is largely responsible for most of these physiologic adaptations in bears, including delayed implantation, split parturition, and pseudopregnancy.

The bear ovary is covered by a complete bursa (Erickson et al. 1964), which likely reduces egg loss and increases chances of successful fertilization. Bear ovaries frequently release multiple eggs at once and, except in the giant panda, also can ovulate sequentially during the mating season. Unlike most other polyestrous mammals, their ovaries continue to ovulate when the female is pregnant (during early embryonic diapause) and/or has already ovulated that mating season (Himelright et al. 2014). This helps to explain the relatively high proportion (15–20%) of multipaternity litters observed in high-density, free-ranging bear populations (Bellemain et al. 2006b; Ombrello et al. 2016; Shimozuru et al. 2019). Multipaternity, in turn, likely serves to increase genetic diversity, mitigate risks of inbreeding depression, and may provide a paternity confusion strategy to cope with infanticidal males (Steyaert et al. 2012).

Interestingly, the corpora lutea (CL) of ovaries in all eight bear species remain relatively inactive until 0.5–6.5 months after ovulation, depending on the species, at which point all CL in a female's ovaries synchronously reactivate, regardless of when individual ovulations occurred that season (Spady et al. 2007). This CL dormancy feature is unique to carnivores with delayed implantation (bears, wolverines, badgers, etc.). In bears specifically, the length of delay until implantation is highly variable within and between individuals. Among seasonally breeding bear species, however, the timing of implantation within the year is consistent and associated with latitude (i.e. late October/early November in brown and American black bears in the Temperate Zone; Spady et al. 2007). At the species level, the duration of delay is similarly long between five of the seven species examined to date (10–27 weeks; Spady et al. 2007). In contrast, the delay in giant pandas is slightly shorter on average (7–15.5 weeks; Zhang et al. 2009). Whether sun bears exhibit delayed implantation remains unclear. Although past studies concluded that sun bears lack delayed implantation, data indicate that a period of at least 2–4 weeks of CL dormancy after ovulation occurs (see Schwarzenberger et al. 2004; Frederick et al. 2012). Because implantation is preceded by active luteal function, sun bear embryos may exhibit a short period of diapause lasting at least 2 weeks. Both long and short delayed implantation have been hypothesized as mechanisms for enhancing reproductive fitness via increased multipaternity and cryptic female choice (Birkhead & Møller 1993). Hence, even the aseasonal sun bears would benefit from this.

The fertilized bear egg grows and develops normally after conception until it reaches the preimplantation blastocyst stage, at which point growth drastically slows and development is arrested (Himelright et al. 2014). The end of delay begins with CL reactivation, resulting in increased progestins, which initiates the uterine secretory phase. Only then do the diapaused blastocysts implant and resume development. Because birth occurs just 6–9 weeks after implantation, depending on the species, cubs are altricial at birth, with giant panda cubs the least developed (Mead 1993; Tsubota & Kanagawa 1993).

Other polyestrous mammals do not have this delay in CL activity; therefore, once a mating results in pregnancy, high progestin levels produced by the active CL block any subsequent follicular development and estruses that same season. Accordingly, sequential ovulations during the mating season of most polyestrous species only occur if the previous estrus did not result in a pregnancy. Hence, without some physiological means to do so, polyestrous species cannot become pregnant following successful fertilization during a previous ovulation. This is not true for bears, however, thus it is theoretically possible for them to become pregnant while they are already pregnant (superfetation). Although superfetation has not yet been proven in bears, it has been documented in related Carnivora with delayed implantation. Superfetation is further supported by the observations that ovulations can occur weeks apart in polyestrous bears (Himelright et al. 2014), and that all CL present reactivate in synchrony after the mating season (Spady et al. 2007). In addition, within 1–2 weeks after CL reactivation, the uterus has achieved the secretory phase, and all diapaused embryos implant and develop in synchrony (Tsubota & Kanagawa 1993; Yamane et al. 2009). In theory, these embryos may have been conceived by different sires at different times in the season, but because they all implant at the same time, they would be born into the same litter and of the same relative developmental age.

Giant pandas and sun bears are spontaneous ovulators (Durrant et al. 2002; Frederick et al. 2013), but in the other six bear species, ovulation is presumed to be induced by physical stimulation of vaginal nerves during coitus (Boone et al. 1998). However, subsequent studies (e.g. Okano et al. 2006) support the conclusion that spontaneous ovulation, or ovulation induced by triggers other than coitus (such as other physical contact, or perhaps olfactory or visual cues), is the primary mode of ovulation in bears. All bear species exhibit pseudopregnancy, which is a consequence of ovulation without subsequent fertilization of ova. Pseudopregnant females share many behavioral and physiological characteristics with pregnant females, such as nest-building behavior, uterine hypertrophy, and mammary development, all in response to prolonged exposure to elevated progestins and other hormones, but without the embryo (Concannon 2009).

Sexually Selected Infanticide (SSI)

SSI is a male reproductive strategy in which an adult male kills (or induces abortion) of unrelated offspring to shorten the time to when he can impregnate the mother, and thus increases his fitness (Hrdy 1979). SSI is common across many taxa. In terrestrial mammals, it has been most clearly studied in rodents, primates, and carnivores. The three requirements for SSI to evolve are simple: (1) the interbirth interval of the female should be shortened if dependent offspring are killed; (2) the probability of a male killing his own offspring should be very low; and (3), the male, or his close relative, should have a high probability of fathering the mother's next offspring. SSI would benefit the fitness of the male, but is very costly to the fitness of the mother, and thus female counterstrategies would also be expected to have evolved (Ebensperger 1998). These counterstrategies primarily include: (1) female aggression and coalitions to defend offspring; (2) mothers avoiding males that are too large to repel; (3) promiscuity to confuse paternity so males tolerate young; and (4) assistance in defending the territory of a family group.

The entire SSI process is extremely difficult to document in bears, as they are generally solitary, have large and overlapping home ranges, and are often difficult to observe. Among bears, SSI has mostly been studied in the brown bear, in which it can induce >30% cub loss per year (Gosselin et al. 2017; Figure 2.6). In general, bears clearly meet the first requirement for SSI, as they have extended interbirth intervals, during which mothers do not enter estrus (lactational anestrus) or mate. Killing cubs, or even yearlings, would return the mother to a breeding condition earlier. In Scandinavian brown bears, for example, females typically enter estrus just a few days after losing an entire litter during the mating season (Steyaert et al. 2014). The second and third requirements are usually not as clear, although some genetic evidence for both requirements exists in brown bears (Bellemain et al. 2006a; Davoli et al. 2018). Unless a male has never mated before, immigrates into an area, or can recognize the females he mated with the previous year, there may be a nontrivial probability that an infanticidal male could kill his own offspring, particularly if there were few other males in the population. However, males can undoubtedly recognize females, as individual recognition is widespread in a great variety of animals (Tibbetts & Dale 2007) and bears, like many mammals, spend considerable effort physically and chemically advertising their presence (Lamb et al. 2017). Furthermore, the female counterstrategy of confusing paternity by mating with several males would only work if males can recognize individual females they have mated with (Bellemain et al. 2006b; Fernández-Gil 2013). The second requirement for SSI is thus likely met in bears, with

Figure 2.6 Male brown bears sometimes consume their infanticide victims, as the cub claw and hair in the feces depicted evidences. Here, a GPS-marked adult male killed and ate a cub-of-the-year of a GPS-marked female during the mating season of 2010 in south-central Sweden. (Photo by S.M.J.G. Steyaert.)

males not killing the cubs of mothers that they mated with the previous year. The third SSI requirement, where the infanticidal male has a high probability of fathering the victimized mother's next offspring, is least clear. Not being territorial, male bears do not have exclusive breeding rights with any female. However, as females can rapidly enter estrus after losing an entire litter, consorting or guarding such females until estrus could give males (even if not dominant) an opportunity to mate and to father the female's next offspring. Because bear density varies by over two orders of magnitude, there are likely areas with a low enough density of adult male bears that even smaller infanticidal males can successfully consort and mate with the female. The third requirement of the SSI hypothesis is thus likely met in some species or populations, particularly for large, dominant males.

Human Impacts on the Mating System of Bears

Bears typically roam over large areas during the mating season (Dahle & Swenson 2003), which can increase the probability of coming into contact with humans or human-related infrastructures, compared to other periods of the year (Penteriani et al. 2018). For instance, road mortality of brown bears in Europe is higher during the mating season (Kaczensky et al. 2003), when bears likely cross roads more often than when they have smaller home ranges. The despotic nature of male bears is most pronounced during the mating season. More vulnerable bears, such as subadults and family groups, may avoid male aggression during the mating season by using areas relatively close to humans, which also make them more prone to become involved in human–bear interactions and conflict (Elfström et al. 2014; Steyaert et al. 2016). Some human activities focus on sensitive places where bears congregate during the mating season (Fernández-Gil et al. 2006) and where females with cubs try to stay away from other bears. That is the case of ecotourism focused on the observation of bears (Penteriani et al. 2017). Bears may leave such important areas or reduce activity during daytime because of human disturbance, which in turn can have negative effects on mating, cub rearing, and, ultimately, fitness (Fortin et al. 2016; Penteriani et al. 2017). It also remains unclear how male–female associations during the mating season occur, to what extent human activities and infrastructures limit roam-to-mate behavior, and how that influences reproductive success. A better understanding of where male and female bears interact and how they move during the mating season would allow the quantification of the influence of anthropogenic disturbances on the mating system of various bear species.

Of importance from a conservation and management point of view is that trophy-hunting of adult males can cause rapid male turnover and home range rearrangement, which may artificially increase SSI and can limit population growth. This mechanism is controversial, as it cannot be generalized among bear species and populations. For example, strong evidence for this mechanism exists in a population in south-central Sweden (Swenson et al. 1997; Gosselin et al. 2015, 2017; Leclerc et al. 2017). Studies in North America, however, have found that removing adult males had no discernible effect on the survival of cubs of mothers that lived within the dead males' home range (McLellan 2005, 2015; Schwartz et al. 2006) and cub survival is lowest in populations where adult male survival is highest (McLellan, unpublished data). Furthermore, an experimental removal of male American black bears did not show a change in cub survival (Czetwertynski et al. 2007). Undoubtedly, demographic parameters (e.g. population density, operational sex ratio, size dimorphism), as well as ecological context (e.g. availability and clustering of food resources), determine the prevalence of SSI in bear species and populations and how male removal impacts cub survival. How and to what extent, however, remains unclear. Depletion of male bears, for example due to overharvest, may also affect their mating system by reducing the encounter rates of estrous females, as has been documented in polar bears (Molnár et al. 2008).

As for many other wildlife species, habitat destruction and fragmentation can have far-reaching effects on behavior and life history in bears. In giant pandas, for example, females rely on dens for birth and early cub development. Due to the fragility of the cub, selection of dens can be very important to provide protection from predators and buffer the cub from inclement weather (Wei et al. 2019). Female pandas select dens in tree hollows and rock dens based on characteristics that will provide a warm, dry, stable environment, such as the size of the opening. Given widespread logging throughout much of the panda's range, the quantity and quality of tree hollows for rearing cubs has been diminished, and may explain why pandas favor old-growth to secondary-growth forests (Zhang et al. 2011).

Thus, a variety of human activities can have negative effects on bears during the mating season, particularly in human-dominated, fragmented landscapes. In addition, climate change (Derocher et al. 2004; Hertel et al. 2018) and other human-induced disturbances of ecosystems (e.g. pollution) can affect mating systems and strategies. To what extent human impacts affect the mating systems and strategies of bears remains largely unknown. Improving our knowledge about how humans impact mating systems and strategies of bears and wildlife in general is paramount for effective management and conservation, especially for small and endangered populations.

References

Adams, M. (1980). Odour-producing organs of mammals. *Symposia of the Zoological Society of London* 45: 57–86.

Amstrup, S. C. & DeMaster, D. P. (2003). Polar bear, *Ursus maritimus*. *Wild Mammals of North America: Biology, Management, and Conservation* 2: 587–610.

Andersson, M. (1994). *Sexual selection*. Princeton, NJ: Princeton University Press.

Appleton, R. D., Van Horn, R. C., Noyce, K. V., et al. (2018). Phenotypic plasticity in the timing of reproduction in Andean bears. *Journal of Zoology* 305: 196–202.

Bellemain, E., Swenson, J. E. & Taberlet, P. (2006a). Mating strategies in relation to sexually selected infanticide in a non-social carnivore: the brown bear. *Ethology* 112: 238–246.

Bellemain, E., Zedrosser, A., Manel, S., et al. (2006b). The dilemma of female mate selection in the brown bear, a species with sexually selected infanticide. *Proceedings of the Royal Society B* 273: 283–291.

Birkhead, T. R. & Møller, A. P. (1993). Sexual selection and the temporal separation of reproductive events: sperm storage data from reptiles, birds and mammals. *Biological Journal of the Linnean Society* 50(4): 295–311.

Boone, W. R., Catlin, J. C., Casey, K. J., et al. (1998). Bears as induced ovulators: a preliminary study. *Ursus* 10: 503–505.

Boone, W. R., Keck, B. B., Catlin, J. C., et al. (2004). Evidence that bears are induced ovulators. *Theriogenology* 61: 1163–1169.

Bromaghin, J. F., McDonald, T. L., Stirling, I., et al. (2015). Polar bear population dynamics in the southern Beaufort Sea during a period of sea ice decline. *Ecological Applications* 25: 634–651.

Castellanos, A. & Vasquez, D. (2017). Kinship relations in a multi-generational Andean bear (*Tremarctos ornatus*) family in North Ecuador. *International Bear News* 26: 24–25.

Castellanos, A., Ascanta, M. & Jackson, D. (2018). Does Rebecca, a seasoned Andean bear mother, show seasonal birthing patterns? *International Bear News* 27: 57–58.

Clapham, M., Nevin, O. T., Ramsey, A. D. & Rosell, F. (2012). A hypothetico-deductive approach to assessing the social function of chemical signalling in a non-territorial solitary carnivore. *PLoS ONE* 7: e35404.

Clapham, M., Nevin, O. T., Ramsey, A. D. & Rosell, F. (2014). Scent-marking investment and motor patterns are affected by the age and sex of wild brown bears. *Animal Behaviour* 94: 107–116.

Concannon, P. (2009). Endocrinologic control of normal canine ovarian function. *Reproduction in Domestic Animals* 44: 3–15.

Connor, T., Hull, V. & Liu, J. (2016). Telemetry research on elusive wildlife: a synthesis of studies on giant pandas. *Integrative Zoology* 11: 295–307.

Costello, C. M., Creel, S. R., Kalinowski, S. T., Vu, N. V. & Quigley H. B. (2009). Determinants of male reproductive success in American black bears. *Behavioral Ecology and Sociobiology* 64: 125.

Craighead, J. J., Sumner, J. S. & Mitchell, J. A. (1995). *The grizzly bears of Yellowstone: Their ecology in the Yellowstone Ecosystem, 1959–1992*. Washington, DC: Island Press.

Crudge, B., Lees, C., Hunt, M., et al. (2019). *Sun bears: Global status review and conservation action plan, 2019–2028*. IUCN SSC Bear Specialist Group.

Czetwertynski, S. M., Boyce, M. S. & Schmiegelow, F. K. (2007). Effects of hunting on demographic parameters of American black bears. *Ursus* 18: 1–18.

Dahle, B. & Swenson, J. E. (2003). Seasonal range size in relation to reproductive strategies in brown bears *Ursus arctos*. *Journal of Animal Ecology* 72: 660–667.

Davoli, F., Cozzo, M., Angeli, F., Groff, C. & Randi, E. (2018). Infanticide in brown bear: a case-study in the Italian Alps – genetic identification of perpetrator and implications in small populations. *Nature Conservation* 25: 55–75.

Derocher, A. E. & Stirling, I. (1994). Age-specific reproductive performance of female polar bears. *Journal of Zoology* 234: 527–536.

Derocher, A. E. & Stirling, I. (1996). Aspects of survival in juvenile polar bears. *Canadian Journal of Zoology* 74: 1246–1252.

Derocher, A. E. & Wiig, Ø. (1999a). Infanticide and cannibalism of juvenile polar bears (*Ursus maritimus*) in Svalbard. *Arctic* 52: 307–310.

Derocher, A. E. & Wiig, Ø. (1999b). Observation of adoption in polar bears (*Ursus maritimus*). *Arctic* 52: 413–415.

Derocher, A. E., Lunn, N. J. & Stirling, I. (2004). Polar bears in a warming climate. *Integrative and Comparative Biology* 44: 163–176.

Derocher, A. E., Andersen, M., Wiig, Ø. & Aars J. (2010). Sexual dimorphism and the mating ecology of polar bears (*Ursus maritimus*) at Svalbard. *Behavioral Ecology and Sociobiology* 64: 939–946.

Durrant, B., Czekala, N., Olson, M., et al. (2002). Papanicolaou staining of exfoliated vaginal epithelial cells facilitates the prediction of ovulation in the giant panda. *Theriogenology* 57: 1855–1864.

Ebensperger, L. A. (1998). Strategies and counterstrategies to infanticide in mammals. *Biological Reviews* 73: 321–346.

Edwards, M. A. & Derocher, A. E. (2015). Mating-related behaviour of grizzly bears inhabiting marginal habitat at the periphery of their North American range. *Behavioural Processes* 111: 75–83.

Elfström, M., Zedrosser, A., Jerina, K., et al. (2014). Does despotic behavior or food search explain the occurrence of problem bears in Europe? *Journal of Wildlife Management* 78(5): 881–893.

Emlen, S. T. & Oring, L. W. (1977). Ecology, sexual selection and the evolution of mating systems. *Science* 197: 215–223.

Enciso, M. A. & Guimarães, M. A. B. V. (2013). Knowing the reproductive endocrinology in the female Andean bear through non-invasive methods. *International Bear News* 22: 33–34.

Erickson, A. W., Nellor, J. & Petrides, G. A. (1964). The black bear in Michigan.

Michigan State University Agriculture Experiment Station East Lansing Research Bulletin 4: 1–102.

Fernández-Gil, A. (2013). Behavior and conservation of large carnivores in human-dominated landscapes. Brown bears and wolves in the Cantabrian Mountains. PhD thesis, University of Oviedo, Spain.

Fernández-Gil, A., Naves, J. & Delibes, M. (2006). Courtship of brown bears *Ursus arctos* in northern Spain: phenology, weather, habitat and durable mating areas. *Wildlife Biology* 12: 367–373.

Filipczyková, E., Heitkönig, I. M. A., Castellanos, A., Hantson, W. & Steyaert, S. M. J. G. (2017). Marking behavior of Andean bears in an Ecuadorian cloud forest: a pilot study. *Ursus* 27(2): 122–128.

Fortin, J. K., Rode, K. D., Hilderbrand, G. V., et al. (2016). Impacts of human recreation on brown bears (*Ursus arctos*): a review and new management tool. *PLoS ONE* 11: e0141983.

Frederick, C., Kyes, R., Hunt, K., et al. (2010). Methods of estrus detection and correlates of the reproductive cycle in the sun bear (*Helarctos malayanus*). *Theriogenology* 74: 1121–1135.

Frederick, C., Hunt, K. E., Kyes, R., Collins, D. & Wasser, S. K. (2012). Reproductive timing and aseasonality in the sun bear (*Helarctos malayanus*). *Journal of Mammalogy* 93: 522–531.

Frederick, C., Hunt, K., Kyes, R., et al. (2013). Social influences on the estrous cycle of the captive sun bear (*Helarctos Malayanus*). *Zoo Biology* 32: 581–591.

Fredriksson, G. M. & Wich, S. A. (2006). Frugivory in sun bears (*Helarctos malayanus*) is linked to El Niño-related fluctuations in fruiting phenology, East Kalimantan, Indonesia. *Biological Journal of the Linnean Society* 89: 489–508.

Fredriksson, G. M., Danielsen, L. S. & Swenson, J. E. (2007). Impacts of El Niño related drought and forest fires on sun bear fruit resources in lowland dipterocarp forest of East Borneo. *Biodiversity and Conservation* 16: 1823–1838.

Friebe, A., Evans, A. L., Arnemo, J. M., et al. (2014). Factors affecting date of implantation, parturition, and den entry estimated from activity and body temperature in free-ranging brown bears. *PLoS ONE* 9(7): e101410.

Garcia-Rangel, S. (2012). Andean bear *Tremarctos ornatus* natural history and conservation. *Mammal Review* 42: 85–119.

Garshelis, D. L. (2009). Family Ursidae (bears). In: Wilson, D. E. & Mittermeier, R.A. (Eds.), *Handbook of the mammals of the world, Vol. 1 – Carnivores* (pp. 448–497). Barcelona: Lynx Edicions.

Garshelis, D. L. & Hellgren, E. C. (1994). Variation in reproductive biology of male black bears. *Journal of Mammalogy* 75: 175–188.

Garshelis, D. L., Joshi, A. R., Smith, J. L. & Rice C. G. (1999). *Sloth bear conservation action plan in bears: Status survey and conservation action plan* (pp. 225–240). Gland: International Union for the Conservation of Nature and Natural Resources.

Gittleman, J. L. (1994). Are the pandas successful specialists or evolutionary failures? *BioScience* 44: 456–464.

Gopal, R. (1991). Ethological observations on the sloth bear (*Melursus ursinus*). *Indian Forester* 117: 915–920.

Gosselin, J., Zedrosser, A., Swenson, J. E. & Pelletier, F. (2015). The relative importance of direct and indirect effects of hunting mortality on the population dynamics of brown bears. *Proceedings of the Royal Society B: Biological Sciences* 282: 20141840.

Gosselin, J., Leclerc, M., Zedrosser, A., et al. (2017). Hunting promotes sexual conflict in brown bears. *Journal of Animal Ecology* 86: 35–42.

Hamer, D. & Herrero, S. (1990). Courtship and use of mating areas by grizzly bears in the Front Ranges of Banff National Park, Alberta. *Canadian Journal of Zoology* 68: 2695–2697.

Hertel, A. G., Bischof, R., Langval, O., et al. (2018). Berry production drives bottom-up effects on body mass and reproductive success in an omnivore. *Oikos* 127(2): 197–207.

Himelright, B. M., Moore, J. M., Gonzales, R. L., et al. (2014). Sequential ovulation and fertility of polyoestrus in American black bears (*Ursus americanus*). *Conservation Physiology* 2: cou051.

Hrdy, S. B. (1979). Infanticide among animals: a review, classification, and examination of the implications for the reproductive strategies of females. *Ethology and Sociobiology* 1: 13–40.

Hunter, L. & P. Barrett. (2019). *Carnivores of the world*. Second edition. Princeton, NJ: Princeton University Press.

Iibuchi, R., Nakano, N., Nakamura, T., et al. (2009). Change in body weight of mothers and neonates and in milk composition during denning period in captive Japanese black bears (*Ursus thibetanus japonicus*). *Japanese Journal of Veterinary Research* 57: 13–22.

Jonkel, C. & Cowan, I. (1971). The black bear in the spruce-fir forest. *Wildlife Monographs* 27: 3–57.

Joshi, A. R. (1996). The home range, feeding habits, and social organization of sloth bears (*Melursus ursinus*) in Royal Chitwan National Park, Nepal. PhD thesis, University of Minnesota, Minneapolis, USA.

Joshi, A. R., Smith, J. L. D. & Garshelis, D. L. (1999). Sociobiology of the myrmecophagous sloth bear in Nepal. *Canadian Journal of Zoology* 77(11): 1690–1704.

Kaczensky, P., Knauer, F., Krze, B., et al. (2003). The impact of high speed, high volume traffic axes on brown bears in Slovenia. *Biological Conservation* 111: 191–204.

Katayama, A., Tsubota, T., Yamada, F., Kita, I. & Tiba, T. (1996). Reproductive evaluation of Japanese black bears (*Selenarctos thiberanus japonicus*) by observation of the ovary and uterus. *Japanese Journal of Zoo Wildlife Medicine* 1: 26–32.

Kleiner, J. D., Van Horn R. C., Swenson J. E. & Steyaert, S. M. J. G. (2018). Rub-tree selection by Andean bears in the Peruvian dry forest. *Ursus* 29(1): 58–66.

Kolenosky, G. B. (1990). Reproductive biology of black bears in east-central Ontario. *Bears: Their Biology and Management* 8: 385–392.

Komatsu, T., Tsubota, T., Kishimoto, M., Amasaki, S. & Tiba, T. (1994). Puberty and stem cell for the initiation and resumption of spermatogenesis in the male Japanese black bear (*Selenarctos thibetanus japonicus*). *Journal of Reproduction and Development* 40: j65–j71.

Kovach, A. I. & Powell, R. A. (2003). Effects of body size on male mating tactics and paternity in black bears, *Ursus americanus*. *Canadian Journal of Zoology* 81: 1257–1268.

Kozakai, C., Yamazaki, K., Nemoto, Y., et al. (2013). Fluctuation of daily activity time budgets of Japanese black bears: relationship to sex, reproductive status, and hardmast availability. *Journal of Mammalogy* 94: 351–360.

Krofel, M., Filacorda, S. & Jerina, K. (2010). Mating-related movements of male brown bears on the periphery of an expanding population. *Ursus* 21: 23–30.

Kurt, F. & Jayasuriya, A. (1968). Notes on a dead bear. *Loris* 11: 182–183.

Lamb, C. T., Mowat, G., Gilbert, S. L., et al. (2017). Density-dependent signaling: an alternative hypothesis on the function of chemical signaling in a non-territorial solitary carnivore. *PLoS ONE* 12: e0184176.

Larivière, S. (2001). *Ursus americanus. Mammalian Species* 647: 1–11.

Laurie, A. & Seidensticker, J. (1977). Behavioural ecology of the sloth bear (*Melursus ursinus*). *Journal of Zoology* 182 (2): 187–204.

Leclerc, M., Frank, S. C., Zedrosser, A., Swenson, J. E. & Pelletier, F. (2017). Hunting promotes spatial reorganization and sexually selected infanticide. *Scientific Reports* 7: 45222.

LeCount, A. L. (1987). Causes of black bear cub mortality. *Bears: Their Biology and Management* 7: 75–82.

Liu, D., Wei, R., Zhang, G., et al. (2008). Male panda (*Ailuropoda melanoleuca*) urine contains kinship information. *Chinese Science Bulletin* 53: 2793–2800.

Martin-Wintle, M. S., Shepherdson, D., Zhang, G., et al. (2015). Free mate choice enhances conservation breeding in the endangered giant panda. *Nature Communications* 6: 10125.

Martin-Wintle, M. S., Shepherdson, D., Zhang, G., et al. (2017). Do opposites attract? Effects of personality matching in breeding pairs of captive giant pandas on reproductive success. *Biological Conservation* 207: 27–37.

Mauritzen, M., Derocher, A. E. & Wiig, Ø. (2001). Space-use strategies of female polar bears in a dynamic sea ice habitat. *Canadian Journal of Zoology* 79: 1704–1713.

McLellan, B. N. (2005). Sexually selected infanticide in grizzly bears: the effects of hunting on cub survival. *Ursus* 16(2): 141–156.

McLellan, B. N. (2015). Some mechanisms underlying variation in vital rates of grizzly bears on a multiple use landscape. *Journal of Wildlife Management* 79: 749–765.

McLoughlin, P. D., Ferguson, S. H. & Messier, F. (2000). Intraspecific variation in home range overlap with habitat quality: a comparison among brown bear populations. *Evolutionary Ecology* 14: 39–60.

Mead, R. A. (1993). Embryonic diapause in vertebrates. *Journal of Experimental Zoology* 266: 629–641.

Molnár, P. K., Derocher, A. E., Lewis, M. A. & Taylor, M. K. (2008). Modelling the mating system of polar bears: a mechanistic approach to the Allee effect. *Proceedings of the Royal Society B – Biological Sciences* 275: 217–226.

Moore, J. A., Xu, R., Frank, K., Draheim, H. & Scribner, K. T. (2015). Social network analysis of mating patterns in American black bears (*Ursus americanus*). *Molecular Ecology* 24: 4010–4022.

Muller-Schwarze, D. (2006). *Chemical ecology of vertebrates.* Cambridge: Cambridge University Press.

Nakamura, S., Yokoyama, M., Katayama, A., Morimitsu, Y. & Saita, E. (2011). Growth pattern and its characteristics of outer morphology of Asiatic black bear. *Wildlife Monograph of Hyogo* 3: 107–116.

Ngoprasert, D., Reed, D. H., Steinmetz, R. & Gale, G. A. (2012). Density estimation of Asian bears using photographic capture-recapture sampling based on chest marks. *Ursus* 23(2): 117–133.

Nie, Y.-G., Zhang, Z.-J., Swaisgood, R. R. & Wei F.-W. (2012a). Effects of season and social interaction on fecal testosterone metabolites in wild male giant pandas: implications for energetics and mating strategies. *European Journal of Wildlife Research* 58: 235–241.

Nie, Y., Swaisgood, R. R., Zhang, Z., Liu, X. & Wie, F. (2012b). Reproductive competition and fecal testosterone in wild male giant pandas (*Ailuropoda melanoleuca*). *Behavioral Ecology and Sociobiology* 66: 721–730.

Nie, Y., Speakman, J. R., Wu, Q., et al. (2015). Exceptionally low daily energy expenditure in the bamboo-eating giant panda. *Science* 349: 171–174.

Norris, T. (1969). Ceylon sloth bear. *International Wildlife* 12: 300–303.

Oi, T., Ohnishi, N., Yamada, F. & Kitahara, E. (2008). Characteristics of sex-age composition of Asian black bears killed as nuisances in Kyoto. *Mammalian Science* 48: 17–24.

Okano, T., Nakamura, S., Nakashita, R., et al. (2006). Incidence of ovulation without coital stimuli in captive Japanese black bears (*Ursus thibetanus japonicus*) based on serum progesterone profiles. *Journal of Veterinary Medical Science* 68: 1133–1137.

Ombrello, T. A., Chinnici, N. L. & Huffman, J.E. (2016). Multiple paternities in American black bears from New Jersey. *Journal of the Pennsylvania Academy of Science* 90: 21–24.

Owen, M. A., Swaisgood, R. R., Slocomb, C., et al. (2015). An experimental investigation of chemical communication in the polar bear. *Journal of Zoology* 295: 36–43.

Pan, W. (2014). *A chance for lasting survival: Ecology and behavior of wild giant pandas.* Washington, DC: Smithsonian Institution Scholarly Press.

Penteriani, V., López-Bao, J. V., Bettega, C., et al. (2017). Consequences of brown bear viewing tourism: a review. *Biological Conservation* 206: 169–180.

Penteriani, V., Delgado, M. D. M., Krofel, M., et al. (2018). Evolutionary and ecological traps for brown bears *Ursus arctos* in human-modified landscapes. *Mammal Review* 48: 180–193.

Peyton, B. (1980). Ecology, distribution, and food habits of spectacled bears, *Tremarctos ornatus*, in Peru. *Journal of Mammalogy* 61: 639–652.

Phillips, W. W. A. (1984). *Manual of the mammals of Sri Lanka* (pp. 290–296). Wildlife and Nature Protection Society of Sri Lanka.

Prater S. H. (1965). *The book of Indian animals.* Bombay: Bombay Natural History Society.

Puschmann, V. W., Schuppel, K. F. & Kronberger, H. (1977). Detection of blastocyst in uterine lumen of Indian bear (*Melursus u. ursinus*). In: Ippen, R. & Schrader, H. D. (Eds.) *Sickness in zoos* (pp. 389–391). Berlin: Akademie Verlag.

Ramsay, M. A. & Stirling, I. (1986). On the mating system of polar bears. *Canadian Journal of Zoology* 64: 2142–2151.

Ramsay, M. A. & Stirling, I. (1988). Reproductive biology and ecology of female polar bears (*Ursus maritimus*). *Journal of Zoology* 214: 601–634.

Ratnayeke, S., van Manen, F. T. & Padmalal, U. (2007). Home ranges and habitat use of sloth bears *Melursus ursinus inornatus* in Wasgomuwa National Park, Sri Lanka. *Wildlife Biology* 13: 272–284.

Richardson, E. S. (2014). The mating system and life history of the polar bear. PhD dissertation, University of Alberta, Edmonton.

Rosell, F., Jojola, S. M., Ingdal, K., et al. (2011). Brown bears possess anal sacs and secretions may code for sex. *Journal of Zoology* 283: 143–152.

Rosenthal, M. & Haggerty, C. (1989). *Proceedings of the first international symposium on the spectacled bear.* Chicago: Lincoln Park Zoo.

Rosenthal, R. A. (1999). *International studbook for the spectacled bear (*Tremarctos ornatus*).* Chicago: The Lincoln Park Zoological Society.

Rosing-Asvid, A., Born, E. W. & Kingsley, M. C. S. (2002). Age at sexual maturity of males and timing of the mating season of polar bears (*Ursus maritimus*) in Greenland. *Polar Biology* 25: 878–883.

Schaller, G. B., Hu, J., Pan, W. & Zhu, J. (1985). *The giant pandas of wolong.* Chicago: University of Chicago Press.

Schwartz, C. C., Keating, K. A., Reynolds, H. V., et al. (2003). Reproductive maturation and senescence in the female brown bear. *Ursus* 14(2): 109–119.

Schwartz, C. C., Haroldson, M. A., White, G. C., et al. (2006). Temporal, spatial, and environmental influences on the demographics of grizzly bears in the greater Yellowstone Ecosystem. *Wildlife Monographs* 161: 1–68.

Schwarzenberger, F., Fredriksson, G., Schaller, K. & Kolter, L. (2004). Fecal steroid analysis for monitoring reproduction in the sun bear (*Helarctos malayanus*). *Theriogenology* 62: 1677–1692.

Sergiel, A., Naves, J., Kujawski, P., et al. (2017). Histological, chemical and behavioural evidence of pedal communication in brown bears. *Scientific Reports* 7: 1052.

Shimozuru, M., Shirane, Y., Tsuruga H., et al. (2019). Incidence of multiple paternity and inbreeding in high-density brown bear populations on the Shiretoko Peninsula, Hokkaido, Japan. *Journal of Heredity* 110(3): 321–331.

Smith, T. G. & Aars, J. (2015). Polar bears (*Ursus maritimus*) mating during late June on the pack ice of northern Svalbard, Norway. *Polar Research* 34: 25786.

Spady, T. J., Lindburg, D. G. & Durrant, B. S. (2007). Evolution of reproductive seasonality in bears. *Mammal Review* 37: 21–53.

Stenhouse, G., Boulanger, J., Lee, J., et al. (2005). Grizzly bear associations along the eastern slopes of Alberta. *Ursus* 16(1): 31–40.

Steyaert, S. M. J. G., Endrestøl, A., Hackländer, K., Swenson J. E. & Zedrosser, A. (2012). The mating system of the brown bear *Ursus arctos*. *Mammal Review* 42: 12–34.

Steyaert, S. M. J. G., Swenson, J. E. & Zedrosser, A. (2014). Litter loss triggers estrus in a nonsocial seasonal breeder. *Ecology and Evolution* 4: 300–310.

Steyaert, S. M. J. G., Leclerc M., Pelletier, F., et al. (2016). Human shields mediate sexual conflict in a top predator. *Proceedings of the Royal Society B: Biological Sciences* 283: 20160906.

Stirling, I., Spencer, C. & Andriashek, D. (2016). Behavior and activity budgets of wild breeding polar bears (*Ursus maritimus*). *Marine Mammal Science* 32: 13–37.

Stone, I. R. & Derocher, A. E. (2007). An incident of polar bear infanticide and cannibalism on Phippsoya, Svalbard. *Polar Record* 43: 171–173.

Swaisgood, R. R., Lindburg, D. G. & Zhang, H. (2002). Discrimination of oestrous status in giant pandas (*Ailuropoda melanoleuca*) via chemical cues in urine. *Journal of Zoology* 257: 381–386.

Swenson, J. E., Sandegren, F., Söderberg, A., et al. (1997). Infanticide caused by hunting of male bears. *Nature* 386: 450–451.

Tattoni, C., Bragalanti, N., Groff, C. & Rovero, F. (2015). Patterns in the use of rub trees by the Eurasian brown bear. *Hystrix, the Italian Journal of Mammalogy* 26: 118–124.

Tibbetts, E. A. & Dale, J. (2007). Individual recognition: it is good to be different. *Trends in Ecology and Evolution* 22: 529–537.

Trivers, R. L. (1972). Parental investment and sexual selection. In: Campbell B. G. (Ed.), *Sexual selection and the descent of man 1871–1971* (pp. 136–179). Chicago: Aldine.

Tsubota, T. & Kanagawa, H. (1993). Morphological characteristics of the ovary, uterus and embryo during the delayed implantation period in the Hokkaido brown bear (*Ursus arctos yesoensis*). *Journal of Reproduction and Development* 39: 325–331.

Tsubota, T., Howell-Skalla, L., Boone, W. R., Garshelis, D. L. & Bahr, J. M. (1998). Serum progesterone, oestradiol, luteinizing hormone and prolactin profiles in the female black bear (*Ursus americanus*). *Animal Reproduction Science* 53: 107–118.

Vaisefeld, M. A. & Chestin, I. E. (1993). *Bears: brown bear, polar bear, Asian black bear. Distribution, ecology, use and protection.* Moscow: Nauka.

Wei, W., Swaisgood, R. R., Owen, M. A., et al. (2019). The role of den quality in giant panda conservation. *Biological Conservation* 231: 1892196.

Wiig, Ø., Gjertz, I., Hansson, R. & Thomassen, J. (1992). Breeding behaviour of polar bears in Hornsund, Svalbard. *Polar Record* 28: 157–159.

Wong, S., Servheen, C., Ambu, L. & Norhayati, A. (2005). Impacts of fruit production cycles on Malayan sun bears and bearded pigs in lowland tropical forest of Sabah, Malaysian Borneo. *Journal of Tropical Ecology* 21: 627–639.

Yamamoto, K., Tsubota, T. & Kita, I. (1998). Observation of sexual behavior of captive Japanese black bears, *Ursus tibetanus japonicus. Journal of Reproduction and Development* 44: j13–j18.

Yamamoto, T., Tamatani, H., Tanaka, J., et al. (2012). Multiple paternity in Asian black bear *Ursus thibetanus* (Ursidae, Carnivora) determined by microsatellite analysis. *Mammalia* 77: 215–217.

Yamane, M., Yamamoto, Y., Tsujimoto, T. & Osawa, T. (2009). Relationship between uterine morphology and peripheral concentrations of sex steroid hormone in wild Japanese black bears (*Ursus thibetanus japonicus*). *Animal Reproduction Science* 113: 251–262.

Yamazaki, K. (2017). *Challenges in management and conservation of Asian black bears in Japan.* Tokyo: University of Tokyo Press.

Yoganand, K. (2005). Behavioural ecology of sloth bear (*Melursus ursinus*) in Panna National Park, Central India. PhD Thesis. Saurashtra University, India.

Yoganand, K., Rice, C. G. & Johnsingh, A. J. T. (2013). Sloth bear: *Melursus ursinus*. In: Johnsingh, A. J. T. & Manjrekar, N. (Eds.), *Mammals of South Asia* (pp. 438–456). Telangana, India: Orient Blackswan pvt. Ltd.

Zedrosser, A., Bellemain, E., Taberlet, P. & Swenson, J. E. (2007). Genetic estimates of annual reproductive success in male brown bears: the effects of body size, age, internal relatedness and population density. *Journal of Animal Ecology* 76: 368–375.

Zeyl, E., Aars, J., Ehrich, D., Bachmann L. & Wiig, Ø. (2009). The mating system of polar bears: a genetic approach. *Canadian Journal of Zoology* 87: 1195–1209.

Zhang, H., Li, D., Wang, C. & Hull, V. (2009). Delayed implantation in giant pandas: the first comprehensive empirical evidence. *Reproduction* 138: 979–986.

Zhang, Z., Swaisgood, R. R., Zhang, S., et al. (2011). Old-growth forest is what giant pandas really need. *Biology Letters* 7(3): 403–406.

Zhang, Z., Sheppard, J. K., Swaisgood, R. R., et al. (2014). Ecological scale and seasonal heterogeneity in the spatial behaviors of giant pandas. *Integrative Zoology* 9: 46–60.

Zhu, L., Wu, Q., Dai, J., Zhang, S. & Wei, F. (2011). Evidence of cellulose metabolism by the giant panda gut microbiome. *Proceedings of the National Academy of Sciences* 108: 17714–17719.

Zhu, X., Lindburg, D. G., Pan, W., Forney, K. A. & Wang, D. (2001). The reproductive strategy of giant pandas: infant growth and development and mother-infant relationships. *Journal of Zoology* 253: 141–155.

Chapter 3

Interspecific Interactions between Brown Bears, Ungulates, and Other Large Carnivores

Andrés Ordiz, Miha Krofel, Cyril Milleret, Ivan Seryodkin, Aimee Tallian, Ole-Gunnar Støen, Therese Ramberg Sivertsen, Jonas Kindberg, Petter Wabakken, Håkan Sand, and Jon E. Swenson

Introduction

Large carnivores, such as brown bears (*Ursus arctos*), wolves (*Canis lupus*), and tigers (*Panthera tigris*), can play a key ecological role from their apex position in trophic systems (Ordiz et al. 2013). Within the overall context of bottom-up and top-down regulation of ecosystems (Sinclair & Krebs 2002), predation by large carnivores often induces demographic and behavioral changes in prey species. These vertical interactions between different trophic levels are important regulatory mechanisms in nature. On the other hand, competitive interactions between species, or horizontal interactions within the same trophic level, are also common. Interspecific interactions between large carnivores are widespread in many ecosystems and can play an important role in community structure and stability. For instance, competition between carnivores can reduce the population size of their respective competitor, which in turn affects lower trophic levels (Palomares & Caro 1999; Caro & Stoner 2003).

Competition and other interactions among carnivores can lead to spatial avoidance, shifts in habitat use, altered predation patterns, and other behavioral changes. For example, kleptoparasitism or the theft of kills can force affected species to additional hunting (Krofel et al. 2012; Elbroch et al. 2015). Indirect competition for shared resources, such as prey or habitat, can also affect population dynamics, predation patterns, and habitat selection (e.g. Ordiz et al. 2015). Such interactions can either weaken or strengthen the top-down effects of carnivores by altering predator densities or predation patterns. For instance, scavenging, with kleptoparasitism between carnivores as one of its major mechanisms, is a key ecological process involved in energy flow in ecosystems (DeVault et al. 2003). Interspecific interactions can cause the extirpation of one species, but more often result in resource partitioning and, thus, coexistence (Apps et al. 2006). The outcome of interactions may be mediated by the environmental context and the effects of humans, which can have an overwhelming impact on the population dynamics of the interacting species, their distribution, and the effects of competition (Apps et al. 2006; Krofel & Jerina 2016). In fact, this may be the norm for large carnivores inhabiting human-dominated landscapes, given the long history of human persecution and the general avoidance of people by large carnivores (see Chapter 18). Overall, predator–prey relationships and other interspecific interactions are important topics for ecology and management, because they can influence the distribution and abundance of species (Schoener 1983; Wisz et al. 2013).

Because interactions between apex predators can have cascading effects on lower trophic levels, research and management would benefit from a multispecies or ecosystem-focused approach, rather than focusing efforts on single species (Ordiz et al. 2015; Krofel & Jerina 2016). In addition, large carnivores generate considerable public interest and are a central issue in human–wildlife conflicts. Predation is the mechanism driving apex predators' function in nature, but it is also a source of conflict with different stakeholders, e.g. hunters and livestock owners, when predation affects wild game, domestic or semi-domestic species (depredation). This situation is challenging when trying to secure long-term carnivore conservation and coexistence with people in the human-dominated landscapes that currently characterize most of our planet.

Brown bears are very efficient predators of newborn ungulates. Predation by brown bears often results in additive neonate ungulate mortality (e.g. Griffin et al. 2011, see below), which is of particular interest in ecosystems where bears are sympatric with other large carnivores, such as canids and felids, that also prey on ungulates. This is especially relevant given the great extent to which brown bears co-occur with other carnivores in the Northern Hemisphere (Figure 3.1).

In this chapter, we summarize the effects of brown bear predation on both wild and domestic ungulate populations and present the current state of knowledge on interactions between brown bears and other large carnivores, focusing on felids and canids (in particular, wolves). Besides the inherent interest of this topic from an ecological perspective, it also has direct implications for the conservation and management of predators and their ungulate prey; e.g. the latter are both commonly exposed to human management and predation by large carnivores (Jonzén et al. 2013).

Bears and Wild Ungulate Populations

All bear species have the potential to kill wild ungulates, but because bears are omnivores and opportunistic predators, they have relatively loose predator–prey relationships with ungulates. This means that bear population densities are probably

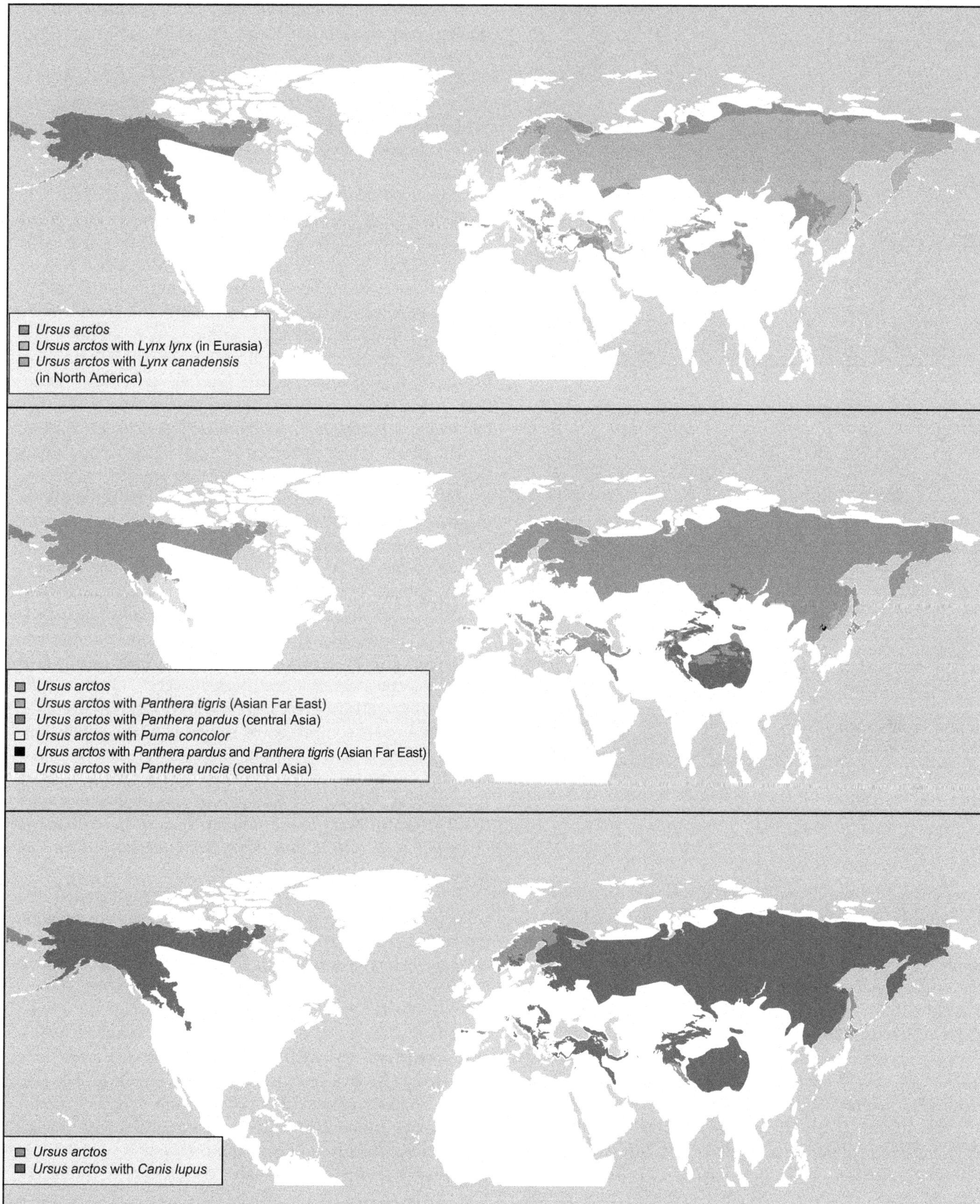

Figure 3.1 Worldwide distribution of brown bears and its overlap with the distribution of several felids (top and central panel) and wolves (bottom panel) in the Northern Hemisphere. (*Source*: IUCN 2019. The IUCN Red List of Threatened Species, Version 6.2, www.iucnredlist.org. Downloaded March 16, 2019.) (A black and white version of this figure will appear in some formats. For the color version, please refer to the plate section.)

Figure 3.2 Moose calf (top panel) and reindeer calf (bottom panel) killed by brown bears in Sweden. (Photos by A. Ordiz.)

more affected by the abundance of other food sources than ungulates, i.e. bear densities may fluctuate independently of ungulate densities. However, the effect of bear predation on an ungulate population may primarily vary with changes in the relative density of the bear and ungulate species, i.e. the effect of bears is likely stronger as their relative density increases (Zager & Beecham 2006).

Brown bears are efficient predators on neonatal moose (*Alces alces*), elk (*Cervus canadensis*), caribou (reindeer in Europe; *Rangifer tarandus*), and muskoxen (*Ovibos moschatus*) in North America (Zager & Beecham 2006; Arthur & Vecchio 2017) and neonatal moose and reindeer in Europe (Swenson et al. 2007; Karlsson et al. 2012; Figure 3.2). Although less common, bears can also prey on adult moose, elk, caribou, and muskoxen in North America (Zager & Beecham 2006; Arthur & Vecchio 2017) and on adult moose, red deer (*Cervus elaphus*), and wild boar (*Sus scrofa*) in Europe (Dahle et al. 2013; Krofel et al., unpublished data). Bears apparently do not regularly prey on American bison (*Bison bison*) (Wyman 2002) or sika deer (*Cervus nippon*) (Sato et al. 2005).

It is worth mentioning that other bear species are also predatory. American black bears (*Ursus americanus*) are efficient predators of neonatal moose, elk, caribou, white-tailed deer (*Odocoileus virginianus*), and mule deer (*Odocoileus hemionus*), but very rarely prey on adult ungulates (Zager & Beecham 2006; Monteith et al. 2014). In Taiwan, the relatively small-sized adult Formosan muntjac (*Muntiacus reevesi*) and Formosan serow (*Naemorhedus swinhoei*) are part of the diet of Asiatic black bears (Hwang et al. 2002), which also prey on Japanese serow (*Capricornis crispus*) (Huygens et al. 2003). Although Andean bears have been reported to attack woolly tapir (*Tapirus pinchaque*) and prey on white-tailed deer, guanaco (*Lama guanicoe*), and vicuña (*Vicugna vicugna*), there are few supporting quantitative data (Stirling & Derocher 1990). Polar bears (*Ursus maritimus*) scavenge carcasses when they find them, and occasionally kill reindeer (Derocher et al. 2000).

The effects of bear predation on wild ungulate populations vary enormously, but predation generally appears to be additive at low ungulate densities and more compensatory as ungulate population densities approach carrying capacity (Zager & Beecham 2006). Compensatory mortality means that instead of causing additional mortality to the population, as would be the case with additive mortality, the occurrence of predation does not cause higher total mortality rates. Additionally, the overall impact of bear predation on ungulate population growth is dependent on the targeted segment of the prey population, e.g. variation in age- and sex-specific survival differentially affect population growth (Gervasi et al. 2012). In seven of eight reviewed studies of moose populations in Alaska, brown and American black bear predation dominated calf mortality and this mortality was primarily additive (Boertje et al. 2010). These findings are consistent with studies in other brown bear–moose systems, such as in Scandinavia (Swenson et al. 2007), and other bear–ungulate systems in North America (Griffin et al. 2011).

Bear predation on neonates, especially in combination with wolf predation, has probably caused negative population growth in several populations of moose and elk (Garrott et al. 2009; Boertje et al. 2010). Brown bear predation on adult ungulates is seemingly compensatory to other sources of mortality; however, exceptions have been observed in several systems (e.g. Zager & Beecham 2006). For example, brown bear predation on adult and calf muskoxen apparently increased in an area of Alaska, perhaps because caribou and moose populations had declined, causing a decline in the muskox population (Arthur & Vecchio 2017). Nevertheless, Keech et al. (2000) in Alaska and Swenson et al. (2007) in Scandinavia showed the existence of reproductive compensation in moose females following loss of calves. Female moose can have higher reproductive rates the year after predation occurred, i.e. managers should expect partial compensation of large carnivore predation through higher reproduction the year after a moose female lost her calves, especially if this happened early in the calves' lives (Swenson et al. 2007).

Bears and Domestic Ungulate Populations: A Case Study from Scandinavia

Brown bears prey on domestic and semi-domestic livestock in many parts of their range, including sheep, cows, goats, and reindeer (e.g. Knight & Judd 1979; Kaczensky 1999). Here, we highlight the case of brown bear predation of semi-domestic reindeer in northern Scandinavia as a case study of human–bear conflict. Brown bears can be efficient depredators of semi-domestic reindeer, especially during calving in spring, when neonates in their first weeks of life are easy prey for bears (Karlsson et al. 2012; Sivertsen 2017). Large losses of reindeer calves to bear predation frequently lead to conflict with reindeer owners. Pastoral reindeer husbandry, with more than 1200 years of tradition in Fennoscandia (Ruong 1982), is an essential part of the culture and livelihoods of the Sámi people (Zabel & Holm-Müller 2008). Traditionally, the reindeer owners have protected their reindeer by killing bears on their calving grounds. However, the core areas of traditional Sámi reindeer herding were also the main areas where bears avoided extirpation in Sweden during the government-driven large carnivore extermination in the nineteenth century. In 1930, only ~130 individual bears survived in mountainous areas in northern and central Sweden and one area in Norway (Swenson et al. 1995).

After the abolishment of bounties in 1893 and legal protection in 1986, the bear population increased and reached ~3000 bears in Sweden in 2008 (Kindberg et al. 2011; Swenson et al. 2017). This bear population is once again decreasing, due to increased harvest quotas set by management agencies, mostly driven by conflicts with reindeer husbandry and moose hunters, in order to stop the population growth. Although reindeer herders claim that losses are substantial, the absence of reliable data on brown bear predation has made it difficult to reach an agreement on compensation schemes and preventive interventions safeguarding both Sámi reindeer herding and a viable brown bear population. In order to address the need for reliable data, the Swedish government initiated a study on brown bear kill rate on reindeer in forested calving grounds in 2010–2012 (Karlsson et al. 2012). The results showed that brown bears killed calves almost exclusively during, and shortly after, peak calving, i.e. in the three last weeks of May. The estimated predation rate indicated that brown bears may have caused a considerable proportion (39% and 67%) of the total calf losses observed within the two reindeer herding districts. Average annual calf mortality in the herding districts was approximately 43% and 41%, and the estimated annual bear-caused calf mortality was around 29% and 16% of the total number of calves born annually within the two studied reindeer herding districts (Sivertsen 2017). A follow-up study on interventions in 2013–2016 showed that calving in corrals prevents most losses to bears, but was too expensive, increased the risk of diseases, and was not compatible with traditional Sámi reindeer herding traditions (Frank et al. 2017). Reduction of the bear density in calving grounds, either as management killing or area-differentiated license hunting, was considered the most efficient way to solve the problem (Frank et al. 2017). Nevertheless, a zoning system, prioritizing carnivore conservation and reindeer herding in different areas, might also help reduce conflict (Ordiz et al. 2017).

The recent increase in hunting quotas has given hunters the opportunity to reduce the bear population, but increased bear harvest does not seem to have reduced the conflict in the reindeer husbandry areas. This is because calving areas are remote, far from roads, and less available for brown bear hunters. Therefore, most of the removal of bears in calving areas has been carried out as authorized spring management culls (i.e. not as part of the regular hunting season), often targeting females with cubs and using helicopters. This has raised questions about the ethics of management removals and caused conflict among stakeholders, including the general public, people interested in bear conservation, and hunter organizations.

Brown Bear Interactions with Other Large Carnivores

Brown bears interact with many other large carnivores, including other bear species (e.g. Schwartz et al. 2010), felids (e.g. Krofel et al. 2012), and canids (e.g. Ordiz et al. 2015). There is extensive overlap in the distribution of brown bears, several felids, and wolves, the largest canid species, throughout the Northern Hemisphere (Figure 3.1). In particular, this section focuses on brown bear–felid and brown bear–wolf interactions, while acknowledging that interactions with several bear species and other canids and felids also occur where their distributions overlap.

Much of the existing literature on interactions between large carnivores reports on direct and indirect interactions at the individual level. Yet, we still lack knowledge about the effects of interspecific interactions beyond individuals, i.e. at the population level. Scarcity of such information has been specifically highlighted for the interactions between wolves and bears (Ballard et al. 2003) and, more broadly, for large carnivores in general (Périquet et al. 2015). In fact, interspecific competition between free-ranging apex predators is challenging to quantify, because such species are elusive and logistically complex and expensive to study.

Interactions between bears and felids or canids include all the forms of interspecific interactions described earlier in the chapter, i.e. they use the kills of each other via kleptoparasitism, they can displace each other and, in agonistic situations, they can kill each other (see below). As the largest terrestrial scavengers with excellent olfactory abilities, brown bears are indeed one of the most important dominant scavengers and kleptoparasites in the Holarctic region (Krofel et al. 2012), and kleptoparasitism is likely the most common form of interaction between bears and other carnivores.

Brown Bears and Felids

Throughout their range in the Northern Hemisphere, brown bears overlap with multiple species of wild cats. However, the available literature on interspecific interactions is limited to tiger, mountain lion (*Puma concolor*), Eurasian lynx (*Lynx lynx*), common leopard (*Panthera pardus*), and snow leopard (*Panthera uncia*) (Figures 3.1 and 3.3). All of them prey primarily on wild ungulates, and kleptoparasitism is the most frequently reported interaction between brown bears and felids (Krofel et al. 2012). Smaller felids predominantly hunt small prey (i.e. smaller than their body size), which are quickly consumed, and this probably explains their low level of interactions with bears. On the other hand, larger felids mainly hunt ungulates, which may take them several days to consume. Extended handling time increases the potential for carcasses to be usurped by the scavenging bears (Krofel et al. 2012; Allen et al. 2015). It has also been suggested that scavenging by bears may have been an important factor in shaping felid patterns of prey selection toward smaller prey and results in consequent morphological and behavioral adaptations of predators during the evolution of some felids, like the genus *Lynx* (Krofel et al. 2012).

In the Rocky Mountains of North America, brown bears and American black bears frequently scavenge on mountain lion kills, providing them with an important source of energy (Murphy et al. 1998; Elbroch et al. 2015). In that system, bears displaced mountain lions from 7% of their kills in winter and 12% of kills between spring and autumn (Murphy et al. 1998). Similarly, brown bears usurp ungulate carcasses killed by Eurasian lynx in Asia and Europe (Zhiryakov & Baidavletov 2003; Mattisson et al. 2011; Krofel et al. 2012), sometimes by following lynx tracks in the snow (Kostoglod 1981; Krofel & Kos 2007). Kleptoparasitic interactions between brown bears and Eurasian lynx were studied in detail in the Dinaric Mountains of Slovenia and Croatia, where the predicted proportion of all lynx kills found by bears throughout the year was 32% and lynx were estimated to lose 15% of the biomass of their ungulate kills. In response, lynx increased ungulate kill rate by 23%, but due to long searching times needed to capture new prey, they were only able to compensate for 59% of the lost biomass (Krofel et al. 2012). The frequency of bear scavenging on lynx kills was strongly dependent on bear density and activity patterns. Theft of kills was also highest during lynx pregnancy and lactation periods, suggesting that kleptoparasitism by bears could affect lynx reproductive success (Krofel et al. 2012). In Slovenia, these interactions were strongly influenced by bear management practices. Kleptoparasitism intensity differed almost threefold between different bear management zones (e.g. interaction increased in zones with higher bear protection level and more intensive artificial feeding) and the presence of an artificial bear feeding site locally increased the odds of lynx losing kills fivefold (Krofel & Jerina 2016). A bear was reported consuming a lynx in Kamchatka, but also a lynx was observed feeding on a bear carcass in the same area (Mosolov & Valentsev 2003). However, such observations could also represent scavenging rather than predation events.

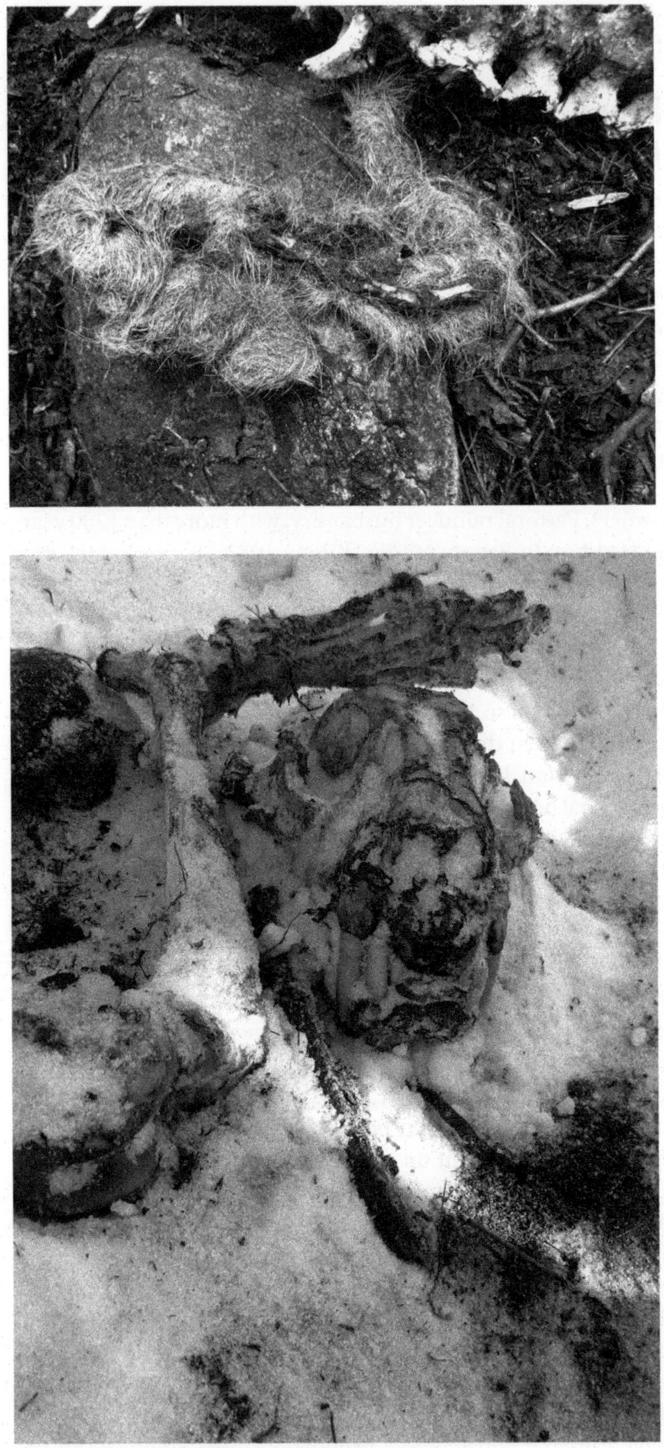

Figure 3.3 Scat of brown bear with tail and hair of tiger (top) and tiger carcass partially consumed by a brown bear (bottom). (Photos by I. Seryodkin.)

Although brown bears are typically dominant in interactions with lynx and mountain lions, the situation is less

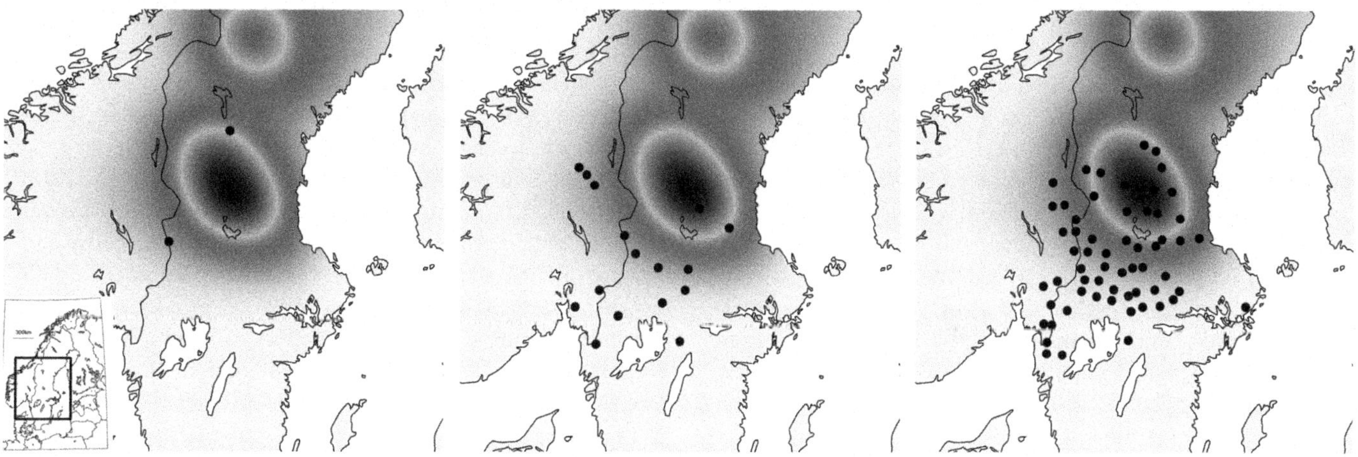

Figure 3.4 Location of wolf pairs (black dots) and brown bear density (grayscale shadows show the densest bear areas) in south-central Scandinavia, northern Europe, during the ongoing recovery phase of the wolf population, in 1985 (left), 2000 (central), and 2010 (right panel). Modified from Ordiz et al. (2015).

predictable in interactions with tigers. Among 45 documented cases of aggressive encounters between tigers and brown bears, the fight ended with the death of the bear in 51% of the cases, of the tiger in 27%, while both animals survived in 22% (Seryodkin et al. 2018). According to tiger diet analyses, brown bears can account for up to 5% of a tigers' diet, although Asiatic black bears are more commonly consumed (Tkachenko 2012). Consumption does not necessarily mean predation, as tigers also scavenge on bears that died from other causes. Although tigers kill most bears during their active period, tigers also predate on hibernating bears in their dens. Most of the brown bears killed by tigers are young animals, or females, and are generally killed by male tigers (Seryodkin et al. 2018). Predation on tigers by brown bears is probably more intensive in years with lower food availability, when bears were occasionally reported to chase tigers for long distances (Kostoglod 1981). Predation by brown bears also contributes to tiger cub mortality (Gorokhov 1997).

Despite the risk, bears will also consume the remains of tiger-killed prey. In the Russian Far East, signs of kleptoparasitism by brown or Asiatic black bears were recorded at 17% of tiger kills. In cases where brown bears were present at tiger kills, bears either used the carcass after the tigers already left (42%), usurped the kill from the tiger (25%), or both species shared the kill by alternately feeding on it (17%). For the remaining 17% of the cases, the order of use was unknown. Most of the tigers in these cases were females (Seryodkin et al. 2018). On rare occasions, tigers scavenge on bear-killed prey. For example, Rakov (1965) reported that a bear killed an adult tigress while it was eating a wild boar killed by that bear.

Finally, interactions between brown bears and common and snow leopards remain mostly unstudied. Pikunov and Korkishko (1992) reported a case from the Far East where a brown bear followed the tracks of a leopard and usurped its kill. Heptner and Sludskii (1972) mentioned the single case of a two-year-old brown bear killed and partly eaten by a snow leopard in the Tien Shan range.

Brown Bears and Canids

Brown bear populations overlap with several canid species throughout the Northern Hemisphere. For instance, coyotes (*Canis latrans*) overlap the entire brown bear distribution in North America, and bears overlap with golden jackals (*Canis aureus*) in different parts of Eurasia. However, the red fox (*Vulpes vulpes*) and the wolf have the greatest overlap with brown bears; both are sympatric with bears across most of their range in North America and Eurasia. Here, we focus on interactions between bears and wolves (Figure 3.1), which have received greater attention than interactions between bears and smaller canids.

Similar to bear–felid interactions, bears often kleptoparasitize wolf kills, both in North America (Ballard et al. 2003; Smith et al. 2003) and Europe (Milleret 2011; Ordiz et al. 2015). The outcome of interactions between bears and wolves at carcasses varies, but in general, bears often dominate and limit the access of wolves to kills (Boertje et al. 1988; MacNulty et al. 2001). For instance, Ballard et al. (2003) compiled 108 direct bear–wolf interactions in North America, outside Yellowstone National Park, and 57% of them occurred at kill sites, with bears dominating in most of those encounters (Ballard et al. 2003). In Europe, brown bears were recorded at 20% of wolf-killed prey in Slovenia, and 50–60% of wolf kills in Scandinavian studies, but wolves were never recorded at bear-killed prey (authors' unpublished data). Nevertheless, simultaneous scavenging by both species has also been reported (Smith et al. 2003; Lewis & Lafferty 2014).

Beyond interactions at specific locations, such as kill sites or dens of brown bears and wolves (Ballard et al. 2003), the potential effect of interspecific interactions between these large carnivores has been studied in Scandinavia at larger scale in recent years. The recent and ongoing recovery of the wolf population in areas where brown bears have been both continually present and absent in Sweden and Norway, in combination with long-term monitoring efforts of both species, offered a great opportunity to understand competition and coexistence between these two apex predators. Bears and wolves heavily rely on a common prey resource, i.e. seasonally available moose calves (Swenson et al. 2007; Sand et al. 2008), which sets the stage for competition. For these reasons, much of what we know about the nature of interactions between wolves and bears comes from Scandinavia, where researchers are studying bear–wolf interactions with two complementary approaches, in terms of habitat selection and kill rates.

At the landscape level and during the ongoing demographic recovery and expansion of the wolf population in Scandinavia, the establishment of wolf pairs was negatively affected by the presence of bears and humans (Ordiz et al. 2015; Figure 3.4). At the finer scale of habitat selection, inside home ranges of radio-collared bears and wolves and within the area where both species overlap in central Scandinavia, bears and wolves segregated more than expected by chance (Milleret et al. 2018). Wolves tended to select for moose occurrence, young forests, and rugged terrain more than bears, which likely reflected the different requirements of an obligate carnivore and an omnivore. However, both species generally avoided human-related habitats during daytime (Milleret et al. 2018).

Regarding predation rates, it was previously suggested that wolf kill rates would be higher where they coexisted with brown bears, because they would be forced to hunt more often to compensate for the loss of food caused by bear kleptoparasitism (e.g. Ballard et al. 2003). However, recent research suggests that wolves sympatric with bear populations may actually kill less often. A study comparing wolf kill rates with respect to bear presence–absence in Scandinavia and Yellowstone National Park found that bear presence was negatively correlated with wolf kill rate in both systems (Tallian et al. 2017). The mechanism behind this pattern is under investigation, but it could involve either interference competition, i.e. a species directly alters the resource-attaining behavior of another species, or exploitation competition, i.e. species interact indirectly competing for common resources, such as territory or food, or both. Bear presence may increase wolf handling time (e.g. wolves stay longer at a kill to defend or gain access to the carcass), search time (e.g. wolf and bear predation both reduce the availability of shared prey, making them more difficult to find), or both. Regardless of the mechanism explaining these results, they suggest that the total impact of wolves and brown bears on prey may be less than the sum of their individual impacts. If so, interactions between wolves and bears may partially dampen the effect of the other carnivore on ungulate mortality (Tallian et al. 2017).

Habitat- and predation-related studies (Ordiz et al. 2015; Tallian et al. 2017; Sanz-Pérez et al. 2018) suggest that brown bears have a negative impact on wolf habitat selection and foraging ability, such that wolf populations that coexist with brown bears may suffer fitness consequences. However, wolf presence may have the opposite effect on bear populations. Wolf kills are an important source of protein for scavenger communities, including brown and American black bears (Wilmers et al. 2003). Wolves kill ungulates consistently throughout the course of a year, providing a spatiotemporal stable source of meat for bears and other scavengers (Wikenros et al. 2013). Consistent access to wolf kills may even provide a buffer to climate change-induced reductions in winter-killed ungulates. Shorter winters may lower ungulate mortality, with fewer carcasses available for scavengers, and this is an important source of post-denning protein for bears in some systems (Wilmers & Getz 2005).

Ultimately, determining the energetic costs of interactions between bears and other large carnivores, and ascertaining their consequences for predator population dynamics, will help us understand the costs of sympatry among apex predator populations. Understanding the costs of sympatry will allow us to clarify the effect of carnivore coexistence on prey populations and lower trophic levels. Gaining detailed information on coexisting apex predators will also help tune the adaptive management of economically important ungulate game species, such as moose, elk or red deer, and other deer species. In this sense, the current recovery of some large carnivores in Europe (Chapron et al. 2014) and North America (Bruskotter & Shelby 2010), along with the expanded use of GPS-based telemetry, should provide opportunities to study interspecific competition among sympatric apex predators at both the individual and population level across multiple ecosystems. Brown bears co-occur with multiple large carnivores (Figure 3.1) and ungulate species throughout their current distribution in the Northern Hemisphere, which highlights the importance of understanding the outcomes of interspecific interactions among carnivores and with their prey populations.

References

Allen, M. L., Elbroch, L. M., Wilmers, C. C. & Wittmer, H. U. (2015). The comparative effects of large carnivores on the acquisition of carrion by scavengers. *The American Naturalist* 185: 822–833.

Apps, C. D., McLellan, B. N. & Woods, J. G. (2006). Landscape partitioning and spatial inferences of competition between black and grizzly bears. *Ecography* 29: 561–572.

Arthur, S.M. & Vecchio, P.A. Del. (2017). Effects of grizzly bear predation on muskoxen in northeastern Alaska. *Ursus* **28**: 81–91.

Ballard, W., Carbyn, L. N. & Smith, D. (2003). Wolf interactions with non-prey. In: Mech, L. D. & Boitani, L. (Eds.), *Wolves: Behavior, ecology, and conservation* (pp. 259–271). Chicago, IL: University of Chicago Press.

Boertje, R., Keech, M. & Paragi, T. (2010). Science and values influencing predator control for Alaska moose management. *Journal of Wildlife Management* 74: 917–928.

Boertje, R. D., Gasaway, W. C., Grangaard, D. V. & Kelleyhouse, D. G. (1988). Predation on moose and caribou by radio-collared grizzly bears in east central Alaska. *Canadian Journal of Zoology* 66: 2492–2499.

Bruskotter, J. T. & Shelby, L. B. (2010). Human dimensions of large carnivore conservation and management. *Human Dimensions of Wildlife* 15: 311–314.

Caro, T. M. & Stoner, C .J. (2003). The potential for interspecific competition among African carnivores. *Biological Conservation* 110: 67–75.

Chapron, G., Kaczensky, P., von Arx, M., et al. (2014). Recovery of large carnivores in Europe's modern human-dominated landscapes. *Science* 346: 1517–1519.

Dahle, B., Wallin, K., Cederlund, G., et al. (2013). Predation on adult moose *Alces alces* by European brown bears. *Wildlife Biology* 19: 165–169.

Derocher, A. E., Wiig, Ø. & Bangjord, G. (2000). Predation of Svalbard reindeer by polar bears. *Polar Biology* 23: 675–678.

DeVault, T. L., Rhodes, O. E. & Shivik, J. A. (2003). Scavenging by vertebrates: behavioral, ecological, and evolutionary perspectives on an important energy transfer pathway in terrestrial ecosystems. *Oikos* 102: 225–234.

Elbroch, L. M., Lendrum, P. E., Allen, M. L. & Wittmer, H. U. (2015). Nowhere to hide: pumas, black bears, and competition refuges. *Behavioral Ecology* 26: 247–254.

Frank, J., Støen, O. G., Segerström, P., et al. (2017). *Kalvning i hägn och områdesriktad jakt på björn som åtgärder för att minska björnars predation på ren.* Rapport från Viltskadecenter 7. Sveriges Lantbruksuniversitet SLU, Grimsø forskningsstation, Riddarhyttan, Sweden. [In Swedish.] https://pub.epsilon.slu.se/15369/7/frank_j_et_al_180425.pdf.

Garrott, R. A., White, P. J. & Watson, F. G. R. (2009). *The ecology of large mammals in central Yellowstone: Sixteen years of integrated field studies.* Terrestrial Ecology Series Volume 3. San Diego, CA: Academic Press.

Gervasi, V., Nilsen, E. B., Sand, H., et al. (2012). Predicting the potential demographic impact of predators on their prey: a comparative analysis of two carnivore–ungulate systems in Scandinavia. *Journal of Animal Ecology* 81: 443–454.

Gorokhov, G. F. (1997). The abundance and structure of the Amur tiger population in southern Sikhote-Alin, in Redkie vidy mlekopitayushchikh i ikh okhrana. In: *Proceedings 2nd All-Union Conf. "Rare Species of Mammals and Their Conservation* (pp. 119–120). Moscow: Nauka.

Griffin, K. A., Hebblewhite, M., Robinson, H. S., et al. (2011). Neonatal mortality of elk driven by climate, predator phenology and predator community composition. *Journal of Animal Ecology* 80: 1246–1257.

Heptner, V. G. & Sludskii, A. A. (1972). Mlekopitayushchiye Sovetskogo Soyuza. In *Mammals of the Soviet Union, Carnivora.* Moscow: Vysshaya Shkola/

Huygens, O. C., Miyashita, T., Dahle, B., Carr, M. & Izumiyama, S. (2003). Diet and feeding habits of Asiatic black bears in the Northern Japanese Alps. *Ursus* 14: 236–245.

Hwang, M.-H., Garshelis, D. L. & Wang, Y. (2002). Diets of Asiatic black bears in Taiwan, with methodological and geographical comparisons. *Ursus* 13: 111–125.

Jonzén, N., Sand, H., Wabakken, P., et al. (2013). Sharing the bounty – adjusting harvest to predator return in the Scandinavian human–wolf–bear–moose system. *Ecological Modelling* 265: 140–148.

Kaczensky, P. (1999). Large carnivore depredation on livestock in Europe. *Ursus* 11: 59–72.

Karlsson, J., Støen, O. G., Segerström, P., et al. (2012). *Björnpredation på ren och potentiella effekter av tre förebyggande åtgärder.* Rapport från Viltskadecenter 6. Sveriges Lantbruksuniversitet SLU, Grimsø forskningsstation, Riddarhyttan, Sweden. [In Swedish.]

Keech, M. A., Bowyer, R. T., Hoef, J. M. Ver, et al. (2000). Life-history consequences of maternal condition in Alaskan moose. *The Journal of Wildlife Management* 64: 450–462.

Kindberg, J., Swenson, J. E., Ericsson, G., et al. (2011). Estimating population size and trends of the Swedish brown bear *Ursus arctos* population. *Wildlife Biology* 17: 114–123.

Knight, R. R. & Judd, S. (1979). Grizzly bears that kill livestock. *International Conference on Bear Research and Management* 5: 186–190.

Kostoglod V. E. (1981). The experience of prolonged tracking of the brown bear in Sikhote-Alin. *Bulletin of Moscow Society of Naturalists. Biological series* 86: 3–12.

Krofel, M. & Jerina, K. (2016). Mind the cat: conservation management of a protected dominant scavenger indirectly affects an endangered apex predator. *Biological Conservation* 197: 40–46.

Krofel, M. & Kos, I. (2007). Evidence of the brown bear (*Ursus arctos*) tracking the Eurasian lynx (*Lynx lynx*) on the Snežnik plateau, Slovenia. *Natura Sloveniae* 9: 45–46.

Krofel, M., Kos, I. & Jerina, K. (2012). The noble cats and the big bad scavengers: effects of dominant scavengers on solitary predators. *Behavioral Ecology and Sociobiology* 66: 1297–1304.

Lewis, T. M. & Lafferty, D. J. R. (2014). Brown bears and wolves scavenge humpback whale carcass in Alaska. *Ursus* 25: 8–13.

MacNulty, D., Varley, N. & Smith, D. (2001). Grizzly bear, *Ursus arctos*, usurps bison calf, *Bison bison*, captured by wolves, *Canis lupus*, in Yellowstone National Park, Wyoming. *Canadian Journal of Zoology* 115: 495–498.

Mattisson, J., Persson, J., Andrén, H. & Segerström, P. (2011). Temporal and spatial interactions between an obligate predator, the Eurasian lynx (*Lynx lynx*), and a facultative scavenger, the wolverine (*Gulo gulo*). *Canadian Journal of Zoology* 89: 79–89.

Milleret, C. (2011). Estimating wolves (*Canis lupus*) and brown bear (*Ursus arctos*) interactions in Central Sweden. Does the emergence of brown bears affect wolf predation patterns? Masters thesis, Université Joseph Fourier-Grenoble.

Milleret, C., Ordiz, A., Chapron, G., et al. (2018). Habitat segregation between brown bears and gray wolves in a human-dominated landscape. *Ecology and Evolution* 8: 1–17.

Monteith, K. L., Bleich, V. C., Stephenson, T. R. & Pierce, B. M. (2014). Life-history characteristics of mule deer: effects of nutrition in a variable environment. *Wildlife Monographs* 186: 1–64.

Mosolov, V. & Valentsev, A. (2003). Lynx of Eurasia. The Kamchatka. In: *The lynx. Regional features of ecology, use and protection* (pp. 408–423). Moscow: Nauka.

Murphy, K., Felzien, G., Hornocker, M. & Ruth, T. (1998). Encounter competition between bears and cougars: some ecological implications. *Ursus* 10: 55–60.

Ordiz, A., Bischof, R. & Swenson, J. E. (2013). Saving large carnivores, but losing the apex predator? *Biological Conservation* 168: 128–133.

Ordiz, A., Milleret, C., Kindberg, J., et al. (2015). Wolves, people, and brown bears influence the expansion of the recolonizing wolf population in Scandinavia. *Ecosphere* 6: 1–14.

Ordiz, A., Sæbø, S., Kindberg, J., Swenson, J. E. & Støen, O. G. (2017). Seasonality and human disturbance alter brown bear activity patterns: implications for circumpolar carnivore conservation? *Animal Conservation* 20: 51–60.

Palomares, F. & Caro, T. M. (1999). Interspecific killing among mammalian carnivores. *The American Naturalist* 153: 492–508.

Périquet, S., Fritz, H. & Revilla, E. (2015). The lion king and the hyaena queen: large carnivore interactions and coexistence. *Biological Reviews* 90: 1197–1214.

Pikunov, D. G. & Korkishko, V. G. (1992). *Leopard of Far East.* Moscow: Nauka [in Russian].

Rakov, N. V. (1965). The current distribution of the tiger in the Amur-Ussuri region. *Zoologicheskii Zhurnal* 44: 433–441.

Ruong, I. (1982). *Samerna i historien och nutiden.* Stockholm: BonnierFakta.

Sand, H., Wabakken, P., Zimmermann, B., et al. (2008). Summer kill rates and predation pattern in a wolf–moose system: can we rely on winter estimates? *Oecologia* 156: 53–64.

Sanz-Pérez, A., Ordiz, A., Sand, H., et al. (2018). No place like home? A test of the natal habitat-biased dispersal hypothesis in Scandinavian wolves. *Royal Society Open Science* 5: 181379.

Sato, Y., Mano, T. & Takatsuki, S. (2005). Stomach contents of brown bears (*Ursus arctos*) in Hokkaido, Japan. *Wildlife Biology* 11: 133–144.

Schoener, T. W. (1983). Field experiments on interspecific competition. *The American Naturalist* 122: 240–285.

Schwartz, C. C., Cain, S. L., Podruzny, S., Cherry, S. & Frattaroli, L. (2010). Contrasting activity patterns of sympatric and allopatric black and grizzly bears. *Journal of Wildlife Management* 74: 1628–1638.

Seryodkin, I. V., Miquelle, D. G., Goodrich, J. M., Kostyria, A. V. & Petrunenko, Y. K. (2018). Interspecific relationships between the Amur tiger (*Panthera tigris altaica*) and brown (*Ursus arctos*) and Asiatic black (*Ursus thibetanus*) bears. *Biology Bulletin* 45: 853–864.

Sinclair, A. R. E. & Krebs, C. J. (2002). Complex numerical responses to top-down and bottom-up processes in vertebrate populations. *Philosophical Transactions of the Royal Society of London B* 357: 1221–1231.

Sivertsen, T. R. (2017). Risk of brown bear predation on semi-domesticated reindeer calves interactions and landscape heterogeneity. Doctoral thesis. Swedish University of Agricultural Sciences, Uppsala, Sweden.

Smith, D. W., Peterson, R. O. & Houston, D. B. (2003). Yellowstone after wolves. *BioScience* 53: 330–340.

Stirling, I. & Derocher, A. (1990). Factors affecting the evolution and behavioral ecology of the modern bears. *Bears: Their Biology and Management* 8: 189–204.

Swenson, J. E., Wabakken, P., Sandegren, F., et al. (1995). The near extinction and recovery of brown bears in Scandinavia in relation to the bear management policies of Norway and Sweden. *Wildlife Biology* 1: 11–25.

Swenson, J., Dahle, B., Busk, H., et al. (2007). Predation on moose calves by European brown bears. *Journal of Wildlife Management* 71: 1993–1997.

Swenson, J. E., Schneider, M., Zedrosser, A., et al. (2017). Challenges of managing a European brown bear population; lessons from Sweden, 1943–2013. *Wildlife Biology* 1: wlb.00251.

Tallian, A., Ordiz, A., Metz, M. C., et al. (2017). Competition between apex predators? Brown bears decrease wolf kill rate on two continents. *Proceedings of the Royal Society B: Biological Sciences* 284: 20162368.

Tkachenko, K. N. (2012). Specific features of feeding of the Amur tiger *Panthera tigris altaica* (Carnivora, Felidae) in a densely populated locality (with reference to Bol'shekhekhtsirskii Reserve and its environs). *Biology Bulletin* 39: 279–287.

Wikenros, C., Sand, H., Ahlqvist, P. & Liberg, O. (2013). Biomass flow and scavengers use of carcasses after re-colonization of an apex predator. *PLoS One* 8: e77373.

Wilmers, C. C. & Getz, W. M. (2005). Gray wolves as climate change buffers in Yellowstone. *PLoS Biology* 3: e92.

Wilmers, C. C., Crabtree, R. L., Smith, D. W., Murphy, K. M. & Wayne, M. (2003). Trophic facilitation by introduced top predators: grey wolf subsidies to scavengers in Yellowstone National Park. *Journal of Animal Ecology* 72: 909–916.

Wisz, M. S., Pottier, J., Kissling, W. D., et al. (2013). The role of biotic interactions in shaping distributions and realised assemblages of species: implications for species distribution modelling. *Biological Reviews* 88: 15–30.

Wyman, C. (2002). Grizzly on a bull bison in Yellowstone National Park. *Ursus* 13: 375–377.

Zabel, A. & Holm-Müller, K. H. (2008). Conservation performance payments for carnivore conservation in Sweden. *Conservation Biology* 22: 247–251.

Zager, P. & Beecham, J. (2006). The role of American black bears and brown bears as predators on ungulates in North America. *Ursus* 17: 95–108.

Zhiryakov, V. A. & Baidavletov, R. Z. (2003). Lynx of Eurasia. Kazakhstan. In: *The lynx. Regional features of ecology, use and protection* (pp. 344–367). Moscow: Nauka.

Chapter 4

Adaptations and Competitive Interactions of Tropical Asian Bear Species Define Their Biogeography: Past, Present, and Future

Robert Steinmetz, David L. Garshelis, and Anwaruddin Choudhury

Three potentially competing bear species inhabit tropical Asia: the sloth bear (*Melursus ursinus*), sun bear (*Helarctos malayanus*), and Asiatic black bear (henceforth black bear; *Ursus thibetanus*) (Figure 4.1). Sun bears (30–80 kg), the smallest species of bear in the world, are about half the size of black bears (65–150 kg) and sloth bears (55–145 kg). Sloth bears occur in the Indian subcontinent (Nepal, India) and Sri Lanka. Black bears range from the Russian Far East to South-East Asia and west to Iran, crossing northern and north-eastern India. Sun bears occupy mainland South-East Asia and the two large islands of Sumatra and Borneo, but only extend west to north-eastern India and north to Yunnan, China. Sloth bears and sun bears are wholly tropical–subtropical species, whereas black bears range widely into temperate regions of Asia.

Geographic ranges of the black bear and sun bear overlap extensively. Ranges of the sloth bear and black bear abut, and slightly overlap, along the edges of their respective ranges in the Terai arc of southern Nepal and northern India, and may have once overlapped more in north-eastern India. The ranges of all three species converge in north-east India (Figure 4.2).

What factors generate the separation of sloth bears geographically from black and sun bears? What factors facilitate the extensive sympatry of black bears and sun bears? How are these patterns structured by evolutionary history and competition between bear species, and what mechanisms facilitate their coexistence or maintain their separation? Has current forest loss and degradation benefited one species over another? If so, has interspecific competition played a part? These questions are the focus of this chapter.

The spatial distribution of a species is determined by three factors: (1) evolutionary history and dispersal capacities; (2) the distribution of resources and environmental conditions favorable to survival and reproduction; and (3) the biotic environment involving predators, pathogens, and competitors (Wiens 2011; Louthan et al. 2015). The necessary resources and conditions (factor 2) constitute the fundamental niche of a species. Species distributions can be constrained by competitors (factor 3), causing realized niches to be smaller than fundamental niches (Holt 2009). In terms of biogeography this means that places outside the observed distribution of a species might be suitable (in terms of resources, climate, etc.), but are occupied by a competitor that is better able to use and monopolize the resources, or physically exclude occupation by another similar species.

Feeding Habits of Tropical Asian Bears

All three tropical bear species are omnivorous, yet there are distinct differences in their diets. Each consumes both fruits and insects, although in different proportions. The sloth bear possesses a unique suite of morphological adaptations specifically for myrmecophagy (ant- and termite-eating): the absence of two upper incisors, a long snout with a vaulted palate, mobile protrusible lips, and closable nostrils (Garshelis et al. 1999; Sacco & Van Valkenburgh 2004). These features enable sloth bears to efficiently suck in large quantities of termites and ants (see also Chapter 9). This capability is so impressive (and social insects abundant in its habitat; Figure 4.3) that the sloth bear is able to subsist almost solely on a diet of insects, violating an otherwise general pattern in carnivore energetics that constrains invertebrate-feeding species to body masses less than 22 kg (Carbone et al. 1999). Fruits, when abundant, may comprise up to 90% of the sloth bear diet; however, termites, ants, and other insects comprise >80% of the diet during non-fruiting periods (Seidensticker et al. 2011), and may dominate the diet year-round in some areas (Joshi et al. 1997).

The sun bear too possesses morphological features facilitating insectivory (see also Chapter 8). Proportionate to its body size, the sun bear has especially large, strong canines (Christiansen 2008). The skull is wide, supporting large masseter muscles, giving the sun bear an exceptionally powerful bite force. Together, these features might be functional adaptations for breaking into hard termite colonies and for tearing into the trunks of tropical hardwood trees to access nests of stingless bees (Fredriksson 2012). Alternately, these adaptations might have evolved primarily for self-defense, as the sun bear is small-bodied and must contend with potentially dangerous predators including tigers, leopards, dholes, and black bears; nevertheless, their teeth and skull certainly function well to enhance insect-feeding. When fruit is scarce, as occurs periodically between masting events in Indonesia and Malaysia, insects are the dietary mainstay for sun bears; but when fruits are sufficiently available, sun bears are almost

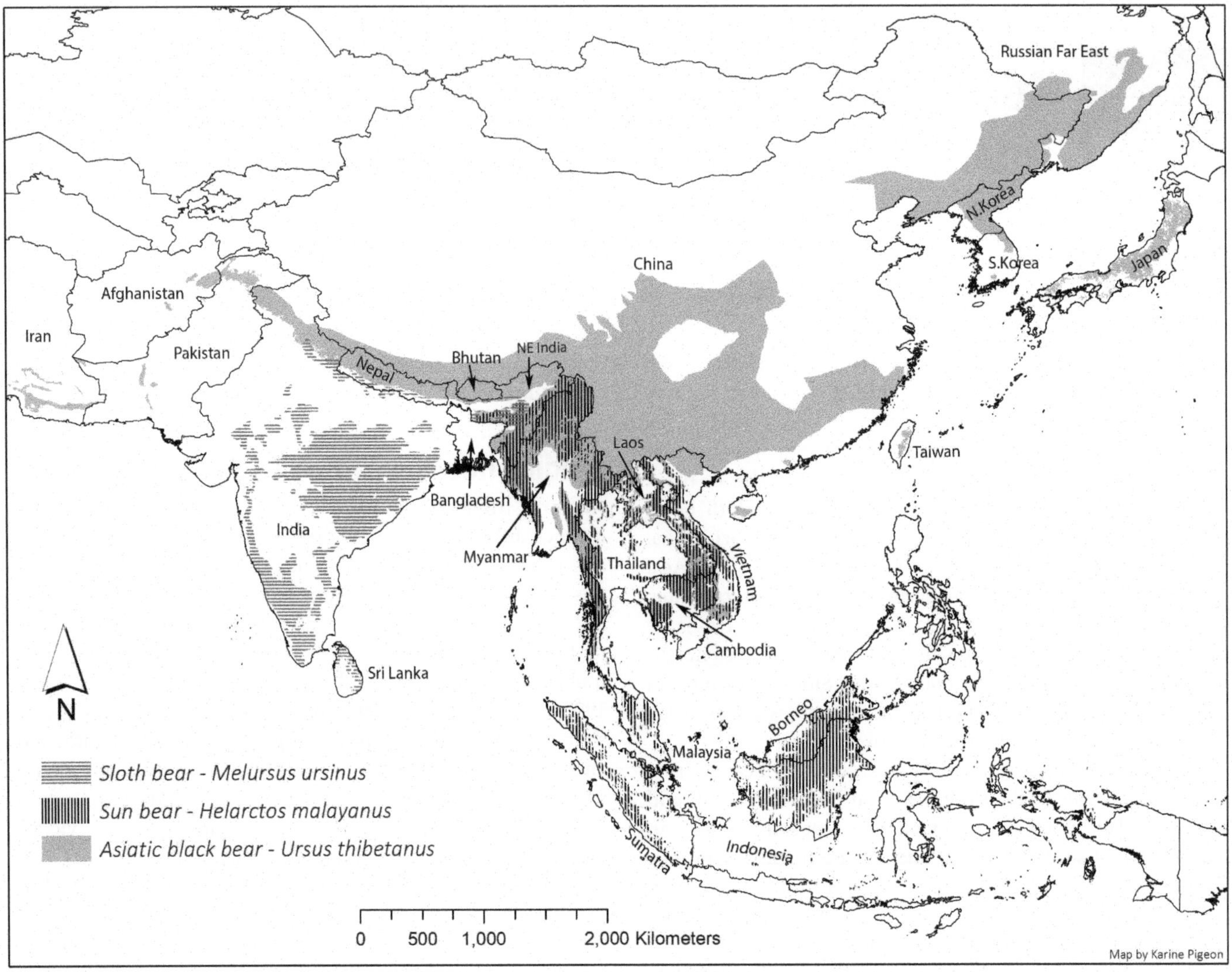

Figure 4.1 Global distributions of Asiatic black bears, sloth bears, and sun bears. Maps are from the Bear Specialist Group of IUCN (International Union for Conservation of Nature), combining definite and possible range.

completely frugivorous (Wong et al. 2002; Fredriksson et al. 2006; Steinmetz et al. 2011).

Black bears feed mainly on fruit (Steinmetz et al. 2011). Insects generally constitute a small proportion of their diet (0–4% relative volume) throughout their range (Hwang et al. 2002; Huygens et al. 2003; Garshelis, 2004; see also Chapter 10). Thus, both sun bears and black bears feed predominantly on fruit when available, with sun bears in particular falling back on insects when fruit is scarce (Hwang et al. 2002; Wong et al. 2002; Fredriksson et al. 2006).

A key difference between these species is their ability to subsist for an extended time on insects. Sloth bears clearly can. Although sun bears are highly insectivorous when necessary, they become emaciated and some even starve during lengthy fruiting failures (Wong et al. 2005; Fredriksson 2012). Black bears also regularly eat insects during seasons with low fruit availability (Koike 2010), but they cannot meet their energy demands on an exclusively insect diet (Yamazaki et al. 2012). Thus, black bears and sun bears depend on fruits, whereas sloth bears can efficiently exploit fruit-poor habitats with an abundance of insects. These characteristics of morphology, feeding habits, and niches have largely dictated the geographical distributions and ecological overlap of the three species.

Range Limits, Overlap, and Separation

The sloth bear is restricted to the Indian subcontinent, with at least 90% of its range in India. There is no evidence, presently or historically, that this species ranged east into Myanmar. The black bear and sun bear both occur in Myanmar and across South-East Asia, but the black bear does not range south into Malaysia, where rainforests replace seasonal tropical forests. Sun bears continue south into the Sundaic region, which includes the southern Malay Peninsula and the islands of Sumatra and Borneo, but northward they do not cross the Red River in northern Vietnam or the Brahmaputra in Assam, India.

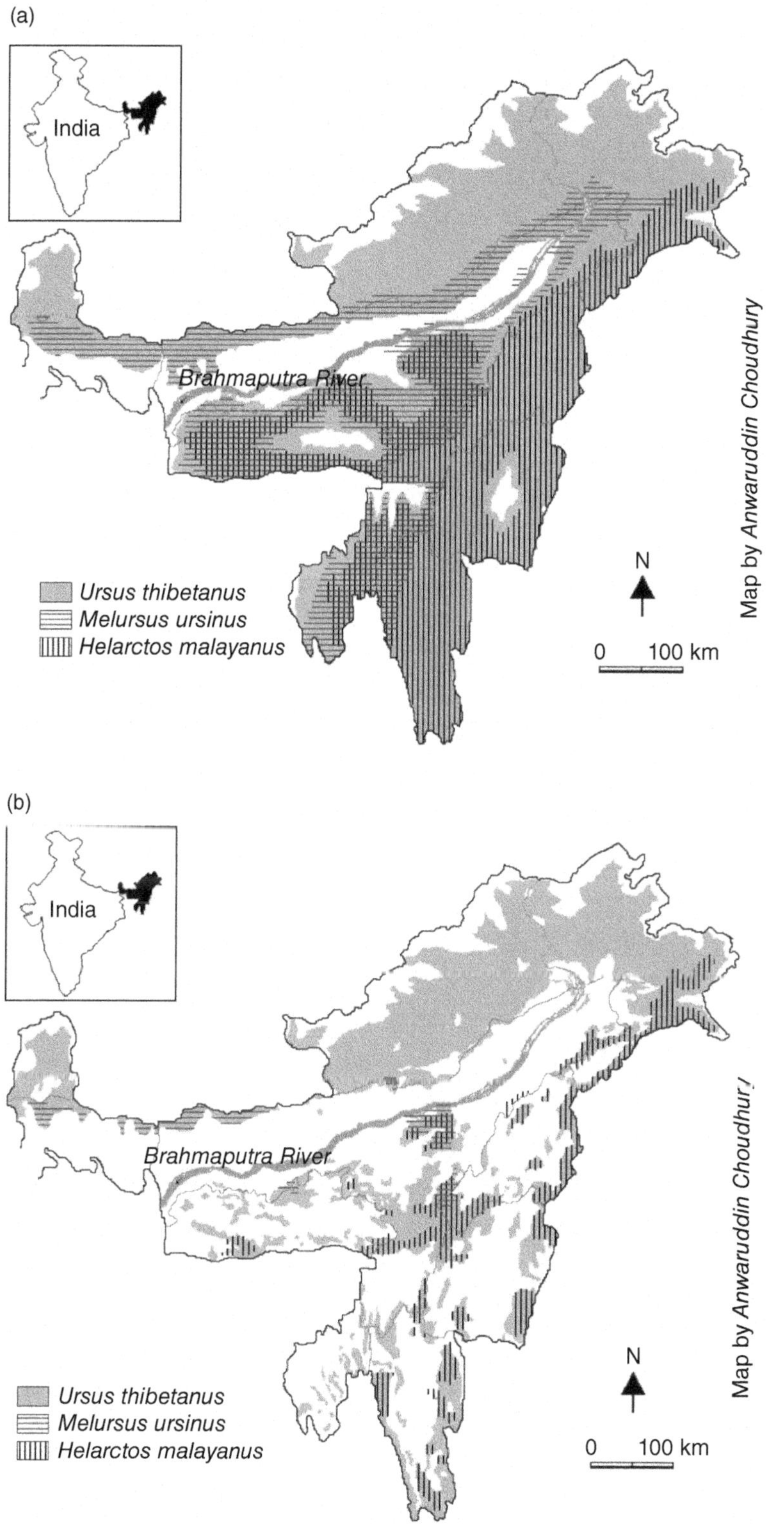

Figure 4.2 Distribution and overlap of Asiatic black bears, sloth bears, and sun bears in north-east India, (a) historically, *ca.* 1950, and (b) currently. This region of convergence represents the eastern extent of sloth bears and western extent of sun bears, and the only place where all three tropical Asian bears once overlapped. The data are gleaned from various sources of literature, interviews, photographs, and specimens, and point locations were connected to form estimated ranges.

Sloth bear and black bear distributions overlap in one small area in the Indian state of Uttarakhand (including Corbett and Rajaji National Parks) within the Terai arc, although on a fine scale the actual overlap may be quite limited (Seidensticker et al. 2011). In Nepal, the ranges of these two species are mainly allopatric, with black bears in the Middle Hills and sloth bears in the lowland Terai, but recently a few black bears have been camera-trapped in two Terai national parks, Bardia and Banke (Yadav et al. 2017). Moving eastward, their ranges diverge such that only sloth bears occur in Chitwan National Park, Nepal,

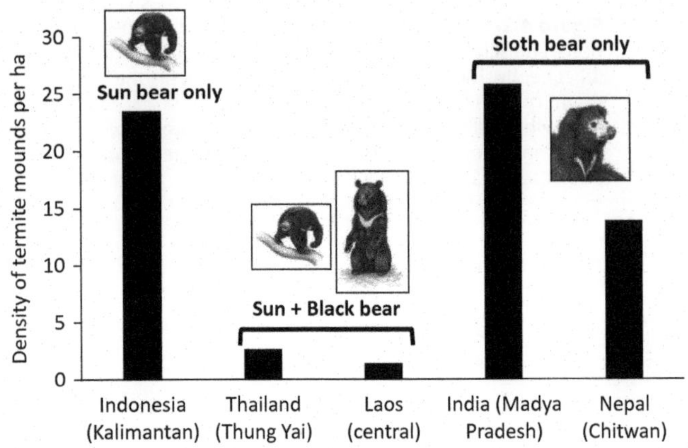

Figure 4.3 Differences in availability of epigeal (mound-building) termites in different geographic zones inhabited by the tropical Asian bears. Termite data are from: Indonesia (Mathisen 2003), Thailand (Steinmetz et al. 2011), Laos (Miyagawa et al. 2011), central India (Akhtar et al. 2004), and Nepal (Joshi et al. 1997).

and only black bears occur in Bhutan. Just south of Bhutan, in Manas National Park, India, the two species co-occur.

Within north-eastern India there is a patchwork (Figure 4.2b), with some areas occupied by only one species (e.g. only sloth bears in Kaziranga National Park, with one record of a black bear), and some occupied by two species (black bears with either sloth bears or sun bears). Historically, there appears to be a limited degree of overlap between sloth bears and sun bears (Figure 4.2a). However, the distribution of each of these species in north-east India remains somewhat uncertain because they are often confused, by local people, by government authorities, and even in camera trap photos. This makes many reports of species overlap suspect, both presently and historically. It is unclear whether there is (or was) any area where all three species share(d) the same portion of forest. There is tentative evidence that all three species may occur in northern parts of Karbi Anglong in Assam, during the rainy season when sloth bears purportedly move there from flooded areas in Kaziranga (Choudhury 2001, 2011).

The distributions of black bears and sun bears in mainland South-East Asia (Myanmar, Thailand, Laos, Cambodia, and Vietnam) are better known. Within this area, which spans millions of square kilometers, sun bears and black bears commonly co-occur within most habitats.

Black bear range mainly coincides with broadleaved forests. They are limited northward (in the Russian Far East; Figure 4.1) by boreal coniferous forest (occupied by brown bears, *Ursus arctos*), and limited elevationally by treeline. They occur up to 4300 m in India, and possibly even higher in Nepal. In mainland South-East Asia, black bears are commonly found in montane habitats; conversely, sun bears are more commonly found at lower elevations but tend to be rare at high elevations (Steinmetz 2011; Steinmetz et al. 2011). However, in Borneo and Sumatra, where black bears are absent, sun bears reach to at least 2000 m (Tumbelaka & Fredriksson 2006); they were frequently camera-trapped at 700–1940 m in Sumatra (Linkie et al. 2007). The sloth bear is primarily a lowland species, occurring mostly below 500 m. This elevation range corresponds to termite and ant availability, which declines sharply with increasing elevation in the tropics (Collins 1980).

Fossils Shed Light on Past Ranges

Are the geographic ranges of tropical Asian bears recent arrangements, or have they been this way for millennia? Sloth bears, black bears, and sun bears originated from a common ancestor in Eurasia during the Miocene to Pliocene transition, 5–6 million years ago (Krause et al. 2008). This period was characterized by the global climate becoming cooler and drier, which caused the fragmentation of forests and expansion of savanna and grassland (Zachos et al. 2001; Woodburne 2004). Savanna ecosystems are often rich in insect abundance but poor in fruit abundance; they can support extremely high densities of social insects like termites (Ferrar 1982). A food this small yet so abundant would be an evolutionary pressure promoting feeding specializations for rapid consumption, and other adaptations for living on a diet low in carbohydrates. The evolution of ecological specialization tends to be favored by the abundance (as opposed to scarcity) of a particular environmental state or resource (Futuyma 1988). It may be that this new niche, afforded by a habitat that was rich in insect availability but likely poor in fruit abundance, promoted development of the unique feeding adaptations of the sloth bear. These specialized feeding habits may have also limited the sloth bear to the region where the food to which they were adapted was sufficient to sustain them.

Few sloth bear fossils have been found, and, with one possible exception, all occur within the current range of this species. A fossil bear named *Ursus theobaldi*, identified only from a few damaged teeth, might actually be a sloth bear. Its location in north-western India (Erdbrink 1953), well beyond the current distribution, indicates their distribution might have been wider historically, and perhaps overlapped more extensively with black bears, which currently inhabit that area and presumably did so in the past.

Whereas sloth bears specialized, and accordingly thrived only in the region where termites and ants were abundant, black bears remained generalists, little evolved from the prototypic form. Fossil remains of this species are difficult to distinguish from *Ursus minimus*, their presumed ancestor, and have led to many apparent misinterpretations of their former range (Wagner et al. 2011). Purported *U. thibetanus* fossils have been found across a very broad area, as far north as the Ural Mountains (Russia) and Germany, and as far west as France, dating from the early Pliocene to late Pleistocene; however, it now seems likely that this species originally evolved in Asia and spread to Europe for a time in the early Pleistocene. Black bear fossils from the Middle and Late Pleistocene are widespread in Asia, occurring in Vietnam, China, and Thailand (Tougard 2001), but do not extend south into Malaysia or Indonesia, or into the Indian subcontinent.

In contrast to the black bear, sun bear fossils are scarce in mainland Asia. However, there are three locales where both sun bear and black bear fossils have been found: two in northern Vietnam (Tougard 2001; Bacon et al. 2008), and one in Shanxi province, northern China (well north of the presumed Pleistocene range for this species, creating some doubt as to the location of this find; Erdbrink 1953). Late Pleistocene sun bear fossils are far more widespread in Sumatra, Borneo, and Java (where they went extinct less than 10,000 years ago) (Bacon et al. 2008).

This fossil history imparts the following conclusions relevant to species interactions. First, there is no evidence that ranges of black bears and sloth bears ever broadly overlapped. That is, their current separation, with marginal overlap in places, does not seem to result from range contraction, but rather sloth bears evolving to take advantage of an area where black bears could not live. Second, black bears and sun bears have coexisted in mainland South-East Asia for hundreds of thousands of years. They have similar ecological needs, yet one does not outcompete and exclude the other, at least on a broad scale. Third, sun bears never ranged west beyond north-east India (claims of their occurrence in Nepal were in error). The western ecological limits for this species remain unclear, but suffice to say that north-east India is the western limit of sun bears, the eastern limit of sloth bears, and central within the range of black bears, making it an intriguing area of interest (Figure 4.2).

Why are Sloth Bears Disjunct From Black Bears and Sun Bears?

The black bear occupies a wide range of habitats, from an arid, sparsely vegetated landscape in Iran to tropical forests of South-East Asia to high elevations in the Himalayas. Likewise, the sun bear occupies habitats as diverse as dry forest with sparse tree cover in Laos, to dense rainforest with towering canopies of dipterocarps in Borneo. It would seem that the fundamental niches of black bears and possibly sun bears would include forested habitats of peninsular India, and that they have had sufficient time historically to reach the area – so why don't they occur there? We believe there are two reasons: (1) low abundance and diversity of fruits, and (2) competitive exclusion by sloth bears. Sloth bear habitats in the Indian subcontinent appear to be particularly poor in fruit availability, yet relatively rich in termites (Figure 4.3), a food to which the species is uniquely adapted. For example, average density of fruit trees in a sloth bear-occupied sal forest in India was 38 per ha, or just 19% of overall tree density (Akhtar et al. 2004), whereas at a site in Thailand (where sloth bears are absent but black and sun bears coexist) there were 158–223 fruit trees per ha, comprising 40–53% of overall tree density (Steinmetz et al. 2013). In contrast, termites are 6–20 times more abundant in sloth bear habitat than outside sloth bear range (Figure 4.3).

Over evolutionary time scales, the prior adaptation and occupation by one species can preclude another species from evolving to occupy the same habitat (Wiens 2011). This process, known as niche pre-emption, integrates abiotic and biotic influences on species distribution (Wiens 2011). When the black bear arrived at the boundary of sloth bear-occupied territory, it was excluded from invading this region by the existing occupation of a more specialized species that was well-adapted for efficient exploitation of the resources there (Futuyma 1988). Unlike in mainland South-East Asia, where abundant fruit, combined with the sun bear's ability to supplement its diet with less-preferred insects (see below) facilitates coexistence of black bears and sun bears, sloth bear range is mainly rich in insects, and fruit abundance is too scanty to support another frugivorous bear species (Figure 4.4).

Another example of niche pre-emption occurred in Japan. Black bears dispersed into Japan from continental Asia during the mid-Pleistocene, but remained within a small region of Honshu Island, even after tens of thousands of years, seemingly constrained by the larger-bodied brown bear, which already occupied much of the region (Wu et al. 2015). Fossil evidence and mtDNA indicate that the brown bear went extinct on Honshu Island during the late Pleistocene, after which the black bear expanded its range (but never occupied Hokkaido Island, where brown bears still exist). The extinction of the brown bear, probably a major potential competitor, on Honshu apparently opened niche space for the black bear, which until then had been excluded from otherwise suitable habitat.

Niche pre-emption does not require the winning species to be larger. For example, among the many islands off the coast of Alaska, some are occupied by brown bears and some (mainly smaller ones) by American black bears (*U. americanus*), and there is evidence that very high densities of the latter, smaller species preclude colonization by brown bears (Mattson et al. 2005). In India, we posit that whereas sloth bears and black bears are similar in size, sloth bears gained an advantage by colonizing the subcontinent earlier, and likely reached high densities there, as they require a relatively small amount of space (Garshelis et al. 1999). They also maintained a competitive advantage in their ability to thrive in this fruit-poor, insect-rich environment. Both of these advantages precluded occupancy by black bears (and possibly sun bears).

We now turn to the opposite question: what limits sloth bears to the north (where black bears occur) and east (where both black bears and sun bears occur)? The answer might again be related to food supply in relation to competitive advantages. The uplands and mountains to the north and east of sloth bear range appear to have much lower termite abundance than the Indian subcontinent (Garshelis et al. 2015; Figure 4.3). Thus, the sloth bear's advantage in termite-rich regions wanes outside of its current range, and this is perhaps exacerbated further by competition from black bears and sun bears.

Why do Black Bears and Sun Bears Broadly Overlap?

Studies in Thailand have shown that black bears and sun bears use the same habitats and both rely mostly on fruit, tracking the same fruiting tree species through time (Ngoprasert et al. 2011;

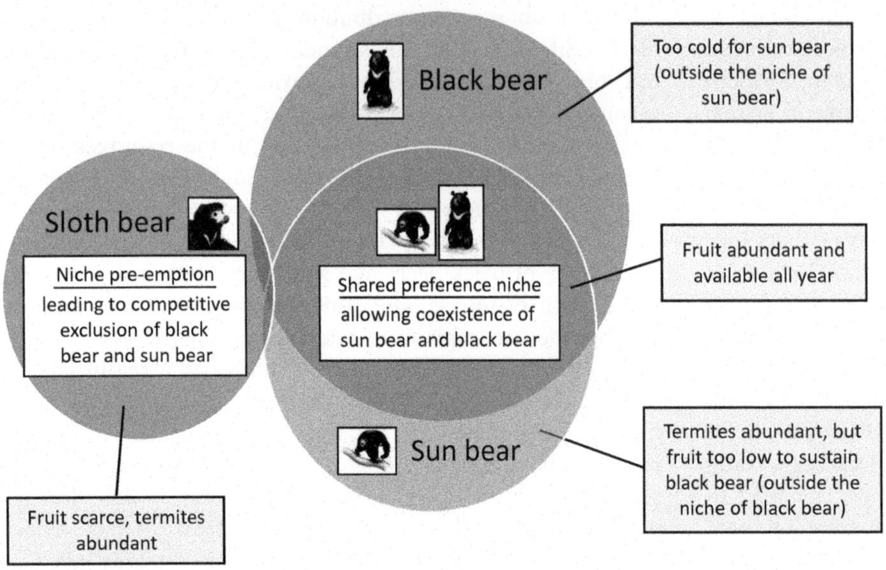

Figure 4.4 Model of geographic overlap of tropical Asian bear species, as a function of evolutionary history, ecological niches, and resource abundance. Size of circles is roughly proportionate to global range size. Overlap among circles depicts amount of geographic overlap: Asiatic black bears and sun bears overlap extensively, whereas sloth bears overlap little with the others. White boxes explain major mechanisms of ecological separation or overlap. Gray boxes describe niche conditions in different regions of overlap or non-overlap.

Steinmetz et al. 2013). However, some important differences were also found: sun bears were the main consumer of insects, and black bears predominated in fruit-rich montane forest (Steinmetz et al. 2011, 2013). Yet in the Sundaic region, where black bears are absent, sun bears are relatively common in montane forest, suggesting that their low density in montane habitats on the mainland may be due to competitive exclusion by the larger black bears. Sun bears, because of their smaller size and lower absolute food requirements, may be better able than black bears to subsist on insects as a fallback food. In north-east India, sun bears consume an almost equal proportion of insects and plants year-round (Sethy & Chauhan 2018).

Niche partitioning can arise under such circumstances, whereby species share preferences for a resource or habitat, but differ in their tolerances or competitive abilities, so become differentially distributed along an environmental gradient (Rosenzweig 1991). The coexistence of black bears and sun bears appears to be structured this way, with overlapping fundamental niches combined with asymmetric competition, known as the "shared preference niche" (Rosenzweig 1991). Extensive niche overlap between these two species may also reflect an abundance of food: they share resources without really competing (Wiens 1993).

Black bears may be limited from occupying the Sundaic region not because of sun bears, but because of fruiting cycles (Figure 4.4). Black bears reach their southernmost limit at about 8°52′N in Thailand. This location, 200 km south of the Isthmus of Kra, marks a phytogeographical ecotone between seasonal tropical forests to the north and aseasonal rainforests to the south (Woodruff 2003). South of this point most tree species fruit synchronously during masting events, every 3–7 years, followed by extended intermast periods with very little fruit (Primack & Corlett 2005). Sun bears survive these cycles by periodically switching to insects. In contrast, the seasonal climate of mainland South-East Asia, with wet seasons and dry seasons, promotes asynchronous fruiting among tree species, yielding seasonally changing fruit supplies, but no extended periods without fruit. For example, in a primary dipterocarp forest in Indonesian Borneo, an average of 13 trees/ha provided fruit for sun bears during mast fruiting months, whereas only 0.6 trees/ha provided fruit during an intermasting 2-year span (Fredriksson et al. 2006). In contrast, in evergreen and deciduous forests in Thailand, where sun bears and black bears coexist, density of trees with fruit exceeded 10 trees/ha in the rainy season and 6 trees/ha even in the relatively fruit-poor dry season (Steinmetz et al. 2013).

Species Interactions in a Changing World

Major differences in the abundance and relative availability of key bear foods across Asia, combined with differences in the species' evolutionary histories and ecological niches, have produced the biogeographic patterns of distribution of black bears, sun bears, and sloth bears that we see today (Figure 4.4). However, each of these species is under threat from combinations of habitat loss or degradation, poaching for parts, or conflict-related killing. These species may have different vulnerabilities to these threats. Black bears are most sought by poachers for their body parts (gall bladder and paws), but also seem most tolerant of habitat alteration.

The south Asian region where the three species intersect is near the center of the geographic range of black bears, but the fringe of the ranges of sloth and sun bears. It is therefore unsurprising that along these fringes, the latter two species have been most vulnerable to habitat change. For example, Yunnan Province in southern China has suffered heavy loss of low and mid-elevational rainforests, which have been converted into plantations of banana, rubber, and tea; sun bears once ranged across a significant portion of this province, but today there is just a single confirmed record of a sun bear in China, <1 km from the Myanmar border. In Bhutan, once thought to harbor sloth bears in grasslands along the Indian border, just one sloth bear has been documented in over a decade of camera trapping, 2 km from this border (Garshelis et al. 2015); incursion of the grassland on the Bhutan side with shrubs and trees may have reduced termite abundance. Also, whereas both sun bears and sloth bears previously occupied large portions of Bangladesh, sloth bears have been extirpated, and only a few sun bears persist. In a forest patch in Assam, India, bordering Bangladesh, local people assert that sun bears disappeared within the past decade (Talukdar & Choudhury 2017). In a park recently surveyed in Meghalaya, north-east India, local people reported the existence of three bear species decades ago, but there has been no recent evidence of sloth bears or sun bears. Meanwhile, in all of these areas (Yunnan, China; Bhutan; Bangladesh; north-east India), black bears remain reasonably common.

Under modern changing conditions – deteriorating or shrinking habitat, and population reductions from direct killing – the balance of these bear species within the matrix of habitats is being disrupted, particularly towards the edges of their ranges. Rapidly changing pressures from humans have replaced the ecological pressures that these species became adapted to over millennia. Shrinking habitats could lead to greater competition and reduced sympatry between black bears and sun bears, and increased extirpation for all three species. It is yet unknown whether interactions among these species will affect their persistence over the broad region where they co-occur, but certainly our present knowledge of their ecology should be helpful for looking into the future, and for guiding conservation strategies that take species interactions into account.

Acknowledgments

Thanks to Karine Pigeon and Pantipa Jaikaew for creating maps, and thanks to WWF Thailand and the Minnesota Department of Natural Resources.

References

Akhtar, N., Singh Bargali, H. & Chauhan, N. P. S. (2004). Sloth bear habitat use in disturbed and unprotected areas of Madhya Pradesh, India. *Ursus* 15(2): 203–211.

Bacon, A. M., Demeter, F., Duringer, P., et al. (2008). The Late Pleistocene Duoi U'Oi cave in northern Vietnam: palaeontology, sedimentology, taphonomy and palaeoenvironments. *Quaternary Science Reviews* 27(15–16): 1627–1654.

Carbone, C., Mace, G. M., Roberts, S. C. & Macdonald, D. W. (1999). Energetic constraints on the diet of terrestrial carnivores. *Nature* 402(6759): 286–288.

Choudhury, A. U. (2001). Devastating flood in Kaziranga National Park. *Tiger Paper* 28(3): 24–26.

Choudhury, A. U. (2011). Records of Sloth Bear and Malayan Sun Bear in North East India. Final report to International Association for Bear Research & Management (IBA). The Rhino Foundation for Nature in NE India, Guwahati, Assam, India.

Collins, N. M. (1980). The distribution of soil macrofauna on the west ridge of Gunung (Mount) Mulu, Sarawak. *Oecologia* 44: 263–275.

Christiansen, P. (2008). Feeding ecology and morphology of the upper canines in bears (Carnivora: Ursidae). *Journal of Morphology* 269(7): 896–908.

Erdbrink, D. P. (1953). *A review of fossil and recent bears of the Old World: with remarks on their phylogeny, based upon their dentition (Vol. 1)*. Deventer: Drukkerij Jan de De Lange.

Ferrar, P. (1982). Termites of a South African savanna. II. Densities and populations of smaller mounds, and seasonality of breeding. *Oecologia* 52(1): 133–138.

Fredriksson, G. M. (2012). Effects of El Niño and large-scale forest fires on the ecology and conservation of Malayan sun bears (*Helarctos malayanus*) in East Kalimantan, Indonesian Borneo. Doctoral dissertation, University of Amsterdam.

Fredriksson, G. M., Wich, S. A. & Trisno. (2006). Frugivory in sun bears (*Helarctos malayanus*) is linked to El Niño-related fluctuations in fruiting phenology, East Kalimantan, Indonesia. *Biological Journal of the Linnean Society* 89(3): 489–508.

Futuyma, D. J. & Moreno, G. (1988). The evolution of ecological specialization. *Annual Review of Ecology and Systematics* 19(1): 207–233.

Garshelis, D. L. (2004). Variation in ursid life histories: is there an outlier? In: *Giant pandas: Biology and conservation* (pp. 53–73). Berkeley, CA: University of California Press.

Garshelis, D. L., Joshi A. R., Smith, J. L. D. & Rice, C. G. (1999) Conservation action plan for sloth bears. In: C. Servheen, C., Herrero & S. Peyton, B. (Eds.), *Bears: Status survey and conservation action plan* (pp. 225–240). Gland, Switzerland and Cambridge, UK: IUCN/SSC Bear and Polar Bear Specialist Groups.

Garshelis, D. L., Dhariaya, N. A., Sharp, T. R., et al. (2015). Sloth bears at the northern edge of their range: status of the transboundary population linking northeastern India to Bhutan. Final Report to International Association for Bear Research and Management.

Holt, R. D. (2009). Bringing the Hutchinsonian niche into the 21st century: ecological and evolutionary perspectives. *Proceedings of the National Academy of Sciences* 106(Supplement 2): 19,659–19,665.

Huygens, O. C., Miyashita, T., Dahle, B., et al. (2003). Diet and feeding habits of Asiatic black bears in the Northern Japanese Alps. *Ursus* 14(2): 236–245.

Hwang, M. H., Garshelis, D. L. & Wang, Y. (2002). Diets of Asiatic black bears in Taiwan, with methodological and geographical comparisons. *Ursus* 13: 111–125.

Joshi, A. R., Garshelis, D. L. & Smith, J. L. (1997). Seasonal and habitat-related diets of sloth bears in Nepal. *Journal of Mammalogy* 78(2): 584–597.

Koike, S. (2010). Long-term trends in food habits of Asiatic black bears in the Misaka Mountains on the Pacific coast of central Japan. *Mammalian Biology* 75(1): 17–28.

Krause, J., Unger, T., Noçon, A., et al. (2008). Mitochondrial genomes reveal an explosive radiation of extinct and extant bears near the Miocene–Pliocene boundary. *BMC Evolutionary Biology* 8(1): 220, DOI:10.1186/1471-2148-8-220.

Linkie, M., Dinata, Y., Nugroho, A. & Haidir, I.A. (2007). Estimating occupancy of a data deficient mammalian species living in tropical rainforests: sun bears in the Kerinci Seblat region, Sumatra. *Biological Conservation* 137: 20–27.

Louthan, A. M., Doak, D. F. & Angert, A. L. (2015). Where and when do species interactions set range limits? *Trends in Ecology & Evolution* 30(12): 780–792.

Mathisen, K. M. (2003). Effects of forest fires on termites (Isoptera) and availability of food for the sun bear (*Ursus malayanus*) in a dipterocarp forest of East Kalimantan, Indonesia. MSc thesis, Agriculture University of Norway.

Mattson, D. J., Herrero, S. & Merrill, T. (2005). Are black bears a factor in the restoration of North American grizzly bear populations? *Ursus* 16(1): 11–31.

Miyagawa, S., Koyama, Y., Kokubo, M., et al. (2011). Indigenous utilization of termite mounds and their sustainability in a rice growing village of the central plain of Laos. *Journal of Ethnobiology and Ethnomedicine* 7(1): 24–29.

Ngoprasert, D., Steinmetz, R., Reed, D. H., Savini, T. & Gale, G. A. (2011). Influence of fruit on habitat selection of Asian bears in a tropical forest. *The Journal of Wildlife Management* 75(3): 588–595.

Primack, R. B. & Corlett, R. (2005). *Tropical rain forests: An ecological and biogeographical comparison*. Hoboken, NJ: Blackwell Publishing.

Rosenzweig, M. L. (1991). Habitat selection and population interactions: the search for mechanism. *The American Naturalist* 137: S5–S28.

Sacco, T. & Van Valkenburgh, B. (2004). Ecomorphological indicators of feeding behaviour in the bears (Carnivora: Ursidae). *Journal of Zoology* 263(1): 41–54.

Seidensticker, J., Yoganand, K. & Johnsingh, A. J. T. (2011). Sloth bears living in seasonally dry tropical and moist broadleaf forests and their conservation. In: McShea, W. J., Davies, S. J. & Bhumpakphan, N. (Eds.), *Dry forests of Asia: Conservation and ecology* (pp. 217–236). Washington, DC: Smithsonian Institution Press.

Sethy, J. & Chauhan, N. P. S. (2018). Dietary preference of Malayan sun bear *Helarctos malayanus* in Namdapha Tiger Reserve, Arunachal Pradesh, India. *Wildlife Biology* 2018(1): wlb.00351.

Steinmetz, R. (2011). Ecology and distribution of sympatric Asiatic black bears and sun bears in the tropical dry forest ecosystem of Southeast Asia. In: McShea, W. J., Davies, S. J. & Bhumpakphan, N. (Eds.), *Dry forests of Asia: Conservation and ecology* (pp. 249–274). Washington, DC: Smithsonian Institution Press.

Steinmetz, R., Garshelis, D. L., Chutipong, W. & Seuaturien, N. (2011). The shared preference niche of sympatric Asiatic black bears and sun bears in a tropical forest mosaic. *PLoS One* 6(1): e14509.

Steinmetz, R., Garshelis, D. L., Chutipong, W. & Seuaturien, N. (2013). Foraging ecology and coexistence of Asiatic black bears and sun bears in a seasonal tropical forest in Southeast Asia. *Journal of Mammalogy* 94(1): 1–18.

Talukdar, N. R. & Choudhury, P. (2017). Conserving wildlife wealth of Patharia Hills Reserve Forest, Assam, India: a critical analysis. *Global Ecology and Conservation* 10: 126–138.

Tougard, C. (2001). Biogeography and migration routes of large mammal faunas in South-East Asia during the Late Middle Pleistocene: focus on the fossil and extant faunas from Thailand. *Palaeogeography, Palaeoclimatology, Palaeoecology* 168(3–4): 337–358.

Tumbelaka, L. & Fredriksson, G. M. (2006). The status of sun bears in Indonesia. In: *Understanding Asian bears to secure their future* (pp. 73–78). Ibaraki: Japan Bear Network.

Wagner, J., Čermák, S. & Horáček, I. (2011). The presence of *Ursus ex gr. minimus-thibetanus* in the Late Villányian and its position among the Pliocene and Pleistocene black bears in Europe. *Quaternaire* 4: 39–58.

Wiens, J. A. (1993). Fat times, lean times and competition among predators. *Trends in Ecology & Evolution* 8(10): 348–349.

Wiens, J. J. (2011). The niche, biogeography and species interactions. *Philosophical Transactions of the Royal Society B: Biological Sciences* 366(1576): 2336–2350.

Wong, S. T., Servheen, C. W. & Ambu, L. (2002). Food habits of Malayan sun bears in lowland tropical forests of Borneo. *Ursus* 13: 127–136.

Wong, S. T., Servheen, C., Ambu, L. & Norhayati, A. (2005). Impacts of fruit production cycles on Malayan sun bears and bearded pigs in lowland tropical forest of Sabah, Malaysian Borneo. *Journal of Tropical Ecology* 21(6): 627–639.

Woodburne, M. O. (2004). *Global events and the North American mammalian biochronology*. New York, NY: Columbia University Press.

Woodruff, D. S. (2003). Neogene marine transgressions, palaeogeography and biogeographic transitions on the Thai–Malay Peninsula. *Journal of Biogeography* 30(4): 551–567.

Wu, J., Kohno, N., Mano, S., et al. (2015). Phylogeographic and demographic analysis of the Asian Black Bear (*Ursus thibetanus*) based on mitochondrial DNA. *PLoS One* 10(9): e0136398.

Yadav, S. K., Lamichhane, B. R., Subedi, N., et al. (2017). Himalayan black bear discovered in Babai Valley of Bardia National Park, Nepal, co-occurring with sloth bears. *International Bear News* 26 (3): 23–25.

Yamazaki, K., Kozakai, C., Koike, S., et al. (2012). Myrmecophagy of Japanese black bears in the grasslands of the Ashio area, Nikko National Park, Japan. *Ursus* 23(1): 52–65.

Zachos, J., Pagani, M., Sloan, L., Thomas, E. & Billups, K. (2001). Trends, rhythms, and aberrations in global climate 65 Ma to present. *Science* 292: 686–693.

Chapter

5 Remarkable Adaptations of the American Black Bear Help Explain Why it is the Most Common Bear: A Long-Term Study from the Center of its Range

David L. Garshelis, Karen V. Noyce, Mark A. Ditmer, Pamela L. Coy, Andrew N. Tri, Timothy G. Laske, and Paul A. Iaizzo

American black bears (*Ursus americanus*, hereafter black bears) are by far the most abundant species of bear, numbering more than twice that of all other bear species combined. Here we explore the reasons for their commonness, using a long-term case study from near the geographic center of this species' range: Minnesota, USA.

Minnesota is a Midwestern state that borders extensive boreal forest landscapes in Canada to the north, broad-leaved forests of Wisconsin–Michigan to the east, and vast agricultural landscapes of the Dakotas and Iowa to the west and south, respectively. These agricultural landscapes were formerly prairie, and thus generally not favorable habitat to this forest-dwelling bear. However, black bears once commonly occurred along forested riverine corridors transecting the prairies. In the Red River Valley of northwestern Minnesota and eastern North Dakota, a fur trader in 1800 remarked: "Bears make prodigious ravages in the brush and willows; the plum trees are torn to pieces, and every tree that bears fruit has shared the same fate . . . their dung lies about in the woods as plentiful as that of the buffalo in the meadow" (Henry & Thompson 1897, pp. 101–102).

Forested landscapes, more typical bear habitat, were historically more extensive in Minnesota than today, stretching continuously from the north-central to the south-eastern part of the state. Logging and conversion to agriculture and settlement caused the forest, and hence the black bear range, to retreat northward. Compounding the loss of habitat, bears were widely regarded as vermin, and were readily killed to protect crops and livestock, or simply out of fear. Schorger (1949, p. 151) noted that, for the early settlers of Wisconsin, "the presence of the black bear . . . was second only to the approach of an Indian war party in its power to raise . . . a high pitch of excitement."

From the 1940s through the mid-1960s, government agencies in Minnesota paid a bounty to encourage the killing of bears. As populations rapidly dwindled, concern mounted that bears would totally disappear. In 1971 their status was elevated to big game animal, meaning that bear hunting required a license and was limited to a designated season. In the following decades, hunting regulations became more restrictive, and efforts were made to reduce the killing of bears that caused damage or were perceived as a threat. Today the species is highly regarded as one of the most charismatic large mammals of Minnesota. They have recolonized north-western Minnesota and are frequently sighted south of the forested primary range.

Many US states share this history of black bear decline and resurgence, and today have burgeoning bear populations. To a large extent, the comeback of this species has been a consequence of restrictions on killing, and a fundamental change in how the public perceives and reacts to black bears. However, the success of this species is also due to its biological adaptiveness – its ability to live in a vast array of habitats, to adapt to radically variable food conditions, and to tolerate the presence of people and the changes they have imposed on the landscape. This chapter highlights the adaptability of the black bear using an extensive and diverse data set spanning 38 years.

Study Sites

About 35% of Minnesota is forested, and 90% of that forested area lies north of 46°N latitude. Aspen, northern hardwoods, and black spruce are the dominant forest types. Forested lands are owned privately (36%), by the state (24%), counties (16%), federal government (17%), or industries (paper, timber, 7%).

A key feature of Minnesota forests is that food production for bears varies drastically by season and by year (Garshelis & Noyce 2008). From collections of bear scats we learned that bears in Minnesota consume primarily green vegetation in spring, mainly ants and pupae in June, various sorts of berries as they ripen during July–August, and some late-season berries but primarily hazelnuts and oak acorns in September, until they den in September or October. Bears must search within different habitats to locate different foods, as they become available. During some years, many kinds of foods are available and abundant; in other years, the variety of foods is less, and in some years, fruiting is so poor that bears face a food scarcity.

We chose study sites that varied in extent and type of forest habitat, land ownership, road density, and bear hunting. We established our first study site in 1981, within the Chippewa National Forest (CNF) and surrounding state- and county-owned forests. The main purpose was to better understand the impact of hunting on the bear population. The study site was selected because it was near the center of the bear range

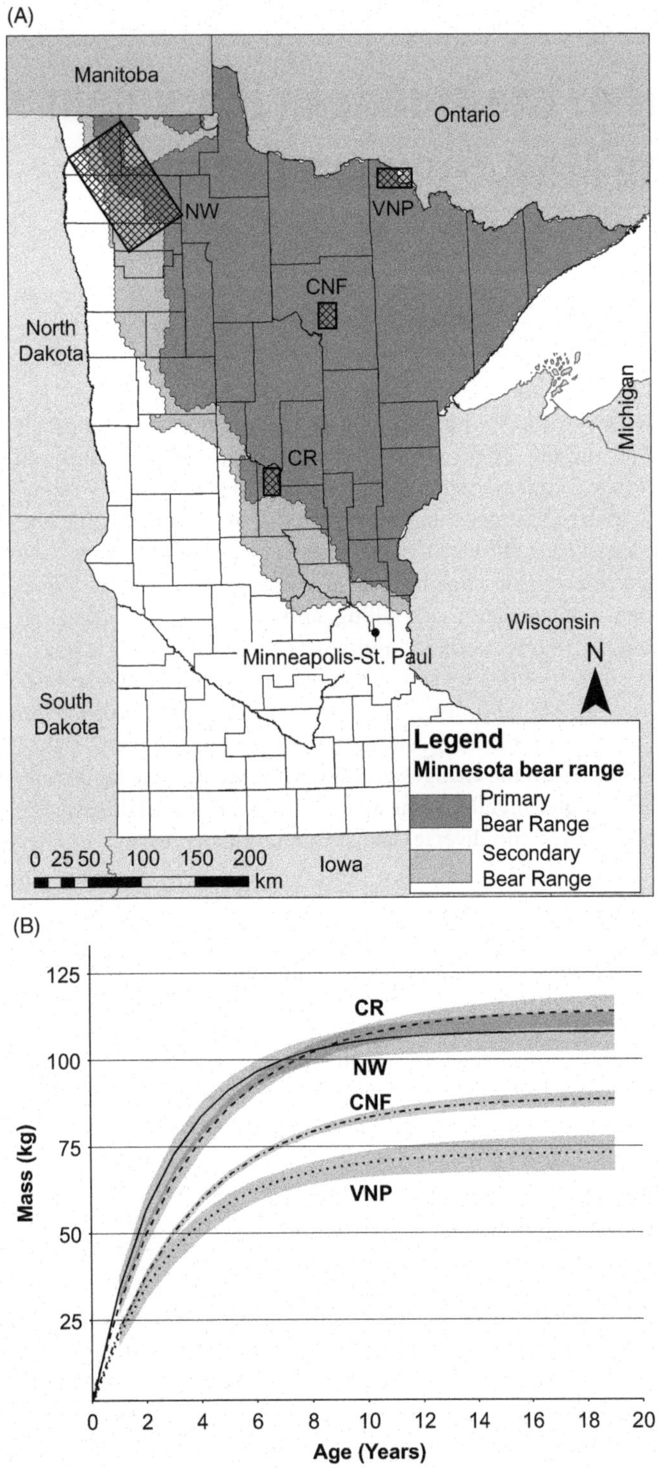

Figure 5.1 (A) American black bears were studied in four areas, representing different parts of their range in Minnesota. Minnesota bear range was differentiated as primary or secondary, based mainly on continuity of forested habitat and overall bear density. Bears near the western and southern fringes of the bear range (NW and CR) had access to more abundant foods, especially in fall, which enabled them to grow more quickly and reach higher maximum weights than bears near the center of the primary range (CNF). Bears along the northern edge of their range in Minnesota (VNP) had the poorest foods, and were smallest. (B) Graphs are nonlinear regressions (±95% confidence interval) of age-specific weight data from denning female bears during February–March. Differences in rates of growth led to earlier maturation of CR and NW females, and very delayed maturation in VNP.

(Figure 5.1A) and was known to be heavily hunted due to extensive public land and forest roads that provided easy access to bear habitat. Also, active logging yielded a complex matrix of variously aged forest stands, providing opportunities to investigate bears' associations with their habitat. The forests are a patchwork of uplands, dominated by aspen, and forested lowlands.

In 1991 our study expanded to include Camp Ripley Training Center, a National Guard facility at the southern periphery of the bear range. Bear hunting is prohibited on Camp Ripley (CR), but bears may be hunted if they range outside. Oaks are plentiful. The 210-km^2 area is long and narrow (6–10 km wide), and bordered by highways with fast-moving traffic and cornfields.

In 1997 we added Voyageurs National Park (VNP), located along the northern edge of the Minnesota bear range (bordering Canada). The study site is a 300-km^2 roadless peninsula bounded by three large lakes. Like CR, hunting is prohibited, but bears are exposed to hunting if they leave the park.

In 2007, we initiated work in a fourth site, at the north-western edge of Minnesota's bear range (NW). This area is largely agricultural, although only 2% of the land area is planted with crops consumed by bears. Forested land, which comprises less than 20% of the area, is patchily distributed in small, privately owned woodlots and state-owned Wildlife Management Areas, which are open to hunting. The area is heavily transected by roads. A low density of bears occupied this area until 1995, when poor natural food in neighboring areas spurred an influx of bears attracted to corn and sun-flowers. Unexpectedly, this wave of bears stayed to reside permanently.

We hypothesized that bears on these four study sites would have different rates of reproduction and different sources of mortality. CNF had the most diverse foods, due to the mixture of forest types and ages, so we thought it would support high reproduction, including large litters. VNP had the poorest and shallowest soils, no timber cutting, and a largely coniferous forest, so provided the least food for bears; we therefore suspected reproduction would be lowest. Although CR had plentiful oaks, and NW had abundant crops, we thought reproduction would be somewhat limited by lack of forest diversity in CR and patchiness of forests in NW. We thought that hunting would be the primary source of mortality in the CNF, collisions with vehicles the main mortality in CR, depredation-related killing the main cause in the NW, and natural mortality dominant in VNP. We theorized that bears in the NW would be most stressed by people, due to the extensive system of farm roads, paucity of forest cover, and their reliance on crops for food.

Research Methods

We captured bears in baited traps made of 55-gallon barrels. We measured and weighed them, took temperature and blood samples, and attached a radiocollar. For the first 25 years we used VHF-transmitting collars, and tracked them from an airplane. We switched to GPS collars as they became available and affordable. Recent GPS collars provide near real-time data by sending locations via a link with a satellite. Collared bears could be killed legally by hunters, but beginning in 2001, we asked hunters to voluntarily avoid shooting them.

We visited all bears in their winter dens each year to reassess their weight and condition, and to quantify the cubs born and the previous year's cubs that survived to yearlings (denned with their mother). We fit collars to yearling bears, and readjusted the fit of all other radiocollars at each den visit. Beginning in 2015, we set trail cameras in front of dens to gain information about when bears exited, and possibly what caused them to do so.

From 1981 to 2019 we radiocollared nearly 700 individuals. We monitored most of them until they died. These study subjects provided the data reported below.

In 2009, we began using cardiac monitors, which were implanted subcutaneously near the heart (Laske et al. 2018). We used these to better understand hibernation physiology, and also to measure potential stress in different circumstances, which we discerned by matching heart rates to GPS locations.

Physical Growth, a Function of Food

Bears on the four study sites grew at different rates and reached different maximum weights (Figure 5.1B), in accordance with abundance of food, especially fall food. Females in CR and NW grew most quickly: on average they reached 100 kg by 8–10 years old, and 90% of this weight by age 5–6 years. In contrast, average maximum weight for CNF females was 85 kg and VNP females only 75 kg, which they reached by their early teens (90% by age 8–9 years). Males also grew most quickly and to heavier maximum weights in CR and NW (one individual reaching 250 kg), but male sizes were more variable within each area than those of females.

Habitat-related differences in the growth of bears were also evident on a finer scale. In the NW, bears of both sexes were heavier and fatter (more prepared for hibernation) if they consumed crops. Frequent crop use also increased the physical stature of males, which likely enhanced their social dominance (Ditmer et al. 2016).

Within the CNF, a sharp habitat division existed between predominantly uplands on the east side versus lowlands in the west; bear foods were more diverse and abundant in the uplands. Accordingly, the skull length (an easily measured index of physical stature) of bears that lived mainly in upland habitats reached maximum size 2 years earlier than for bears that lived principally in lowlands; lowland bears eventually caught up in skull size, but not weight.

Reproductive Variability

Since our first den visits in 1982, we observed 315 litters with 819 cubs. Litters contained 1–5 cubs, with 3 being most common (54% overall; Figure 5.2; average litter size = 2.6). First-time mothers tended to have smaller-than-average litters, as did bears in VNP, but otherwise, litter size did not increase with mother's weight (Noyce & Garshelis 1994).

Mother's weight also did not affect the normal 2-year birthing interval. Bears typically bred in June or July, gave birth in January, and remained with their litter for about 17

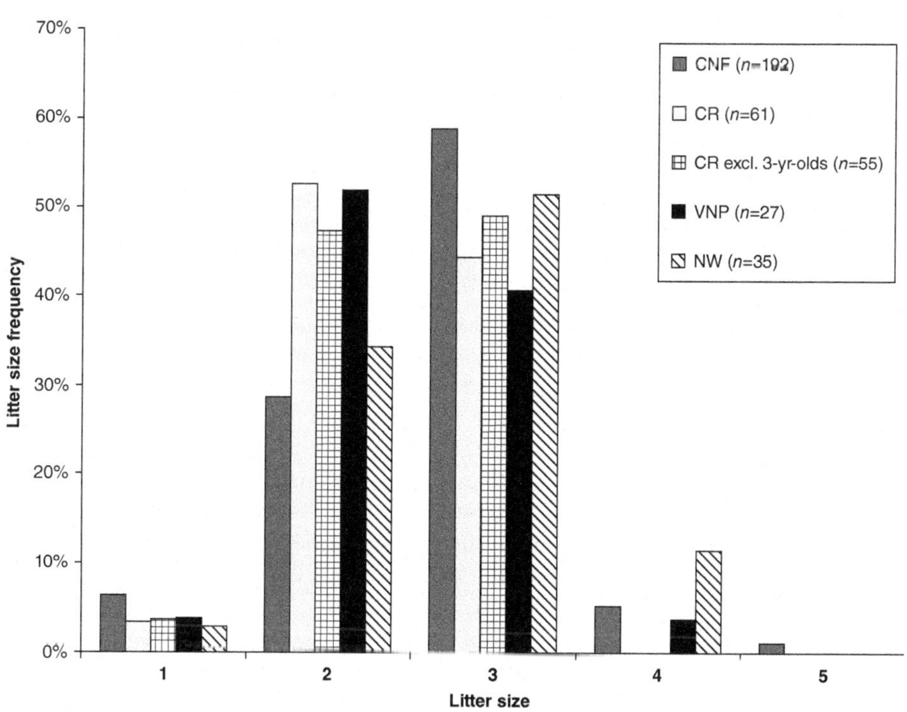

Figure 5.2 Litter size of black bears in Minnesota ranges from one to five cubs, but 90% contain two or three. Three-cub litters are most common in CNF and NW study sites. In CR, 3-year-old mothers have smaller litters than older bears. In VNP, two-cub litters are most common (*n* = number of observed litters).

months, denning together through the cubs' first birthday. Family break-up was prompted by the next breeding activity. Of 195 birthing intervals, all were 2 years, except seven where a whole litter died and the mother reproduced again the next year, and seven with a 3-year gap between births (although there may have been an undetected litter that died soon after being born). We observed no relationship between non-surviving or missed litters and the mother's body condition.

The age that females had their first litter (age of primiparity) was the most variable reproductive parameter, ranging from 3 to 10 years. Typically it was 4 or 5 years, but production of a litter with at least one surviving cub by age 4 varied from 0% of females in VNP to 38% in CNF, to 86% in CR and NW, in accordance with body weight. Only females that weighed at least 41 kg in March (during hibernation) produced their first cubs the following year. Within the CNF, bears living in lowland habitats with less food began reproducing about 2 years later than their heavier counterparts in the adjacent uplands. The weight and growth of cubs during their first year corresponded with the weight of their mother (Noyce & Garshelis 1994) and strongly affected when they matured. Bears that were small as yearlings tended to stay small, and thus reproduced later (Garshelis & Noyce 2008). One anecdote from VNP exemplifies this point: two sisters were the same size as 8-week old cubs, but one was 20% heavier by the time we saw them again as yearlings, still with their mother. This gap continued to widen with age, and ultimately the heavier one produced cubs a year earlier than her sister.

In our study, only two females lived long enough to reach reproductive senescence. One was 25 years old when first captured, and did not produce cubs that year or the next. The other produced 11 litters from age 5 to 25, but none for the next 14 years. If most females lived this long, a year or two difference in when they produced their first litter would have little effect on their lifetime reproductive output. However, because of hunting, females rarely reach reproductive senescence, and in fact, many never reach reproductive age; of the females who survive their cub year, about half die before reaching age 4. Therefore, even a 1-year delay in the age of primiparity might determine whether a bear produces cubs in her lifetime. The downside of maturing early is that young mothers are less experienced in all it takes to be a bear, and so are more apt to lose a whole litter than older first-time mothers.

Male reproduction also is affected by age and weight, because males compete for access to females. Through radiotelemetry we observed male–female pairs during the mating season, and also measured serum testosterone of captured bears. Males that were 6–8 years old were heaviest and had the highest testosterone (Garshelis & Hellgren 1994). Three-year-old males were capable of breeding (and had frequent facial wounds from fighting during the breeding season), but rarely paired with estrous females; when they did, it tended to be late in the season. We speculate that prime breeding males may tire, as we have evidence that they fast during the breeding season (Coy & Garshelis 1992).

Primary Causes of Mortality

Mortality during a cub's first year of life is relatively low in Minnesota (19%, but 25% in VNP). The mortality of male cubs (23%) exceeds that of females (15%). We do not know what causes cubs to die, as we did not radiocollar them until they were 1 year old. However, on a few occasions we observed cubs that fell out of trees while playing or sleeping, scampered carelessly across roads while trailing behind their mother, and in three cases became separated from their mother and never rejoined her. We suspect that male cubs were more adventurous, accounting for their higher mortality. The sex ratio of cubs at 8 weeks old (when we first observed them in the den) was 51% male. By yearling age, it shifted to 52% female.

Bears in Minnesota almost never died in their dens, and never of starvation. Even very skinny yearlings made it through the winter, though some (all weighing <10 kg) died of malnutrition after den emergence in spring. By 15 months of age, natural mortality was a rare event in this population. No bears died of disease. We knew of only two that were killed by wolves, even though wolves were common in all study sites. A few bears also were killed by other bears, and one undernourished 2-year-old starved after biting a porcupine. Of 387 deaths that we documented, only one bear died of natural causes related to old age. She was 39.5 years old, the oldest known wild individual of any species of bear. The average lifespan for a bear in Minnesota is about 4 years for males and 6 years for females (although females in CR and VNP lived, on average, to 7 and 9 years, respectively).

The highest proportion of natural deaths occurred, as we had hypothesized, in the two unhunted populations, VNP and CR. However, VNP and CR bears were hunted when they left the park or military reserve to seek better foods; consequently, even in these two areas, legal hunting was the primary source of mortality, as it was in NW and CNF (Figure 5.3). In CNF we used black-colored radiocollars for most years, purposefully so hunters could not distinguish collared bears, and found that hunting constituted 81% of deaths (excluding cubs, which were not legally hunted). We observed that upland bears were subject to higher hunting mortality because few people hunted the lowland forests, which had poor road access.

In Minnesota, most hunters use bait to lure a bear close enough to shoot. Because bears are so food-driven in fall, baiting is a highly successful hunting technique. Hunting success varies inversely with natural food conditions: in years when fall foods are plentiful, bears are less apt to come to baits than when foods are scarce. Females, which are more wary of baits than males, are disproportionately less inclined to feed at baits than males when natural foods are adequate (Noyce & Garshelis 1997). The greater wariness of females is an adaptive trait that, in a hunted population like Minnesota, yields perpetually male-biased harvests and hence greater potential for population growth.

Hunters' baits often include attractive and nutritious foods like fruits and nuts (as well as sweets), which may enhance

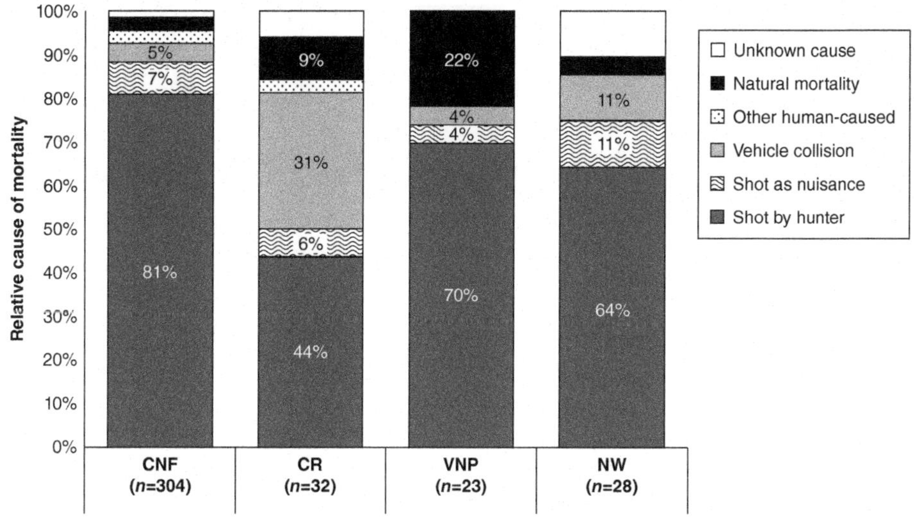

Figure 5.3 Legal hunting is the leading cause of death for bears across Minnesota. CR and VNP are unhunted, but bears are vulnerable to hunting when they leave on foraging forays. Vehicle collisions just outside CR are also relatively common. Natural mortality is common only in VNP. (*n* = number of collared bears monitored until death).

reproduction by hastening growth and maturity; on the other hand, bait consumption is a mortality risk. Individual bears must weigh the attraction of the concentrated but foreign food against the risks they discern from human smells at the site. One of our study bears with a heart monitor showed an elevated heart rate for 3 hours before being shot at a hunter's bait (Laske et al. 2011). Wary or shy bears that avoid baits or only feed at bait sites nocturnally are more likely to survive, thus potentially altering bear behavior (e.g. attraction to human foods) on a population level.

The two other main sources of mortality for bears in Minnesota are collisions with a vehicle and being shot as a nuisance. As we had postulated, road kills were most common for CR, which is bordered by highways, and depredation kills were most common in NW (Figure 5.3), where bears are surrounded by agriculture, including fields of corn and sunflowers. Even so, hunting was still the dominant source of mortality in these areas. This was despite our requests since 2001 (i.e. the entire span of our study in NW) for hunters not to shoot collared bears, which we marked with large, colorful ear tags. Hence, collared bears likely underrepresented hunting as a source of mortality.

Home Ranges and Migrations Relate to Food

During the 1980s, when our study was limited to the CNF and all bears were tracked from an airplane, we were struck by the number of bears that made excursions well outside their normal summer home range during August–September, most returning to den about 6 weeks later. Due to the speed and apparent directness in their travel, and the long distances covered, it was nearly impossible to track these bears enroute. Nevertheless, significant trends emerged from these data (Noyce & Garshelis 2011, 2014), which were later corroborated with more detailed data from GPS collared bears (Figure 5.4). Bears generally traveled south and/or west to areas with better food resources (oaks and agriculture). Such travel was common (~40% of bear-years) and included both sexes and all ages, although males tended to travel farther (median = 26 km) than females (median = 10 km). Sometimes bears from different parts of the study area ended up together at distant locations; and they seemed to use similar travel routes. Taken together, these are characteristics of a migration, albeit not the familiar en-masse migration that occurs in birds or ungulates.

Early accounts of naturalists in this region mention distinct, well-used trails that "could easily be followed for 20 or 30 miles" and "droves of bears" travelling together on a migration (Schorger 1949). Minnesota hunters in the mid-1800s waited for migrating bears to cross a certain predictable "bear pass" along a major river. These anecdotes have been largely dismissed by scientists as exaggerations. Possibly the old migration routes became less predictable with changing landscapes and declining bear populations. Nevertheless, our data clearly showed that many bears followed similar routes and ended up close together. It seems that a network of migratory travelways really does exist. We found that these were not governed by environmental features, nor were they learned from their mother. We surmised that bears cue off the scent trails of each other, with routes varying somewhat year to year.

We expected that bears that migrated in fall would be particularly inclined to visit hunters' baits along the way, because the anticipation of finding a concentrated food source is what motivated their migration. However, we observed just the reverse: hunting-related mortality of migrating bears was less than for bears that remained in their home ranges, suggesting that bears were more wary of people while traveling outside their home ranges (Noyce & Garshelis 2011).

Also unexpectedly, we found that bears migrated most often not when food supplies in their home range were scarce, but when food was plentiful. In years of food scarcity, it could be risky to expend energy migrating, as point sources of more abundant food might be difficult to find. Conversely, in good food years, there would likely be many patches of superabundance, which could significantly bolster weight gain. This

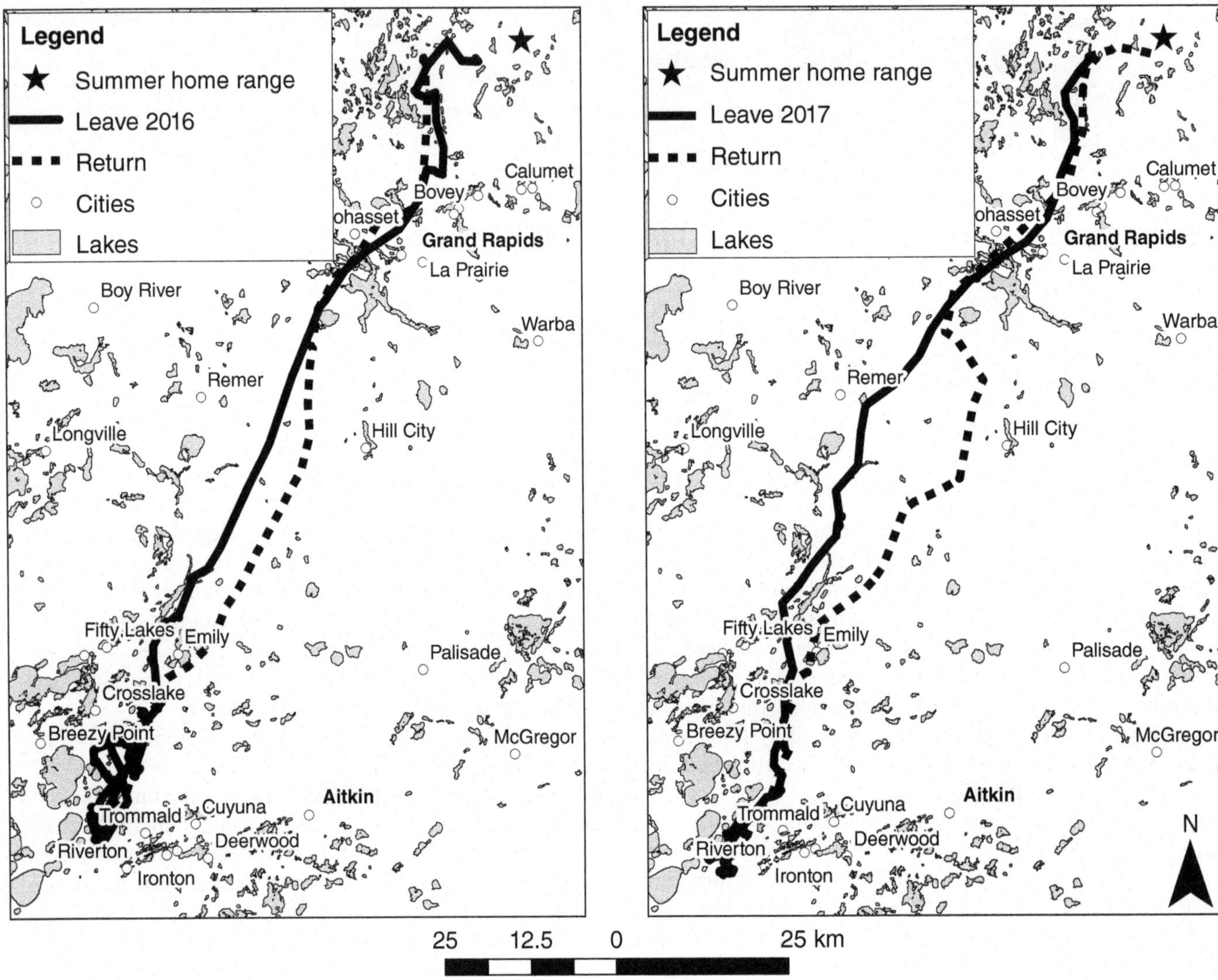

Figure 5.4 Bears from the CNF study site, near the center of Minnesota's bear range, frequently migrate south in late summer or early fall to locate areas of richer food (typically an oak stand or cornfield). This was discovered initially with VHF-collared bears, located weekly from an airplane. GPS collars, taking locations every 2 hours, highlighted the directness of the travel routes, there and back. Shown here is an adult male with a GPS collar who travelled virtually the same migration route in consecutive years (left 2016, right 2017).

could make the difference in whether females produced their first litter, the weight of their cubs, and when these cubs reached maturity. Lowland females on the CNF, whose growth and reproduction were suppressed, migrated at twice the rate of upland females.

VNP bears also migrated southward. This area typically had low food abundance in fall. In some years bears swam 3–5 km across a large lake to reach an area where both natural and human-related foods were more abundant. In one particular year, when wild plums along this south shore were especially plentiful, most of the VNP bears migrated outside the park. The bears had either detected the scent of these fruits wafting across the lake, or they simply left the park because they were hungry and happened upon this bounty of food just outside.

CR and NW bears did not show the same migratory patterns. Most CR bears left the reserve at some time, but we could not discern a particular pattern of directional movement, possibly because CR is already at the southern fringe of the range. NW bears often expanded their fall home ranges to include crops or oak stands, but as these were scattered about the landscape, they also showed no consistent migratory pattern.

The patchiness of the NW landscape required bears to cross agricultural fields that lacked food or cover. During these crossings, their heart rates accelerated beyond what was expected for their hastened movement across the field, indicating a stress response (Ditmer et al. 2015b). Traversing this landscape, which was mainly non-habitat embedded with patches of forest, inflated their home range size. Whereas weekly home ranges of males averaged about 130 km^2, their weekly ranges shifted spatially during the year, creating yearly home ranges that encompassed over 800 km^2 – the largest known for this species (Ditmer et al. 2018a). These expansive home ranges were partly the result of males searching for a suitable crop field, where they would feast for some time, apparently (based on their slow heart rates) feeling unthreatened. This was similar to the CNF bears in terms of long travels to find concentrated sources of food, but differed in not being directional on a population level. Also unlike CNF, females in NW travelled outside their home ranges to use crops mainly as a fallback when natural food conditions were poor (Ditmer et al. 2016). By studying captive bears that had never eaten corn or sunflowers, we found that females were inherently less apt than males to try these unfamiliar foods (Ditmer et al. 2015a). Females with cubs were especially hesitant to visit crop fields, likely trying to avoid infanticide by males (one female that ventured into a cornfield had one of her cubs attacked by a male). The sprawling wanderings of males and their use of crops led us to predict that some will continue to expand their range further west, into even more treeless habitat; however, it is unlikely that females, which are more bound to forests and natural foods, will follow (Ditmer et al. 2018a).

Roads seem to have little impact on the movements of Minnesota bears. Heart monitors showed that bears were wary when they approached and crossed roads, which occurred about every 3 or 4 days (but sometimes multiple times per day); however, their slightly elevated heart rates indicated that it was a low-stress activity (Ditmer et al. 2018b). They crossed high-traffic roads mainly at night, but otherwise showed no avoidance of roads.

Denning and Hibernation – Surviving Half the Year Without Food

Minnesota bears typically entered dens during September–October. Pregnant females denned the earliest: following summer breeding, the blastocyst (not yet an embryo) does not implant until November, but by denning time, progesterone is noticeably elevated, so females are physiologically aware of their impending pregnancy.

Bears used a wide variety of dens, some of which seem (to a human observer) unappealing for a 6–7-month hibernaculum in one of the coldest places within the range of this species. Many individuals used different types of dens year to year, but they did not become choosier in their den selection as they got older. Also, they rarely reused a den in a subsequent year. The seeming indifference to the quality of their den was one of the most surprising findings of our study, as we tend to think of hibernation as a critical and vulnerable time for bears.

Bears commonly dug dens into the ground. Females excavated dens more than males. Both sexes also used existing structures, like brush piles from former logging operations or natural hollows under a root mass of a tipped-over tree (Figure 5.5). Classic rock caves were only available in VNP. Dens became more insulated as snow accumulated; sometimes just a small hole permitted air exchange (and some had no air hole) (Figure 5.6). On the opposite end of the spectrum were nest dens, comprised of a pile of branches, boughs, or grass collected from the immediate vicinity (Figure 5.5D). Some nests were situated under thick conifer cover, whereas others had no overhead cover, exposing the bear to snow and wind, and providing no protection against radiative cooling on clear winter nights [temperatures dropped below −18°C (0°F) for an average of 50 days each winter]. Nest dens tended to be more common among males, but also were used by females, including some giving birth.

Not uncommonly, we observed wolf tracks around dens. A bear in a nest, especially a female with newborn cubs, would seem like an easy target, yet only once did we find a denning bear (a young male) injured by wolves (and once observed sign of a scuffle between a female with cubs and wolves, but no injuries to the bear). Oddly, whereas some bears in nests remained vigilant to sounds, with some leaving as we approached, others, with their head tucked under their chest, exhibited no overt reaction to our presence, even to within 2 m.

Many nest dens were in relatively remote areas. One quarter of CNF males made a northward fall migration averaging 50 km (some having just returned from a southward migration to forage) to den in a vast roadless bog (Noyce & Garshelis 2011). In NW, many bears denned in marshes where they created a huge woven nest of cattails, open to the sky. It is difficult to explain the paradox of bears moving to remote areas to den, possibly to avoid wolves or people, yet then often denning in an open nest. It is also difficult to explain movements to remote habitats by some bears while others denned just tens of meters from roads, railroads, snowmobile trails, or houses, seemingly unconcerned about noise and human proximity.

We surmise that bears rely more on their remarkable physiological adaptations than choosing an optimal den for their well-being during winter. One of their most notable adaptations is for "fight or flight" during hibernation. Rectal temperature of Minnesota bears drops only a few degrees from 37°C in summer to typically 32–36°C while hibernating, enabling them to awaken and respond quickly, if needed. This distinguishes them from other hibernators, which are small enough to fully escape predation risks by hibernating in deep burrows or crevices; this security allows them to drop body temperature drastically, thus saving energy but losing the ability to respond quickly.

We learned that part of the bear's suite of special adaptations includes modulation of their heart rate. Implanted heart monitors revealed heart rates slowing from daytime averages of 85–90 bpm during summer to 20–25 bpm in winter, and becoming synchronized with their breathing, which slows to 2–4 per minute. This respiratory sinus arrhythmia allows the heart to pause and rest (for 5–30 seconds), then pump 3–5 times after an inhalation to circulate oxygenated blood (Laske et al. 2010, 2011). Because the blood can be stagnant between breaths, bears avoid strokes or heart attacks by tripling clotting time during hibernation (Iles et al. 2017). However, this

(A) (B) (C) (D)

Figure 5.5 Bears in Minnesota use various types of dens: (A) excavated (here dug into a mound along the edge of a farm field); (B) under the roots of a tipped-over tree; (C) under a brush pile (from logging or storms); and (D) in an open nest that they construct. Individuals often switch den types year to year, and rarely reuse dens. Type of den does not affect reproduction or survival; even cubs born in open nests survive normally. Other animals (e.g. squirrels, rabbits, wolves) commonly approach dens, as in panel (A). (Photos by D. Garshelis.) (A black and white version of figure 5.5D will appear in some formats. For the color version, please refer to the plate section.)

adaptation involves only clotting within their vessels. Hibernating bears retain or enhance their ability to heal external injuries that might be incurred just before or during denning (Iaizzo et al. 2012).

Respiratory sinus arrhythmia is so characteristic of hibernation that we could tell, by disruptions in this pattern, when females gave birth (early–mid January), and when physiological hibernation ended in spring (Laske et al. 2017). The trail cameras we set at dens revealed a transitional process from denning to activity. Bears spent days to weeks exiting and re-entering their den (generally beginning mid-March) before leaving the site (late March or early April). Females with cubs delayed their departure from den sites by an average of 3 weeks, allowing cubs time to gain strength and skills for walking and tree climbing. Many bears remained at their den site after they first emerged to collect more bedding material, especially if it had become saturated by melting snow. One benefit of nest dens may be that they stayed drier, because no snow dripped from above, and they were typically thicker than the beds bears made inside more structured dens (Figure 5.5d).

Adaptability, the Key to Success

The black bear has rebounded across North America to become one of the continent's true conservation success stories. This success relied on people's growing tolerance for a species that might sometimes cause property damage, but otherwise presents little danger to human safety. Restrictions on hunting were also paramount to this success. We learned in this study that natural mortality is very low, and legal hunting

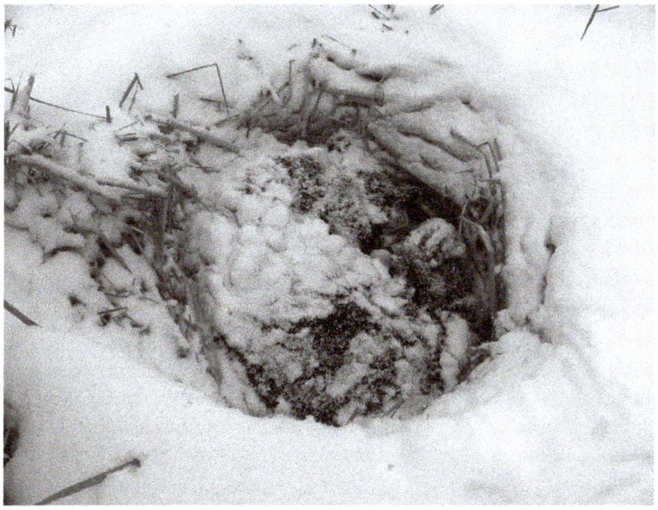

Figure 5.6 American black bears in northern Minnesota – one of the coldest places within their range – commonly spent the winter in unprotected dens that they construct. Even some females that gave birth in mid-winter, like the bear pictured here, used dens with little or no cover. Due to the insulatory properties of their thick layer of fat and dense winter coat, and reduced blood supply to the skin, snow that fell on their backs sublimated or was shaken off, rather than melting into their fur. (Photo by D. Garshelis.) (A black and white version of this figure will appear in some formats. For the color version, please refer to the plate section.)

the driving cause of mortality; even in unhunted areas, bears commonly venture out and are eventually killed by hunters. Thus, management agencies have direct control over the size of black bear populations, unlike species whose mortality is affected largely by environmental factors.

Food conditions, though, had significant effects on reproduction. Baseline food conditions within our four study sites differed greatly, producing different-sized bears with different reproductive rates, due mostly to different rates of maturity. Reproduction is constrained to what the land will support. Poor fruiting years reduce the chance that young females reach the threshold weight for first giving birth, causing lower population-wide cub production some years (Garshelis & Noyce 2008). However, once females began reproducing, poor food years did not disrupt their normal 2-year cycle or cause higher cub mortality. Experienced mothers found ways of caring for cubs in the face of periodic adversity.

Black bears are masters of contending with food variability. They are able to take advantage of far-flung hotspots of bountiful food; when normal foods are lacking, they use substitute foods, sometimes from human-related sources. Meanwhile, their tolerance and resilience to undernutrition is profound. The smallest hibernating yearlings (5–6 kg) weighed about the same as the largest 2-month-old cubs; the heaviest hibernating yearlings (52–65 kg) were 10 times larger, and were more than double the weight of the smallest hibernating 2-year-olds (19–22 kg). Some juveniles barely gained weight one year, but doubled or nearly tripled their weight the next.

Minnesota was an ideal place to study the adaptiveness of this species. The food supply here is particularly variable spatially and temporally, yielding natural experiments in how bears cope. We were fortunate to have been able to conduct our studies over a very long time frame (approaching four decades), enabling us to witness large swings in food abundance. Also, Minnesota is noted for its long, cold winters: food is entirely absent for half the year. Certainly, the most remarkable adaptation of this species is its ability to withstand such a long period without food or water, and extreme cold (–40°C); moreover, during this period, females give birth, lactate, and nurture altricial cubs.

The bears' ability to store fat is key to their ability to survive a half-year without eating. Thus, fall is often considered a critical time for bears, as fall foods dictate how much fat they can amass. Extensive fall migration by Minnesota bears is on par with hibernation as an adaptation for surviving prolonged food deprivation. Unexpectedly, this otherwise solitary animal is not only generally tolerant of multiple individuals sharing a rich fall feeding area, but also reliant on social cues to navigate to such sites.

Often neglected in considering the seasonal adaptations of bears is the springtime. In Minnesota, bears emerge from dens having lost 15–50% of their body weight. Particularly for young, skinny bears, this is the toughest time of year. The ground is still frozen and partly snow-covered; late spring snowstorms are not unusual. Shoots of green vegetation begin to poke up about 2 weeks after den emergence. Fruits are not available for 3.5 months. Bears are relatively inactive during this period, becoming increasingly active as the spring progresses (Laske et al. 2011, 2017). It was previously thought that, before the fruiting season, bears continued to lose weight feeding exclusively on green vegetation, and later insects. However, we discovered that most bears not only gained weight, but grew structurally during spring – another surprising finding (Noyce & Garshelis 1998).

Despite the many studies of this species across its range, each year we learn more. Many of our early well-reasoned hypotheses turned out to be incorrect. Long-term field studies employing constantly improving techniques and technology remain essential for gaining a deeper understanding of how the black bear has been able to thrive in a landscape increasingly dominated by people. While it is appealing for nature conservation to promote the image of bears, generally, as icons of the wilderness, that characterization distracts from the real reason why this species has been so successful: its adaptability to change.

Acknowledgments

This work was funded by the Minnesota Department of Natural Resources with additional support from the Minnesota Army National Guard, Voyageurs National Park, Agassiz National Wildlife Refuge, and University of Minnesota's Institute for Engineering in Medicine and Department of Surgery. Medtronic plc donated the cardiac devices. We graciously thank the many people who assisted with data collection.

References

Coy, P. L. & Garshelis, D. L. (1992). Reconstructing reproductive histories of black bears from the incremental layering in dental cementum. *Canadian Journal of Zoology* 70: 2150–2160.

Ditmer, M. A., Burk, T. E. & Garshelis, D. L. (2015a). Do innate food preferences and learning affect crop raiding by American black bears? *Ursus* 26: 40–52.

Ditmer, M. A., Garshelis, D. L., Noyce, K. V., et al. (2015b). Behavioral and physiological responses of American black bears to landscape features in an agricultural region. *Ecosphere* 6(3): 28. http://dx.doi.org/10.1890/ES14-00199.1

Ditmer, M. A., Garshelis, D. L., Noyce, K. V., Haveles, A. W. & Fieberg, J. R. (2016). Are American black bears in an agricultural landscape being sustained by crops? *Journal of Mammalogy* 97: 54–67.

Ditmer M. A., Garshelis, D. L., Noyce, K. V. & Fieberg, J. R. (2018a). Delineating the ecological and geographic edge of an opportunist: the American black bear exploiting an agricultural landscape. *Ecological Modelling* 387: 205–219.

Ditmer, M. A., Rettler, S. J., Fieberg, J. R., et al. (2018b). American black bears perceive the risks of crossing roads. *Behavioral Ecology* 29: 667–675.

Garshelis, D. L. & Hellgren, E. C. (1994). Variation in reproductive biology of male black bears. *Journal of Mammalogy* 75: 175–188.

Garshelis, D. L. & Noyce, K. V. (2008). Seeing the world through the nose of a bear – diversity of foods fosters behavioral and demographic stability. In: Fulbright T. & Hewitt, D. (Eds.), *Frontiers in wildlife science: Linking ecological theory and management applications* (pp. 139–163). Boca Raton, FL: CRC Press.

Henry, A. & Thompson, D. (1897). *New light on the early history of the greater Northwest. The manuscript journals of Alexander Henry and of David Thompson 1799–1814. Exploration and adventure among the Indians on the Red, Saskatchewan, Missouri, and Columbia Rivers.* Coues, E. (Ed.), vol. 1. New York, NY: Harper.

Iaizzo, P. A., Laske, T. G., Harlow, H. J., McClay, C. B. & Garshelis, D. L. (2012). Wound healing during hibernation by black bears (*Ursus americanus*) in the wild: elicitation of reduced scar formation. *Integrative Zoology* 7: 77–89.

Iles, T. L., Laske, T. G., Garshelis, D. L. & Iaizzo, P. A. (2017). Blood clotting behavior is innately modulated in *Ursus americanus* during early and late denning relative to summer months. *Journal of Experimental Biology* 220: 455–459.

Laske, T. G., Harlow, H. J., Garshelis, D. L. & Iaizzo, P. A. (2010). Extreme respiratory sinus arrhythmia enables hibernating black bear survival – physiological insights and applications to human medicine. *Journal of Cardiovascular Translational Research* 3: 559–569.

Laske, T. G., Garshelis, D. L. & Iaizzo, P. A. (2011). Monitoring the wild black bear's reaction to human and environmental stressors. *BMC Physiology* 11: 13. doi:10.1186/1472-6793-11-13

Laske, T. G., Iaizzo, P. A. & Garshelis, D. L. (2017). Six years in the life of a mother bear – the longest continuous heart rate recordings from a free-ranging mammal. *Scientific Reports* 7:40732. doi: 10.1038/srep40732.

Laske, T. G., Evans, A. L., Arnemo, J. M., et al. (2018). Development and utilization of implantable cardiac monitors in free-ranging American black and Eurasian brown bears: system evolution and lessons learned. *Animal Biotelemetry* 6: 13. doi.org/10.1186/s40317-018-0157-z

Noyce, K. V. & Garshelis, D. L. (1994). Body size and blood characteristics as indicators of condition and reproductive performance in black bears. *International Conference on Bear Research and Management* 9: 481–496.

Noyce, K. V. & Garshelis, D. L. (1997). Influence of natural food abundance on black bear harvests in Minnesota. *Journal of Wildlife Management* 61: 1067–1074.

Noyce, K. V. & Garshelis, D. L. (1998). Spring weight changes in black bears in northcentral Minnesota: the negative foraging period revisited. *Ursus* 10: 521–531.

Noyce, K. V. & Garshelis, D. L. (2011). Seasonal migrations of black bears (*Ursus americanus*): causes and consequences. *Behavioral Ecology and Sociobiology* 65: 823–835.

Noyce, K. V. & Garshelis, D. L. (2014). Follow the leader: social cues help guide landscape-level movements of American black bears (*Ursus americanus*). *Canadian Journal of Zoology* 92: 1005–1017.

Schorger, A. W. (1949). The black bear in early Wisconsin. *Transactions of the Wisconsin Academy of Sciences, Arts and Letters* 39: 151–194.

Chapter 6

Giant Panda (*Ailuropoda melanoleuca*)

Ronald R. Swaisgood, William M. McShea, David Wildt, Vanessa Hull, Jindong Zhang, Megan A. Owen, Zejun Zhang, Zachary Dvornicky-Raymond, Marc Valitutto, Dihua Li, Zhang Hemin, Jenny Santiestevan, and Fuwen Wei

Names

Genus: *Ailuropoda*

Species: Giant panda *Ailuropoda melanoleuca* (David, 1869)

Names in other languages: **Chinese:** Da xióngmāo; **Spanish:** Panda gigante; **French:** Panda géant, Ours panda; **Italian**: Panda, Panda gigante; **German**: Bambusbär.

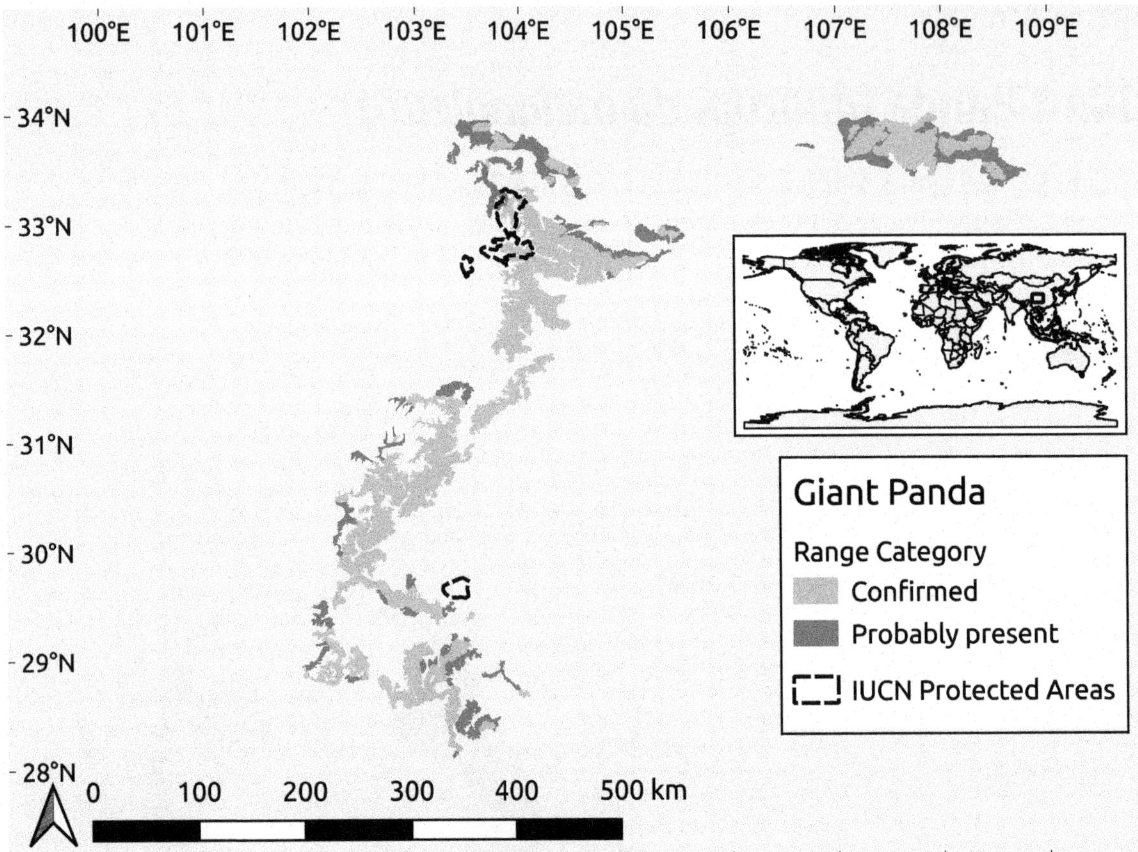

Figure 6.1 Current distributional range of the giant panda. The giant panda once occupied a wide range throughout southern and central China and parts of South-East Asia, but today occupies a few isolated mountain ranges in Sichuan, Gansu, and Shaanxi provinces only. (*Source*: IUCN Red List of Threatened Species.)

Introduction

The giant panda *Ailuropoda melanoleuca* holds an iconic position among endangered wildlife, occupying that pantheon of most recognized and adored species. Docile, black and white, with appealing features, the panda has won the hearts of many worldwide. The panda, however, also shares its habitat with the world's most populous country, China, and its population numbers have declined for millennia until today the panda occupies just a small portion of its former range (Figure 6.1). Some may dispute the panda's place in humanity's collective conscience; some have even argued that they are so undeserving and poorly adapted that we should let them slide into inevitable extinction. If there is one message to learn from this chapter, it is that the panda is exquisitely adapted to its environment and that its current status is the result of millennia of anthropogenic impacts, not any inherent deficiencies in the species' biology. Even so, is the panda worth it? Have we allocated too many resources to saving the panda, funds that might have been better allocated to other species? A recent analysis suggests not (F. Wei et al. 2018). If one calculates the economic value to humans, we find that the ecosystem services provided by the nearly 70 reserves set aside for panda protection generate value in excess of 10-fold the cost of their establishment and maintenance. If one factors in the cultural value of pandas to people (a difficult and crude estimate), that return on investment rises to 27 to 1. Further, pandas occupy some of the most biodiverse forests in the temperate world, and panda reserves provide shelter for a great variety of plant and animal life (Li & Pimm 2016). Finally, this investment has benefited the panda itself: habitat loss has been arrested and panda numbers have stabilized and begun to increase, spurring the IUCN to downlist the panda from Endangered to Vulnerable (Swaisgood et al. 2016, 2017).

Taxonomy

The giant panda (*Ailuropoda melanoleuca*), or Da xióngmāo in the Chinese language, is a member of the family Ursidae in the order Carnivora. Although its classification was at one time in doubt due to several morphological and ecological affinities with the red – or lesser – panda (*Ailurus fulgens*), genetic evidence now conclusively places the giant panda with bears and the red panda with Ailuridae (O'Brien et al. 1985; F. Wei &

Zhang 2009; see also Chapter 1). The resemblance between the two species is the product of convergent evolution as a result of occupying a similar niche. The giant panda is somewhat distinctive from other bears and this may be due to its basal location in ursid phylogeny. The panda's closest living relative is the Andean bear (*Tremarctos ornatus*; see also Chapter 7). The giant panda's ancestry dates back to the Miocene ~8 million years ago, as indicated by fossil evidence of an extinct species, *Ailurarctos lufengensis*.

Significant genetic divergence among populations has been demonstrated, but none significant enough to call into question designation as a single species. Recent techniques utilizing whole-genome sequencing clarified earlier confusion about population divergence and possible subspecies designation (Zhao et al. 2013). The northern population in the Qinling diverged from the southern populations about 0.3 million years ago, associated with glaciation-driven habitat change. The southern population subsequently diverged into two population clusters, the Minshang and the Qionglai–Daxiangling–Xiaoxiangling–Liangshan populations, about 2800 years ago, driven by anthropogenic habitat loss and habitat fragmentation. These natural and anthropogenic processes are responsible for the three distinct genetic population clusters observed today.

Despite a long history including two major population expansions and bottlenecks, and recent dramatic anthropogenically driven decline (Zhao et al. 2013), the panda retains a high level of genetic diversity. Early studies reported low diversity compromising evolutionary potential and potentially contributing to their endangered status, but later application of microsatellite markers revealed much higher genetic diversity (F. Wei et al. 2012). These findings were later confirmed with genomic approaches (Li et al. 2010; Zhao et al. 2013). An interesting and surprising recent development has called into question the panda's long evolutionary history as a bamboo specialist (Han et al. 2019). Stable-isotope analysis of ancient pandas indicates that ancient pandas had a much broader dietary niche, consuming a higher proportion of non-bamboo food items as recently as the mid-Holocene.

Descriptive Notes

Giant pandas are a medium-sized ursid with adults weighing about 70–150 kg. When standing on all paws they are about 100 cm tall at the shoulder and 1.5 m long, with a short tail (10–15 cm). Their tooth formula is incisors 3/3, canines 1/1, premolars 4/4, and molars 2/3 (Huang 1993). The teeth are substantial and include the canines and carnassial teeth typical of bears. The molars are broad, and extensive musculature allows a powerful bite force for grinding bamboo stems. A unique adaptation is the extension of a wrist bone on the heel of the front paws which allows the grasping of food with a single paw. Pandas are generally terrestrial, but their claws have sufficient grasping power for adults to climb trees during the mating season and young to climb trees for rest or to escape danger (Schaller et al. 1985; Pan 2014). The panda's pelage is dense with a characteristic black and white pattern: mostly white with black pelage across the ears, eyes, muzzle, limbs and shoulders (Caro et al. 2017; Figure 6.2). There are usually subtle individual differences in the width of the black along the shoulder and rump (Zheng et al. 2016). Adult males are generally 10–15% larger than females of the same age, with no differences in pelage. In general, giant pandas in the Qinling Mountains are smaller than other populations.

Figure 6.2 Giant panda by Francesco Rinaldi, with permission. (A black and white version of this figure will appear in some formats. For the color version, please refer to the plate section.)

The young of both sexes reach sexual maturity at about 4–6 years of age in captivity but may mature as late as 7–8 years of age in the wild. In the wild, they can reproduce until at least 15 years of age, depending on body condition, while in captivity than can reproduce to age 20. Giant pandas live up to 26 years in the wild and 38 years in captivity.

Habitat

The current reduced range of giant pandas is not indicative of the total environmental niche they can occupy (Swaisgood et al. 2016). Our knowledge of giant panda habitat needs is dependent on habitat selection models derived within this reduced range. We do know that giant pandas need to consume large quantities of bamboo daily and that adult females need denning shelter for nursing and protecting their altricial young (Schaller et al. 1985; Zhang et al. 2007; Pan 2014; W. Wei et al. 2019). Pandas are found at the elevations where preferred bamboo species are prevalent, and they shift elevation and forest type with shifts in plant phenology (Nie et al. 2015b; Hull et al. 2016; Li et al. 2017c), occupying elevation ranges from 1100–2600 m in the northern Qinling Mountain range to 1200–3600 m in the more southern ranges (Swaisgood et al. 2016). Giant pandas are confined to forests in mountainous areas (Figure 6.3), but are found more often on the flatter, warmer slopes within these mountains (Hull et al. 2014; W. Wei et al. 2018).

Figure 6.3 The iconic giant panda caught on a motion-activated camera. Famous for their black and white coloration, these markings have been nominated as one of the factors contributing to the "cute factor." New evidence suggests that the black and white coloration of the giant panda plays a role in crypsis against snowy backgrounds. (Photo by Jindong Zhang.) (A black and white version of this figure will appear in some formats. For the color version, please refer to the plate section.)

Bamboo cover consistently ranks as one of the most important variables pandas use in selecting habit across ecological scales (Schaller et al. 1985; Linderman et al. 2005; Zhang et al. 2011; Hull et al. 2014; W. Wei et al. 2018). Bamboo is most prevalent in mixed-deciduous/conifer and conifer forest types. Although pandas readily use secondary forests and secondary forests are known to support panda populations (Hull et al. 2014), studies conducted on a range-wide scale indicate that pandas preferentially occupy primary forests (Zhang et al. 2011; W. Wei et al. 2018). Climate change will also impact bamboo distributions in ways that probably require movement of pandas to suitable habitat (Tuanmu et al. 2013; Yang et al. 2017; Wang et al. 2018) or natural dispersal or assisted migration of bamboo species adapted to changing climatic conditions (F. Wei et al. 2015b).

Repeated studies have found that giant pandas are negatively associated with various forms of human disturbance (W. Wei et al. 2018), such as farming and settlements, roads, and livestock grazing in panda habitat (Wang et al. 2015). Whether avoidance is due to the lack of needed resources near humans or disturbance by humans and domestic animals is unknown. Giant pandas have shifted their distribution to higher elevations in recent years despite improved habitat at lower elevations; this distributional shift may be due to human disturbance and livestock grazing at lower elevations (W. Wei et al. 2018).

Although mature forest with abundant bamboo is the preferred habitat of giant pandas, they will disperse across bamboo-free areas above the tree line, through some planted forests, and agricultural areas (Pan 2014). Dispersal corridors are an essential landscape feature (Wang et al. 2014; F. Wei et al. 2015b), not only for genetic connectivity, but because bamboo populations engage in mass flowering events at intervals of 20–100 years. Giant pandas also live in one of the most tectonically active regions of the world, and steep slopes result in frequent rock and mud slides removing habitat. These infrequent but catastrophic disturbances require a network of dispersal corridors that pandas will traverse to access other patches of suitable habitat (F. Wei et al. 2015b; Swaisgood et al. 2017; Xu et al. 2017).

Overall, suitable habitat for pandas has increased in recent years, following large habitat losses in the 1970s and 1980s (Xu et al. 2017; Wang et al. 2018; W. Wei et al. 2018), and these positive habitat trends have contributed to panda population stabilization and increase (Swaisgood et al. 2016, 2017).

Movements and Home Range

Giant pandas occupy yearly home ranges varying from 2.8 to 29 km^2 in size (mean ~ 7 km^2) (Connor et al. 2016). This small home range relative to most other bears is probably related to their stable, abundant, and replenishing bamboo food source (Schaller et al. 1985; Pan 2014). Pandas do not defend territories. Males have 1.5 times larger home ranges than females (Connor et al. 2016), likely due to the male mating strategy of monitoring multiple females to maximize access to mates (Schaller et al. 1985; Nie et al. 2012b; Pan 2014). Males also exhibit a dispersed space-use pattern compared to the centralized space-use pattern seen in females, a difference related both to the male mating strategy and the need for mothers to establish a centralized home base to rear young (Schaller et al. 1985; Pan 2014). Pandas have a number of core areas within the home range (0.7–1.3 km^2 in size), a pattern indicative of their foraging strategy of spending several days eating within a single localized bamboo patch before moving to other nearby patches in succession (Zhang et al. 2014; Hull et al. 2015; W. Wei et al. 2015). Home ranges overlap by 13–34% among neighboring pandas (Connor et al, 2016). Some studies suggest significant attraction among neighboring pandas (Hull et al. 2015), whereas other work shows neither attraction nor avoidance (Zhang et al. 2014). Due to constraints on the number of animals fitted with GPS satellite collars in these studies, the degree of spatial attraction among pandas in overlapping ranges is difficult to ascertain. Pandas may shift home ranges across years, while home ranges also expand as they mature but shrink when mothers rear young (Pan 2014).

Giant pandas move around 300–600 m per day (Connor et al. 2016). They display tortuosity in their movement paths, and exhibit area-restricted searches in food patches at spatial scales <700 m (Zhang et al. 2014). Pandas also display directed movement patterns to return to the same patches after lag times of several months, suggesting good spatial memory (Schaller et al. 1985; Hull et al. 2015). Daily distance moved is higher in spring than other times of year, possibly due to both searching for mates and foraging widely for succulent bamboo shoots in the growing season (Schaller et al. 1985; Pan 2014).

Giant pandas exhibit female-biased natal dispersal (Zhan et al. 2007; Zhang et al. 2014). Home ranges of juvenile and subadult pandas are adjacent to their mothers (Pan 2014), but as female subadults enter adulthood, they travel 5–34 km away from their mothers to establish their own range, often engaging in short long-distance movements prior to moving permanently. In contrast, most males exhibit philopatry and stay closer to their mother's range.

Giant pandas undergo seasonal elevational migration corresponding with bamboo growing seasons and temperature shifts, a pattern that varies across regions. For instance, pandas in Wolong Nature Reserve in the Qionglai mountains shift from high elevation (2600–3600 m) to low elevation (2400–2800 m) for roughly two months in spring to forage on high-nutrient umbrella bamboo (*Fargesia robusta*) shoots (Schaller et al. 1985). In the Qinling mountains, pandas shift in the other direction, spending most of the year at a lower elevation (900–2000 m) and shifting to higher elevation (2000–3200 m) in summer for the *Fargesia* bamboo growing season (Pan 2014). A one-way migration is typically completed over a period of a few days to a few months.

Activity Patterns

Unlike other ursids, pandas do not hibernate and are active throughout the year (Schaller et al. 1985; Zhang et al. 2015). Pandas exhibit seasonality in activity patterns, with the highest activity in spring and winter and the lowest activity in summer–autumn (Pan 2014; Zhang et al. 2017a), attributed to seasonality of reproduction and foraging strategy. Actually, they exhibit higher-frequency activity cycles during the winter when resources are relatively poor, as well as during the spring mating season when more travel is required. Pandas exhibit a unique daily activity pattern, as they are active for around 13–14 hours per day with activity peaks occurring in the morning, afternoon, and around midnight (Schaller et al. 1985; Zhang et al. 2015). Most active time is spent foraging, with resting occurring in roughly 2–4-hour bouts in between peak foraging times. Daily activity patterns also relate to weather, as pandas are more active when solar radiation is higher (Zhang et al. 2015), their activity being synchronized with both air temperature and solar radiation (Zhang et al. 2017a).

Feeding Ecology

Although belonging to the order Carnivora, the giant panda has evolved to be reliant almost exclusively on low-quality but constantly available bamboo as a foraging resource (Schaller et al. 1985; Nie et al. 2015a, 2015b). They are known to consume more than 40 species of bamboo in the wild, and readily consume species from outside their native range in captivity. In each mountain range, there are usually several bamboo species distributed along elevational gradients. For example, pandas in the Qinling mountains commonly feed on *Bashania fargesii* (<1900 m) from November to April and on *Fargesia qinlingensis* (>1900 m) during the remainder of the year. In the Qionglai mountains, *B. fangiana*, *F. robusta* and *Yushania brevipaniculata* are staple food bamboos.

Although pandas are occasionally found to ingest other plants such as *Zea mays*, *Cucurbita moschata*, *Heracleum candicans*, *Houttuynia cordata*, *Equisetum hiemale*, *Angelica* sp., and even animal carcasses, bamboo constitutes >98% of their diet (Schaller et al. 1985). Numerous adaptations make the panda exquisitely adapted to a life foraging on bamboo (F. Wei et al. 2015a). These famously include the panda's "thumb" (a modified radial sesamoid bone), specialized dentition, and large zygomatic arches and musculature for masticating such a tough, fibrous diet. However, pandas retain the simple digestive tract of their carnivore ancestors, and their genome does not code for enzymes needed to digest cellulose (F. Wei et al. 2012). Metagenomic analysis revealed a number of gut microbial symbionts that aid in cellulose digestion (Zhu et al. 2011). Even so, the panda's digestive efficiency is far less than that of other herbivorous mammals. To counter these digestive challenges, pandas consume large volumes of bamboo and allocate 14 hours of foraging to every 24-hour cycle. They thus ingest in excess of 12 kg of bamboo daily and defecate about 50 times each day.

The panda's foraging strategy is thus characterized by low efficiency, high throughput, and low rates of energy acquisition. Energetically, one might rightly argue that they live on a knife's edge. This strategy affects many other aspects of panda ecology. Signs of energy conservation are detected in many aspects of their biology, including metabolic rate (Nie et al. 2015a), reproduction (Nie et al. 2012c, 2015b), maternal care (Zhu et al. 2001), and communication (Nie et al. 2012a). They do not put on enough body fat to hibernate, as seen in sympatric Asiatic black bears. Pandas remain active, foraging year-round. Their movement ecology is constrained to favor more gentle slopes (W. Wei et al. 2015; Connor et al. 2016), which require less energy than navigating steep slopes.

Panda foraging strategies are fine-tuned to optimize acquisition of nutrients and energy. Although difficult to observe directly in the wild, pandas leave many telltale signs of their feeding decisions, both in their prolific production of feces and in the bamboo "stumps" they leave behind indicating their diet selection. From these signs, much has been learned about panda foraging. They are selective with regard to bamboo age and the parts of the plant (Schaller et al. 1985; Li et al. 2017c), preferring younger, more nutritious bamboo and typically favoring leaves, which have a higher protein : cellulose ratio than culms. Underlying nutritional composition governs these decisions, guiding pandas to select the most nutritious bamboo during the optimal season, based on key nutrients such as calcium, phosphorus, and nitrogen (Nie et al. 2015b). Acquisition of these nutrients also guides seasonal patterns of elevational migration, bringing pandas to the most nutritious bamboos for the time of year. Pandas move through the environment efficiently between bamboo patches, then adopt more tortuous paths in areas of suitable bamboo (Zhang et al. 2014), stopping to forage most intensively at the most nutritious, protein-rich, cellulose-poor patches (W. Wei et al. 2015).

Pandas do not maintain the same foraging strategy year-round, but alter how they select bamboo and where they forage as the nutritional quality of bamboo changes with seasons (Schaller et al. 1985; Nie et al. 2015b; Li et al. 2017c). For

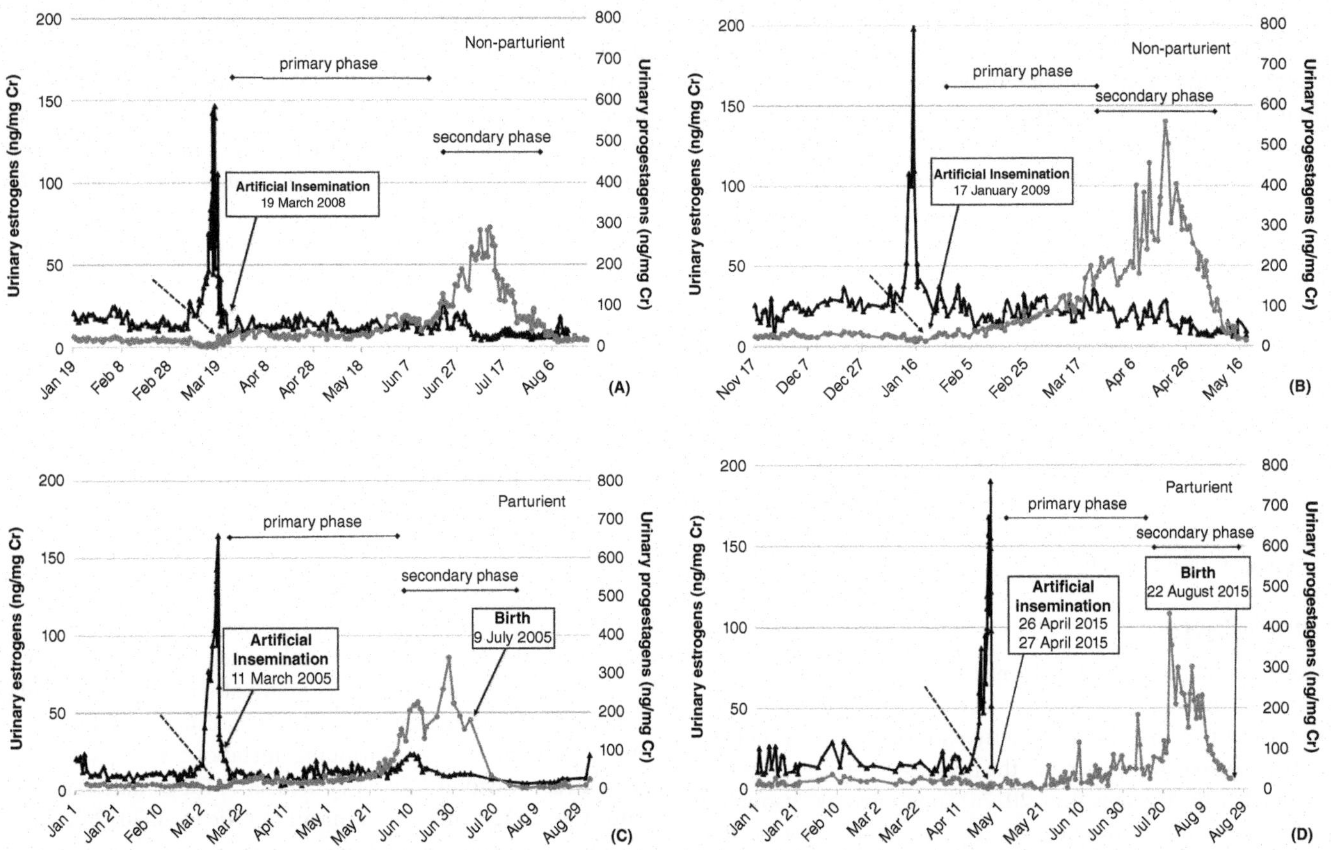

Figure 6.4 Urinary estrogen and progestagen metabolite (adjusted for water content using concentration of creatinine, Cr) profiles in a single giant panda female during four different years. Artificial insemination with fresh/thawed sperm was performed at the time of falling estrogen (black arrow) and the slight rise in peri-ovulatory progestagen (gray arrow), with panels A and B being non-parturient (no offspring) and panels C and D resulting in live, surviving young.

example, studies in Foping Nature Reserve have shown that pandas there show a strong preference for eating leaves throughout most of the year, but as the nutritional quality of those leaves is compromised as the season progresses to winter and the leaves whither, pandas switch to eating culms more often (Li et al. 2017c). They also prefer younger bamboo, but as availability of young bamboo decreases due to preferential consumption, pandas begin to eat older bamboo more. This means they are most nutritionally challenged at the end of winter, just as the mating season begins and pandas need to expend more energy searching and competing for mates. In fact, these energetic constraints may explain why male pandas appear unable to sustain high levels of metabolically expensive testosterone throughout the mating season (Nie et al. 2012c). This period is also characterized by high mortality rates, suggesting that bamboo, although abundant year-round, may be a seasonally limiting resource that regulates population size. New bamboo shoots also appear essential for the giant panda nutritional strategy, across all populations studied (Schaller et al. 1985; Pan 2014; Nie et al. 2015b; Li et al. 2017c). Following the end of the most limiting period of winter, new bamboo growth attracts pandas, and they eat bamboo shoots almost exclusively for a month or two. This resource thereby allows them to recover from the annual period of deprivation, and thus disturbance or degradation of this habitat could have disastrous consequences.

Reproduction and Growth

Reproduction in the giant panda is well established due to field observations (Schaller et al. 1985; Nie et al. 2012b; Pan 2014) and conservation breeding studies integrating behavioral, endocrine, and semen evaluations (Wildt et al. 2006; see also Chapter 2).

Reproductive traits appear remarkably consistent between animals in the wild versus captivity. Pandas display seasonal reproduction, with the female exhibiting a single estrus between February and June, with a preponderance in March/April. Because estrus is only 24–72 h, the female devotes <1% of its annual lifespan to sexual activity. Indices of estrus are increased scent-marking, vocalizations, decreased appetite, tail flagging, backward walking, and, eventually, lordosis (Swaisgood et al. 2003; Owen, Swaisgood & Blumstein 2016). There is a gradual 1-2-week increase in urinary estrogen metabolite excretion resulting in a peak and then rapid fall (Figure 6.4) signaling ovulation (Steinman et al. 2006). Estrogen dynamics at this time are reflected in vaginal cytological changes. There also is peri-ovulatory luteinization of the ovarian follicle(s) revealed by a slight subbaseline decrease and then increase in progestogen that coincides with falling estrogen (Kersey et al. 2011). Peak and

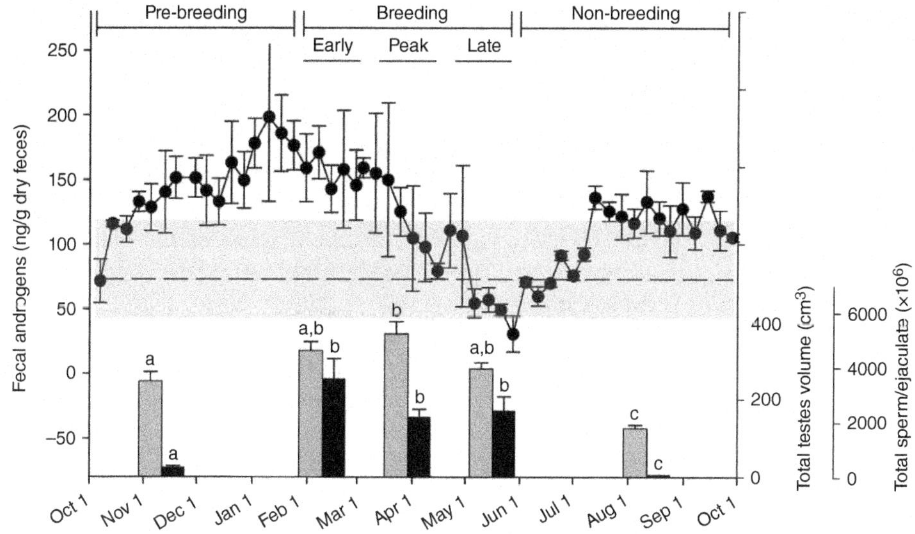

Figure 6.5 Fecal androgen metabolites, testicular volume, and total sperm per ejaculate in giant pandas (*ex situ* in China). Mean (± standard error of the mean, SEM) androgens (solid circles) during pre-breeding (October–January), early (February–early March, peak in mid-March–mid-April), and late (mid-April–May) breeding seasons, as well as the non-breeding (June–September) season. Baseline (± SEM) androgen in the dashed, shaded area. Mean (± SEM) values for testicular volume are presented by gray bars and total sperm per ejaculate by black bars (a,b,c within a row, values with different superscripts denote differences among seasons, $P < 0.05$). Reprinted with permission of Aitken-Palmer et al. (2012) and *Biology of Reproduction*.

declining estrogen are associated with tail-up and walking backward behaviors, but not to a certain threshold in steroid production. Excreted estrogen concentration varies among individuals and within the same female over successive years.

The post-ovulation luteal phase has two components, the "primary" where the progestagen metabolite is excreted at modest levels for 61–122 days (Figure 6.4) and the secondary phase of ~45–55 days characterized by a ~ 20-fold higher progestagen rise, likely associated with implantation (Kersey et al. 2011). Delayed nidation (with a free-floating conceptus) explains the biphasic progestagen profile and variable gestation length, and perhaps avoids seasonally late births. Adrenal glucocorticoids also are increased during peri-estrus and late gestation, facilitating energetic demands (Kersey et al. 2011). An ovulating, mated giant panda becomes parturient or non-parturient, with indistinguishable estrogen and progestagen profiles (Kersey et al. 2010). In either case, there can be decreased appetite, nest building, cradling behaviors, vulvar swelling, mammary gland enlargement, and lethargy.

Adult males produce prodigious numbers of sperm (averaging $>4500 \times 10^6$/ejaculate) that are highly motile (>65%) and structurally normal (≥65%) (Wildt et al. 2006). In captivity, there is protracted male reproductive seasonality (Aitken-Palmer et al. 2012), ensuring sustained fertility for the forthcoming, abrupt female cycle. Fecal androgen excretion, testes size, sperm density (Figure 6.5), and frequency of reproductive behaviors (locomotion, scent marking, vocalizations) increase from the pre-breeding to early breeding season. Excreted adrenal glucocorticoids rise with breeding season, likely to enhance metabolism, maximize energy stores, and provide a competitive edge for mating (Kersey et al. 2011). Testes volume and sperm concentration are maximal in peak season with optimal semen quality occurring 2–3 months after peak androgen (Aitken-Palmer et al. 2012). There are basal testes size, aspermic ejaculates, and absence of reproductive behaviors from June to September. There is no relationship between androgen excretion concentration/patterns and breeding success (Kersey et al. 2010). From the limited information on reproductive endocrinology available from wild pandas, there seem to be some patterns that are divergent from captivity. Male giant pandas do not appear to maintain elevated androgen levels throughout the breeding season, instead demonstrating elevated androgen only when estrous females are present (Nie et al. 2012b). This finding is thought to be due to the greater energy constraints facing wild pandas, which may be unable to maintain high metabolic rates associated with high androgen levels.

Successful reproduction in nature requires plentiful resources and absence of human disturbance (Zhu et al. 2001; Pan 2014). An estrual female in the wild can attract up to five males that can travel long distances and assemble in the female's territory to compete (Schaller et al. 1985; Nie et al. 2012b). Mating is by the most competitive male. Successful breeding in captivity is associated with a biologically relevant, enriched environment with adequate space and complexity to minimize animal stress and allow performance of evolved behaviors (Swaisgood et al. 2006). This includes managing animals in individual enclosures and providing visual, olfactory, and acoustic contact. A strong bout of sexual interaction in protected contact is a prerequisite to introducing the designated pair that can result in 61% breeding success (Li et al. 2017b).

There is no definitive test for pregnancy, although a urinary decrease in the acute-phase protein ceruloplasmin (Willis et al. 2011) or increased prostaglandin $F_{2\alpha}$ metabolite (Roberts et al. 2018) has been associated with late gestation. Gestational duration (first copulation or artificial insemination to birth) ranges from 84 to 273 days, the widest variance of all ursids (Zhu et al. 2001). Although there is wide distribution in estrus onset, 75% of births occur in August and September, coincident with excreted progestagen returning to baseline. A pregnant female gives birth in a rock cave or old-growth tree cavity. Half of all litters have two cubs, although one is typically abandoned to die. The prevailing hypothesis is that twinning in the giant panda is an ancient Ursidae trait where the norm is two or more cubs (Zhu et al. 2001).

In the wild, the interbirth interval is 2–3 years with a maximal lifetime contribution of 7–8 cubs by reproductive senescence at 19 or 20 years of age (Zhu et al. 2001; Pan 2014). Offspring production can be more than twice as high in the breeding centers because of the practice of twin-swapping to accommodating dams and improved husbandry that has increased cub survival to >90%.

Development

Average neonate birth weight is 112 g. Individual cubs average only 0.12% of the dam's body weight, the smallest size proportion of any other eutherian mammal (Zhu et al. 2001). Neonates have sparse short, white hair covering the entire body. This altriciality mandates extraordinary maternal investment in care-giving behaviors, including holding the vulnerable neonate continuously for the first 2–3 weeks of life to prevent hypothermia (Zhu et al. 2001; Snyder et al. 2003, 2016). This also explains the dam's post-partum fast and the cub's frequent vocalizations during the same time period, probably seeking nipple contact.

As in other bears, there is relatively slow maturation of the young (Zhu et al. 2001). The neonate's skin in the area where black hair will eventually grow turns gray by 8–10 days of age. The typical black and white marking of the adult are evident by 23–25 days when cubs are gaining about 60 g/day. Eyes begin opening from 40 to 49 days and are open by 75 days. The denning interval in the wild is ~4 months, with emergence in early winter. By 6 months of age, cubs gain 110 g daily and are beginning to transition to eating bamboo leaves. The weaning norm in the wild is 18–24 months.

Behavior

Due to the altriciality of panda cubs, maternal behavior is notably active, and females engage in a range of behaviors that promote cub suckling, warmth, and comfort. Maternal behaviors are often initiated in response to the plaintive vocalizations cubs emit; however, maternal responsiveness declines over time (Snyder et al. 2003). Birth and initial cub development occur in dens, typically inside tree cavities or in rock caves (Zhu et al. 2001; Zhang et al. 2007; W. Wei et al. 2019). Pandas do not hibernate, and these dens are only used for rearing cubs. Maternal females are highly selective in choosing den sites, preferring cavities of a certain size and shape that protect cubs from the elements and predators. Preferred den sites have a higher capacity to buffer against extremes of temperature and humidity outside, and provide stable microclimatic conditions for rearing vulnerable cubs.

Foraging behavior is of the utmost importance to allow pandas to subsist on their plentiful, but low-nutrient diet of bamboo (see Feeding Ecology). The selection, processing, and consumption of bamboo consists of a fairly stereotyped repertoire of behaviors that serve to efficiently exploit bamboo as a food resource (Swaisgood et al. 2018). Most aspects of these specialized behaviors have been found to be consistent between wild-born and captive-born giant pandas, suggesting that maternal mentorship may not drive the acquisition of these skills. While the complex motor patterns associated with efficient processing and consumption of bamboo appear developmentally stable in different rearing environments, pandas' ability to select the most nutritious parts of the plant (see Feeding Ecology) appears to be compromised during development in captivity.

Communication and reproduction are the best-studied aspects of panda behavior. As a solitary-living animal, olfactory (or chemical) communication is the primary mode of communication throughout most of the year (Schaller et al. 1985; Swaisgood et al. 2004). Pandas deposit scent in the form of urine and specialized anogenital gland secretions on trees at traditional sites that serve as a sort of community bulletin board. Pandas are highly selective of their scent deposition sites, preferring some types of trees over others, selecting relatively open microhabitat around scent trees, and selecting certain bark characteristics based on roughness of texture (Nie et al. 2012a). These selective behaviors serve to increase signal range, increase signal duration, and maximize the probability that the signal will be detected by conspecifics. Pandas also deploy four distinct postures to deposit scents, including a handstand posture used by adult males to convey competitive dominance (White et al. 2002). A series of investigations on captive pandas have demonstrated that pandas can extract a great deal of information from investigating scent (Swaisgood et al. 2004). Pandas can discriminate conspecific odors to determine gender, reproductive condition, age class, kin, individual identity, and other important social information. Semiochemical investigations using gas chromatography have shown that panda chemical signals are composed of a complex mixture of chemical constituents that differ between the sexes and change from the breeding to non-breeding seasons (Yuan et al. 2004; Zhou et al. 2019).

Acoustic communication, by contrast, is primarily utilized during the mating season (and between mother and cub), and wild pandas vocalize infrequently outside of reproductive contexts (vocalization outside the breeding season is more common in captivity, where pandas are housed near conspecifics and also direct vocalizations towards their caretakers). The complexity of information conveyed in vocal signals, however, is no less pronounced than it is for chemical signals (Charlton et al. 2009, 2010, 2011, 2018a). Panda vocalizations carry information regarding identity, male testosterone levels, and female reproduction condition, and play an instrumental role in courtship and copulation. Among their vocal repertoire, pandas emit high-pitched bleats and chirps, which are closely tied to sexual and affiliative motivation, and lower-pitched moans and barks, which are associated with aggressive intent or perhaps ambivalence between attraction and avoidance. Panda audition (hearing) can be described as typical of large carnivores (Owen, Keating, & Denes 2016). However, giant panda hearing sensitivity diverges significantly from that of the polar bear in the upper frequencies, with functional hearing retained into the ultrasonic range. This functionality may support effective intraspecific acoustic communication in the dense bamboo forests that giant pandas inhabit, and may be especially relevant to the transmission of the bleat vocalization through this habitat (Charlton et al. 2018b).

Visual communication is less prominent and less well-studied in pandas. However, some behaviors associated with estrus and leading up to mating are visual signals, such as backward walking, pink vulva, and tail up in females (Owen, Swaisgood, & Blumstein 2016) and footscraping in males. The black and white pelage markings are a prominent feature of pandas, and are thought to play a role in visual communication. While the body markings are likely involved in camouflage, the facial markings may convey information about identity or aggressive intent (Caro et al. 2017). Panda vision is not well studied, but it seems reasonably acute and they possess color vision typical for carnivores (Keller et al. 2006). Presumably, color vision aids in selection of nutritious bamboo, but could also be used in social communication, such as assessment of female reproductive condition via vulva color.

Pandas display a large variety of specialized behaviors associated with reproduction (Owen, Swaisgood, & Blumstein 2016) (see Reproduction). These behaviors aid in localizing potential mates, reducing aggression, overcoming a tendency to avoid conspecifics, and supporting the courtship process (Swaisgood et al. 2006); they also allow managers of captive pandas to evaluate the estrous status of females and reproductive readiness. In the wild, when a female comes into estrus she disperses her scent widely, which serves to attract several male suitors that compete for reproductive access (Schaller et al. 1985; Nie et al. 2012b). This intermale competition can last several days, but typically dominance is established in the first 2 days and escalated and injurious aggression is replaced with the use of visual and acoustic signals, and conflict is mediated by threat and avoidance. The largest male typically prevails and only a single male mates with the female. In captivity, both male and female choice have been shown to be important (Martin-Wintle et al. 2015). Much higher rates of copulation and fertilization are achieved when partners are allowed to choose their own mates than when they are paired with randomly selected partners that are less preferred. Personality also plays a role in these captive parings, with certain combinations of partners having better reproductive performance than other personality pairings (Martin-Wintle et al. 2017). These findings are extremely important for guiding reproductive management in conservation breeding programs, but their relevance to mating behavior in wild populations remains unknown.

Parasites and Diseases

Parasitism is prevalent in wild panda populations and may be a cause for concern (Zhang et al. 2008). *Baylisascaris schroederi* is the most common and harmful parasite in the panda. Estimated prevalence in captive and wild pandas ranges from 50% to 100% (Zhang et al. 2008). Although the life cycle of *B. schroederi* remains unknown, ova are passed in the host feces and into the environment, where they can persist in soil for years, from which they can be ingested (Zhang et al. 2008). Visceral larval migrans (VLM) refers to disease caused by aberrant migration of nematode larvae, which can be profoundly inflammatory. Diagnosis has historically relied on identification of adult worms or eggs in the feces of infected individuals. Recent studies have shown that polymerase chain reaction (PCR) can be used to detect *B. schroederi* eggs with much higher sensitivity and specificity (Wang et al. 2013). Other species of nematode have been identified in pandas, including *Ancylostoma ailuropodae*, *Ogmocotyle sikae*, *Toxocaris seleactis*, and *Strongyloides* sp., but their clinical significance remains unknown.

Over 12 hard tick species have been described infesting pandas, including ticks from the genera *Ixodes*, *Haemaphysalis*, and *Dermacentor* (Cheng et al. 2013). Although it is currently unknown whether these species act as a vector for diseases afflicting pandas, tick infestation alone can cause morbidity, such as dermatitis, or in extreme infestations may result in anemia, weight loss, and death. Giant pandas can also be infested by *Chorioptes* mites, which have been recently confirmed to be a unique species, *Chorioptes panda* (Wang et al. 2012). This mite has been isolated from both wild and captive individuals. Acariasis caused by *C. panda* presents as erythema, crusting, and alopecia (due to pruritus).

Of the four protozoal parasites reported in pandas, *Toxoplasma gondii* is the only one known to have resulted in clinical morbidity and eventual mortality (Ma et al. 2015). The microsporidian pathogen *Enterocytozoon bieneusi* is an opportunistic enteric parasite that primarily resides in the small intestine of the host, but no clinical signs have been described in pandas.

Status in the Wild

The giant panda is considered Vulnerable by the International Union for the Conservation of Nature (IUCN), although it was listed as Endangered until 2016 (Swaisgood et al. 2016). The Chinese government has also recently downlisted the giant panda to Vulnerable using similar criteria (Zhang et al. 2017b). These conservation status assessments incorporate the best available evidence regarding population size and trends, the amount and trends of supporting habitat, and the degree of population fragmentation threatening a species, among other factors. A Vulnerable species under IUCN criteria still faces high risk of extinction in the wild, so the downlisting of the giant panda does not indicate that further conservation efforts are not warranted. On the contrary, ongoing conservation action is required to address the small population size and continuing and emerging threats facing giant pandas (Kang & Li 2016; Swaisgood et al. 2016, 2017; Xu et al. 2017). The giant panda is also protected against international trade by the Convention on International Trade in Endangered Species of Wild Fauna and Flora (CITES), Appendix 1 (the highest level of protection).

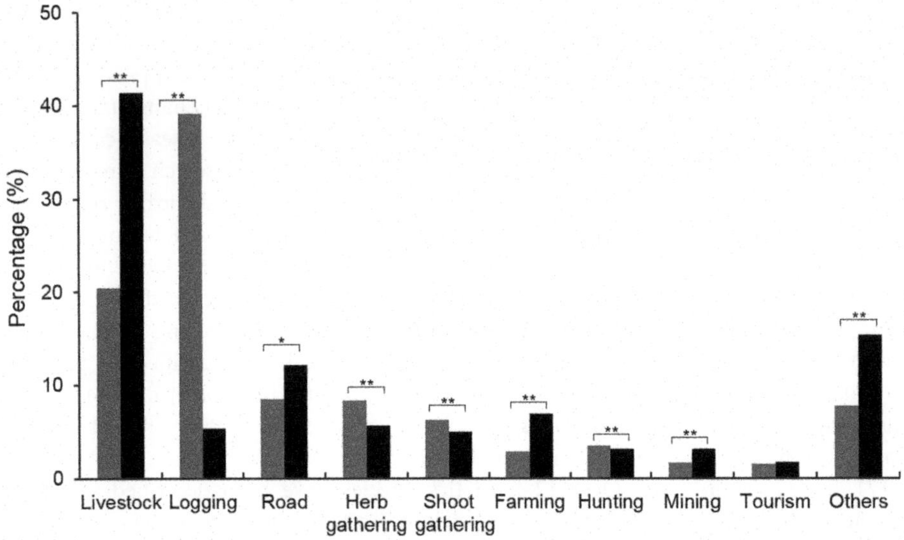

Figure 6.6 Plot-level occurrence of human disturbance in the third (gray; 1999–2003) and fourth (black; 2011–2014) Giant Panda National Surveys in Sichuan Province, China. Proportions of human disturbance and biological covariates that differ between the third and fourth national surveys are indicated by $*P < 0.05$, $**P < 0.01$. Reprinted with permission of W. Wei et al. (2018b) and *Conservation Letters.*

The panda's distribution today (Figure 6.1) represents a small proportion of its historical range. Pandas once occupied much of southern China and were found as far north as Beijing. Range retraction was associated first with a warming climate toward the end of the Pleistocene (Han et al. 2019), but was greatly expedited during recent history due to rapidly expanding human populations, conversion of habitat for agricultural expansion, and, later, industrialization and urbanization (Zhu et al. 2013; Li et al. 2015b). Remnants of panda populations today are found in the Qinling, Minshan, Qionglai, Liangshan, Daxiangling, and Xiaoxiangling mountain ranges.

The status of the giant panda is better known than for most bear species – indeed, than most species of wildlife – because the Chinese government invests in an intensive rangewide national survey about once a decade (SFA 2015). Hundreds of workers spend tens of thousands of worker-days conducting transects through all suitable panda habitat. Pandas are notoriously difficult to find in their dense bamboo habitat, so the census relies on signs, primarily feces. Although DNA census techniques have been developed for pandas (Zhan et al. 2006; Wei et al. 2012), they have not been widely applied in the national survey, and population numbers are estimated using fecal remains. This necessarily crude estimate may underestimate population size, as suggested by comparisons with molecular census methods (Zhan et al. 2006), yet still provides useful information on population trends.

Examining the four surveys conducted to date, beginning in the 1980s, reveals a trend of rapid decline, followed by stabilization and a gradual increase in population numbers. The most recent survey, completed in 2014, estimated 1864 individual pandas (with cubs <2 years old excluded). This encouraging trend for pandas runs counter to continued decline for ~ 90% of species listed on the International Union for the Conservation of Nature (IUCN) Red List of Threatened Species (Hoffmann et al. 2010).

Behind this emerging success story are huge investments in resources and political will. China's State Forestry Administration and other government agencies have implemented a number of policies that have been beneficial to panda conservation (F. Wei et al. 2015b; Liu et al. 2016). Heavy penalties for poaching pandas were soon followed by the establishment of a rapidly growing protected area system designed to arrest the decline in panda populations. China now has 67 panda reserves protecting two thirds of the panda population and half of all suitable habitat (SFA 2015), and has plans to establish a large Giant Panda National Park providing protection for much of the remaining habitat. One of the most significant remaining threats to pandas is severe habitat fragmentation leading to population isolation, and the national park will establish corridors to help connect many of these populations. In addition to the reserve system, two large ecocompensation programs have incentivized local farmers to conserve and restore forests in the panda's range. The National Forest Conservation Program bans logging and the Grain to Green Program pays farmers to plant forests instead of crops in many mountainous areas. Together, these efforts have contributed to many positive trends (Figure 6.6), including increasing forest cover, increasing suitability of secondary forests, improved financing of panda reserves, a reduction in the human agricultural population in the panda's range, expanding panda range, and increasing population size (Vina et al. 2007; F. Wei et al. 2015b; Swaisgood et al. 2017; W. Wei et al. 2018; Xu et al. 2017).

Despite these successes, some threats continue and new ones are emerging (Figures 6.6 and 6.7). Climate change is one of the most prominent emerging threats, although its impacts on pandas remain uncertain. Several early models produced dire forecasts, with loss of panda habitat estimated at 37–100% during this century (Tuanmu et al. 2013; Li et al. 2015a) and associated reductions in panda range and numbers. However, these models may not contain biologically realistic assumptions; for example, assuming that bamboo species in

Figure 6.7 Emerging threats facing giant panda conservation: (A) livestock is encroaching at increasing rates into protected and unprotected panda habitat, with unknown consequences; (B) agricultural expansion (facilitated by climate change) eats away at panda habitat at the lower elevations; (C) bamboo, the principal food source of pandas, is predicted to experience distributional shifts under climate change; (D) remaining old-growth forests, an important ecological factor providing den sites and associated with panda occupancy, will need protection; (E) roads further fragment an already subdivided panda population; and (F) ecotourism is increasing in China, which may be a source of disturbance and habitat degradation. Reprinted with permission of Swaisgood et al. (2017) and *Conservation Letters*.

the pandas' current range are the only species they can depend on in the future (F. Wei et al. 2015b). Historically, pandas occupied lower-elevation, warmer climates and foraged on bamboo species adapted to different bioclimatic envelopes, so a warming climate may be expected to facilitate migration of these species into the pandas' range, replacing those on which they currently depend. More recently, modification of assumptions, different modeling techniques, and inclusion of additional factors such as biotic interactions has led to much more hopeful results, with net habitat loss estimate at 16% (Wang et al. 2018). Thus, our current state of knowledge indicates we must acknowledge uncertainty, but should expect more modest negative impacts on panda populations from climate change. Potential mitigation strategies to offset these effects include assisted migration strategies for bamboo species to ensure species impacted by climate change are replaced with species adapted to the forecast climate patterns for current panda-occupied habitat (F. Wei et al. 2015b).

A second major emerging threat with uncertain impacts on pandas is livestock grazing, which may be a source of disturbance or may compromise panda habitat. The presence of livestock is increasing rangewide in panda habitat, both inside and outside protected areas (W. Wei et al. 2018), and there is evidence that pandas avoid areas used by livestock (Wang et al. 2015; Li et al. 2017a; W. Wei et al. 2018), reducing the habitat available to pandas. Socioeconomic factors, such as an increase in beef consumption in China and restrictions on other uses of panda habitat such as logging and firewood collection (Liu et al. 2016; Xu et al. 2017), must be addressed to mitigate these impacts. Ecocompensation programs, which have been empirically shown to improve panda habitat and enhance panda protections (Vina et al. 2007; Liu et al. 2016), are now in jeopardy from reform of collective forest policies impacting 15% of remaining panda habitat (Yang et al. 2015). Continued economic development in China provides the impetus for a variety of new infrastructure projects such as roads, hydroelectric dams, and mining that further threaten panda habitat (SFA 2015). Fragmentation and domestic animals also threaten to introduce novel pathogens that threaten panda health.

Thus, despite many recent gains in panda conservation, the landscape in which pandas live continues to come under threat from a barrage of human activities that will challenge the panda's future. The downlisting to Vulnerable was the result of concerted conservation efforts on the part of the Chinese government and NGOs, yet emerging threats still threaten panda populations. The giant panda will remain a conservation-dependent species requiring protection and management for the foreseeable future.

Status in Captivity

Giant pandas were once notorious for their poor breeding performance in captivity and the population was not self-sustaining. However, numerous improvements in husbandry, nutrition, and behavioral management led to dramatic increases in reproductive success, especially through natural mating (Li et al. 2017b). The captive population began to grow rapidly, increasing from <200 to >500 individuals in less than two decades. The captive population is currently self-sustaining and produces surplus individuals for reintroduction (for more details see Chapter 23).

Authors' note

We would like to dedicate this chapter to David Wildt, friend, colleague and leading voice in reproductive biology and conservation. His pioneering work has revealed many of the physiological processes underlying giant panda reproduction and led the way for the development of assisted reproduction techniques. The same could be said for many other species which have benefited from his contributions. We will miss his leadership and collegiality.

References

Aitken-Palmer, C., Hou, R., Burrell, C., et al. (2012). Protracted reproductive seasonality in the male giant panda (*Ailuropoda melanoleuca*) reflected by patterns in androgen profiles, ejaculate characteristics, and selected behaviors. *Biology of Reproduction* 86(6): 195.

Caro, T., Walker, H., Rossman, Z., Hendrix, M. & Stankowich, T. (2017). Why is the giant panda black and white? *Behavioral Ecology* 28(3): 657–667.

Charlton, B. D., Huang, Y. & Swaisgood, R. R. (2009). Vocal discrimination of potential mates by female giant pandas (*Ailuropoda melanoleuca*). *Biology Letters* 5: 597–599.

Charlton, B. D., Keating, J. L., Li, R., Huang, Y. & Swaisgood, R. R. (2010). Female giant panda (*Ailuropoda melanoleuca*) chirps advertise the caller's fertile phase. *Proceedings of the Royal Society B* 277: 1101–1106.

Charlton, B. D., Keating, J. L., Kersey, D., et al. (2011). Vocal cues to male androgen levels in giant pandas. *Biology Letters* 7: 71–74.

Charlton, B. D., Martin-Wintle, M. S., Owen, M. A., Zhang, H. & Swaisgood, R. R. (2018a). Vocal behaviour predicts mating success in giant pandas. *Royal Society Open Science* 5(10): 181323.

Charlton, B. D., Owen, M. A., Keating, J. L., et al. (2018b). Sound transmission in a bamboo forest and its implications for information transfer in giant panda (*Ailuropoda melanoleuca*) bleats. *Scientific Reports* 8(1): 12754.

Cheng, W. Y., Zhao, G. H., Jia, Y. Q., et al. (2013). Characterization of *Haemaphysalis flava* (Acari: Ixodidae) from Qingling subspecies of giant panda (*Ailuropoda melanoleuca qinlingensis*) in Qinling Mountains (Central China) by morphology and molecular markers. *PLoS ONE* 8(7): e69793.

Connor, T., Hull, V. & Liu, J. (2016). Telemetry research on elusive wildlife: a synthesis of studies on giant pandas. *Integrative Zoology* 11(4): 295–307.

Han, H., Wei, W., Hu, Y., et al. (2019). Diet evolution and habitat contraction of giant pandas via stable isotope analysis. *Current Biology* 29(4): 664–669.e662.

Hoffmann, M., Hilton-Taylor, C., Angulo, A., et al. (2010). The impact of conservation on the status of the world's vertebrates. *Science* 330: 1503–1509.

Huang, W. (1993). The skull, mandible and dentition of giant pandas (Ailuropoda): morphological characters and their evolutionary implications. *Vertebrata Palasiatica* 31: 191–207.

Hull, V., Roloff, G., Zhang, J., et al. (2014). A synthesis of giant panda habitat selection. *Ursus* 25(2): 148–162.

Hull, V., Zhang, J., Zhou, S., et al. (2015). Space use by endangered giant pandas. *Journal of Mammalogy* 96(1): 230–236.

Hull, V., Zhang, J., Huang, J., et al. (2016). Habitat use and selection by giant pandas. *PLoS ONE* 11(9): e0162266.

Kang, D. & Li, J. (2016). Premature downgrade of panda's status. *Science* 354: 295.

Keller, A. S., Snyder, R. J., Marr, M. J., et al. (2006). Color vision in the giant panda (*Ailuropoda melanoleuca*). *Learning and Behavior* 3: 154–161.

Kersey, D. C., Wildt, D. E., Brown, J. L., et al. (2010). Unique biphasic progestagen profile in parturient and non-parturient giant pandas (*Ailuropoda melanoleuca*) as determined by faecal hormone monitoring. *Reproduction* 140: 183–193.

Kersey, D. C., Wildt, D. E., Brown, J. L., et al. (2011). Rising fecal glucocorticoid concentrations track reproductive activity in the female giant panda (*Ailuropoda melanoleuca*). *General and Comparative Endocrinology* 173(2): 364–370.

Li, B. V. & Pimm, S. L. (2016). China's endemic vertebrates sheltering under the protective umbrella of the giant panda. *Conservation Biology* 30(2): 329–339.

Li, B. V., Pimm, S. L., Li, S., Zhao, L. & Luo, C. (2017a). Free-ranging livestock threaten the long-term survival of giant pandas. *Biological Conservation* 216: 18–25.

Li, D., Wintle, N. J. P., Zhang, G., et al. (2017b). Analyzing the past to understand the future: natural mating yields better reproductive rates than artificial insemination in the giant panda. *Biological Conservation* 216: 10–17.

Li, R., Fan, W., Tian, G., et al. (2010). The sequence and de nova assembly of the giant panda genome. *Nature* 463: 311–317.

Li, R., Xu, M., Wong, M. H. G., et al. (2015a). Climate change threatens giant panda protection in the 21st century. *Biological Conservation* 182: 93–101.

Li, X., Jiang, G., Tian, H., et al. (2015b). Human impact and climate cooling caused range contraction of large mammals in China over the past two millennia. *Ecography* 38(1): 74–82.

Li, Y., Swaisgood, R. R., Wei, W., et al. (2017c). Withered on the stem: is bamboo a seasonally limiting resource for giant pandas? *Environmental Science and Pollution Research* 24(11): 10,537–10,546.

Linderman, M., Bearer, S., An, L., et al. (2005). The effects of understory bamboo on broad-scale estimates of giant panda habitat. *Biological Conservation* 121: 383–390.

Liu, J., Hull, V., Yang, W., et al. (2016). *Pandas and people: Coupling human and natural systems for sustainability*. Oxford: Oxford University Press.

Ma, H., Wang, Z., Wang, C., et al. (2015). Fatal *Toxoplasma gondii* infection in the giant panda. *Parasite* 22: 30–30.

Martin-Wintle, M. S., Shepherdson, D., Zhang, G., et al. (2015). Free mate choice enhances conservation breeding in the endangered giant panda. *Nature Communications* 6: 10125.

Martin-Wintle, M. S., Shepherdson, D., Zhang, G., et al. (2017). Do opposites attract? Effects of personality matching in breeding pairs of captive giant pandas on reproductive success. *Biological Conservation* 207: 27–37.

Nie, Y., Swaisgood, R. R., Zhang, Z., et al. (2012a). Giant panda scent-marking strategies in the wild: role of season, sex and marking surface. *Animal Behaviour* 84(1): 39–44.

Nie, Y., Swaisgood, R. R., Zhang, Z., Liu, X. & Wei, F. (2012b). Reproductive competition and fecal testosterone in wild male giant pandas (*Ailuropoda melanoleuca*). *Behavioral Ecology and Sociobiology* 66(5): 721–730.

Nie, Y.-G., Zhang, Z.-J., Swaisgood, R. R. & Wei, F.-W. (2012c). Effects of season and social interaction on fecal testosterone metabolites in wild male giant pandas: implications for energetics and mating strategies. *European Journal of Wildlife Research* 58(1): 235–241.

Nie, Y., Speakman, J. R., Wu, Q., et al. (2015a). Exceptionally low daily energy expenditure in the bamboo-eating giant panda. *Science* 349(6244): 171–174.

Nie, Y., Zhang, Z., Raubenheimer, D., et al. (2015b). Obligate herbivory in an ancestrally carnivorous lineage: the giant panda and bamboo from the perspective of nutritional geometry. *Functional Ecology* 29(1): 26–34.

O'Brien, S. J., Nash, W. G., Wildt, D. E., Bush, M. E. & Benveniste, R. E. (1985). A molecular solution to the riddle of the giant panda's phylogeny. *Nature* 317 (6033): 140–144.

Owen, M. A., Keating, J. L., Denes, S. K., et al. (2016). Hearing sensitivity in context: conservation implications for a highly vocal endangered species. *Global Ecology and Conservation* 6: 121–131.

Owen, M. A., Swaisgood, R. R. & Blumstein, D. T. (2016). Signalling behaviour is influenced by transient social context in a spontaneously ovulating mammal. *Animal Behaviour* 111: 157–165.

Pan, W. (2014). *A chance for lasting survival: Ecology and behavior of wild giant pandas*. Washington, DC: Smithsonian Institution Scholarly Press.

Roberts, B. M., Brown, J. L., Kersey, D. C., et al. (2018). Use of urinary 13,14, dihydro-15-keto-prostaglandin F2α (PGFM) concentrations to diagnose pregnancy and predict parturition in the giant panda (*Ailuropoda melanolecua*). *PLoS ONE* 13(5): e0195599.

Schaller, G. B., Hu, J., Pan, W. & Zhu, J. (1985). *The giant pandas of Wolong*. Chicago, IL: University of Chicago Press.

SFA. (2015). *Report of the Fourth National Giant Panda Survey*. Beijing: Science Publishing House.

Snyder, R. J., Zhang, A. J., Zhang, Z. H., et al. (2003). Behavioral and developmental consequences of early rearing experience for captive giant pandas (*Ailuropoda melanoleuca*). *Journal of Comparative Psychology* 117: 235–245.

Snyder, R. J., Perdue, B. M., Zhang, Z., Maple, T. L. & Charlton, B. D. (2016). Giant panda maternal care: a test of the experience constraint hypothesis. *Scientific Reports* 6: 27509.

Steinman, K. J., Monfort, S. L., McGeehan, L., et al. (2006). Endocrinology of the giant panda and application of hormone techology to species management. In: Wildt, D. E., Zhang, A. J., Zhang, H., Janssen, D. & Ellis, S. (Eds.), *Giant pandas: Biology, veterinary medicine and management* (pp. 198–230). Cambridge: Cambridge University Press.

Swaisgood, R. R., Zhou, X., Zhang, G., Lindburg, D. G. & Zhang, H. (2003). Application of behavioral knowledge to giant panda conservation. *International Journal of Comparative Psychology* 16: 65–84.

Swaisgood, R. R., Lindburg, D., White, A. M., Zhang, H. & Zhou, X. (2004). Chemical communication in giant pandas: experimentation and application. In: Lindburg, D. & Baragona, K. (Eds.), *Giant pandas: Biology and conservation* (pp. 106–120). Berkeley, CA: University of California Press.

Swaisgood, R. R., Zhang, G., Zhou, X. & Zhang, H. (2006). The science of behavioural management: creating biologically relevant living environments in captivity. In: Wildt, D. E., Zhang, A. J., Zhang, H., Janssen, D. & Ellis, S. (Eds.), *Giant pandas: Biology, veterinary medicine and management* (pp. 274–298). Cambridge: Cambridge University Press.

Swaisgood, R. R., Wang, D. & Wei, F. (2016). *Ailuropoda melanoleuca*. IUCN Red List of Threatened Species. www.iucnredlist.org/details/712/0.

Swaisgood, R. R., Wang, D. & Wei, F. (2017). Panda downlisted but not out of the woods. *Conservation Letters* 11: e12355.

Swaisgood, R. R., Martin-Wintle, M. S., Owen, M. A., Zhou, X. & Zhang, H. (2018). Developmental stability of foraging behavior: evaluating suitability of captive giant pandas for translocation. *Animal Conservation* 21: 474–482.

Tuanmu, M.-N., Viña, A., Winkler, J. A., et al. (2013). Climate-change impacts on understorey bamboo species and giant pandas in China's Qinling Mountains. *Nature Climate Change* 3(3): 249–253.

Vina, A., Bearer, S., Chen, X., et al. (2007). Temporal changes in giant panda habitat connectivity across boundaries of Wolong Nature Reserve, China. *Ecological Applications* 17(4): 1019–1030.

Wang, F., McShea, W. J., Wang, D., et al. (2014). Evaluating landscape options for corridor restoration between giant panda reserves. *PLoS ONE* 9(8): e105086.

Wang, F., McShea, W. J., Wang, D. & Li, S. (2015). Shared resources between giant panda and sympatric wild and domestic mammals. *Biological Conservation* 186: 319–325.

Wang, F., Zhao, Q., McShea, W. J., et al. (2018). Incorporating biotic interactions reveals potential climate tolerance of giant pandas. *Conservation Letters* 11(6): e12592.

Wang, N., Li, D.-S., Zhou, X., et al. (2013). A sensitive and specific PCR assay for the detection of *Baylisascaris schroederi* eggs in giant panda feces. *Parasitology International* 62(5): 435–436.

Wang, S., Gu, X., Fu, Y., et al. (2012). Molecular taxonomic relationships of *Psoroptes* and *Chorioptes* mites from China based on COI and 18S rDNA gene sequences. *Veterinary Parasitology* 184 (2): 392–397.

Wei, F. & Zhang, Z. (2009). Family Ailuridae (red panda). In: Wilson, D. E. & Mittermeier, R.A. (Eds.), *Handbook of the mammals of the world. Vol. 1. Carnivores* (pp. 498–503). Barcelona: Lynx Edicions.

Wei, F., Hu, Y., Zhu, L., et al. (2012). Black and white and read all over: the past, present and future of giant panda genetics. *Molecular Ecology* 21(23): 5660–5674.

Wei, F., Hu, Y., Yan, L., et al. (2015a). Giant pandas are not an evolutionary cul-de-sac: evidence from multidisciplinary research. *Molecular Biology and Evolution* 32(1): 4–12.

Wei, F., Swaisgood, R., Hu, Y., et al. (2015b). Progress in the ecology and conservation of giant pandas. *Conservation Biology* 29: 1497–1507.

Wei, F., Costanza, R., Dai, Q., et al. (2018). The value of ecosystem services from giant panda reserves. *Current Biology* 28: 2174–2180.

Wei, W., Nie, Y., Zhang, Z., et al. (2015). Hunting bamboo: foraging patch selection and utilization by giant pandas and implications for conservation. *Biological Conservation* 186: 260–267.

Wei, W., Swaisgood, R. R., Dai, Q., et al. (2018). Giant panda distributional and habitat-use shifts in a changing landscape. *Conservation Letters* 11: e12575.

Wei, W., Swaisgood, R. R., Owen, M. A., et al. (2019). The role of den quality in giant panda conservation. *Biological Conservation* 231: 189–196.

White, A. M., Swaisgood, R. R. & Zhang, H. (2002). The highs and lows of chemical communication in giant pandas (*Ailuropoda melanoleuca*): effect of scent deposition height on signal discrimination. *Behavioral Ecology and Sociobiology* 51: 519–529.

Wildt, D. E., Zhang, A. J., Zhang, H., Janssen, D. & Ellis, S. (2006). *Giant pandas: Biology, veterinary medicine and management*. Cambridge: Cambridge University Press.

Willis, E. L., Kersey, D. C., Durrant, B. S. & Kouba, A. J. (2011). The acute phase protein ceruloplasmin as a non-invasive marker of pseudopregnancy, pregnancy, and pregnancy loss in the giant panda. *PLoS ONE* 6(7): e21159.

Xu, W., Viña, A., Kong, L., et al. (2017). Reassessing the conservation status of the giant panda using remote sensing. *Nature Ecology & Evolution* 1: 1635–1638.

Yang, B., Busch, J., Zhang, L., et al. (2015). China's collective forest tenure reform and the future of the giant panda. *Conservation Letters* 8(4): 251–261.

Yang, H., Viña, A., Tang, Y., et al. (2017). Range-wide evaluation of wildlife habitat change: a demonstration using giant pandas. *Biological Conservation* 213: 203–209.

Yuan, H., Liu, D., Sun, L., et al. (2004). Anogenital gland secretions code for sex and age in the giant panda, *Ailuropoda melanoleuca*. *Canadian Journal of Zoology* 82(10): 1596–1604.

Zhan, X., Li, M., Zhang, Z., et al. (2006). Molecular censusing doubles giant panda population estimate in a key nature reserve. *Current Biology* 16(12): 451–452.

Zhan, X., Zhang, Z., Wu, H., et al. (2007). Molecular analysis of dispersal in giant pandas. *Molecular Ecology* 16(18): 3792–3800.

Zhang, J., Hull, V., Huang, J., et al. (2015). Activity patterns of the giant panda (*Ailuropoda melanoleuca*). *Journal of Mammalogy* 96(6): 1116–1127.

Zhang, J., Hull, V., Ouyang, Z., et al. (2017a). Modeling activity patterns of wildlife using time-series analysis. *Ecology and Evolution* 7(8): 2575–2584.

Zhang, J.-S., Daszak, P., Huang, H.-L., et al. (2008). Parasite threat to panda conservation. *Ecohealth* 5(1): 6–9.

Zhang, L., Luo, Z., Mallon, D., Li, C. & Jiang, Z. (2017b). Biodiversity conservation status in China's growing protected areas. *Biological Conservation* 210: 89–100.

Zhang, Z., Swaisgood, R. R., Wu, H., et al. (2007). Factors predictiing den use by maternal giant pandas. *Journal of Wildlife Management* 71: 2694–2698.

Zhang, Z., Swaisgood, R. R., Zhang, S., et al. (2011). Old-growth forest is what giant pandas really need. *Biology Letters* 7: 403–406.

Zhang, Z., Sheppard, J. K., Swaisgood, R. R., et al. (2014). Ecological scale and seasonal heterogeneity in the spatial behaviors of giant pandas. *Integrative Zoology* 9(1): 46–60.

Zhao, S., Zheng, P., Dong, S., et al. (2013). Whole-genome sequencing of giant pandas provides insights into demographic history and local adaptation. *Nature Genetics* 45(1): 67–71.

Zheng, X., Owen, M. A., Nie, Y., et al. (2016). Individual identification of wild giant pandas from camera trap photos – a systematic and hierarchical approach. *Journal of Zoology* 300(4): 247–256.

Zhou, W., Nie, Y., Hu, Y., et al. (2019). Seasonal and reproductive variation in chemical constituents of scent signals in wild giant pandas. *Science China Life Sciences* 62: 648–660.

Zhu, L., Wu, Q., Dai, J., Zhang, S. & Wei, F. (2011). Evidence of cellulose metabolism by the giant panda gut microbiome. *Proceedings of the National Academy of Sciences* 108(43): 17,714–17,719.

Zhu, L., Hu, Y., Qi, D., *et al.* (2013). Genetic consequences of historical anthropogenic and ecological events on giant pandas. *Ecology* 94(10): 2346–2357.

Zhu, X., Lindburg, D. G., Pan, W., Forney, K. A. & Wang, D. (2001). The reproductive strategy of giant pandas: infant growth and development and mother–infant relationships. *Journal of Zoology, London* 253: 141–155.

Chapter 7

Andean Bear (*Tremarctos ornatus*)

Ximena Velez-Liendo, David Jackson, Manuel Ruiz-García, Armando Castellanos, Santiago Espinosa, and Andrés Laguna

Names

Genus: *Tremarctos* Gervais, 1855

Species: *Tremarctos ornatus* Cuvier, 1825

Names in other languages: **Spanish** Oso Andino, Oso Frontino, Oso de Anteojos; **English** Andean bear, spectacled bear; **French** Ours Andin, Ours à lunettes; **Italian** Orso Andino; **Quechua** Jukumari, Ucumari. **Other common names**: Iznachi, Manaba, Juco, Ucucu, Uco, Uca, Mashiramo, Puca Mate, Yanapuma, Yuramateo.

Taxonomy

The Andean bear is the only extant bear species in South America. Frederick Cuvier suggested the species name "*ornatus*" in 1825 (Cuvier 1825), and it was Gervais in 1855 who created the genus *Tremarctos*. Later, Kurtén (1966) and Thenius (1976) placed *T. ornatus* within the subfamily Tremarctinae together with other short-faced fossil bear species.

Phylogenetic and molecular genetics studies carried out by Ruiz-García et al. (2018) supported the existence of two Evolutionary Significant Units (ESUs; Moritz 1994). The first (Northern Andean Clade, NAC) comprises all the bears from Venezuela, Colombia, Ecuador, and north-central Peru, whereas the second (Southern Andean Clade, SAC) comprises the bears from southern Peru and the northern and central Bolivian Andes. The temporal split between the ESUs was estimated to have occurred around 500,000 ya (years ago). Different historical demographic analyses showed signatures of significant population expansions for the species as a whole and in each one of the ESUs found. These population expansions began between 690,000 and 50,000 ya. Coupled with the fact that the SAC has a slightly higher level of genetic diversity than the NAC, these results suggest a slightly earlier origin of the SAC. A recent work is supportive of the hypothesis that the southern Andean bear population was the original (Mitchell et al. 2016). The study revealed that the fossil *Arctotherium* (ancient DNA) was more closely related to the extant *T. ornatus* than to the North American short-faced bears. They concluded that a striking convergent evolution existed between the South American and North American short-faced bears. This finding substantiates the results of Ruiz-García et al. (2018), indicating that *Tremarctos* may not have arrived in South America in a recent immigration event. Contrarily, Andean bears appear to have originated in South America from a common ancestor with *Arctotherium*. This should mean that *Tremarctos* (or its ancestor) could have been living in South America much earlier than traditionally thought and the expansion of this species was from south to north, contrary to the traditional theory.

Subspecies and Distribution

The Andean bear is endemic of the Tropical Andes, and the only remaining species – with no subspecies – of the Tremarctinae family (Garshelis 2009). Its long and narrow distribution from Venezuela to Bolivia encompasses 260,000 km^2 (Figure 7.1) of a wide variety of Tropical Andean ecosystems (García-Rangel 2012). The species is currently present in five main areas: (1) Sierra de Perijá and Cordillera de Mérida in Venezuela; (2) the

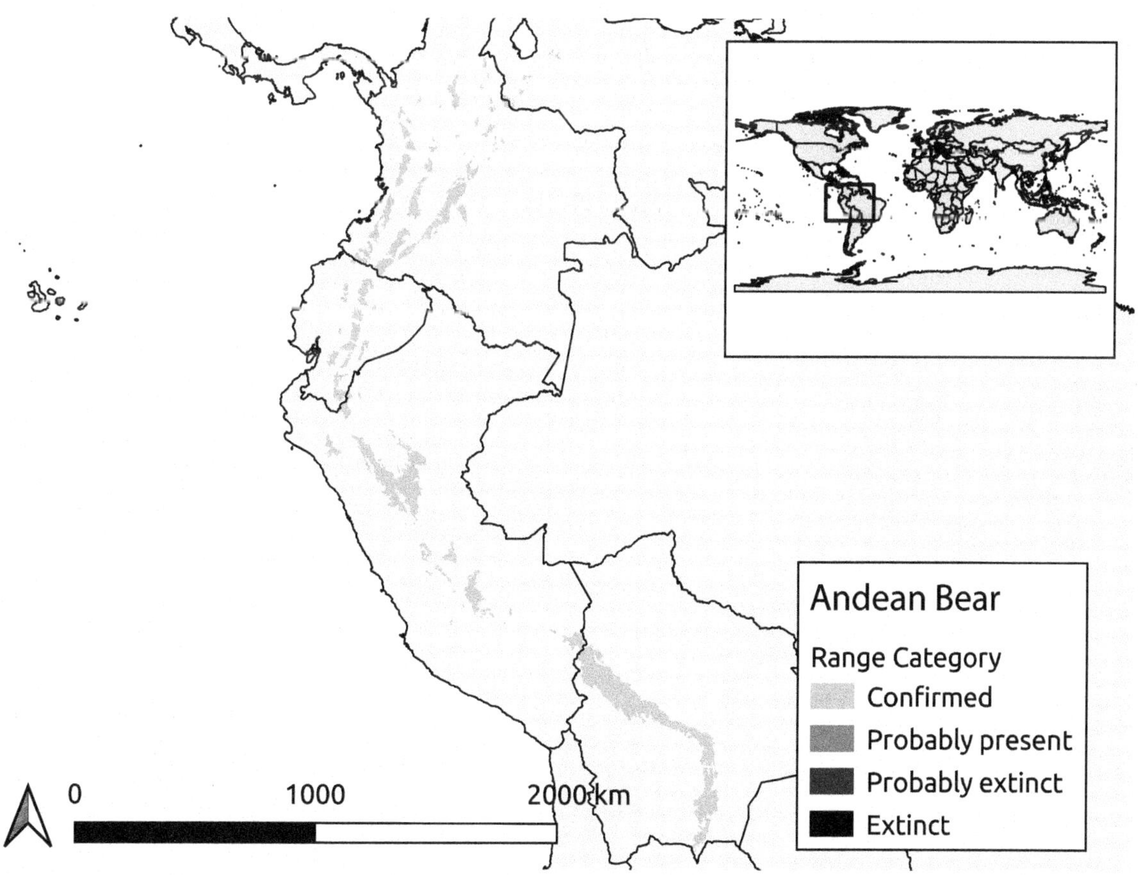

Figure 7.1 The distribution of the Andean bear is long (~ 4600 km) and narrow (~ 200–650 km). From north to south, Andean bears are found in Sierra de Perijá and Cordillera de Mérida in Venezuela; the Occidental, Central, and Oriental Andean mountain ranges of Colombia; both eastern and western slopes of the Ecuadorian Andes; across the three Peruvian Andean mountain ranges, including a portion of the North Pacific coastal desert; and in the eastern slope of the Tropical Andes in Bolivia. (*Source*: IUCN, Red List of Threatened Species 2017.)

Occidental, Central, and Oriental Andean mountain ranges of Colombia; (3) the eastern and western slopes of the Ecuadorian Andes; (4) the three Peruvian Andean mountain ranges; and (5) on the eastern slope of the Tropical Andes in Bolivia (Velez-Liendo & García-Rangel 2017). The presence of Andean bears in Argentina is still under debate (Rumiz et al. 2012; Cosse et al. 2014). Although the occurrence of the species in Panama was previously mentioned by Hershkovitz (1957), Goldstein et al. (2008) did not find recent evidence to support this claim.

Descriptive Notes

Andean bears are short-faced, medium-sized bears (Figure 7.2) Adult weight ranges from 30 to 200 kg and body length from 0.88 to 2.22 m (Servheen et al. 1999; Macdonald 2001; Paisley 2001), with adult female bears a third smaller than male bears (Garcia-Rangel 2012). As characteristic of the subfamily Tremarctinae, the Andean bear's skull is massive, deep, and convex (Mondolfi 1989), and it is the shortest in length among extant bear species (Christiansen 2007). Other specific features include a more pronounced masseteric crest, an extended zygomatico-mandibularis muscle, an epicondylar humeral foramen situated above the internal epicondyle and 13 pairs of ribs – one fewer than all other ursids (Domico & Newman 1988; García-Rangel 2012).

Figure 7.2 Facial markings of the first Andean bear, M1, caught and fitted with a motion-sensitive radio collar in April 1998 by Dr Susanna Paisley. The information collected in the consecutive years showed that the Andean bear, once assumed to be a nocturnal species, is a diurnal species; Andean bears do not hibernate, but night resting periods are longer during the dry season (April–September).

Due to their arboreal activity, Andean bears have longer fore-limbs than hind limbs and strong, sharp, curved, and laterally flattened non-retractable claws. Andean bears share the ursid dental formula of 42 teeth (i 3/3, c 1/1, p 4/4, m 2/3), but, unlike other bear species: (a) the fourth premolar has blunt lophs, three pulp cavities, and three roots (Peyton 1999); (b) canines are small and blade-shaped; (c) premolars are reduced, carnassials are modified; and (d) their large lips are free from their gums. The mandible's structure responds to an "omnivorous" diet that follows a mainly frugivorous/folivorous pattern with occasional consumption of animal protein (Sacco & Van Valkenburgh 2004; Christiansen & Wroe 2007). Like giant pandas (*Ailuropoda melanoleuca*; Chapter 6), Andean bears have a false thumb used in food manipulation, as well as facilitating arboreal activity (Salesa et al. 2006). The tail is rudimentary and is hidden in the fur of the bear's posterior (García-Rangel 2012; Castellanos et al. 2016). Their coat is black; however, dark-red-brown individuals have been photographed (Emmons & Feer 1997), including a "golden spectacled bear," an individual with apparently a degree of albinism, in the Northern Andes range of Peru.

Distinctive white to light-yellow marks can be present around the eyes, down to the cheeks and muzzle, under the chin and can extend across the throat onto the chest (Emmons & Feer 1997). These marks are the most distinctive phenotypic features of the species, from individuals with no marks at all to those with perfect "spectacles" (Figure 7.2). Both facial markings and fur color can change drastically as bears get older, with the first variation in appearance occurring at around the age of two, and then another change when they are 7–9 years old (Van Horn et al. 2015).

Hematological and serum biochemical values appear to demonstrate significantly higher levels of serum triglycerides in comparison to other bear species (Castellanos et al. 2010). This could be indicative of a physiological mechanism to cope with the periodic starvation that occurs at times of year when food sources are limited. Also, Andean bears exhibit low levels of blood urea nitrogen and alkaline phosphatase similar to those of the giant panda, which is likely to be a reflection of their predominantly vegetarian diet (Castellanos et al. 2010). Male bears have significantly higher blood protein levels than females, which is likely a direct correlation to sexual dimorphism and the larger body size of male bears.

Habitat

Due to their adaptive plasticity, Andean bears inhabit a variety of habitats and occupy a wide range of altitudes between 200 and 4900 m (Castellanos et al. 2016; Velez-Liendo & García Rangel 2017; Castellanos et al. 2019). Actually, the Andean bear habitat covers a great variety of ecosystems along the Tropical Andes, including tropical dry forests, tropical moist lowland and montane forests, tropical dry and moist shrublands, and tropical high-altitude shrubland and grasslands (Velez-Liendo & García Rangel 2017). The Andean bear's evolutionary feeding adaptations to hard, vegetative matter like bromeliads, in particular species of the genus *Puya*, and their capacity to survive in regions where water and food resources are limited suggest that these bears were originally a dry tropical forest species (García-Rangel 2012; Appleton et al. 2013).

Movements and Home Range

Although Andean bears are generally solitary, strong social bonds exist between mothers and their cubs (Stirling & Derocher 1990). Subadult females often establish home ranges in or near their natal range, while subadult males disperse widely in order to establish their own home ranges far from their mother's home range. Jackson et al. (2017) reported a multi-generational Andean bear family group in which a female bear cub from a previous litter remained close and interacted with her mother and younger siblings over a period of over 5 years.

Andean bear home range, habitat use, and movement patterns contrast significantly between the sexes (Castellanos et al. 2016). Although there are few studies on Andean bear home range (Paisley 2001; Castellanos 2005), a significant sexual disparity in home range size between males and females have been reported by Castellanos et al. (2016) in northern Ecuador. Males ($n = 3$) showed an average home range size of approximately 150 km^2, whereas female home ranges were smaller (34 km^2; $n = 5$). Extensive home range overlap exists between males and females, which may suggest a lack of territorial dominance between the sexes (Paisley 2001; Castellanos et al. 2016). Daily movements also show great individual and local variations. For example, Paisley (2001) reported the median linear movement of two male radio-tracked bears was 800 m for an 18–60-h period, and a maximum daily movement of at least 6 km. In contrast, Castellanos et al. (2016) observed male bears travelling up to 18.75 km per day (e.g. when moving between core areas). Such large territorial displacements realized by male bears appear to be largely motivated by the pursuit of females and searching for food (Castellanos et al. 2016). Female Andean bears have much smaller home ranges than males and tend to concentrate their movements within relatively small areas where food is abundant. However, in areas where corn plantations exist, females are known to travel longer distances to obtain readily available and nutritious sustenance, and have been observed to travel approximately 6 km in a day (Castellanos 2005; Castellanos et al. 2016).

Andean bears present seasonal migrations across different habitats, which are likely determined by the phenology of plants they use as food (Mondolfi 1989; Paisley 2001). There are exceptions, however, with individuals observed to remain in the cloud forest habitat of northern Ecuador throughout their entire life cycle. This appears to be a result of the presence of perennial *Guzmania* and *Greigia* bromeliads and *Chusquea* bamboo, which are nutritious and abundant food items for bears in the cloud forest environment (Castellanos 2011; Castellanos et al. 2016).

Activity Patterns

Andean bears are mainly diurnal with variation of their activity patterns depending on season and geographic areas (Paisley & Garshelis 2006). In Bolivia and Ecuador, Andean bears showed a bimodal activity rhythm, with peaks of activity at mid-morning and mid-afternoon, and little activity during hours of darkness (Paisley 2001; Castellanos et al. 2016). Andean bears do not hibernate, but activity patterns and time budget differences between dry and wet seasons were observed (Paisley 2001; Garshelis 2009). Night resting periods tended to be longer during dry seasons, while midday rests were longer during the rainy season. According to data on reintroduced individuals, feeding accounts for about 70% of the species activity budget, similar to giant pandas and American black bears (A. Castellanos, personal communication 2009). *In situ* observations of 30 individuals (males, females, subadults, and cubs) revealed that the most frequent activities were those related to feeding and displacement (Laguna & Jurrius 2018).

Feeding Ecology

The Andean bear has a feeding strategy that consists of an extensive and diverse variety of plant, fungi, and animal food resources (García-Rangel 2012; Castellanos et al. 2016). Anatomical adaptations such as a large zygomatico-mandibularis muscle, stiff jaw, large molar area, and thick skull bones allow for hard and fibrous vegetative matter to be ground effectively (Sacco & Van Valkenburgh 2004). Throughout most of its range, the Andean bear feeds primarily on species from the Bromeliaceae, Arecaceae, Poaceae, Ericaceae, and Lauraceae families (García-Rangel 2012; Castellanos et al. 2016). On a seasonal basis, fruits are a key food source, particularly to females with cubs (Figueroa & Stucchi 2009). For example, in Ecuador, the main cloud forest fruits consumed by bears are higueron wild figs (*Ficus* spp.), wild avocados (*Nectandra* spp., *Ocotea* spp.), and myrtle (*Myrcianthes* spp.) (Castellanos et al. 2016). Feeding behavior studies are limited to *Puya* spp., with bears selectively feeding on only a few plants within each patch (Peyton 1986; Goldstein & Salas 1993).

Andean bears also have the capacity to hunt animal prey such as wild pigs, mountain tapirs, deer, vicuña, and even livestock (Paisley 2001; Garcia-Rangel 2012, Castellanos et al. 2016). Bear attacks on cattle can be related to a lack of native food items, although Andean bears have been recorded to naturally hunt similar-sized mountain tapirs in paramo habitat (Castellanos 2014). Depredation generally occurs in grasslands adjacent to forested areas, and usually involves the active pursuit of prey, followed by clawing and biting in the scapula region in order to immobilize prey using their powerful paws and claws. After a successful hunt, Andean bears drag the carcass into the forest, often covering distances of up to 1 km. Depending on the size, Andean bears consume their prey in a period of between 2 and 10 days (Laguna & Jurrius 2018). In northern Ecuador it is reported than one bear can kill around 70–80 cattle per year and for certain individuals, meat consumption represents the greater percentage of their diet (Laguna 2013; Laguna & Jurrius 2018). The Andean bear's predatory behavior has increased negative perceptions toward the species in areas where bears predate on cattle, and can jeopardize the success of Andean bear conservation programs (Espinosa & Jacobson 2012).

Widespread deforestation of natural habitat by humans has increased cattle attacks and crop raiding by bears, resulting in retaliatory killings of "problem" bears (Figueroa & Stucchi 2009; García-Rangel 2012; see also Chapter 15). Among the main non-native food sources exploited by Andean bears are corn (*Zea mays*), granadilla (*Passiflora ligularis*), banana (*Musa* spp.), sugar cane (*Saccharum officinarum*), avocado (*Persea americana*), and tree tomato (*Solanum betaceum*). There are also sporadic reports of Andean bears fishing for rainbow trout (*Oncorhyncus mykiss*) in Colombia and Ecuador, especially in artificially stocked lakes and fish farms (Castellanos et al. 2016).

Reproduction and Growth

Information on Andean bear reproduction has been primarily obtained from captive animals. In captivity, Andean bear females reach sexual maturity at around 4 years of age, while in the wild it is thought to occur earlier, when females are 2–3 years old. It is likely that the age of sexual maturity may vary depending on food availability and weight (Rodríguez-Clark & Sánchez-Mercado 2006; Garshelis 2009). In zoos, courtship and reproductive behavior lasts between 1 and 5 days, and involves non-aggressive fighting, frolicking, and reciprocal high-pitched grumbles (Castellanos et al. 2016). In the wild, bears are polygamous, with males copulating with multiple mates throughout their home ranges. Courtship behavior is largely unknown, although females have been observed with assemblages of up to four disputing males stalking their trail (Castellanos 2015).

The Andean bear is polyestrous, a facultative seasonal breeder and, like most bear species, is capable of delayed implantation (Claro-Hergueta et al. 2007). Females show three to four estrous cycles during a single breeding season with no seasonal ovarian activity (Enciso & Guimarães 2013). Differences in timing of estrus, mating, and parturition of a species can be considered an adaptation to wide geographic distribution across latitudes, altitudes, and habitats, and accompanying variation in food availability and other environmental factors (Spady et al. 2007). In the wild, Andean bear mating behavior has been reported to occur throughout the year, but most activity has been recorded between May and August in the southern part of its distribution, i.e. part of Ecuador, Peru, and Bolivia, while in its northern distribution (Venezuela, Colombia, and part of Ecuador), a larger period of mating and courtship activity, between March and October, has been recorded. As with other bear species, the length of gestation is short, highly variable, and difficult to calculate due to delayed embryonic implantation. In captivity, the gestation period has been observed to vary between 160 and 257 days (García-Rangel 2012). The only accurate gestation period in the wild was calculated to be 120–125 days, which is the shortest recorded for the species (Castellanos 2015; Castellanos & Jackson 2018). Andean bears' birthing also varies with latitude and has been reported to occur predominantly between February and September (Garshelis 2004; Claro-Hergueta et al. 2007). Litter size can vary from one to four, with twins being most common (Claro-Hergueta et al. 2007; García-Rangel 2012). Field observations in Bolivia suggested that births occur 2–3 months before the peak of the fruit season, perhaps an adaptation to allow mothers to leave their den when fruits are plentiful (Velez 1999). Likewise, in the cloud forest region of northern Ecuador, birthing appears to coincide with the corn ripening cycle, in order for mothers to easily access food while cubs are lactating (Castellanos et al. 2016).

Cubs measure approximately 18 cm and weigh between 300 and 500 g at birth. They open their eyes when they are *ca.* 42 days old, and at around 3 months of age are able to follow their mother for short distances (García-Rangel 2012). Cubs are generally believed to lactate for approximately 1 year (García-Rangel 2012), although in Ecuador, cubs have recently been observed lactating until at least the age of 14 months (Castellanos & Jackson 2018). Andean bear mothers construct maternal dens in the days prior to parturition, where they remain with their cubs during their development until they are approximately 3 months old. During this time, females may become severely malnourished due to their selfless dedication to their cubs (Castellanos et al. 2016). In Ecuador it has been observed that female bears remain in the den site throughout the entire initial 3-month maternal period, yet they have also been observed to leave lactating cubs alone in the nest for periods of up to 48 hours while in search of food (Castellanos 2015; Castellanos et al. 2016). Maternal care is provided for approximately 2–3 years before mother and cubs separate (Paisley 2001; García-Rangel 2012). If cubs die after birth or are stillborn, females can become receptive within 4–6 weeks (Claro-Hergueta et al. 2007). Generally, males grow faster and live longer than females (Rodríguez-Clark & Sánchez-Mercado 2006).

Behavior

Andean bears are generally solitary animals that only interact to mate and during periods of parental care (Peyton 1999; Jackson et al. 2017). However, home range studies have shown significant inter- and intrasexual home range overlaps, probably indicating the presence of intraspecific territorial tolerance and low levels of territorial behavior, similar to other ursids (Jackson et al. 2017). However, when food is abundant bears have been reported to congregate with conspecifics (Paisley 2001; Jackson et al. 2017).

Andean bears are elusive, and aggressiveness only occurs when under threat, hurt, and/or with cubs (Paisley 2001). Andean bears in the wild prefer to avoid interactions with humans and, if they detect potential danger, they generally flee or climb the nearest tree in order to avoid confrontation (Castellanos et al. 2016). Andean bears are excellent climbers, not only of trees where they build platforms or nests (Figure 7.3), but also on vertical rock walls (Castellanos et al. 2016). These tree platforms are made of broken or bent

Figure 7.3 A tree nest built by an Andean bear in Carrasco National Park (Bolivia). Tree nests are assemblages of bent branches resembling large platforms strong enough to sustain the weight of a full-grown bear. Tree nests are used primarily for feeding purposes rather than resting and are associated with cattle carcass feeding sites, possibly to secure the prey from other carnivores, i.e. jaguars and foxes. (Photo by Ximena Velez-Liendo.)

branches and serve multiple behavioral purposes, including feeding, resting, protection, and escape (Torres 2008; Castellanos et al. 2016). Andean bears are also very good swimmers and have been observed swimming across large lakes (Castellanos et al. 2016).

Communication is largely achieved through scent-marking using special exocrine glands on the skin (Garshelis 2009; García-Rangel 2012). The main communication method involves rubbing against specific tree species to strip epiphytic moss and bark, followed by claw marking, scent marking, and the deposition of hormonal secretions. The principal purpose of this behavior is generally to announce one's presence to potential mates. Despite the apparent lack of territorial behavior between Andean bears, claw and scent marking might also play a role in determining dominance and position in the social dominance hierarchical system (Castellanos et al. 2016). Vocal communication, although rare in Andean bears, is used frequently during the maternal period for mother and cub interactions (Castellanos et al. 2001; Castellanos 2008). Seven different vocalizations have been identified for Andean bears (Castellanos et al. 2016).

Parasites and Diseases

The most commonly reported internal parasites in captive individuals were ascarid nematodes (Wolff 1989; Schaul 2006), while in wild populations, nematode (Strongyloidea, Ascarididae, and Ancylostomatidae), trematode (Paragonomidae), and protozoa species including *Blastocystis* sp., *Cryptosporidium* sp., and *Giardia* sp. were identified in fecal samples collected in Venezuela and Peru (Goldstein 1989; Figueroa 2015). The only ectoparasites found on wild Andean bears were *Trichodectes ferrisi* lice collected in Venezuela (Cardozo-de-Almeida et al. 2003), and *Ixodes fuscipes* ticks from a rescued individual (D. Jackson, personal communication September 2019).

Diseases such as canine distemper and rabies can affect wild Andean bear populations, but there is no evidence of wild individuals infected by these viruses. On the other hand, ex-situ diseases commonly associated with this species are alopecia, immune-related, and gastrointestinal diseases. Recently, alopecia syndrome, a commonly associated disease of adult females, was successfully treated (Drake et al. 2017).

Status in the Wild

Mortality and Wild Offtake Factors

According to the latest assessment on the Andean bear's status and conservation, the expansion of the agricultural frontier together with inadequate farming practices have been the main drivers of habitat loss and degradation (Velez-Liendo & García-Rangel 2017). A landscape assessment of habitat suitability and connectivity carried out for the Red List assessment revealed a reduction of 28.5% of key habitat patches or areas large enough to sustain a viable population during the past 20 years (Velez-Liendo & García-Rangel 2017). At national level, Venezuela showed the greatest loss of key patches (70%), with only two of these key patches available to sustain its bear population. Peru, Colombia, and Ecuador lost around a third each (31%, 29%, and 27%, respectively), and Bolivia 19%.

Illegal killing is an important yet globally underestimated threat for Andean bears. In some areas, such as the central Valley of Tarija, Bolivia, the Andean bear is presumed extinct due to illegal killing (X. Velez-Liendo, personal communication March 2019). Bears are killed either by opportunistic encounters, sport hunting, or as retaliation after being caught damaging maize crops or preying on livestock (Figure 7.4) (Goldstein 1991; Morales-Vargas 2003; Torres 2008).

Global projections of climate change effects show a general tendency toward upslope displacement of the mountain biome, suggesting that the Tropical Andes is among the most vulnerable regions (Malcolm et al. 2006; Beaumont et al. 2011). Consequently, changes in climate regimes are considered a growing threat for Andean bears in the wild (Velez-Liendo & García-Rangel, 2017). With an increment of +0.7°C in the last century, and a projected increase of 4.3 ± 0.7°C by 2100 (IPCC 2013), extensive changes in the bear's habitat are expected. Actually, the tropical high-altitude grassland, the most fragile ecosystem of all, is likely to lose 30% of its area (Tovar et al. 2013), and projected reduction in annual rainfall is likely to affect epiphytic bromeliads (Colwell et al. 2008). Tropical dry and moist shrublands are likely to lose 24% of their area (Tovar et al. 2013), due to a significant variation in the number of dry months (IPCC 2013), while tropical moist lowland forest, montane forests, and tropical dry forests are likely to suffer a loss of around 18% in area due to upslope displacements.

The Andean bear has been listed as Vulnerable by the IUCN since 1982 (Velez-Liendo & García-Rangel 2017) and

Figure 7.4 Widespread deforestation of natural habitat by humans has increased cattle attacks and crop raiding by bears and other large carnivores. As a result, retaliatory killing of bears is one of the main threats to this species' global conservation. Shown here is one of the last bears in Tarija, Bolivia, the southernmost area where bears inhabit worldwide. (Photo by Ximena Velez-Liendo.)

it has been listed in CITES Appendix I since 1975 (Anonymous 2009). Wild populations are believed to be in decline mainly due to habitat loss and fragmentation, illegal killing, and effects of climate change on habitat quality (Velez-Liendo & García-Rangel 2017). National assessment of the bear population estimated 1100–1600 bears in Venezuela (Ruiz-García 2003), 3000–6000 in Colombia (Ruiz-García 2003), 1200–2000 in Ecuador (Cuesta & Suárez 2001; Viteri Espinel 2007), *ca.* 5000 bears in Peru (Peyton 1999), and *ca.* 3000 bears in Bolivia (Velez-Liendo 2010; Figure 7.5). Although there is a network of 58 protected areas across the Andean bear's distribution, only a few of them have the adequate budget and staff necessary to monitor and protect bear populations (García-Rangel 2012). Despite efforts to establish, maintain, and connect these protected areas through corridors, e.g. Vilcabamba–Amboro between Peru and Bolivia, these areas have received little interest and large portions of the bear's habitat are still unprotected (Kattan et al. 2004; Surkin et al. 2010; Yerena & García-Rangel 2010; Hoffman et al. 2011).

Figure 7.5 Camera-trap picture of an Andean bear in the dry forests of Tarija, Bolivia. Once believed extinct from this ecosystem, this 2017 photo not only shows the most notorious features of this short-faced bear species, but a hope for bear conservation. Courtesy Andean Bears and People Project.

Studies on population genetics of the Andean bear have been important in determining the status of some wild populations. For instance, in the context of conservation policies for the Andean bear in Colombia, Ruiz-García et al. (2019a) genetically analyzed this species in nine geographical regions from Colombia, identifying the two most differentiated populations, which may qualify them as different Management Units (MUs). Likewise, DNA analyses in Ecuador concluded there was no genetic differentiation of populations in this country, suggesting the Andes did not represent an obstacle for dispersion and individuals in this country could be treated as a unique MU (Ruiz-García et al. 2019b).

Status in Captivity

Zoos have undergone fundamental changes in their function and structure. Besides keeping their role as an "attraction," zoos have emphasized their contribution to the conservation of biodiversity. Andean bears in captivity have generated vital information on this species' biology, *ex situ* actions including education programs, and captive breeding programs. Today there are 225 Andean bear individuals in captivity in 163 institutions around the world (Hall 2017). During the past years, zoos and *ex situ* institutions have been transforming from traditional animal display institutions, to conservation centers. A Species Survival Plan (SSP), a breeding program designed to enhance the genetic viability and demographic stability of animal populations in zoos and aquariums through the Association of Zoos and Aquariums (AZA), is in place for this charismatic bear species. There are currently only 39 Andean bears in AZA-accredited zoos and only six potentially viable breeding pairs in the SSP population (www.aza.org/).

References

Anonymous. (2009). *Tremarctos ornatus.* UNEP-WCMC Species Database: CITES-Listed Species. www.unep-wcmc-apps.org/isdb/CITES/Taxonomy/tax-common-result.cfm?source=animals anddisplaylanguage=engand Common=25966

Appleton, R., Tobler, M. & Van Horn, R. (2013). *A comparison of Andean bear (*Tremarctos ornatus*) densities using camera traps at waterholes and bear trails in the tropical dry forest of Northwestern Peru.* Provo, UT: International Association for Bear Research and Management.

Beaumont, L. J., Pitman, A., Perkins, S., et al. (2011). Impacts of climate change on the world's most exceptional ecoregions. *Proceedings of the National Academy of Sciences* 108(6): 2306–2311.

Cardozo-de-Almeida, M., Linardi, P. M. & Costa, J. (2003). The type specimens of chewing lice (Insecta, Mallophaga) deposited in the entomological collection of Instituto Oswaldo Cruz, Rio de Janeiro, RJ, Brazil. *Memórias do Instituto Oswaldo Cruz* 98(2): 233–240.

Castellanos, A. (2005). Preliminary results of the three-year telemetry study of Andean bear in the Intag Region, Ecuador. In *16th International Conference on Bear Research and Management.* Trentino, Italy: International Association for Bear Research and Management.

Castellanos, A. (2008). *Guía para la rehabilitación, liberación y seguimiento de oso andino, Tremarctos ornatus, en Ecuador.* Quito, Ecuador: Fundación Espíritu del Bosque.

Castellanos, A. (2011). Home ranges of Andean bears in Intag region, Ecuador. *Ursus* 22: 65–73.

Castellanos, A. (2014). Co-occurrence of Andean bear and mountain tapir in the Papallacta region, Cayambe Coca National Park, Ecuador: A brief description. *International Bear News* Summer 23(2).

Castellanos, A. (2015). Maternal behaviour of a female Andean bear in the paramo of Cayambe Coca National Park, Ecuador. 10.13140/RG.2.1.4185.0405.

Castellanos, A. & Jackson, D. (2018). Biological Research: Does Rebecca, a seasoned Andean bear mother, show seasonal birthing patterns? *International Bear News* Fall 27(3).

Castellanos, A., Altamirano, M. & Tapia, G. (2001). Ecology and behaviour of reintroduced Andean bears in the Biological Reserve Maquipucuna, Ecuador: implications in conservation. *Ukuku, Boletín Informativo sobre la Conservación del Oso Andino* 3: 23–26.

Castellanos, A., Arias, L., Jackson, D. & Castellanos, R. (2010). Hematological and serum biochemical values of Andean bears in Ecuador. *Ursus* 21: 115–120.

Castellanos, A., Jackson, D. & Arias, L. (2016). *Guidelines for the rescue, rehabilitation, release and post-release monitoring of Andean bears.* Quito, Ecudador: Andean Bear Foundation.

Castellanos, A., Jackson, D. & Ascanta, M. (2019). Are reports of cub abandonment in Andean bears a result of increasing human encroachment? *International Bear News* Spring 28(1): 14–15.

Christiansen, P. (2007). Evolutionary implications of bite mechanics and feeding ecology in bears. *Journal of Zoology* 272: 423–443.

Christiansen, P. & Wroe, S. (2007). Bite forces and evolutionary adaptations to feeding ecology in carnivores. *Ecology* 88: 347–358.

Claro-Hergueta, F., Dollinger, P., Göltenboth, R., et al. (2007) *EAZA Ursid husbandry guidelines.* Köln, Germany: Kölner Zoo.

Colwell, R. K., Brehm, G., Cardelús, C. L., Gilman, A. C. & Longino, J. T. (2008). Global warming, elevational range shifts, and lowland biotic attrition in the wet tropics. *Science* 322(5899): 258–261.

Cosse, M., Fernando Del Moral Sachetti, J., Mannise, N. & Acosta, M. (2014). Genetic evidence confirms presence of Andean bears in Argentina. *Ursus* 25(2): 163–172.

Cuesta, F. & Suárez, L. (2001). *Oso de anteojos (*Tremarctos ornatus*).* Quito, Ecuador: SIMBIO/EcoCiencia/Ministerio del Ambiente/IUCN.

Cuvier, F. (1825). *Ours des cordilières du chili. En: Histoire Naturelle des Mammifères,* Tomo V (E. Geoffroy-Saint-Hilaire & F.G. Cuvier). Paris: Belin Éditeur.

Drake, G. J., Nuttall, T., López, J., et al. (2017). Treatment success in three Andean bears (*Tremarctos ornatus*) with alopecia syndrome using oclacitinib maleate (Apoquel®). *Journal of Zoo and Wildlife Medicine* 48(3): 818–828.

Domico, T. & Newman, M. (1988). *Bears of the world.* New York, NY: Facts on File.

Emmons, L. & Feer, F. (1997). *Neotropical rainforest mammals: A field guide.* Chicago, IL: University of Chicago Press.

Enciso, M. A. & Guimarães, M. A. B. V. (2013). Knowing the reproductive endocrinology in the female Andean bear through non-invasive methods. *International Bear News* 22: 33–34.

Espinosa, S. & Jacobson, S. K. (2012). Human–wildlife conflict and environmental education: evaluating a community program to protect the Andean bear in Ecuador. *The Journal of Environmental Education* 43: 155–165.

Figueroa, J. (2015). New records of parasites in free-ranging Andean bears from Peru. *Ursus* 26: 21–27.

Figueroa, J. & Stucchi, M. (2009). *El oso andino: alcance sobre su historia natural.* Lima, Peru: Asociación para la Investigación y Conservación de la Biodiversidad.

García-Rangel, S. (2012). Andean bear *Tremarctos ornatus* natural history and conservation. *Mammal Review* 42(2): 85–119.

Garshelis, D. L. (2004). Variation in ursid life histories: is there an outlier? In: Lindburg, D. & Baragona, K. (Eds.), *Giant pandas. Biology and conservation* (pp. 53–73). Berkeley, CA: University of California Press.

Garshelis, D. L. (2009). Family Ursidae. In: Wilson, D. E. & Mittermeier, R. A. (Eds.), *Handbook of the mammals of the world: Volume 1. Carnivores* (pp. 448–497). Barcelona: Lynx Edicions.

Goldstein, I. (1989). *Distribution, habitat use, and diet of spectacled bears (*Tremarctos ornatus*) in Venezuela.* In Rosenthal, M. A. (Ed.), *Proceedings of the First International Symposium on the Spectacled Bear.* Chicago, IL: Lincoln Park Zoological Gardens.

Goldstein, I. (1991). Spectacled bear predation and feeding behaviour on livestock in Venezuela. *Studies on Neotropical Fauna and Environment* 26: 231–235.

Goldstein, I. & Salas, L. (1993). Foraging pattern on *Puya* sp. (Bromeliaceae) by *Tremarctos ornatus* (Ursidae) at Páramo El Tambor, Venezuela. *Ecotrópicos* 6: 24–29.

Goldstein, I., Guerrero, V. & Moreno, R. (2008). Are there Andean bears in Panamá? *Ursus* 19: 185–189.

Hall, A. L. (2017). *Spectacled bear studbook,* Tremarctos ornatus. St Helier, Jersey: Durrell Wildlife Conservation Trust.

Hershkovitz, P. (1957). On the possible occurrence of the spectacled bear *Tremarctos ornatus* (F. Cuvier, 1825) in Panama. *Säugetierkundliche Mitteilungen* 5: 122–123.

Hoffman, D., Oetting, I., Arnillas, C. A. & Ulloa, R. (2011). *Cambio climático y áreas protegidas en los Andes Tropicales.* In: Herzog, S. K., Martínez, R., Jørgensen, P. M. & Tiessen, H. (Eds.), *Climate change and biodiversity in the Tropical Andes.* Säo José dos Campos: Inter-American Institute for Global Change Research (IAI) and Paris, France: Scientific Committee on Problems of the Environment (SCOPE).

IPCC. (2013). *Summary for Policymakers. Climate Change 2013: The Physical Science Basis. Contribution of Working Group I to the Fifth Assessment Report of the Intergovernmental Panel on Climate Change, Intergovernmental Panel on Climate Change.* New York, NY: Cambridge University Press.

Jackson, D., Castellanos, A. & Vasquez, D. (2017). Kinship relations in a multi-generational Andean bear (*Tremarctos ornatus*) family in north Ecuador. *International Bear News* Spring 26(1): 24.

Kattan, G., Hernandez, O. L., Goldstein, I., et al. (2004). Range fragmentation in the spectacled bear *Tremarctos ornatus* in the northern Andes. *Oryx* 38: 155–163.

Kurten, B. (1966). Pleistocene bears of North America. 1. Genus *Tremarctos*, spectacled bear. *Acta Zoologica Fennica* 115: 1–120.

Laguna, A. (2013). Estudio del conflicto hombre-oso en el norte de Ecuador. *Actas del I Simposio Latinoamericano de Tapires & II Congreso Ecuatoriano de Mastozoología*, Puyo, Ecuador.

Laguna, A. & Jurrius, I. (2018). *Manual de Atención y Prevención de Ataques por Oso Andino (*Tremarctos ornatus*) al Ganado en Ecuador.* Quito, Ecuador: Ministerio del Ambiente.

Macdonald, D. (2001). *The new encyclopedia of mammals.* Oxford, UK: Oxford University Press.

Malcolm, J. R., Liu, C., Neilson, R. P., Hansen, L. & Hannah, L. (2006). Global warming and extinctions of endemic species from biodiversity hotspots. *Conservation Biology* 20: 538–548.

Mitchell, K. J., Bray, S. C., Bover, P., et al. (2016). Ancient mitochondrial DNA reveals convergent evolution of giant short-faced bears (Tremarctinae) in North and South America. *Biology Letters* 12: 1–4.

Mondolfi, E. (1989). Notes on the distribution, habitat, food habits, status and conservation of the spectacled bear (*Tremarctos ornatus*) in Venezuela. *Mammalia* 53: 525–544.

Morales-Vargas, A. M. (2003). Evaluación de Daños Causados por Vertebrados Silvestres en Maizales de Pajan, K'Apna y Wayrapata (ANMIN Apolobamba, La Paz - Bolivia). BSc dissertation, Universidad Mayor de San Andrés, La Paz, Bolivia.

Moritz, C. (1994). Defining "evolutionarily significant units" for conservation. *Trends in Ecology and Evolution* 9: 373–375.

Paisley, S. (2001). Andean bear and people in Apolobamba, Bolivia: culture, conflict and conservation. PhD thesis, University of Kent, Kent, UK.

Paisley, S. & Garshelis, D. L. (2006). Activity patterns and time budgets of Andean bears (*Tremarctos ornatus*) in the Apolobamba Range of Bolivia. *Journal of Zoology (London)* 268: 25–34.

Peyton, B. (1986). A method for determining habitat components of the spectacled bear (*Tremarctos ornatus*). *Vida Silvestre Neotropical* 1: 68–78.

Peyton, B. (1999). Spectacled bear conservation action plan. In: Servheen, C., Herrero, S. & Peyton, B. (Eds.), *Bears: Status survey and conservation action plan* (pp. 157–198). Gland, Switzerland: IUCN/SSC Bear and Polar Bear Specialist Groups.

Rodríguez-Clark, K. M. & Sánchez-Mercado, A. (2006). Population management of threatened taxa in captivity within their natural ranges: lessons from Andean bears (*Tremarctos ornatus*) in Venezuela. *Biological Conservation* 129: 134–148.

Ruiz-García, M. (2003). Molecular population genetic analysis of the spectacled bear (*Tremarctos ornatus*) in the northern Andean area. *Hereditas* 138: 81–93.

Ruiz-García, M., Arias, J. Y., Castellanos, A., Kolter, L. & Shostell, J. M. (2018). Molecular evolution (mitochondrial and nuclear microsatellites markers) in the Andean bear (*Tremarctos ornatus*; Ursidae, Carnivora): How many ESUs are there? In: Ortega, J. & Maldonado, J. E. (Eds.), *Conservation genetics in mammals.* Cham: Springer-Verlag.

Ruiz-García, M., Arias, J. Y., Restrepo, H., Cáceres-Martínez, C. & Shostell, J. M. (2019a). The genetic structure of the Andean bear (*Tremarctos ornatus*; Ursidae, Carnivora) in Colombia by means of mitochondrial and microsatellite markers. *Journal of Mammalogy* (in press).

Ruiz-García, M., Castellanos, A., Arias, J.Y., Shostell, J.M. (2019b). Genetics of the Andean bear (*Tremarctos ornatus*; Ursidae, Carnivora) in Ecuador: when the Andean Cordilleras are not an obstacle. *Mammalian Biology* (in press).

Rumiz, D. I., Brown, A. D., Perovic, P. G., et al. (2012). El Ucumar (*Tremarctos ornatus*), mito y realidad de su presencia en la Argentina. *Mastozoología neotropical* 19(2).

Sacco, T. & Van Valkenburgh, B. (2004). Ecomorphological indicators of feeding behaviour in the bears (Carnivora: Ursidae). *Journal of Zoology* 263: 41–54.

Salesa, M. J., Siliceo, G., Antón, M., et al. (2006). Anatomy of the 'false thumb' of *Tremarctos ornatus* (Carnivora, Ursidae, Tremarctinae): phylogenetic and functional implications. *Estudios Geológicos* 62: 389–394.

Schaul, J. C. (2006). *Baylisascaris transfuga* in captive and free-ranging populations of bears (Family: Ursidae). Dissertation, The Ohio State University, Ohio, USA.

Servheen, C., Herrero, H. & Peyton, B. (1999). *Bears: Status survey and conservation action plan.* Gland, Switzerland: IUCN.

Spady, T. J., Lindburg, D. G. & Durrant, B. S. (2007). Evolution of reproductive seasonality in bears. *Mammal Review* 42: 84–119.

Stirling, I. & Derocher, A. E. (1990). Factors affecting the evolution and behavioral ecology of the modern bears. In: *Bears: Their biology and management* (pp. 189–204). Missoula, MT: International Association for Bear Research and Management.

Surkin, J., Flores, M., Ledezma, J. C., et al. (2010). Integrating protected areas and landscapes: Lessons from the Vilcabamba–Amboró conservation corridor (Bolivia–Perú). In: Worboys G. L., Francis, W. & Lockwood, M. (Eds.), *Connectivity conservation management: A global guide* (pp. 199–211). London, UK: Earthscan.

Thenius, E. (1976). Zur stammesgeschichtlichen Herkunft von Tremarctos (Ursidae, Mammalia). *Z. Säugetierk* 41: 109–114.

Torres, D. (2008). Caracterización de conflictos socio-espaciales entre la ganadería y los grandes mamíferos en el sector cuenca del río Nuestra Señora. Parque Nacional Sierra Nevada, Venezuela. Universidad de Los Andes, Mérida, Venezuela.

Tovar, C., Arnillas, C. A., Cuesta. F. & Buytaert, W. (2013). Diverging responses of tropical Andean biomes under future climate conditions. *PLoS ONE* 8(5).

Van Horn, R. C., Zug, B., Appleton, R. D., et al. (2015). Photos provide information on age, but not kinship, of Andean bear. *PeerJ* 3: e1042.

Velez, X. (1999). Caracterización y uso de hábitat por el oso Andino en la cuenca alta del Río San Jacinto, Cochabamba. Thesis, Universidad de San Simón.

Velez-Liendo, X. (2010). *Conservation of Andean bears (*Tremarctos ornatus*) in a fragmented landscape – Habitat models, potential distribution and patch connectivity.* Antwerp, Belgium: University of Antwerp.

Velez-Liendo, X. & García-Rangel, S. (2017). *Tremarctos ornatus.* The IUCN Red List of Threatened Species 2017.

Viteri Espinel M. P. (2007). Conservation genetics of Andean bears (*Tremarctos ornatus*) in northeastern Ecuador: molecular tools, genetic diversity and population size. MSc dissertation, University of Idaho, Moscow, ID, USA.

Wolff, P. (1989). Selected medical aspects of the spectacled bear. In: Rosenthal, M. A. (Ed.), *Proceedings of the First International Symposium on the Spectacled Bear* (pp. 313–318). Chicago, IL: Lincoln Park Zoological Garden.

Yerena, E. & García-Rangel, S. (2010). The implementation of an interconnected system of protected areas in the Venezuelan Andes. In: Worboys, G. L., Francis, W. & Lockwood, M. (Eds.), *Connectivity conservation management: A global guide* (pp. 233–244). London: Earthscan.

Chapter 8

Sun Bear (*Helarctos malayanus*)

Lorraine Scotson, Cheryl Frederick, Kirsty Officer, and Wai-Ming Wong

Names

Genus: *Helarctos*

Species: Sun bear *Helarctos malayanus* (Raffles, 1821)

Subspecies: *Helarctos malayanus malayanus* Raffles, 1822; *Helarctos malayanus euryspilus* Horsfield, 1825

Names in other languages: **French:** Ours Malais, Ours des cocotiers; **Spanish:** Oso Malayo, Oso de Sol; **German**: Malaienbär; **Italian:** Orso Malese; **Thai and Lao:** "Mi Maa"; **Indonesian and Malaysian:** Beruang Madu

The English name, the sun bear, is commonly thought to refer to its often crescent-shaped chest mark. Due to their small, stocky build and tendency to raid bees' nests for honey, sun bears are locally known as the dog bear (e.g. "Mi Maa" in Thai and Lao) and the honey bear (e.g. "Beruang Madu" in Indonesian).

Taxonomy

Evolutionary theory suggests that historic changes in Asia's land connectivity, caused by the rise and fall of seaways, led to the allopatric divergence of many tropical species, including several of the Asian bears (Hughes et al. 2003; Krause et al. 2008; Kutschera et al. 2017). The phylogenic placement of the sun bear within Ursidae remains unclear. The sun bear likely evolved in South-East Asia from the main lineage of ursine bears, the Auvergne bear *Ursus minimus*, about 1 million years after the sloth bear line branched, the first modern ursine lineage to emerge (McLellan & Reiner 1994). MtDNA research suggests that the sun bear and the American black bear *Ursus americanus* diverged as sister taxa after the sloth bear *Melursus ursinus*, followed by the Asiatic black bear (Zhang & Ryder 1994). The IUCN recognizes the sun bear as *Helarctos malayanus* in the most recent species account (Scotson et al. 2017a). However, opinions vary on whether to maintain the sun bear as a distinct genus, *Helarctos*, or to include it in the genus *Ursus*, with four other bear species (Meijaard 2004; see also Chapter 1).

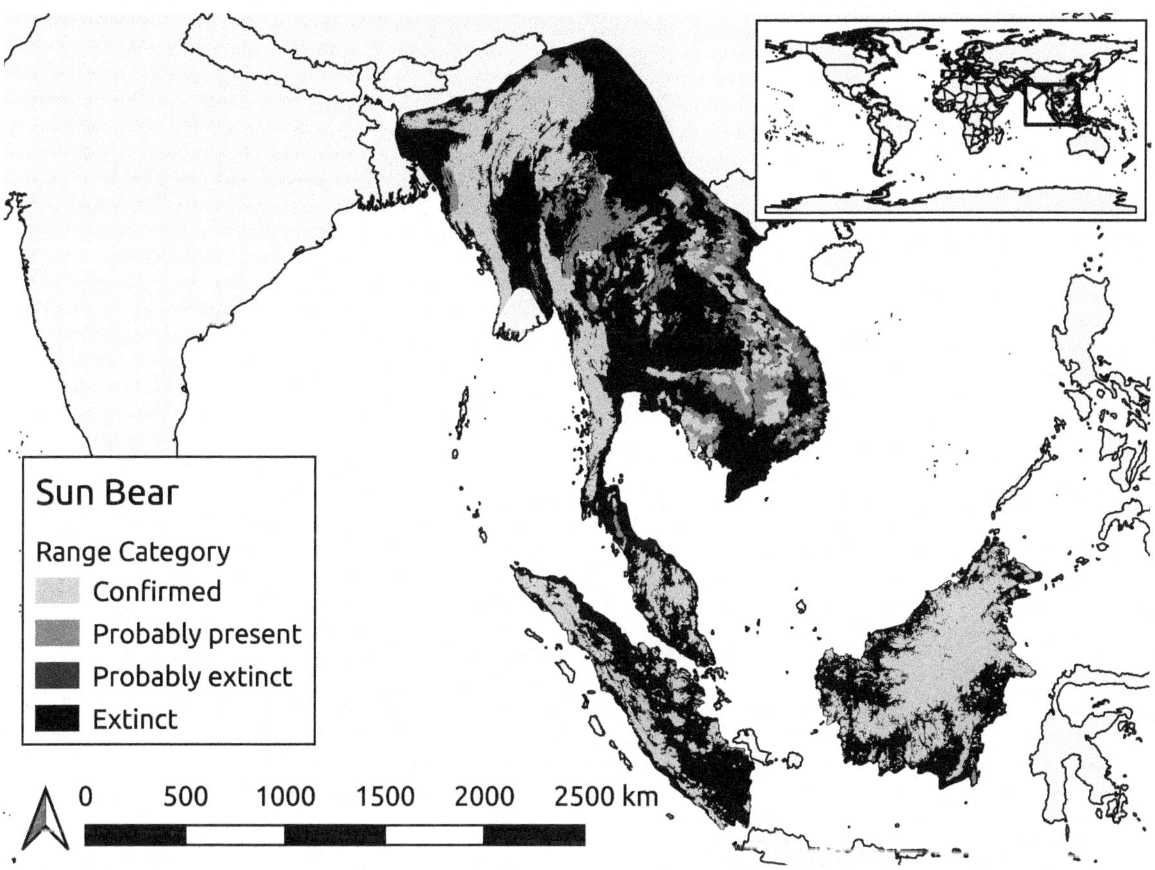

Figure 8.1 Sun bear range in 2016. Range is generally concurrent with areas under natural forest (Scotson et al. 2017a; *Source*: IUCN Red List of Threatened Species).

Subspecies and Distribution

Currently, there are two recognized subspecies of sun bear: *Helarctos malayanus malayanus* occurring in mainland South-East Asia, small parts of adjacent China, India and Bangladesh, and the island of Sumatra. *Helarctos malayanus euryspilus* occurs only on Borneo (Meijaard 2004). There is one case of a wild Asiatic black bear *Ursus thibetanus*–sun bear hybrid recorded in Cambodia (Galbreath et al. 2008).

The sun bear is distributed throughout mainland and insular South-East Asia. The northern range extends to Bangladesh, north-east India, and the Red River in north-east Vietnam. Sun bear range once extended at least into Yunnan Province of China – where they may now be locally extirpated – and fossil records from the Pleistocene have been found much further north (Erdbrink 1953). Assam, in north-east India, marks the north-western confirmed range limit (Wroughton 1916; Higgins 1932). Sun bear range may be limited in the north-west by competition with the sloth bear (Steinmetz 2011). The southern range extends to the islands of Borneo and Sumatra (Erdbrink 1953; Scotson et al. 2017a; Figure 8.1). There are no records in the last 500 years of sun bear occurring further east than Borneo. Records exist from the southern island of Java from the middle–late Pleistocene (Erdbrink 1953).

The sun bear present-day range occurs within seven potential subpopulations. Two are created by natural oceanic barriers between the islands of Sumatra and Borneo. On mainland South-East Asia, five others are a result of progressive human-made fragmentation: (1) northern mainland (Bangladesh, north-east India, northern Myanmar), with a possible fracture occurring in north-east Myanmar; (2) central Myanmar (a small patch of isolated forest area surrounded by areas of heavy human influence); (3) central South-East Asia (southern Myanmar, Lao PDR, Vietnam, Thailand, northern Cambodia); (4) south-central South-East Asia (southern Cambodia, south-west Thailand); and (5) the Thai–Malay Peninsula (Scotson 2019).

Descriptive Notes

H. m. malayanus body measurements: **shoulder height**, 70–72 cm; **head and body length**, 100–140 cm; **body mass**, 27–65 kg; **tail length**, 3–7 cm (Pocock 1941; Lekagul & McNeely 1977; Chakraborty 2003). *H. m. euryspilus* body measurements: **shoulder height**, 41–67cm (Wong, unpublished data); **head and body length**, 112.5–126 cm; **body mass**, 48–63 kg; **tail length**, 3–9 cm (Payne et al. 1989). Dental formula: I 3/3, C 1/1, P 3–4/3, M 2/3 (×2) 38–39 (Allen 1938; Pocock 1941; Lekagul & McNeely 1977).

The sun bear is distinctive for its diminutive size, being the smallest of the extant bears (Stirling & Derocher 1990). The sun bear is a medium-sized mammal with a stocky build,

Figure 8.2 Sun bear by Alberto Mattia Perronace, with permission.

Figure 8.3 Sun bear by Francesco Rinaldi, with permission. (A black and white version of this figure will appear in some formats. For the color version, please refer to the plate section.)

short black pelt, small ears, and a pale-colored muzzle and chin (Figures 8.2 and 8.3). The Bornean subspecies *H. m. euryspilus*, classified based on cranial variation in specimens from Borneo compared with other parts of South-East Asia, has a lower maximum body length than *H. m. malayanus* found on mainland Asia and Sumatra (Payne et al. 1989; Meijaard 2004). Dathe (1975) noted mainland-origin captive bear weights of ~ 65 kg for males and ~50 kg for females (Lekagul & McNeely 1977). Wild *H. m. euryspilus* from Borneo have been recorded at lower weights: 30–60 kg for males and 20–40 kg for females (Normua et al. 2004; Fredriksson 2005).

The sun bear has a yellowish crescent shape on its chest, although this patch is highly variable in shape, extent and color, ranging from pale whitish to deep orange and occasionally speckled (Meijaard 1999). The patch contrasts with the typically jet black, short pelt, which is thickened with underwool. Their skin is loose, sometimes forming wrinkles on the forehead and folds in the low back when standing. Sun bear have a silvery gray to pale tan color on the muzzle that may extend to surround the eyes (Fetherstonhaugh 1940; Pocock 1941). Their ears are round and small, measuring 4–6 cm, and there are whorls of hair on their foreheads and behind the ears, which occasionally cover shoulder and ventral areas of their compact bodies (Allen 1938; Pocock 1941; Lekagul & McNeely 1977).

A striking aspect of the sun bear is the large cranium, which is only slightly longer than it is wide, giving them a "thick-headed" look (Dathe 1975; Meijaard 2004; Figure 8.4). Pocock (1932) observed that male skulls were wider and longer than females'. Meijaard (2004) found that *H. m. euryspilus* skulls were significantly smaller than *malayanus*.

The sun bear's powerful jaw has stout robust canines and is missing an upper and lower premolar (Allen 1938; Meijaard 2004). The rhinarium of the short snout extends to the upper lip, allowing for protractible lips and tongue (Pocock 1917). The tongue is the longest of any bear species, measuring up to a remarkable 46 cm. Long and narrow, the tongue is well adapted for extracting insects and honey from tree crevices (Figure 8.5).

This species may have evolved to fill a primate-like niche, being physiologically adapted to forage both terrestrially and arboreally (Fredriksson et al. 2006). Many of their physical

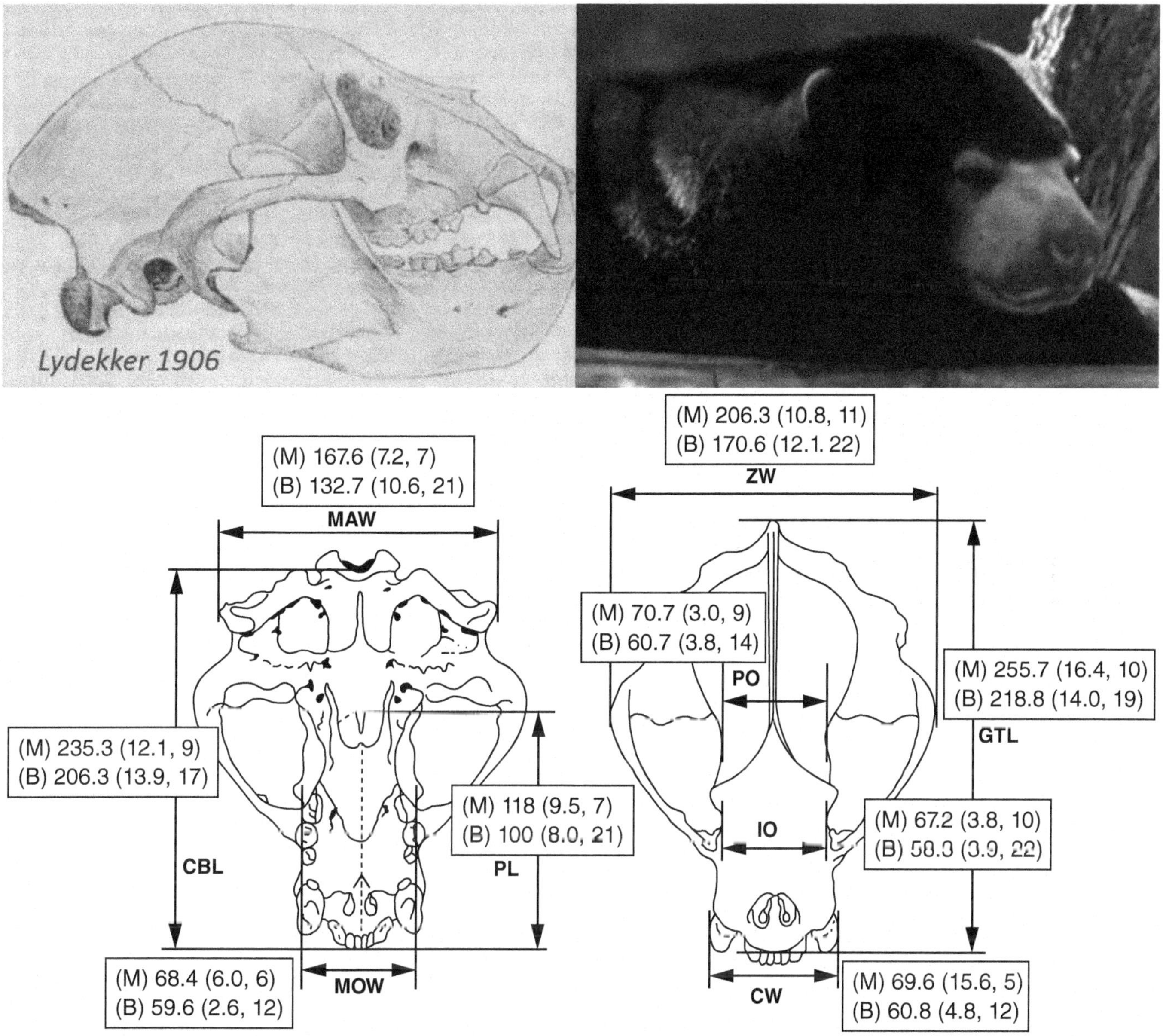

Figure 8.4. Sun bear cranium and comparative measurements of the skulls of *Helarctos malayanus* subspecies. Skull illustration adapted from Pocock (1932) in Meijaard (2004), with average measurements added (mm, SD, and sample number [*n*]) for mainland Asia (M) and Bornean (B) subspecies of both sexes. GTL, greatest total length; CBL, condylobasal length; ZW, zygomatic width; IO, interorbital width; CW, canine width; PO, postorbital width; PL, palate length; MAW, mastoid width; MOW, molar width. (Photo by Cheryl Frederick.)

features facilitate skilful climbing (Meijaard 1999; Fredriksson 2005). The fore limbs are slightly bowed inward due to the arrangements of the ankle joint and muscles (Pocock 1932; Sasaki et al. 2005). The paws are long and wide relative to body size, with bare soles and long, strongly curved, sickle-shaped claws (Allen 1938; Dathe 1975; Chakraborty 2003), which leave conspicuous claw marks gouged in the trees they climb (Steinmetz et al. 2011). The large paws are adept at breaking into logs and digging up termites and ants (Fredriksson et al. 2006). This combination of features makes the sun bear able to exploit a variety of resources at different feeding strata (Pocock 1932; Sasaki et al. 2005).

Habitat

The sun bear occurs in lowland tropical forests, montane forests, and degraded secondary forests. In mainland South-East Asia, this species inhabits seasonal ecosystems with a long dry season (3–7 months), during which rainfall is <100 mm per month. These seasonal forests are usually spread among a mosaic of forest types that includes semi-evergreen, mixed deciduous, dry dipterocarp, and montane evergreen forest. Here, the range of the sun bear overlaps with that of the Asiatic black bear in this seasonal forest mosaic (Steinmetz et al. 2011). In the south-eastern extreme of their range, in Borneo, Sumatra, and peninsular Malaysia, tropical evergreen

Figure 8.5 Sun bear characteristics. (Photos by Gari Weinraub top left and Free the Bears top right, bottom left and right.)

rainforest comprises the main habitat. Climate is generally constant across this habitat, with high annual rainfall that is relatively evenly distributed throughout the year. Tropical evergreen rainforest includes a wide diversity of forest types used by sun bears, including lowland dipterocarp, peat swamp, freshwater swamp, limestone/karst hills, hill dipterocarp, and lower montane forest (Linkie et al. 2007; Wong et al. 2013). Although present across a wide range of habitats, sun bears preferentially use habitats supporting high densities of fruiting trees (Steinmetz et al. 2011). Therefore, sun bears favor lowland and hill forests, where fruit productivity and other food availability (e.g. insect nests) are expected to be higher. As lowland forest habitats become increasingly reduced due to resource extraction and human settlements, sun bear populations in much of their remaining ranges are now fragmented and, in many cases, isolated (Scotson 2019).

Movements and Home Range

Home range, recorded only for the insular region, is small compared to other bear species (e.g. Blanchard & Knight 1991; Hwang & Garshelis 2007). Home range, and associated density of sun bears, probably varies across the two ecologically distinct categories of tropical forests that comprise their natural range. Male home ranges have been recorded from 6.5 to 27.5 km^2 (Normua et al. 2004; Wong et al. 2004; Cheah 2013) and females from 1.2 to 5.1 km^2 (Normua et al. 2004; Fredriksson 2012). Sun bears do not appear to do seasonal migrations, but may alter range across seasons and years according to food availability, during local and seasonal fruiting events and mast fruiting or when shifting to alternative food sources during low fruiting years (Fredriksson 2012; Cheah 2013; Steinmetz et al. 2013). Within the mainland (i.e. all range north of the Isthmus of Kra, southern Thailand), the home range is unknown, but could be larger due to interspecific competition with the sympatric Asiatic black bear, which dominates in areas of high fruit availability (Steinmetz et al. 2011). Sun bear south of the Isthmus of Kra in aseasonal tropical forests may occur at four or five times the density of bears on the mainland (Scotson et al. 2017a).

Activity Patterns

Sun bear average daily movements range between 1.5 and 5.6 km (Wong et al. 2004; Fredriksson 2012; Cheah 2013). Activity patterns may be nocturnal or diurnal, depending on local conditions. In general, sun bears are more nocturnal in disturbed areas, whereas sun bear in pristine forests are more diurnal and crepuscular (Griffiths & Schaik 1993). Indeed, when an area is heavily disturbed by humans, sun bears may shift from predominantly diurnal to nocturnal rhythms of activity to avoid encounters with humans (Wong et al. 2004). In Sumatra, camera traps detected sun bear activity throughout the day and detections generally peaked during dawn and dusk, indicating a crepuscular activity pattern (Linkie et al. 2007). Likewise, in East Kalimantan, in Indonesian Borneo, five reintroduced sun bears were predominately diurnal and crepuscular (Fredriksson 2012).

Feeding Ecology

Sun bears are generalist omnivores, able to travel large distances to exploit varied food resources (including human foods). They feed primarily on termites, ants, beetle larvae, stingless bee larvae and honey, a large variety of fruits, and sometimes palms and flowers (McConkey & Galetti 1999; Augeri 2005; Tumbelaka & Fredriksson 2006). The sun bear may be more insectivorous in areas of range overlap with the Asiatic black bear (i.e. north of the Isthmus of Kra), which dominates in areas of high fruit availability (Steinmetz 2011). Sun bears also feed in, and travel through, agricultural areas close to the forest edge (Normua et al. 2004; Fredriksson 2005; Cheah 2013; Scotson et al. 2014). Plantations and agricultural fields may supplement wild diet in times of food shortage, but there is no evidence that bears can survive without access to natural forest.

Reproduction and Growth

Little is known about reproduction in wild sun bears, with much documented knowledge obtained from observations of captive individuals in zoos. Sun bears have four nipples, two pectoral and two abdominal, and birth one or sometimes two cubs; but even in zoos, two cubs are rare (Garshelis 2004; Schwarzenberger et al. 2004; Frederick 2008). Sun bear differ from other ursids that have defined breeding seasons and delayed implantation (Spady et al. 2007). Captive studies of sun bear demonstrate that estrus can occur in any month and that both sexes have aseasonal hormone profiles (Hesterman et al. 2005; Frederick et al. 2012). Births occur year-round in zoos and in the wild (Schwarzenberger et al. 2004; Spady et al. 2007; Frederick et al. 2012). In East Kalimantan, captive births peak in the wet season, whereas in North America, they peak during fall and winter (Schwarzenberger et al. 2004; Frederick et al. 2012). These patterns likely reflect climatic conditions favorable to fecundity. Sun bear are polyestrous, with up to three breeding cycles a year. If mating is unsuccessful, estrus can recur in 3.5–4 months (Spady et al. 2007; Frederick et al. 2010). Estrus can also occur post-partum, shortly after cub death or removal (Spady et al. 2007; Frederick et al. 2012). Sun bears ovulate spontaneously without copulation although exposure to a male may influence whether and how often they cycle (Frederick et al. 2010, 2013). Elevated estrogen and physical and behavioral changes associated with estrus may last from a day to over a week (Domico 1988; Johnston et al. 1994; Schwarzenberger et al. 2004; Frederick et al. 2010).

Mating takes place over 1–4 days (Normua et al. 2004; Schwarzenberger et al. 2004; Frederick et al. 2010). Wild females reach adult size at 1.5–2 years (Fredriksson, personal communication). Onset of estrus can occur at 2 years, and sun bears normally breed by their third year (Dathe 1970; Schwarzenberger et al. 2004; Frederick et al. 2010). In zoos, sun bear live into their early 30s and studbook records indicate age spans of 4–27 years for successfully siring and giving birth (Marti et al. 2012; Frederick et al. 2013). In the wild, sun bears may use tree cavities or fallen hollowed logs as denning sites (Wong et al. 2004).

Gestation is between 96 and 110 days, with no evidence of delayed implantation (Spady et al. 2007; Frederick et al. 2012). Sun bears exhibit obligate pseudopregnancies during non-conceptive cycles, even without access to males (Schwarzenberger et al. 1998; Onuma et al. 2001; Frederick et al. 2012). Pseudopregnancies can include lactation and behavioral changes, such as excavating a den and increased appetite (Onuma et al. 2001; Frederick et al. 2012).

Like other ursids, cubs are blind and highly altricial, weighing 300–325 g at birth (Dathe 1962, 1970; Hall & Swaisgood 2009). Cub development is variable, with rearing conditions playing a role (Hall & Swaisgood 2009). Eyes open between 14 and 50 days old (Weber 1969; Hall & Swaisgood 2009). The upper canines erupt first by day 42 or earlier, followed by upper incisors, with full adult dentition achieved by 18 months (Fetherstonhaugh 1940; Weber 1969). Cubs crawl by day 40 and can follow the mother out of the den at ~2.5 months (Poglayen-Neuwell 1986; Hall & Swaisgood 2009). Sun bear are especially attentive mothers, showing continuous contact and fasting for the first few days post-partum (Hall & Swaisgood 2009). Care includes spending considerable time holding, responding to vocal signals, and interacting with the cub (Hall & Swaisgood 2009; Owen et al. 2014).

Behavior

The sun bear lives a terrestrial and arboreal lifestyle, foraging for insects and fallen fruits on the ground. Paws are highly dextrous. The back paws hold food items in place while the front paws are used to manipulate them. The tongue laps up what falls on their bellies (a behavior observed when sun bears feed from termite mounds; Nowak 1991; Frederick, personal observation; Wong, personal communication).

Small body size might be advantageous in escaping predators by climbing (Stirling & Derocher 1990). Small body size also facilitates climbing and clinging to trees for long periods of time, to break into bees' nests and to forage in outlying branches of tall trees. Sun bear use branches of tall trees to build nests, and hollow logs and cavities beneath trees as resting or bedding sites (Wong et al. 2004). Tree nests are made by breaking surrounding branches toward them and constructing a V-shaped structure, usually from large branches near to the main trunk of the tree (Fredriksson 2005). Tree nests may afford some protection from humans and dogs by providing cover and a higher vantage point. Zoo-housed bears build nests during the day or night by gathering, carrying, placing, and even intertwining diverse materials to achieve the desired effect (Frederick, personal observation).

There is little information about wild social behaviors. Sun bears are considered solitary, with the most common social grouping being a female with cub(s). Limited field data suggest adults form only brief associations for mating (Normua et al. 2004), with infrequent reports of three adult bears occurring together at feeding sites (Fetherstonhaugh 1940; Fredriksson 2005).

Sun bears tend to avoid humans and are generally shy and inoffensive (Fetherstonhaugh 1940). However, if cornered or surprised, sun bears may become aggressive (especially mothers with cubs), and bark and charge. Barking may be preceded by showing the ventral chest patch while standing and stretching the neck skin, giving the bear a larger appearance (Fredriksson, personal observation cited in Meijaard 1999). Agonistic vocalizations include hoarse grunts, barks, growling, and loud roars (Payne et al. 1989; Fredriksson 2005). In rural farming communities, where bears raid crop fields and humans gather forest resources in bear habitat (including hunting bears), sun bear have attacked and injured farmers and hunters (Scotson et al. 2014; see also Chapters 15 and 17).

Sun bears are one of several bear species with chest markings that augment a threat posture while rearing up (Lekagul & McNeely 1977; Stirling & Derocher 1990). During interactions, bears tend to approach one another head on, and chest and face markings likely serve to identify individuals, and the positioning of ears, mouth, teeth, and overall body posture as communication signals (Stirling & Derocher 1990).

Parasites and Disease

There is little recorded on parasites and disease in wild sun bears, with most knowledge from captive bears. Published studies report conditions caused by gastrointestinal parasites, fungi, and bacteria, and a variety of other infectious and non-infectious diseases. Until more is known about the incidence of disease in wild sun bears, the influence of the captive environment on disease susceptibility and manifestation, while suspected to be significant, remains unknown.

Five parasites have been reported from wild sun bears in South-East Asia: the mosquito-born *Plasmodium* spp., a cestode *Pentorchis arkteios*, a hookworm *Ancylostoma malayanum*, and three ticks *Haemaphysalis semermis*, *H. leachi*, and *H. hystricis* (Rogers & Rogers 1976; Siew Te Wong, personal communication). Starvation of wild sun bears caused by fruiting failures (Wong et al. 2005) may be exacerbated by parasite load. In captive bears, recurrent roundworm infestations are common (e.g. *Baylisascaris transfuga*; Rogers & Rogers 1976). Other gastrointestinal parasites reported include *Trichuris* and *Ancylostoma* spp. (Jenantika et al. 2019).

Bacterial and fungal conditions recorded in captive bears include the potentially fatal salmon poisoning disease, caused by the bacterium *Neorickettsia helminthoeca* in live trout (Gai & Marks 2008), and skin disease associated with the fungus *Microsporum canis* (Groves 1969).

Infectious conditions probably include viral spillover from other species, with sun bears showing signs of disease after infection with canine adenovirus 1 (Goodnight & Emanuelson 2012) and foot-and-mouth disease (Officer et al. 2014). A novel gammaherpes virus, ursid herpesvirus 1, has been detected in sun bears in apparent association with oral squamous cell carcinoma (Lam et al. 2013).

Non-infectious conditions include neoplasia (particularly of the gastrointestinal and hepatobiliary systems), dental disease, and degenerative joint disease, perhaps reflecting the longevity of captive individuals (Canfield et al. 1990; Blake & Collins 2002; Kitchener 2004). There are multiple reports of squamous cell carcinoma, at oral, ocular, and mandibular sites (Blake & Collins 2002; Mylniczenko et al. 2005; Lam et al. 2013) and a single report of mesothelial meningioma (Chien et al. 2013). Sun bears appear more prone to cystic alterations of the uterine cervix compared to other bear species (Goeritz et al. 2006) and there are several reports of gastro-intestinal accidents, including foreign body obstruction and gastric dilation and volvulus (Montali et al. 1981; Blake & Collins 2002).

Degenerative joint disease is common in captive bears and old-age-related osteoarthritis, particularly of the elbow joints, is frequent (Bourne et al. 2010). A high prevalence of spondyloarthropathy has been recorded in sun bear skeletons, indicative of inflammatory and infectious etiologies and biomechanical stress (Kompanje et al. 2000; Nunn et al. 2007).

Dental disease is almost ubiquitous in captive bears with a high prevalence of calculus deposition and periodontal disease (Bourne et al. 2010). Enamel erosion and crown fractures of the canine teeth, exposing the pulp cavity and giving rise to apical infections, are also recorded (Kompanje et al. 2000; Fleming & Burn 2014). Canine tooth crown loss has been observed in two wild sun bears in Kalimantan, Indonesia (Fredriksson, personal communication).

Eye conditions, including chronic keratitis, cataracts, lens luxation, hyalitis, retinal degeneration, retinal vascular changes (dilation, spasm, aneurysm), and hemorrhage have been observed in captive sun bears, as well as age-related changes such as lenticular sclerosis (opacity of the lens) (Blake & Collins 2002; Hartley, personal communication).

Status in the Wild

The sun bear is categorized as Vulnerable on the IUCN Red List, with population change, estimated by expert opinion, projected to decline globally by >30% in the next 30+ years (Scotson et al. 2017a). Some populations may meet the IUCN criteria for Endangered based on declining population trend. Commercial trade is considered one of the biggest threats to wild populations, with trade occurring both domestically and internationally with China, Thailand, and Vietnam (Foley et al. 2011; Scotson et al. 2017a). Hunting bears is prohibited throughout most of South-East Asia, but poor law enforcement capacity coupled with the high value of bear gall bladders, paws, and cubs creates a high incentive for poaching (Nooren & Claridge 2001; Scotson & Brocklehurst 2013). The sun bear is classed as CITES Appendix I, which restricts trade, unless under exceptional circumstances. Import permits can be issued if specimens are not to be used for commercial purposes, or if the import purposes are not deemed to be detrimental to the survival of the species. Nevertheless, the

Figure 8.6 Sun bear cub rescued from poachers in Cambodia. This was the 184th bear (sun bears and Asiatic black bears together) to be rescued by Free the Bears in Cambodia, here pictured with the Cambodian Program Manager, Chuon Vuthy. Rescue carried out by Cambodia's Wildlife Rapid Rescue Team. (Photo by Free the Bears/Peter Yen.)

Figure 8.7 A rescued sun bear in Cambodia. Bears rescued by Free the Bears enjoy life in large semi-natural enclosures within sanctuaries throughout South-East Asia. (Photo by Free the Bears/Peter Yen.) (A black and white version of this figure will appear in some formats. For the color version, please refer to the plate section.)

sun bear was one of the most commonly seized species of bears across Asia between 2000 and 2011 (Burgess et al. 2014; Figure 8.6).

Sun bear are poached mainly for their gall bladders, which have been used in Chinese traditional medicine for thousands of years (Foley et al. 2011; see also Chapter 26). China is the world's biggest consumer of bear bile and is presumed to be the main driver of international demand. Wild bear products are imported for domestic consumption, by Chinese international tourists, and by expatriate Chinese communities living in other countries (Livingstone et al. 2018). Traditionally, bile was sourced from the gall bladders of bears hunted and killed in the wild. In more recent years, since the 1980s, several Asian countries have established the business of farming live bears for their bile (Servheen et al. 1999). It is unknown how farming practices are affecting sun bear populations, but it is widely assumed by conservation scientists that farming has a negative impact, largely due to the demand for this product (Dutton et al. 2011; IUCN 2012; Livingstone & Shepherd 2014; Garshelis & Steinmetz 2016; Scotson et al. 2017a). An IUCN Recommendation, WCC-2012-Rec-139, urges range state governments of sun bear and Asiatic black bear to phase out bear farming (IUCN 2012). However, bear farms prevalently hold other bear species in greater numbers (Animals Asia 2011; Livingstone & Shepherd 2014; BANCA 2017; Education for Nature 2017). Low numbers of sun bears in bear farms may be in part due to their rarity in bear farming nations, which are more concentrated in the north of the sun bear range, and also due to sun bears having a low concentration of ursodeoxycolic acid in their gall bladders (the active medicinal component of bear bile) compared to the other species (Vang et al. 2016).

Sun bears are also threatened by high rates of deforestation (Sodhi et al. 2010; Miettinen et al. 2011). Declines based on habitat loss data over a 14-year period (2000–2014) were 22% on Sumatra and Borneo and ~9% on the mainland. Projected over a 30-year period, population declines in the insular region might surpass 50%, meeting the IUCN criteria for listing this subpopulation as Endangered, if sun bears were listed in the same way as brown bears, *Ursus arctos*, on the population level (McLellan et al. 2017; Scotson et al. 2017b). Likewise, based on these projections, the Borneo subspecies *H. m. euryspilus* meets the IUCN criteria for the Endangered category (IUCN 2016).

Status in Captivity

Sun bears are kept in captivity in non-range state zoos and animal collections, in rescue centers within range countries and, to a lesser extent, in bear farms in South-East Asia. The number of captive sun bears in non-range countries (including Australia and Japan) is less than 100. Within range countries, members of the South East Asian Zoo Association hold more than double this, with yet more held within sanctuaries across South-East Asia (Crudge et al. 2019; see also Chapter 23). Rescue centers throughout sun bear range face major challenges in maintaining enough space and resources for displaced, confiscated, rescued, injured, orphaned, nuisance, and ex-pet sun bears (Figure 8.7).

In North America, the captive population has low breeding potential (i.e. an aging population with female bias) and the Association of Zoos and Aquariums' Bear Taxon Advisory Group has designated the sun bear as a "phase out" species, still considered important for conservation, research, and education (Snyder & Thompson 2015). European zoos, while facing similar problems, have fared better, and the European Association of Zoos and Aquaria program for sun bears remains active (Kolter, personal communication).

References

Allen, G. M. (1938). *The mammals of China and Mongolia.* Natural history of central Asia. New York, NY: The American Museum of Natural History.

Animals Asia. (2011). *Bear farming industry in China.* Chengdu, China: Animals Asia Foundation.

Augeri, D. M. (2005). *On the biogeographic ecology of the Malayan Sun Bear.* Cambridge, UK: University of Cambridge.

BANCA. (2017). *The bear trade and the bear farms in cross-border regions in eastern Myanmar.* Yangon, Myanmar: Biodiversity and Nature Conservation Association.

Blake, C. N. & Collins, D. (2002). Captive ursids: results and selected findings of a multi-institutional study. In: *Proceedings from the Annual Conference of the American Association of Zoo Veterinarians,* Milwaukee, Wisconsin, October 5–10 (pp. 21–26). Schaumberg, IL: American Association of Zoo Veterinarians.

Blanchard, B. M. & Knight, R .R. (1991). Movements of Yellowstone grizzly bears. *Biological Conservation* 58: 41–67.

Bourne, D. C., Cracknell, J. M. & Bacon, H. J. (2010). Veterinary issues related to bears (Ursidae). *International Zoo Yearbook* 44: 16–32.

Burgess, E. A., Stoner, S. S. & Foley, K. E. (2014). *Brought to bear: An analysis of seizures across Asia (2000–2011).* Petaling Jaya, Selangor, Malaysia: TRAFFIC Southeast Asia.

Canfield, P., Bellamy, T., Blyde, D., et al. (1990). Pancreatic lesions and hepatobiliary neoplasia in captive bears. *Journal of Zoo and Wildlife Medicine* 21: 471–475.

Chakraborty, R. (2003). A catalogue of mammalian exhibits of zoological gallaries of the Indian Museum. *Records of the Zoological Survey of India. Occasional Paper* No. 219: 1–99.

Cheah, C. (2013). The ecology of Malayan sun bears (*Helarctos malayanus*) at the Krau Wildlife Reserve, Pahang, Malaysia and adjacent plantations. PhD dissertation, University Putra Malaysia, Malaysia.

Chien, Y. C., Lien, C. Y., Guo, J. C., et al. (2013). Meningothelial meningioma in a Malayan sun bear (*Helarctos malayanus*). *Journal of Veterinary Diagnostic Investigation* 25: 636–640.

Crudge, B., Lees, C., Hunt, M., et al. (2019). Sun bears: Global status review & conservation action plan, 2019–2028. IUCN SSC Bear Specialist Group/IUCN SSC Conservation Planning Specialist Group/Free the Bears/TRAFFIC. Available from https://portals.iucn.org/library/sites/library/files/documents/2019-041-En.pdf

Dathe, H. (1962). Breeding the Malayan bear *Helarctos malayanus. International Zoo Yearbook* 3: 92.

Dathe, H. (1970). A second generation birth of captive sun bears: at East Berlin Zoo. *International Zoo Yearbook* 10: 79–79.

Dathe, H. (1975). *Malayan sun bears.* In: *Grzimek's animal life encyclopedia* (pp. 141–142). New York, NY: Van Nostrand Reinhold Company.

Domico, T. (1988). *Bears of the world,* 3rd edn. New York, NY: Facts on File.

Dutton, A. J., Hepburn, C. & Macdonald, D. W. (2011). A stated preference investigation into the Chinese demand for farmed vs. wild bear bile. *PLoS ONE* 6: e21243.

Education for Nature. (2017). Progress tackling wildlife crime illuminates challenges ahead. *Wildlife Crime* 1: 4.

Erdbrink, D. P. (1953). *A review of the fossil and recent bears of the Old World with remarks on their philogenese based upon their dentition.* Deventer: Drukkerij Jan de Lange.

Fetherstonhaugh, A. H. (1940). Some notes on Malayan bears. *The Malayan Nature Journal* 1: 15–19.

Fleming, M. & Burn, C. C. (2014). Behavioural assessment of dental pain in captive Malayan sun bears (*Helarctos malayanus*). *Animal Welfare* 23: 131–140.

Foley, K. E., Stengel, C. J. & Shepherd, C. R. (2011). *Pills, powders, vials and flakes: The bear bile trade in Asia.* Petaling Jaya; Cambridge: Traffic Southeast Asia.

Frederick, C. (2008). *The reproductive biology and behavior of the sun bear (*Ursus malayanus*).* Seattle, WA: University of Washington.

Frederick, C., Kyes, R., Hunt, K., et al. (2010). Methods of estrus detection and correlates of the reproductive cycle in the sun bear (*Helarctos malayanus*). *Theriogenology* 74: 1121–1135.

Frederick, C., Hunt, K. E., Kyes, R., Collins, D. & Wasser, S. K. (2012). Reproductive timing and aseasonality in the sun bear (*Helarctos malayanus*). *Journal of Mammalogy* 93: 522–531.

Frederick, C., Hunt, K., Kyes, R., et al. (2013). Social influences on the estrous cycle of the captive sun bear (*Helarctos malayanus*). *Zoo Biology* 32: 581–591.

Fredriksson, G. (2005). Human–sun bear conflicts in East Kalimantan, Indonesian Borneo. *Ursus* 16: 130–137.

Fredriksson, G. (2012). Effects of El Niño and large-scale forest fires on the ecology and conservation of Malayan sun bears (*Helarctos malayanus*) in East Kalimantan. Institute for Biodiversity and Ecosystem Dynamics. PhD dissertation, University of Amsterdam, Netherlands.

Fredriksson, G. M., Wich, S. A. & Trisno. (2006). Frugivory in sun bears (*Helarctos malayanus*) is linked to El Niño-related fluctuations in fruiting phenology, East Kalimantan, Indonesia. *Biological Journal of the Linnean Society* 89: 489–508.

Gai, J. J. & Marks, S. L. (2008). Salmon poisoning disease in two Malayan sun bears. *Journal of the American Veterinary Medical Association* 232: 586–588.

Galbreath, G. J., Hunt, M., Clements, T. & Waits, L. P. (2008). An apparent hybrid wild bear from Cambodia. *Ursus* 19: 85–86.

Garshelis, D. L. (2004). *Variation in ursid life histories: is there an outlier?* In: Lindburg, D. & Baragona, K. (Eds.), *Giant pandas: Biology and conservation* (pp. 53–73). Los Angeles, CA: University of California Press.

Garshelis, D. & Steinmetz, R. (2016). *Ursus thibetanus*, Asiatic black bear. IUCN Red List of Threatened Species. www.iucnredlist.org

Goodnight, A. L. & Emanuelson, K. (2012). Two cases of canine adenovirus type 1 infection in Malayan sunbears (*Helarctos malayanus*) at the Oakland Zoo, California. In: *Proceedings from the Annual Conference of the American Association of Zoo Veterinarians,* Oakland, California, October 21–26 (p. 77). Schaumberg, IL: American Association of Zoo Veterinarians.

Goeritz, F., Hermes, R., Jewgenow, K., et al. (2006). High incidence of cysts of the cervix uteri in captive Malayan sun bears.

In: *Proceedings from the Annual Conference of the American Association of Zoo Veterinarians*, Tampa, Florida, September 19–24 (pp. 114–115). Schaumberg, IL: American Association of Zoo Veterinarians.

Griffiths, M. & Schaik, C. P. (1993). The impact of human traffic on the abundance and activity periods of Sumatran rain forest wildlife. *Conservation Biology* 7: 623–626.

Groves, M. G. (1969). Griseofulvin treatment of *Microsporum canis* infection in Malayan sun bears (*Helarctos malayanus*). *Journal of the American Veterinary Medical Association* 155: 1090–1092.

Hall, S. S. & Swaisgood, R. R. (2009). Maternal care and cub development in the sun bear. *Ursus* 20: 143–151.

Hesterman, H., Wasser, S. K. & Cockrem, J. F. (2005). Longitudinal monitoring of fecal testosterone in male Malayan sun bears (*U. malayanus*). *Zoo Biology* 24: 403–417.

Higgins, J. C. (1932). The Malay bear. *Journal of the Bombay Natural History Society* 35: 673–674.

Hughes, J. B., Round, P. D. & Woodruff, D. S. (2003). The Indochinese–Sundaic faunal transition at the Isthmus of Kra: an analysis of resident forest bird species distributions. *Journal of Biogeography* 30: 569–580.

Hwang, M. H. & Garshelis, D. L. (2007). Activity patterns of Asiatic black bears (*Ursus thibetanus*) in the Central Mountains of Taiwan. *Journal of Zoology* 271: 203–209.

IUCN. (2012). *Resolutions and Recommendations World Conservation Congress*. Jeju, Republic of Korea, September 6–15. Gland, Switzerland: IUCN.

IUCN. (2016). *The IUCN Red List of Threatened Species*. Gland, Switzerland: IUCN.

Jenantika, P. U., Fahrimal, Y. & Sayuti, A. (2019). Identifikasi parasit gastrointestinal pada beruang madu (*Helarctos malayanus*) di Taman Margasatwa medan [Identification of gastrointestinal parasites in sun bear (*Helarctos malayanus*) in Taman Margasatwa]. *Jurnal Ilmiah Mahasiswa Veteriner* 3: 142–148.

Johnston, L. A., Donoghue, A. M., Igo, W., et al. (1994). Oocyte recovery and maturation in the American black bear (*Ursus americanus*): a model for endangered ursids. *Journal of Experimental Zoology* 269: 53–61.

Kitchener, A. C. (2004). The problems of old bears in zoos. *International Zoo News* 51: 282–293.

Kompanje, E. J. O., Klaver, P. S. J. & de Vries, G. Th. (2000). Spondyloarthropathy and osteoarthosis in three Indomalayan bears: *Ursus ursinus* Cuvier, 1823, *Ursus thibetanus* Raffles, 1821, and *Ursus malayanus* Shaw & Nodder, 1791 (Mammalia: Carnivora: Ursidae). *Contributions to Zoology* 69: 259–269.

Krause, J., Unger, T., Noçon, A., *et al.* (2008). Mitochondrial genomes reveal an explosive radiation of extinct and extant bears near the Miocene–Pliocene boundary. *BMC Evolutionary Biology* 8: 220. doi: 10.1186/1471-2148-8-220.

Kutschera, V. E., Bidon, T., Hailer, F., et al. (2017). Bears in a forest of gene trees: phylogenetic inference is complicated by incomplete lineage sorting and gene flow. *Molecular Biology and Evolution* 31: 2004–2017.

Lam, L., Garner, M. M., Miller, C. L., et al. (2013). A novel gammaherpesvirus found in oral squamous cell carcinomas in sun bears (*Helarctos malayanus*). *Journal of Veterinary Diagnostic Investigation* 25: 99–106.

Lekagul, B. & McNeely, J. A. (1977). *Mammals of Thailand*. Bangkok, Thailand: Association for the Conservation of Wildlife.

Linkie, M., Dinata, Y., Nugroho, A. & Haidir, I.A. (2007). Estimating occupancy of a data deficient mammalian species living in tropical rainforests: sun bears in the Kerinci Seblat region, Sumatra. *Biological Conservation* 137: 20–27.

Livingstone, E. & Shepherd, C. R. (2014). Bear farms in Lao PDR expand illegally and fail to conserve wild bears. *Oryx* 50: 1–9.

Livingstone, E., Gomez, L. & Bouhuys, J. (2018). A review of bear farming and bear trade in Lao People's Democratic Republic. *Global Ecology and Conservation* 13: e00380.

Marti, K., Theis, M., Lindsey, S., et al. (2012). *Mainland sun bear (*Helarctos malayanus malayanus*) and Bornean sun bear (*Helarctos malayanus euryspilus*)*. AZA Animal Program Population Viability Analysis Report. Chicago, IL: American Association of Zoo Veterinarians.

McConkey, K. & Galetti, M. (1999). Seed dispersal by the sun bear *Helarctos malayanus* in Central Borneo. *Journal of Tropical Ecology* 15: 237–241.

McLellan, B. & Reiner, D. C. (1994). *A Review of Bear Evolution. Bears: Their Biology and Management. Vol. 9, Part 1: A Selection of Papers from the Ninth International Conference on Bear Research and Management*, Missoula, Montana, February 23–28 (pp. 85–96). Anchorage, AK: International Association for Bear Research and Management.

McLellan, B. N., Proctor, M. F., Huber, D. & Michel, S. (2017). *Ursus arctos*, Brown bear. IUCN Red List of Threatened Species. www.iucnredlist.org.

Meijaard, E. (1999). Human-imposed threats to sun bears in Borneo. *Ursus* 11: 185–192.

Meijaard, E. (2004). Craniometric differences among Malayan sun bears (*Ursus malayanus*); evolutionary and taxonomic implications. *Raffles Bulletin of Zoology* 52: 665–672.

Miettinen, J., Shi, C. & Liew, S. C. (2011). Deforestation rates in insular Southeast Asia between 2000 and 2010. *Global Change Biology* 17: 2261–2270.

Montali, R. J., Hoopes, P. J. & Bush, M. (1981). Extrahepatic biliary carcinomas in Asiatic bears. *Journal of the National Cancer Institute* 66: 603–608.

Mylniczenko, N. D., Manharth, A. L., Clayton, L. A., Feinmehl, R. & Robbins, M. (2005). Successful treatment of mandibular squamous cell carcinoma in a Malayan sun bear (*Helarctos malayanus*). *Journal of Zoo and Wildlife Medicine* 36: 346–348.

Nooren, H. & Claridge, G. (2001). *Wildlife trade in Laos: The end of the game*. Netherlands: International Union for Conservation of Nature.

Normua, F., Higashi, S., Ambu, L. & Mohamed, M. (2004). Notes on oil palm plantation use and seasonal spatial relationships of sun bears in Sabah, Malaysia. *Ursus* 15: 227–231.

Nowak, R. M. (1991). *Walker's mammals of the world*. Choice Reviews Online. Fifth edition. Baltimore, MD: John Hopkins University Press.

Nunn, C. L., Rothschild, B. & Gittleman, J. L. (2007). Why are some species more commonly afflicted by arthritis than others? A comparative study of spondyloarthropathy in primates and carnivores. *Journal of Evolutionary Biology* 20: 460–470.

Officer, K., Lan, N. T., Wicker, L., et al. (2014). Foot-and-mouth disease in Asiatic black bears (*Ursus thibetanus*). *Journal of Veterinary Diagnostic Investigation* 26: 705–713.

Onuma, M., Suzuki, M. & Ohtaishi, N. (2001). Reproductive pattern of the sun bear (*Helarctos malayanus*) in Sarawak. *Journal of Veterninary Medicine Science* 63: 293–297.

Owen, M. A., Hall, S., Bryant, L. & Swaisgood, R. R. (2014). The influence of ambient noise on maternal behavior in a Bornean sun bear (*Helarctos malayanus euryspilus*). *Zoo Biology* 33: 49–53.

Payne, J., Francis, C. M. & Philipps, K. (1989). *A field guide to the mammals of Borneo*. Sabah, Malaysia: The Sabah Society with World Wildlife Fund Malaysia.

Pocock, R. (1917). A new genus of Ursidae. *Annals and Magazine of Natural History* 8: 128–129.

Pocock, R. I. (1932). The black and brown bears of Europe and Asia. Part II. *Journal of the Bombay Natural History Society* 36: 101–138.

Pocock, R. I. (1941). *The fauna of British India, including Ceylon and Burma. Mammalia*. London, UK: Taylor & Francis.

Poglayen-Neuwell, I. (1986). An unusual method of transport of young sun bears by their mothers. *Der Zoologische Garten* 56, 437–438.

Rogers, L. L. & Rogers, S. M. (1976). Parasites of bears: a review. *Proceedings of the Third International Conference on Bears* Paper 42: 411–430.

Sasaki, M., Endo, H., Wiig, Ø., et al. (2005). Adaptation of the hindlimbs for climbing in bears. *Annals of Anatomy* 187: 153–160.

Schwarzenberger, F., Schaller, K., Chaduc, Y., Pagan, O. & Kolter, L. (1998). Faecal steroid analysis for monitoring ovarian function and the effect of PZP (porcine zona pellucida protein) in the sun bear (*Helarctos malayanus*). *Proceedings of the European Association of Zoo and Wildlife Veterinarians* 2: 387–395.

Schwarzenberger, F., Fredriksson, G., Schaller, K. & Kolter, L. (2004). Fecal steroid analysis for monitoring reproduction in the sun bear (*Helarctos malayanus*). *Theriogenology* 62: 1677–1692.

Scotson, L. (2019). Exploring potential range connectivity of sun bear (Carnivora: Ursidae: Ursinae). *Raffles Bulletin of Zoology* 67: 67–76.

Scotson, L. & Brocklehurst, M. (2013). Bear poaching in Lao PDR is exposed as an increasing threat to wild populations. *International Bear News* 22: 22–23.

Scotson, L., Vannachomchan, K. & Sharp, T. (2014). More valuable dead than deterred? Crop-raiding bears in Lao PDR. *Wildlife Society Bulletin* 38.

Scotson, L., Fredriksson, G., Augeri, D., et al. (2017a). *Helarctos malayanus*, sun bear. IUCN Red List of Threatened Species. www.iucnredlist.org

Scotson, L., Fredriksson, G., Ngoprasert, D., Wong, W. M. & Fieberg, J. (2017b). Projecting range-wide sun bear population trends using tree cover and camera-trap bycatch data. *PLoS ONE* 12: doi: 10.1371/journal.pone.0185336.

Servheen, C., Herrero, S., Peyton, B., et al. (1999). *Bears: Status survey and conservation action plan*. Gland, Switzerland and Cambridge, UK: IUCN/SSC Bear and Polar Bear Specialist Goups.

Snyder, R. & Thompson, D. (2015). *Bear TAG Regional Collection Plan*, 4th edition. Maryland: Bear Taxon Advisory Group, Association of Zoos and Aquaria.

Sodhi, N. S., Posa, M. R. C., Lee, T. M., et al. (2010). The state and conservation of Southeast Asian biodiversity. *Biodiversity and Conservation* 19: 317–328.

Spady, T. J., Lindburg, D. G. & Durrant, B. S. (2007). Evolution of reproductive seasonality in bears. *Mammal Review* 37: 21–53.

Steinmetz, R. (2011). *Ecology and distribution of sympatric Asiatic black bears and sun bears in the seasonally dry forests of Southeast Asia. The ecology and conservation of seasonally dry forests in Asia*. Washington DC: Smithsonian Institution Scholarly Press.

Steinmetz, R., Garshelis, D. L., Chutipong, W. & Seuaturien, N. (2011). The shared preference niche of sympatric Asiatic black bears and sun bears in a tropical forest mosaic. *PLoS ONE* 6: e14509.

Steinmetz, R., Garshelis, D. L., Chutipong, W. & Seuaturien, N. (2013). Foraging ecology and coexistence of Asiatic black bears and sun bears in a seasonal tropical forest in Southeast Asia. *Journal of Mammalogy* 94: 1–18.

Stirling, I. & Derocher, A. E. (1990). Factors affecting the evolution and behavioral ecology of the modern bears. *Bears: Their Biology and Management* 8: 189–204.

Tumbelaka, L. & Fredriksson, G. M. (2006). The status of sun bears in Indonesia. In: *Understanding Asian bears to secure their future* (pp. 73–78). Ibaraki: Japan Bear Network.

Vang, S., Longley, K., Steer, C. J., et al. (2016). Assessing the illegal bear trade in Myanmar through conversations with poachers: topology, perceptions, and trade links to China. *PLoS ONE* 11: 1–4.

Weber, E. (1969). Notes on hand rearing a Malayan sun bear *Helarctos malayanus* at Melbourne Zoo. *International Zoo Yearbook* 9: 163–163.

Wong, S. T., Servheen, C. W. & Ambu, L. (2004). Home range, movement and activity patterns, and bedding sites of Malayan sun bears *Helarctos malayanus* in the rainforest of Borneo. *Biological Conservation* 119: 169–181.

Wong, S. T., Servheen, C., Ambu, L. & Norhayati, A. (2005). Impacts of fruit production cycles on Malayan sun bears and bearded pigs in lowland tropical forest of Sabah, Malaysian Borneo. *Journal of Tropical Ecology* 21: 627–639.

Wong, W. M., Leader-Williams, N. & Linkie, M. (2013). Quantifying changes in sun bear distribution and their forest habitat in Sumatra. *Animal Conservation* 16: 216–223.

Wroughton, R. C. (1916). Bombay Natural History Society's mammal survey of India, Burma and Ceylon. *Journal of the Bombay Natural History Society* 24: 773–782.

Zhang, Y. P. & Ryder, O. A. (1994). Phylogenetic relationships of bears (the Ursidae) inferred from mitochondrial DNA sequences. *Molecular Phylogenetics and Evolution* 3: 351–359.

Chapter 9

Sloth Bear (*Melursus ursinus*)

Harendra Singh Bargali, Karine E. Pigeon, Naim Akhtar, Thomas Sharp, and Kajal K. Jadav

Names: Sloth bear, honey bear, lip bear

Genus: *Melursus,* Meyer, 1793

Species: *Melursus ursinus* (Shaw, 1791)

Subspecies: *Melursus ursinus ursinus*, India; *Melursus ursinus inornatus*, Sri Lanka

Names in other languages

Italian: orso labiato; **French**: ours lippu de l'Inde, ours prochile lippu; **German**: Lippenbär; **Spanish**: oso bezudo; **Indian**: bhalu; **Nepal**: Kathe Bhalu; **Malay**: beruang sloth; **Tamil/Malayalam**: Karadi

Synonym(s): *Bradypus ursinus* Shaw, 1791; *Melursus lybius* Meyer, 1793

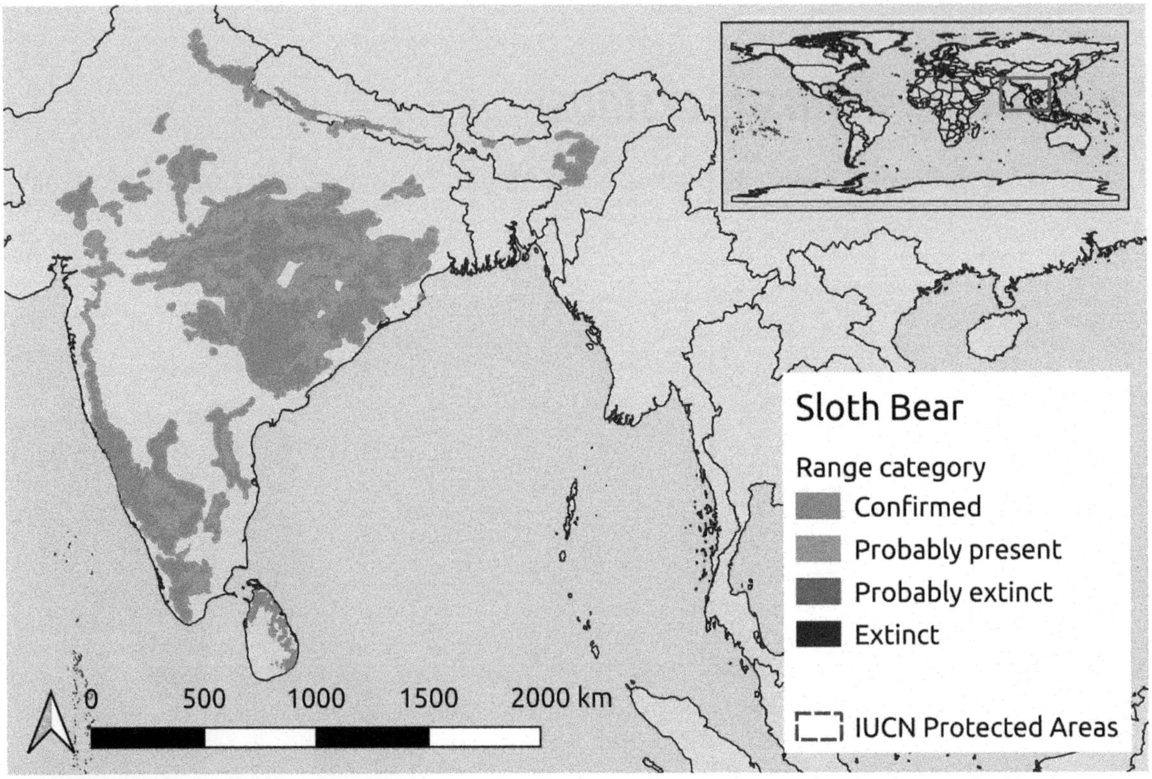

Figure 9.1 Sloth bear distribution. (*Source*: IUCN Red List of Threatened Species.)

Taxonomy

Evidence suggests that sloth bears (*Melursus ursinus*) evolved roughly 5 million years ago and are potentially the oldest of the currently extent species of the Ursidae (Krause et al. 2008; Pagès et al. 2008; Kumar et al. 2017; see also Chapter 1). The exact position of sloth bears on the evolutionary tree is still debated, and while mitochondrial DNA (mtDNA) analyses and morphological studies place the speciation of sloth bears earlier than all other ursine bears, nuclear gene analyses indicate a position much closer to sun bears (Krause et al. 2008; Abella et al. 2012; Kutschera et al. 2014; Kumar et al. 2017). George Shaw (1971) first described sloth bears as *Ursine bradypus*, a "bear-like sloth," and thought that the species was related to South American sloths (*Bradypus* spp.) because of its long, thick claws and because it lacks the two upper incisors that are common to most mammals. Although the name "sloth bear" still persists, phylogenetic studies have confirmed sloth bears as a tropical bear species with no direct relationship to South American sloths. Considering the complex evolutionary history of bears, there are still current debates on whether sloth bears should be kept in their own genus, *Melursus*, or included in the *Ursus* genus (Krause et al. 2008; Pagès et al. 2008; Kitchener 2010; see Chapter 1).

Subspecies and Distribution

Two subspecies of sloth bears have been recognized, *Melursus ursinus ursinus* from the Indian peninsula and a physically smaller and shorter subspecies, *Melursus ursinus inornatus*, endemic to Sri Lanka (Pocock 1941a; Corbet & Hill 1992; Kitchener 2010).

At least 90% of the current sloth bear range is restricted to India, with the only current range countries being India, Nepal, and Sri Lanka (Dharayia et al. 2016; see Figure 9.1). Historically, sloth bears were also present in Bangladesh, and possibly Bhutan and Myanmar. Although the historical presence of sloth bears in Myanmar is doubtful considering the mountainous habitat of the region, the species was recently extirpated from Bangladesh (Islam et al. 2013). Moreover, investigations in Bhutan suggest that the species was either never present in the country (other than the odd migrant) or only occurred in very low numbers (Garshelis et al. 2015).

Although sloth bears are widely distributed in India, their current distribution is patchy and fragmented. Estimates on areas occupied by sloth bears vary from 200,000 km^2 (Johnsingh 2003) to 400,000 km^2 (Sathyakumar et al. 2012). In the north, sloth bear range partly overlaps with the range of Asiatic black bears (*Ursus thibetanus*; Bargali 2012). In the north-eastern part of sloth bear range (i.e. north-east India), sloth bears overlap with Asiatic black bears and sun bears (*Helarctos malayanus*) (Choudhury 2011; see also Chapter 4).

Descriptive Notes

Sloth bears are medium-sized bears characterized by a distinctively shaggy dusty-black coat and a common, but not consistently present, broad cream colored U- or V-shaped chest blaze

Figure 9.2 A standing sloth bear, showing its typical white chest mark. (Photo by Ansar Khan.)

Figure 9.3 Sloth bear by Francesco Rinaldi, with permission. (A black and white version of this figure will appear in some formats. For the color version, please refer to the plate section.)

(Figures 9.2 and 9.3; Laurie & Seidensticker 1977; Prater 1980). Some fully brown or whitish individuals have also been observed (Pocock 1941b; Brander 1982; Phillips 1984). Sloth bear hairs are especially long around the neck and the back of the head, often hiding their ears altogether. Sloth bears are the only species of bear with autapomorphic craniodental morphology adapted to myrmecophagy (Erdbrink 1953; Abella et al. 2012). Sloth bears can be characterized by an absence of the two upper incisors, an elongated concave palate, well-developed lips, flanges on the nose, long front claws, a long shaggy coat, a nearly naked snout, and a low metabolic rate (Pocock 1941a; McNab 1992). Adult males generally weigh 80–145 kg while females generally weigh between 60 and 100 kg. Sloth bears stand 65–85 cm at shoulder and are 140–170 cm long (Figure 9.4; Prater 1980; Garshelis et al. 1999a).

Habitat

Across their distribution range, sloth bears occupy a variety of habitats. Sloth bears are primarily a lowland species and usually occur in habitats below 1000 m elevation in India and Nepal (Garshelis et al. 1999a; Johnsingh 2003) and below 300 m in Sri Lanka (Ratnayeke et al. 2007). However, sloth bear scats have been collected from higher elevations as well. On the Indian mainland, sloth bears occur in semi-arid forested habitats (Johnsingh 2003; Akhtar et al. 2004), montane forests of the Western Ghats (Johnsingh 2003), moist evergreen forests, dry deciduous forests, moist deciduous forests, degraded scrub forests, and tall grasslands (Desai et al. 1997). Moreover, sloth bears are also found in riparian forests and tall grass areas on the floodplains of Nepal (Joshi et al. 1995) and in the Bhrahmaputra valley of Assam (Brander 1982). Presently, sloth bears are widely distributed across the tropical forest of the Indian subcontinent (Yoganand et al. 2006). Sloth bear ranges overlap with Asiatic black bear in north India in Uttarakhand and preferentially live in dry deciduous forests (Bargali 2012), whereas in north-east India, sloth bear range overlaps with the ranges of Asiatic black bear and sun bear (Choudhury 2011). According to Jhala et al. (2008), sloth bears occupy 32% of India's total forested area. Sloth bears occur frequently in moist and dry deciduous forest (42% and 33%, respectively) and less frequently in wet evergreen (13%) and dry scrub (6%) forests (Yoganand et al. 2006).

In Sri Lanka, sloth bears were once abundant in forests of the dry zone lowlands (Phillips 1984); however, increasing

Figure 9.4 The sloth bear is characterized by a shaggy, dusty-black coat. (Photo by Ansar Khan.) (A black and white version of this figure will appear in some formats. For the color version, please refer to the plate section.)

agricultural and human settlements have resulted in rapid loss of forest cover and fragmentation of sloth bear habitat (Santiapillai & Santiapillai 1990). Sloth bear range constitutes about 17% of Sri Lanka's total land area (Ratnayeke et al. 2007) and remains primarily in dry zone lowlands with substantial monsoon forest, where human population density and activity are relatively low.

In Nepal, sloth bears exhibit seasonal home range shifts from tall grasslands and riverine forests (alluvial floodplains with mean elevation 150 m) to upland sal (*Shorea robusta*) forests (maximum 700 m) with the onset of the monsoon (Joshi et al. 1995).

Sloth bears use dens both for breeding and for shelter (Akhtar et al. 2007; Baskaran et al. 2015). Depending on habitat, dens can be located in nullah beds, rocky areas both on hills and on flat areas, under roots of trees, hillocks with big boulders, escarpment habitat, and lantana shrub thickets (Laurie & Seidensticker 1977; Akhtar et al. 2007; Yoganand et al. 2005). In the case of bouldery hillocks, it has been observed that dens located in the same hillock were used by sloth bear, leopard, hyena, and jackal without any competition (Akhtar et al. 2007).

Sloth bears are also reported to have disappeared from many forest areas of their former range (Krishnan 1972). The species has been extirpated from Bangladesh and is extremely rare in Bhutan.

Movements and Home Range

While only a few studies have investigated movement patterns for sloth bears, e.g. Panna National Park (Yoganand et al. 2005), Chitwan National Park (Laurie & Seidensticker 1977; Joshi et al. 1995), Mudumalai wildlife sanctuary (Desai et al. 1997), and North Bilaspur Forest Division (NBFD), Chhattisgarh (Akhtar 2004), sloth bear movements appear to vary based on both external (availability of food in various habitat types, availability of den sites and shelters, competition for natural resources) and internal (sex, reproductive status, and age of individuals) factors.

Males generally have larger home ranges than females: in NBFD, India, two males showed a home range of 20 and 22 km^2, whereas the home range of one female was 8 km^2 (Akhtar 2004). In Royal Chitwan National Park, Nepal, Joshi et al. (1995) found that sloth bears used relatively small areas as home ranges: minimum, maximum, and average home ranges were 5, 27, and 14 km^2, respectively, for males, and 4, 23, and 9 km^2, respectively, for females. In Mudumalai Wildlife Sanctuary, India, the home range of two subadult males moving together was reported as 5 km^2, while the home range of a female with two subadult cubs was reported as 19 km^2 (Desai et al. 1997). However, Yoganand et al. (2005) recorded home range sizes in Panna National Park ranging from 12 km^2 for a female to 85 km^2 for male, a much larger area compared to other studies. Although further research is needed to understand factors influencing home range sizes across sloth bear distribution, availability and abundance of food resources (Joshi et al. 1997; Yoganand et al. 2005), and perhaps competition with other individuals, are likely key factors.

Home ranges vary seasonally, with bears generally preferring shrublands, sal forests, and habitats near water, as well as croplands, plantations, and open habitats (Akhtar et al. 2004; Yoganand et al. 2005). In Royal Chitwan National Park, Nepal, sloth bears exhibited seasonal home range shifts from grassland and riverine forest of alluvial floodplains to upland sal forests (Joshi et al. 1995). Such a seasonal shift was likely prompted by the availability of termites and hardness of termite mounds (Joshi et al. 1995, 1997).

Activity Patterns

In human-dominated landscapes, sloth bears are known to forage during night time and return to dens or daytime resting areas in the early morning (Bargali et al. 2005; Samad & Hosetti 2017). Most records of crop raiding seem to occur around night time, likely because sloth bears tend to avoid humans (Laurie & Seidensticker 1977), although most attacks on humans by sloth bears occur during the day (Bargali et al. 2005; Garcia et al. 2016; Dhamorikar et al. 2017; see also Chapter 17). By being active at night or during evenings and mornings, sloth bears could also be reducing their daytime activity to avoid intense heat by remaining under the shade of large boulders and dense undergrowth, especially during the summer (Bargali et al. 2005; Yoganand et al. 2005; Ramesh et al. 2013). Year-round activity patterns of sloth bears can be described according to food availability. In and around the Terai of Nepal, sloth bears mostly use low-lying grasslands, riverine, and sal (*Shorea robusta*) forests during the dry season (November–April) and the adjacent upland sal forests during the remainder of the year (Joshi et al. 1997). Movement between lowland and upland areas has been attributed to the accessibility of termites. Sloth bears prey on termites in underground colonies or in mounds, and mounds are presumably preferred during the monsoon rains because the rain softens the mounds, making access to termites easier (Laurie & Seidensticker 1977; Joshi et al. 1995, 1997; Bargali et al. 2004).

Feeding Ecology

As opportunistic omnivores, sloth bear diet varies seasonally and geographically across their range (Baskaran et al. 1997; Joshi et al. 1997; Sreekumar & Balakrishnan 2002; Bargali et al. 2004; Mewada & Dharaiya 2010; Sukhadiya et al. 2013).

While out foraging, sloth bears walk slowly, frequently sniffing the ground, and when foraging for insects, they can be heard scratching the ground with their forepaws, and loudly sucking and blowing.

Insects, especially ants and termites (*Odontotermes*, *Macrotermes*, *Microcerotermes*, *Hypotermes*, and *Reticulitermes* spp.), and fruit are their main food source, and sloth bears are known to follow the availability of termites and ants within their home range, as well as the phenology of fruiting species such as *Ficus* spp., *Cassia fistula*, *Zizyphus* spp., and many other species (Laurie & Seidensticker 1977; Joshi et al. 1997; Bargali et al. 2004).

In line with their myrmecophagous adaptations, sloth bear diet typically includes a large proportion of insects. However, during the fruiting season (May–August) fruit can be the major component of their diet. Although some fruits (*Zizyphus* spp.) are locally available during winter, termites and ants generally constitute the majority of sloth bear diet during months other than the fruiting season (Laurie & Seidensticker 1977; Joshi et al. 1997; Bargali et al. 2004). Although sloth bears can feed on an extensive list of native species including *Ficus bengalensis*, *F. virens*, *F. religiosa*, *F. racemosa*, *Zizyphus mauritiana*, *Z. oenoplia*, *Z. nummularia*, *Aegle marmelos*, *Briedelia squamosa*, *Diospyros melanoxylon*, *Buchanania lanzan*, *Schleichera oleosa*, *Syzygium cumini*, *Cassia fistula*, and *Madhuca indica*, they also raid orchards and crops to feed on species such as mango (*Mangifera* spp.), guava (*Psidium* spp.), mulberry (*Morus indica*), yams (*Dioscorea* spp.), sweet potato (*Ipomoea* spp.), ground nut (*Arachis hypogaea*), papaya (*Carica papaya*), and maize (*Zea mays*) at the fringe of forested areas (Laurie & Seidensticker 1977; Joshi et al. 1997; Bargali et al. 2004; Samad & Hosetti 2017). In degraded habitats, fruits of *Ficus* spp. can serve as crucial components of sloth bear diet during the dry summer (Laurie & Seidensticker 1977; Bargali et al. 2004).

Reproduction and Growth

Information on the reproductive behavior of sloth bears varies from study to study. In India and Nepal, sloth bears typically breed during summer (May–July) (Laurie & Seidensticker 1977; Joshi et al. 1999), whereas in Sri Lanka, such seasonality does not appear to exist (Norris 1969; Laurie & Seidensticker 1977; Phillips 1984; Gopal 1991). Females attain maturity and begin mating at about 3 years of age and can breed with multiple males (Joshi et al. 1999). During the mating period, noisy interaction and fighting between males has been observed (Laurie & Seidensticker 1977; Joshi et al. 1999). The gestation period generally lasts 6–7 months, and depending on the timing of breeding, cubs are born from September to January (Jacobi 1975; Laurie & Seidensticker 1977; Joshi et al. 1999), after a period of delayed implantation (Puschmann et al. 1977). However, breeding and birthing have also been

Figure 9.5 Mother sloth bear (on the left) with two subadult cubs. (Photo by Ikrar Khan.)

reported during other times of the year (Norris 1969; Laurie & Seidensticker 1977; Phillips 1984; Gopal 1991). Usually, female sloth bears search for a concealed place such as protected dens or ground shelters to give birth (see Denning). A litter of two is common (Laurie & Seidensticker 1977; Phillips 1984; Gopal 1991; Joshi et al. 1999), but litters can range from one to three cubs (Norris 1969; Brander 1982; Heath & Mellen 1983). Females remain in partial or complete seclusion for a period of 3–6 weeks after giving birth, and rarely come out of hiding, even to forage. Females therefore mostly live on fat reserves and metabolic water (Jacobi 1975; Joshi et al. 1999). The cubs are born blind and remain blind for about 3 weeks, and they can leave dens in about 4–5 weeks (Norris 1969). Females generally begin to emerge from dens at nighttime, 1–2 weeks before bringing their cubs out of the den. Mothers carry young cubs on their back for about 6–9 months, a behavior unique to sloth bears among ursids, and nurse for about 12–14 months (Joshi et al. 1999). Mothers likely carry cubs on their back as a way to prevent attacks on cubs by predators (Laurie & Seidensticker 1977; Gopal 1991; Joshi et al. 1999). In addition, females also aggressively defend their cubs by charging other animals. Cubs begin descending from their mother's back at about 3 months of age, but once frightened, they can rapidly climb again onto their mother's back. Cubs generally stay with their mother for 1.5–2.5 years (Figure 9.5; Joshi et al. 1999). After separating from their mother, siblings may stay together for months, and this behavior could be a helpful defense against larger bears or other predators. Little is known about the longevity of sloth bears in the wild; however, in captivity, sloth bears have lived up to 40 years (Ward & Kynaston 1995).

Behavior

Sloth bears are usually solitary, but their non-aggressive behavior toward the presence of other bears suggests that they are not territorial (Laurie & Seidensticker 1977; Joshi et al. 1999; Bargali et al. 2012; Garcia et al. 2016). Passive congregations of sloth bears have been observed near fruiting or flowering trees (Prater 1980; Brander 1982; Baskaran 1990). Still, some

Figure 9.6 A bouldery hillock in central India, which typically provides den sites to sloth bears. (Photo by Harendra Singh Bargali.)

Figure 9.7 A typical sloth bear habitat in central India, showing bouldery hillocks with dens, ponds and human settlements. (Photo by Harendra Singh Bargali.)

evidence suggests that females with cubs and small males stay away from potentially more favorable habitats such as upland areas during the monsoon to avoid males, and in line with these findings, Joshi et al. (1999) observed one female moving upland while solitary, but remaining in the lowlands while she had cubs (Joshi et al. 1995, 1999). Sloth bears, like other bear species, communicate by marking trees (i.e. rubbing, biting, or clawing) throughout the year (Laurie & Seidensticker 1977; Joshi et al. 1999). Although sloth bears can be active throughout any period of the day, they are generally considered to be nocturnal and mostly remain active during the evening, night, and morning hours (Laurie & Seidensticker 1977; Joshi et al. 1995; Yoganand et al. 2006; Bargali et al. 2012; Ramesh et al. 2013). However, females with cubs and subadults have been observed to be more active during daylight hours and less active at night, likely to avoid both adult males and nocturnal predators such as tigers and leopards (Joshi et al. 1999).

Denning

Denning plays an important role in sloth bear ecology, and as a tropical species, sloth bear dens differ significantly from those of temperate bear species. Sloth bears make use of two types of dens: resting and maternal (Eisenberg & Lockhart 1972; Laurie & Seidensticker 1977; Desai et al. 1997; Garshelis et al. 1999b; Akhtar et al. 2007). Resting dens are shelters used during daylight hours when sloth bears are least active, while maternal dens are shelters used by female sloth bears exclusively to give birth and raise cubs.

Resting dens are used by sloth bears irrespective of season or sex and these dens vary greatly across sloth bear range depending on available habitat. The dens are in the form of caves, overhangs, crevices between big boulders, the spaces between roots of trees or fallen trees and bushes, and even loose mud (Eisenberg & Lockhart 1972; Laurie & Seidensticker 1977; Desai et al. 1997; Garshelis et al. 1999b; Akhtar et al. 2007). Resting dens offer protection from environmental factors such as the heat of the day (Ahktar et al. 2007) and potential predators (Figure 9.6). In central India, including areas where caves may not be abundant, resting dens are generally located near forest edges close to villages, presumably because of the availability of crop fields and water (Figure 9.7; Ahktar et al. 2004). Resting dens in the rocky scrub jungle of eastern Karnataka are always made up of naturally occurring caves and overhangs. These dens vary a great deal from one another in terms of their physical structure, namely their entrance size, height, width, depth, and general complexity of the inner cave. Several individual bears have been observed using the same den at different times. Adult males have been observed extensively scent-marking the entrance to resting dens (Swaminathan et al. 2017). Resting dens may be an important place where bears communicate and learn about other bears in the area (Swaminathan et al. 2017). In NBFD, Madhya Pradesh, India, collared male bears used various den sites while females showed strong den fidelity (Akhtar 2004).

Maternal dens are either dug by sloth bears preparing to give birth or are in the form of naturally occurring caves (Joshi et al. 1999; Arun et al. 2017). A few excavated maternal dens have been documented in Royal Chitwan National Park, Nepal (Joshi et al. 1999) and in the Wildlife SOS Bannerghatta Bear Rescue Center (BBRC) in Karnataka, India (Arun et al. 2017). Within the rocky scrub jungle habitat of the Deccan Plateau in eastern Karnataka, only naturally occurring caves have been observed as maternal dens. Differences in the use of natural caves versus excavated dens are likely due to variations in the abundance and complexity of naturally occurring caves and in the availability of adequate soil to excavate dens across sloth bear range. However, when sloth bears make use of natural caves or hollows for maternal dens, they appear to choose them with extreme care and select complex caves with at least two or more chambers (Seshamani & Satyanarayan 1997). In addition to the safety of the cubs, the location of the den relative to resources, namely water and food, appears to be of prime importance.

Human–Sloth Bear Interactions

Considering the nature and severity of injuries inflicted by sloth bears, and considering the number of human fatalities resulting from sloth bear attacks, these bears are one of the most dangerous wild animals across their distribution range (see Chapter 17). Sloth bears are widely distributed in India, southern Nepal, and eastern Sri Lanka, with substantial spatial and temporal variation in the nature and intensity of human–sloth bear conflicts (see Chapter 15). Especially in India, but also in Nepal, degradation and collection of non-timber forest products (NTFPs) by communities living in forest fringe villages results in direct competition for food resources between sloth bears and humans (Rajpurohit & Chauhan 1996; Rajpurohit & Krausman 2000; Bargali et al. 2004; Debata et al. 2017). Sloth bears raid agricultural fields for crops such as maize, ground nut, potatoes, sweet potatoes, Bengal gram (*Cicer arietinum*), and sugarcane (*Saccharum officinarum*) (Palita et al. 2014), and also orchards for mulberry (*Morus* spp.) (Samad & Hosetti 2017), guava, and mango (Laurie & Seidensticker 1977; Bargali et al. 2005; Samad & Hosetti 2017). In addition, food items such as Tendu fruit (*Diospyros melanoxylon*), mahua (*Madhuca indica*) flowers and seeds, and chironji (*Buchanania lanzan*) that are preferred by sloth bears are also collected by local communities in varying quantities (Bargali et al. 2004). Moreover, small-scale collection of mohline leaves (*Bauhinia vahlii*), sal leaves (*Shorea robusta*), bamboo (*Dendrocalamus strictus*), chironji (*Buchanania lanzan*), wild mushrooms, honey, and other forest products from areas inhabited by sloth bears not only increases the chances of encounters between humans and sloth bears but also results in shortage of food resources for sloth bears, therefore significantly contributing to human–sloth bear conflicts across their range (Bargali et al. 2005; Dhamorikar et al. 2017). In addition, stone quarrying poses a potential threat to sloth bears because of the resulting habitat loss in already degraded sloth bear habitats (Mardaraj 2015; Debata et al. 2017). Human–sloth bear conflicts have had direct negative consequences for sloth bear populations and incidents of retaliatory killings have been recorded across most of their range (Bargali et al. 2005; Mardaraj, 2015; Debata et al. 2017).

In Sri Lanka, owing to the availability of a large extent of contiguous forest, sloth bears rarely enter human settlements or raid crops (Ratnayeke et al. 2014); however, incidents of human casualties by sloth bears occur in unprotected forest areas (Ratnayeke et al. 2006). In Nepal, sloth bear populations exist mainly within protected areas of the Terai, and occasions of sloth bears raiding crops or attacking humans outside of natural habitats are considered rare (Joshi et al. 1995). Sloth bears in Nepal have been responsible for some crop damage (Laurie & Seidensticker 1977), although more recent studies have not observed this to be an issue (Joshi et al. 1995).

Parasites and Diseases

Most diseases have been documented in captivity and very few have been reported in free-ranging sloth bears. Both infectious and non-infectious diseases have been reported (Table 9.1).

Sloth bears are more susceptible to worm infections because of their digging and feeding behavior that predispose

Table 9.1 List of some of the most common diseases of sloth bears.

Infectious diseases			Non-infectious diseases
Bacterial	Viral	Fungal	
Tuberculosis	Rabies	Candidiasis	Dental infections involving canine tooth
Leptospirosis	Infectious canine hepatitis		Periapical abscess
Anthrax (rare)	Canine distemper		Auricular hematoma
Colibacillosis	Canine parvo virus		Gigantiform cementoma
Salmonellosis			Scrotal hernia
			Cataract
			Fractures
			Adenocarcinoma
			Cholangiocarcinoma
			Lymphoid leukemia
			Seminoma
			Cholelithiasis
			Gastroesophageal reflux disease (GRED)
			Idiopathic epilepsy
			Hemangioma

Figure 9.8 Scat of a sloth bear infected by roundworms. (Photo by Kajal K. Jadav.)

Figure 9.9 Sloth bear sighting during a wildlife watching safari in a Tiger Reserve in India. (Photo by Ikrar Khan.)

them to exposure to saprophytic pathogens more than any other species of bears (Schaul 2006). However, the bear roundworm *Baylisascaris* is an ascaroid that has been reported in all species of bears (Schaul 2006). *Baylisascaris* infection is common in bears in captivity, but resulting mortality is rare (Figure 9.8; Klos & Lang 1982). Ectoparasites are usually present in areas such as the nasal bridge, face, ears, and belly region where the coat is less dense.

Hepatozoon ursi has been detected in captive sloth bears by Pawar et al. (2018). Sloth bears are infected through ingestion of hematophagus arthopods such as ticks which are definitive hosts and contains sporulated oocytes (Baneth et al. 1998). Such infections remain subclinical, but concomitant infection of bacterial or viral origin potentially weakens the animal's immune system and in such immune-compromised sloth bears *Hepatozoon* species may be a potential pathogen.

Status in the Wild

There are only a few reliable estimates of sloth bear density and population numbers across their range (Garshelis et al. 1999a, 1999b; Ratneyeke et al. 2007), and most studies are based on questionnaire surveys. From questionnaires, Yoganand et al. (1999) estimated a population of 9000–13,000 sloth bears in India. In Sri Lanka, Santiapillai and Santiapillai (1990) estimated 300–600 sloth bears. However, based on home ranges of radio-collared bears, Ratneyeke et al. (2007) estimated more than 100–150 sloth bears in Wasgomuwa National Park alone (about 5% of sloth bear range in Sri Lanka), which indicates that sloth bear population likely exceeds the estimated number reported by Santiapillai and Santiapillai (1990). In Nepal, Garshelis et al. (1999b) used mark–recapture models to estimate density and reported a population of 200–250 sloth bears in Royal Chitwan National Park. Extrapolating these findings to suitable habitats within the country, Garshelis et al. (1999a) estimated the population to be <1000, and possibly <500 at that time. Using available literature, Garshelis et al. (1999a) also estimated the total world population to be 10,000–25,000, and current estimates lie between <10,000 and 20,000 depending on methods and data employed (Dharaiya et al. 2016). However, reliable estimates across the entire sloth bear range are still lacking, and the historical and current worldwide population estimates are far from actual reliability. Although Garshelis et al. (1999a) acknowledged the lack of reliable population estimates for this species two decades ago, little improvement has been made on this issue within the last 20 years (Dharaiya et al. 2016).

Threats to Sloth Bears

All sloth bear range countries are developing countries with increasing human populations. Human demands and development associated with developing countries therefore exert strong pressure on available natural resources across sloth bear range. The biggest threat to sloth bear conservation is the removal of habitat and the degradation of available habitat (Cowan 1972; Servheen 1990; Garshelis et al. 1999a; Kemf et al. 1999). In addition, poaching for trade in body parts, especially for gall bladders, is a key threat (Servheen 1990; Garshelis et al. 1999a). In Bangladesh, sloth bears were extirpated as a result of habitat loss and over-hunting (Islam et al. 2013). In Bhutan, the species is either extremely rare, or only vagrants remain (Garshelis et al. 2015). An important aspect of sloth bear conservation is the threat to sloth bear populations outside of protected areas. In India, even though there are more than 700 protected areas, most of these protected areas target conservation of other flagship species such as tigers while claiming to support sloth bear conservation as well. Sloth bears commonly use areas outside of protected areas, and most of these areas are interspersed with human settlements, agricultural fields, and a range of human activities. Almost all such human-dominated landscapes are witnessing rapid changes due to human-induced activities (Mardaraj 2015; Debata et al. 2017). Such landscapes are crucial not only for the survival of sloth bears but also for other important wildlife, including flagship species such as tigers and elephants (Figure 9.9; Dhamorikar et al. 2017). Degradation or loss of such critical areas not only disrupts connectivity between populations, but also isolates populations in the remaining forest patches. Individuals using isolated patches can often come into direct conflict with the surrounding human population that share common food resources (Rajpurohit & Chauhan 1996; Rajpurohit & Krausman 2000; Bargali et al. 2004; Debata et al. 2017). Collection of NTFPs from sloth bear habitat, both at the commercial and the community level, not only results in shortage of food resources but also further degrades habitats, increasing chances of human–sloth bear conflict (Bargali et al. 2005; Dhamorikar et al. 2017).

Historically, one of the main threats to sloth bear populations was the capture of cubs to use as dancing bears that provided revenue from shows in streets and venues of villages, especially for tourists. The history of keeping sloth bears in captivity and using them as dancing bears dates back to the thirteenth century and the pre-Mughal era. The Kalandars, who practiced the tradition of capturing sloth bears, were once common in the many towns in India and used to travel from house to house to advertise their bears' performances. Officers in British India often kept sloth bears as pets. Despite a ban on the practice that was enacted under Wildlife (Protection) Act, 1972, as many as 800 dancing bears were performing in the streets of India during the latter part of the twentieth century, particularly on the highway between Delhi, Agra, and Jaipur. Sloth bear cubs used to be poached from forests and trained to dance and follow commands through coercive stimuli and starvation. The Government of India also imposed a ban on dancing bears but success in eliminating this practice came after much effort from government agencies, and national and international organizations resulting in confiscation of almost all dancing bears in 2009. Morever, with the help of the Indian government and movements launched by non-government conservation organizations such practices have been largely controlled across their range (D'Cruze et al. 2011).

Status in Captivity

In India, sloth bears are charismatic and interesting species for zoo visitors. In particular, zoos at Bhopal, Bilaspur, Raipur, and Lucknow have been famous for keeping white (albino) sloth bears that look like polar bears from a distance. Currently, zoos at Lucknow and Raipur have albino sloth bears in their collection. Most of the bears in Indian zoos are of wild origin, have a high degree of heterogeneity, and can be an asset if being used for potential planned conservation breeding. In India, the Central Zoo Authority (CZA) records indicate that sloth bears used to be kept in more than 70 zoos but are now restricted to approximately 48 zoos located in 18 states. Sloth bears in these zoos consist of 449 individuals including 217 males, 221 females, and 11 unknown individuals (Anonymous 2018a).

Internationally, the first sloth bear reported in North America was at the Philadelphia zoo, in 1889. Sloth bears have been exhibited in zoos since the beginning of the 1900s, and according to Vineyard (2013), the *ex situ* sloth bear population consisted of 47 individuals (20 male, 21 female, and six unknown) at 21 zoos associated with the American Zoos Association (AZA). Of note, in 1979, the San Diego zoo (California, USA) became the first North American zoo with resident sloth bears from Sri Lanka (subspecies *inornatus*). All other sloth bears in North America originated from India. The lack of ideal housing and genetic variability are major factors behind the very low reproduction rate in captivity; thus, *ex situ* population viability becomes a major problem (Marti et al. 2012). In addition, biliary adenocarcinoma was the leading cause of death for adult sloth bears housed in various zoological institutions in the United States (Anderson et al. 2018). As per the most recent data available in 2018, the World Association of Zoos and Aquariums (WAZA) studbook (current to February 2016), there were 1168 individuals in captivity, including 490 individuals living at 50 institutions (Anonymous 2018b).

References

Abella, J., Alba, D. M., Robles, J. M., et al. (2012). *Kretzoiarctos gen. nov.*, the oldest member of the giant panda clade. *PLoS ONE* 7(11): e48985. doi:10.1371/journal.pone.0048985

Akhtar, N. (2004). Habitat use and ranging pattern of sloth bear (*Melursus ursinus*) in North Bilaspur Forest Division, Madhya Pradesh. Doctorate Degree at Saurastra University Rajkot, Gujarat.

Akhtar, N., Bargali, H. S. & Chauhan, N. P. S. (2004). Sloth bear habitat use in disturbed and unprotected areas of Madhya Pradesh, India. *Ursus* 15: 203–211.

Akhtar, N., Bargali, H. S. & Chauhan, N. P. S. (2007). Characteristics of sloth bear day dens and use in disturbed and unprotected habitat of North Bilaspur Forest Division, Chhattisgarh, Central India. *Ursus* 18: 203–208.

Anderson, K., Garner, M. M. & Dennis, P. M. (2018). Causes of mortality in sloth bears (*Melursus ursinus*) housed in U.S. zoos. *The Journal of Zoo and Aquarium Research* 6(1).

Anonymous. (2018a). Annual Inventory reports of zoo animals. http://cza.nic.in/page/en/inventory-of-animals-in-zoos

Anonymous. (2018b). International studbooks for rare species of wild animals in captivity. *International Zoo Yearbook* 52: 448–478. DOI:10.1111/izy.12194.

Arun, A. S., Sharp, T., Pillai, H., et al. (2017). Sloth bear (*Melursus ursinus*) maternity denning at the Wildlife SOS Bannerghatta Bear Rescue Centre, India. *International Zoo Yearbook* 52: 1–10. DOI:10.1111/izy.12179.

Baneth, G., Aroch, I., Tai, N., et al. (1998). *Hepatozoon* species infection in domestic cats: a retrospective study. *Veterinary Parasitology* 79: 123–133.

Bargali, H. S. (2012). Distribution of different species of bears and status of human–bear conflict in the State of Uttarakhand, India. *Advances in Biological Research* 6(3): 121–127.

Bargali, H. S., Akhtar, N. & Chauhan, N. P. S. (2004). Feeding ecology of sloth bears (*Melursus ursinus*) in a disturbed area in central India. *Ursus* 15(2): 212–217.

Bargali, H. S., Akhtar, N. & Chauhan, N. P. S. (2005). Characteristics of sloth bear attacks and human casualties in North Bilaspur Forest Division, Chhattisgarh, India. *Ursus* 16(2): 263–267.

Baskaran, N. (1990). An ecological investigation on the dietary composition and habitat utilization of sloth bear (*Melursus ursinus*) at Mudumalai wildlife sanctuary, Tamil Nadu (South India). MPhil thesis, A.V.C. College, Mannambandal, Tamil Nadu, India.

Baskaran, N., Sivanagesan, N. & Krishnamoorthy, J. (1997). Food habits of the sloth bear at Mudumalai Wildlife Sanctuary, Tamil Nadu, South India. *Journal of the Bombay Natural History Society* 94: 1–9.

Baskaran, N., Sivanagesan, N., Krishnamoorthy, J., et al. (2015). On the behavioural ecology of sloth bear (*Melursus ursinus* Shaw 1791) in Mudumalai Wildlife Sanctuary, Western Ghats, India. In Gupta, V. K. & Verma, A. K. (Eds.), *Animal diversity, natural history and conservation* (Vol. 5, pp. 313–333). New Delhi: Daya Publishing House.

Brander, A. A. D. (1982). *Wild animals in central India.* Dehradun: Natraj Publishers.

Choudhury, A. U. (2011). *Records of sloth bear and Malayan sun bear in North East India.* Final report to International Association for Bear Research & Management (IBA). Guwahati, Assam, India: The Rhino Foundation for Nature in NE India.

Corbet, G. B. & Hill, J. E. (1992). *The mammals of the Indomalayan region: A systematic review.* London: Natural History Museum Publications, and Oxford: Oxford University Press.

Cowan, I. McT. (1972). The status and conservation of bears (Ursidae) of the world –1970. *International Conference on Bear Research and Management* 2: 343–367.

D'Cruze, N., Sarma, U. K., Mookerjee, A., et al. (2011). Dancing bears in India: a sloth bear status report. *Ursus* 22(2): 99–105.

Debata, S., Swain, K. K., Sahu, H. K., et al. (2017). Human–sloth bear conflict in a human-dominated landscape of northern Odisha, India. *Ursus* 27(2):90–98.

Desai, A. A., Bhaskaran, N. & Ventaktesh, S. (1997). Behavioural ecology of the sloth bear in Mudumalai Wildlife Sanctuary and National Park. Tamil Nadu and Bombay Natural History Society collaborative project. Mumbai, India.

Dhamorikar, A. H., Mehta, P., Bargali, H., et al. (2017). Characteristics of human–sloth bear (*Melursus ursinus*) encounters and the resulting human casualties in the Kanha–Pench corridor, Madhya Pradesh, India. *PLoS ONE* 12(4): e0176612. https://doi.org/10.1371/journal.pone.0176612

Dharaiya, N., Bargali, H. S. & Sharp, T. (2016). *Melursus ursinus.* The IUCN Red List of Threatened Species 2016. e. T13143A45033815.http://dx.doi.org/10.2305/IUCN.UK.2016-3.RLTS.T13143A45033815.en

Eisenberg, J. F. & Lockhart, M.C. (1972). An ecological reconnaissance of Wilpattu National Park, Ceylon. *Smithsonian Contributions to Zoology* 101: 1–119.

Erdbrink, D. P. (1953). *A review of fossil and recent bears of the Old World.* Deventer: Drukkerij Jan de Lange.

Garcia, K. C., Joshi, H. M. & Dharaiya, N. (2016). Assessment of human–sloth bear conflicts in North Gujarat, India. *Ursus* 27(1): 5–10.

Garshelis, D. L., Joshi, A. R., Smith, J. L. D., et al. (1999a). Sloth bear conservation action plan. In: Servheen, C., Herrero S. & Peyton, B. (Eds.), *Bears: Status survey and conservation action plan* (pp. 225–240). Gland, Switzerland and Cambridge, UK: IUCN/SSC Bear and Polar Bear Specialist Groups.

Garshelis, D., Joshi, A. & Smith, D. (1999b). Estimating density and relative abundance of sloth bears. *Ursus* 11: 87–98.

Garshelis, D. L., Dhariaya, N. A., Sharp, T. R., et al. (2015). Sloth bears at the northern edge of their range: Status of the transboundary population linking northeastern India to Bhutan. Final report to International Association for Bear Research and Management.

Gopal, R. (1991). Ethological observations on the sloth bear (*Melursus ursinus*). *Indian Forester* 117: 915–920.

Heath, D. C. & Mellen, J. D. (1983). Development of maternally reared sloth bear cubs in captivity. Unpublished Sixth International Conference on Bear Research and Management.

Islam, M. A., Uddin, M., Aziz, M. A., et al. (2013). Status of bears in Bangladesh: Going, going, gone? *Ursus* 24: 83–90.

Jacobi, E. F. (1975). Breeding sloth bears in Amsterdam zoo. In: Martin, R. D. (Ed.), *Breeding endangered species in captivity* (pp. 351–356). London: Academic Press.

Jhala, Y. V., Gopal, R. & Qureshi, Q. (Eds.). (2008). *Status of tigers, co-predators and prey in India.* Dehradun: National Tiger Conservation Authority, New Delhi, and Wildlife Institute of India.

Johnsingh, A. J. T. (2003). Bear conservation in India. *The Journal of the Bombay Natural History Society* 100: 190–201.

Joshi, A. R., Garshelis, D. L. & Smith J. L. D. (1995). Home ranges of sloth bears in Nepal: implications for conservation. *Journal of Wildlife Management* 59: 204–213.

Joshi, A. R., Garshelis, D. L. & Smith J. L. D. (1997). Seasonal and habitat-related diets of sloth bears in Nepal. *Journal of Mammalogy* 78: 584–597.

Joshi, A. R., Smith, J. L. D. & Garshelis, D. L. (1999). Sociobiology of the myrmecophagous sloth bear in Nepal. *Canadian Journal of Zoology* 77: 1690–1704.

Kemf, E., Wilson, A. & Servheen, C. (1999). *Bears in the Wild.* 1999 WWF Species Status Report. Gland, Switzerland: World Wide Fund For Nature.

Kitchener, A. C. (2010). Taxonomic issues in bears: impacts on conservation in zoos and the wild, and gaps in current knowledge. *International Zoo Yearbook*, 44: 33–46.

Klos, H-G. & Lang, E. M. (1982). *Handbook of Zoo Medicine.* New York, NY: Van Nostrand Reinhold Co.

Krause, J., Unger, T., Nocon, A., et al. (2008). Mitochondrial genomes reveal an explosive radiation of extinct and extant bears near the Miocene–Pliocene boundary. *BMC Evolutionary Biology* 8: 220. DOI:10.1186/1471-2148-8-220.

Krishnan, M. (1972). An ecological survey of the large mammals of peninsular India. *The Journal of the Bombay Natural History Society* 69: 47–49.

Kumar, V., Lammers, F., Bidon, T., et al. (2017). The evolutionary history of bears is characterized by gene flow across species. *Scientific Reports* 7: 46487. DOI:10.1038/srep46487.

Kutschera, V. E., Bidon, T., Hailer, F., et al. (2014). Bears in a forest of gene trees: phylogenetic inference is complicated by incomplete lineage sorting and gene flow. *Molecular Biology and Evolution*, 31(8): 2004–2017. DOI:10.1093/molbev/msu186

Laurie, A. & Seidensticker, J. (1977). Behavioural ecology of the sloth bear

(*Melursus ursinus*). *Journal of Zoology, London* 182: 187–204.

Mardaraj, P. C. (2015). Identifying key issues for conservation of sloth bear (*Melursus ursinus*) in Rajnilgiri, Odisha, Eastern India. Report submitted to The Rufford Small Grant Foundation, United Kingdom.

Marti, K., Theis, M., Thompson, D., et al. (2012). *Sloth bear (*Melursus ursinus*) AZA Animal Program Population Viability Analysis Report.* Chicago, IL: Lincoln Park Zoo.

McNab, B. K. (1992). Rate of metabolism in the termite eating sloth bear (*Ursus ursinus*). *Journal of Mammalogy* 73: 168–172.

Mewada, T. & Dharaiya, N. (2010). Seasonal dietary composition of sloth bear in the reserve forest of Vijaynagar forest, North Gujarat, India. *Tigerpaper* 37(2): 8–13.

Norris, T. (1969). Ceylon sloth bear. *International Wildlife* 12: 300–303.

Pages, M., Calvignac, S., Klein, C., et al. (2008). Combined analysis of fourteen nuclear genes refines the Ursidae phylogeny. *Molecular Phylogenetics and Evolution* 47(1): 73–83. DOI:10.1016/j.ympev.2007.10.019.

Palita, S. K., Kar, T. & Debata, S. (2014). Human–sloth bear interactions: preliminary survey from Semiliguda range of Koraput Forest Division, Southern Odisha, India. *Pranikee-Journal of Zoological Society of Orissa*, XXVI: 13–21.

Pawar, R. M., Poornachandar, A. & Arun, A.S. (2018). Molecular prevalence and characterization of infection in Indian sloth bears (*Melursus ursinus*). *Wildlife SOS Compendium of Publications* 1: 195.

Phillips, W. W. A. (1984). *Manual of the mammals of Sri Lanka.* Part III, 2nd edition. Bataramulla: Wildlife and Nature Protection Society of Sri Lanka.

Pocock, R. I. (1941a). *The Fauna of British India, including Ceylon and Burma. Mammalia. Vol II: Carnivora (suborders Aeluroidae (part) and Arctoidae).* London: Taylor and Francis.

Pocock, R. I. (1941b). The black and brown bears of Europe and Asia. Part II. The sloth bear (*Melursus*) and the Asian black bear (*Selenarctos*) and the Malayan sun bear (*Helarctos*). *Journal of Bombay Natural History Society* 36: 101–138.

Prater, S. H. (1980). *The book of Indian animals.* Bombay: Bombay Natural History Society.

Puschmann, V. W., Schuppel, K. F. & Kronberger, H. (1977). Detection of blastocyst in uterine lumen of Indian bear (*Melursus ursinus*). In: Ippen, R. & Schrader, H. D. (Eds.), *Sickness in zoos* (pp. 389–391). Berlin: Akademie Verlag [in German with English abstract].

Rajpurohit, K. S. & Chauhan, N. P. S. (1996). *Study of animal damage problems in and around protected areas and managed forest in India. Phase I: Madhya Pradesh, Bihar and Orissa.* Dehradun: Wildlife Institute of India.

Rajpurohit, K. S. & Krausman, P. R. (2000). Human–sloth-bear conflicts in Madhya Pradesh, India. *Wildlife Society Bulletin* 28: 393–399.

Ramesh, T., Kalle, R., Sankar, K., et al. (2013). Activity pattern of sloth bear *Melursusursinus* (Mammalia: Ursidae) in Mudumalai Tiger Reserve, Western Ghats, India. *Journal of Threatened Taxa* 5(5): 3989–3992. DOI:10.11609/JoTT.o3071.3989-92.

Ratnayeke, S., Wijeyamohan, S. & Santiapillai, C. (2006). The status of sloth bears in Sri Lanka. In: Oi, T., Mano, T., Yamazaki, K., et al. (Eds.), *Understanding Asian bears to secure their future* (pp. 35–40). Gifu: Japan Bear Network.

Ratnayeke, S., van Manen, F. T. & Padmalal, U. K. G. K. (2007). Landscape characteristics of sloth bear range in Sri Lanka. *Ursus* 18: 189–202.

Ratnayeke, S., Van Manen, F. T., Pieris, R., et al. (2014). Challenges of large carnivore conservation: sloth bear attacks in Sri Lanka. *Human Ecology* 42: 467–479.

Samad, K. S. A. & Hosetti, B. B. (2017). Sloth bear *Melursus ursinus*–human conflict: a case study of unprotected bear habitat in Kudligi taluk, Ballari district, Karnataka. *International Journal of Zoology Studies* 2 (6): 255–260.

Santiapillai, A. & Santiapillai, C. (1990). Status, distribution and conservation of the sloth bear (*Melursus ursinus*) in Sri Lanka. *Tiger Paper* 17(1): 13–15.

Sathyakumar, S., Kaul, R., Ashraf, N. V. K., et al. (2012). *National Bear Conservation and Welfare Action Plan.* Ministry of Environment and Forests, Wildlife Institute of India, and Wildlife Trust of India.

Schaul, J. C. (2006). *Baylisascaris transfuga* in captive and free-ranging populations of bears (family: Ursidae). Dissertation, Ohio State University.

Servheen, C. (1990). The status and conservation of the bears of the world. *Proceedings of the International Conference on Bear Research and Management, Monograph Series* 2: 1–32.

Seshamani, G. & Satyanarayan, K. (1997). *The dancing bears of India.* London, UK: The World Society for the Protection of Animals.

Sreekumar, P. G. & Balakrishnan, M. (2002). Seed dispersal by the sloth bear (*Melursus ursinus*) in South India. *Biotropica* 34: 474–477.

Sukhadiya, D., Joshi, J. V. & Dharaiya, N. (2013). Feeding ecology and habitat use of sloth bear in Jassore wildlife sanctuary, Gujarat, India. *Indian Journal of Ecology* 40(1): 14–18.

Swaminathan, S., Arun, A. S., Sharp, T., et al. (2017). Sloth bear pede-marking caught on video. *International Bear News* 26(2): 27–28.

Vineyard, T. (2013). North American regional studbook, sloth bear (*Melursus ursinus*). Cleveland Metroparks Zoo, Cleveland, OH, 1–20. https://issuu.com/vineyardt/docs/sloth_bear_studbook_2013

Ward, P. & Kynaston, S. (1995). *Bears of the world.* London, UK: Blandford.

Yoganand, K., Johnsingh, A. J. T. & Rice, C. G. (1999). *Annual technical report (October 1998 to September 1999) of the project 'Evaluating Panna National Park with special reference to the ecology of sloth bear'.* Dehradun: Wildlife Institute of India.

Yoganand, K., Rice, C. G. & Johnsingh, A. J. T. (2005). *Evaluating Panna National Park with special reference to the ecology of sloth bear.* Final Project Report. Dehradun: Wildlife Institute of India.

Yoganand, K., Rice, C. G. & Johnsingh, A. J. T. (2006). Is the sloth bear in India secure? A preliminary report on distribution, threats and conservation requirements. *Journal of the Bombay Natural History Society* 103: 2–3.

Chapter 10

Asiatic Black Bear (*Ursus thibetanus*)

Chinatsu Kozakai, Ivan Seryodkin, Karine E. Pigeon, Koji Yamazaki, Sangay Wangchuk, Shinsuke Koike, Toshio Tsubota, and Yonten Jamtsho

Names

Genus: *Ursus* Linnaeus, 1758

Species: Asiatic black bear *Ursus thibetanus* G. [Baron] Cuvier, 1823

Subspecies: *formosanus* Swinhoe, 1864; *gedrosianus* Blanford, 1877; *japonicus* Schlegel, 1857; *laniger* (Pocock, 1932); *mupinensis* (Heude, 1901); *ussuricus* (Heude, 1901)

Other common names: Asian black bear, Himalayan black bear, moon bear

Names in other languages: Urudu: Kaala Reech or Bhaloo; **Khowar**: Ukhch; **Punjabi**: Reech; **Balochi**: Mum; **Bhutan**: Dhom; **Vietnam**: Gau ngua; **Chinese**: 黑熊; **Russia**: гималайский медведь or белогрудый медведь; **Korean**: 반달가슴곰; **Japan**: ツキノワグマ (tsukinowaguma); **French**: Ours à collier; **German**: Kragenbär; **Spanish**: Oso negro asiatico; **Italian**: Orso nero asiatico; Orso tibetano

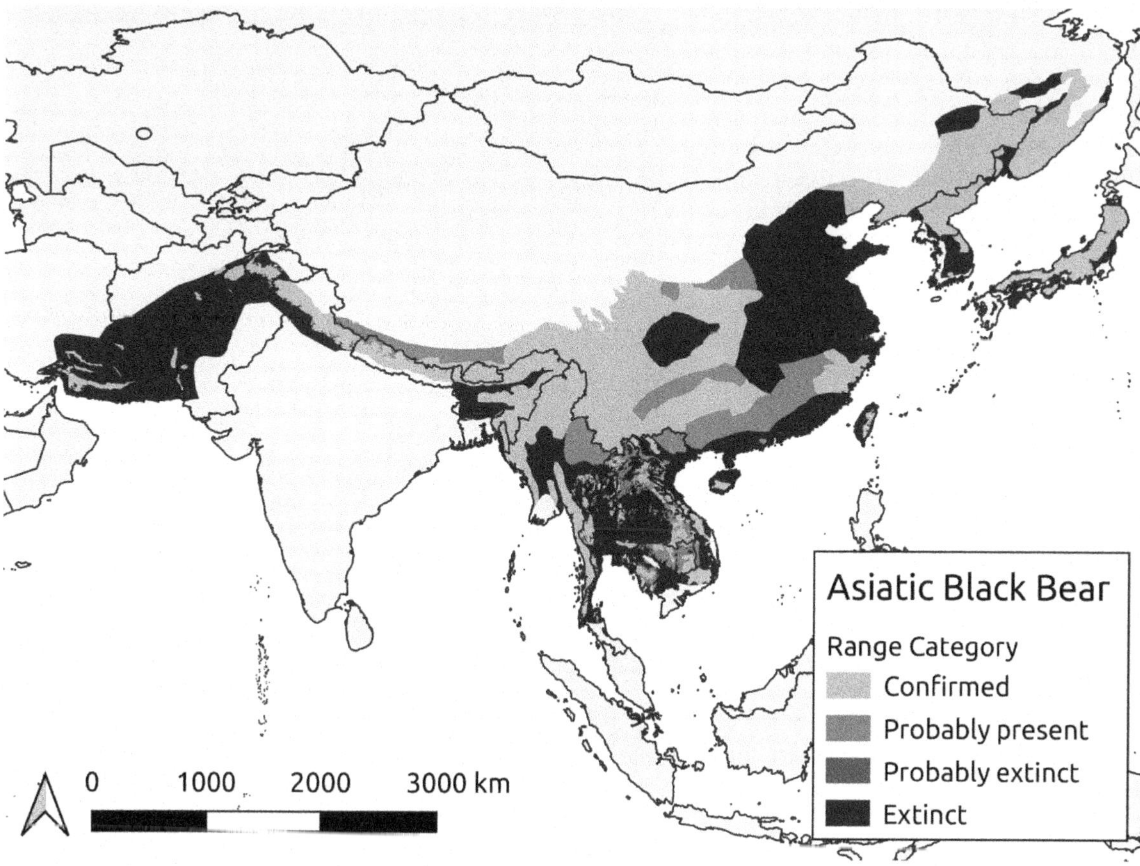

Figure 10.1 Geographic range of Asiatic black bear. (*Source*: IUCN 2016. Red List of Threatened Species.)

Taxonomy

The Asiatic black bear is a mammal from the order Carnivora and the family Ursidae (see also Chapter 1). The type locality is from the Sylhet region, presently divided between Assam in India and Bangladesh. In the past, this species was included in the genus *Selenarctos* (meaning moon bear).

The tooth morphology of the Asiatic black bear is most closely related to *U. minimus*, which appeared about 5–6 million years ago and is the first identifiable representative of the ursine bears (Garshelis 2009). Evidence from the fossil records, morphology, and mitochondrial phylogeny suggests a closer relationship between the Asiatic black bear and the American black bear *U. americanus* (Krause et al. 2008; Wu et al. 2015). On the other hand, autosomal and Y-chromosomal sequences support a grouping with the American black bear, being sister group to the brown bear *U. arctos* and polar bear *U. maritimus* clade (Bidon et al. 2014; Kutschera et al. 2014). A natural Asiatic black bear–sun bear *H. malayanus* hybrid has been reported in Cambodia (Galbreath et al. 2008), and a captive hybrid between the Andean bear *Tremarctos ornatus* and the Asiatic black bear also has been reported (Mondolfi & Boede 1981). A recent genome-scale study suggests a complex process of bear species evolution, including the fact that gene flow is not necessarily limited to closely related species pairs. In fact, there is strong ancestral gene flow between the Asiatic black bear and the ancestors of the brown bear, polar bear, and American black bear (Kumar et al. 2017).

Subspecies and Distribution

The fossil record suggests that Asiatic black bears were more broadly distributed and occurred as far west as Europe (Baryshnikov 2010; Baryshnikov & Zakharov 2013). Their current distribution edge is now south-eastern Iran, and they now only occur in Asian countries (listed from west to east): Iran, Afghanistan, Pakistan, India, Nepal, Bhutan, Bangladesh, Myanmar, China, Thailand, Laos, Cambodia, Vietnam, North and South Korea, Taiwan, Russia, and Japan (Figure 10.1).

At least seven extant subspecies have been recognized (Wozencraft 2005; see also Chapter 1). Although the distribution and genetic characteristics of most subspecies are still poorly understood, subspecies populations in the Balochistan region of southern Pakistan and Iran (*U. t. gedrosianus*), Taiwan (*U. t. formosanus*), and Japan (*U. t. japonicus*) appear to be valid (Garshelis 2009). Our current understanding of extant distributions is as follows (Garshelis & Steinmetz 2016; Kadariya et al. 2018): (1) at the western range limit, *U. t. gedrosianus* is one of the most threatened populations and inhabits a long-isolated arid thorn forest (the Baluchistan/Balochistan region of Iran and southern Pakistan); (2) *U. t. laniger* also occurs in small populations in the western Himalayas (northern Pakistan, north-west and north-east India, and Nepal, and may also occur in Afghanistan); (3) *U. t. thibetanus* has a broad distribution across the eastern Himalayas and south-east Asia (Nepal, north-east India, Bhutan, Bangladesh, Myanmar, south-western China, Laos, Thailand, Cambodia, and Vietnam). However, a recent study (Kadariya et al. 2018) suggests that the distribution of *U. t. laniger* and *U. t. thibetanus* may overlap in the eastern Himalayas (Nepal, India, Bhutan, and Myanmar). At this point, the distribution of the Asiatic black bear is wider and includes low-elevation sites in the hill states of north-eastern India, and continues across mainland South-East Asia, stretching south into Myanmar and Thailand to ~200 km north of the Malaysian border. Population isolation and fragmentation is extensive in East and South-East Asia, and in China. There are no records of the Asiatic black bear in Malaysia; (4) *U. t. mupinensis* inhabits central and south-western China (possibly including Hainan island), and is the most widespread subspecies in China; (5) *U. t. ussuricus* occurs in north-eastern China (mainly in Heilongjiang Province), south-east Russia, and North Korea. A small isolated population also exists in the Jiri Mountains, in southern South Korea. This population was reintroduced from individuals of the Russian Far East and North Korea since 2004 by the Korean Government (Kim et al. 2011). It is believed that there were about 40 individual bears of this subspecies in North Korea in 2015; (6) Formosan black bears (*U. t. formosanus*) occur only in Taiwan; and, finally, (7) the Japanese black bear (*U. t. japonicus*) occurs on the Honshu and Shikoku islands of Japan.

In the Japanese populations, genetic characteristics vary within a relatively narrow geographical range (Ohnishi et al. 2009). For example, in the Tohoku region (see also Chapter 25) there are substantial differences in skull morphology between two populations in the Iwate Prefecture (Amano et al. 2004). This large variation in genetic characteristics suggests that gene flow has been limited between these distributional ranges despite the fact that the two regions are only 5 km apart at the narrowest point of the river basin. As previously mentioned, we still have a poor understanding of the phylogeography of Asiatic black bears in most regions of Asia.

The Asiatic black bear is replaced by other bear species in forested areas at the edge of their range, but there are regions where Asiatic black bears are sympatric with brown bears, sun bears, and giant panda (*Ailuropoda melanoleuca*) (Latham et al. 2012; Garshelis & Steinmetz 2016; Yadav et al. 2017). Notably, Asiatic black bears overlap extensively with sun bears in South-East Asia (see Chapter 4), with brown bears in the northern parts of Asiatic black bear range (i.e. northern Pakistan, India, Nepal, south-eastern Russia, and north-western and eastern China), and with giant panda (China). Sloth bears and Asiatic black bears, however, are mostly allopatric. Recent evidence shows that these species overlap at least partly at the edge of their respective ranges in the lowlands of the Terai Arc of Nepal and India, but while these lowlands are extensively used by sloth bears, their use by Asiatic black bears is limited (Pigeon 2018). It is possible that the Asiatic black bear, sloth bear, and sun bear overlap in small pockets of north-east India, but these species are easily confused and historical or current overlap of these three species has yet to be confirmed (see Chapter 4).

Figure 10.2 Asiatic black bear by Francesco Rinaldi, with permission. (A black and white version of this figure will appear in some formats. For the color version, please refer to the plate section.)

Descriptive Notes

Body measurements: **head and body length**: 110–190 cm; **tail length**: less than 12 cm; **body mass**: 60–200 kg (rarely 250 kg) for adult males, and 35–140 kg (rarely 170 kg) for adult females. The maximum weights occur after accumulating fat in autumn, prior to winter hibernation; **dental formula**: I 3/3, C 1/1, P 4/4, M 2/3 (some of P1–3 often missing); **chromosome**: $2n = 74$.

The principal coat color phase is black. Rare dark/reddish brown phases are also known in *U. t. gedrosianus* of Pakistan (Sheikh 2006), and a rare blond color phase was discovered in Cambodia, Thailand, and Laos (Galbreath et al. 2001). On the chest, there is generally, but not always, a white or cream/yellow mark, often in the shape of a V or crescent moon (Figure 10.2). The common name "moon bear" was derived from this chest mark. The chest marking is sometimes small or absent, but it is possible to identify unique individuals from interindividual variations in these chest markings (Higashide et al. 2012, 2013). Some individuals also have a white chin patch. Although muzzle color varies from light to dark, it is generally dark brown, and the color does not extend past the eyes. All subspecies are characterized by large, rounded ears that are easily distinguishable from the mane. Asiatic black bears also have short, dark claws that allow them to climb trees with ease. Most subspecies also have a large mane around the neck, although it is not distinctly recognized in some subspecies (e.g. *U. t. japonicus*). Overall hair length and thickness also vary regionally but the hair is relatively short (except for the mane) and well kept. For example, *U. t. laniger* has longer and thicker fur with abundant under-wool, and chest marks that are usually smaller and whiter than in *U. t. thibetanus*.

Habitat

The distribution of Asiatic black bears roughly coincides with the forest distribution throughout Asia. These bears dominate temperate broadleaved deciduous forests (Figure 10.3A), and are potentially effective seed dispersers among mammals (Koike et al. 2011; Tsuji & Su 2018). Indeed, because this species can range from mountain tops to lowland following plant phenology and disperse seeds toward the mountain tops, these bears potentially are helping the plant species to escape from global warming (Naoe et al. 2016). Asiatic black bears also occur in temperate conifer forests, tropical forests, and boreal forests (Figure 10.3B). In addition, they occur seasonally in alpine areas above the tree line in the Himalayan countries, Taiwan, and Japan. These bears also use agricultural land (Honda 2009; Liu et al. 2009; Takahata et al. 2017) and raid crops (see also Chapter 15). Habitat loss due to logging and expansion of agriculture and plantation is a major threat to these bears (Garshelis & Steinmetz 2016). At least 10% of the potential distributional area for Asiatic black bear is not currently suitable because of urban or suburban encroachment (Escobar et al. 2015). More details about habitats in each country/region are as follows.

Western south Asia: At the edge of their western distribution, in Iran and southern Pakistan, *U. t. gedrosianus* inhabits

Figure 10.3 Different Asiatic black bear habitats: (A) broadleaved deciduous (oak) forest in Japan (photo by Shinsuke Koike); (B) Korean pine forest in Russia (photo by Ivan Seryodkin); (C) mountain habitats in Nepal (photo by Toshio Tsubota); (D) subtropical broadleaved forest, siwaliks, Nepal (photo by Karine Pigeon).

subtropical thorn forests (hot and arid environments with scattered woodlands), primarily in mountainous regions (Fahimi et al. 2011; Ghadirian et al. 2017). In this region, connectivity among bear habitats is thought to be critical to access seasonal foods and to provide gene flow among local populations (Almasieh et al. 2016).

Western and eastern Himalayas (south Asia): In northern Pakistan, *U. t. laniger* are found in hills/valleys in temperate forests at 1500–4000 m and these bears prefer steep, high-altitude habitats with dense forests (Abbas et al. 2015; Ali et al. 2018). In Afghanistan, Asiatic black bear habitat is historically characterized by extended temperate (oak, *Quercus* spp.) forests up to 2500 m, where they are displaced by coniferous forests (Stevens et al. 2011). Asiatic black bears in western Himalayas inhabit forested hills ranging from 1200 to 3300 m (Sathyakumar & Choudhury 2007) and prefer temperate–riverine habitats where rich food resources can be found (Sathyakumar et al. 2015). In the Annapurna Conservation Area of central Nepal (Figure 10.3C), *U. t. laniger* also occurs (Kadariya et al. 2018), and these bears are distributed in *Quercus* spp. forests between 1900 and 3100 m (Bista & Aryal 2013).

In Makalu-Barun National Park in eastern Nepal, conifer forests are the most suitable habitats for Asiatic black bears when compared to broadleaved forests (Bista et al. 2018). Asiatic black bears in the eastern states of northern India occur between 70 and 4300 m (Sathyakumar & Choudhury 2007). Khangchendzonga, India is an important area that connects the high-elevation landscape of India, Nepal, China, and Bhutan. In this area, the percentage of acorn-producing trees in the temperate zone was recently determined to be the most important factor in bear utilization during the pre-hibernation period (Bista et al. 2018). The Wangchuck Centennial National Park in Bhutan has records of Asiatic black bears between 2500 m (broadleaf forest) and 4500 m (alpine meadows; Figure 10.4) (Jamtsho & Wangchuk 2016).

In south-western China, Asiatic black bear occurrence is strongly related to forest cover throughout the province (Liu et al. 2009) and the highest occupancy of Asiatic black bears occurs primarily in broadleaved forests at higher elevation (Wang et al. 2015). More than half of the total range of Asiatic black bears is found in China. Forested areas in China have been increasing, but most of those reforestation areas are in poor habitat for Asiatic black bears (Garshelis & Steinmetz 2016).

The Asiatic black bear is also found in the lowland tropical forest of the eastern Himalayas (Dasgupta et al. 2015) (Figure 10.3D). In Bangladesh, the forest cover is now <7% of the land area and Asiatic black bears inhabit small forest patches that are mainly covered with mixed evergreen and deciduous forests dominated by sal *Shorea robusta* (Islam et al. 2013). Asiatic black bears do not occur in the tropical, mainly dry forests, scrub and thorn woodland habitats of central and southern India and Sri Lanka where sloth bears are found (Garshelis & Steinmetz 2016).

Figure 10.4 Asiatic black bear at an elevation of 3500 m in Wangchuck Centennial National Park, Bhutan. (Photo by UWICER.)

South-East Asia: In this region, Asiatic black bears inhabit a mosaic of tropical forests (evergreen and deciduous forests) along with sun bears. In Thailand, Asiatic black bears use semi-evergreen, mixed deciduous, and montane evergreen forests (Steinmetz et al. 2011), and fruit abundance is reported to be higher in locations where the Asiatic black bears are present compared to areas where they are absent (Ngoprasert et al. 2011). In Laos, Asiatic black bears depend on tropical forests and areas of high elevation and high tree density away from roads (Scotson 2017). Asiatic black bears live in evergreen, semi-evergreen, and deciduous forests in Vietnam (Xuan 2006), while in Cambodia, they are recorded in evergreen forest at 480–530 m (Gray et al. 2012). There are no records of Asiatic black bears in Malaysia where sun bears occur (Garshelis & Steinmetz 2016).

East Asia: In Japan, *U. t. japonicus* mainly live in temperate deciduous oak forests on the islands of Honshu and Shikoku. Seasonal and annual differences in habitat use are observed: Asiatic black bears especially prefer deciduous forests during autumn compared to other seasons (Izumiyama & Shiraishi 2004; Takahata et al. 2014) and their patterns of habitat selection change in response to hard mast productivity (Arimoto et al. 2011). During summer, Asiatic black bears in the Alps of Northern Japan use subalpine and alpine zones at high elevation (Izumiyama & Shiraishi 2004). Asiatic black bears in the Central Japan Alps are known to use the narrow fringes of human-dominated lands (the edge of a red pine forest in the foothills) that provides good food conditions during summer (Takahata et al. 2013, 2014, 2017). In the Ashio-Nikko Mountain of central Honshu, Asiatic black bears use grasslands and feed on ants (Yamazaki et al. 2012). In addition, in suburban areas of northern Honshu, these bears use mountainous habitat during daytime and primarily move around the village at night (Sakamoto et al. 2009). In Taiwan, *U. t. formosanus* inhabits areas ranging from 300 to >3500 m (mainly below 2000 m) and prefers broadleaf and mixed broadleaf–coniferous forests

Figure 10.5 Asiatic black bear in Russia taken by a camera trap. (Photo by Ivan Seryodkin.) (A black and white version of this figure will appear in some formats. For the color version, please refer to the plate section.)

(Hwang et al. 2010). In Russia, *U. t. ussuricus* (Figure 10.5) occurs primarily in mountain broadleaved (*Quercus mongolica*) forests and mixed broadleaf–Korean pine *Pinus koraiensis* from the sea coast to 800 m (Seryodkin et al. 2003; Aramilev 2006; Latham et al. 2012).

Movements and Home Range

In temperate forests in Japan, during spring and summer, female Asiatic black bears show high site fidelity over the years (Yamamoto et al. 2012; Kozakai et al. 2017), and kin-related spatial structure and female-biased natal philopatry have been observed/suggested (Ishibashi & Saitoh 2004; Ohnishi & Osawa 2014). During autumn, their home range size changes according to availability of hard mast (Kozakai et al. 2011). During poor mast events (autumn), movements of both sexes increase, and females seem to travel longer distances compared to males (Koike et al. 2012). Although the kin-related spatial structure among males and females temporarily dissolves in a poor mast year (Ohnishi et al. 2011), matrilineal site fidelity appears to re-form by the end of the autumn when females return to sites near their primary range before den entry (Kozakai et al. 2017). Similarly, in Taiwan, movements of bears in autumn are related to the availability of acorns (Hwang et al. 2010). On the other hand, in the Karuizawa area of central Honshu, Japan, females do not migrate long distances seasonally or annually (Yamamoto et al. 2012).

Home range sizes are not affected by reproductive status and age among females (Yamamoto et al. 2012). Although home range overlaps among both sexes (Izumiyama & Shiraishi 2004), exclusion of (or avoidance by) females and smaller males from prime feeding areas occupied by larger males was suggested in some areas (Huygens & Hayashi 2001; Hwang et al. 2010). The sex ratio of nuisance individuals that damage corn crops is extremely biased to males (Saito et al. 2008).

Activity Patterns

Activity time budgets, daily activity rhythms, and timing of den entry/emergence have been studied by radio-telemetry through monitoring VHF beacons (Reid et al. 1991; Hwang & Garshelis 2007) and activity sensors built into GPS collars (Kozakai et al. 2008; Yamazaki et al. 2008). Although bears are diurnal during spring and summer in the tropical zone of Taiwan and temperate zone of Japan, nocturnal activity occurs during good mast falls in both countries (Hwang & Garshelis 2007; Kozakai et al. 2009). In addition, in suburban areas of Japan, Asiatic black bears shift to nocturnal behavior quite rapidly when moving toward agricultural and residential areas (Sakamoto et al. 2009). A remote camera-trap study conducted in Iran to investigate the use of caves by *U. t. gedrosianus* (Fahimi et al. 2011) revealed that these bears use rock caves as shelters during daytime and feed at night, possibly to avoid extreme heat from the sun or to avoid encounters with humans.

Bears are active 54–57% of the day throughout the year in Taiwan (Hwang & Garshelis 2007), but activity time budgets have been found to vary by sex and reproductive status in Japan (Kozakai et al. 2013). In addition, autumn is a critical part of the yearly cycle when bears accumulate fat reserves before hibernation (hyperphagia); thus, time budgets during that time also vary according to the availability of hard mast. Daily time budgets in poor mast years rapidly decrease after October and den entry is earlier than during years of good/moderate mast (Kozakai et al. 2013).

Feeding Ecology

Asiatic black bears feed on a wide range of foods and their diet is mainly composed of fruit and green vegetation (Koike 2010; Steinmetz et al. 2013). Asiatic black bears also feed on insects such as bees and ants, and also kill and scavenge livestock and wild animals for meat (Koike 2010; Seryodkin 2015b; Koike et al. 2016).

In temperate forest environments, bears tend to consume fresh leaves with high crude protein and low fiber content during spring (Furusaka et al. 2017). During late summer and autumn, Asiatic black bears mostly consume fresh fruit, and tend to prefer large fruit growing in tall trees (Koike & Masaki 2019). In the Ashio Nikko Mountains of Japan, where the landscape is characterized by open grasslands resulting from past mining activities, Asiatic black bears spent 7–8 h/day feeding on ants and alter their foraging behavior (time foraging per ant nest) based on ant nest composition during summer, i.e. the availability of pupae may trigger ant feeding by bears. Because queen pupae were heavier than members of other castes, calories per individual were higher. "Time foraging per ant nest" (TPN) varied during the study period (early summer). There was a negative relationship between frequency of occurrence of pupae in ant nests and TPN; because pupae cannot move by themselves, bears could consume them effectively and quickly (Fujiwara et al. 2013).

However, this behavior has been found to result in insufficient energy intake (Yamazaki et al. 2012; Fujiwara et al. 2013). Bears may have used ants for essential amino acids that they are unable to produce themselves. In temperate forests during autumn, Asiatic black bears particularly rely on hard mast from the Fagaceae family to attain sufficient fat reserves for hibernation. However, plants from the Fagaceae family have a reproductive strategy characterized by many synchronous mast crops, followed by few additional mast crops only. Thus, the autumn mast yield results in food scarcity that can cause Asiatic black bears to adjust their feeding habits (Koike et al. 2012) and movement patterns (see below). In the Russian Far East, up to 98% of Asiatic black bear diet can be composed of pine nuts and oak acorns during autumn (Seryodkin 2015b).

Asiatic black bears are arboreal feeders, and when feeding on hard mast, these bears often break branches and pile them up in the canopy, forming what appears to be a platform or nest. This behavior is specific to Asiatic black bears and these arboreal feeding site signs seem to be used more in trees with large mast energy values and during poor mast years (Tochigi et al. 2018a). Asiatic black bears are opportunistic feeders and target a variety of abundant food sources to maximize energy intake based on food types available on the landscape. The amount of meat that results from kills or carrion, as well as the amount of fruit in their diet, has been observed to vary according to availability (Koike et al. 2013).

Asiatic black bears also feed on anthropogenic foods such as crops (corn and oats), cultivated fruits in orchards, apiaries, fish farms, and livestock throughout Asia (Japan Bear Network 2006; Can et al. 2014; Ali et al. 2018). In the hot and arid habitats of Iran, Asiatic black bears feed on human-related foods more than observed in other regions (Ghadirian et al. 2017), and a high occurrence of agricultural crops in diets of Asiatic black bears has also been reported in north-western Pakistan (Ali et al. 2018). In Bhutan, where animal husbandry is the mainstay of the country's economy, Asiatic black bears mainly kill sheep and yaks during September and October (Sangay & Vernes 2008). Asiatic black bears strip the bark of conifer trees to feed in plantations, and this behavior results in management issues in Japan (Kobashikawa & Koike 2016) (see also Chapter 15). Genetic relationship among the bark-stripping bears is higher than that among the other bear species (Kitamura & Ohnishi 2011). Stable isotope analysis using tissues has been increasingly used to determine the variety of food types, and the amount of human-related foods, in diets of different Asiatic black bear populations (Mizukami et al. 2005a, 2005b; Koike et al. 2013).

Reproduction and Growth

The mating season of Asiatic black bears occurs between May and August (Yamamoto et al. 1998; Spady et al. 2007). In males, spermatogenesis is initiated during hibernation in March and peaks in May and June to correspond with estrus in females (Komatsu et al. 1997). Females are induced ovulators: they ovulate in response to stimuli from males and without coitus (Okano et al. 2006a). Although fertilization occurs within the reproductive tract soon after breeding, the fertilized egg ceases to develop at the blastocyst stage for several months, a reproductive strategy referred to as delayed implantation or embryonic diapause (Tsubota et al. 2001). Implantation may occur between late November and early December, when bears begin hibernation (Sato et al. 2001). Pregnant females give birth between late January and early February (Iibuchi et al. 2009), and maintain homeothermic conditions and supply nutrients to the fetus by stimulating thermogenesis, gluconeogenesis, and urea recycling during hibernation (Shimozuru et al. 2013; see also Chapter 2).

Based on skull measurement, growth appears to cease around 5 years of age in males and 4 years of age in females (Amano et al. 2004). The body weight of single cubs is significantly higher than that of twin cubs and the daily weight gain of single cubs (32.7 g/day) is approximately 1.65 times greater than that of twins (19.8 g/day) (Iibuchi et al. 2009). As observed by spermatogenesis in the testis, males attain reproductive maturity at 2–4 years of age (Komatsu et al. 1994). Females attain reproductive maturity at 4 years of age, as observed from ovulation scars (corpus luteum or corpus albicans; Katayama et al. 1996). A recent study reports that the reproductive history of individuals is reflected in the cementum annuli width of Asiatic black bears as observed with other bear species (Tochigi et al. 2018b): the annular cementum growth of females with cubs in or after August is significantly smaller than during years when females are with yearlings or without offspring. The relationship between age and tooth wear is estimated by counting tooth cementum annuli (Shimoinaba & Oi 2015).

Behavior

From the Himalayan foothills southward (including Taiwan), where food is available during all months and not covered by snow, Asiatic black bears do not hibernate excepted for parturient females. On the other hand, in temperate zone and more northern latitudes, Asiatic black bears can hibernate from October to May in East Asia (Reid et al. 1991; Seryodkin et al. 2003; Koike & Hazumi 2008; Kozakai et al. 2013; Yamamoto et al. 2016). Early den entry during poor mast years is therefore likely related to the lack of available food. In central Honshu, Japan, hard mast production in autumn has been linked to the timing of den exit (springtime) for males, and temperatures in November and March have been linked to the timing of den entry and emergence, respectively (Yamamoto et al. 2016). As observed for other bear species, parturient females in Russia and Japan emerge from dens later than females without cubs and males (Seryodkin et al. 2003; Yamamoto et al. 2016). In Japan, Asiatic black bears mainly use tree cavities, tree roots, and rock crevices (Figure 10.6) as dens, and den sites are commonly located in protected areas or on remote steep slopes, presumably to escape hunting pressure

Figure 10.6 Asiatic black bear on a steep slope of rock in the Ashio-Nikko Mountains, Japan. (Photo by Sinh Futagami.)

(Huygens et al. 2001; Koike & Hazumi 2008). In Russia, Asiatic black bears also use hollow trees, ground nests, and caves/ under rocks for denning (Seryodkin et al. 2003).

Interspecific feeding association has been reported between Asiatic black bears and Himalayan langurs *Semnopithecus schistaceus*: bears feed on acorns that are incidentally dropped by langurs feeding in trees above (Nautiyal & Huffman 2018). In Russia, during denning, Asiatic black bears can be predated by tigers *Panthera tigris* and brown bears and, in some cases, have been observed to avoid predation by climbing inside hollow trees (Seryodkin et al. 2003).

Parasites and Diseases

Meat and feces of Asiatic black bears contain and hold zoonotic pathogens (Cantlay et al. 2017), and bears also disperse arthropods such as ticks that latch on to their bodies (Yumoto et al. 2010). Q fever, a zoonosis, is caused by *Coxiella burnetii.* In Japan, Asiatic black bears have a high probability of testing positive for antibody titers to *C. burnetii* (Ejercito et al. 1993). In humans, outbreaks of parasitic infection caused by trichinellosis have been caused by eating raw meat of Asiatic black bears and brown bears (Yamaguchi 1991). In addition, the genus *Hepatozoon* includes apicomplexan parasites that infect terrestrial vertebrates as intermediate hosts and hematophagous arthropods (e.g. ticks) as final hosts. While influences on human health are unknown, *Hepatozoon ursi* is observed in the lungs of Asiatic black bears and mature *Hepatozoon* oocysts have been detected in two tick species, *Haemaphysalis japonica* and *Haemaphysalis flava*, collected on the bears infected with *H. ursi* (Uni et al. 2003; Kubo et al. 2008). In Russia, trichinosis has also been observed in Asiatic black bears (Seryodkin 2015a).

Status in the Wild

Asiatic black bears are designated as Vulnerable in the IUCN Red List (Garshelis & Steinmetz 2016), and have been listed in Appendix I of CITES since 1979. The Baluchistan subspecies *U. t. gedrosianus* is listed as Critically Endangered in the 1996 IUCN Red List.

The main threat to this species is habitat loss due to the expansion of increasing human populations leading to activities such as commercial logging, farming, roadways, and dam construction, as well as the use of various natural resources by communities bordering Asiatic black bear habitats. Commercial poaching for trade of bear body parts (especially gall bladders and paws) is also a serious threat, especially because bear bile has been an important part of traditional medicine in Eastern Asia. Although all international trade of bear body parts including bile farming is illegal under CITES, illegal domestic and international trading of bear parts is common (Garshelis & Steinmetz 2016). Because they often raid and damage crops and timber, and also attack humans (see Chapters 15 and 17), considerable numbers of Asiatic black bears are killed annually. Incidental hunting (by snare and box trap originally set for ungulates), legal sports hunting (which is only legal in Japan except some Prefectures and Russia), capture/ trade of cubs for bear baiting (i.e. illegal fights with dogs), and for dancing bears (Japan Bear Network 2006) also occur (see Chapter 26). The threat of climate change to Asiatic black bears worldwide is uncertain and needs to be monitored at local and international levels.

Although there is a pervasive lack of rigorous population estimates sufficient to accurately document trends in most populations of Asiatic black bears, the smallest populations are found in Iran and in South Korea; Japanese populations are stable or increasing; populations in Bhutan and Thailand are stable; and populations in other counties are decreasing (Garshelis & Steinmetz 2016). Throughout Japan, except for Shikoku island, the current range of Asiatic black bears has been expanding into suburban areas that contain recovering forests (see also Chapter 25). This range expansion is a result of diminishing rural human populations, but still causes intrusions into human-populated areas (Japan Bear Network 2014). Until the 2000s, the focus of bear conservation and management in Japan had been the endangered populations in the western region of the country. Since the 2000s, the objective of the Japanese management plans is now to simultaneously reduce human–bear conflicts while maintaining numbers of Asiatic black bears in most of Japan (Kozakai et al. 2015). In Bhutan and probably many more countries, human–bear conflicts have been increasing: crop raiding, livestock predation, and bear attacks on humans are a serious problem even in the National Parks (Jamtsho & Wangchuk 2016). In these regions, high numbers of bears are killed in towns or agricultural areas.

Status in Captivity

As mentioned above, captive bear farming to extract bile is conducted legally in China, and illegally in other countries (Garshelis & Steinmetz 2016). Bear farming has raised concerns for animal welfare and conservation. Captive breeding programs are an important method to conserve endangered species, and in Asiatic black bears, techniques of semen collection and cryopreservation have been established (Okano et al. 2006b). Fecal estradiol and progesterone analyses have been used to estimate the endocrine profile for reproductive cycles (Chang et al. 2011). Other studies on annual changes in body weight (Hashimoto & Yasutake 1999), body fat ratio (Nakamura et al. 2008), and serum leptin concentrations (Tsubota et al. 2008), and seasonal changes in intravenous glucose tolerance test (Kamine et al. 2012) have been conducted, and can help improve our understanding of the biology of this species.

References

Abbas, F., Bhatti, Z., Haider, J. & Mian, A. (2015). Bears in Pakistan: distribution, population biology and human conflicts. *Journal of Bioresource Management* 2(2): 1–13.

Ali, A., Waseem, M., Teng, M., et al. (2018). Human–Asiatic black bear (*Ursus thibetanus*) interactions in the Kaghan Valley, Pakistan. *Ethology Ecology & Evolution* 30(5): 399–415.

Almasieh, K., Kaboli, M. & Beier, P. (2016). Identifying habitat cores and corridors for the Iranian black bear in Iran. *Ursus* 27(1): 18–30.

Amano, M., Oi, T. & Hayano, A. (2004). Morphological differentiation between adjacent populations of Asiatic black bears, *Ursus thibetanus japonicus*, in northern Japan. *Journal of Mammalogy* 85(2): 311–315.

Aramilev, V. V. (2006). The conservation status of Asiatic black bears in the Russian Far East. In: *Understanding Asian bears to secure their future* (pp. 86–89). Ibaraki: Japan Bear Network.

Arimoto, I., Goto, Y., Nagai, C. & Furubayashi, K. (2011). Autumn food habits and home-range elevations of Japanese black bears in relation to hard mast production in the beech family in Toyama Prefecture. *Mammal Study* **36**(4): 199–208.

Baryshnikov, G. F. (2010). Middle Pleistocene *Ursus thibetanus* (Mammalia, Carnivora) from Kudaro caves in the Caucasus. *Proceedings of the Zoological Institute RAS* 314(1): 67–79.

Baryshnikov, G. F. & Zakharov, D. S. (2013). Early Pliocene bear *Ursus thibetanus* (Mammalia, Carnivora) from Priozernoe locality in the Dniester basin (Moldova Republic). *Proceedings of the Zoological Institute RAS* 317(19): 3–10.

Bidon, T., Janke, A., Fain, S. R., et al. (2014). Brown and polar bear Y chromosomes reveal extensive male-biased gene flow within brother lineages. *Molecular Biology and Evolution* 31(6): 1353–1363.

Bista, R. & Aryal, A. (2013). Status of the Asiatic black bear *Ursus thibetanus* in the southeastern region of the Annapurna Conservation Area, Nepal. *Zoology and Ecology* 23(1): 83–87.

Bista, M., Panthi, S. & Weiskopf, S. R. (2018). Habitat overlap between asiatic black bear *Ursus thibetanus* and red panda *Ailurus fulgens* in Himalaya. *PLoS ONE* 13(9): 1–12.

Can, Ö. E., D'Cruze, N., Garshelis, D. L., Beecham, J. & Macdonald, D. W. (2014). Resolving human–bear conflict: a global survey of countries, experts, and key factors. *Conservation Letters* 7(6): 501–513.

Cantlay, J. C., Ingram, D. J. & Meredith, A. L. (2017). A review of zoonotic infection risks associated with the wild meat trade in Malaysia. *EcoHealth* 14(2): 361–388.

Chang, G.-R., Yang, C.-C., Hsu, S.-H. et al. (2011). Fecal reproductive steroid profiles for monitoring reproductive patterns in female formosan black bears (*Ursus thibetanus formosanus*). *Annales Zoologici Fennici* 48(5): 275–286.

Dasgupta, S., Choudhury, P., Ashraf, N. V. K., Bhattacharjee, P. C. & Kyarong, S. (2015). Food preference of rehabilitated Asiatic black bear cubs in lowland tropical forests of northeast India. *Asian Journal of Conservation Biology* 4(1): 20–25.

Ejercito, C. L. A., Cai, L., Htwe, K. K. et al. (1993). Serological evidence of *Coxiella burnetii* infection in wild animals in Japan. *Journal of Wildlife Diseases* 29(3): 481–484.

Escobar, L. E., Awan, M. N. & Qiao, H. (2015). Anthropogenic disturbance and habitat loss for the red-listed Asiatic black bear (*Ursus thibetanus*): using ecological niche modeling and nighttime light satellite imagery. *Biological Conservation* 191: 400–407.

Fahimi, H., Yusefi, G. H., Madjdzadeh, S. M., et al. (2011). Camera traps reveal use of caves by Asiatic black bears (*Ursus thibetanus gedrosianus*) (Mammalia: Ursidae) in southeastern Iran. *Journal of Natural History* 45(37–38): 2363–2373.

Fujiwara, S., Koike, S., Yamazaki, K., Kozakai, C. & Kaji, K. (2013). Direct observation of bear myrmecophagy: relationship between bears' feeding habits and ant phenology. *Mammalian Biology* 78: 34–40.

Furusaka, S., Kozakai, C., Nemoto, Y., et al. (2017). The selection by the Asiatic black bear (*Ursus thibetanus*) of spring plant food items according to their nutritional values. *ZooKeys* 2017(672). doi:10.3897/zookeys.672.10078

Galbreath, G. J., Hean, S. & Montgomery, S. M. (2001). A new color phase of *Ursus thibetanus* (Mammalia: Ursidae) from Southeast Asia. *Natural History Bulletin of the Siam Society* 49: 107–111.

Galbreath, G. J., Hunt, M., Clements, T. & Waits, L. P. (2008). An apparent hybrid wild bear from Cambodia. *Ursus* 19(1): 85–86.

Garshelis, D. L. (2009). Family Ursidae (bears). In: Wilson, D. & Mittermeier, R. (Eds.), *Handbook of the mammals of the world. Vol. 1. Carnivores* (pp. 448–497). Barcelona: Lynx Edicions.

Garshelis, D. & Steinmetz, R. (2016). *Ursus thibetanus* (errata version published in 2017). The IUCN Red List of Threatened Species 2016: e.T22824A114252336.

Ghadirian, T., Qashqaei, A. T., Soofi, M., Abolghasemi, H. & Ghoddousi, A. (2017). Diet of Asiatic black bear in its westernmost distribution range, southern Iran. *Ursus* 28(1): 15–19.

Gray, T. N. E., Rattanak, O., Keavuth, H., Chanrattana, P. & Maxwell, A. L. (2012). The status of large mammals in eastern Cambodia: a review of camera trapping data 1999–2007. *Cambodian Journal of Natural History* 1: 42–55.

Hashimoto, Y. & Yasutake, A. (1999). Seasonal changes in body weight of female Asiatic black bears under captivity. *Mammal Study* 24(1): 1–6.

Higashide, D., Miura, S. & Miguchi, H. (2012). Are chest marks unique to Asiatic black bear individuals? *Journal of Zoology*, 288(3): 199–206.

Higashide, D., Miura, S. & Miguchi, H. (2013). Evaluation of camera-trap designs for photographing chest marks of the free-ranging Asiatic black bear, *Ursus thibetanus*. *Mammal Study* 38(1): 35–39.

Honda, T. (2009). Environmental factors affecting the distribution of the wild boar, sika deer, Asiatic black bear and Japanese macaque in central Japan, with implications for human–wildlife conflict. *Mammal Study* 34(2): 107–116.

Huygens, O. C. & Hayashi, H. (2001). Use of stone pine seeds and oak acorns by Asiatic black bears in central Japan. *Ursus* 12: 47–50.

Huygens, O., Goto, M., Izumiyama, S., Hayashi, H. & Yoshida, T. (2001). Denning ecology of two populations of Asiatic black bears in Nagano prefecture, Japan. *Mammalia* 65(4): 417–428.

Hwang, M. H. & Garshelis, D. L. (2007). Activity patterns of Asiatic black bears (*Ursus thibetanus*) in the central mountains of Taiwan. *Journal of Zoology* 271(2): 203–209.

Hwang, M. H., Garshelis, D. L., Wu, Y. H. & Wang, Y. (2010). Home ranges of Asiatic black bears in the central mountains of Taiwan: gauging whether a reserve is big enough. *Ursus* 21(1): 81–96.

Iibuchi, R., Nakano, N., Nakamura, T., et al. (2009). Change in body weight of mothers and neonates and in milk composition during denning period in captive Japanese black bears (*Ursus thibetanus japonicus*). *Japanese Journal of Veterinary Research* 57(1): 13–22.

Ishibashi, Y. & Saitoh, T. (2004). Phylogenetic relationships among fragmented Asian black bear (*Ursus thibetanus*) populations in Western Japan. *Conservation Genetics* 5(3): 311–323.

Islam, M. A., Uddin, M., Aziz, M. A. et al. (2013). Status of bears in Bangladesh: going, going, gone? *Ursus* 24(1): 83–90.

Izumiyama, S. & Shiraishi, T. (2004). Seasonal changes in elevation and habitat use of the Asiatic black bear (*Ursus thibetanus*) in the Northern Japan Alps. *Mammal Study* 29(1): 1–8.

Jamtsho, Y. & Wangchuk, S. (2016). Assessing patterns of human–Asiatic black bear interaction in and around Wangchuck Centennial National Park, Bhutan. *Global Ecology and Conservation* 8: 183–189.

Japan Bear Network. (2006). *Understanding Asian bears to secure their future*. Ibaraki: Japan Bear Network.

Japan Bear Network. (2014). *Changes in distribution of Asiatic black bears and brown bears in Japan*. Ibaraki: Japan Bear Network.

Kadariya, R., Shimozuru, M., Maldonado, J. E., et al. (2018). High genetic diversity and distinct ancient lineage of Asiatic black bears revealed by non-invasive surveys in the Annapurna Conservation Area, Nepal. *PLoS ONE* 13(12): e0207662.

Kamine, A., Shimozuru, M., Shibata, H. & Tsubota, T. (2012). Changes in blood glucose and insulin responses to intravenous glucose tolerance tests and blood biochemical values in adult female Japanese black bears (*Ursus thibetanus japonicus*). *Japanese Journal of Veterinary Research* 60: 5–13.

Katayama, A., Tsubota, T., Yamada, F., Kita, I. & Tiba, T. (1996). Reproductive evaluation of Japanese black bears (*Selenarctos thibetanus japonicus*) by observation of the ovary and uterus. *Japanese Journal of Zoo and Wildlife Medicine* 1(1): 26–32.

Kim, Y.-J., Hong, Y.-J., Min, M.-S., et al. (2011). Genetic status of Asiatic black bear (*Ursus thibetanus*) reintroduced into South Korea based on mitochondrial DNA and microsatellite loci analysis. *Journal of Heredity* 102(2): 165–174.

Kitamura, F. & Ohnishi, N. (2011). Characteristics of Asian black bears stripping bark from coniferous trees. *Acta Theriologica* 56(3): 267–273.

Kobashikawa, S. & Koike, S. (2016). Spatiotemporal factors affecting bark stripping of conifer trees by Asiatic black bears (*Ursus thibetanus*) in Japan. *Forest Ecology and Management* 380: 100–106.

Koike, S. (2010). Long-term trends in food habits of Asiatic black bears in the Misaka Mountains on the Pacific coast of central Japan. *Mammalian Biology* 75(1): 17–28.

Koike, S. & Hazumi, T. (2008). Notes on Asiatic black bears denning habits in the Misaka Mountains, central Japan. *Ursus* 19(1): 80–84.

Koike, S. & Masaki, T. (2019). Characteristics of fruits consumed by mammalian frugivores in Japanese temperate forest. *Ecological Research* 34: 246–254.

Koike, S., Masaki, T., Nemoto, Y., et al. (2011). Estimate of the seed shadow created by the Asiatic black bear *Ursus thibetanus* and its characteristics as a seed disperser in Japanese cool-temperate forest. *Oikos* 120(2): 280–290.

Koike, S., Kozakai, C., Nemoto, Y., et al. (2012). Effect of hard mast production on foraging and sex-specific behavior of the effect of hard mast production on foraging and sex-specific behavior of the Asiatic black bear (*Ursus thibetanus*). *Mammal Study* 37(1): 21–28.

Koike, S., Soga, M., Enari, H., Kozakai, C. & Nemoto, Y. (2013). Seasonal changes and altitudinal variation in deer fecal pellet decay. *European Journal of Wildlife Research* 59(5): 765–768.

Koike, S., Nakashita, R., Kozakai, C., et al. (2016). Baseline characterization of the diet and stable isotope signatures of bears that consume natural foods in central Japan. *European Journal of Wildlife Research* 62(1). doi:10.1007/s10344-015-0969-6

Komatsu, T., Tsubota, T., Kishimoto, M., Hamasaki, S. & Tiba, T. (1994). Puberty and stem cell for the initiation and resumption of spermatogenesis in the male Japanese black bear (*Selenarctos thibetanus japonicus*). *Journal of Reproduction and Development* 40(6): j65–j71.

Komatsu, T., Yamamoto, Y., Atoji, Y., Tsubota, T. & Suzuki, Y. (1997). Seasonal changes in subcellular structures of Leydig and Sertoli cells in the Japanese black bear, *Ursus thibetanus japonicus*. *Archives of Histology and Cytology* 60(3): 225–234.

Kozakai, C., Koike, S., Yamazaki, K. & Furubayashi, K. (2008). Examination of captive Japanese black bear activity using activity sensors. *Mammal Study* 33(3): 115–119.

Kozakai, C., Yamazaki, K., Nemoto, Y., et al. (2009). Behavioural study of free-ranging Japanese black bears II – How does a bear manage in a year of food shortage? In: Oi, T., Ohnishi, N., Koizumi, T. & Okochi, I. (Eds.), *FFPRI Scientific Meeting Report 4 Biology of Bear Intrusions – Proceedings of an International Workshop on "The Mechanism of the Intrusion of Bears into Residential Areas"* (pp. 64–66). Ibaraki: Forestry and Forest Products Research Institute Matunosato.

Kozakai, C., Yamazaki, K., Nemoto, Y., et al. (2011). Effect of mast production on home range use of Japanese black bears. *Journal of Wildlife Management* 75(4). doi:10.1002/jwmg.122

Kozakai, C., Yamazaki, K., Nemoto, Y., et al. (2013). Fluctuation of daily activity time budgets of Japanese black bears: relationship to sex, reproductive status, and hard-mast availability. *Journal of Mammalogy* 94(2). doi:10.1644/11-MAMM-A-246.1

Kozakai, C., Kondo, M., Arimoto, I., et al. (2015). I. Background and legal structure on conservation and management of bears. *Honyurui Kagaku (Mammalian Science)* 55(2): 219–239.

Kozakai, C., Nemoto, Y., Nakajima, A., et al. (2017). Influence of food availability on matrilineal site fidelity of female Asian black bears. *Mammal Study* 42(4). doi:10.3106/041.042.0404

Krause, J., Unger, T., Noçon, A., et al. (2008). Mitochondrial genomes reveal an explosive radiation of extinct and extant bears near the Miocene–Pliocene boundary. *BMC Evolutionary Biology* 8(1): 220.

Kubo, M., Uni, S., Agatsuma, T. et al. (2008). *Hepatozoon ursi* n. sp. (Apicomplexa: Hepatozoidae) in Japanese black bear (*Ursus thibetanus japonicus*). *Parasitology International* 57(3): 287–294.

Kumar, V., Lammers, F., Bidon, T., et al. (2017). The evolutionary history of bears is characterized by gene flow across species. *Scientific Reports* 7(1). doi:10.1038/srep46487

Kutschera, V. E., Bidon, T., Hailer, F., et al. (2014). Bears in a forest of gene trees: phylogenetic inference is complicated by incomplete lineage sorting and gene flow. *Molecular Biology and Evolution* 31(8): 2004–2017.

Latham, E., Stetz, J. B., Seryodkin, I., et al. (2012). Non-invasive genetic sampling of brown bears and Asiatic black bears in the Russian Far East. *Ursus* 23(2): 145–158.

Liu, F., McShea, W., Garshelis, D., et al. (2009). Spatial distribution as a measure of conservation needs: an example with Asiatic black bears in south-western China. *Diversity and Distributions* 15(4): 649–659.

Mizukami, R. N., Goto, M. & Izumiyama, S. (2005a). Estimation of feeding history by measuring carbon and nitrogen stable isotope ratios in hair of Asiatic black bears. *Ursus* 16(1): 93–101.

Mizukami, R. N., Goto, M., Izumiyama, S., et al. (2005b). Temporal diet changes recorded by stable isotopes in Asiatic black bear (*Ursus thibetanus*) hair. *Isotopes in Environmental and Health Studies* 41(1): 87–94.

Mondolfi, E. & Boede, E. O. (1981). A hybrid of the spectacled bear (*Tremarctos ornatus*) and the Asiatic black bear (*Selenarctos thibetanus*) born at the Maracay Zoological Park, Venezuela. *Memoria de La Sociedad de Ciencias Naturales La Salle* 41(December): 143–148.

Nakamura, S., Okano, T., Shibata, H., et al. (2008). Relationships among changes of serum leptin concentration, leptin mRNA expression in white adipose tissue (WAT), and WAT fat-cell size in female Japanese black bears (*Ursus thibetanus japonicus*). *Canadian Journal of Zoology-Revue Canadienne De Zoologie*, 86(9): 1042–1049.

Naoe, S., Tayasu, I., Sakai, Y., et al. (2016). Mountain-climbing bears protect cherry species from global warming through vertical seed dispersal. *Current Biology* 26(8): R315–R316.

Nautiyal, H. & Huffman, M. A. (2018). Interspecific feeding association between central Himalayan langurs (*Semnopithecus schistaceus*) and Himalayan black bears (*Ursus thibetanus*), in a temperate forest of the western Indian Himalayas. *Mammal Study* 43(1): 1–6.

Ngoprasert, D., Steinmetz, R., Reed, D. H., Savini, T. & Gale, G. A. (2011). Influence of fruit on habitat selection of Asian bears in a tropical forest. *Journal of Wildlife Management* 75(3): 588–595.

Ohnishi, N. & Osawa, T. (2014). A difference in the genetic distribution pattern between the sexes in the Asian black bear. *Mammal Study* 39(1): 11–16.

Ohnishi, N., Uno, R., Ishibashi, Y., Tamate, H. B. & Oi, T. (2009). The influence of climatic oscillations during the Quaternary Era on the genetic structure of Asian black bears in Japan. *Heredity* 102(6): 579–589.

Ohnishi, N., Yuasa, T., Morimitsu, Y. & Oi, T. (2011). Mass-intrusion-induced temporary shift in the genetic structure of an Asian black bear population. *Mammal Study* 71: 67–71.

Okano, T., Nakamura, S., Komatsu, T., et al. (2006a). Characteristics of frozen–thawed spermatozoa cryopreserved with different concentrations of glycerol in captive Japanese black bears (*Ursus thibetanus japonicus*). *The Journal of Veterinary Medical Science* 68(10): 1101–1104.

Okano, T., Nakamura, S., Nakashita, R., et al. (2006b). Incidence of ovulation without coital stimuli in captive Japanese black bears (*Ursus thibetanus japonicus*) based on serum progesterone profiles. *The Journal of Veterinary Medical Science* 68(10): 1133–1137.

Pigeon, K. (2018). What is it about the Terai of Nepal that favors sloth bears over Asiatic black bears? Biological research. *International Bear News* 27(3): 49–51.

Reid, D., Jiang, M., Teng, Q., Qin, Z. & Hu, J. (1991). Ecology of the Asiatic black bear (*Ursus thibetanus*) in Sichuan, China. *Mammalia* 55(2): 221–237.

Saito, M., Yamauchi, K., Aoi, T., et al. (2008). Individual identification of Asiatic black bears using extracted DNA from damaged crops. *Ursus* 19(2): 162–167.

Sakamoto, Y., Kunisaki, T., Sawaguchi, I., et al. (2009). A note on daily movement patterns of a female Asiatic black bear (*Ursus thibetanus*) in a suburban area of Iwate Prefecture, northeastern Japan. *Mammal Study* 34(3): 165–170.

Sangay, T. & Vernes, K. (2008). Human–wildlife conflict in the Kingdom of Bhutan: patterns of livestock predation by large mammalian carnivores. *Biological Conservation* 141(5): 1272–1282.

Sathyakumar, S. & Choudhury, A. (2007). Distribution and status of the Asiatic black bear in India. *Journal of the Bombay Natural History Society* 104: 316–323.

Sathyakumar, S., Sharma, L. K. & Charoo, S. A. (2015). Ecology of Asiatic black bear (*Ursus thibetanus*) in Dachigam National Park, Kashmir, India. Final project report, Dehradun.

Sato, M., Tsubota, T., Komatsu, T., et al. (2001). Changes in sex steroids, gonadotropins, prolactin, and inhibin in pregnant and nonpregnant Japanese black bears (*Ursus thibetanus japonicus*). *Biology of Reproduction* 65(4): 1006–1013.

Scotson, L. (2017). Distribution, range connectivity, and trends of bear populations in Southeast Asia. Doctoral thesis, University of Minnesota.

Seryodkin, I. V. (2015a). Trichinosis of brown bear and Asiatic black bear in the Russian Far East. In *Bulletin of KrasGAU* 12: 168–173.

Seryodkin, I. V. (2015b). Diet composition of brown bear and Asiatic black bear in the Middle Sikhote-Alin (Russian Far East): comparative study. *The Bulletin of Irkutsk State University* 12: 32–38.

Seryodkin, I. V, Kostyria, A. V, Goodrich, J. M. et al. (2003). Denning ecology of brown bears and Asiatic black bears in the Russian Far East. *Ursus* 14(2): 153–161.

Sheikh, K. M. (2006). The status of bears in Pakistan. In: *Understanding Asian bears to secure their future* (pp. 1–6). Ibaraki: Japan Bear Network.

Shimoinaba, S. & Oi, T. (2015). Relationship between tooth wear and age in the Japanese black bear in Hiroshima Prefecture, Japan. *Mammal Study* 40(1): 53–60.

Shimozuru, M., Iibuchi, R., Yoshimoto, T., et al. (2013). Pregnancy during hibernation in Japanese black bears: effects on body temperature and blood biochemical profiles. *Journal of Mammalogy* 94(3): 618–627.

Spady, T. J., Lindburg, D. G. & Durrant, B. S. (2007). Evolution of reproductive seasonality in bears. *Mammal Review* 37(1): 21–53.

Steinmetz, R., Garshelis, D. L., Chutipong, W. & Seuaturien, N. (2011). The shared preference niche of sympatric Asiatic black bears and sun bears in a tropical forest mosaic. *PLoS ONE* 6(1). doi:10.1371/journal.pone.0014509

Steinmetz, R., Garshelis, D. L., Chutipong, W. & Seuaturien, N. (2013). Foraging ecology and coexistence of Asiatic black bears and sun bears in a seasonal tropical forest in Southeast Asia. *Journal of Mammalogy* 94(1): 1–18.

Stevens, K., Dehgan, A., Karlstetter, M., et al. (2011). Large mammals surviving conflict in the eastern forests of Afghanistan. *Oryx* 45(2): 265–271.

Takahata, C., Nishino, S., Kido, K. & Izumiyama, S. (2013). An evaluation of habitat selection of Asiatic black bears in a season of prevalent conflicts. *Ursus* 24(1): 16–26.

Takahata, C., Nielsen, S. E., Takii, A. & Izumiyama, S. (2014). Habitat selection of a large carnivore along human–wildlife boundaries in a highly modified landscape. *PLoS ONE* 9(1): e86181.

Takahata, C., Takii, A. & Izumiyama, S. (2017). Season-specific habitat restriction in Asiatic black bears, Japan. *Journal of Wildlife Management* 81(7): 1254–1265.

Tochigi, K., Masaki, T., Nakajima, A., et al. (2018a). Detection of arboreal feeding signs by Asiatic black bears: effects of hard mast production at individual tree and regional scales. *Journal of Zoology* 305(4): 223–231.

Tochigi, K., Tamatani, H., Kozakai, C., et al. (2018b). Reproductive histories of Asian black bears can be determined by cementum annuli width. *Mammal Study* 43(4): 261.

Tsubota, T., Taki, S., Nakayama, S., et al. (2001). Immunolocalization of steroidogenic enzymes in the corpus luteum and the placenta of the Japanese black bear, *Ursus thibetanus japonicus*, during pregnancy. *Reproduction* 121(4): 587–594.

Tsubota, T., Sato, M., Okano, T., et al. (2008). Annual changes in serum leptin concentrations in the adult female Japanese black bear (*Ursus thibetanus japonicus*). *Journal of Veterinary Medical Science* 70(12): 1399–1403.

Tsuji, Y. & Su, H. H. (2018). Macaques as seed dispersal agents in Asian forests: a review. *International Journal of Primatology* 39(3): 356–376.

Umemura, Y., Koike, S., Kozakai, C., et al. (2018). Using a novel method of potential available energy to determine masting condition influence on sex-specific habitat selection by Asiatic black bears. *Mammalia* 82(3): 288–297.

Uni, S., Matsubayashi, M., Ikeda, E. & Suzuki, Y. (2003). Characteristics of a hepatozoonosis in lungs of Japanese black bears (*Ursus thibetanus japonicus*). *Journal of Veterinary Medical Science* 65(3): 385–388.

Wang, F., McShea, W. J., Wang, D. & Li, S. (2015). Shared resources between giant panda and sympatric wild and domestic mammals. *Biological Conservation* 186: 319–325.

Wozencraft, W. C. (2005). Order carnivora. In: Wilson, D. E. & Reede, D. M. (Eds.), *Mammal species of the world: A taxonomic and geographic reference. Vol.1*, 3rd edition (pp. 532–628). Baltimore, MD: Johns Hopkins University Press.

Wu, J., Kohno, N., Mano, S., et al. (2015). Phylogeographic and demographic analysis of the Asian black bear (*Ursus thibetanus*) based on mitochondrial DNA. *PLoS ONE* 10(9): 1–19.

Xuan, D. (2006). The current status and conservation of bears in Vietnam. In *Understanding Asian bears to secure their future* (pp. 61–65). Hokkaido: Japan Bear Network.

Yadav, S., Lamichhane, B., Subedi, N., et al. (2017). Himalayan black bear discovered in Babai valley of Bardia National Park, Nepal, co-occurring with sloth bears. *International Bear News* 26(3): 23–25.

Yamaguchi, T. (1991). Present status of trichinellosis in Japan. *Southeast Asian Journal of Tropical Medicine and Public Health* 22(Suppl): 295–301.

Yamamoto, K., Tsubota, T. & Kita, I. (1998). Observation of sexual behavior of captive Japanese black bears, *Ursus thibetanus japonicus*. *Journal of Reproduction and Development* 44(5): j13–18.

Yamamoto, T., Oka, T., Ohnishi, N., et al. (2012). Genetic characterization of northernmost isolated population of Asian black bear (*Ursus thibetanus*) in Japan. *Mammal Study* 37(2): 85–91.

Yamamoto, T., Tamatani, H., Tanaka, J., et al. (2016). Abiotic and biotic factors affecting the denning behaviors in Asiatic black bears *Ursus thibetanus*. *Journal of Mammalogy* 97: 128–134.

Yamazaki, K., Kozakai, C., Kasai, S., et al. (2008). A preliminary evaluation of activity sensing GPS collars for estimating daily activity patterns of Japanese black bears. *Ursus* 19: 154–161.

Yamazaki, K., Kozakai, C., Koike, S., et al. (2012). Myrmecophagy of Japanese black bears in the grasslands of the Ashio area, Nikko National Park, Japan. *Ursus* 23(1). doi:10.2192/URSUS-D-10-00012.1

Yumoto, K., Yamazaki, K., Koike, S., et al. (2010). Ixodid ticks collected from Japanese black bears in the Northern Kanto District, Central Japan (Arachnida, Acarina). *Bulletin of Ibaraki Nature Museum* 13: 81–84.

Chapter 11

American Black Bear (*Ursus americanus*)

Joseph D. Clark, Jon P. Beckmann, Mark S. Boyce, Bruce D. Leopold, Anne E. Loosen, and Michael R. Pelton

Names

Genus: *Ursus* (Linnaeus, 1758)

Species: *Ursus americanus* (Linnaeus, 1758)

Subspecies: *Ursus americanus altifrontalis* (Elliot 1903); *Ursus americanus amblyceps* (Baird 1859); *Ursus americanus americanus* (Pallas 1779); *Ursus americanus californiensis* (Miller 1900); *Ursus americanus carlottae* (Osgood 1901); *Ursus americanus cinnamomum* (Audubon and Bachman 1854); *Ursus americanus emmonsii* (Dall 1895); *Ursus americanus eremicus* (Merriam 1904); *Ursus americanus floridanus* (Merriam 1904); *Ursus americanus hamiltoni* (Cameron 1957); *Ursus americanus kermodei* (Hornaday 1905); *Ursus americanus luteolus* (Griffith 1821); *Ursus americanus machetes* (Elliot 1903); *Ursus americanus perniger* (Allen 1910); *Ursus americanus pugnax* (Swarth 1911); and *Ursus americanus vancouveri* (Hall 1928).

Names in other languages: **French:** Ours noir américain; **Spanish:** Oso negro americano; **Italian**: orso nero; **German:** Amerikanischer Schwarzbär; **Cherokee:** ᎠᏫᏆᏍ ᏋᏂᏍ ᏊᎾ; **Lakota:** Mato

Introduction

American black bears (*Ursus americanus*) are endemic to North America, having speciated from other ursids some 1.2–1.8 million years ago (Kurtén & Anderson 1994). During that time, black bears came to occupy nearly all of the forested areas of the North American continent. Historically, black bears were one of the most important mammals to indigenous peoples of North America by providing food, fat, hides, and tools (Raybourne 1987). Bears also played important roles in indigenous culture as symbols of strength, hard work, and love (see Rockwell 1991). In the eighteenth and nineteenth centuries in the US, Canada, and Mexico, European settlers with firearms reduced black bear numbers (Williams 1930), but the axe and plow had equally negative effects on bear populations as forests gave way to agriculture and livestock. Black bears also faced intense pressure from unregulated market hunting, poaching, and predator control programs. Black bears disappeared from large portions of their North American range, being relegated to swamps, thickets, rugged mountains, and other areas deemed too inaccessible, unsuitable, or undesirable to be occupied by humans (Pelton & van Manen 1994). In the early twentieth century, Canadian and US governments established national parks, forests, and refuges, and state and provincial legislatures created wildlife management agencies and enacted regulations to manage wildlife populations. By the late twentieth century, the combination of public land infrastructure, habitat improvements, harvest regulations, law enforcement, and information–education programs resulted in significant increases in black bear numbers and range expansions in the US and Canada (Pelton & van Manen 1994). Their recovery was aided in no small part by the remarkable adaptability of this animal (Pelton 2003; Scheick & McCown 2014). Today, the American black bear is one of the most iconic, abundant, and economically important bear species in the world. Unfortunately, black bear populations in Mexico have not recovered similarly to those in the US and Canada, and the status of the species there is uncertain.

Taxonomy

American black bears are mammals from the order Carnivora and family Ursidae. American black bears are in the genus *Ursus*, which also includes grizzly (*Ursus arctos*) and polar bears (*Ursus maritimus*). Evolution of the black bear was shaped by cyclical glacial events during the Pleistocene that resulted in population isolation, expansion, recolonization, and hybridization. Mitochondrial DNA analyses suggest that black bears diverged from grizzly and polar bear predecessors about 3.3 million years ago (Kumar et al. 2017). This evolutionary pathway is a relatively recent finding, with earlier accounts suggesting that American black bears were more closely related to Asiatic black bears (*Ursus thibetanus*, McLellan & Reiner 1994). The taxonomic legacy of these early ursids in North America was further complicated by frequent hybridization among bear species (Kumar et al. 2017).

Subspecies and Distribution

Hall (1981) compiled 16 recognized subspecies of the American black bear (Figure 11.1). Two of the subspecies (*U. a. luteolus* and *U. a. floridanus*) were originally classified as nearly extirpated, but the population status of both has since greatly improved (Laufenberg et al. 2016; Florida Fish and Wildlife Conservation Commission 2011). The current and historical distribution of *U. a. machetes* is data-deficient, thus its conservation status in Mexico is unknown (Delfín-Alfonso et al. 2012). Subspecies classifications such as those from Hall (1981) were based primarily on skull morphology and other physical characteristics. A more recent analysis of black bear taxonomy performed by Puckett et al. (2015) using nuclear and mitochondrial DNA found that black bear lineages arose from three subcontinental nuclear clusters (i.e. Alaskan, Eastern, and Western; Figure 11.1). Puckett et al. (2015) concluded that *U. a. pugnax* was comprised of two distinct lineages (Alaska and Western). Further, they recommended combining the northern range of *U. a. pugnax* with *U. a. emmonsii* and *U. a. perniger*, as these regions formed an admixed cluster in the analysis. *U. a. cinnamomum* also contained admixture from the Eastern (*U. a. americanus*) lineage. *U. a. floridanus* and *U. a. luteolus* had not diverged from *U. a. americanus*; thus, they suggested that all bears in the eastern range should be described as *U. a. americanus*. This was supported by an analysis of clustering that accounted for isolation-by-distance (Bradburd et al. 2018). While the Puckett et al. (2015) genomic results do not include south Texas, mitochondrial haplotype data

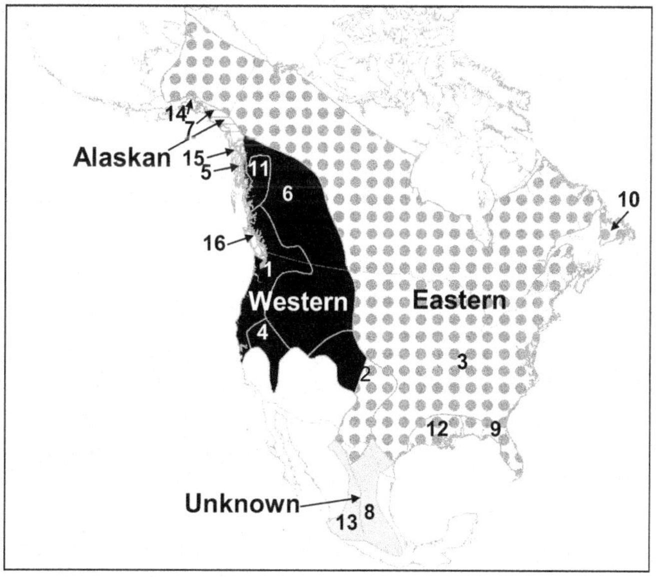

Figure 11.1 Subspecies designations for American black bear adapted from Hall (1981) and Larivière (2001) are (1) *U. a. altifrontalis*, (2) *U. a. amblyceps*, (3) *U. a. americanus*, (4) *U. a. californensis*, (5) *U. a. carlottae*, (6) *U. a. cinnamomum*, (7) *U. a. emmonsii*, (8) *U. a. eremicus*, (9) *U. a. floridanus*, (10) *U. a. hamiltoni*, (11) *U. a. kermodei*, (12) *U. a. luteolus*, (13) *U. a. machetes*, (14) *U. a perniger*, (15) *U. a. pugnax*, and (16) *U. a. vancouveri*. Three subcontinental nuclear clusters based on Puckett et al. (2015) are Eastern (polka dot), Western (black), and Alaskan (horizontal line).

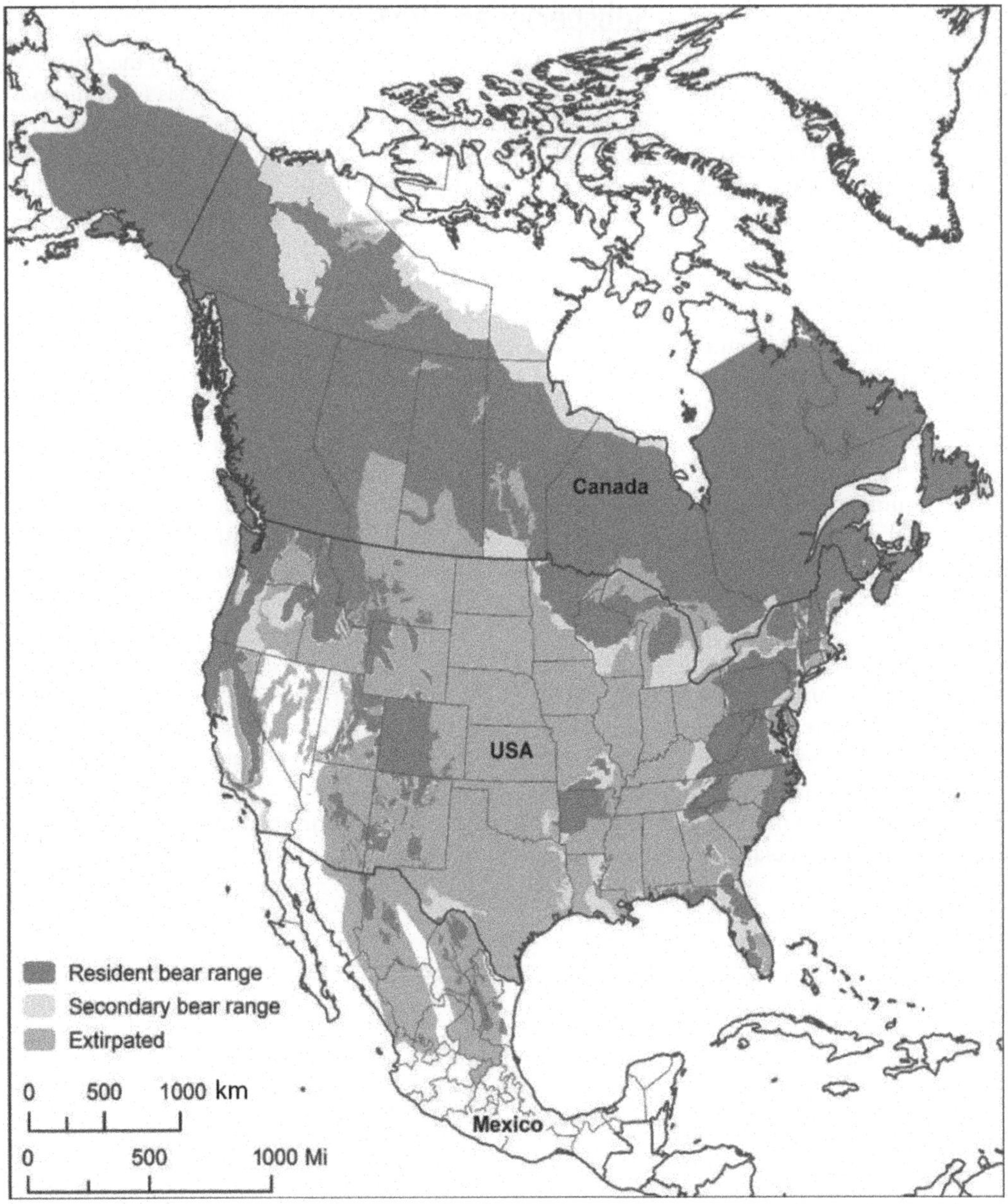

Figure 11.2 Range of the American black bear including resident, secondary, and extirpated range (adapted from Scheick & McCown 2014).

(Onorato et al. 2004) and high microsatellite divergence (Onorato et al. 2007) from south-western populations suggest that the currently recognized range of *U. a. amblyceps* should be divided where populations in New Mexico are included in the western genomic lineage and populations in Texas in the eastern genomic lineage. Despite these new findings, most jurisdictions continue to rely on Hall's taxonomic designations.

The American black bear is the most widely distributed of the three bear species in North America, extending from Yukon Territory, Canada, through Alaska, USA, and south to central Mexico (Figure 11.2). The latitudinal distribution of black bears ranges from 20 to 60°N and their longitudinal distribution ranges from 70 to 155°W (Maehr et al. 2001). The species can be found from 0 to 3500 m above sea level. Black bears are present in all provinces and territories of Canada except Prince Edward Island, where they are now extinct. Black bears are distributed throughout the US except the south-western states, where their distribution is patchy, and the central US, where agriculture is intensive and forest cover limited. In Mexico, the black bear is found only in the Sierra Madre Occidental and Sierra Madre Oriental mountain ranges.

Descriptive Notes

The American black bear is the smallest of the three ursids occurring in North America. Its body morphology is stocky with a thick neck, massive legs, strong shoulder musculature, and short tail (Figure 11.3). Black bears are plantigrade with strong, recurved, non-retractile claws that they use to climb trees. The black bear is agile and quick, and can run up to

50 km/h for short distances (Pelton 2003). Black bears have flexible lips and a prehensile-like tongue that allows them to pick fruits and berries and consume small prey (e.g. ants, grubs). Coat color can be black, various shades of brown, bluish grey (Rogers 1980), and a rare white phase occurs on the Kermode Islands off the Pacific coast of British Columbia, Canada (Hornaday 1905). The black coat color is more frequently found in boreal, montane, and temperate rain forests, whereas non-black phases are more frequent in open forests of western North America or in desert biomes (Rounds 1987; Figure 11.4). The dental formula is 3/3, 1/1, 4/4, 2/3, resulting in 42 teeth (Pelton 2003). The canines are prominent, yet the carnassial molars are reduced, presumably as an adaptation for omnivory. Their smaller body size, arboreal behavior, and dark coloration may make black bears better adapted for living in forested environments than grizzly bears (Herrero 1978).

Figure 11.3 American black bear by Francesco Rinaldi, with permission. (A black and white version of this figure will appear in some formats. For the color version, please refer to the plate section.)

The body mass of American black bears varies by age, sex, season, and region. Black bears are sexually dimorphic, with males being about 20% greater in mass than females (Pelton 2003). Seasonal variation in body mass is driven by preparation for and recovery from hibernation. Masses are greatest in autumn before denning and lowest in early spring after individuals emerge from winter dens, although body sizes in extreme southern latitudes may not vary seasonally (Maehr et al. 2001). Mass also varies by region, with bears from northern and eastern North America generally being larger than bears from western interior regions. There are many exceptions to these regional trends, because access to anthropogenic food sources, either in the form of agriculture or garbage, and access to spawning salmon can greatly contribute to body mass in bears (Alt 1980; Maehr et al. 2001). Average black bear masses have ranged from 86 kg for adult males in California, USA (Larivière 2001) and Great Smoky Mountains National Park, Tennessee, USA (McLean & Pelton 1990) to 135 kg in New York, USA (New York State Department of Environmental Conservation 2007), although adult males greater than 365 kg have been reported. Average masses of females have been reported to range from 50 kg in Great Smoky Mountains National Park (McLean & Pelton 1990) to 74 kg in New York (New York State Department of Environmental Conservation 2007). Head-and-body lengths of adults typically range from 120 to 200 cm and shoulder heights range from 70 to 105 cm. The skull has a large cranium, strong jaws, and long rostrum filled with turbinate bones that aid in olfaction.

(A)

(B)

Figure 11.4 Black (A, photo by Clint Turnage) and brown (B, photo by Spencer Rettler) color phases of the American black bear.

Habitat

Black bears are highly adaptable mammals, occupying a wide array of environments ranging from Arctic tundra to subtropical forest (Karelus et al. 2016, 2018) and from deserts (Hellgren et al. 2005) to deciduous temperate forests. Black bear habitat selection varies seasonally and regionally (Hellgren et al. 1991; Clark et al. 1994). Black bears have a noteworthy ability to take advantage of seasonally available food resources over wide spatial extents, often traversing great distances and elevational gradients to do so (Noyce & Garshelis 2014). Thus, food is a primary driver for habitat selection (Rogers 1976; Lindzey & Meslow 1977a; Landers et al. 1979; Costello & Sage 1994; Vaughan 2002; Humm et al. 2017).

Dens are important for hibernation and reproduction and, where available, female black bears prefer elevated tree cavities (Pelton et al. 1980; Weaver & Pelton 1994; Oli et al. 1997). Bears will also den on the ground in thickets, rotten logs, rocky outcroppings, excavations, or even in agricultural fields (Ditmer et al. 2018). Bears in exurban settings will sometimes den in crawl spaces underneath houses. Research has shown that tree dens are thermodynamically superior to other den types (Johnson et al. 1978). Secure dens are critical for black bears sympatric with grizzly bears because predation on adults or cubs by grizzlies has been documented (Mattson et al. 1992).

Black bears have had to adapt their behaviors and habitat use in response to landscape alterations by humans. Bears can be sensitive to disturbance by humans and riparian zones are often selected as escape cover (Karelus et al. 2016; Shakeri et al. 2018). Roads and highways pose threats to bears by increasing direct mortality from vehicle collisions and increasing access by hunters and poachers (van Manen et al. 2012). Roads and other forms of human activity can also displace bears into less-favored habitats (Brody & Pelton 1989; Kasworm & Manley 1990; Clark et al. 1994; Cushman & Lewis 2010). In south-west Alberta, Canada, densities of female black bears were greater on private land relative to public lands where recreational use was high (Loosen et al. 2019). Such human-influenced changes in black bear habitat utilization can result in changes to population genetic structure caused by isolation and barriers to gene flow (Draheim et al. 2018; Proctor et al. 2020). Genetic structural changes precipitated by habitat loss can lead to reduced genetic diversity and inbreeding, resulting in even greater stressors to population dynamics than habitat losses alone. Nevertheless, bears are becoming increasingly accustomed to sharing the landscape with humans, often relying on agriculture (Ferreira et al. 2018) in rural landscapes and garbage in urban and exurban areas (Beckmann & Berger 2003a; Don Carlos et al. 2009; Wynn-Grant et al. 2018).

Movements and Home Range

Movement rates and home range sizes for black bears vary depending on sex, reproductive status, season, and the productivity of the local environment. Male black bears, as is true with most mammals, have larger home ranges and greater rates of movement than females (Garshelis et al. 1983). In an analysis of home range data from Michigan, Missouri, and Mississippi, USA, the annual home range size (all home ranges are 95% fixed kernel estimates) for males were 89.9 km^2 compared with 18.7 km^2 for females (Gantchoff et al. 2018). Solitary females there had home ranges 6.9 times greater than those of females with cubs in Maryland, USA (Jones et al. 2015), but a different study in Michigan found no difference in home range sizes between solitary females and those with cubs (Norton et al. 2018).

The size of home ranges also varies by season. Home ranges for yearlings in Maryland averaged 9 km^2 during spring, 15 km^2 in summer, and 20 km^2 in fall (Jones et al. 2015). Likewise, bears in Florida, USA, had larger home ranges in autumn compared with summer (Moyer et al. 2007). Black bears typically have smaller home ranges in landscapes that are more productive. For example, home ranges in Michigan, Missouri, and Mississippi were smaller in areas with more abundant vegetation, even when forested areas were fragmented with agriculture (Gantchoff et al. 2018). In an area in Minnesota, USA that was predominantly agricultural and forest cover averaged only 17%, home ranges were extremely large, averaging 834 km^2 for males and 91 km^2 for females (Ditmer et al. 2018).

Dispersal is a mechanism that enables bears to avoid inbreeding and to colonize unexploited habitats. Yet, it is one of the least studied demographic parameters (Clevenger & Pelton 1990; Schwartz & Franzmann 1992; Lee & Vaughan 2003). The inbreeding avoidance hypothesis in mammals suggests that only one sex needs to disperse (Greenwood 1980) and in black bears, males are the dispersing sex (Costello et al. 2008). Yearling males typically disperse from their natal range in early summer before the mating season (Clevenger & Pelton 1990). Dispersal distances may depend on habitat quality or bear density (Costello et al. 2008). Rogers (1987) reported mean dispersal distances of yearling males in Minnesota of 61 km, but Lee and Vaughan (2003) found that only three of 11 yearling males dispersed greater than 20 km in Virginia, USA. Philopatry of yearling females is high (Jones et al. 2015). Most yearling females do not disperse from the natal home range and occupy home ranges that overlap with those of their mothers (Rogers 1987; Elowe & Dodge 1989; Lee & Vaughan 2003).

Bears have overlapping home ranges and segregate themselves mostly by mutual avoidance, probably mediated by chemical (i.e. scent) communication (Burst & Pelton 1983; Horner & Powell 1990; Powell et al. 1997; Taylor et al. 2015). The most dominant males tend to have home ranges that overlap the highest-quality habitats (Garshelis & Pelton 1981; Koehler & Pierce 2003; Noyce & Garshelis 2014). Male black bears occasionally kill cubs, which might force females with cubs to select home ranges away from areas occupied by adult males (Powell et al. 1997). In south-west Alberta, Canada, female black bears were found more frequently on private lands where they might be avoiding the high density of males found on adjacent public lands (Loosen et al. 2019). However, Czetwertynski et al. (2007) found that home ranges for females with cubs were not positioned to avoid males in north-western Alberta and north-eastern Saskatchewan. Likewise, Norton et al. (2018) failed to find evidence that home ranges of female

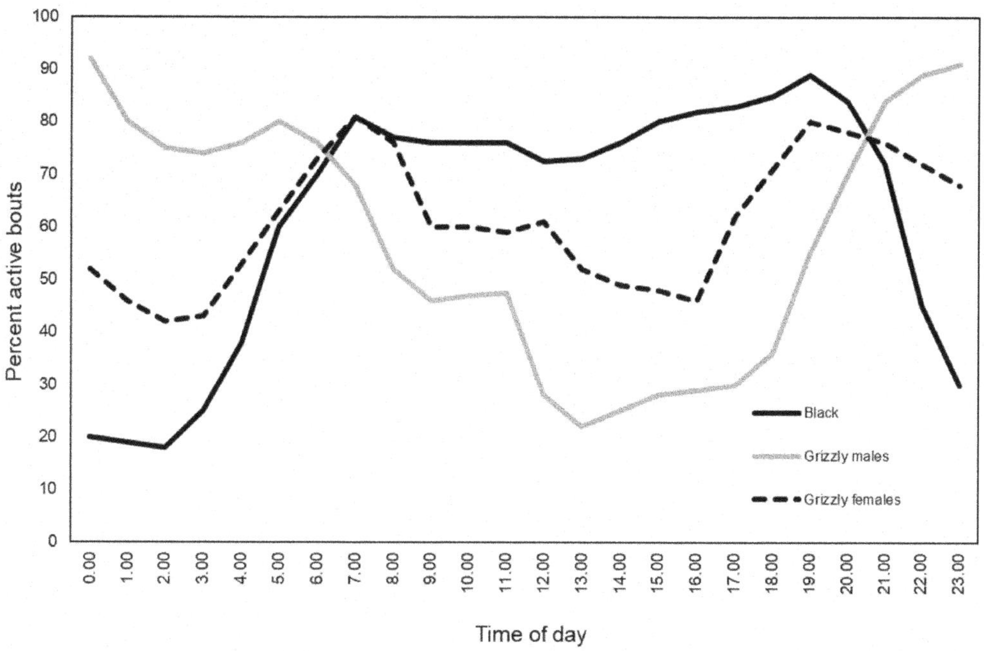

Figure 11.5 Diel activity patterns of black and grizzly bears in north-west Wyoming, June–September, 1995–1996 (adapted from Holm et al. 1999).

black bears with cubs in Michigan were located in areas with less risk of encountering males. Clearly, more work is needed to fully understand the spatial relationships between male and female bears.

Activity Patterns

Activity patterns of American black bears vary depending on a variety of factors including time of year, food availability, age, sex, reproductive status, and level of anthropogenic influence (e.g. Beckmann & Berger 2003b; Bridges et al. 2004). Several studies on black bears demonstrate that they are crepuscular, being most active during dawn and dusk, moderately active during the day, and least active at night (Amstrup & Beecham 1976; Lewis & Rachlow 2011). The evening peak of activity may be more pronounced than the early morning period (Bridges et al. 2004).

Across seasons, black bears can shift their daily activity patterns and, in some instances, bears have been found to be more active at night during spring and autumn compared with summer (Amstrup & Beecham 1976). In eastern North American populations of black bears, such as those found in Great Smoky Mountain National Park, Quigley (1982) ascertained that bears exhibited a crepuscular daily pattern of activity, but that activity was modified seasonally. Further, activity tended to be greatest during the day in all seasons except autumn, when nocturnal activity was greatest. This same pattern holds for western black bear populations, such as those in California and Washington, USA (Lindzey & Meslow 1977a; Ayres et al. 1986). Most seasonal variation in activity patterns was attributed to changes in foraging patterns; monthly activity levels tended to be least in the post denning months of April to early June and pre-denning months of November and December. Activity levels were greater in July and August, during the breeding season.

Activity patterns of black bears are driven to some extent by sex and age class of the bear, with subadult and yearling females generally having the greatest daytime activity levels and adult males being the least day-active. The presence of cubs reduced adult female activity patterns in some studies (Ayres et al. 1986; Garshelis & Pelton 1980) but had no effect in others (Lindzey & Meslow 1977b; Bridges et al. 2004). The presence of sympatric grizzly bears may also influence black bear activity patterns (Holm et al. 1999; Figure 11.5). Fortin et al. (2013a, 2013b) found that black bears remained active during the day and for longer periods of time when grizzly bears were present in the Greater Yellowstone Ecosystem, USA. This suggests that black bears may resort to eating less-digestible food when faced with grizzly bear predation, thus requiring more foraging time. The presence of humans can also affect activity patterns of black bears. This is particularly true at the wildland–urban interface, where bears shifted to more nocturnal activity patterns compared with wildland conspecifics (Reimchen 1998; Beckmann & Berger 2003a, 2003b; Merkle et al. 2013).

Feeding Ecology

Although American black bears are members of order Carnivora, their diet is the most omnivorous of the three North American ursids. Black bears have a relatively inefficient system for digesting plant material (Pritchard & Robbins 1990; McLellan 2011) so, to compensate, much of their lives is spent searching for food. The extent to which bears rely on plant material versus meat in their diet is not only a function of the habitat where they live, but also of season. Bears have a primarily plant-based diet, but seasonal abundance of neonate ungulates and anadromous salmon (*Oncorhynchus* spp.) can lead to periods of increased meat consumption (see below).

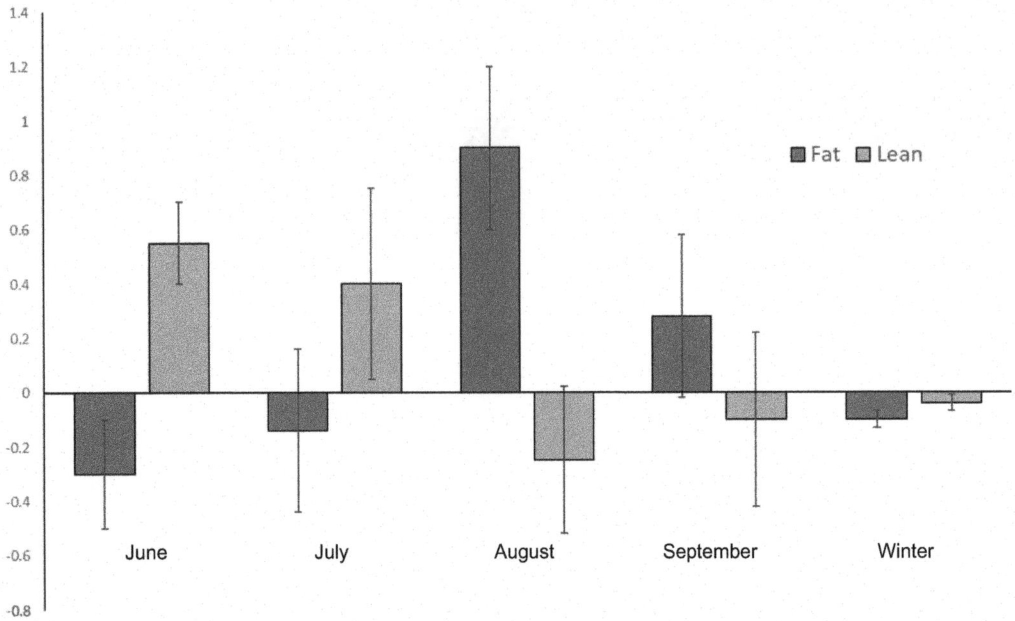

Figure 11.6 Mean rates of fat and lean tissue deposition on six black bears in south-eastern British Columbia, 1991–1994, showing the higher amount of fat acquired by bears in autumn, the hyperphagia period (adapted from McLellan 2011).

Black bears are well adapted to take advantage of temporally and spatially heterogeneous food. They have the mental capacity to remember patchy food resources from previous years and may use chemical cues from other bears to exploit those areas (Noyce & Garshelis 2014). Black bears may suddenly leave summer ranges to travel great distances to seasonally available food patches, where they may join other bears that have done the same.

Although the variety of food items in black bear diets, particularly plants, is extremely diverse, some general patterns emerge. Following denning, black bears consume herbaceous matter such as grasses and forbs. In many eastern deciduous forests, the parasitic plant squawroot (*Conopholis americana*) is an important spring-time food (Beeman & Pelton 1980). Spring is often referred to as the "negative foraging period" because bears are thought to lose body mass following den emergence (Poelker & Hartwell 1973). While spring is typically a time of low food availability, bears may make more efficient use of plant proteins than originally presumed, and young bears in particular can put on weight between den emergence and early summer (Noyce & Garshelis 1998). In late spring and early summer in drier climates, bears may shift to eating neonate ungulates. In the Greater Yellowstone Ecosystem, spring black bear diets were 23% meat (Fortin et al. 2013b). In Newfoundland, black bears exploit caribou calves during the spring calving season, but quickly switch to other foods as calves get larger and become more difficult to capture (Rayl et al. 2018). Salmon is also an important seasonal food source for coastal black bears. This depends, however, on the presence of grizzly bears and the abundance of spawning salmon (Reimchen 2000; Fortin et al. 2007). Belant et al. (2006) found that black bears did not exploit salmon during below-average spawning runs, likely due to interspecific exclusion by grizzly bears, but did so in above-average years. During autumn, black bears shift to eating hard mast (Maehr & Brady 1984; Land 1994). This hard mast consumption drives many aspects of black bear ecology and affects their body condition throughout winter and into the following spring (Clark 2004). This period of intense food consumption prior to denning is called hyperphagia, when black bears can consume up to 20,000 kcal per day (Nelson et al. 1983). Because bears intensify caloric consumption and storage into just a few weeks, autumn is a critical period in the annual cycle for black bears (Figure 11.6). In North American forests, the American chestnut (*Castanea dentata*) was an extremely important food for bears before it was essentially extirpated by an exotic fungal pathogen, *Cryphonectria parasitica*, in the early to mid-1900s. Since then, black bears in eastern forests have relied more heavily on other types of hard mast such as acorns (*Quercus* spp.), beechnuts (*Fagus* spp.), and hickory nuts (*Carya* spp.) (Costello & Sage 1994; Vaughan 2002). Acorns of the black oak group (i.e. *Erythrobalanus*) have greater crude fat levels (18.1–36.9%) than the white oak group (i.e. *Lepidobalanus*, 4.1–9.6%), but white oak acorns are more palatable (Short 1976).

Boreal forests are the largest continuous habitat type occupied by black bears (Young & Ruff 1982), but conditions in those areas can be challenging for these animals (Mosnier et al. 2008). Diversity and productivity of soft mast (i.e. berries and soft fruits) producing species tend to be lower in coniferous compared with deciduous forests (Jonkel & Cowan 1971; Clark et al. 1994), and conifer tree stands do not provide as much fatty hard mast (i.e. nuts; Lindzey & Meslow 1977a; Rogers 1987). However, boreal areas contain berry-producing species high in sugars that bears can then convert to fat. Bears adapt to ripening fruit on the landscape by making elevational and spatial adjustments to consume soft mast species such as buffaloberry (*Shepherdia canadensis*), crowberry (*Empetrum nigrum*), bearberry (*Arctostaphylus uva-ursi*), and blueberry (*Vaccinium* spp.; Pelchat & Ruff 1986; MacHutchon 1989; Raine & Kansas 1990). Such high-quality food for black bears is often found in recently burned or logged areas (Costello & Sage 1994; Brodeur et al. 2008). Black bears in alpine environments often compensate for the lack of hard mast by preying on the young of moose (*Alces alces*), white-tailed deer (*Odocoileus virginianus*), elk (*Cervus elaphus*), and caribou (*Rangifer tarandus*; Schlegel 1976; Mathews & Porter 1988; Schwartz & Franzmann 1991; Linnell et al. 1995; Latham et al. 2011). Although acorns also occur in many areas in western regions, bears also consume pine nuts (*Pinus* spp.) in autumn. Some black bears rely solely on soft mast species, particularly in areas where hard mast species are declining, e.g. whitebark pine (*Pinus albicaulis*).

Black bears are opportunists at the edges of their range such as the subtropics of Florida where they consume saw palmetto (*Serenoa repens*) fruits and cabbage palmetto (*Sabal palmetto*) "hearts" (Maehr & Brady 1984). Similarly, bears in northern Mexico consume madrone (*Arbutus xalapensis*) berries and prickly pear (*Opuntia* spp.) fruits (Doan-Crider & Hewitt 2007). This dietary range is one of the reasons black bears have such a wide distribution and are so adaptable to altered landscapes (see also Chapter 5).

As with other behaviors, diets and feeding ecology of black bears are heavily influenced by the presence of humans and anthropogenic foods. Several studies (Beckmann & Berger 2003a, 2003b; Merkle et al. 2013; Johnson et al. 2015) have demonstrated that black bears living near human areas shifted to primarily foraging on garbage and other anthropogenic foods. Black bears also rely on anthropogenic foods more often during periods when natural foods are scarce, such as during a drought or a mast crop failure (Baruch-Mordo et al. 2014), or in drier and less-productive systems (Johnson et al. 2015). This shift toward human foods has altered black bear behavior including changes in daily activity patterns, denning chronology, demographics, movement ecology, fitness, and elevated levels of human–bear conflicts (e.g. see Baruch-Mordo et al. 2014; Beckmann & Lackey 2018).

Reproduction and Growth

Black bears are polygamous and males have home ranges that are large enough to encompass the home ranges of many females. A few dominant males will typically sire most of the cubs in a particular year (Kovach & Powell 2003; Moore et al. 2015), but litters born to individual females with multiple paternities (i.e. polyandry) have also been documented (Schenk & Kovacs 1995; Ombrello et al. 2016). Bears are induced ovulators whereby egg release is triggered by copulation (Boone et al. 2004). This reproductive trait may be advantageous because mates do not have to copulate during the short time window of a fixed ovulation. This may be particularly advantageous for solitary species with large home ranges that, like bears, may not come into frequent contact with other individuals (Greenwald 1956).

Black bears exhibit delayed implantation, a physiological trait common to over 100 mammal species (Heldstab et al. 2018). Delayed implantation is a reproductive process whereby a fertilized egg develops into a small cohesion of cells known as a blastocyst, at which time development is suspended. Instead of immediately attaching to the uterine wall, implantation of the blastocyst occurs later, effectively lengthening the gestation period. It has been hypothesized that delayed implantation is an adaptation to bridge long winters with little food (Heldstab et al. 2018). Delayed implantation may also give bears the flexibility to vary timing of implantation and subsequent birth or prevent implantation altogether depending on the physiological status of the mother (Bunnell & Tait 1981; Rogers 1987; Robbins et al. 2012).

Female black bears generally enter their winter dens from mid-November to late January in southern latitudes in North America and from late September to late November in the north and west. Birth dates range from late December to mid-February, peaking in mid-January. The true gestation period (i.e. implantation to parturition) is only about 60–70 days (Kolenosky & Strathearn 1987; Hellgren et al. 1990), and cubs are altricial, typically weighing between 0.3 and 0.4 kg at birth (Oftedal et al. 1993). Cubs nurse and remain in winter dens with their mothers until March or April in southern latitudes to April or May in the north, when they exit the dens. Cubs remain with their mothers during spring and summer and re-den with her the following autumn as yearlings. After den emergence the following spring, the female separates herself from the yearlings, and she enters estrus again. The breeding season for black bears typically occurs from late May to August, peaking in July, and is relatively consistent across latitudes.

Bears have evolved a reproductive strategy that maximizes cub survival. Cubs remain with their mothers for more than a year and during this time the young cubs learn about food resources, dangers, and social communication, all of which are important to their overall survival. Litter sizes of black bears vary from one to four cubs, with five cubs being possible but rare (Alt 1989). Reproductive rates vary widely by geographic region. In the western US and Canada, with reported mean litter sizes ranging from 1.6 to 2.1, mean intervals between litters have ranged from 2.0 to 4.4 years, and mean age of primiparity from 5.9 to 8.4 years (Jonkel & Cowan 1971; Kasworm & Thier 1994; Miller 1994; Garshelis et al. 2005). Reproductive rates generally are greater in their eastern North American distribution, with mean litter sizes ranging from 2.1 to 2.9 cubs, birth intervals from 2.0 to 2.3 years, and age of primiparity from 3.0 to 6.3 years (Kordek & Lindzey 1980; Alt 1982; Rogers 1987; Elowe & Dodge 1989; Hellgren & Vaughan 1989; Kolenosky 1990; McDonald & Fuller 2001). In a meta-analysis of black bear demographics across North America, Beston (2011) found that fecundity (average number of female offspring produced per adult female per year)

Table 11.1 Posterior mean vital rates and 95% credible intervals (CrI) for American black bears from a Bayesian meta-analysis of demographic studies in eastern and western North America, 1959–2007 (adapted from Beston 2011).

		Cub survival	Yearling survival	Subadult survival	Adult survival	Fecundity
East						
	Mean	0.65	0.74	0.77	0.82	0.58
	95% CrI	0.60–0.70	0.65–0.81	0.69–0.83	0.77–0.86	0.54–0.62
West						
	Mean	0.54	0.72	0.77	0.88	0.45
	95% CrI	0.43–0.65	0.57–0.83	0.69–0.83	0.83–0.92	0.36–0.54

averaged 0.58 in the east and 0.45 in western US and Canada (Table 11.1). Cub sex ratios are typically near parity.

Reproductive rates of black bears are greatly influenced by food availability (Elowe & Dodge 1989; Schwartz & Franzmann 1991). Black bears with high-carbohydrate diets in the autumn are more likely to produce cubs than bears on low-carbohydrate diets (Elowe & Dodge 1989). In the boreal forest, black bear productivity is closely tied to moose calf consumption (Schwartz & Franzmann 1991; Miller 1994). Food shortages can result in reproductive synchrony, which are reproductive failures followed by alternating large and small annual cub cohorts (Free & McCaffery 1972; Lindzey & Meslow 1977b; LeCount 1983; McLaughlin et al. 1994).

Mortality rates for adult black bears typically are low, with only minor differences between eastern and western ranges (Beston 2011, Table 11.1). Consequently, most of the variation in population growth comes from reproduction. Reproduction and cub survival may be density-dependent in some black bear populations and killing of cubs by adult male bears has been documented (Jonkel & Cowan 1971; LeCount 1982; Lindzey et al. 1983; Czetwertynski et al. 2007; Garrison et al. 2007; Laufenberg et al. 2016). Sexually selected infanticide, an evolutionary strategy whereby adult males kill offspring of rival males in the breeding population, is suspected but has not been conclusively documented in black bears (LeCount 1987; Garrison et al. 2007). Most cub mortality occurs between 1.5 and 5 months of age (Elowe & Dodge 1989; Garrison et al. 2007). Population growth in black bears is variable and is typically low, but Bunnell and Tait (1981) suggested that annual growth rates nearing 24% are possible. Black bear populations reintroduced to high-quality habitat have been shown to grow at an annual rate of 18% for a sustained period of time (Murphy et al. 2015).

Behavior

American black bears are extremely intelligent animals (Vonk et al. 2012). They have excellent hearing, but it seems that they may have poor vision, with limited ability to perceive color (Bacon & Burghardt 1976). Black bears have a keen sense of smell, which they use to find food, detect other bears, and find mates (Kilham & Gray 2002; Gonzales et al. 2013). In many areas, black bears mark trees, signs, and other objects by biting, clawing, and rubbing. Marking is thought to be a form of chemical communication and is most often practiced by males during the breeding season (Taylor et al. 2015), but females and cubs commonly rub as well. Bears vocalize to communicate with cubs, to show aggression, and to issue warnings. Bears will slap the ground, sway their heads, lay their ears back, or narrow their rostrums when displaying aggression.

In spring, following den emergence, black bears turn over rocks and rip into logs in search of insects and use their long tongues effectively for this purpose. Black bears are adept at climbing trees and often consume nuts and berries before these fruits fall to the ground. Black bears will sometimes make "nests" of branches in trees when consuming mast. If food resources are abundant, many black bears will feed close to one another. Black bears commonly travel along "bear trails," which appear as a staggered pattern of shallow depressions in the forest floor. Bears have the habit of placing their feet in the same places time after time along these trails, with other bears following suit, resulting in this unusual pattern.

Black bears are usually solitary, although cubs or yearlings will often accompany females, and males and females can be found together during the mating season. Male black bears exhibit a dominance hierarchy, but females also may display dominance. Subadult males, which are often forced by adult males to place home ranges in marginal sites, are more likely to come into conflict with humans. Such conflicts include consuming unsecured garbage, damaging personal property, raiding bird feeders, or destroying bee yards (White et al. 1995). Bear predation on livestock, such as cattle (primarily calves), sheep, and pigs, has also been documented (Horstman & Gunson 1982; Morehouse & Boyce 2017). Bears consuming the cambium layer of trees are a significant problem for timber growers in the north-western US and western Canada, as this reduces the economic value of the growing forest stands (see also Chapter 15). The best and most common-sense practice is to eliminate the possibility for such behaviors to develop by securing garbage and other attractants. Diversionary feeding programs have been used to dissuade bears from consuming cambium. Black bears have attacked and killed humans, although it rarely occurs (see also Chapter 17). However, predatory attacks, whereby the bear perceives humans as food, are more common for black bears than for any other bear species (Herrero et al. 2011).

Parasites and Diseases

American black bears are habitat generalists and, as such, are exposed to a variety of disease agents through their foraging behavior. Bears are often found around farms, ranches, and residential areas, which may increase their exposure to diseases of humans and domestic animals. In general, however, infectious diseases are not a major mortality or morbidity factor in black bears. This is probably related to bears' solitary behavior and occurrence at relatively low densities. It should be noted, however, that disease monitoring in wild black bear populations is sporadic. Although disease prevalence is thought to be low, some black bear populations exist in small, isolated enclaves. This insularization increases the chances of disease transmission among conspecifics and these small populations become vulnerable to any additive mortality factor, including disease.

Based on diagnoses in diseased animals or detection of serum antibodies, viral diseases reported in free-ranging black bears include rabies, canine distemper, parvovirus, and gammaherpesvirus, and West Nile virus (Walroth et al. 1996; Farajollahi et al. 2003; Cottrell et al. 2013; Stephenson et al. 2015; Black et al. 2019). Bacterial diseases include bubonic plague, tularemia, and brucellosis (Binninger et al. 1980; Clover et al. 1989). Only one protozoan disease (toxoplasmosis) has been reported in black bears (Binninger et al. 1980; Chambers et al. 2012).

Parasites are common, but usually do not impact black bears at the population level. A variety of endoparasites has been reported in wild bears including *Macracanthorhynchus ingens*, *Gongylonema pulchrum*, *Baylisascaris transfuga*, *Strongyloides* sp., *Capillaria putorii*, *Trichinella spiralis*, and *Dirofilaria ursi* (Crum et al. 1978; Manville 1978; Duffy et al. 1994). Bears are particularly susceptible to *Trichinella spiralis*, which can be transmitted to humans by eating undercooked, infected bear meat. In areas with salmon, bears can ingest the fluke *Nanophyetus salmincola*, resulting in "salmon poisoning" disease (Schlegel et al. 1968). Ticks are prevalent on black bear coats. There are six common tick species found on black bears: *Dermacentor variabilis*, *D. albipictus*, *Ixodes scapularis*, *I. affinis*, *Amblyomma americanum*, and *A. maculatum* (Manville 1978; Addison et al. 1978; Yabsley et al. 2009). Mange from *Sarcoptes scabei* and *Demodex* sp. has been reported in black bears, which can result in death (Schmitt et al. 1987; Forrester et al. 1993; Fitzgerald et al. 2008; Niedringhaus et al. 2019).

Status in the Wild

Population recovery was an important management goal from the 1960s through the early 1990s, when many habitat improvement efforts and conservative hunting regulations were initiated. Black bear populations throughout North America have increased in abundance and distribution in recent decades, and black bears are now classified as a game species in most jurisdictions. Consequently, black bear populations throughout North America have increased in abundance and distribution in recent decades (Garshelis & Hristienko 2006; Lackey et al. 2013; Scheick & McCown 2014). In a survey of wildlife managers in USA and Canada with black bear populations in 1988 and 2001, 32 jurisdictions reported population increases during that time, 10 jurisdictions reported stable populations, and two (i.e. Alabama and Montana, USA) reported declines (Hristienko & McDonald 2007). Today, the American black bear is the most common of the eight extant species in the family Ursidae (Servheen 1990). It is estimated that more than 750,000 black bears roam the North American continent (Garshelis & Hristienko 2006). Moreover, black bears have returned to several US states in recent decades where they were once extirpated (e.g. Alabama, Arkansas, Ohio, Kentucky, Maryland, Missouri, Mississippi, Nevada, New Jersey, and Oklahoma). Forty states in the US, 12 Canadian provinces and territories, and six states in northern Mexico now have black bears (Scheick & McCown 2014).

Concomitant with increasing and expanding bear populations, numbers of bears struck by motor vehicles and human–bear conflicts have increased (Hristienko & McDonald 2007). Accordingly, black bear management in many portions of North America has gradually shifted from population recovery to increasing harvest opportunities and reducing human–bear conflicts (Kellert 1992; Bowman et al. 2004; see also Chapter 15). Public sport hunting is the primary mechanism employed to maintain bear populations at levels acceptable to the public, or so-called "social carrying capacity," and many jurisdictions have initiated liberal hunting regulations. Most hunts are in the autumn, but some states and provinces allow spring hunting. Hunting with dogs or baiting is allowed by many jurisdictions. Hunting is a significant source of revenue for wildlife conservation and management activities and is the primary funding mechanism for bear management in most areas. However, hunting participation and revenues are declining across most regions of North America (US Census Bureau 2010). Increasing and expanding black bear populations, combined with decreasing hunter participation and revenue, present substantial challenges to successful bear management. With human development continuing to encroach on wildlife habitat and many black bear diets subsidized by human foods (e.g. fruit trees, crops, garbage) leading to increased levels of human–bear conflicts (see also Chapter 15), the twenty-first century has brought new tensions between the public, wildlife managers, and bears.

Despite the overall success of black bear population recovery in North America, conservation concerns persist in some portions of its historical range. In the south-eastern US, for example, many bear populations are small and insular. These isolated populations are susceptible to inbreeding depression, disease, road mortality, and habitat loss, which can interact to create a so-called "extinction vortex" (Gilpin & Soulé 1986). Cryptorchidism (the failure of one or both testes to descend) has been reported in such isolated populations in Florida (Dunbar et al. 1996). Cryptorchidism has been linked to low genetic diversity in Florida panthers (*Puma concolor*, Barone et al. 1994), and inbreeding is a suspected cause of cryptorchidism in black bears. Conservation efforts to aid in the recovery of these small populations include habitat protection; restoration of travel corridors to link small isolated populations; protection from hunting; reduced vehicle speed limits to minimize road mortalities; and reintroductions to restore bears to recovered habitats, establish "stepping-stone" populations to increase connectivity, and restore genetic diversity (Wear et al. 2005; Clark et al. 2015; Murphy et al. 2019). Small populations may require active management for many years because of the threat of continued human expansion. In Mexico, primary threats are the result of an increasing human population (Valdez et al. 2006) and deforestation, where an estimated 450,000–1,500,00 ha of forest are removed annually (Landa et al. 1997). Recent modeling efforts, however, suggest that potential black bear habitat exists in Mexico, particularly in mountainous terrain with temperate climatic conditions (Delfín-Alfonso et al. 2012). Areas where bears are found in Mexico can be in public, private, or communal ownership (Valdez et al. 2006), and conflicting interests among land stewards have made wildlife management in Mexico difficult, even in protected areas. Whereas the sale of black bear parts is prohibited by international treaty, black bear parts are still commonly sold in open markets in Mexico. Moreover, limited financial capacity, lack of incentives, and lack of expertise has hampered management efforts for Mexico's bears and other wildlife (Valdez et al. 2006). As a consequence, Mexico is the only country where American black bears are considered endangered, where they have been listed since 1986 (Doan-Crider & Hellgren 1996).

Today, the public demands transparent, science-based black bear management programs. Wildlife management agencies need reliable information and tools to respond to changing management circumstances without putting the long-term viability of bear populations at risk. Not only are managers of bear populations under increasing public scrutiny, but agencies faced with increasing black bear populations need reliable information on how best to reduce human–bear conflicts. The increasingly urban public often demands alternatives to lethal control of black bears involved in conflicts. In addition, population-monitoring methods, which allow agencies to evaluate management actions, are often insufficient. Some population monitoring methods lack the power to detect small but meaningful changes in population parameters, may not be feasible at spatial and temporal scales most beneficial to managers, or may provide little information on underlying population processes and the mechanisms driving these processes. Garshelis and Hristienko (2006) called for wildlife agencies to use different or complementary research methods to improve data quality for science-based decision-making. Further, extensive regional approaches to research and monitoring, rather than fine-scaled approaches at local levels, may better meet the shared objective of sustainable bear populations.

Although the restoration of the American black bear is one of our greatest conservation success stories, the future status of the species is far from certain. Continued human population growth, urbanization, agriculture, and roads pose significant threats if unchecked. Invasive species, particularly plants or plant diseases, have the potential to affect bears. Aging oak–hickory forests in eastern North America are threatened with replacement by tree species that may not be as favorable to bears. Climate change may change plant communities upon which bears rely. Bears are becoming an increasingly common attribute of urban and suburban landscapes and are more frequently coming into close contact with humans. Climate changes that may lead to more frequent, more intense, and longer droughts likely will exacerbate this issue in the future. Management agencies need better tools to monitor black bears, and the long-term future of this animal in North America rests with the ability of management agencies to reliably detect changes in bear numbers or bear habitats. Just as importantly, the public will have to develop a better tolerance and willingness to coexist with a species that is becoming increasingly common around people (Bowman et al. 2001). American black bears have shown the ability to co-occupy the landscape with humans; the question of whether the opposite is also true remains to be answered.

Status in Captivity

American black bears are a popular zoological exhibit with 370 individuals (175 male, 182 female, 13 unknown) reported in 105 institutions worldwide (Zoo Information Management System data, compiled by J. Beckmann, August 1, 2019). Most are in North America, with 83 institutions reporting 252 black bears (120 male, 119 female, 13 unknown) followed by Europe with 17 institutions reporting 104 bears (52 males, 52 females), Asia with three institutions reporting five bears (five females), Africa with one institution reporting five bears (one male, four females), and South America with one institution reporting one bear (one male) (see also Chapter 23). Like all bear species, American black bears are challenging to manage in a captive environment and, without proper enrichment, can develop fixed and repetitive behavior patterns termed stereotypies (van Keulen-Kromhout 1978). Environmental enrichment activities that simulate natural foraging may help reduce these stereotypic behaviors (Carlstead et al. 1991; Law & Reid 2010). Because of the large number of bears in captivity, and the relatively stable conservation status of the species, the role of zoos in American black bear conservation is not as important as it may be for some other bear species. However, a number of specialized facilities are devoted to the rehabilitation and release of American black bears that have been injured or orphaned as cubs. These facilities attempt to house and care for bears in a semi-natural environment to avoid habituation with humans. These animals are typically fed natural foods and contact with humans is kept to a minimum. Once the bears are physiologically ready to be released, they typically are transported to a relocation site and released, either prior to denning or in the spring following hibernation. Recent research suggests that survival rates of these rehabilitated bears is similar to wild bears and the incidence of human–bear conflict following rehabilitation is low (Beecham et al. 2015; Blair et al. 2020).

References

Addison, E. M., Pybus, M. J. & Rietveld, H. J. (1978). Helminth and arthropod parasites of black bear, *Ursus americanus*, in central Ontario. *Canadian Journal of Zoology* 56: 2122–2126.

Allen, J. A. (1910). The black bear of Labrador. *Bulletin of the American Museum of Natural History* 28: 1–6.

Alt, G. L. (1980). Rate of growth and size of Pennsylvania black bears. *Pennsylvania Game News* 51(12): 7–17.

Alt, G. L. (1982). Reproductive biology of Pennsylvania's black bear. *Pennsylvania Game News* 53(2): 9–15.

Alt, G. L. (1989). Reproductive biology of female black bears and early growth and development of cubs in northeastern Pennsylvania. PhD dissertation, University of West Virginia.

Amstrup, S. C. & Beecham, J. (1976). Activity patterns of radio-collared black bears in Idaho. *Journal of Wildlife Management* 40: 340–348.

Audubon, J. J. & Bachman, J. (1854). *The viviparous quadrupeds of North America*. New York, NY: V. G. Audubon.

Ayres, L. A., Chow, L. S. & Graber, D. M. (1986). Black bear activity patterns and human induced modifications in Sequoia National Park. *Ursus* 6: 151–154.

Bacon, E. S. & Burghardt, G. M. (1976). Learning and color discrimination in the American black bear. *Ursus* 3: 27–36.

Baird, S. F. (1859). Mammals of the boundary. *United States and Mexican Boundary Survey* 2: 1–62.

Barone, M. A., Roelke M. E., Howard, J., et al. (1994). Reproductive characteristics of male Florida panthers: comparative studies from Florida, Texas, Colorado, Latin America, and North American zoos. *Journal of Mammalogy* 75: 150–162.

Baruch-Mordo, S., Wilson, K. R., Lewis, D. L., et al. (2014). Stochasticity in natural forage production affects use of urban areas by black bears: implications to management of human–bear conflicts. *PLoS ONE* 9: e85122.

Beckmann, J. P. & Berger, J. (2003a). Using black bears (*Ursus americanus*) to test ideal-free distribution models experimentally. *Journal of Mammalogy* 84: 594–606.

Beckmann, J. P. & Berger, J. (2003b). Rapid ecological and behavioural changes in carnivores: the responses of black bears (*Ursus americanus*) to altered food. *Journal of Zoology* 261: 207–212.

Beckmann, J. P. & Lackey, C. W. (2018). Lessons learned from a 20-year collaborative study on American black bears. *Human–Wildlife Interactions* 12: 396–404.

Beecham, J. J., Hernando, M. D. G., Karamanlidis, A. A., et al. (2015). Management implications for releasing orphaned, captive-reared bears back to the wild. *Journal of Wildlife Management* 79: 1327–1336.

Beeman, L. E. & Pelton, M. R. (1980). Seasonal foods and feeding ecology of black bears in the Smoky Mountains. *Ursus* 4: 141–147.

Belant, J. L., Kielland, K., Follmann, E. H., et al. (2006). Interspecific resource partitioning in sympatric ursids. *Ecological Applications* 16: 2333–2343.

Beston, J. A. (2011). Variation in life history and demography of the American black bear. *Journal of Wildlife Management* 75: 1588–1596.

Binninger, C. E., Beecham, J. J., Thomas, L. A., et al. (1980). A serologic survey for selected infectious diseases of black bears in Idaho. *Journal of Wildlife Diseases* 16: 423–430.

Black, W., Troyer, R. M., Coutu, J., et al. (2019). Identification of gammaherpesvirus infection in free-ranging black bears (*Ursus americanus*). *Virus Research* 259: 46–53.

Blair, C. D., Muller, L. I., Clark, J. D., et al. (2020). Survival and conflict behavior of American black bears after rehabilitation. *Journal of Wildlife Management* 84: 75–86.

Boone, W. R., Keck, B. B., Catlin, J. C., et al. (2004). Evidence that bears are induced ovulators. *Theriogenology* 61: 1163–1169.

Bowman, J. L., Leopold, B. D., Vilella, F. J., et al. (2001). Attitudes of landowners towards American black bears compared between areas of high and low bear populations. *Ursus* 12: 153–160.

Bowman, J. L., Leopold, B. D., Vilella, F. J., et al. (2004). A spatially explicit model, derived from demographic variables, to predict attitudes towards black bear restoration. *Journal of Wildlife Management* 68: 223–232.

Bradburd, G. S., Coop, G. M. & Ralph, P. L. (2018). Inferring continuous and discrete population genetic structure across space. *Genetics* 210: 33–52.

Bridges, A. S., Vaughn, M. R. & Klenzendorf, S. (2004). Seasonal variation in American black bear *Ursus americanus* activity patterns: quantification via remote photography. *Wildlife Biology* 10: 277–284.

Brodeur, V., Ouellet, J. P., Courtois, R., et al. (2008). Habitat selection by black bears in an intensively logged boreal forest. *Canadian Journal of Zoology* 86: 1307–1316.

Brody, A. J. & Pelton, M. R. (1989). Effects of roads on black bear movements in western North Carolina. *Wildlife Society Bulletin* 17: 5–10.

Bunnell, F. L. & Tait, D. E. N. (1981). Population dynamics of bears – implications. In: Smith, T. D. & Fowler, C. (Eds.), *Dynamics of large mammal populations* (pp. 75–98). New York, NY: John Wiley and Sons, Inc.

Burst, T. L. & Pelton, M. R. (1983). Black bear mark trees in the Smoky Mountains. *Ursus* 5: 45–53.

Cameron, A. W. (1957). A new black bear from Newfoundland. *Journal of Mammalogy* 37:538–540.

Carlstead, K., Seidensticker, J. & Baldwin, R. (1991). Environmental enrichment for zoo bears. *Zoobiology* 10: 3–16.

Chambers, D. L., Ulrey, W. A., Guthrie, J. M., et al. (2012). Seroprevalence of *Toxoplasma gondii* from free-ranging black bears (*Ursus americanus*) from Florida. *Journal of Parasitology* 98: 674–675.

Clark, J. D. (2004). Oak–black bear relationships in southeastern uplands. In: Spetich, M. (Ed.), *Upland Oak Ecology Symposium: History, current conditions, and sustainability* (pp. 116–119). Technical Report SRS-73. Asheville, NC: General US Department of Agriculture, Forest Service, Southern Research Station.

Clark, J. D., Clapp, D. L., Smith, K. G., et al. (1994). Black bear habitat use in relation to food availability in the Interior Highlands of Arkansas. *Ursus* 9(1): 309–318.

Clark, J. D., Laufenberg, J. S., Davidson, M., et al. (2015). Connectivity among subpopulations of Louisiana black bears as estimated by a step-selection function. *Journal of Wildlife Management* 79: 1347–1360.

Clevenger, A. P. & Pelton, M. R. (1990). Pre and post breakup movements and space use of black bear family groups in Cherokee National Forest, Tennessee. *Ursus* 8: 289–295.

Clover, J. R., Hofstra, T. D., Kuluris, B. G., et al. (1989). Serologic evidence of *Yersinia pestis* infection in small mammals and bears from a temperate rainforest of north coastal California. *Journal of Wildlife Diseases* 25: 52–60.

Costello, C. M. & Sage, R. W., Jr. (1994). Predicting black bear habitat selection from food abundance under 3 forest management systems. *Ursus* 9(1): 375–387.

Costello, C. M., Creel, S. R., Kalinowski, S. T., et al. (2008). Sex-biased natal dispersal and inbreeding avoidance in American black bears as revealed by spatial genetic analyses. *Molecular Ecology* 17: 4713–4723.

Cottrell, W. O., Keel, M. K., Brooks, J. W., et al. (2013). First report of clinical disease associated with canine distemper virus infection in a wild black bear (*Ursus americanus*). *Journal of Wildlife Diseases* 49: 1024–1027.

Crum, J. M., Nettles, V. F. & Davidson, W. R. (1978). Studies on endoparasites of the black bear (*Ursus americanus*) in the southeastern United States. *Journal of Wildlife Diseases* 14: 178–186.

Cushman, S. A. & Lewis, J. S. (2010). Movement behavior explains genetic differentiation in American black bears. *Landscape Ecology* 25: 1613–1625.

Czetwertynski, S. M., Boyce, M. S. & Schmiegelow, F. K. (2007). Effects of hunting on demographic parameters of American black bears. *Ursus* 18: 1–18.

Dall, W. H. (1895). The St. Elias bear. *Science* 2: 87.

Delfín-Alfonso, C. A., López-González, C. A. & Equihua, M. (2012). Potential distribution of American black bears in northwest Mexico and implications for their conservation. *Ursus* 23: 65–77.

Ditmer, M. A., Noyce, K. V., Fieberg, J. R., et al. (2018). Delineating the ecological and geographic edge of an opportunist: the American black bear exploiting an agricultural landscape. *Ecological Modelling* 387: 205–219.

Doan-Crider, D. & Hellgren, E. C. (1996). Population characteristics and winter ecology of black bears in Coahuila, Mexico. *Journal of Wildlife Management* 60: 398–407.

Doan-Crider, D. & Hewitt, D. G. (2007). Metapopulations, food, and people: bear management in northern Mexico. In: Fulbright, T. E. & Hewitt, D. G. (Eds.), *Wildlife science: Linking ecological theory and management applications* (pp. 177–194). Boca Raton, FL: CRC Press.

Don Carlos, A. W., Bright, A. D., Teel, T. L., et al. (2009). Human–black bear conflict in urban areas: an integrated approach to management response. *Human Dimensions of Wildlife* 14: 174–184.

Draheim, H. M., Moore, J. A., Fortin, M. J., et al. (2018). Beyond the snapshot: landscape genetic analysis of time series data reveal responses of American black bears to landscape change. *Evolutionary Applications* 11: 1219–1230.

Duffy, M. S., Greaves, T. A. & Burt, M. D. B. (1994). Helminths of the black bear, *Ursus americanus*, in New Brunswick. *Journal of Parasitology* 80: 478–480.

Dunbar, M. R., Cunningham, M. W., Wooding, J. B., et al. (1996). Cryptorchidism and delayed testicular descent in Florida black bears. *Journal of Wildlife Diseases* 32: 661–664.

Elliot, D. G. (1903). Descriptions of apparently new species of the genera *Heteromys* and *Ursus* from Mexico and Washington. *Field Columbian Museum Publications, Zoology Series 3* 80: 233–237.

Elowe, K. D. & Dodge, W. E. (1989). Factors affecting black bear reproductive success and cub survival. *Journal of Wildlife Management* 53: 962–968.

Farajollahi, A., Panella, N. A., Carr, P., et al. (2003). Serologic evidence of West Nile virus infection in black bears (*Ursus americanus*) from New Jersey. *Journal of Wildlife Diseases* 39: 894–896.

Ferreira, A. S., Peres, C. A., Bogoni, J. A., et al. (2018). Use of agroecosystem matrix habitats by mammalian carnivores (Carnivora): a global-scale analysis. *Mammal Review* 48: 312–327.

Fitzgerald, S. D., Cooley, T. M. & Cosgrove, M. K. (2008). Sarcoptic mange and *Pelodera* dermatitis in an American black bear (*Ursus americanus*). *Journal of Zoo and Wildlife Medicine* 39: 257–259.

Florida Fish and Wildlife Conservation Commission. (2011). Florida black bear status review report. https://myfwc.com/media/1960/florida-black-bear-bsr.pdf. Accessed January 28, 2019.

Forrester, D. J., Spalding, M. G. & Wooding, J. B. (1993). Demodicosis in black bears (*Ursus americanus*) from Florida. *Journal of Wildlife Diseases* 29: 136–138.

Fortin, J. K., Farley, S. D., Rode, K. D., et al. (2007). Dietary and spatial overlap between sympatric ursids relative to salmon use. *Ursus* 18(1): 19–30.

Fortin, J. K., Ware, J. V., Jansen, H. T., et al. (2013a). Temporal niche switching by grizzly bears but not American black bears in Yellowstone National Park. *Journal of Mammalogy* 94: 833–844.

Fortin, J. K., Schwartz, C. C., Gunther, K. A., et al. (2013b). Dietary adjustability of grizzly bears and American black bears in Yellowstone National Park. *Journal of Wildlife Management* 77: 270–281.

Free, S. L. & McCaffrey, E. (1972). Reproductive synchrony in the female black bear. *Ursus* 2: 199–206

Gantchoff, M., Wang, G. M., Beyer, D., et al. 2018. Scale-dependent home range optimality for a solitary omnivore. *Ecology and Evolution* 8: 12,271–12,282.

Garrison, E. P., McCown, J. W. & Oli, M. K. (2007). Reproductive ecology and cub survival of Florida black bears. *Journal of Wildlife Management* 71: 720–727.

Garshelis, D. L. & Hristienko, H. (2006). State and provincial estimates of American black bear numbers versus assessments of population trend. *Ursus* 17: 1–7.

Garshelis, D. L. & Pelton, M. R. (1980). Activity of black bears in the Great Smoky Mountains National Park. *Journal of Mammalogy* 61: 8–19.

Garshelis, D. L. & Pelton, M. R. (1981). Movements of black bears in the Great Smoky Mountains National Park. *Journal of Wildlife Management* 45: 912–925.

Garshelis, D. L., Quigley, H. B., Villarubia, C. R., et al. (1983). Diel movements of black bears in the southern Appalachians. *Ursus* 5: 11–19.

Garshelis, D. L., Gibeau, M. L. & Herrero, S. (2005). Grizzly bear demographics in and around Banff National Park and Kananaskis country, Alberta. *Journal of Wildlife Management* 69: 277–297.

Gilpin, M. E. & Soulé, M. E. (1986). Minimum viable populations: processes of species extinction. In: Soulé, M. E. (Ed.), *Conservation biology: The science of scarcity and diversity* (pp. 19–34). Sunderland, MA: Sinauer Associates.

Gonzales, R. L., Mendoza, A. V., Himelright, B. M., et al. (2013). American black bear mating behavior and chemosensation of estrus. *Ursus* 24(2): 139–147.

Greenwald, G. S. (1956). The reproductive cycle of the field mouse, *Microtus californicus*. *Journal of Mammalogy* 37: 213–222.

Greenwood, P. J. (1980). Mating systems, philopatry and dispersal in birds and mammals. *Animal Behaviour* 28: 1140–1162.

Griffith, E. (1821). *General and particular descriptions of the vertebrated animals arranged conformably to the modern discoveries and improvements in zoology. Order Carnivora*. London, UK: Baldwin, Cradock and Joy.

Hall, E. R. (1928). A new race of black bear from Vancouver Island, British Columbia, with remarks on other northwest coast forms of *Euarctos*. *University of California Publications of Zoology* 30: 231–242.

Hall, E. R. (1981). *The mammals of North America*, 2nd edn. New York, NY: John Wiley and Sons.

Heldstab, S. A., Müller, D. W., Graber, S. M., et al. (2018). Geographical origin, delayed implantation, and induced ovulation explain reproductive seasonality in the Carnivora. *Journal of Biological Rhythms* 33: 402–419.

Hellgren, E. C. & Vaughan, M. R. (1989). Denning ecology of black bears in a southeastern wetland. *Journal of Wildlife Management* 53: 347–353.

Hellgren, E. C., Vaughan, M. R., Gwazdauskas, F. C., et al. (1990). Endocrine and electrophoretic profiles during pregnancy and nonpregnancy in captive female black bears. *Canadian Journal of Zoology* 69: 892–898.

Hellgren, E. C., Vaughan, M. R. & Stauffer, D. F. (1991). Macrohabitat use by black bears in a southeastern wetland. *Journal of Wildlife Management* 55: 442–448.

Hellgren, E. C., Onorato, D. P. & Skiles, J. R. (2005). Dynamics of a black bear population within a desert metapopulation. *Biological Conservation* 122: 131–140.

Herrero, S. M. (1978). A comparison of some features of the evolution, ecology, and behavior of black and grizzly/brown bears. *Carnivore* 1: 7–17.

Herrero, S., Higgins, A., Cardoza, J. E., et al. (2011). Fatal attacks by American black bear on people: 1900–2009. *Journal of Wildlife Management* 75: 596–603.

Holm, G., Lindzey, F. & Moody, D. (1999). Interactions of sympatric black and grizzly bears in northwest Wyoming. *Ursus* 11: 99–108.

Hornaday, W. T. (1905). A new white bear from British Columbia. *Annual Report of the New York Zoological Society* 9: 81–86.

Horner, M. A. & Powell, R. A. (1990). Internal structure of home ranges of black bears and analyses of home-range overlap. *Journal of Mammalogy* 71: 402–410.

Horstman, L. P. & Gunson, J. R. (1982). Black bear predation on livestock in Alberta. *Wildlife Society Bulletin* 10: 34–39.

Hristienko, H. & McDonald, J. E., Jr. (2007). Going into the 21st century: a perspective on trends and controversies in the management of the American black bear. *Ursus* 18: 72–88.

Humm, J. M., McCown, J. W., Scheick, B. K., et al. (2017). Spatially explicit population estimates for black bears based on cluster sampling. *Journal of Wildlife Management* 81: 1187–1201.

Johnson, H. E., Breck, S. W., Baruch-Mordo, S., et al. (2015). Shifting perceptions of risk and reward: dynamic selection for human development by black bears in the western United States. *Biological Conservation* 187: 164–172.

Johnson, K. G., Johnson, D. O. & Pelton, M. R. (1978). Simulation of winter heat loss for a black bear in a closed tree den. *Proceedings of the Eastern Black Bear Workshop* 4: 155–156.

Jones, M. D., Tri, A. N., Edwards, J. W., et al. (2015). Home-range dynamics of female *Ursus americanus* (Pallas) (American black bear) in a recovering population in western Maryland. *Northeastern Naturalist* 22: 830–841.

Jonkel, C. J. & Cowan, I. McT. (1971). The black bear in the spruce-fir forest. *Wildlife Monographs* 27.

Karelus, D. L., McCown, J. W., Scheick, B. K., et al. (2016). Home ranges and habitat selection by black bears in a newly colonized population in Florida. *Southeastern Naturalist* 15: 346–364.

Karelus, D. L., McCown, J. W., Scheick, B. K., et al. (2018). Microhabitat features influencing habitat use by Florida black bears. *Global Ecology and Conservation* 13: e00367.

Kasworm, W. F. & Manley, T. L. (1990). Road and trail influences on grizzly bears and black bears in northwest Montana. *Ursus* 8: 79–84.

Kasworm, W. F. & Their, T. J. (1994). Adult black bear reproduction, survival, and mortality sources in northwest Montana. *Ursus* 9(1): 223–230.

Kellert, S. R. (1992). Public attitudes towards bears and their conservation. *Ursus* 9(1): 43–50.

Kilham, B. & Gray, E. (2002). *Among the bears: Raising orphan cubs in the wild*. New York, NY: Henry Holt and Company.

Koehler, G. M. & Pierce, D. J. (2003). Black bear home-range sizes in Washington: climatic, vegetative, and social influences. *Journal of Mammalogy* 84: 81–91.

Kolenosky, G. B. (1990). Reproductive biology of black bears in east-central Ontario. *Ursus* 8: 385–392.

Kolenosky, G. B. & Strathearn, S. M. (1987). Winter denning of black bears in east-central Ontario. *Ursus* 7: 305–316.

Kordek, W. S. & Lindzey, J. S. (1980). Preliminary analysis of female reproductive tracts from Pennsylvania black bears. *Ursus* 4: 159–161.

Kovach, A. I. & Powell, R. A. (2003). Effects of body size on male mating tactics and paternity in black bears, *Ursus americanus*. *Canadian Journal of Zoology* 81: 1257–1268.

Kumar, V., Lammers, F., Bidon, T., et al. (2017). The evolutionary history of bears is characterized by gene flow across species. *Scientific Reports* 7: 46487.

Kurtén, B. & Anderson, E. (1994). *Pleistocene mammals of North America*. New York, NY: Columbia University Press.

Lackey, C. W., Beckmann, J. P. & Sedinger, J. (2013). Bear historical ranges revisited: documenting the increase of a once extirpated population in Nevada. *Journal of Wildlife Management* 77: 812–820.

Land, E. D. (1994). *Southwest Florida black bear habitat use, distribution, movements, and conservation strategy*. Final Report W-41-32. Tallahassee, FL: Florida Game and Fresh Water Fish Commission.

Landa, R., Meave, J. & Carabias, J. (1997). Environmental deterioration in rural Mexico: an examination of the concept. *Ecological Applications* 7: 316–329.

Landers, J. L., Hamilton, R. J., Johnson, A. S., et al. (1979). Foods and habitat of black bears in southeastern North Carolina. *Journal of Wildlife Management* 43: 143–153.

Larivière, S. (2001). *Ursus americanus*. *Mammalian Species* 647: 1–11.

Latham, A. D. M., Latham, M. C. & Boyce, M. S. (2011). Habitat selection and spatial relationships of black bears (*Ursus americanus*) with woodland caribou (*Rangifer tarandus caribou*) in northeastern Alberta. *Canadian Journal of Zoology* 89: 267–277.

Laufenberg, J. S., Clark, J. D., Hooker, M. J., et al. (2016). Demographic rates and population viability of black bears in Louisiana. *Wildlife Monographs* 194.

Law, G. & Reid, A. (2010). Enriching the lives of bears in zoos. *International Zoo Yearbook* 44: 65–74.

LeCount, A. L. (1982). Characteristics of a central Arizona black bear population. *Journal of Wildlife Management* 46: 861–868.

LeCount, A. L. (1983). Denning ecology of black bears in central Arizona. *Ursus* 5: 71–78.

LeCount, A. L. (1987). Causes of black bear cub mortality. *Ursus* 7: 75–82.

Lee, D. J. & Vaughan, M. R. (2003). Dispersal movements by subadult American black bears in Virginia. *Ursus* 14(2): 162–170.

Lewis, J. S. & Rachlow, J. L. (2011). Activity patterns of black bears in relation to sex, season, and daily movement rates. *Western North American Naturalist* 71: 388–396.

Lindzey, F. G. & Meslow, C. E. (1977a). Home range and habitat use by black bears in southwestern Washington. *Journal of Wildlife Management* 41: 413–425.

Lindzey, F. G. & Meslow, E. C. (1977b). Population characteristics of black bears on an island in Washington. *Journal of Wildlife Management* 41: 408–412.

Lindzey, J. S., Alt, G. L., McLaughlin, C. R., et al. (1983). Population response of Pennsylvania black bears to hunting. *Ursus* 5: 34–39.

Linnaeus, C. (1758). *Systema naturae* (Vol. 1). Stockholm: Laurentii Salvii.

Linnell, J. D., Aanes, R. & Andersen, R. (1995). Who killed Bambi? The role of predation in the neonatal mortality of temperate ungulates. *Wildlife Biology* 1: 209–224.

Loosen, A., Morehouse, A. T. & Boyce, M. S. (2019). Land tenure shapes black bear density and abundance on a multi-use landscape. *Ecology and Evolution* 9: 73–89.

MacHutchon, A. G. (1989). Spring and summer food habits of black bears in the Pelly River Valley, Yukon. *Northwest Science* 63: 116–118.

Maehr, D. S. & Brady, J. R. (1984). Food habits of Florida black bears. *Journal of Wildlife Management* 48: 230–235.

Maehr, D. S., Hellgren, E. C., Bingham, R. L., et al. (2001). Body mass of American black bears from Florida and Mexico. *Southwestern Naturalist* 46: 129–133.

Manville, A. M. (1978). Ecto- and endoparasites of the black bear in northern Wisconsin. *Journal of Wildlife Diseases* 14: 97–101.

Mathews, N. E. & Porter, W. F. (1988). Black bear predation of white-tailed deer neonates in the central Adirondacks. *Canadian Journal of Zoology* 66: 1241–1242.

Mattson, D. J., Knight, R. R. & Blanchard, B. M. (1992). Cannibalism and predation on black bears by grizzly bears in the Yellowstone ecosystem, 1975–1990. *Journal of Mammalogy* 73: 422–425.

McDonald, J. E. & Fuller, T. K. (2001). Prediction of litter size in American black bears. *Ursus* 12: 93–102.

McLaughlin, C. R., Matula, G. J., Jr & O'Connor, R. J. (1994). Synchronous reproduction by Maine black bears. *Ursus* 9(1): 471–479.

McLean, P. K. & Pelton, M. R. (1990). Some demographic comparisons of wild and panhandler bears in the Smoky Mountains. *Ursus* 8: 105–112.

McLellan, B. N. (2011). Implications of a high-energy and low-protein diet on the body composition, fitness, and competitive abilities of black (*Ursus americanus*) and grizzly (*Ursus arctos*) bears. *Canadian Journal of Zoology* 89: 546–558.

McLellan, B. & Reiner, D. C. (1994). A review of bear evolution. *Ursus* 9(1): 85–96.

Merkle, J. A., Robinson, H. S., Krausman, P.R., et al. (2013). Food availability and foraging near human developments by black bears. *Journal of Mammalogy* 94: 378–385.

Merriam, C. H. (1904). Four new bears from North America. *Proceedings of the Biological Society of Washington* 17: 153–156.

Miller, J. (1900). *True bear stories*. Chicago, IL: Rand McNally and Company.

Miller, S. D. (1994). Black bear reproduction and cub survivorship in south-central Alaska. *Ursus* 9(1): 263–273.

Moore, J. A., Xu, R., Frank, K., et al. (2015). Social network analysis of mating patterns in American black bears (*Ursus americanus*). *Molecular Ecology* 24: 4010–4022.

Morehouse, A. T. & Boyce, M. S. (2017). Troublemaking carnivores: conflicts with humans in a diverse assemblage of large carnivores. *Ecology and Society* 22: 4.

Mosnier, A., Ouellet, J. P. & Courtois, R. (2008). Black bear adaptation to low productivity in the boreal forest. *Ecoscience* 15: 485–497.

Moyer, M. A., McCown, J. W. & Oli, M. K. (2007). Factors influencing home-range size of female Florida black bears. *Journal of Mammalogy* 88: 468–476.

Murphy, S. M., Cox, J. J., Clark, J. D., et al. (2015). Rapid growth and genetic diversity retention in an isolated reintroduced black bear population in the Central Appalachians. *Journal of Wildlife Management* 79: 807–818.

Murphy, S. M., Hast, J. T., Augustine, B. C., et al. (2019). Early genetic outcomes of American black bear reintroductions in the Central Appalachians, USA. *Ursus* 29 (2): 119–133.

Nelson, R. A., Folk, G. E., Jr., Pfeiffer, E. W., et al. (1983). Behavior, bio-chemistry, and hibernation in black, grizzly, and polar bears. *Ursus* 5: 284–290.

New York State Department of Environmental Conservation. (2007). *Black bears in New York: Natural history, range, and interactions with people*, 2nd edition. Albany, NY: New York State Department of Environmental Conservation.

Niedringhaus, K. D., Brown, J. D., Ternent, M., et al. (2019). The emergence and expansion of sarcoptic mange in American black bears (*Ursus americanus*) in the United States. *Veterinary*

Parasitology: Regional Studies and Reports 17: 100303.

Norton, D. C., Belant, J. L., Bruggink, J. G., et al. (2018). Female American black bears do not alter space use or movements to reduce infanticide risk. *PLoS ONE* 13(9): e0203651.

Noyce, K. V. & Garshelis, D. L. (1998). Spring weight changes in black bears in northcentral Minnesota: the negative foraging period revisited. *Ursus* 10: 521–531.

Noyce, K. V. & Garshelis, D. L. (2014). Follow the leader: social cues help guide landscape-level movements of American black bears (*Ursus americanus*). *Canadian Journal of Zoology* 92: 1005–1017.

Oftedal, O. T., Alt, G. L., Widdowson, E. M., et al. (1993). Nutrition and growth of suckling black bears (*Ursus americanus*) during their mothers' winter fast. *British Journal of Nutrition* 70: 59–79.

Oli, M. K., Jacobson, H. A. & Leopold, B. D. (1997). Denning ecology of black bears in the White River National Wildlife Refuge, Arkansas. *Journal of Wildlife Management* 61: 700–706.

Ombrello, T. A., Chinnici, N. L. & Huffman, J. E. (2016). Multiple paternities in American black bears from New Jersey. *Journal of the Pennsylvania Academy of Science* 90: 21–24.

Onorato, D. P., Hellgren, E. C., Van Den Bussche, R. A., et al. (2004). Phylogeographic patterns within a metapopulation of black bears (*Ursus americanus*) in the American Southwest. *Journal of Mammalogy* 85: 140–147.

Onorato, D. P., Hellgren, E. C., Van Den Bussche, R. A., et al. (2007). Genetic structure of American black bears in the desert southwest of North America: conservation implications for recolonization. *Conservation Genetics* 8: 565–576.

Osgood, W. H. (1901). Natural history of the Queen Charlotte Islands, British Columbia and natural history of the Cook Inlet region, Alaska. *North American Fauna* 21: 1–87.

Pallas, P. S. (1779). Spicilegia Zoologica, quibus novae imprimis et obcurae animalium species iconibus. Berolini, G.A. Lange, 1774–1780. *Fasicle* 14: 5.

Pelchat, B. O. & Ruff, R. L. (1986). Habitat and spatial relationships of black bears in boreal mixedwood forest of Alberta. *Ursus* 6: 81–92.

Pelton, M. R. (2003). Black bear. In: Feldhamer, G. A., Thompson, B. C. & Chapman, J. A. (Eds.), *Wild mammals of North America: Biology, management, and economics* (pp. 504–514). Baltimore, MD: Johns Hopkins University Press.

Pelton, M. R. & van Manen, F. (1994). Distribution of black bears in North America. *Proceedings of the Eastern Workshop on Black Bear Research and Management* 12: 133–138.

Pelton, M. R., Beeman, L. E. & Eagar, D. C. (1980). Den selection by black bears in the Great Smoky Mountains National Park. *Ursus* 4: 149–151.

Poelker, R. J. & Hartwell, H. D. (1973). *Black bear of Washington*. Seattle, WA: Washington State Game Department Biological Bulletin No. 14.

Powell, R. A., Zimmerman, J. W. & Seaman, D. E. (1997). *Ecology and behaviour of North American black bears: Home ranges, habitat and social organization.* New York, NY: Chapman and Hall.

Pritchard, G. T. & Robbins, C. T. (1990). Digestive and metabolic efficiencies of grizzly and black bears. *Canadian Journal of Zoology* 68: 1645–1651.

Proctor, M. F., Kasworm, W., Teisberg, J., et al. (2020). American black bear population fragmentation detected with pedigrees in the trans-border Canada–United States region. *Ursus* 31(e1).

Puckett, E. E., Etter, P. D., Johnson, E. A., et al. (2015). Phylogeographic analyses of American black bears (*Ursus americanus*) suggest four glacial refugia and complex patterns of postglacial admixture. *Molecular Biology and Evolution* 32: 2338–2350.

Quigley, H. B. (1982). Activity patterns, movement ecology and habitat utilization of black bears in the Great Smoky Mountains National Park, Tennessee. Masters thesis, University of Tennessee.

Raine, R. M. & Kansas, J. L. (1990). Black bear seasonal food habits and distribution by elevation in Banff National Park, Alberta. *Ursus* 8: 297–304.

Raybourne, J. W. (1987). The black bear: home in the highlands. In: Kallman, H., Agee, C. P., Goforth, W. R., et al. (Eds.), *Restoring America's wildlife: 1937–1987* (pp. 105–117). Washington, DC: U. S. Fish and Wildlife Service, U. S. Government Printing Office.

Rayl, N. D., Bastille-Rousseau, G., Organ, J. F., et al. (2018). Spatiotemporal heterogeneity in prey abundance and vulnerability shapes the foraging tactics of an omnivore. *Journal of Animal Ecology* 87: 874–887.

Reimchen, T. E. (1998). Nocturnal foraging behaviour of black bears, *Ursus americanus*, on Moresby Island, British Columbia. *Canadian Field-Naturalist* 112: 446–450.

Reimchen, T. E. (2000). Some ecological and evolutionary aspects of bear–salmon interactions in coastal British Columbia. *Canadian Journal of Zoology* 78: 448–457.

Robbins, C. T., Ben-David, M., Fortin, J. K., et al. (2012). Maternal condition determines birth date and growth of newborn bear cubs. *Journal of Mammalogy* 93: 540–546.

Rockwell, D. (1991). *Giving voice to bear: North American Indian myths, rituals, and images of the bear.* Lanham, MD: Roberts Rinehart Publishers.

Rogers, L. L. (1976). Effects of mast and berry crop failures on survival, growth and reproductive success of black bears. *Transactions of the North American Wildlife and Natural Resources Conference* 41: 431–438.

Rogers, L. L. (1980). Inheritance of coat color and changes in pelage coloration in black bears in northeastern Minnesota. *Journal of Mammalogy* 61: 324–327.

Rogers, L. L. (1987). Effects of food supply and kinship on social behavior, movements, and population growth of black bears in northeastern Minnesota. *Wildlife Monographs* 97.

Rounds, R. C. (1987). Distribution and analysis of colourmorphs of the black bear (*Ursus americanus*). *Journal of Biogeography* 14: 521–538.

Scheick, B. & McCown, W. (2014). Geographic distribution of American black bears in North America. *Ursus* 25: 24–33.

Schenk, A. & Kovacs, K. M. (1995). Multiple mating between black bears revealed by DNA fingerprinting. *Animal Behavior* 50: 1483–1490.

Schlegel, M. (1976). Factors affecting elk calf survival in north central Idaho: a progress report. *Proceedings of the Annual Conference of the Western Association of State Game and Fish Commissioners* 56: 342–355.

Schlegel, M. W., Knapp, S. E. & Millemann, R. E. (1968). 'Salmon poisoning' disease. V. Definitive hosts of the trematode vector, *Nanophyetus salmincola*. *Journal of Parasitology* 54: 770–774.

Schmitt, S. M., Cooley, T.M., Friedrich, P.D., et al. (1987). Clinical mange of the black bear (*Ursus americanus*) caused by *Sarcoptes scabiei* (Acarina, Sarcoptidae). *Journal of Wildlife Diseases* 23: 162–165.

Schwartz, C. C. & Franzmann, A.W. (1991). Interrelationships of black bears to moose and forest succession in the northern coniferous forest. *Wildlife Monographs* 113.

Schwartz, C. C. & Franzmann, A. W. (1992). Dispersal and survival of subadult black bears from the Kenai Peninsula, Alaska. *Journal of Wildlife Management* 56: 426–431.

Servheen, C. (1990). The status and conservation of the bears of the world. International Conference Bear Research and Management. Monograph Series No. 2.

Shakeri, Y. N., White, K. S. & Levi, T. (2018). Salmon-supported bears, seed dispersal, and extensive resource subsidies to granivores. *Ecosphere* 9: e02297.

Short, H. L. (1976). Composition and squirrel use of acorns of black and white oak groups. *Journal of Wildlife Management* 40: 479–483.

Stephenson, N., Higley, J. M., Sajecki, J. L., et al. (2015). Demographic characteristics and infectious diseases of a population of American black bears in Humboldt County, California. *Vector-Borne and Zoonotic Diseases* 15: 116–123.

Swarth, H. S. (1911). Birds and mammals of the 1909 Alexander Alaskan Expedition. *University of California Publications of Zoology* 7: 9–172.

Taylor, A. P., Allen, M. L. & Gunther, M. S. (2015). Black bear marking behaviour at rub trees during the breeding season in northern California. *Behaviour* 152: 1097–1111.

US Census Bureau. (2010). *United States census 2010*. Washington, DC: US Census Bureau, http://2010.census.gov/2010census/data/index.php. Accessed April 1, 2012.

Valdez, R., Guzmán-Aranda, J. C., Abarca, F. J., et al. (2006). Wildlife conservation and management in Mexico. *Wildlife Society Bulletin* 34: 270-282

van Keulen-Kromhout, G. (1978). Zoo enclosures for bears: their influence on captive behavior and reproduction. *International Zoo Yearbook* 18: 177–186.

van Manen, F. T., McCollister, M. F., Nicholson, J.M., et al. (2012). Short-term impacts of a 4-lane highway on American black bears in eastern North Carolina. *Wildlife Monographs* 181.

Vaughan, M. R. (2002). Oak trees, acorns, and bears. In: McShea, W. J. & Healy, W. M. (Eds.), *Oak forest ecosystems: Ecology and management for wildlife* (pp. 224–240). Baltimore, MD: The Johns Hopkins Press.

Vonk, J., Jett, S. E. & Mosteller, K. W. (2012). Concept formation in American black bears, *Ursus americanus*. *Animal Behaviour* 84: 953–964.

Walroth, R., Brown, N., Wandeler, A., et al. (1996). Rabid black bears in Ontario. *Canadian Veterinary Journal* 37: 492.

Wear, B. J., Eastridge, R. & Clark, J. D. (2005). Factors affecting settling, survival, and viability of black bears reintroduced to Felsenthal National Wildlife Refuge, Arkansas. *Wildlife Society Bulletin* 33: 1363–1374.

Weaver, K. M. & Pelton, M. R. (1994). Denning ecology of black bears in the Tensas River basin of Louisiana. *Ursus* 9(1): 427–433.

White, T. H., Jr., Shropshire, C. S. & Staten, M. (1995). Black bear damage in the Mississippi Alluvial Valley. *Eastern Wildlife Damage Conference* 7: 109–117.

Williams, S. C. (1930). *Adair's history of the American Indians*. New York, NY: Promontory Press.

Wynn-Grant, R., Ginsberg, J. R., Lackey, C. W., et al. (2018). Risky business: modeling mortality risk near the urban-wildland interface for a large carnivore. *Global Ecology and Conservation* 16: e00443.

Yabsley, M. J., Nims, T. N., Savage, M. Y., et al. (2009). Ticks and tick-borne pathogens and putative symbionts of black bears (*Ursus americanus floridanus*) from Georgia and Florida. *Journal of Parasitology* 95: 1125–1128.

Young, B. F. & Ruff, R. L. (1982). Population dynamics and movements of black bears in east central Alberta. *Journal of Wildlife Management* 46: 845–860.

Chapter 12

Brown Bear (*Ursus arctos*; Eurasia)

Jon E. Swenson, Hüseyin Ambarlı, Jon M. Arnemo, Leonid Baskin, Paolo Ciucci, Pjotr I. Danilov, Miguel Delibes, Marcus Elfström, Alina L. Evans, Claudio Groff, Anne G. Hertel, Djuro Huber, Klemen Jerina, Alexandros A. Karamanlidis, Jonas Kindberg, Ilpo Kojola, Miha Krofel, Josip Kusak, Tsutomu Mano, Mario Melletti, Yorgos Mertzanis, Andrés Ordiz, Santiago Palazón, Jamshid Parchizadeh, Vincenzo Penteriani, Pierre-Yves Quenette, Agnieszka Sergiel, Nuria Selva, Ivan Seryodkin, Michaela Skuban, Sam M.J.G. Steyaert, Ole-Gunnar Støen, Konstantin F. Tirronen, and Andreas Zedrosser

Names

Genus: *Ursus* (Linnaeus, 1758)

Species: *Ursus arctos* Linnaeus, 1758

Subspecies: *U a. arctos*; *U. a. isabellinus*; *U. a. syriacus*; *U. a. pruinosus*; *U. a. lasiotus*; *U. a. collaris*; *U. a. beringianus/piscator*

Names in other languages:

Albanian, ariu; **Arabic**, دب بنى (daba bunaa) or دب أسمر (daba asmar); **Armenian**, արջ (arj); **Basque**, hartz; **Belarussian**, мядзведзь (mjadzvedz'); **Breton** (Brezhoneg), arzh gell; **Bulgarian**, мечка (mečka); **Catalan**, ós bru; **Chinese**, 棕熊 (zōngxióng); **Croatian**, smeđi medvjed; **Czech**, mědved hnědý; **Danish**, brunbjørn; **English**, brown bear; **Esperanto**, urso; **Estonian**, karu;

Faroese, bjørn; **Farsi** (Iranian), خرس قهوه ای (khers-e qahvehei /ghahvehei); **Finnish**, karhu; **French**, ours brun; **Georgian**, დათვი (datvi), **German**, Braunbär; **Greek**, καφέ αρκούδα (kafe arkouda); **Hindi**, भालू (bhalu); **Hungarian**, barna medve; **Icelandic**, björn; **Irish** (Gaeilge), béar donn; **Italian**, orso bruno; **Japanese**, ヒグマ(higuma); **Korean**, 갈색 곰 (galsaeg gom); **Kurdish**, hirç; **Latvian**, brūnais lācis; **Lithuanian**, rudoji meška; **Macedonian**, мечка (mečka); **Maltese**, debb=ors; **Mongolian**, ᠬᠦᠷᠡᠨ ᠪᠠᠭᠠᠪᠭᠠᠢ or хүрэн баавгай (khuren baavgai); **Norwegian**, brunbjørn; **Polish**, niedźwiedź brunatny; **Portuguese**, urso; **Romanian**, urs; **Romansh**, urs; **Romany**, rish (m.) rishni (f.); **Russian**, бурый медведь (buryi medved'); **Sámi**, guovža; **Scottish** (Gaelic), mathan; **Serbian**, мрки медвед (mrki medved); **Slovak**, medveď hnedý; **Slovenian**, rjavi medved; **Spanish**, oso pardo; **Swedish**, brunbjörn; **Turkish**, bozayı; **Ukrainian**, бурий ведмідь (buryi vedmid'); **Welsh** (Cymraeg), arth.

Taxonomy

Phylogeography

For the systematics, evolution, and genetics of bears, see Chapter 1. The brown bear and cave bear (*Ursus spelaeus*) diverged in Eurasia approximately 1.2–1.4 million years ago (Loreille et al. 2001) and brown bears subsequently inhabited most of the continent. Here we review how climatic fluctuations during and after the Pleistocene shaped the genetic relationships within Eurasian brown bears, especially focusing on the effects of the Last Glacial Maximum (LGM, 26,000–19,000 years ago), when most of northern Eurasia was covered with continental ice sheets (Saarma et al. 2007).

In the mid-1990s, Randi et al. (1994), Taberlet and Bouvet (1994), and Kohn et al. (1995) studied the genetics of several small, endangered European brown bear populations, based on sequences of a portion of the mitochondrial DNA (mtDNA) control region, to assess their genetic status and to determine appropriate donor populations for possible augmentation attempts. They reported two main lineages, the western and eastern, which diverged approximately 514,000 years ago (Anijalg et al. 2018). This laid the groundwork for many studies on the phylogeography of Eurasian brown bears, making it a model species in studying post-Pleistocene migrations (Swenson et al. 2011). Based on additional mtDNA studies, these two lineages have been subdivided into clades and subclades, which apparently arose when the bears were confined to glacial refugia (Saarma et al. 2007; Davison et al. 2011; Anijalg et al. 2018).

The phylogeography of Eurasian brown bears is best documented for the western lineage in Europe. It encompasses the bears presently found in Western Europe and extinct populations in North Africa (Algeria), Lebanon (Calvignac et al. 2008, 2009), and in Anatolia, Turkey (Çilingir et al. 2016). This lineage consists of two clades (1a and 1b), which diverged approximately 100,000 years ago (Davison et al. 2011). The important glacial refugium for clade 1a bears was the Iberian Peninsula, from which bears migrated northwards and recolonized central and northern Europe, Britain, Ireland, and southern Scandinavia. The glaciated Alps acted as a barrier for some of the clade 1b bears, which were in refugia on the Italian Peninsula, in the Balkans, and in the Carpathian Mountains and Anatolia, Turkey, where clade 3a bears also occur (Sommer & Benecke 2005; Bray et al. 2013; Çilingir et al. 2016). Clade 1b bears moved northward after most of Europe had been colonized by clade 1a bears, and were generally excluded from Western Europe, although a 5300-year-old clade 1b bear has been found in northern Denmark (Bray et al. 2013). In addition to these refugia, some brown bears also survived in the cold tundra–steppes of central Europe during the LGM, which was probably similar to the arctic habitats inhabited by brown bears in North America today (Valdiosera et al. 2007). It is not known if the now-extinct 1a brown bears in Algeria came there naturally from Spain or if they were introduced by Romans or Carthaginians (Calvignac et al. 2008). Subfossil samples of brown bears from an extinct population in Morocco are not closely aligned with any known clade of living brown bears and may have diverged while migrating to western North Africa from the Middle East (Calvignac et al. 2008).

The eastern lineage in Europe (clade 3a) is found in northern Scandinavia, Russia and adjacent countries (Finland, Estonia, Belarus), part of Romania (Taberlet & Bouvet 1994), and Anatolia, Syria, and Georgia (Çilingir et al. 2016). Saarma et al. (2007) suggested that clade 3a bears recolonized most of northern Europe from a refugium in the western Carpathians (Slovakia) and perhaps the adjacent Great Hungarian Plain. There is a sharp delineation between these bears and clade 1a bears at the zone of contact in Scandinavia (Taberlet et al. 1995), but the lineages are sympatric in eastern Romania and Turkey (Kohn et al. 1995; Çilingir et al. 2016).

The phylogeography of brown bears in Asia is less studied and appears to be more complicated. Prior to a climatic cooling period approximately 54,000 years ago (Anijalg et al. 2018), 3a clade bears occurred throughout Eurasia, extending from Romania and eastward to the border of Canada, 12,000 km away (Korsten et al. 2009), and into Turkey and Syria (Calvignac et al. 2009). Korsten et al. (2009) concluded that the founder population of this clade underwent a severe demographic bottleneck, followed by a sudden demographic expansion throughout most of continental Eurasia from a single glacial refugium after the LGM. Later, important Pleistocene refugia were probably the Carpathian Mountains in Eastern Europe, the Altai-Sayan Mountains in Central Asia (Anijalg et al. 2018), and perhaps the Mediterranean region of Turkey (Çilingir et al. 2016). Brown bears on Hokkaido belong to three allopatric clades, 3a, 3b, and 4, the latter two of which have been reported only from Hokkaido and may have reached the island at different times when

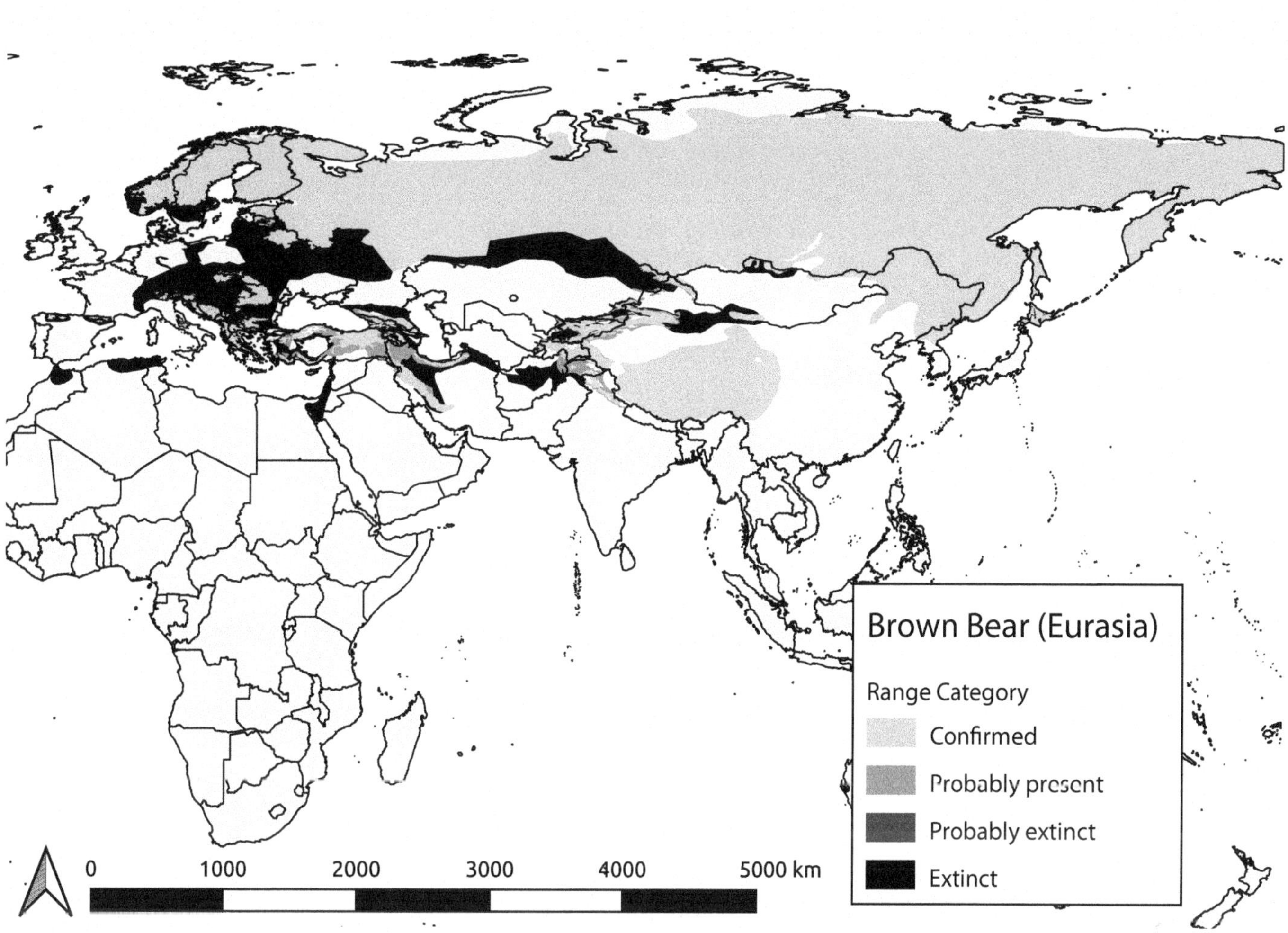

Figure 12.1 Eurasian brown bear distribution. (*Source:* IUCN 2017, Red List of Threatened Species.) IUCN only maps areas where species became extinct after about 1500 AD, so not all former distributional areas for brown bears are shown here. Please refer to Table 12.1.

Hokkaido was connected to the mainland (Hirata et al. 2013). Another Asian clade, clade 5, is known only from Tibet. It diverged approximately 343,000 years ago, suggesting that some brown bears dispersed to the Tibetan Plateau and thereafter remained isolated from the source population (Hirata et al. 2013). A clade 7 has also been identified in Iran and eastern Turkey (Calvignac et al. 2009; Çilingir et al. 2016).

Subspecies and Distribution

Only brown bear populations, not subspecies, are in the IUCN Red List. A list of the currently recognized subspecies of brown bears is provided in Chapter 1.

The brown bear is the most widespread bear species (Figure 12.1) and is found over portions of North America, Europe, and Asia (Table 12.1). The species is extinct in North Africa, where it disappeared from Egypt in the 1500s (Manlius 1998) and Algeria and Morocco in the 1800s (Hamdine et al. 1998).

In Europe, the brown bear has disappeared from many countries in the last centuries, due to high human pressure, habitat loss, and climate change (Albrecht et al. 2017). It occupies about 1,200,000 km^2 on the continent, excluding Russia, with an estimated population of 17,000 individuals in 10 populations (Chapron et al. 2014; McLellan et al. 2016; Huber 2018; Table 12.1). Russia has the largest brown bear population, estimated at over 200,000 individuals (www.ohotcontrol.ru/resource/number/; see also Table 12.1), yet this estimation and its trend over recent decades varies greatly among sources (Kudrenko 2018). Several small and isolated populations occur in southern and western Europe (Alpine and Abruzzo in Italy, Cantabrian in Spain, and Pyrenean in Spain–France; Ciucci et al. 2015; Groff et al. 2018).

The status of brown bear populations is poorly known in Asia outside Russia. A large brown bear population (>2000) in the Lesser Caucasus in Turkey is connected to populations in Georgia, Iran, and northern Iraq (Ambarlı et al. 2016, 2018). Bears also occur in northern Georgia, Armenia, and Azerbaijan (McLellan et al. 2017) and in northern (Alborz Mountains), north-western (Arasbaran Biosphere Reserve), and western (Zagros Mountains) Iran (Etemad 1985). Brown bears occur in scattered and isolated populations in Himalayan India, Pakistan, and Nepal (Nawaz 2007; Aryal et al. 2010;

Table 12.1 Main distribution, population estimates, trends, and threats of brown bear in Eurasia.

Occurrence	Population estimate (n. ind.)	Trend	Threats	References
Afghanistan	Isolated population	?	?	McLellan et al. (2017)
Albania	180–200	Stable to decreasing	Illegal killing/capturing Habitat destruction	Bego (2007), Kaczensky et al. (2013)
Algeria	Extinct			McLellan et al. (2017)
Armenia	250	Decreasing	Poaching and capturing, human–bear conflict, habitat destruction	McLellan et al. (2017), Burton et al. (2018)
Austria	5	Decreasing	Illegal killing	Kruckenhauser et al. (2009)
Belgium	Extinct			McLellan et al. (2017)
Bosnia-Herzegovina	1000	Increasing	Poor management Livestock depredation	Kaczensky et al. (2013), Huber (2018)
Bulgaria	Central Balkan 100–130 Rio-Rhodopean 430–460	Stable	Forest plantation Illegal killing	Ministry of Environment and Waters (MEW) (2008), Zlatanova et al. (2009)
China	West ~6000 North-east ~1000	?	?	Gong & Harris (2006)
Croatia	937 (846–1072)	Increasing	Artificial food sources	Huber (2018)
Czech Republic	Cross-border individuals		Poaching Habitat fragmentation	(Skuban, personal communication)
Denmark	Extinct			McLellan et al. (2017)
Egypt	Extinct			McLellan et al. (2017)
Estonia	700	Increasing	None	Männil & Kont (2012)
Finland	1600–1800	Stable	None	Kojola & Heikkinen (2006)
France	Pyrenees Mountains (extinct but reintroduced) 22 (min. 46 in France + Spain)	Increasing	Illegal killing Hunting accidents Small population	Decaluwe et al. (2011) Sentilles et al. (2018)
Georgia	450	Decreasing	Habitat destruction Illegal killing	Lortkipanidze (2010)
Germany	Extinct (vagrant)	?		Huber (2018)
Greece	450–500	Stable/ increasing	Illegal killing Habitat loss/degradation Traffic fatalities	Mertzanis et al. (2009), Karamanlidis et al. (2015)
Hungary	Extinct Few records	?	?	Huber (2018)
India	Himalaya sector 130–220 (population shared with Pakistan)	?	?	Sathyakumar et al. (2012)
Iran	The Alborz Mountains 1300. The Arasbaran Biosphere Reserve, 100. The Zagros Mountains, <100	Unknown	Human–bear conflicts in the forms of illegal shooting and stoning, vehicle collision, habitat loss, degradation, and destruction	Gutleb et al. (2002), Esmaeeli (2007), McLellan et al. (2017)
Iraq	<100	Decreasing	Habitat destruction, illegal killing and capturing	Garshelis & McLellan (2011)
Ireland	Extinct			McLellan et al. (2017)
Israel	Extinct			McLellan et al. (2017)

Table 12.1 *(cont.)*

Occurrence	Population estimate (n. ind.)	Trend	Threats	References
Italy	Central Apennine 50 Alps 52–63	Stable Increasing	Illegal killing Small population Local acceptance	Gervasi et al. (2012), Ciucci et al. (2015), Groff et al. (2018)
Japan	~2200	?	?	Mano (2006)
Jordan	Extinct			McLellan et al. (2017)
Kyrgyzstan	?		Small population	McLellan et al. (2017)
Latvia	10–15	Stable	Human disturbance Human infrastructure	Pilāts & Ozoliņš (2003)
Lebanon	Origin uncertain			McLellan et al. (2017)
Liechtenstein	Extinct			McLellan et al. (2017)
Lithuania	Extinct			McLellan et al. (2017)
Luxembourg	Extinct			McLellan et al. (2017)
Moldova	Extinct			McLellan et al. (2017)
Monaco	Extinct			McLellan et al. (2017)
Mongolia	Altai Mountains? Gobi Desert 21–29	?	Isolated population (Gobi Desert)	Tumendemberel et al. (2015)
Montenegro	378	?	?	Huber (2018)
Morocco	Extinct			McLellan et al. (2017)
Netherlands	Extinct			McLellan et al. (2017)
North Macedonia	160–200	Stable	Illegal killing Low acceptance	Kaczensky et al. (2013)
Norway	150	Increasing	Illegal killing Low acceptance Disturbance	Chapron et al. (2014)
Pakistan	Hindu Kush 15–20 Himalaya sector 130–220 (shared with India)	?	?	Nawaz (personal communication, 2014 in Abbas et al. 2015), McLellan et al. (2017)
Palestine	Extinct			McLellan et al. (2017)
Poland	80–100	Stable	Habitat fragmentation/loss Urbanization Recreation	Jakubiec (2008), Selva et al. (2011), Huber (2018)
Portugal	Extinct			McLellan et al. (2017)
Romania	5850–6300	Stable	None	Ionescu (2002), Huber (2018)
Russian Federation	225,100	Increasing	Forest logging Illegal killing	www.ohotcontrol.ru/resource/number/
Serbia	Carpathian and East Balkan 8 Dinaric-Pindos 60	Stable Increasing	Habitat loss Illegal killing	Paunović et al. (2008)
Slovakia	1034–1439	Increasing	Vehicle collisions Low acceptance Habitat destruction Illegal killing	Huber (2018)
Slovenia	875–1130	Increasing	Human acceptance Habitat fragmentation Vehicle collisions	Jerina et al. (2018)

Table 12.1 *(cont.)*

Occurrence	Population estimate (n. ind.)	Trend	Threats	References
Spain	Cantabrian Mountains 183–278 Pyrenees 22–27	Increasing	Small population Illegal killing Low acceptance Habitat disturbance	Palazón et al. (2011), Pérez et al. (2014)
Sweden	2800–3300	Increasing/ stable	Illegal killing Local acceptance	Kindberg et al. (2011), Swenson et al. (2017)
Switzerland	0–2	Fluctuating	Vehicle collisions Human–bear encounters Predation on livestock	BAFU (2009), Huber (2018)
Syrian Arab Republic	Origin uncertain			McLellan et al. (2017)
Tajikistan	?		Small population	McLellan et al. (2017)
Tunisia	Extinct			McLellan et al. (2017)
Turkey	3400–4000	Stable or Increasing	Habitat destruction, Illegal killing, human–bear conflict, vehicle collisions	Ambarlı et al. (2016)
Ukraine	320		?	Huber (2018)
United Kingdom	Extinct			McLellan et al. (2017)
Uzbekistan	?		Small population	McLellan et al. (2017)

Sathyakumar et al. 2012; Abbas et al. 2015) and the southern Pamir Mountains in Afghanistan, Kyrgyzstan, Tajikistan, and Uzbekistan, where populations seem to be connected. Probably the most isolated population occurs in the Gobi Desert in southern Mongolia (Tumendemberel et al. 2015). Mongolia also hosts brown bears in the Altai Mountains, most probably connected with populations in China, Russia, and Kazakhstan (Groff, personal communication). Finally, a large population occurs in Hokkaido, Japan, but population estimates are not reliable (Mano 2006).

Figure 12.2 Brown bear by Giulia Bombieri, with permission.

Descriptive Notes

Patitschniak-Arts (1993) provided a detailed description of the brown bear. The species is characterized by a disk-shaped facial profile, heavy body with a distinctive shoulder hump, and long claws, but shorter than those in North American grizzly bears (Figure 12.2). The brown bear has a massive head with a prominent nose and, in common with many other bear species, it has small, rounded ears, small eyes, and a short tail. It has a massive and powerful body and great strength. Colors vary from the usual brown to many shades lighter to blonde and many shades darker to almost black (Figures 12.3 and 12.4).

On average, the brown bear is the second largest bear after the polar bear (*Ursus maritimus*), although the largest brown bears are as large as the largest polar bears (Derocher 2012). The size of brown bears is closely related to the amount of meat they eat, particularly salmon (*Oncorhynchus* spp.; Hilderbrand et al. 1999b). Thus, the largest brown bears in Eurasia are found on Kamchatka (Kurtén 1973), where they have access to salmon.

A comparative study found no significant difference in brown bear body mass in northern (Sweden and Norway) and southern (Croatia and Slovenia) Europe. Asymptotic body masses were 115 kg in spring and 141 kg in autumn for southern females, 248 kg and 243 kg for southern males, 96 kg and 158 kg for northern females, and 201 kg and 273 kg for northern males, respectively. Thus, males had about twice the mass of females in the spring and about 1.7 times as much in autumn. Northern bears gained more body mass before hibernation and lost more during hibernation than southern bears, probably because hibernation was twice as long in the north (Swenson

(A)

(B)

Figure 12.3(A,B) The Gobi bear, which lives in the mountainous desert of southern Mongolia, is the only desert-dwelling brown bear in Eurasia. (Photo by Jon Swenson.) (A black and white version of this figure will appear in some formats. For the color version, please refer to the plate section.)

et al. 2007). Population mean masses of adult females in nine Eurasian populations ranged from 94 to 142 kg, but this does not include any of the largest bears from the Russian Pacific Coast (Zedrosser et al. 2011). Some 400-kg males have been reported in autumn in Europe (Karamanlidis, personal communication).

Habitat

Brown bears inhabit a wide range of habitats within their vast distribution. An ideal habitat is characterized by areas of high-quality food and shelter (forest cover and rugged landscapes) for resting, rearing young, and hibernation (Nawaz et al. 2014). Brown bears often live in large, continuous forests, ranging from virgin and semi-natural forests to managed woodlands, with a variety of tree species compositions and bioclimatic zones (Zedrosser et al. 2001). However, bears also can utilize open landscapes, such as steppe ecosystems and agricultural landscapes, including grain fields with corn, wheat, oat, barley, vineyards, and orchards (Skuban et al. 2017).

Figure 12.4 Eurasian brown bear by Francesco Rinaldi, with permission. (A black and white version of this figure will appear in some formats. For the color version, please refer to the plate section.)

This brings them into direct contact with people and their infrastructure, which often leads to increased human–bear conflicts (Skuban et al. 2017; see also Chapter 15).

Most brown bears in southern latitudes live in temperate broadleaved and mixed forests with a moderate climate and distinct warm and cold seasons (Zedrosser et al. 2001). Dominating tree species are spruce (*Picea* spp.), fir (*Abies* spp.), and pine (*Pinus* spp.), with beech (*Fagus* spp.) and oak (*Quercus* spp.) providing important food items in autumn (European Environment Agency 2006). Although bears in Italy, Spain, Greece, Turkey, Bulgaria, Iraq, and Syria can also inhabit Mediterranean forests interspersed by grassland and shrubs, brown bears in southern countries mainly inhabit forests with coniferous species and beech, oak, hazelnut (*Corylus* spp.), and birch (*Betula* spp.) trees (Figure 12.5). Bears from southern countries are also adapted to a landscape with a mosaic character with a high human footprint and dry summer conditions (Malcom et al. 2002).

In higher altitudes in Europe, Kyrgyzstan, Tajikistan, Uzbekistan, Kazakhstan, China, and Japan, as well as in some parts of India, Nepal, Mongolia, and Turkey, brown bears occupy

Figure 12.5 A brown bear climbing up a tree in northern Iran. (Photo by Hamed Tizrooyan.) (A black and white version of this figure will appear in some formats. For the color version, please refer to the plate section.)

Figure 12.6 Standing up on its hind legs allows a brown bear to get more information from its senses of smell, sight, and hearing. (Photo by Mario Melletti.) (A black and white version of this figure will appear in some formats. For the color version, please refer to the plate section.)

Figure 12.7 During autumn hyperphagia, brown bears eat large amounts of food and accumulate fat for hibernation and reproduction. (Photo by Vincenzo Penteriani.) (A black and white version of this figure will appear in some formats. For the color version, please refer to the plate section.)

temperate coniferous forests (Olson et al. 2000). Brown bears also occur occasionally in temperate forests in coastal regions near the northern parts of the Black (Vaisfeld & Chestin 1993) and Caspian seas (Gutleb et al. 2002). The taiga or boreal forest is typical for Scandinavia and northern Russia (Olson et al. 2000), and is characterized by spruce, pine, and birch and a forest understory of ericaceous shrubs (e.g. *Vaccinium* spp.; Elfström et al. 2008) (Figures 12.6 and 12.7).

Japan and Asian Russia harbor populations of coastal bears with access to abundant and nutrient-rich foods, such as spawning salmon and whale carcasses (Yudin 1993; Figure 12.8).

(A)

(B)

Figure 12.8 Populations of coastal brown bears from the easternmost areas of the species distribution have access to spawning salmon (A) and whale carcasses (B). (Photos by Ivan Seryodkin.)

Due to this availability of highly energetic food, these are the largest brown bears in Eurasia.

Brown bears also inhabit Palaearctic temperate grasslands, savannas, and shrublands, for example in the western Tian Shan and the Altai steppe regions of Kazakhstan, Uzbekistan, Kyrgyzstan, and Tajikistan (Olson et al. 2000). The Gissano-Altai open woodland covers the Turkestan region, Gissar, Zeravshan, and the Altai Mountains. Some parts of Gissano-Altai are characterized by patches of wild fruits and nut-bearing trees, including walnut (*Juglans regia*), pistachio (*Pistacia vera*), and almonds (*Amygdalus* spp.), from which bears obtain substantial amounts of high-caloric seeds during hyperphagia (Vaisfeld & Chestin 1993). Brown bears also occupy Palaearctic montane grasslands and shrublands, including high mountainous plateaus. They occur in the Himalaya, Karakoram, and Hindu Kush mountains in Pakistan, India, Afghanistan, Nepal, and Bhutan (Nawaz et al. 2014). Brown bears inhabit alpine steppes on the central and western Tibetan Plateau in China. Brown bears have been recorded in the Himalaya Mountains up to 5500 m (Nawaz et al. 2014). On the Kola Peninsula, bears penetrate into the tundra along river valleys up to the Barents Sea coast (Danilov & Tirronen 2011).

The Gobi bear is the only desert-dwelling brown bear in Eurasia. These critically endangered bears live in the mountainous desert of southern Mongolia.

Human impacts on ecosystems include habitat alteration and fragmentation due to agricultural activities, forestry, urbanization, ski stations/resorts, hydroelectric power plants, and transportation infrastructure. In many parts of Eurasia, natural habitats and undisturbed areas are shrinking. Additionally, bears must cope with human disturbances posed by outdoor activities in many formerly untouched areas. Climate change also threatens brown bears in some areas, especially in dry and barren habitats (Penteriani et al. 2019) and lower elevations (Albrecht et al. 2017).

Movements and Home Range

Brown bears generally restrict their movements to a fixed area all year round. This home range is defined as the area regularly traversed and used by an individual to find food resources, mating partners, and to produce offspring. Bears are solitary but not territorial, i.e. two individuals may occupy the same range, but avoid each other in time and space. Adult female brown bears maintain stable home ranges over multiple years. Home ranges of females are generally smaller than those of males and have a greater home range overlap with other females, particularly related ones (Frank et al. 2018). Daughters are often philopatric and establish their home range in parts of their mother's range, which may lead to spatial matrilineal assemblages of related females (Støen et al. 2005). Related females are more tolerant of each other than unrelated females and tend to exclude unrelated females from these matrilineal assemblages (Støen et al. 2005).

Figure 12.9 Brown bear mating generally occurs during spring, but can also extend into the summer. (Photo by Ivan Seryodkin.)

There is considerable variation in average home range size and overlap among populations in Eurasia, mostly as a function of habitat quality, food availability, and local bear density. Where food is locally abundant, such as salmon in Kamchatka, bears are more tolerant of each other, maintain smaller home ranges (6–15 km^2), and occur in high population densities (Seryodkin et al. 2017a). At the other extreme, home ranges span up to 2400 km^2 in the Gobi Desert of Mongolia, where forage is extremely sparse and widely interspersed and bears travel large distances between oases to find food and mating opportunities (Luvsamjamba et al. 2016). Brown bears in Europe maintain average home range sizes ranging from 160–640 km^2 in the Scandinavian boreal forest (Hertel et al. 2019b), to 58–362 km^2 in Greece (Mertzanis et al. 2011), and to 14–83 km^2 in mixed evergreen coniferous forest and oak stands in Turkey (Ambarlı 2012). Females with cubs move much shorter distances in southern than in northern Europe (Ordiz et al. 2007), likely reflecting smaller bear home ranges. Like home range sizes, movement distances vary seasonally, depending on the availability and spatial distribution of resources, such as food, bed sites, mating partners, etc. Average daily movement distances of brown bears may range from 1.5 to 3.5 km (Huber & Roth 1993; Seryodkin et al. 2014). During the mating season, solitary females and males show high rates of movements and scent marking to find a mating partner (Steyaert et al. 2012). During this time, males may exhibit long-distance excursions outside their usual range in order to find a mating partner (Dahle & Swenson 2003b; Krofel et al. 2010; Figure 12.9). Females accompanied by cubs display the lowest levels of movement and daily displacement (Hertel et al. 2019b). Especially during the mating season, females with cubs avoid potentially infanticidal males by selecting less productive habitat (Steyaert et al. 2016). In the late summer and autumn, food search governs movements and movement patterns depend largely on the spatial distribution of food resources. Where nutritious foods are spatially clustered, e.g. supplementary feeding sites, garbage dumps, or agricultural fields, bears often revisit sites (Selva et al. 2017). However, where food is distributed more regularly and at lower densities, such as berries, pine nuts, and acorns, bears exhibit continuous, slow, meandering movements (Hertel et al. 2016). In some populations, bears show clear diurnal movements between foraging and resting sites (Martin et al. 2010). Even a seasonal migration of more than 100 km to search for food has been documented (Cozzi et al. 2016).

Depending on the predominant diel activity pattern (see section below for more details), bears rest once or twice a day. Bears almost never use the same bed twice and hence direct their movements to find a suitable resting site. Bears select sites of dense vertical and horizontal vegetation cover for resting to reduce their risk of detection by humans (Ordiz et al. 2011). Because brown bears maintain stable multiannual home ranges, they certainly have good knowledge and memory about the spatial distribution of resources and habitat types within their home range.

Activity Patterns

Except for winter, when brown bears are typically hibernating in dens, bears are active for approximately 50% of the day (Ordiz et al. 2011; Seryodkin et al. 2013). Brown bears display mostly crepuscular and nocturnal activity in human-dominated landscapes (Kaczensky et al. 2006; Seryodkin et al. 2013), are more active at daytime in remote areas or when they rely on some specific foods (Klinka & Reimchen 2002), and tend to be more diurnal when days are becoming shorter (Ordiz et al. 2013). Brown bears seem to be more nocturnal in Europe than in North America, which may reflect the longer history of persecution in Europe (Kaczensky et al. 2006; Ordiz et al. 2011). Nevertheless, human tolerance can be high in some areas where bears and people have coexisted for a long time (Zarzo-Arias et al. 2018).

Daily activity patterns vary among bear sex and age classes (Hertel et al. 2019b). Females with cubs are more active during daytime than other bears, especially in spring, likely to avoid infanticide of cubs by males that try to access females (Steyaert et al. 2012). The different diurnal activity of females with cubs, compared to other bears, is a well-documented pattern (e.g. Ordiz et al. 2007). Subadults can display an intermediate behavior, being more active during daytime than adult bears (Kaczensky et al. 2006; Hertel et al. 2019b). This may allow subadult bears to access food resources at times when adults are less active, likely to reduce the risks of intraspecific aggression (Klinka & Reimchen 2002). In human-dominated landscapes, greater subadult activity during the daytime may also reflect a lack of experience in encounters with humans or human activities, which might be a major reason why adult bears are mostly active at night (Kaczensky et al. 2006).

Food availability, the breeding season, and human activity seem to be the most important factors influencing the typically crepuscular and nocturnal activity patterns of Eurasian brown bears. This applies to small, native bear populations, such as in the Cantabrian Mountains, Spain (Clevenger et al. 1990) and Rodopi, Greece (Mertzanis et al. 2005); reintroduced bears, such as in Trentino, Italy (Tattoni et al. 2015), and larger populations, such as in northern Europe (Ordiz et al. 2013). Nevertheless, within populations, bears also show individual variation in daily activity patterns, which is in part a function of differences in density of human infrastructure, density of conspecifics, and foraging opportunities (Martin et al. 2010; Hertel et al. 2017; Penteriani et al. 2017). Beyond that, variation in diel activity preferences seems to be highly conserved within an individual and persists throughout its lifetime (Hertel et al. 2019a).

Feeding Ecology

Brown bears are opportunistic omnivores with an amazingly diverse diet and, because of their size and winter hibernation, they must consume large quantities of food and/or select high-nutritional food items to satisfy their energetic demands (Bojarska & Selva 2012). Their choice of foods depends on the availability of potential food resources in time and space. Brown bear diets in Eurasia include plant material, soft and hard fruits, fungi, earthworms, ants, salmon, small mammals, and large ungulates. Livestock, bees and honey, and human-provided food can be important in some areas (Karamanlidis et al. 2011; Bojarska & Selva 2012; Niedziałkowska et al. 2019). Brown bear populations in temperate forest biomes have the most diverse diet (Bojarska & Selva 2012).

Although brown bears are among the largest terrestrial carnivores, they feed mostly on plants, which generally is more than half of their diet. The bear's typical carnivore digestive tract means that they digest meat efficiently, but plants poorly. They consume meat, as predators or scavengers, sometimes using kills of other large predators (Krofel & Jerina 2016; see Chapter 3 on interspecific interactions). Populations in locations with deeper snow cover, lower temperatures, and lower productivity consume significantly more vertebrates, fewer invertebrates, and less hard mast. The consumption of vertebrates was highest in the higher-latitude tundra and lowest in temperate forest biomes, whereas fruits and hard mast show the opposite pattern (Vulla et al. 2009; Bojarska & Selva 2012). Also, brown bears in Europe feed more on ungulates (mean 10.5%) than in Asia (mean 6.8%) (Niedziałkowska et al. 2019), although this can vary regionally. In north-western Russia, ungulates comprised 21% of the diet in the Murmansk Region (Nasimovich & Semenov-Tyan-Shanskiy 1951), 13% in Karelia, and 4–5% in the Leningrad and Pskov regions (Danilov et al. 1979).

Seasonality is also an inherent characteristic of brown bear diet and relates to both the species' energetic and nutritional needs after (hypophagia) and before (hyperphagia) hibernation, and the temporal availability of foods. The need for protein after the hibernation period is reflected in the peak in ungulate consumption in early spring (Ciucci et al. 2014; Tirronen et al. 2016; Niedziałkowska et al. 2019), mostly preyed on as young animals or scavenged. The most important large wild mammal prey across Eurasia is moose (*Alces alces*), but bears also consume reindeer (*Rangifer tarandus*), wild yak (*Bos grunniens*), European bison (*Bison bonasus*), Tibetan antelope (*Pantholops hodgsonii*), wild boar (*Sus scrofa*), musk deer (*Moschus moschiferus*), red deer (*Cervus elaphus*), snow sheep (*Ovis nivicola*), and wild goats (*Capra* spp.). Smaller mammals can be important prey in some areas, such as the long-tailed ground squirrel (*Spermophilus undulatus*) in north-eastern Asia, plateau pika (*Ochotona curzoniae*) in Tibet, and golden marmot (*Marmota caudata*) in Pakistan (Egorov 1971; Mordosov 1993; Chernyavsky & Krechmar 2001; Aichun et al. 2006; Zyryanov 2012; Nawaz et al. 2014). Spawning salmon provide a major food resource along the Pacific Coast (Krechmar 1995; Seryodkin 2006). Reptiles (tortoises) have been found in summer diets in southern Europe (Paralikidis et al. 2010).

Insects are usually consumed in spring–summer (Ciucci et al. 2014), particularly ants in boreal forests (Swenson et al. 1999) and, although to a lesser extent, also in temperate forests (Tosoni et al. 2018). In warmer locations, other groups of insects (mainly bees and wasps) are more important. At Lake Baikal, bears eat swarming caddisflies (Ustinov 1993).

In spring and summer, succulent foliage is also an important food item. Particularly in the spring, bears eat green grasses and sedges, leaves, and young shoots and in some areas of Eurasia, they also dig and eat starch-rich rhizomes, tubers, and roots (Danilov et al. 1979; Chernyavsky & Krechmar 2001; Ciucci et al. 2014; Stenset et al. 2016). In spring, bears can strip the bark of some trees, especially conifers, to forage on sapwood (Zyśk-Gorczyńska et al. 2014).

Brown bears track pulses and waves of food resources, which are typically available in large quantities for a limited period. This mostly occurs during autumn hyperphagia, when bears eat large amounts of food and accumulate fat for hibernation and reproduction. These food waves include fleshy fruits and hard mast over most of the bear range (and also salmon on the Pacific Coast), and influence bear reproductive success (Hertel et al. 2017). Berries are particularly important in this regard, especially bilberry (*Vaccinium myrtillus*). Crowberry (*Empetrum* spp.), lingonberry (*V. vitis-idaea*), cranberries (*V. oxycoccos*), bearberry (*Arctostaphylos uva-ursi*), rowan (*Sorbus* spp.), dogwoods (*Cornus* spp.), bird cherry (*Prunus padus*), raspberries and blackberries (*Rubus* spp.), buckthorns (*Rhamnus* spp.), and currants (*Ribes* spp.) are among the fleshy fruits most consumed by bears. Hard mast, abundant during masting years, includes seeds of beech, oaks, walnut, hazelnut (*Corylus avellana*), and chestnut (*Castanea sativa*) and is important autumn food in southern latitudes, including Central and Southern Europe, the Caucasus, Central Asia, and the Far East (Ciucci et al. 2014). The seeds of pines (*Pinus sibirica* throughout Siberia, *P. koraiensis* and *P. pumila* in the Far East) are a crucial food in the area from the Ural Mountains to the Pacific Ocean. During years of mast failure, bears may experience food shortage during fall, unless natural productivity allows for the availability of alternative high-quality foods, such as berries and other fleshy fruits (Ciucci et al. 2014).

Nowadays, most of the bear range is under the influence of human activities, which can alter bear diet. For instance, a decline in the consumption of marine and terrestrial animals has been documented in recent centuries in Japan (Matsubayashi et al. 2015). Dietary shifts are probably most pronounced when bears are, intentionally or unintentionally, provided with foods of human origin. These comprise an important or even major component of brown bear diet in some parts of their Eurasian range (Bojarska & Selva 2012). Intentional artificial feeding of bears with corn, oats, livestock carcasses, and other anthropogenic foods is a controversial but widespread management practice used in some countries for various reasons, including hunting, ecotourism, and conflict prevention (Steyaert et al. 2014; Penteriani et al. 2017; Figure 12.10). In addition bears: (a) readily visit developed areas, orchards, and fields to feed on garbage, slaughter remains, fruits and crops, including apples (*Malus* spp.), cherries (*Prunus* spp.), pears (*Pyrus communis*), and several kinds of cereal grains (e.g. Karamanlidis et al. 2011; Skuban et al. 2016); and (b) can predate on free-ranging livestock, sheep, and goats. However, bears' diet may also remain the same near settlements compared to in remote areas, suggesting other reasons than food accessibility behind bear occurrence near people (Elfström et al. 2014). Beside access to anthropogenic food, bear food habits can be influenced by human disturbance, human use of natural resources, and climate change through seasonally reduced accessibility or availability of key food sources (Wilmers & Post 2006; Rodríguez et al. 2007).

Reproduction and Growth

Body size, mass, and growth are important life-history traits that influence behavioral, physiological, and reproductive characteristics of organisms (Stearns 1992). Brown bear cubs are born during hibernation in late January and the beginning of February (Robbins et al. 2012; Friebe et al. 2013), and the number, survival, and size of cubs depend on maternal conditions prior to entering the den (Robbins et al. 2012). Cubs are born with a mass of ~650 g, and both sexes have approximately the same mass in the first few months of life (Robbins et al. 2012; see also Chapter 2). Sexual size dimorphism becomes apparent at the yearling stage, when male yearlings are about 3% larger and 8% heavier than females (Dahle et al. 2006). Female Eurasian brown bears attain 90% asymptotic body mass at approximately 5–6 years of age (Zedrosser et al. 2006; Swenson et al. 2007), whereas males reach 90% asymptotic mass at approximately 8–9 years of age (Zedrosser et al. 2007). In accordance with life-history theory (Stearns 1992), size and growth in bears are correlated positively with good environmental conditions and negatively with population density (Dahle et al. 2006; Zedrosser et al. 2006). Mean adult mass in European and Japanese populations ranges from approximately 90 to 140 kg for females (Swenson et al. 2007; Zedrosser et al. 2011) and 200–245 kg for males in fall

Figure 12.10 Brown bears at artificial feeding points. (A) A brown bear obtaining corn from an automatic feeder in Dinaric forest in Slovenia (photo by Miha Krofel). (B) Adult and subadult brown bears at an artificial feeding point in the Finnish Karelia (photo by Vincenzo Penteriani). (C) Female eating corn on Rakitna plateau, Slovenia (photo by Miha Krofel).

(Swenson et al. 2007; Moriwaki et al. 2018), although males up to 400 kg have been reported (Moriwaki et al. 2018; Karamanlidis, personal communication). The body size of bears living in marginal habitats in Central Asia can be considerably smaller; for example, adult female masses of 73 kg (mean) in Deosai National Park, Pakistan (Nawaz et al. 2008) and 51–78 kg in the Gobi Desert, Mongolia (Reynolds et al. 2010).

In mammals, females reach sexual maturity at a threshold body mass, after which energy is allocated primarily to reproduction (Gaillard et al. 2000). Most female brown bears in Eurasia reach primiparity at 4–5 years of age (Zedrosser et al. 2009; Knott et al. 2014; Shimozuru et al. 2017), 2–5 years younger than in North America (e.g. Zedrosser et al. 2011; McLellan 2015). Three-year-old European females have been observed to give birth (Frkovic et al. 2001; Zedrosser et al. 2004).

In mammals, primiparous females have a poorer reproductive performance than multiparous females (Stearns 1992). This also applies to brown bears, where primiparous mothers have on average smaller litters and a higher cub loss than multiparous mothers (Zedrosser et al. 2009). Average litter sizes in Eurasia range from 1.3 in marginal habitats in Central Asia (Nawaz et al. 2008), 1.6 on Hokkaido, Japan, to 2.3–2.5 in Europe (Zedrosser et al. 2011). Also, the interlitter interval is highly variable in Eurasian brown bear populations; about 2 years in most European populations (Knott et al. 2014; Van de Walle et al. 2018), 2.4 years on Hokkaido (Shimozuru et al. 2017), and up to 5.7 years in marginal populations in Central Asia (Nawaz et al. 2008). Female–female interactions, probably caused by resource competition, can also affect reproductive performance, as an adult female's probability of having cubs in a given year is influenced by whether her nearest neighboring adult female had cubs (Ordiz et al. 2008) and philopatric females that have not dispersed from their mother's home range show delayed primiparity (Støen et al. 2006a).

In iteroparous mammals, conditions experienced early in life may have long-lasting effects on lifetime reproductive success (Stearns 1992). Both yearling and adult body size are positively associated with measures of lifetime reproductive success and individual fitness in both sexes of brown bears (Bellemain et al. 2006; Zedrosser et al. 2007, 2013). Larger and older males produce more offspring, likely because large body size directly benefits male–male competition and is advantageous in endurance competition for access to females (Bellemain et al. 2006; Zedrosser et al. 2007). Mean adult female mass is positively related to mean litter size in brown bears (Hilderbrand et al. 1999b; Zedrosser et al. 2011). Populations with a long-term persecution history by humans have, on average, larger litter sizes in relation to adult female body mass than populations with a short-term persecution history, which suggests that humans may affect bear life history traits through harvesting (Zedrosser et al. 2011; Zedrosser et al. 2013; Van de Walle et al. 2018).

Behavior

General Behavior

Brown bears are generally considered a solitary and nonterritorial species (Dahle & Swenson 2003b), displaying a dominance hierarchical social system. The family group (mother with her cubs) represents the most stable social entity, as it lasts between 1.5 and 5.5 years in Eurasian populations (Dahle & Swenson 2003a; Nawaz et al. 2008). Exceptionally, mixed litters, with young born in different years, have been observed (Swenson & Haroldson 2008). Direct social interactions between adults occur mainly during the mating season. As adult males are dominant and may kill subadults and females with dependent offspring, a spatial or temporal segregation occurs, based on sex, age, and reproductive class (Steyaert et al. 2013). Occasionally in some populations (e.g. Cantabrian Mountains, Alps, White Sea), several bears may aggregate at seasonally high concentrations of food (Tirronen et al. 2015; Ruiz-Villar et al. 2019; Groff, personal communication). In such situations, a clear dominance hierarchy may be established.

Bears communicate through scent-marking, often deposited by rubbing on trees or other objects, thereby signaling their presence and social and reproductive status to neighboring, or passing, conspecifics (Rosell et al. 2011; Sergiel et al. 2017). Rubbing behavior as an "indirect social interaction" is well documented in several populations (Karamanlidis et al. 2007; Puchkovskiy 2009; Sato et al. 2013; Figure 12.11).

Natal dispersal is sex-biased, with males dispersing further from the natal range than females, probably to avoid encounters with resident conspecifics and inbreeding (Støen et al. 2006b; Zedrosser et al. 2007; Jerina & Adamic 2008). Frequent road crossings and elevated encounter risk with humans or conspecifics increase mortality rates for young bears during dispersal. The maximum dispersal recorded in Europe was a male that moved 467 km (Bartoń et al. 2019).

The period when cubs are with their mother is likely important for the behavioral development of the cubs. The cognitive capacities and the behavioral plasticity (Gittleman 1986; Deecke 2012) of brown bears suggest that social learning by the cubs can occur by observing the mother (Morehouse et al. 2016). This behavioral plasticity of bears also has been shown in personality studies of older bears (Hertel et al. 2019a). Brown bears show correlations between repeatable behaviors, including habitat selection (Leclerc et al. 2016). Individuals that travel over longer distances also have an overall greater displacement and are more diurnal (Hertel et al. 2019a). Such individual behavioral patterns may have important implications for the adaptability and persistence of brown bears in human-dominated landscapes (Hertel et al. 2019a).

Hibernation and Denning Behavior

Denning behavior, phenology, and physiology are important components of brown bear life history, as they may spend as

(A) (B)

Figure 12.11 Brown bear scent-marking. (A) Rubbing trees in a mixed forest in the Cantabrian Mountains, north-western Spain (photo by Vincenzo Penteriani). (B) Female brown bear smelling a rubbing tree in the Dinaric Mountains in Slovenia (photo by Miha Krofel).

Figure 12.12 Two different types of brown bear dens: a natural cave in Slovenia (A, photo by Miha Krofel) and an excavated hole in Russia (B, photo by Ivan Seryodkin).

much as half of their life in winter dens, including some important events; gestation, birth, and lactation. A meta-analysis of brown bear denning periods across the species' range indicated consistent patterns in the relationship between latitude and denning duration (Krofel et al. 2017). Also, the availability of a continuous supply of food during winter can reduce denning duration (Krofel et al. 2017). The duration of denning also depends on the reproductive status and the age of the individuals; pregnant females exhibit the longest denning period and adult males emerge from their dens first (Manchi & Swenson 2005). Smaller bears have both a longer denning duration and a lower body temperature, but body size does not affect den entrance date (Evans 2016).

During the denning period, brown bears generally avoid areas with increased human disturbance, with adult males often choosing den sites in more remote areas than younger bears and females with offspring (Elfström et al. 2008; Elfström & Swenson 2009). Dens are also most often situated in steep terrain with few roads, making them less accessible to people (Petram et al. 2004; Elfström et al. 2008).

The diversity of den types reflects the local habitat, e.g. availability of cavities, and the behavioral plasticity of brown bears (Figure 12.12). Adult male bears seem to have fewer requirements for cover and protection during winter denning (Elfström & Swenson 2009). Possibly this reflects different physiological requirements (Hellgren 1998). Depending on availability, bears use dens in natural cavities underneath rocks or boulders, ground nests under or without conifer canopy cover, in cavities under roots of large standing trees, or dens dug in the soil or anthills (Camarra 1997; Huber & Roth 1997; Elfström & Swenson 2009). A study of the frequency of covered, semi-covered, and open dens in 37 regions of the former Soviet Union showed that open and semi-covered dens prevailed in the southern taiga and mixed and deciduous mountain forests, where continuous snow cover did not exceed 160 days. In these areas, the denning period ranged from 45 to 160 days (Danilov 1991).

Hibernating brown bears undergo drastic changes in all aspects of their biology and physiology during winter dormancy (Nelson et al. 1973). Although bear physiology has been studied for decades (primarily in American black bears, *Ursus americanus*), many aspects of their basic biology remain a mystery and are subjects of ongoing research. Recent advances in biologging have allowed some of the knowledge gaps about basic denning ecology to be filled. Well established aspects of bear hibernation physiology include a decrease in body temperature, heart rate, respiratory rate, and metabolism (Folk et al. 1974; Tøien et al. 2011, 2015) during hibernation and an increase in fat storage prior to hibernation (Hilderbrand et al. 1999a). Due to these changes, physiology varies dramatically throughout the year, with hematological and biochemical variables showing the involvement of nearly all organ systems (Huber et al. 1997; Kusak et al. 2005; Græsli et al. 2014).

Climatic factors also modulate brown bear hibernation phenology (Evans et al. 2016a; Delgado et al. 2018). Evans et al. (2016a) assessed the drivers of den entry and exit using GPS-collared bears with heart rate and body temperature loggers. They found that the high variability in den entry and exit dates between years was explained by variation in ambient temperature between years, with the warmer years associated with later entry and shorter hibernation. Bears enter the den when snow arrives and when ambient temperature reaches 0°C. Heart rate variability, a proxy of sympathetic nervous system activity, drops dramatically once bears enter their dens, indirectly suggesting that denning is coupled tightly to metabolic suppression. Body temperature began increasing 2 months before den exit and was driven by ambient temperature. Heart rate variability increases only 3 weeks before exit, likely indicating activation of the sympathetic nervous system to restore euthermic metabolism. Bears exit the den when ambient temperature reaches the bears' presumed upper critical temperature, perhaps because they were uncomfortably warm. This is

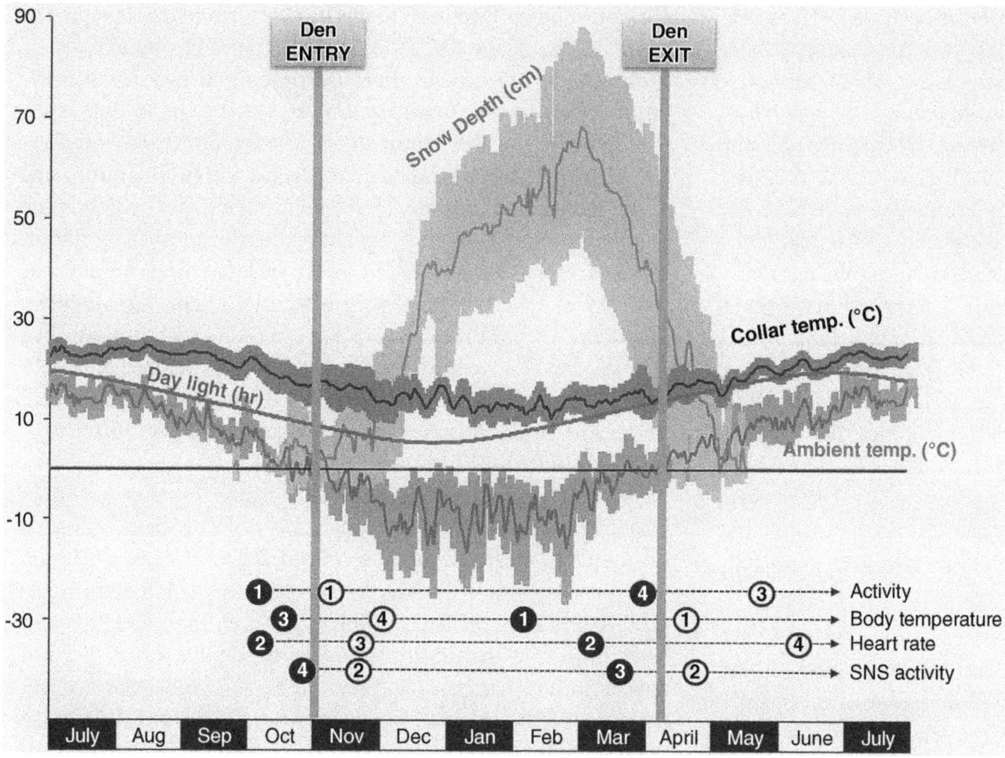

Figure 12.13 Sequence of environmental and physiological events in relation to the entry into and exit from hibernation by brown bears in Sweden. Heart rate variability was used as a proxy for nervous system activity (Evans et al. 2016a; https://frontiersinzoology.biomedcentral.com/articles/10.1186/s12983-016-0140-6; https://creativecommons.org/licenses/by/4.0/).

consistent with a study based on tracks from den exits, where higher spring temperatures were strongly associated with earlier den exit (Delgado et al. 2018). Therefore, hibernation is initiated primarily by environmental cues and terminated by physiological constraints (Figure 12.13). Smaller bears have up to 1.5°C lower body temperature than larger bears in winter, probably because smaller animals have a lower surface-to-volume ratio, resulting in more thermal conductance and poorer heat conservation (Evans 2016).

Capturing Eurasian brown bears during winter presents a significant disturbance, as it takes 15–20 days from capture before body temperature and heart rate return to hibernation levels. However, no effect on duration or magnitude of hibernation has been detected (Evans et al. 2016b).

Parasites and Diseases

Brown bears are susceptible to most of the pathological conditions that affect other mammalian species. As wild bears have a more limited lifespan and diseased individuals are eliminated efficiently, one normally sees healthy bears in a wild population. A disease is defined as a particular abnormal condition that negatively affects the structure or function of part or all of an organism. This is difficult to document in wild individuals, so only a few individual cases have been reported anecdotally. Also, the presence of parasites is usually only a normal host–parasite system (Huber & Ehrlich 1981), as is the evidence of exposure to viral and bacterial pathogens (Madić et al. 1993; Modrić & Huber 1993). Many organic diseases, like tumors, usually develop at an age that most free-living bears do not reach. Typically, the longer lives of captive bears, plus their husbandry conditions, allow for more diverse and severe pathological changes, which are also monitored more closely and described in the literature more frequently. The most frequent causes of death in most wild brown bear populations are not diseases, but the traumas caused by gunshots or traffic accidents (Frkovic et al. 1987; Bischof et al. 2008). Other causes are rarely reported in free-living bears, such as poisoning (Reljić et al. 2012) or tumors (Hohšteter et al. 2018). Additionally, brown bears may be exposed to various pollutants (like heavy metals and pesticides) (Lazarus et al. 2014, 2018a, 2018b; Herceg Romanić et al. 2015). Increased stress, or loss of genetic diversity due to isolation, may affect the immune response. The major histocompatibility complex may be one of the indicators of the state of a population (Kuduk et al. 2012).

Zooanthroponoses, diseases that can infect both bears and humans, such as trichinosis or rabies, receive special attention in research on brown bear health issues. For example, Seryodkin (2015) reported that 85% of cases of human trichinosis in Russia were the result of eating brown bear meat. In the Russian Far East, the causative agent of trichinosis in bears is *Trichinella nativa*, with *Trichinella* prevalence varying from 2.4% in the Sakhalinsky Region to 30.6% in the Kamchatsky Region. Thus, knowledge about the spread of trichinosis in bears and the life cycle of the disease in bear populations is necessary for the prevention of helminthiasis in humans.

The parasites and infectious diseases of brown bears have been well studied, especially in the Russian Far East. The following number of species have been recorded: 13 nematodes, one trematode, three cestodes, two Acanthocephalans, four ticks, one flea, one Hippoboscidae, and 11 causative agents of infectious diseases, including rabies, *Simplexvirus*, *Toxoplasma gondii*, canine morbillivirus, *Chlamydia* sp., and *Mycoplasma* sp. (Seryodkin 2010, 2015; Esaulova et al. 2012; Seryodkin et al. 2012, 2016, 2017b; Tranbenkova et al. 2012; Shchelkanov et al. 2016; Medvedev & Seredkin 2018; Naidenko et al. 2018; Panova et al. 2018; Seryodkin, unpublished data). In the Murmansk Region, brown bear scats contained eggs of the following helminths: *Dicrocoelium* sp., *Diphyllobothrium* sp., Anoplocephalidae, Capillariidae, *Baylisascaris* sp., and two types of Strongylida (Bugmyrin et al. 2017). The most widespread intestinal nematode in bears is apparently *Baylisascaris transfuga* (De Ambrogi et al. 2011), which has never been documented in humans. Other endoparasites (Aghazadeh et al. 2015; Papadopoulos et al. 2017) and ectoparasites, like chewing lice (Esteruelas et al. 2016), also have been reported.

Status in the Wild

The International Union for Conservation of Nature and Natural Resources (IUCN) has listed the conservation status of the brown bear globally to be of Least Concern (LC), as it is one of the world's most widely distributed terrestrial mammals and is not declining significantly globally (McLellan et al. 2017). The global population exceeds 200,000, with more than half of these in Eurasia. The IUCN Red List (2012) lists 44 populations of brown bears globally (Table 12.1). Because many populations are small, isolated, and in need of conservation attention, they have been assessed individually using the IUCN Red List Criteria for isolated populations (McLellan et al. 2017).

Status in Captivity

The brown bear is a species that is commonly kept in captivity. Species360 (formerly International Species Information System/ISIS, www.species360.org) and its Zoo Information Management System (ZIMS) include data on 800 individual brown bears (359 males, 365 females, and 76 "other") kept in 215 institutions around the world. However, the vast majority of the data originate from Europe (137 institutions with 510 individuals) and North America (57 institutions with 171 individuals), whereas Asian, South American, and African zoos have fewer reported captive brown bears; 23 institutions with 103 individuals, five institutions with 11 individuals, and two institutions with four individuals, respectively.

Captive brown bears are commonly recorded by their subspecies or origin. "*Ursus arctos*" of unknown and/or mixed origin (463 individuals, of which 421 are reported by 95 European and 74 by 12 Asian institutions) is the most common subspecies/origin recorded in captivity. Of six subspecies traditionally recognized in zoo records in Eurasia, confirmed *Ursus arctos arctos* is reported by 40 European and three Asian institutions, with 119 and five individuals, respectively. Seven institutions (three Asian and four European) report having 14 individuals of *U. a. isabellinus*. Two individuals of *U. a. pruinosus* are reported to be in the Kobe Oji Zoo in Japan and 11 individuals of *U. a. lasiotus* are kept in three European and two Asian institutions. *Ursus arctos beringianus* is reported to be in only eight zoos in Europe, with 15 individuals (Species360 2019). Additionally, captive brown bear data are collected within the framework of the European Studbook (ESB) for Brown Bears, which is the population management program of the European Association of Zoos and Aquaria (EAZA) and includes only animals in captivity in EAZA membership institutions. The vast majority of individuals in European zoos have been born in captivity, but over the last few decades, 2–3 orphaned cubs or problem bears have been moved from the wild to zoos each year. The zoological gardens of North America probably receive a higher number of wild-born individuals (Kolter & Zee 2008).

Several European institutions keep a large number of brown bears: Parque de la Naturaleza de Carbaceno in Spain, 76 bears; Orsa Predator Park in Sweden, 15 bears; Tbilisi Zoo in Georgia and the Foundation for the Preservation of Wildlife and Cultural Assets in Armenia, 13 individuals each (Species360 2019). Nevertheless, a significant number of captive bears do not appear in any database, as they are kept in zoos that are not members of zoo associations, such as Species360 or EAZA, or in poorly controlled circuses, sanctuaries, rehabilitation centers, mini-zoos, or privately. At least 19 bear sanctuaries exist in Europe, with the goal of receiving bears from places where they have been kept in poor or abusive conditions, have been confiscated from illegal trade, orphaned wild bear cubs that were hand-raised by people, or even bears that were captured in the wild due to problematic behavior and placed in a sanctuary. For example, 92 brown bears presently are kept in a brown bear sanctuary in Romania. The total number of bears in European sanctuaries probably exceeds 300 individuals (Weegenaar, Cuyten & Sergiel, unpublished data). Some sanctuaries carry out efforts to rehabilitate and reintroduce brown bears to the wild (Beecham et al. 2015), such as the Orphan Bears Rescue Center in Russia (Sergei Pazhetnov, personal communication).

Nationwide studies on the welfare of captive bears have been conducted, for example in Albania, Armenia, Bosnia and Herzegovina, Georgia, Germany, and Poland. Additional studies have assessed specific problems with captive bears, such as the origin and welfare status of bears displayed at restaurants in Albania (Karamanlidis et al. 2014). Such bears have often been found to be in very poor physical condition (e.g. locomotory disabilities, skeletal deformities, dental problems) due to inappropriate holding conditions and inadequate care. Also, several types of behavioral disorders have been observed in these animals, from pacing to severe self-mutilation (e.g. Sergiel & Maślak 2014).

Within the European Union, the Zoo Directive (Council Directive 1999/22/EC) requires licensed zoological gardens to accommodate their animals under conditions that aim to satisfy the biological and conservation requirements of the individual species. For example, recommended conditions for the appropriate accommodation of bears are that enclosures should be larger than 4000 m^2 or 1 ha of enriched environment

(Sergiel & Maślak 2014). However, regulations implementing the directive differ among countries and often fail to secure the welfare of captive bears. For example, in Austria at least 300 m^2 is required per two individuals, whereas only 100 m^2 is required in Poland.

Acknowledgments

Thanks to Maja Szymańska for valuable assistance in acquiring Zoo Information Management System data on captive brown bear populations.

References

Abbas, F., Bhatti, Z. I., Haider, J. & Mian, A. (2015). Bears in Pakistan: distribution, population biology and human conflicts. *Journal of Bioresource Management* 2: 1–13.

Aghazadeh, M., Elson-Riggins, J., Reljić, S., et al. (2015). Gastrointestinal parasites and the first report of *Giardia* spp. in a wild population of European brown bears (*Ursus arctos*) in Croatia. *Veterinarski Arhiv* 85: 201–210.

Aichun, X., Zhigang, J., Chunwang, L., et al. (2006). Summer food habits of brown bears in Kekexili Nature Reserve, Qinghai–Tibetan Plateau, China. *Ursus* 17: 132–137.

Albrecht, J., Bartoń, K.A., Selva, N., et al. (2017). How humans and climate change propelled the 12,000-year decline of the largest terrestrial carnivore. *Scientific Reports* 7: 10399.

Ambarlı, H. (2012). Spatio-temporal ecology, habitat use and population size of brown bears (*Ursus arctos*) in Yusufeli, Turkey. PhD thesis, Middle East Technical University.

Ambarlı, H., Erturk, A. & Soyumert, A. (2016). Current status, distribution, and conservation of brown bear (Ursidae) and wild canids (gray wolf, golden jackal, and red fox; Canidae) in Turkey. *Turkish Journal of Zoology* 40: 944–956.

Ambarlı, H., Mengüllüoğlu, D., Fickel, J. & Förster, D.W. (2018). Population genetics of the main population of brown bears in southwest Asia. *PeerJ* 6: e5660.

Anijalg, P., Ho, S. Y. W., Davison, J., et al. (2018). Large-scale migrations of brown bears in Eurasia and to North America during the Late Pleistocene. *Journal of Biogeography* 45: 394–405.

Aryal, A., Sathyakumar, S. & Schwartz, C. (2010). Current status of brown bears in the Manasalu Conservation Area, Nepal. *Ursus* 21: 109–114.

BAFU. (2009). Konzept Bär. Managementplan für den Braunbären in der Schweiz. 1–23. Available from www.bafu.admin.ch/bafu/de/home/themen/biodiversitaet/publikationen-studien/publikationen/konzept-bar–managementplan-fuer-den-braunbaeren-in-der-schweiz.html

Bartoń, K. A., Zwijacz-Kozica, T., Zięba, F., Sergiel, A. & Selva, N. (2019). Bears without borders: long-distance movement in human-dominated landscapes. *Global Ecology and Conservation* 17: e00541.

Beecham, J. J., De Gabriel Hernando, M., Karamanlidis, A. A., et al. (2015). Management implications for releasing orphaned, captive-reared bears back to the wild. *Journal of Wildlife Management* 79: 1327–1336.

Bego, F. (2007). The Brown Bear (*Ursus arctos*) Action Plan. The Republic of Albania, Ministry of Environment, Forest and Water Administration. Report prepared under the GEF Biodiversity Enabling Activity, pp. 1–60.

Bellemain, E., Zedrosser, A., Manel, S., et al. (2006). The dilemma of female mate selection in the brown bear, a species with sexually selected infanticide. *Proceedings of the Royal Society* 273: 283–291.

Bischof, R., Fujita, R., Zedrosser, A., Soderberg, A. & Swenson, J.E. (2008). Hunting patterns, the ban on baiting, and harvest demographics of brown bears in Sweden. *Journal of Wildlife Management* 72: 79–88.

Bojarska, K. & Selva, N. (2012). Spatial patterns in brown bear *Ursus arctos* diet: the role of geographical and environmental factors. *Mammal Review* 42: 120–143.

Bray, S. C., Austin, J. J., Metcalf, J. L., et al. (2013). Ancient DNA identifies post-glacial recolonisation, not recent bottlenecks, as the primary driver of contemporary mtDNA phylogeography and diversity in Scandinavian brown bears. *Diversity and Distributions* 19: 245–256.

Bugmyrin, S. V., Tirronen, K. F., Panchenko, D. V., et al. (2017). Helminths of brown bears (*Ursus arctos*) in the Kola Peninsula. *Parasitology Research* 116: 1755–1760.

Burton, A. C., Fisher, J. T., Adriaens, P., et al. (2018). Density and distribution of a brown bear (*Ursus arctos*) population within the Caucasus biodiversity hotspot. *Journal of Mammalogy* 99: 1249–1260.

Calvignac, S., Hughes, S., Tougard, C., et al. (2008). Ancient DNA evidence for the loss of a highly divergent brown bear clade during historical times. *Molecular Ecology* 17: 1962–1970.

Calvignac, S., Hughes, S. & Hänni, C. (2009). Genetic diversity of endangered brown bear (*Ursus arctos*) populations at the crossroads of Europe, Asia and Africa. *Diversity and Distributions* 15: 742–750.

Camarra, J. J. (1997). Caractéristiques et utilisation de tanières hivernales d'ours brun (*Ursus arctos*) dans les Pyrénées. *Gibier Faune Sauvage* 4: 391–405.

Chapron, G., Kaczensky, P., Linnell, J. D. C., et al. (2014). Recovery of large carnivores in Europe's modern human-dominated landscapes. *Science* 346: 1517–1519.

Chernyavsky, F. B. & Krechmar, M. A. (2001). *Brown bear (*Ursus arctos*) in north-east of Siberia*. Magadan: Institute of Biological Problems of the North.

Çilingir, F. G., Akın Pekşen, Ç., Ambarlı, H., Beerli, P. & Bilgin, C. C. (2016). Exceptional maternal lineage diversity in brown bears (*Ursus arctos*) from Turkey. *Zoological Journal of the Linnean Society* 176: 463–477.

Ciucci, P., Tosoni, E., Di Domenico, G., Quattrociocchi, F. & Boitani, L. (2014). Seasonal and annual variation in the food habits of Apennine brown bears, central Italy. *Journal of Mammalogy* 95: 572–586.

Ciucci, P., Gervasi, V., Boitani, L., et al. (2015). Estimating abundance of the remnant Apennine brown bear population using multiple noninvasive genetic data sources. *Journal of Mammalogy* 96: 206–220.

Clevenger, A. P., Purroy, F. J. & Pelton, M. R. (1990). Movement and activity patterns of a European brown bear in the Cantabrian mountains, Spain. *International Conference on Bear Research and Management* 8: 205–211.

Cozzi, G., Chynoweth, M., Kusak, J., et al. (2016). Anthropogenic food resources foster the coexistence of distinct life history strategies: year-round sedentary and migratory brown bears. *Journal of Zoology* 300: 142–150.

Dahle, B. & Swenson, J. E. (2003a). Family breakup in brown bears: are young forced to leave? *Journal of Mammalogy* 84: 536–540.

Dahle, B. & Swenson, J. E. (2003b). Seasonal range size in relation to reproductive strategies in brown bears *Ursus arctos*. *Journal of Animal Ecology* 72: 660–667.

Dahle, B., Zedrosser, A. & Swenson, J. E. (2006). Correlates with body size and mass in yearling brown bears (*Ursus arctos*). *Journal of Zoology* 269: 273–283.

Danilov, P. I. (1991). Brown bear dens in the USSR. In: Pazhetnov, V. S. (Ed.), *Bears of the USSR, state of populations* (pp. 56–70). Moscow: Rzhev.

Danilov, P. I. & Tirronen, K. F. (2011). Large predators in the Russian North-West. *Beiträge zur Jagd- und Wildforschung* 36: 19–25.

Danilov, P. I., Rusakov, O. S. & Tumanov, I. L. (1979). *Brown bear//Carnivore mammals of north-west of the USSR.* Ivanter, E. (Ed). Leningrad: Nauka.

Davison, J., Ho, S. Y. W., Bray, S. C., et al. (2011). Late-Quaternary biogeographic scenarios for the brown bear (*Ursus arctos*), a wild model species. *Quarternary Science Reviews* 30: 418–430.

De Ambrogi, M., Aghazadeh, M., Hermosilla, C., et al. (2011). Occurrence of *Baylisascaris transfuga* in wild populations of European brown bears (*Ursus arctos*) as identified by a new PCR method. *Veterinary Parasitology* 179: 272–276.

Decaluwe, F., Camarra, J. J., Sentilles, J. & Quenette, P. Y. (2011). Le statut de la population d'ours brun dans les Pyrénées: bilan 2010 et actualités 2011. *Faune Sauvage* 292: 48–50.

Deecke, B. V. (2012). Tool-use in the brown bear (*Ursus arctos*). *Animal Cognition* 15: 725–730.

Delgado, M. M., Tikhonov, G., Meyke, E., et al. (2018). The seasonal sensitivity of brown bear denning phenology in response to climatic variability. *Frontiers in Zoology* 15: 41.

Derocher, A. E. (2012). *Polar bears; A complete guide to their biology and behavior.* Baltimore, MD: Johns Hopkins University Press.

Egorov, O. V. (1971). *Ursus arctos L.* (1758) – brown bear. In: Tavrovsky, V. A. (Ed.), *Mammals of Yakutia* (pp. 416–430). Moscow: Nauka.

Elfström, M. & Swenson, J. E. (2009). Effects of sex and age on den site use by Scandinavian brown bears. *Ursus* 20: 85–93.

Elfström, M., Swenson, J. E. & Ball, J. P. (2008). Selection of denning habitats by Scandinavian brown bears *Ursus arctos*. *Wildlife Biology* 14: 176–187.

Elfström, M., Davey, M. L., Zedrosser, A., et al. (2014). Do Scandinavian brown bears approach settlements to obtain high-quality food? *Biological Conservation* 178: 128–135.

Esaulova, N. V., Seryodkin, I. V., Konyaev, S. V., Malkina, A. V. & Borisov, M. Y. (2012). Fauna of bear's helminthes from Sakhalin Island and south of Russian Far East. *Russian Veterinary Journal: Small Domestic and Wildlife Animals* 4: 16–19.

Esmaeeli, H. R. (2007). Study of distribution and status of brown bear *Ursus arctos* (Linneaus 1758): Carnivora Ursidae in Fars Province. Iran's Department of Environmental Conservation Library. Final report.

Esteruelas, N. F., Malmsten, J., Bröjer, C., et al. (2016). Chewing lice *Trichodectes pinguis pinguis* in Scandinavian brown bears (*Ursus arctos*). *International Journal for Parasitology: Parasites and Wildlife* 5: 134–138.

Etemad, E. (1985). *The mammals of Iran.* Tehran: Iran's Department of the Environment Publication.

European Environment Agency. (2006). *European forest types: Categories and types for sustainable forest management reporting and policy.* Luxembourg: Office for Official Publications of the European Communities.

Evans, A. L. (2016). Ecophysiology of brown bears; basic physiology and effects of hibernation, pregnancy, body mass and capture. PhD, Hedmark University of Applied Sciences.

Evans, A. L., Singh, N. J., Friebe, A., et al. (2016a). Drivers of hibernation in the brown bear. *Frontiers in Zoology* 13: 7.

Evans, A. L., Singh, N. J., Fuchs, B., et al. (2016b). Physiological reactions to capture in hibernating brown bears. *Conservation Physiology* 4: cow061.

Folk, G. E., Larson, A. & Folk, M. (1974). Physiology of hibernating bears. *International Conference on Bear Research and Management* 3: 373–380.

Frank, S. C., Leclerc, M., Pelletier, F., et al. (2018). Sociodemographic factors modulate the spatial response of brown bears to vacancies created by hunting. *Journal of Animal Ecology* 87: 247–258.

Friebe, A., Zedrosser, A. & Swenson, J. E. (2013). Detection of pregnancy in a hibernator based on activity data. *European Journal of Wildlife Research* 59: 731–741.

Frkovic, A., Ruff, R. L., Cicnjak, L. & Huber, Đ. (1987). Brown bear mortality during 1946–85 in Gorski Kotar, Yugoslavia. *International Conference on Bear Research and Management* 7: 87–92.

Frkovic, A., Huber, Đ. & Kusak, J. (2001). Brown bear litter size in Croatia. *Ursus* 12: 103–106.

Gaillard, J. M., Festa-Bianchet, M., Yoccoz, N. G., Loison, A. & Toigo, C. (2000). Temporal variation in fitness components and population dynamics of large herbivores. *Annual Review of Ecology and Systematics* 31: 367–393.

Garshelis, D. L. (2009). Family Ursidae. In: Wilson, D. E. & Mittermeier, R. A. (Eds.), *Handbook of the mammals of the world. Vol. 1. Carnivores* (pp. 448–497). Barcelona: Lynx Edicions.

Garshelis, D. & McLellan, B. (2011). What's so special about the Syrian bear? *International Bear News* 21: 5–7.

Gervasi, V., Ciucci, P., Boulanger, J., Randi, E. & Boitani, L. (2012). A multiple data source approach to improve abundance estimates of small populations: the brown bear in the Apennines, Italy. *Biological Conservation* 152: 10–20.

Gittleman, J. L. (1986). Carnivore brain size, behavioral ecology, and phylogeny. *Journal of Mammalogy* 67: 23–36.

Gong, J. & Harris, R.B. (2006). The status of bears in China. In: *Understanding Asian bears to Secure their Future* (pp. 50–56). Ibaraki, Japan: Japan Bear Network.

Græsli, A.R., Fahlman, Å., Evans, A.L., Bertelsen, M.F., Arnemo, J.M. & Nielsen, S.S. (2014). Haematological and biochemical reference intervals for free-ranging brown bears (*Ursus arctos*) in Sweden. *BMC Veterinary Research* **10**: 183.

Groff, C., Angeli, F., Bragalanti, N., et al. (2018). *Large carnivores report.* Trento: Autonomous Province of Trento's Forestry and Wildlife Department.

Gutleb, B., Ghaemi, R. A. & Kusak, J. (2002). Brown bear in Iran. *International Bear News* 11: 20.

Hamdine, W., Thevenot, M. & Michaux, J. (1998). Recent history of the brown bear in the Maghreb. *Comptes Rendus de l'Academie des Sciences - Serie III: Sciences de la Vie* 321: 565–570.

Hellgren, E. C. (1998). Physiology of hibernation in bears. *Ursus* 10: 467–477.

Herceg Romanić, S., Klinčić, D., Kljaković-Gašpić, Z., et al. (2015). Organochlorine pesticides and polychlorinated biphenyl congeners in wild terrestrial mammals from Croatia: interspecies comparison of residue levels and compositions. *Chemosphere* 137: 52–58.

Hertel, A. G., Steyaert, S. M., Zedrosser, A., et al. (2016). Bears and berries: species-specific selective foraging on a patchily distributed food resource in a human-altered landscape. *Behavioral Ecology and Sociobiology* 70: 831–842.

Hertel, A. G., Swenson, J. E. & Bischof, R. (2017). A case for considering individual variation in diel activity patterns. *Behavioral Ecology* 28: 1524–1531.

Hertel, A. G., Leclerc, M., Warren, D., et al. (2019a). Don't poke the bear: using tracking data to quantify behavioural syndromes in elusive wildlife. *Animal Behaviour* 147: 91–104.

Hertel, A. G., Zedrosser, A., Kindberg, J., Langval, O. & Swenson, J. E. (2019b). Fluctuating mast production does not drive Scandinavian brown bear behavior. *Journal of Wildlife Management* https://doi.org/10.1002/jwmg.21619.

Hilderbrand, G. V., Jenkins, S. G., Schwartz, C. C., Hanley, T. A. & Robbins, C. T. (1999a). Effect of seasonal differences in dietary meat intake on changes in body mass and composition in wild and captive brown bears. *Canadian Journal of Zoology* 77: 1623–1630.

Hilderbrand, G. V., Schwartz, C. C., Robbins, C. T., et al. (1999b). The importance of meat, particularly salmon, to body size, population productivity, and conservation of North American brown bears. *Canadian Journal of Zoology* 77: 132–138.

Hirata, D., Mano, T., Abramov, A. V., et al. (2013). Molecular phylogeography of the brown bear (*Ursus arctos*) in northeastern Asia based on analyses of complete mitochondrial DNA sequences. *Molecular Biology and Evolution* 30: 1644–1652.

Hohšteter, M., Šoštarić-Zuckermann, I. C., Reljić, S., et al. (2018). Intestinal adenocarcinoma in a European brown bear (*Ursus arctos*) – a case report. *Veterinarski Arhiv* 88: 569–579.

Huber, Đ. (2018). *Ursus arctos*. The IUCN Red List of Threatened Species 2018e. T41688A133236401. Accessed December 8, 2018.

Huber, Đ. & Ehrlich, I. (1981). A survey of helminthoparasites of wildlife in the Plitvice Lakes National Park: ecological relationships and a stability regulation model in the host–parasite system. *Internationales Symposium über die Erkrankungen der Zootiere* 23: 155–160.

Huber, Đ. & Roth, H. U. (1993). Movements of European brown bears in Croatia. *Acta Theriologica* 38: 151–159.

Huber, D. & Roth, H. U. (1997). Denning of brown bears in Croatia. *International Association on Bear Research and Management* 9: 79–83.

Huber, Đ., Kusak, J., Zvorc, Z. & Rafaj, R. (1997). Effects of sex, age, capturing method, and season on serum chemistry values of brown bears in Croatia. *Journal of Wildlife Diseases* 33: 790–794.

Ionescu, O. (2002). *Bear status and management in Carpathians, Bucharest, Romania*. Bucharest: Forest Research and Management Planning Institute (ICAS).

Jakubiec, Z. (2008). 1354 Niedźwiedź *Ursus arctos*. In *Monitoring gatunków i siedlisk przyrodniczych ze szczególnym uwzględnieniem specjalnych obszarów ochrony siedlisk Natura 2000. Wyniki monitoringu*. Unpublished report for the Główny Inspektorat Ochrony Środowiska, Warszawa.

Jerina, K. & Adamic, M. (2008). Fifty years of brown bear population expansion: effects of sex-biased dispersal on rate of expansion and population structure. *Journal of Mammalogy* 89: 1491–1501.

Jerina, K., Polaina, E., Huber, Đ., et al. (2018). Reconstruction of brown bear population dynamics in Slovenia and Croatia for the period 1998–2018. Prepared within C5 action of LIFE DINALP BEAR Project (LIFE13 NAT/SI/0005).

Kaczensky, P., Huber, Đ., Knauer, F., et al. (2006). Activity patterns of brown bears (*Ursus arctos*) in Slovenia and Croatia. *Journal of Zoology* 269: 474–485.

Kaczensky, P., Chapron, G., Von Arx, M., et al. (2013). Status, management and distribution of large carnivores – bear, lynx, wolf & wolverine – in Europe. Part 1 (pp. 1–72). Available from http://ec.europa.eu/environment/nature/conservation/species/carnivores/pdf/task_1_part1_statusoflcineurope.pdf

Karamanlidis, A. A., Youlatos, D., Sgardelis, S. & Scouras, Z. (2007). Using sign at power poles to document presence of bears in Greece. *Ursus* 18: 54–61.

Karamanlidis, A. A., Sanopoulos, A., Georgiadis, L. & Zedrosser, A. (2011). Structural and economic aspects of human–bear conflicts in Greece. *Ursus* 22: 141–151.

Karamanlidis, A. A., Pllaha, S., Krambokoukis, L., Shore, K. & Zedrosser, A. (2014). Preliminary brown bear survey in southeastern Albania. *Ursus* 25: 1–7.

Karamanlidis, A. A., De Gabriel Hernando, M., Krambokoukis, L. & Gimenez, O. (2015). Evidence of a large carnivore population recovery: counting bears in Greece. *Journal for Nature Conservation* 27: 10–17.

Kindberg, J., Swenson, J. E., Ericsson, G. & Bellemain, E. (2011). Estimating population size and trends of the Swedish brown bear *Ursus arctos* population. *Wildlife Biology* 17: 114–123.

Klinka, D. R. & Reimchen, T. E. (2002). Nocturnal and diurnal foraging behaviour of brown bears (*Ursus arctos*) on a salmon stream in coastal British Columbia. *Canadian Journal of Zoology* 80: 1317–1322.

Knott, E., Bunnefeld, N., Huber, Đ., et al. (2014). The potential impacts of changes in bear hunting policy for hunting organisations in Croatia. *European Journal of Wildlife Research* 60: 85–97.

Kohn, M., Knauer, F., Stoffella, A., Schroder, W. & Paabo, S. (1995). Conservation genetics of the European brown bear: a study using excremental PCR of nuclear and mitochondrial sequences. *Molecular Ecology* 4: 95–103.

Kojola, I. & Heikkinen, S. (2006). Structure of expanded brown bear population at the edge of the range in Finland. *Annales Zoologici Fennici* 43: 258–262.

Kolter, L. & Zee, J. (2008). Inventory of captive ursids – report from the Captive Bears Expert Team. *International Bear News* 171: 9–11.

Korsten, M., Ho, S. Y. W., Davison, J., et al. (2009). Sudden expansion of a single brown bear maternal lineage across northern continental Eurasia after the last ice age: a general demographic model for mammals? *Molecular Ecology* 18: 1963–1979.

Krechmar, M. A. (1995). Geographical aspects of the feeding of the brown bear (*Ursus arctos* L.) in the extreme northeast of Siberia. *Russian Journal of Ecology* 26: 436–443.

Krofel, M. & Jerina, K. (2016). Mind the cat: conservation management of a protected dominant scavenger indirectly affects an endangered apex predator. *Biological Conservation* 197: 40–46.

Krofel, M., Filacorda, S. & Jerina, K. (2010). Mating-related movements of male brown bears on the periphery of an expanding population. *Ursus* 21: 23–29.

Krofel, M., Spacapan, M. & Jerina, K. (2017). Winter sleep with room service: denning behaviour of brown bear with access to anthropogenic food. *Journal of Zoology* 302: 8–14.

Kruckenhauser, L., Rauer, G., Daeubl, B. & Haring, E. (2009). Genetic monitoring of a founder population of brown bears (*Ursus arctos*) in central Austria. *Conservation Genetics* 10: 1223–1233.

Kudrenko, S. (2018). Factors contributing to human Injuries and fatalities inflicted by brown bears (Ursus arctos) in Russia, 1932–2017. Master thesis, Norwegian University of Life Sciences.

Kuduk, K., Babik, W., Bojarska, K., et al. (2012). Evolution of major histocompatibility complex class I and class II genes in the brown bear. *BMC Evolutionary Biology* 12: 197.

Kurtén, B. (1973). Transberingian relationships of *Ursus arctos* Linne (brown and grizzly bears). *Societas Scientiarum Fennica Commentation Biology* 65: 1–10.

Kusak, J., Rafaj, R. B., Žvorc, Z., et al. (2005). Effects of sex, age, body mass, and capturing method on hematologic values of brown bears in Croatia. *Journal of Wildlife Diseases* 41: 843–847.

Lazarus, M., Sekovanić, A., Reljić, S., et al. (2014). Selenium in brown bears (*Ursus arctos*) from Croatia: relation to cadmium and mercury. *Journal of Environmental Science and Health, Part A* 49: 1392–1401.

Lazarus, M., Orct, T., Reljić, S., et al. (2018a). Trace and macro elements in the femoral bone as indicators of long-term environmental exposure to toxic metals in European brown bear (*Ursus arctos*) from Croatia. *Environmental Science and Pollution Research* 25: 21,656–21,670.

Lazarus, M., Sekovanić, A., Orct, T., et al. (2018b). Sexual maturity and life stage influences toxic metal accumulation in Croatian brown bears. *Archives of Environmental Contamination and Toxicology* 74: 339–348.

Leclerc, M., Vander Wal, E., Zedrosser, A., et al. (2016). Quantifying consistent individual differences in habitat selection. *Oecologia* 180: 697–705.

Loreille, O., Orlando, L., Patou-Mathis, M., et al. (2001). Ancient DNA analysis reveals divergence of the cave bear, *Ursus spelaeus*, and brown bear, *Ursus arctos*, lineages. *Current Biology* 11: 200–203.

Lortkipanidze, B. (2010). Brown bear distribution and status in South Caucasus. *Ursus* 21: 97–103.

Luvsamjamba, A., Reynolds, H., Yansanjav, A., et al. (2016). Review of Gobi bear research (*Ursus arctos gobiensis*, Sokolov and Orlov, 1992). *Arid Ecosystems* 6: 206–212.

Madić, J., Huber, Đ. & Lugović, B. (1993). Serologic survey for selected viral and rickettsial agents of brown bears (*Ursus arctos*) in Croatia. *Journal of Wildlife Diseases* 29: 572–576.

Malcom, J. R., Liu, C., Miller, L. B., Allnutt, T. & Hansen, L. (2002). *Habitats at risk*. Gland, Switzerland: WWF.

Manchi, S. & Swenson, J.E. (2005). Denning behaviour of Scandinavian brown bears *Ursus arctos*. *Wildlife Biology* 11: 123–132.

Manlius, N. (1998). L'ours brun en Egypte. *Ecologie* 29: 565–581.

Männil, P. & Kont, R. (2012). *Action plan for the conservation and management of wolf, lynx and brown bear in Estonia in 2012–2021*. Tallinn: Estonian Ministry of the Environment.

Mano, T. (2006). The status of brown bears in Japan. In: *Understanding Asian bears to secure their future* (pp. 111–121). Ibaraki: Japan Bear Network.

Martin, J., Basille, M., Van Moorter, B., et al. (2010). Coping with human disturbance: spatial and temporal tactics of the brown bear (*Ursus arctos*). *Canadian Journal of Zoology* 88: 875–883.

Matsubayashi, J., Morimoto, J.O., Tayasu, I., et al. (2015). Major decline in marine and terrestrial animal consumption by brown bears (*Ursus arctos*). *Scientific Reports* 17: 9203.

McLellan, B. N. (2015). Some mechanisms underlying variation in vital rates of grizzly bears on a multiple use landscape. *Journal of Wildlife Management* 79: 749–765.

McLellan, B. N., Proctor, M. F., Huber, Đ. & Michel, S. (2016). *Brown bear (*Ursus arctos*) isolated populations (Supplementary material to* Ursus arctos *redlisting account)*. The IUCN Red List of Threatened Species. Accessed December 5, 2018.

McLellan, B. N., Proctor, M. F., Huber, Đ. & Michel, S. (2017). Ursus arctos *(amended version of 2017 assessment)*. The IUCN Red List of Threatened Species 2017: e. T41688A121229971. Accessed December 5, 2018.

Medvedev, S. G. & Seredkin, I. V. (2018). Fleas (Siphonaptera) of carnivores (Mammalia, Carnivora) of the Russian Far East. *Parazitologiya* 52: 315–327.

Mertzanis, Y., Isaak, I., Mavridis, A., Nikolaou, O. & Tragos, A. (2005). Movements, activity patterns and home range of a female brown bear (*Ursus arctos*, L.) in the Rodopi Mountain Range, Greece. *Belgian Journal of Zoology* 135: 217–221.

Mertzanis, G., Giannakopoulos, A. & Pylidis, C. (2009). *Ursus arctos* (Linnaeus, 1758). In: Legakis, A. & Maragou, P. (Eds.), *Red Data Book of the threatened animal species of Greece* (pp. 387–389). Athens: Hellenic Zoological Society.

Mertzanis, G., Mazaris, A., Sgardelis, S., et al. (2011). Telemetry as a tool to study spatial behavior and patterns of brown bears as affected by the newly constructed Egnatia highway – N. Pindos – Greece. In: O. Krejcar (Ed.), *Modern telemetry* (pp. 307–328). IntechOpen, DOI: 10.5772/24958. Available from: www.intechopen.com/books/modern-telemetry/telemetry-as-a-tool-to-study-spatial-behaviour-and-patterns-of-brown-bears-as-affected-by-the-newly.

Ministry of Environment and Waters (MEW). (2008). Action Plan for the Brown Bear in Bulgaria.

Modrić, Z. & Huber, Đ. (1993). Serologic survey for leptospirae in European brown bears (*Ursus arctos*) in Croatia. *Journal of Wildlife Diseases* 29: 608–611.

Mordosov, I. I. (1993). Yakutia. In: Vaisfeld, M. A. & Chestin, I. E. (Eds.), *Bears: Brown bear, polar bear, Asian black bear* (pp. 301–318). Moscow: Nauka.

Morehouse, A. T., Graves, T. A., Mikle, N. & Boyce, M. S. (2016). Nature vs. nurture: evidence for social learning of conflict behaviour in grizzly bears. *PLoS ONE* 11: e0165425.

Moriwaki, J., Omori, R., Shimozuru, M., et al. (2018). Evaluation of body condition using body mass and chest girth in brown bears of Hokkaido, Japan (*Ursus arctos yesoensis*). *Japanese Journal of Veterinary Research* 66: 71–81.

Naidenko, S. V., Hernandez-Blanko, J. A., Seryodkin, I. V., et al. (2018). Serum prevalence of the bears in the Russian Far East to different pathogens. *Zoologicheskii Zhurnal* 97: 1189–1194.

Nasimovich, A. A. & Semenov-Tian-Shansky, O. I. (1951). Brown bear food resources and the assessment of its role as a predator in the Lapland reserve. *Bulleten Moskovskogo obshchestva Ispytateley Prirody, Otdelenie Biologicheskoe* 56(4): 3–12.

Nawaz, M. A. (2007). Status of the brown bear in Pakistan. *Ursus* 18: 89–100.

Nawaz, M. A., Swenson, J. E. & Zakaria, V. (2008). Pragmatic management increases a flagship species, the Himalayan brown bears, in Pakistan's Deosai National Park. *Biological Conservation* 141: 2230–2241.

Nawaz, M. A., Martin, J. & Swenson, J.E. (2014). Identifying key habitats to conserve the threatened brown bear in the Himalaya. *Biological Conservation* 170: 198–206.

Nelson, R. A., Wahner, H. W., Jones, J. D., Ellefson, R. D. & Zollman, P. E. (1973). Metabolism of bears before, during, and after winter sleep. *American Journal of Physiology* 224: 491–496.

Niedziałkowska, M., Hayward, M. W., Borowik, T., Jędrzejewski, W. & Jędrzejewska, B. (2019). A meta-analysis of ungulate predation and prey selection by the brown bear *Ursus arctos* in Eurasia. *Mammal Research* 64: 1–9.

Olson, D. M., Dinerstein, E., Robin, A., et al. (2000). *The Global 2000. A representation approach to conserving the Earth's distinctive ecoregions.* Gland: WWF.

Ordiz, A., Rodriguez, C., Naves, J., et al. (2007). Distance-based criteria to identify minimum number of brown bear females with cubs in Europe. *Ursus* 18: 158–167.

Ordiz, A., Støen, O.-G., Swenson, J. E., Kojola, I. & Bischof, R. (2008). Distance-dependent effect of the nearest neighbour: spatio-temporal patterns in brown bear reproduction. *Ecology* 89: 3327–3335.

Ordiz, A., Støen, O.-G., Delibes, M. & Swenson, J. E. (2011). Predators or prey? Spatio-temporal discrimination of human-derived risk by brown bears. *Oecologia* 166: 59–67.

Ordiz, A., Støen, O.-G., Sæbø, S., et al. (2013). Lasting behavioural responses of brown bears to experimental human encounters. *Journal of Applied Ecology* 50: 306–314.

Palazón, S., Ruiz-Olmo, J. & Batet, A. (2011). El oso en Cataluña: La población se consolida. *Quercus* 304: 16–23.

Panova, O. A., Khrustalev, A. V., Kurnosova, O. P., Seryodkin, I. V. & Kalinkin, Y. N. (2018). Parasitological examination of bears in the north-east of the Altai Mountains and the Middle Sikhote-Alin. *Theory and Practice of Combating Parasitic Diseases* 19: 74–377.

Papadopoulos, E., Komnenou, A., Poutachides, T., et al. (2017). Detection of *Dirofilaria immitis* in a brown bear (*Ursus arctos*) in Greece. *Helminthologia* 54: 257–261.

Paralikidis, N. P., Papageorgiou, N. K., Kontsiotis, V. J. & Tsiompanoudis, A. C. (2010). The dietary habits of the brown bear (*Ursus arctos*) in western Greece. *Mammalian Biology* 75: 29–35.

Patitschniak-Arts, M. (1993). *Ursus arctos. Mammalian Species* 439: 1–10.

Paunović, M., Ćirović, D. & Milenković, M. (2007). *Status and conservation of carnivores in Serbia.* CIC Proceedings of the Symposium "Coexistance of man and carnivores: Threat or Benefit?", May 1, 2007 (pp. 111–117). Belgrade, Serbia.

Penteriani, V., López-Bao, J. V., Bettega, C., et al. (2017). Consequences of brown bear viewing tourism: a review. *Biological Conservation* 206: 169–180.

Penteriani, V., Zarzo-Arias, A., Novo Fernández, A., Bombieri, G. & Lopez-Sanchez, C. A. (2019). Responses of an endangered brown bear population to climate change based on predictable food resource and shelter alterations. *Global Change Biology* 25: 1133–1151.

Pérez, T., Naves, J., Vázquez, J. F., et al. (2014). Estimating the population size of the endangered Cantabrian brown bear through genetic sampling. *Wildlife Biology* 20: 300–309.

Petram, W., Knauer, F. & Kaczensky, P. (2004). Human influence on the choice of winter dens by European brown bears in Slovenia. *Biological Conservation* 119: 129–136.

Pilāts, V. & Ozoliņš, J. (2003). Status of brown bear in Latvia. *Acta Zoologica Lituanica* 13: 65–71.

Puchkovskiy, S. (2009). Selectivity of tree species as activity target of brown bear in taiga. *Contemporary Problems of Ecology* 2: 260–268.

Randi, E., Gentile, L., Boscagli, G., Huber, Đ. & Roth, H. U. (1994). Mitochondrial DNA sequence divergence among some west European brown bear (*Ursus arctos* L.) populations. Lessons for conservation. *Heredity* 73: 480–489.

Reljić, S., Srebočan, E., Huber, Đ., et al. (2012). A case of a brown bear poisoning with carbofuran in Croatia. *Ursus* 23: 86–91.

Reynolds, H. V., Craighead, D. J., Proctor, M., Luvsanjamba, A. & Mijiddorj, B. (2010). *Gobi bear conservation in Mongolia.* Ulaanbaatar, Mongolia: The Gobi Bear Project Team.

Robbins, C. T., Meray, B.-D., Fortin, J. & Nelson, L. O. (2012). Maternal condition determines birth and growth of newborn bear cubs. *Journal of Mammalogy* 93: 540–546.

Rodríguez, C., Naves, J., Fernández-Gil, A., Obeso, J. R. & Delibes, M. (2007). Long-term trends in food habits of a relict brown bear population in northern Spain: the influence of climate and local factors. *Environmental Conservation* 34: 36–44.

Rosell, F., Jojola, S. M., Ingdal, K., et al. (2011). Brown bears possess anal sacs and secretions may code for sex. *Journal of Zoology* 283: 143–152.

Ruiz-Villar, H., Morales-Gonzalez, A., Bombieri, G., Zarzo-Arias, A. & Penteriani, V. (2019). Characterization of a brown bear aggregation during the hyperphagia period in the Cantabrian Mountains, NW Spain. *Ursus* 29: 93–100.

Saarma, U., Ho, S. Y. W., Pybus, O. G., et al. (2007). Mitogenetic structure of brown bears (*Ursus arctos* L.) in northeastern Europe and a new time frame for the formation of European brown bear lineages. *Molecular Ecology* 16: 401–413.

Sathyakumar, S., Kaul, R., Ashraf, N. V., Mookerjee, A. & Menon, V. (2012). *National Bear Conservation and Welfare Action Plan.* Ministry of Environment and Forests, Wildlife Institute of India and Wildlife Trust of India.

Sato, Y., Kamiishi, C., Tokaji, T., et al . (2013). Selection of rub trees by brown bears (*Ursus arctos*) in Hokkaido, Japan. *Acta Theriologica* 59: 129–137.

Selva, N., Zwijacz-Kozica, T., Sergiel, A., Olszanska, A. & Zięba, F. (2011). *Management plan for the brown bear*

Ursus arctos *in Poland.* Warsaw: University of Life Sciences.

Selva, N., Teitelbaum, C. S., Sergiel, A., et al. (2017). Supplementary ungulate feeding affects movement behavior of brown bears. *Basic and Applied Ecology* 24: 68–76.

Sentilles, J., Camarra, J. J., Vanpé, C. & Quenette, P. Y. (2019). *Suivi de l'ours brun dans les Pyrénées françaises, rapport annuel 2018* (pp. 1–50). Paris: ONCFS, Office national de la chasse et de la faune sauvage.

Sergiel, A. & Maślak, R. (2014). The welfare of captive bears. In: Gardocka, T., Gruszczyńska, A., Maślak, R. & Sergiel, A. (Eds.), *The welfare of animals in zoos and EU legal standards* (pp. 129–139). Warsaw: Publishing House ELIPSA.

Sergiel, A., Naves, J., Kujawski, P., et al. (2017). Histological, chemical and behavioural evidence of pedal communication in brown bears. *Scientific Reports* 7: 1052.

Seryodkin, I. V. (2006). The biology and conservation status of brown bears in the Russian Far East. In: *Understanding Asian bears to secure their future* (pp. 1–145). Ibaraki: Japan Bear Network.

Seryodkin, I. V. (2010). Basic types of interspecific relationship of brown bear *Ursus arctos* in the Sikhote-Alin. *Achievements in the Life Sciences* 2: 134–145.

Seryodkin, I. V. (2015). Trichinosis of brown bear and Asiatic black bear in the Russian Far East. *Bulletin of KrasGAU* 12: 167–173.

Seryodkin, I. V., Esaulova, N. V., Mukhacheva, A. S., Petrunenko, Y. K. & Miquelle, D. G. (2012). Endoparasites of large carnivores in Primorskii Krai. In: Seryodkin, I. V. & Miquelle, D. G. (Eds.), *Diseases and parasites of wildlife in Siberia and the Russian Far East* (pp. 127–136). Vladivostok: Dalnauka.

Seryodkin, I. V., Kostyria, A. V., Goodrich, J. M. & Miquelle, D. G. (2013). Daily activity patterns of brown bears (*Ursus arctos*) of the Silkhote-Alin Mountain Range (Primorskiy Krai, Russia). *Russian Journal of Ecology* 44: 50–55.

Seryodkin, I. V., Kostyria, A. V. & Goodrich, J. M. (2014). Daily and seasonal movements of brown bear in the Sikhote-Alin. *Seria "Biologiya i Ecologiya"* 4: 233–240.

Seryodkin, I. V., Esaulova, N., Tranbenkova, N., et al. (2016). Helminth fauna of brown bears and Asiatic black bears of the Russian Far East. *European Multicolloquium of Parasitology* 12: 10.

Seryodkin, I. V., Paczkowski, J., Borisov, M. Y. & Petrunenko, Y. K. (2017a). Home ranges of brown bears on the Kamchatka Peninsula and Sakhalin Island. *Contemporary Problems of Ecology* 10: 599–611.

Seryodkin, I. V., Thomas, L., Birtles, R., et al. (2017b). Tick-borne pathogens in Ixodidae ticks and their large mammalian hosts in the Russian Far East. *Parazitologiya* 51: 239–252.

Shchelkanov, M. Y., Deviatkin, A. A., Ananiev, V. Y., et al. (2016). Complete genome sequence of a rabies virus strain isolated from a brown bear (*Ursus arctos*) in Primorsky Krai, Russia (November 2014). *Genome Announcements* 4: e00642-16.

Shimozuru, M., Yamanaka, M., Nakanishi, M., et al. (2017). Reproductive parameters and cub survival of brown bears in the Rusha area of the Shiretoko Peninsula, Hokkaido, Japan. *PLoS ONE* 12: e0176251.

Skuban, M., Find'o, S. & Kajba, M. (2016). Human impacts on bear feeding habits and habitat selection in the Pol'ana Mountains, Slovakia. *European Journal of Wildlife Research* 62: 353–364.

Skuban, M., Find'o, S. & Kajba, M. (2017). Bears napping nearby: daybed selection by brown bears in a human-dominated landscapes. *Canadian Journal of Zoology* 96: 1–11.

Sommer, R. S. & Benecke, N. (2005). The recolonization of Europe by brown bears *Ursus arctos* Linnaeus, 1758 after the Last Glacial Maximum. *Mammal Review* 35: 156–164.

Species360. (2019). Zoological Information Management System (ZIMS). Available at zims.Species360.org.

Stearns, S. C. (1992). *The evolution of life histories.* New York, NY: Oxford University Press.

Stenset, N. E., Lutnæs, P. N., Bjarnadóttir, V., et al. (2016). Seasonal and annual variation in the diet of brown bears *Ursus arctos* in the boreal forest of southcentral Sweden. *Wildlife Biology* 22: 107–116.

Steyaert, S. M. J. G., Endrestøl, A., Hackländer, K., Swenson, J. E. & Zedrosser, A. (2012). The mating system of the brown bear *Ursus arctos. Mammal Review* 42: 12–34.

Steyaert, S. M. J. G., Kindberg, J., Swenson, J. E. & Zedrosser, A. (2013). Male reproductive strategy explains spatiotemporal segregation in brown bears. *Journal of Animal Ecology* 82: 836–845.

Steyaert, S. M., Kindberg, J., Jerina, K., et al. (2014). Behavioral correlates of supplementary feeding of wildlife: can general conclusions be drawn? *Basic and Applied Ecology* 15: 669–676.

Steyaert, S. M. J. G., Leclerc, M., Pelletier, F., et al. (2016). Human shields mediate sexual conflict in a top predator. *Proceedings of the Royal Society B* 283: 20160906.

Støen, O.-G., Bellemain, E., Saebo, S. & Swenson, J. E. (2005). Kin-related spatial structure in brown bears *Ursus arctos. Behavioral Ecology and Sociobiology* 59: 191–197.

Stoen, O.-G., Zedrosser, A., Wegge, P. & Swenson, J. E. (2006a). Socially induced delayed primiparity in brown bears *Ursus arctos. Behavioral Ecology and Sociobiology* 61: 1–8.

Støen, O.-G., Zedrosser, A., Saebo, S. & Swenson, J. E. (2006b). Inversely density-dependent natal dispersal in brown bears *Ursus arctos. Oecologia* 148: 356–364.

Swenson, J. E. & Haroldson, M. A. (2008). Observations of mixed-aged litters in brown bears. *Ursus* 19: 73–79.

Swenson, J. E., Jansson, A., Riig, R. & Sandegren, F. (1999). Bears and ants: myrmecophagy by brown bears in central Scandinavia. *Canadian Journal of Zoology* 77: 551–561.

Swenson, J. E., Adamic, M., Huber, Đ. & Stokke, S. (2007). Brown bear body mass and growth in northern and southern Europe. *Oecologia* 153: 37–47.

Swenson, J. E., Taberlet, P. & Bellemain, E. (2011). Genetics and conservation of European brown bears *Ursus arctos. Mammal Review* 41: 87–98.

Swenson, J. E., Schneider, M., Zedrosser, A., et al. (2017). Challenges of managing a European brown bear population; lessons from Sweden, 1943–2013. *Wildlife Biology* 2017(4): wlb-00251.

Taberlet, P. & Bouvet, J. (1994). Mitochondrial DNA polymorphism, phylogeography, and conservation genetics of the brown bear *Ursus arctos* in Europe. *Proceedings of the Royal Society of London Series B Biological Sciences* 255: 195–200.

Taberlet, P., Swenson, J. E., Sandegren, F. & Bjaervall, A. (1995). Localization of a contact zone between two highly divergent mitochondrial DNA lineages of the brown bear *Ursus arctos* in Scandinavia. *Conservation Biology* 9: 1255–1261.

Tattoni, C., Bragalanti, N., Fgroff, C. & Rovero, F. (2015). Patterns in the use of rub trees by the Eurasian brown bear. *Hystrix, the Italian Journal of Mammalogy* 26: 118–124.

Tirronen, K. F., Panchenko, D. B. & Kusnecova, D. S. (2015). Brown bear (*Ursus arctos* L.) of the White Sea of the Kola Peninsula. *Vestnik Okhotovedeniya* 12: 125–136.

Tirronen, K. F., Panchenko, D. B. & Kuznetsova, A. S. (2016). New data on the diets of the brown bear (*Ursus arctos* L.) in Karelia and the south of the Kola Peninsula. *Transactions of Karelian Research Centre of the Russian Academy of Sciences* 12: 114–122.

Tøien, Ø., Blake, J., Edgar, D. M., et al. (2011). Hibernation in black bears: independence of metabolic suppression from body temperature. *Science* 331: 906–909.

Tøien, Ø., Blake, J. & Barnes, B. M. (2015). Thermoregulation and energetics in hibernating black bears: metabolic rate and the mystery of multi-day body temperature cycles. *Journal of Comparative Physiology B* 185: 447–461.

Tosoni, E., Mei, M. & Ciucci, P. (2018). Ants as food for Apennine brown bears. *European Zoological Journal* 85: 343–349.

Tranbenkova, N. A., Seryodkin, I. V. & Zhakov, V. V. (2012). Helminths of Kamchatka brown bears. In: Seryodkin, I. V. & Miquelle, D. G. (Eds.), *Diseases and parasites of wildlife in Siberia and the Russian Far East* (pp. 173–178). Vladivostok: Dalnauka.

Tumendemberel, O., Proctor, M., Reynolds, H., et al. (2015). Gobi bear abundance and inter-oases movements, Gobi Desert, Mongolia. *Ursus* 26: 129–142.

Ustinov, S. K. (1993). The Baikal region. In: Vaisfeld, M. A. & Chestin, I. E. (Eds.), *Bears: Brown bear, polar bear, Asian black bear* (pp. 275–301). Moscow: Nauka.

Vaisfeld, M. A. & Chestin, I. E. (1993). *Bears: Distribution, ecology, management, and conservation.* Moscow: Nauka.

Valdiosera, C. E., García, N., Anderung, C., et al. (2007). Staying out in the cold: glacial refugia and mitochondrial DNA phylogeography in ancient European brown bears. *Molecular Ecology* 16: 5140–5148.

Van De Walle, J., Pigeon, G., Zedrosser, A., Swenson, J. E. & Pelletier, F. (2018). Hunting regulation favors slow life histories in a large carnivore. *Nature Communications* 9: 1100.

Vulla, E., Hobson, K. A., Korsten, M., et al. (2009). Carnivory is positively correlated with latitude among omnivorous mammals: evidence from brown bears, badgers and pine martens. *Annales Zoologici Fennici* 46: 395–416.

Wilmers, C. C. & Post, E. (2006). Predicting the influence of wolf-provided carrion on scavenger community dynamics under climate change scenarios. *Global Change Biology* 12: 403–409.

Yudin, V. G. (1993). Aggressive behaviour of brown and the Asian black bear in the coastal areas. In: Pazhetnov, V. S. (Ed.), *Bears in the UDSSR – Status of populations.* Rzhev: Scientific Institut for Nature Conservation and Management of Protected Areas.

Zarzo-Arias, A., Del Mar Delgado, M., Ordiz, A., et al. (2018). Brown bear behaviour in human-modified landscapes: the case of the endangered Cantabrian population, NW Spain. *Global Ecology and Conservation* 16: e00499.

Zedrosser, A., Dahle, B., Swenson, J. E. & Gerstl, N. (2001). Status and management of the brown bear in Europe. *Ursus* 12: 9-20.

Zedrosser, A., Rauer, G. & Kruckenhauser, L. (2004). Early primiparity in brown bears. *Acta Theriologica Sinica* 49: 427–432.

Zedrosser, A., Dahle, B. & Swenson, J. E. (2006). Population density and food conditions determine adult female body size in brown bears. *Journal of Mammalogy* 87: 510–518.

Zedrosser, A., Bellemain, E., Taberlet, P. & Swenson, J. E. (2007). Genetic estimates of annual reproductive success in male brown bears: the effects of body size, age, internal relatedness and population density. *Journal of Animal Ecology* 76: 368–375.

Zedrosser, A., Dahle, B., Stoen, O.-G. & Swenson, J. E. (2009). The effects of primiparity on reproductive performance in the brown bear. *Oecologia* 160: 847–854.

Zedrosser, A., Steyaert, S. M. J. G., Gossow, H. & Swenson, J. E. (2011). Brown bear conservation and the ghost of persecution past. *Biological Conservation* 144: 2163–2170.

Zedrosser, A., Pelletier, F., Bischof, R., Festa-Bianchet, M. & Swenson, J. E. (2013). Determinants of lifetime reproduction in female brown bears: early body mass, longevity, and hunting regulations. *Ecology* 94: 231–240.

Zlatanova, D., Racheva, V. & Fremuth, W. (2009). Habitatverbund für den Braunbären in Bulgarien – Grundlage für die Schaffung Transeuropäischer Wildtiernetze (TEWN) auf dem Balkan. *Naturschutz und Landschaftsplanung* 41: 114–122.

Zyryanov, A. N. (2012). Study of brown bear behavior in natural conditions. In: *Proceedings of Centralno-Sibirsky Nature Reserve* (pp. 155–169). Krasnoyarsk, Russia: Centralno-Sibirsky Nature Reserve.

Zyśk-Gorczyńska, E., Jakubiec, Z. & Wuczyński, A. (2014). Brown bears (*Ursus arctos*) as ecological engineers: the prospective role of trees damaged by bears in forest ecosystems. *Canadian Journal of Zoology* 93: 133–141.

Chapter **13**

Brown Bear (*Ursus arctos*; North America)

Mark A. Haroldson, Melanie Clapham, Cecily C. Costello, Kerry A. Gunther, Katherine C. Kendall, Sterling D. Miller, Karine E. Pigeon, Michael F. Proctor, Karyn D. Rode, Christopher Servheen, Gordon B. Stenhouse, and Frank T. van Manen

Names

Genus: *Ursus* (Linnaeus, 1758)

Species: brown or grizzly bear *Ursus arctos* (Linnaeus, 1758)

Subspecies: brown bear, grizzly bear *Ursus arctos horribilis* (Ord, 1815); Kodiak bear *Ursus arctos middendorffi* (Merriam, 1896)

Names in other languages: **French**: ours brun; **German**: Braunbär; **Spanish**: oso pardo; **Italian**: orso bruno; **Dutch**: bruine beer; **Sioux**: matohota; **Blackfoot**; Kyiyo; **Arapaho**: nonookuneseet; **Shoshone**: bee-yah-ah-gwy; **Salish**: Smx̣eỳčǹ

Other common names: silvertip, Old Ephraim, bruin, golden bear

The scientific name combines the Latin (*ursus*) and Greek (Ἄρκτος [arktos]) words for "bear." North American brown bears are commonly referred to as "grizzly bears" throughout much of their range in the interior of the continent. The name is derived from the bear's "grizzled" appearance caused by its silver-tipped hair. Typically, only populations along the Pacific coast of Canada, Alaska, and the Admiralty, Baranof, and Chichagof islands are referred to as "brown bears." Brown bears occurring on Kodiak Island, the largest representative of the species in North America, are referred to as Kodiak bears. George Ord published the first scientific description of the species in 1815, giving it the scientific name *Ursus horribilis* based on Lewis and Clark's vivid descriptions of the bear's grisly, fear-inspiring behavior (Schullery 2002). Although Lewis and Clark are often credited with being the first to observe the species, grizzly bears were described to them by Native American peoples (Schullery 2002), who revered them as spiritual sources of guidance, medicine, courage, and strength long before Lewis and Clark encountered them (Rockwell 1991). We use the common names brown bear and grizzly bear consistent with their regional usage, recognizing they refer to the same species.

Taxonomy

Brown bears are mammals from the order Carnivora, family Ursidae, and subfamily Ursinae. They belong to the genus *Ursus*, which includes four extant bear species with wide distributions. Besides brown bears, this genus includes the polar bear (*U. maritimus*), American black bear (*U. americanus*), and Asiatic black bear (*U. thibetanus*). The subfamily Ursinae also includes the sloth bear (*Melursus ursinus*) and the sun bear (*Helarctos malayanus*). Two additional subfamilies have one species each: the Tremarctinae are represented by the Andean bear (*Tremarctos ornatus*) and the Ailuropodinae by the giant panda (*Ailuropoda melanoleuca*) (see also Chapter 1).

Bears diverged from early canids approximately 20–25 million years ago (McLellan & Reiner 1994). Of the extant species, the giant panda diverged 12–20 million years ago, followed by the Andean bear about 7–13 million years ago, whereas the six species of Ursinae diverged during the last 5 million years (Kumar et al. 2017). The Ursidae have the distinguishing morphological traits of Carnivora but, except for the polar bear, are not obligate carnivores and have omnivorous or plant-based diets. The generalist omnivore strategy of brown bears allows them to occupy a broad range of habitats in the northern hemisphere (Robbins et al. 2004; Kumar et al. 2017).

Subspecies and Distribution

Subspecies

The subspecies classification of North American brown bears is subject to ongoing debate. Using general geographic distributions, Merriam (1918) proposed over 90 subspecies. Rausch (1963) identified two extant subspecies based on skull measurements, and Kurtén (1973) used these same skull measurements to propose three subspecies. Using cranial and dental measurements, Hall (1984) proposed seven North American subspecies. Current genetics research using maternal mitochondrial DNA (mtDNA) phylogeny does not support the historic taxonomic classifications based on morphological measurements or distribution. Six well-defined mtDNA maternal lineages, defined as clades, have been identified in brown bears across their geographic range, with three of these clades (II, III, and IV) occurring in North America (Miller et al. 2006; Davison et al. 2011). Ashrafzadeh et al. (2016) suggested that brown bears in Iran represent a possible seventh clade. Brown bears from the Admiralty, Baranof, and Chichagof islands of Alaska constitute the oldest and most genetically unique mtDNA clade (i.e. clade II) in the New World and this is a sister taxon to the polar bear (Talbot & Shields 1996; Shields et al. 2000). However, Waits et al. (1998) argued against drastically altering the current taxonomy based on the results from a single mtDNA region. Supporting this notion, Paetkau et al. (1998a) documented that only brown bears on Kodiak Island are genetically distinct from other North American populations, and likely have been isolated since sea levels reached current heights approximately 11,000 years ago. Based on microsatellite DNA, they documented male-mediated geneflow between Admiralty, Baranof, and Chichagof islands and continental populations (Paetkau et al. 1998a). Presently, the classification of Rausch (1963) remains the accepted subspecies classification for North America (Schwartz et al. 2003b). He classified bears from the mainland as *U. arctos horribilis* Ord and those from the Kodiak Island archipelago as *U. a. middendorffi* Merriam.

Colonization

Data from recent studies suggest that brown bears colonized the North American continent on three occasions via the Bering land bridge (Davison et al. 2011). The first colonization, representing clades II, III, and IV, occurred during the late Pleistocene (~70 ky BP) into eastern Beringia (current Alaska) and possibly Admiralty, Baranof, and Chichagof islands. Although brown bears disappeared from Beringia prior to 35 ky BP, some populations had already colonized the mid-continental area of southern Canada (clade IV). Around 21 ky BP, the time of the last glacial maximum, a second colonization occurred representing a subclade of clade III; this clade colonized the area of northern Canada after the continental ice sheets retreated. The final colonization happened at the end of the last glacial maximum (13–9 ky BP), before the Bering Strait again became a barrier to movement due to rising sea levels, resulting in the colonization of Alaska by a second subclade of clade III (Davison et al. 2011).

Historic Range

Following recession of the ice sheets at the end of the Wisconsin Glacial Episode, *Ursus arctos* was widely distributed across North America. The distribution at that time extended from the west coast east to the Canadian province of Ontario and US states of Ohio and Kentucky, but receded in response to unfavorable environmental conditions (Peterson 1965; Guilday 1968). Their southward distribution extended well into northern Mexico (Storer & Tevis 1955).

Prior to European settlement of North America, the brown bear's distribution was expansive, occupying approximately 49% of the continent, primarily in the western half (Storer & Tevis 1955; Craighead & Mitchell 1982; Figure 13.1). According to MacPherson (1965), the eastern extent of the barren-grounds grizzly was near the Thelon, Back, Dubawnt, and Kazan rivers in Northwest Territories. The range in the prairies extended eastward to the great bend of the Missouri River in North Dakota, southward to the Moreau River in South Dakota, and possibly to the Red River region of Texas (Stebler 1972). The southern extent reached into central Mexico (Storer & Tevis 1955). In the grasslands and chaparral of California extending to the Pacific coast, the "golden" grizzly bear was probably abundant (Storer & Tevis 1955).

Current Range

With the advent of European settlement of the North American continent, brown bear range was drastically reduced through active persecution by humans and habitat loss (Mattson et al. 1995; Mattson & Merrill 2002). Reductions were greatest in the southern and eastern portions of the range. Brown bears were extirpated throughout much of their range prior to 1900 and were gone from the plains of Montana by 1885 (Koch 1941). South of the 49th northern parallel that separates US western states from Canada, 31 of 37 remnant brown bear populations present in 1922 were eliminated by 1975 (Servheen 1999), when the species was listed as threatened under the US Endangered Species Act (US Fish and Wildlife Service 1993). There are four extant populations in the conterminous US, three of which are transboundary populations (Proctor et al. 2012). There are no established populations in the North Cascades and Bitterroot Ecosystems in the US, although occasional occurrences have been documented in recent decades. The Greater Yellowstone Ecosystem population is the most southern population and is genetically isolated (Haroldson et al. 2010). The occupied range of these populations represents approximately 6% of historic range within the conterminous US (Figure 13.1).

Brown bears currently occupy about 60% of their historic range in Canada (Figure 13.1; Miller et al. 2013). Brown bears have been extirpated from portions of their historical range in Manitoba; Saskatchewan; the prairies and boreal plains of eastern Alberta; and the hot, dry, and populated south-central plateaus of British Columbia. In the Yukon and Nunavut territories, brown bears are relatively numerous except in areas where people have settled (McLellan & Banci 1999).

Alaska has the largest population of brown bears of any jurisdiction in North America and their distribution has remained relatively unchanged since the mid-1700s (Miller 1993; Miller & Schoen 1999). They occur throughout Alaska except on islands south of Frederick Sound in south-east Alaska, west of Unimak in the Aleutian Chain, the delta of the Yukon and Kuskokwim Rivers, and Bering Sea islands.

Descriptive Notes

Morphology

The skeleton of North American brown bears is larger and heavier than that of most other ursids, but its structure is similar to American black bears. Front and hind legs are approximately equal in length and taper to large plantigrade feet. Plantigrade morphology allows brown bears to stand on their hind feet, which they do frequently to increase their sensory (i.e. olfactory and visual) acuity (Figure 13.2). There are five toes on each foot with non-retractable claws. The claws on the front feet are straight and can be as long as 8 cm, preventing these bears from being proficient tree climbers. The long front claws and the large scapular muscle mass enable them to dig for roots, small mammals, and other foods in tundra and grassland environments (Herrero 1972). The claws on their hind feet are much shorter. Morphological features most useful in distinguishing North American brown bears from sympatric American black bears are their prominent shoulder hump, long foreclaws, and a more concave profile between brow and nose.

Size and Body Mass

Size and body mass vary widely across sex, age, season, reproductive status (females), and diet. Brown bears are sexually dimorphic with males approximately 1.2–2.2 times larger than females (Stringham 1990; Hilderbrand et al. 1999b). Sexual dimorphism is apparent beginning at ages 2–4 (Troyer & Hensel 1969; Pearson 1975; Blanchard 1987). Bears continue to grow throughout their life; however, mass gain more than triples during their first summer and gains decline continuously thereafter (Kingsley et al. 1983).

Body mass is seasonally dynamic. During hyperphagia in late summer and fall, bears gain mass and body fat rapidly and generally reach their annual peak immediately prior to denning, with body mass gains of 30% or more compared with masses in spring (Nelson 1980; Kingsley et al. 1983; Hilderbrand et al. 2000; Schwartz et al. 2014). Bears do not consume food or water during hibernation, which lasts approximately 5–6 months. As they metabolize fat and protein while hibernating, body mass continuously decreases (Hilderbrand et al. 2000). In many populations, foods available in spring following den emergence are primarily vegetative. Such foods often do not cover energetic requirements and bears may continue to lose mass until more nutritious foods are available later in the year.

Habitat productivity strongly influences bear diets and, in turn, affects body mass, which varies widely across populations (Hilderbrand et al. 1999b). Bears with access to more nutritious foods, such as salmon and ungulates, are generally larger and heavier than those with primarily plant-based diets. Brown bears occupying coastal areas of Alaska and British Columbia are the largest in North America. For example, males and females on the Alaska Peninsula averaged 357 and 226 kg, respectively (Miller & Sellers 1992). In contrast, interior males averaged only 145 kg and females 98 kg in south-western Yukon, Canada (Pearson 1975).

Pelage and Molt

Brown bears vary considerably in pelage color and pattern among individuals and with age and season. Most brown bears are indeed brown, but their pelage can vary from blond through various shades of brown to black. Often, they have medium brown underfur, darker legs, and lighter brown or blond heads and shoulders (Rausch 1953; Pearson 1975; Nagy & Russell 1978; Nagy et al. 1983). The majority, but not all, of interior brown bears have lighter-tipped guard hair (tips without pigment) over their head, shoulder, and back, resulting in a silver-tipped or grizzled appearance. The distribution of silver-tipped hairs can form patterns such as a lighter girth

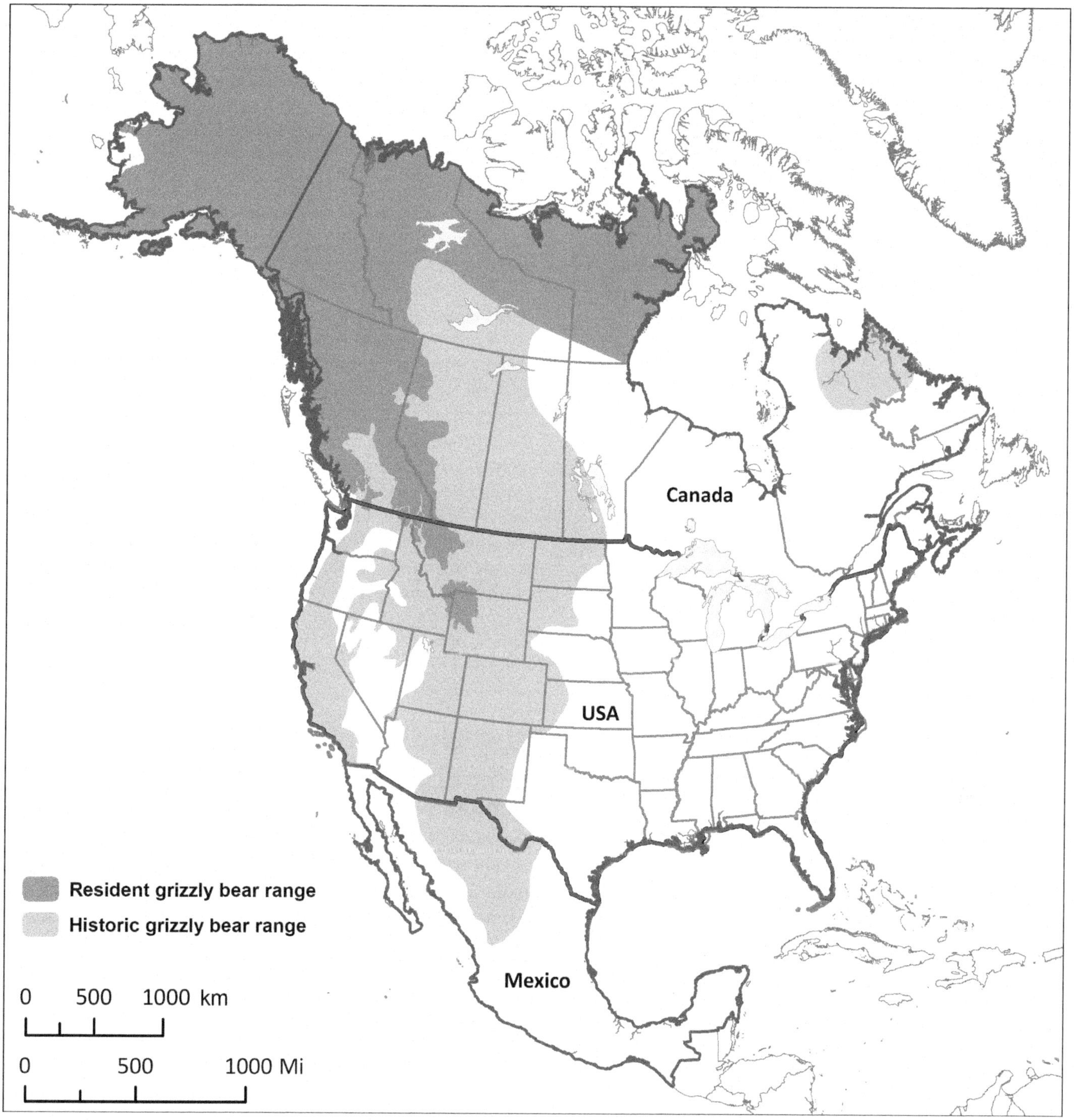

Figure 13.1 Historic and current occupied range of brown bears in North America. *Data sources*: M. Proctor, Birchdale Ecological, Ltd.; International Union for Conservation of Nature (IUCN); Interagency Grizzly Bear Study Team. Map creation: L. Landenburger, US Geological Survey and D. Baisero, Wildlife Conservation Society. (A black and white version of this figure will appear in some formats. For the color version, please refer to the plate section.)

band behind the forelegs. Some interior and most coastal brown bears do not have silver-tipped hair and appear more uniformly brown (Rausch 1953). Grizzled hair can look different depending on light conditions and fur often bleaches during the active season, making it difficult to positively identify individuals over time based on pelage color and pattern. Cubs may have a cream or blond "collar" on the chest and neck, which usually disappears by the time they are 2–4 years old (Rausch 1953; Troyer & Hensel 1969; Knight et al. 1981). Some bears that are brown in the spring gradually bleach to a straw color before becoming brown again after molting (Nagy et al. 1983).

Figure 13.2 Brown bears frequently stand on their hind legs to increase their visual and olfactory acuity. (Photo courtesy of J. Davis.)

The timing of molting varies with latitude, sex, and age. Cubs, hairless at birth, do not molt and can have hair as long as 10 cm by fall (Rausch 1953; Troyer & Hensel 1969). Adult bears molt annually. Generally, mature males begin molting first, followed by younger males and females with dependent young (Quimby & Snarski 1974; Pearson 1975; Nagy et al. 1983). Molting is usually completed in late July through August.

Skull and Dentition

Brown bear skulls are robust and sexually dimorphic (Rausch 1963; Craighead & Mitchell 1982). The skull grows and changes in structure throughout a bear's life (Zavatsky 1976), lengthening from an oval shape at birth to the elongated shape characteristic of the species when they reach sexual maturity. Zygomatic width continues to increase slightly after the skull has reached maximum length (Rausch 1963). Female maximum skull length is attained by 4 years and width by age 8, whereas maximum male skull length and width is reached by age 8 and 10 years, respectively (Rausch 1953). The largest brown bear skull length (45.6 cm) and width (32.5 cm) were recorded in Alaska (Byers & Bettas 1999). Mirroring body size, interior bears have smaller skulls than coastal bears. However, Rausch (1953) emphasized that skull dimensions and shape among individuals are so variable that generalizations for even a local area can rarely be made. For example, skull profiles from brow to nose can be concave and gradually sloping or convex and sloping acutely.

Brown bears generally have 42 teeth with the dental formula I3/3, C1/1, P4/4, M2/3; however, premolars are vestigial and some can be missing (Glass 1974; Pasitschniak-Arts 1993). Although the length of teeth and relative position in the skull are indicators of approximate age, a far more accurate technique to age brown bears is by the number of cementum annuli in teeth (Rausch 1969). Because the small and vestigial premolars are easily extracted, they are typically used to age bears with this method (Matson et al. 1993). Brown bears of both sexes can be distinguished from American black bears based on the crown length (>20.4 mm) or width (>10.5 mm) of the first mandibular molar if the permanent teeth are sufficiently erupted (Gordon 1977).

Genetics

North American brown bears have nearly identical karyotypes to the five other ursine bears, with 74 diploid chromosomes that consist of 72 autosomes (60 acrocentric and 12 metacentric) and two sex chromosomes; a large metacentric X and a small acrocentric Y (Ewer 1973; Pasitschniak-Arts 1993; Waits et al. 1999). As mentioned previously, North American brown bear populations are descendent from three mtDNA clades (Miller et al. 2006; Davison et al. 2011). These clades do not correlate with taxonomic classifications and are hypothesized to have formed before brown bears colonized the North American continent. Current genetic diversity is highest (H_e [expected heterozygosity] = 0.76) in northern populations in the center of the North American distribution and is lowest in insular populations such as the Kodiak Archipelago (H_e = 0.26; Paetkau et al. 1998b). The level of genetic diversity among populations inhabiting relatively undisturbed northern landscapes likely can be maintained only through geneflow on the scale of the current North American distribution (Paetkau et al. 1998b). Among Canada–US transboundary populations, Proctor et al. (2012, p. 16) documented a trend of decreasing heterozygosity at lower latitudes, which they attributed to greater landscape fragmentation associated with anthropogenic influences.

Habitat

Extant brown bear populations in North America are associated with the Rocky, Cascade, Coastal, Alaska, and Brooks mountain ranges, as well as the interior and coastal plains of Alaska and north-western Canada. At the southern extent of currently occupied range, grizzly bear habitat is dominated by open or patchy vegetation associated with mid-elevation montane forest communities. Use of open areas by grizzly bears is common (e.g. avalanche paths, berry shrub fields, riparian areas; Waller & Mace 1997; McLellan & Hovey 2001; Proctor et al. 2015; Figure 13.3), but some selection for edge habitats is generally observed where forest cover is present (Stewart et al. 2013). Above the tree line in mountainous areas, bears also use open areas of alpine grasslands and alpine grass and forb tundra (McLellan & Hovey 2001). Adjacent areas of western prairies are also used by bears and include shortgrass prairie, sagebrush shrublands, and hardwood riparian bottoms at low elevations (Aune et al. 1994). At higher latitudes, where harsh climates limit tree growth, many grizzly bear populations exist in predominantly open areas such as lichen–herbaceous tundra, where resources are dispersed over large expanses (McLoughlin et al. 2002a; MacHutchon & Wellwood 2003; Figure 13.3).

Figure 13.3 Examples of typical brown bear habitats in North America. (A) Brown bears often use open areas but selection for edge habitats is common where forest cover is present; Wyoming, USA. (Photo courtesy of J. Davis.) (B) Shrub fields provide fruits such as huckleberries (*Vaccinium* spp.) during late summer and early fall; Canadian Rocky Mountains. (Photo courtesy of G. MacHutchon.) (C) Avalanche chutes typically remain in early seral stages due to disturbance from snow slides and are used by bears in early spring for emergent vegetal foods and again in late summer as various fruits ripen; Northern Rocky Mountains of Montana, USA. (Photo courtesy of C. Costello.) (D) Avalanche chutes and shrubfields in a glaciated U-shaped valley; Canadian Rocky Mountains. (Photo courtesy of M. Proctor.) (E) Coastal brown bear habitat; Alaska, USA. (Photo courtesy of M. Clapham.) (F) Tundra habitats have low primary productivity, resulting in large home ranges of brown bears; Mackenzie River Delta north of Inuvik, Northwest Territories, Canada. (Photo courtesy of M. Edwards.)

Optimizing efforts to store energy and nutrients for growth, reproduction, and hibernation is an overriding principle in grizzly bear habitat selection (McLellan 2015). A second principle that drives habitat selection is human disturbance that may reduce habitat effectiveness through displacement or risk of mortality (Schwartz et al. 2010b). These forces reflect the paradigm of bottom-up influences of food resources that fuel body condition and reproduction and top-down anthropogenic forces influencing survival, and ultimately fitness and density (McLellan et al. 1999; Nielsen et al. 2004; Schwartz et al. 2010b; Boulanger & Stenhouse 2014; McLellan 2015; Proctor et al. 2017; Lamb et al. 2018).

In the southern portion of the grizzly bear distribution, humans have altered historically productive areas (McLellan 1998). Many rivers no longer support healthy runs of spawning salmon, bison (*Bison bison*) were extirpated on the Great Plains, and human settlement has dominated many productive mountain valleys, excluding grizzly bears. Consequently, grizzly bears in many areas are relegated to using more remote mountainous landscapes where anthropogenic influences are limited (Schwartz et al. 2003b). Within these landscapes, habitat use is driven in large part by the strategy of brown bears to extract energy and nutrients from foods that are relatively easy to digest, such as meat and fish, berries or nuts, selected roots and corms, and young succulent vegetation (Hilderbrand et al. 1999a). Accordingly, vegetation types used by grizzly bear often shift seasonally (Nielsen et al. 2003). In mountainous regions, these shifts follow elevational gradients (Mace & Waller 1997). Habitat use also varies across years, as selection for food items shifts due to variation in availability (Fortin et al. 2013a; Cristescu et al. 2015; Ebinger et al. 2015). Despite these temporal and spatial dynamics, there are some similarities in currently occupied brown bear habitats across North America. For example, bears use vegetation types that contain plentiful ungulates, particularly neonates, wherever it is efficient to hunt them, from Yellowstone National Park to Alaska and northern Canada (Zager & Beecham 2006). Similarly, areas containing rich berry fields are used across North America.

Interior Areas

Spring habitats in the mountainous interior are generally defined by lack of snow cover. Avalanche chutes are prime spring bear habitats and are a good example of high-quality sites for foraging with little or no canopy cover (Mowat et al. 2013; Proctor et al. 2015). The annual disturbance of sliding snow keeps chutes free of trees and soils moist, and provides access to high-protein herbs, shoots, and corms. Riparian areas associated with streams, lakes, or wetlands similarly are productive spring habitats. In many areas, fire acts as a disturbance agent and enhances habitat quality (Hamer & Herrero 1987). In subalpine and alpine areas bears forage on vegetative foods and opportunistically dig for rodents such as ground squirrels (e.g. *Urocitellus columbianus*, *Urocitellus armatus*, *Callospermophilus lateralis*) and northern pocket gophers (*Thomomys talpoides*) and their food caches (McLellan & Hovey 1995; Mattson 2004). There is some evidence that foraging activities involving digging by bears influence the nutrient quality of their foods. For example, glacier lilies (*Erythronium grandiflorum*) form underground corms that are dug extensively by bears in spring and early summer, and these plants produce more seeds and have higher protein and water-soluble carbohydrates than those in previously undisturbed sites (Tardiff & Stanford 1998).

As the active season progresses, various species of berries (e.g. *Vaccinium* spp., *Shepherdia canadensis*, *Arctostaphylos uva-ursi*, *Amelanchier* spp., *Sambucus* spp., *Ribes* spp., *Empetrum nigrum*, *Viburnum edule*) provide productive bear food (McLellan & Hovey 1995; McLellan 2015). These berry fields tend to occur at higher elevations, open canopy sites (typically <30% canopy cover), and often after fire events (McLellan 2015; Proctor et al. 2017; Shores et al. 2019). Although these habitats can be ephemeral, they can remain productive for up to 70 years, provided tree canopy remains intact (Proctor et al. 2018b). In other areas, such as in the Greater Yellowstone Ecosystem and Glacier National Park, Montana, USA, late summer habitat can be in the form of talus slopes in alpine areas that are used by army cutworm moths (*Euxoa auxiliaris*) as thermal refugia during the day; these moths are rich in fats and individual bears have been estimated to consume as many as 40,000 moths in a single day (French et al. 1994; White et al. 1998b). Open grassland and shrublands that produce healthy populations of ungulates such as bison, elk (*Cervus canadensis*), moose (*Alces alces*), or caribou (*Rangifer tarandus*) are used extensively by grizzly bears in the Yellowstone area, in northern British Columbia, Alberta, and Alaska (Boertje et al. 1987; Mattson 1997; Young & McCabe 1997; Hilderbrand et al. 1999a; Zager & Beecham 2006; Miller et al. 2011; Milakovic & Parker 2012; Cristescu et al. 2014; Ebinger et al. 2015; Figure 13.4).

After the period of range contraction (1870–1970), in some locales grizzly bears have been gradually expanding from the remote mountainous areas where they survived over 100 years of persecution. Grizzly bears still avoid heavily settled environments, but where mortality exposure is low, they are moving back into habitats that now contain ranchlands and agricultural areas that provide access to high-calorie anthropogenic foods (Northrup et al. 2012b; Proctor & MacHutchon 2019). Although challenging, managing for coexistence can mediate this shared use of the landscape by humans and grizzly bears (Wilson et al. 2006; Northrup et al. 2012b; Morehouse et al. 2016; Proctor & MacHutchon 2019).

Coastal Areas

Coastal habitats in British Columbia and Alaska are dominated by salmon-bearing rivers and use is influenced by the timing of spawning of each salmon species, ranging from early summer through fall (Hilderbrand et al. 1999b). Consequently, habitat use is a function of the spatial variation of "spawning waves" followed by bears to optimize caloric gains per unit

Figure 13.4 A grizzly bear foraging for blueberries (*Vaccinium* spp.) on the tundra slopes within Denali National Park, Alaska, USA. In early September, blueberries are abundant and the tundra turns red and gold as fall approaches and bears target berries to build fat reserves in preparation for hibernation. (Photo courtesy of J. Schoen.) (A black and white version of this figure will appear in some formats. For the color version, please refer to the plate section.)

effort, with some areas attracting large numbers of bears (Schoen et al. 1994; Deacy et al. 2019). Catchability of salmon by bears is influenced by characteristics of the rivers. For instance, although salmon are prevalent in some rivers of south-western Alaska, such as the Yukon and Kuskokwim deltas, their lower reaches are generally too deep and silt-laden to allow bears to capture salmon moving upstream. When salmon are not available, habitat use largely mimics that of other areas of the continent described previously, but may also involve use of coastal tidal zones for transitional and marine-based food resources such as clams (*Siliqua patula*).

Northern Areas

Home ranges in northern arctic areas are typically large because seasonal food resources are widespread and less concentrated. Non-forested areas include eskers containing overwintered *A. uva-ursi* berries, areas with scavenged ungulates, or caribou for predation, young vegetative growth, and fruit-bearing shrub fields (Gau 1998; McLoughlin 2000; Gau et al. 2002; McLoughlin et al. 2003a; Edwards et al. 2011). Accordingly, habitat use by grizzly bears tends to reflect individual food choices of bears, which can vary substantially (Edwards et al. 2011). In subarctic regions, bears use open dwarf shrublands comprised of willow (*Salix* spp.), birch (*Betula* spp.), and sedges (*Carex* spp.), along with interspersed boreal forest of spruce (*Picea* spp.), fir (*Abies* spp.), poplar (*Populus* spp.), and birch (Darling 1987). In northern coastal areas, bears use sedge marshes and tidal grasslands, salmon streams, and upland areas seasonally (MacHutchon & Wellwood 2003).

Habitat Fragmentation

Grizzly bears show varying degrees of avoidance of human settlements, roads, trails, and site-based human activities, such as mines (Gibeau et al. 2002; Northrup et al. 2012a; Coleman et al. 2013; Cristescu et al. 2016a, 2016b). Avoidance is not always absolute and bears can adapt to human activities to some extent. However, habitat fragmentation can increase the risk of human-caused mortality, thereby reducing permeability of landscapes to bear movement. Using genetic samples from the contiguous US, western Canada, and south-east Alaska, Proctor et al. (2012) documented progressively greater levels of population fragmentation from northern to southern latitudes. Populations that were once interconnected experience diminished demographic connectivity as female movements become severely restricted or non-existent, although genetic connectivity may remain through male immigration. Where habitat fragmentation is greatest and male movements are restricted as well, populations become both demographically and genetically isolated. Habitat fragmentation may also interfere with the ability of grizzly bears to shift habitat use in response to climate change (Chester et al. 2013). Population fragmentation and other potential impacts of habitat fragmentation may be mitigated through habitat linkages. Recent efforts have identified habitat and potential movement corridors in between the Northern Continental Divide and Greater Yellowstone ecosystems primarily in Montana, USA and the southern Canada–USA trans-border region (Proctor et al. 2015; Peck et al. 2017). Those results are enabling managers and organizations to identify and prioritize conservation measures that maintain or enhance grizzly bear dispersal routes. In the trans-border region, substantial management effort has resulted in increased connectivity, including the conservation of valley bottom areas to reconnect remnant patches of grizzly bear habitat with adjacent mountains (Proctor et al. 2018a). In Banff National Park, Alberta, extensive efforts have been invested in highway crossing structures that have successfully connected important habitats on each side of Canada's busiest highway, the Trans-Canada Highway (Ford et al. 2009; Sawaya et al. 2013).

Habitat Management

In southern portions of brown bear range (conterminus states, USA; Alberta and southern British Columbia, Canada), habitat management has historically focused on minimizing habitat displacement from human disturbance and mortality associated with human access to backcountry areas (Mace et al. 1996; Proctor et al. 2018b). Survival of adult female grizzly bears is strongly associated with levels of human access that can tip the balance from population growth (source areas) to decline (sink areas; Schwartz et al. 2010b). Managing for secure habitat standards is a key component of this approach, typically involving thresholds for density of motorized roads and developed sites. Additionally, managers use backcountry food storage orders to reduce attractants at work, recreational, or hunting camps (Mace et al. 1996; Schwartz et al. 2010b; Boulanger & Stenhouse 2014; Proctor et al. 2018b). These actions increase habitat effectiveness and allow bears to access necessary resources for sustainable survival, reproduction, and fitness (Schwartz et al. 2010b; Boulanger & Stenhouse 2014; Proctor et al. 2017). This approach to habitat management likely contributed to recovery of several grizzly bear populations south of Canada, most notably in the Northern Continental Divide and Greater Yellowstone Ecosystems.

Movements and Home Range

Owing to their large body size and higher trophic position, grizzly bears exhibit some of the most extensive movement patterns among North American mammals (McNab 1963; Harestad & Bunnell 1979). Wide-ranging movements allow bears to take advantage of seasonally available foods present in various habitats and across elevational gradients. In most areas, adult grizzly bears show fidelity to a home range or region within which they use their knowledge of local resources and seasonal phenology to acquire not only food, but shelter, den sites, and potential mates (Blanchard & Knight 1991; Edwards et al. 2009; Wirsing et al. 2018). There is a lack of evidence for territorial defense, although subordinate bears may avoid productive food patches being exploited by more dominant individuals (Stonorov & Stokes 1972; Blanchard & Knight 1991; Wielgus & Bunnell 1994a, 1994b; Mace & Waller 1997; Schwartz et al. 2003b; Gende & Quinn 2004).

The spatial and temporal distribution of foods is a primary driver of movement and home-range characteristics. McLoughlin et al. (2000) analyzed data from 30 populations of brown bears in North America to evaluate relationships between home-range characteristics and the relative deviation from mean annual climatic conditions, a surrogate of habitat quality. Home-range size, primarily based on minimum convex polygons, was inversely associated with habitat quality and bear density. Bear foraging movements likely involve habitual visits to known food sources, sampling of local foods, and exploratory forays into novel areas. The magnitude of these foraging strategies varies across North America and movement patterns can be broadly categorized according to three geographic regions: interior, coastal, and barren-ground (i.e. tundra; Craighead & Mitchell 1982; McLoughlin et al. 1999). The traditional view of a home range is most exemplified by interior grizzly bear populations that feed on a wide variety of foods with predictable seasonal phenology, but with annual and spatial variability in abundance. Within interior populations, movements and home ranges vary with annual variation in food availability and conditions. These populations exhibit moderately sized home ranges, ranging from 86 to 408 km^2 for females and 287 to 1918 km^2 for males (McLoughlin et al. 1999, 2000). In the Yellowstone Ecosystem, for example, Blanchard and Knight (1991) found that home-range size increased in drier years and was influenced by availability of a key food, whitebark pine (*Pinus albicaulis*) seeds. Local resources available to individuals can also shape movement patterns. In interior southwest Alaska, Mangipane et al. (2018) found that home-range size increased as landscapes became more homogeneous, likely because bears expanded movements to gain access to a greater variety of foods. In coastal regions, spawning salmon are a predictable and spatially concentrated, high-energy food that commonly dominates the annual diet. In this food-rich environment, smaller home ranges of 24–293 km^2 for females and 115–295 km^2 for males are possible because many individuals can fulfill annual needs within a relatively short distance of spawning streams (McLoughlin et al. 1999, 2000). Salmon-feeding bears track spatial patterns in salmon spawning phenology (Deacy et al. 2016). On Kodiak Island, Barnes (1990) found that female bears visited an average of 2.1 fishing sites per year whereas males visited 3.9 sites. Mean summer ranges were three times larger than either spring or fall ranges because of interdrainage movements between salmon streams. He further noted that traditional annual movements by some individuals varied little across years. In food-limited barren-ground landscapes, bears often range over vast areas to fulfill their annual needs and home ranges are some of the largest recorded for females (210–2434 km^2) and males (645–8171 km^2; McLoughlin et al. 1999, 2000). Edwards et al. (2011) found that movement rates of barren-ground females increased with trophic position, suggesting that search effort increases with increasing levels of carnivory. Additionally, McLoughlin et al. (2003b) found that home ranges of barren-ground grizzly bears increased as the proportional amount of exposed bedrock and other marginal habitats in the environment increased.

These broad relationships between foods and home-range size are also modified by social factors, including population density. Nagy and Haroldson (1989) compared population and home-range characteristics among four populations in Canada and concluded that differences in home-range size were best explained as an inverse function of population density. In the Yellowstone ecosystem, Bjornlie et al. (2014) examined home-range characteristics before and after documented declines in whitebark pine availability. They found that female home ranges were smaller and less variable at higher population

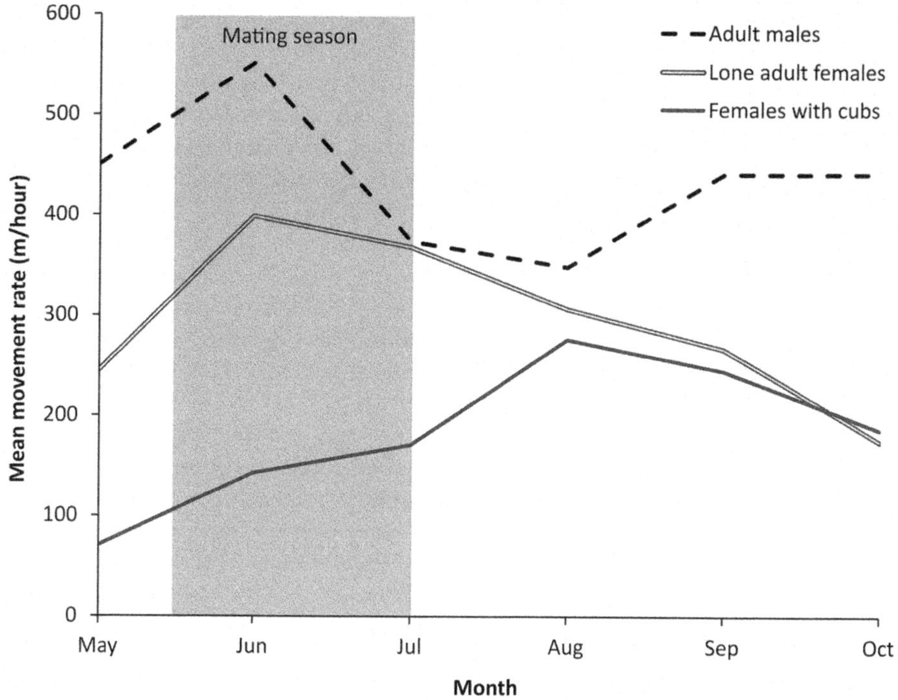

Figure 13.5 Mean movement rate by month for three cohorts of grizzly bears in west-central Alberta, Canada, 1999–2010. The gray rectangle indicates the duration of the breeding season. *Data source*: fRI Research.

densities but did not vary relative to whitebark pine availability.

Males exhibit greater variability of movement and range size compared with females. Among studies reporting home-range sizes in North America, male ranges are typically 2–4 times larger than female ranges (McLoughlin et al. 1999). Males likely use larger areas to gain access to multiple females for mating. This notion is supported by many studies reporting that male movement rates and home ranges are largest during the mating season (Figure 13.5; Glenn & Miller 1980; Graham & Stenhouse 2014). Blanchard and Knight (1991) reported that females exhibited stronger fidelity to seasonal and annual ranges than males and the asymptote in the size of multiannual ranges was more evident in females than in males. In the Canadian Arctic, however, Edwards et al. (2009) revealed lower levels of interannual home-range fidelity among sampled females (24% overlap among annual ranges) and suggested home-range drift was advantageous owing to the small costs of intraspecific interference in the low-density population. Nonetheless, home-range centers were relatively stable, suggesting these bears showed fidelity to a region more than to a specific home range.

Throughout North America, home-range overlap among individuals is generally high. Using weekly telemetry locations of neighboring bears in Montana, Mace and Waller (1997) found that annual home-range overlap of neighboring females ranged from 0% to 94%, averaged 24%, and was generally lower when one or both females had offspring. Mean home-range overlap was 51% among male–male neighbors and 29% among female–male neighbors. In south-eastern British Columbia, McLellan and Hovey (2001) estimated that a typical female home range overlapped those of about 19 males and a male home range overlapped those of about 32 females.

In coastal populations, large numbers of bears congregate around specific fishing sites, resulting in high overlap of seasonal and annual ranges. Barnes (1990) reported peak numbers of 39 bears observed on a single stream during the same day and 86 bears during a 4-day period. McLoughlin (2000) calculated home-range overlap as the product of mean home-range size and bear density and found it was highest for coastal populations, lowest for interior populations, and intermediate for barren-ground populations. Despite their solitary nature, movement analyses also suggest that encounters between individual grizzly bears are relatively frequent. Stenhouse et al. (2005) evaluated spatial associations among neighboring grizzly bears in Alberta and found that only 25% of those involved opposite-sex pairs during the mating season, whereas associations during the berry foraging season were similar for same-sex and opposite-sex associations.

Females alter their movements according to their reproductive status. Combining data on movements and cub survival in interior Alaska, Gardner et al. (2014) suggested that staying near the den longer, limiting movements to a few habitat patches, and reducing activity between midnight and mid-morning appeared effective to reduce cub exposure to potentially infanticidal adult bears, mostly males. They found that individual females reduced their home range by 63–91% when accompanied by cubs. Similar behaviors and contraction of home ranges by females with cubs has been observed in other populations and may be related to limited mobility of cubs (Glenn & Miller 1980; Blanchard & Knight 1991; Collins et al. 2005; Graham & Stenhouse 2014).

Figure 13.6 Along salmon spawning streams such as the McNeil River in Alaska, USA, large male brown bears often congregate and dominate prime fishing locations, whereas smaller females and those with dependent young are relegated to less productive locations. (Photo courtesy of J. Hechtel.)

Ben-David et al. (2004) reported that females with cubs were found further away from salmon streams compared with other cohorts, probably to avoid adult males that sometimes congregate at concentrated food resources (Figure 13.6). A trend toward larger home ranges with increasing litter age has also been observed, and may be a function of increasing nutritional needs of the family group and reduced vulnerability to intraspecific mortality (Blanchard & Knight 1991). On Kodiak Island (Barnes 1990) and the Canadian Arctic, however, range sizes were similar among females of all reproductive classes and an individual-based index of resource availability was a better predictor of home-range size than was reproductive status (Edwards et al. 2013).

After family separation, the process of establishing a home range differs between subadult females and males. Most male bears emigrate from their natal range, undergo a period of transiency that may last several years, and typically settle into a home range coincident with breeding activity. Some subadult males occupy relatively small annual ranges that shift further from natal areas over time (Mace & Waller 1997; McLellan & Hovey 2001), whereas others exhibit long-distance, linear movements that can result in vast range sizes (Glenn & Miller 1980; Blanchard & Knight 1991; Gau et al. 2004). Most females are philopatric and settle in areas near or overlapping their mother's home range (Glenn & Miller 1980; Blanchard & Knight 1991; Mace & Waller 1997; McLellan & Hovey 2001). Movement studies have documented occasional instances of male philopatry and female long-distance dispersal. Male-biased dispersal probably evolved to reduce inbreeding and resource and mate competition (McLellan & Hovey 2001). Based on radiotelemetry monitoring or DNA sampling combined with parentage analysis, mean natal dispersal distances were 30–67 km for males and 10–14 km for females (Glenn & Miller 1980; McLellan & Hovey 2001; Proctor et al. 2004). Maximum dispersal distances were 67–173 km for males and 20–82 km for females. Lacking knowledge of their natal range, other researchers have reported long-distance movements of 95–166 km for subadult males in the tundra of southwest Alaska (Glenn & Miller 1980) and 145–340 km in forests of southern British Columbia (McLellan & Hovey 2001). Directional forays of 201–779 km by subadult males in the Canadian Arctic resulted in extraordinarily large annual home ranges of 2349–32,188 km^2 (Gau et al. 2004).

Home-range fidelity patterns are reflected in analyses of post-translocation movements. Translocation is often used to address human–bear conflict but also for population augmentation or experimental reasons (Miller & Ballard 1982; Blanchard & Knight 1995; Kasworm et al. 2007). Homing behavior is common among translocated bears, but return rates are affected by distance transported, sex, age, and number of prior translocations. Among bears translocated to address human–bear conflicts, frequency of return is generally lowest for subadult females, highest for adult males or adult females, and higher for individuals with previous translocations (Blanchard & Knight 1995; Campbell 1999; Riley et al. 2004). Post-translocation home ranges of bears moved outside of their management unit in Alberta were about three times larger than those of resident bears (Milligan et al. 2018). The success of the population augmentation program in the small Cabinet-Yaak population of Montana–Idaho is likely due to the selection of subadult or young adult individuals from source populations in British Columbia and Montana (Servheen et al. 1987; Kasworm et al. 2007). Among bears alive and located at least one year after translocation, five of eight females and five of five males were still present in the target population area. Pedigree analyses indicate one translocated female produced at least 10 first-generation offspring, 15 second-generation offspring, and one third-generation offspring (Kasworm et al. 2018).

Activity Patterns

Activity patterns of brown bears are commonly viewed as seasonal or annual patterns and diel patterns.

Seasonal Activity

Season or annual activity patterns primarily correspond to the phenology of key food resources and mating activities (Figure 13.7). Following den emergence in spring and into early summer, bear activity increases as they feed on a combination of ungulate carrion, forbs, graminoids, and roots at lower elevations where vegetation green-up occurs first (Mace & Waller 1997; Munro et al. 2006). Females with newborn cubs are an exception to this pattern as they typically remain at higher elevations and close to the den upon emergence and exhibit restricted movements and activities (Haroldson et al. 2002; Graham & Stenhouse 2014). Restricted movements and remaining at higher elevations could both allow cubs to develop and reduce the chance of encountering male bears that may be a source of cub mortality (Rode et al. 2006b). Spring–early summer is the only time period when male and female activity patterns diverge, with male activity and

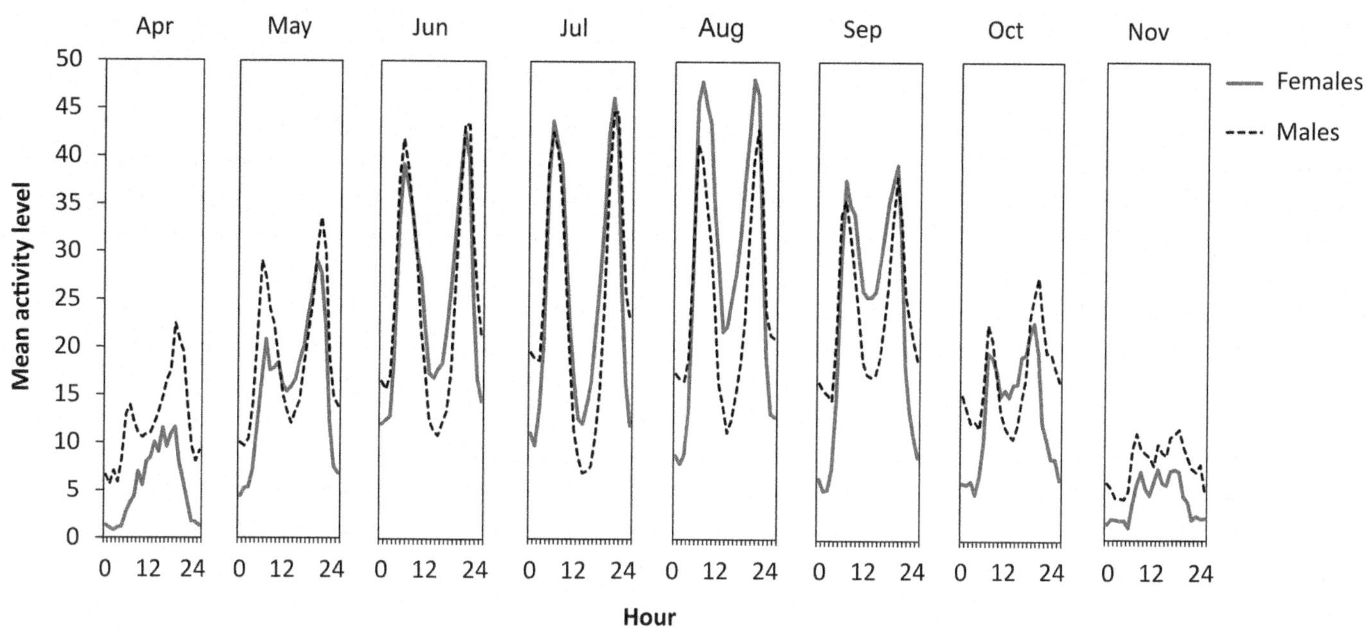

Figure 13.7 Diel activity of 36 female (n = 136,118 values) and 88 male (n = 191,122) grizzly bears by month, Greater Yellowstone Ecosystem, USA, 2002–2012. *Data source*: Interagency Grizzly Bear Study Team.

movements increasing substantially in association with the mating season from May through early July (see Figure 13.5). As the season progresses into summer and early fall, feeding habits increasingly shift to high-calorie food sources that begin the process of accumulating fat reserves to support hibernation (Munro et al. 2006; Robbins et al. 2012). As bears spend more time searching and foraging, overall activity levels reach their annual peak for females and remain high for males (McLellan & McLellan 2015). Fall food resources are generally found at higher elevations on the landscape and bears move to these areas as they follow the phenological advance of preferred foods (McLellan & Hovey 1995; Coogan et al. 2014). Often these resources are spatially concentrated and activities are more focused on foraging than searching, contributing to correspondingly declining activity. With the availability of foods rapidly declining in late fall and early winter, activity levels are lowest prior to den entry.

Diel Activity

Brown bear activity is often concentrated at certain times of the day, known as diel activity. Brown bears in North America are most active during diurnal and crepuscular periods, with bedding occurring most frequently at night (Figure 13.7; J. J. Craighead et al. 1995; Munro et al. 2006; Schwartz et al. 2010a; Graham & Stenhouse 2014). In southern British Columbia, McLellan and McLellan (2015) found that, on average, bears were active during 78% of daytime hours, 77% of civil twilight hours (sun 0–6 degrees below horizon), 47% of nautical twilight hours (sun 6–12 degrees below horizon), and 9% of nighttime hours. Although diel activity patterns can vary seasonally or with specific circumstances, there are generally four distinct activity periods during the non-denning portion of the year, with low levels of activity at night (22:00–05:00), peak activity in the morning (06:00–11:00) and evening (17:00–21:00), and moderate activity in the afternoon (12:00–16:00) (Graham & Stenhouse 2014).

Although most studies indicate that grizzly bears exhibit diurnal or crepuscular activity patterns, bears often alter their diel activity patterns seasonally and in response to human activity (MacHutchon et al. 1998). In the Yellowstone population, Fortin et al. (2013b) observed that mean daily activity levels did not differ between sexes, but male bears were more frequently active at night. Peak hour of activity fluctuated from midday during spring and fall to evening during summer. In coastal British Columbia, Klinka and Reimchen (2002) found that brown bears showed slightly greater efficiency catching salmon with reduced light levels, leading to substantial nocturnal activity. There is also evidence that diel activity patterns and movements can be impacted by human activities, including traffic volumes, such that brown bears become more active at night (Gibeau et al. 2002; Graves et al. 2006; Schwartz et al. 2010a; Northrup et al. 2012b; Kite et al. 2016; Ladle et al. 2019).

Feeding Ecology

Brown bears are omnivorous and feed on a wide variety of plant and animal species, depending on local and seasonal availability (Hilderbrand et al. 1999a; Edwards et al. 2011; Fortin et al. 2013a). Within a single ecosystem, they have been documented to consume as many as 266 different plant and animal species (Gunther et al. 2014). This opportunistic feeding strategy enables brown bears to occupy a wide range of habitats, including Arctic lowlands, inland mountainous terrain, and temperate and coastal regions. Further, they can exhibit substantial dietary variation within a single population

(Edwards et al. 2011; Gunther et al. 2014; Costello et al. 2016a). In some areas and seasons, their diet is nearly or completely plant matter (Rode et al. 2006a; Edwards et al. 2011). Alternatively, dietary meat consumption can be as high as 80% (Hilderbrand et al. 1999a, 1999b). Despite the diversity of brown bear diets, key food resources often play an important role in habitat use and population productivity.

Populations with access to abundant meat resources have been attributed with larger adult female body size (Hilderbrand et al. 1999b; Mowat & Heard 2006). In coastal populations of British Columbia and Alaska, salmon make up a large portion of the diet, which has been associated with larger body size, larger litter size, and higher bear densities compared with populations that lack access to salmon (Hilderbrand et al. 1999b; Ferguson & McLoughlin 2000; Mowat & Heard 2006; Van Daele et al. 2013). However, population productivity is not always related to abundance of meat. Other food resources, such as berries and whitebark pine nuts, can also be important in population productivity (Schwartz et al. 2006; McLellan 2011; Fortin et al. 2013a).

Brown bear diets vary seasonally, in part due to their distribution across highly seasonal, northern habitats, but also due to seasonal variation in nutritional needs associated with the physiological effects of hibernation. Spring diets typically consist of early-growth vegetation that contains high levels of digestible protein compared with later growth (Rode et al. 2006a), including sweet vetch roots (*Hedysarum alpinum*) (Servheen 1983; Rode et al. 2001; Gau et al. 2002; Munro et al. 2006; Costello et al. 2016a). These food resources and meat support recovery of lean body mass and fat reserves used during hibernation (Hilderbrand et al. 1999a, 2000). In autumn, bears become hyperphagic to accumulate significant energy reserves in preparation for hibernation. Carbohydrates, most often in the form of berries (Servheen 1983; Gau et al. 2002; Costello et al. 2016a), have been identified as a key resource for brown bears throughout much of North America because they aid in accumulation of body fat (Rode & Robbins 2000; Pigeon et al. 2016b; Deacy et al. 2017).

Although brown bears consume meat, optimal protein intake falls well below that of true carnivores (Erlenbach et al. 2014). Diets that balance protein and carbohydrate or fat intake maximize metabolic efficiency (Felicetti et al. 2003; Erlenbach et al. 2014; López-Alfaro et al. 2015) and ultimately affect individual fitness (Coogan et al. 2014). The importance of meat resources to brown bear population productivity (Mowat & Heard 2006) is consistent with positive relationships between protein intake and bear body mass (Hilderbrand et al. 1999b; Costello et al. 2016a). However, dietary carbohydrates and fat are more efficiently converted to fat storage (Erlenbach et al. 2014) and observations of bear feeding behavior in the wild support dietary mixing to maximize mass gain, depending on resource availability (Rode & Robbins 2000; Costello et al. 2016a; Deacy et al. 2017). For example, bears have been observed to leave salmon streams to consume berries in autumn (Robbins et al. 2007; Deacy et al. 2017) apparently to avoid overconsumption of nitrogen and increase intake of carbohydrates that can more efficiently be converted for fat storage (Erlenbach et al. 2014). Thus, brown bears must use a range of diets and macronutrient intake to meet their nutritional needs in response to seasonal and annual variation in food availability (Gunther et al. 2014; Costello et al. 2016a).

Figure 13.8 A brown bear fishing for red (sockeye) salmon (*Oncorhynchus nerka*) in the intertidal waters of the Chenik River located in the McNeil River State Game Refuge on the west side of Cook Inlet, Alaska, USA. Once the red salmon enter fresh water, they begin to turn a dark red color and start to lose significant amounts of fat and protein. The bright fish have much higher nutritional value prior to entering fresh water and brown bears often focus their fishing activities near the mouth of rivers in intertidal areas. (Photo courtesy of J. Schoen.) (A black and white version of this figure will appear in some formats. For the color version, please refer to the plate section.)

Due to the diversity of foods consumed by brown bears, a wide range of species have been identified to be of seasonal importance. Salmon and other non-anadromous fish (e.g. cutthroat trout, *Oncorhynchus clarkii*) provide a highly concentrated, energy-rich food source and are an important resource for brown bears where they are available (Felicetti et al. 2004; Mowat & Heard 2006; Figure 13.8). The presence of salmon in streams, however, does not equate to availability. Stream characteristics, including depth and the presence of obstacles that pool salmon, play an important role in whether bears can effectively capture and consume salmon (Quinn et al. 2003; Rode et al. 2006b, 2006c). Ungulates such as moose, elk, bison, and deer (*Odocoileus virginianus*, *O. hemionus*) can be important in bear diets (Gau et al. 2002; Fortin et al. 2013a; Nielsen et al. 2017). Some of the most carnivorous bears occur in areas with caribou (Gau et al. 2002; Mowat & Heard 2006). Brown bears prey primarily on ungulate neonates (Ballard et al. 1981; Munro et al. 2006; Zager & Beecham 2006) but will also scavenge carcasses (Servheen 1983; Mattson 1997), and kill adults (see also Chapter 3). Although bears have been documented to feed on muskoxen (*Ovibos moschatus*), this behavior appears to be a specialization among individuals

Figure 13.9 A North American brown bear in a coastal habitat. (Photo courtesy of M. Clapham.) (A black and white version of this figure will appear in some formats. For the color version, please refer to the plate section.)

within a population (Reynolds et al. 2002). Bears supplement vegetative diets with insects such as army cutworm moths and hymenopterans (White et al. 1998c; Costello et al. 2016a) and small mammals such as Arctic ground squirrels (*Urocitellus parryii*; Hechtel 1985; Gau et al. 2002; Barker et al. 2015), particularly where large mammals are not available. Costello et al. (2016a) estimated insects, mostly ants, to represent 12% of digestible energy consumed by grizzly bears in the Greater Yellowstone Ecosystem. Similarly, in coastal areas, bears will augment diets with clams but use is seasonal and limited to smaller-bodied bears (Smith & Partridge 2004; Rode et al. 2006c).

Although a wide variety of plant species are consumed by brown bears, some commonalities in preferred species are evident across populations. Grasses (graminoids), sedges, and salt marsh vegetation (Figure 13.9), roots (e.g. *Plantago* spp., *Eleocharis* spp., *Triglochin* spp.), and sweet vetch roots are commonly consumed in the spring (Munro et al. 2006; Rode et al. 2006c). Horsetails (*Equisetum arvense*), cow parsnip (*Heracleum sphondylium*), false-truffles (*Rhizopogon* spp.), clover (*Trifolium* spp.), dandelion (*Taraxacum* spp.), and thistle (*Cirsium* spp.) are common spring and summer food items (MacHutchon & Wellwood 2003; Gunther et al. 2014), whereas blueberries (*Vaccinium* spp.), crowberries (*Empetrum nigrum*), russet buffaloberry (*Shepherdia canadensis*; MacHutchon & Wellwood 2003), and whitebark pine nuts are consumed in late summer and autumn as bears prepare for hibernation.

Diet selection by bears depends on seasonal availability but is also affected by body size. Because adult males can be much larger than adult females, their total energetic requirements are greater, and they are able to dominate high-quality food resources (Ben-David et al. 2004; Gende & Quinn 2004; Rode et al. 2006b). Consequently, it is common for diets to vary between males and females and between subadult and adult bears (Mowat & Heard 2006; Rode et al. 2006b).

Denning Ecology

Although most individuals in North American brown bear populations den during winter, in areas such as Kodiak Island, where surrounding warmer ocean waters support a long growing season, more than a quarter of studied males did not den at least 1 year during a 6-year study (Van Daele et al. 1990). It is possible that non-denning was more frequent for more southern, now extirpated, North American populations. Parturient females always den, likely because dens provide a warm and secure environment for altricial, newborn cubs. Typically, bears stay in the same den throughout the denning period, but cases of multiple den use have been documented (Schoen et al. 1987; Miller 1990; Van Daele et al. 1990). Suitable den sites provide adequate soil depth and stability for excavation of a den or a suitable natural cavity so that the den remains dry through the denning period (Van Daele et al. 1990). There is no evidence that suitable den sites are a limiting factor for brown bear populations in North America.

Chronology

Abiotic and biotic factors likely influence the timing of grizzly bear den entry and emergence (McLoughlin et al. 2002b). Difficulties in meeting metabolic requirements associated with unavailability of food towards the end of the growing season are likely the primary triggers for den entry (Schoen et al. 1987; Miller 1990; Pigeon et al. 2016b). The link between food scarcity and the timing of den entry is supported by captive bears that will not hibernate when fed year-round. Additionally, changes in photoperiod may operate on broad temporal scales through endogenous pathways (Folk et al. 1972), whereas specific weather events, such as heavy snowfall, may act as a fine-scale proximal factor for den entry (Craighead & Craighead 1972; McLoughlin et al. 2002b). The earliest dates of den entrance for female brown bears were recorded in the Greater Yellowstone Ecosystem and in central Alaska, with some bears entering during the latter half of September (Miller 1990; Haroldson et al. 2002). In most populations, pregnant females enter dens first, followed by females with cubs, lone females, and adult and subadult males (Schoen et al. 1987; Van Dael et al. 1990; Linnell et al. 2000; Haroldson et al. 2002; Schwartz et al. 2003b; Pigeon et al. 2016b). Prior to den entry, female brown bears may stay in the vicinity of their dens for 1–3 weeks (Harding 1976; Ciarniello et al. 2005). At the northern edge of their range in south-central and north-east Alaska and in the Northwest Territories, brown bears generally enter dens in October (McLoughlin et al. 2002b; Schwartz et al. 2003b). On the eastern slopes of the Northern Rocky Mountains, den entry generally occurs during November (Vroom et al. 1980; Pigeon et al. 2016b).

Warming spring temperatures and the start of the growing season largely trigger den exit (Pigeon et al. 2016b). In addition, unseasonable spring conditions that result in den flooding could trigger early den emergence (Schoen et al. 1987; Van Daele et al. 1990). Adult males are the first to exit dens (Linnell et al. 2000; Schwartz et al. 2003b), whereas females accompanied by newborn cubs consistently exit last (Vroom et al. 1980; Servheen & Klaver 1983; Miller 1990; Haroldson et al. 2002). These females tend to remain near dens for days or even weeks after first emergence (Schoen et al. 1987; Miller 1990; Haroldson et al. 2002), and food intake during this time seems to be minimal (Nelson et al. 1983). Remaining in the vicinity of dens likely reduces intraspecific predation on vulnerable newborn cubs, because previously emerged adult males are already in search of spring vegetative growth in more productive, lowland areas. Following an altitudinal gradient in the onset of the growing season, brown bears in British Columbia that denned at higher elevations typically emerged from dens later than bears that denned at lower elevations at the same latitude (Ciarniello et al. 2005). At the northern edge of their range in south-central and north-east Alaska and the Northwest Territories, brown bears generally emerge from dens during the last week of April or the first half of May (McLoughlin et al. 2002b; Schwartz et al. 2003b). On the eastern slopes of the Northern Rocky Mountains in Alberta and the Yellowstone region, den exit generally occurs during April (Vroom et al. 1980; Haroldson et al. 2002; Pigeon et al. 2016b).

Differences in methodology used in different studies likely account for some of the variations reported in den entry and den exit dates across populations, but research shows consistent evidence that brown bear populations living in harsher environmental conditions (i.e. shorter growing seasons with reduced access to food) den the longest. The denning period is typically 131–189 days for males to 146–204 days for females, and generally increases with latitude (Figure 13.10; e.g. Ciarniello et al. 2005). On Kodiak Island, both sexes have short denning periods given the latitude, likely because of the aforementioned maritime conditions. Pregnant females that gave birth spent 8–39 days longer in dens than other females (Schoen et al. 1987; Miller 1990; Van Daele et al. 1990; Haroldson et al. 2002; Pigeon et al. 2016b). The requirements associated with cub development likely explain the long denning period for females with newborn cubs.

Global warming may result in shallower snow packs and longer growing seasons, thus shortening periods of food scarcity. A longer growing season and less snow will likely lead to shorter hibernation periods, with denning starting later and ending earlier (Schoen et al. 1987; Pigeon et al. 2016b). Haroldson et al. (2002) observed a trend of later den entry dates by males over the time period 1975–1999, which was positively correlated with November minimum monthly temperatures. Climate-induced changes in denning behavior can potentially increase human–bear interactions among some populations because brown bears and humans will be simultaneously active on the landscape for longer periods (Pigeon et al. 2016b).

Den Characteristics

Brown bears in North America usually excavate their dens directly into the ground, but some individuals in Yellowstone National Park and south-west and north-east Alaska also den in natural caves (Reynolds et al. 1976; Judd et al. 1986; Schoen et al. 1987). Bedding materials such as moss and twigs are commonly used. Excavated dens are consistently on steep slopes (~20–60°) and in well-drained soil types, likely reducing the risk of den flooding (Harding 1976; Pigeon et al. 2016a). Excavated dens are generally composed of a narrow entranceway and a tunnel that opens into a chamber, although some dens lack a proper tunnel (Harding 1976; McLoughlin et al. 2002b; Pigeon et al. 2016a). Excavated dens are not often reused, likely because they collapse. However, den reuse has been documented and several authors report fidelity to specific

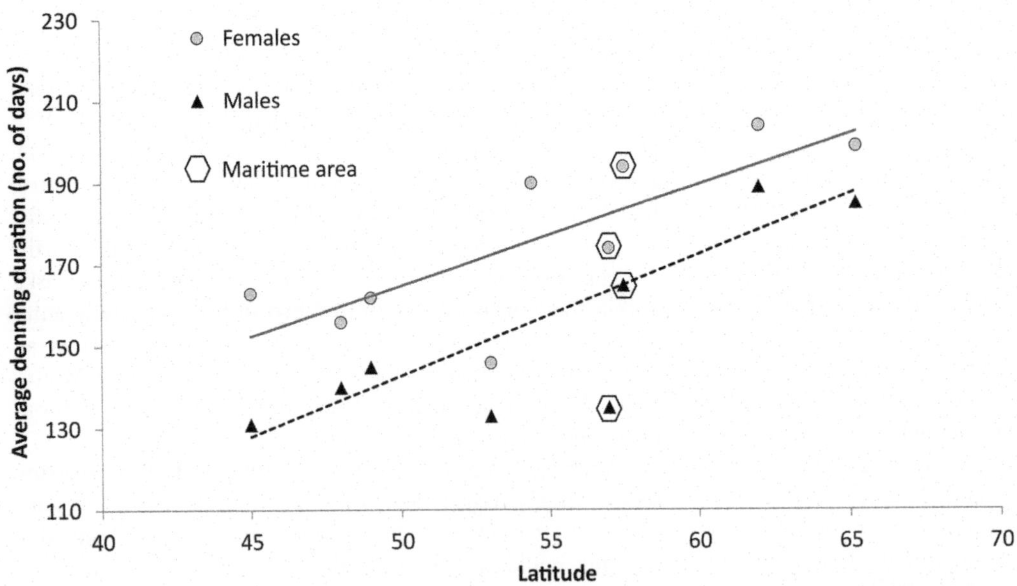

Figure 13.10 Duration (days) of denning period by latitude and sex reported for North American brown bear populations. Linear regression (excludes maritime areas): solid gray line = females (R^2 = 0.64); dashed black line = males (R^2 = 0.85).

areas with multiple dens observed on the same slope in successive years (Vroom et al. 1980; Servheen & Klaver 1983; Miller 1990; Van Daele et al. 1990). Where available, mature trees, shrubs, fallen logs or stumps are used, presumably because these root structures provide den support and stability and hiding cover for the entrance (Harding 1976; McLoughlin et al. 2002b; Pigeon et al. 2016a). Where deep snow accumulation improves the thermal properties of dens, selection for leeward slopes has been reported for several interior grizzly bear populations (Reynolds et al. 1976; Vroom et al. 1980; Miller 1990). In northern regions above the tree line, well-drained, dry, south-facing slopes under tall shrubs were favored (McLoughlin et al. 2002b). However, other researchers found no link between aspect and den locations (Servheen & Klaver 1983; Judd et al. 1986; Van Daele et al. 1990; McLoughlin et al. 2002b; Pigeon et al. 2014). Den insulation likely plays a more important role in cold climates (i.e. the interior and eastern slopes of the Rocky Mountains), but variation in local topography and prevailing wind direction influences snow depths right at dens, and could explain why these studies found no link between aspect and den locations.

Many researchers report brown bears seeking mid- to high-elevation areas for denning, but not all populations have access to an elevation gradient (Vroom et al. 1980; Judd et al. 1986; Miller 1990; Goldstein et al. 2010). Haroldson et al. (2002) found that pregnant females denned at slightly higher elevations than other cohorts in the Yellowstone area. At low elevations, dens are usually excavated on the banks of drainages or small slopes (Harding 1976; Pigeon et al. 2014). Dens in open timbered sites or near the alpine zone are common (Miller 1990; Waller & Mace 1997; Goldstein et al. 2010), but mature forests seem to be preferred when available (Judd et al. 1986; Schoen et al. 1987; MacHutchon et al. 1993). In less-mountainous areas of the plateau of central British Columbia, bears selected den sites under old-growth tree roots and in areas with lower densities of forestry roads (Ciarniello et al. 2005). Dens are generally located away from roads, recreation trails, and other human disturbances (Linnell et al. 2000; Podruzny et al. 2002; Goldstein et al. 2010; Pigeon et al. 2014). Den abandonment due to human disturbance has been documented but is not pervasive, with bears generally tolerating considerable activities near den sites (Linnell et al. 2000).

Reproduction and Growth

Breeding and Copulation

Female brown bears are monoestrous, with breeding periods (courtship, copulation, and post-copulatory association) averaging 49 days (range = 25–92 days) among North American studies (LeFranc et al. 1987). Based on data from 20 different populations, the average breeding season starts around mid-May and ends in mid-July (Schwartz et al. 2003b; Figure 13.11). Only sexually mature females not accompanied by cubs or yearlings enter into estrus. Successful copulations usually are shorter than 25 minutes, but some continue for 60 minutes (Craighead, Sumner & Mitchell 1995). Brown bears are promiscuous: litters can include siblings sired by different fathers and males commonly mate with multiple females during the same breeding season. Males enhance mating opportunities by having large home ranges that overlap multiple female ranges (see Movements and Home Range). They may further enhance their reproductive success by following a female and sequestering her until she enters estrus (Hamer & Herrero 1990). Such pair bonds can last from a few hours to a few weeks. Male competition for mates is high and their breeding success is influenced by male dominance, which generally correlates with body mass and age. Consequently, many males may not successfully breed. For example, genetic evidence showed that about half of breeding-age males in an Arctic population in north-western Alaska did not sire offspring (Craighead, Paetkau, et al. 1995).

Sexual Maturity

Average age of primiparity reported for North American populations is typically 6.5 years, ranging from 5.2 for bears foraging in open-pit garbage dumps in Yellowstone National Park during the 1960s to 10.5 years in the third decade of the Flathead River study in British Columbia, a period characterized by population decline (Table 13.1). This variation is primarily a function of habitat quality, but may also be influenced by population density and whether females disperse. Within each population, there can be substantial variation among individuals. In the Greater Yellowstone Ecosystem, for example, 10% of females produced their first litter at age 4, 30% at age 5, 56% at age 6, and the remaining 4% by age 7 (Schwartz et al. 2006). Female reproductive senescence starts around age 25 (Schwartz et al. 2003a). Males generally become sexually mature at 3.5–7 years (Erickson et al. 1968; Pearson 1975; White et al. 1998a). Reproductive senescence in males occurs at a later age than females but often does not manifest itself because males either do not live long enough or experience loss of body mass and are unable to physically dominate and exclude other males when competing for estrous females (Steyaert et al. 2012).

Delayed Implantation and Litter Size

Like several other members of the Ursidae family, female brown bears experience seasonal delayed implantation, or embryonic diapause (Dittrich & Kronberger 1963). Fertilized ova undergo several cell divisions but remain in the blastocyst stage and implant into the uterine wall in late November or early December. Upon implantation, the gestation period is approximately 60 days, with parturition in late January through early February. The combined period of delayed implantation and gestation is approximately 235 days. Females with greater fat reserves give birth earlier and their cubs grow faster than those nursing from lean mothers. Newborn brown bear cubs are altricial, 230–280 mm in length, and weigh

Figure 13.11 Breeding typically occurs mid-May through mid-July and copulations are generally less than 25 minutes in duration. (Photo courtesy of J. Davis.)

approximately 0.5 kg (Pasitschniak-Arts 1993). Blanchard (1987) found a strong correlation between female body mass and litter size among eight study sites in North America. Given a typical litter size of two cubs and a female body mass of 100–200 kg, the ratio of maternal body mass to that of newborn litters (0.5–1%) is one of the smallest among placental mammals (Oftedal & Gittleman 1989; Sibly & Brown 2009).

North American brown bears typically have litters of 1–3 cubs (Craighead, Sumner & Mitchell 1995), although occasional litters of four have been observed in at least eight different populations inhabiting a variety of habitats (Craighead et al. 1976; Ballard et al. 1991, 1993; Aune et al. 1994; Case & Buckland 1998; McLoughlin et al. 2003b; Miller et al. 2003; Barnes & Van Daele 2006; van Manen & Haroldson 2017). Several litters of five and a litter of six have been observed in Alaska, although adoption could not be ruled out in the latter instance (Modafferi 1984; Wilk et al. 1988). Although rare, adoptions have been documented (Haroldson et al. 2015). Average litter sizes reported for North American populations are comparable to those observed for Eurasian brown bear populations (Table 13.1; see Chapter 12). Average litter size was consistent among 23 study sites in North America, varying between 1.4 and 2.5 cubs (Table 13.1). Litter size is associated with the age of the female producing the litter and food production. Long-term studies indicate that litter size within a population shows little variation over time (Interagency Grizzly Bear Study Team 2012). Young are generally weaned at 2.5 years, thus the expected interval for a successful reproductive cycle is 3 years. However, in areas where habitat productivity is low, such as the North Slope of Alaska, the average weaning age of offspring was 4.5 years (Reynolds 1976). The reproductive cycle is disrupted by the loss of a complete litter, which may delay subsequent reproduction depending on the timing of separation between mother and offspring. Female brown bears show some flexibility in this reproductive sequence, as indicated by rare observations of mixed-age litters composed of cubs and yearlings (Swenson & Haroldson 2008). This phenomenon may occur when a female is temporarily separated from her cubs, mates while still maintaining lactation, and produces another litter in the subsequent year.

Reproductive Rate

Fecundity, or the number of female cubs produced per reproductive-age female per year, is a function of age of primiparity, litter size, cub survival, interbirth interval, and the

Table 13.1 Vital rates and demographic data from brown bear studies conducted in North America during the period 1959–2014.

Study	Location[a]	Time period	Density (bears/1000 km^2)	Annual probability of survival (S): Cubs (<1 year)	Annual probability of survival (S): Yearlings (1–2 years)	Annual probability of survival (S): Subadults (2–4 years)[b]	Annual probability of survival (S): Adults (≥5 years)[b]	Mean litter size	Mean age of primiparity (years)	Mean weaning age (years)	Mean litter interval (years)	Fecundity (m)	Sex ratio M:F[b]	Finite rate of population growth (λ)
Nagy et al. (1983)	Tuktoyaktuk Peninsula and Richards Island; Northwest Territories	1974–1978	3.8–4.7					2.2	5.9	2–3	3.3			
Case and Buckland (1998)	Kugluktuk area; Northwest Territories	1988–1995		0.81	0.76–0.84		F: 0.98	2.3	8.7		3.3			1.026
Ballard et al. (1991, 1993)	Noatak River; Alaska	1986–1990	20	0.87	0.89		M: 0.91, F: 0.94	2.2		2.5			50:50	
Keay et al. (2018)	Denali National Park; Alaska	1991–1998	37	0.34	0.46	M: 0.94, F: 1.00	M: 0.98, F: 0.97	2.1	7.5	2.9	3.1	0.34		0.98
Miller (1990)	Nelchina; Alaska	1978–1989		0.69			M: 0.80, F: 0.92							
Sellers and Aumiller (1994)	McNeil River; Alaska	1963–1991		0.67	0.89		M: 0.94, F: 0.93	2.1					55:45	
Kovach et al. (2006)	South-western Kuskokwim Mountains; Alaska	1993–1997		0.68	0.84[c]		F: 0.94–1.0[d]	2.0	7.2	2.6	2.6	0.214–0.452[d]		1.035–1.047
		1998–2003		0.48	0.72[c]		F: 0.39–1.0[d]			2.9		0.235–0.362[d]		0.963–0.997
McLellan et al. (1999)	Rocky and Columbia Mountains; Alberta, British Columbia, Idaho, Montana, Washington	1975–1997				M: 0.80, F: 0.92	M: 0.88, F: 0.93							
Garshelis et al. (2005)	Bow River Watershed; Alberta	1994–2002		0.79	0.91		M: 0.81–0.84, F: 0.95	1.8	8.4		4.4	0.24		1.040

Table 13.1 *(cont.)*

Study	Location[a]	Time period	Density (bears/ 1000 km²)	Annual probability of survival (*S*): Cubs (<1 year)	Yearlings (1–2 years)	Subadults (2–4 years)[b]	Adults (≥5 years)[b]	Mean litter size	Mean age of primiparity (years)	Mean weaning age (years)	Mean litter interval (years)	Fecundity (*m*)	Sex ratio M:F[b]	Finite rate of population growth (λ)
Boulanger and Stenhouse (2014)	Grand Cache – Yellowhead; Alberta	1999–2012		0.55		M: 0.78, F: 0.88	M: 0.88, F: 0.96	2.1				0.272		
Wielgus and Bunnell (1994)	Kananaskis Park and Bow Crow Forest; Alberta	1980–1984	16	0.78		M: 0.89, F: 0.89–0.93	M: 0.70, F: 0.93	1.4	5.5		3	0.23		0.990–1.010[e]
Hovey and McLellan (1996)	Flathead River; British Columbia and Montana	1979–1994		0.87	0.94	F: 0.93	F: 0.95		6.4			0.422		1.085
McLellan (1989)	South-eastern British Columbia	1979–1987	57–80	0.82	0.88	M: 0.91, F: 0.94	M: 0.92, F: 0.94	2.3	6.0		2.7	0.43		1.084[e]
McLellan (2015)	North Fork Flathead River; British Columbia	1979–1988	16–56[f]	0.84	0.86	M: 0.78, F: 0.96	M: 0.84, F: 0.93	2.4	6.6		2.9	0.374		1.074
		1989–1998		0.86	0.78	M: 0.78, F: 1.00	M: 0.84, F: 0.97	1.8	7.1		3.4	0.257		1.060–1.080
		1999–2010		0.69	1.00	M: 0.78, F: 0.94	M: 0.84, F: 0.89	1.8	10.5		4.2	0.192		0.955–0.980
Wakkinen and Kasworm (2004)	Selkirk Mountains; Idaho, Washington, and British Columbia	1983–2002		0.88	0.78	M: 0.77, F: 0.90	M: 0.91, F: 0.94					0.284		1.018[e]
Wielgus et al. (1994)	Selkirk Mountains; Idaho, Washington, and British Columbia	1985–1990	14–23	0.84		M: 0.90, F: 0.78	M: 0.81, F: 0.96	2.2	7.3		3.0	0.37		1.0[e]
Kendall et al. (2015)	Cabinet-Yaak Ecosystem; Montana	2012	4											
Wakkinen and Kasworm (2004)	Cabinet-Yaak Mountains; Idaho, Montana	1983–2002		0.68	0.88	M: 0.75, F: 0.77	M: 0.85, F: 0.93					0.291		0.964[e]
Aune and Kasworm (1989)	Rocky Mountain East Front; Montana	1976–1987		0.82		M: 0.66, F: 0.92	M: 0.81, F: 0.94	2.2	6.5		2.6			

Aune et al. (1994)	Northern Continental Divide Ecosystem; Montana	1969–1991		0.89	0.86			2.1	5.7	2.7			
Mace and Waller (1998)	Swan Mountains; Montana	1987–1996	16	0.79	0.91	F: 0.825	F: 0.90	1.6	6.0		0.261		0.977
Kendall et al. (2008)	Glacier National Park; Montana	1998 and 2000	30										
Mace et al. (2012)	Northern Continental Divide Ecosystem; Montana	2004–2009		0.61	0.68	F: 0.35	F: 0.95	2.3			0.367		1.031
Costello et al. (2016b)	Northern Continental Divide Ecosystem; Montana	2004–2014		0.55	0.64	M: 0.90, F: 0.95	M: 0.90, F: 0.95	2.1	5.8				1.023
Craighead et al. (1976)	Greater Yellowstone Ecosystem; Wyoming, Montana, and Idaho	1959–1971						2.3	5.2				
Blanchard and Knight (1980)	Greater Yellowstone Ecosystem; Wyoming, Montana, and Idaho	1975–1979						1.9	5.5				
Eberhardt (1995)	Greater Yellowstone Ecosystem; Wyoming, Montana, and Idaho	1981–1994		0.84	0.80		F: 0.94				0.35		1.053
Schwartz et al. (2006)	Greater Yellowstone Ecosystem; Wyoming, Montana, and Idaho	1983–2002	21	0.64	0.82	0.92–0.95[g]	M: 0.88–0.87,[g] F: 0.92–0.95	2.0	5.8	2.78	0.318–0.362	39:61	1.042–1.076
Interagency Grizzly Bear Study Team (2012)	Greater Yellowstone Ecosystem; Wyoming, Montana, and Idaho	2002–2011		0.55	0.54	0.89–0.95[g]	M: 0.94–0.95,[g] F: 0.94–0.95	2.1–2.5			0.286–0.336	50:50	1.003–1.022

[a] Populations ordered from northern to southern latitudes.
[b] F = female, M = male.
[c] Includes 2-year-olds.
[d] Range represent age classes 5–9 years, 10–14 years, 15–19 years, and 20–29 years.
[e] Derived from estimate of intrinsic rate of increase (*r*).
[f] Excludes independent males.
[g] Range represents animals with unresolved fates being censored versus assumed dead.

age of reproductive senescence. In turn, these factors are influenced by habitat productivity, or food supply, and population density. Schwartz et al. (2003a) documented fecundity for 20 brown bear populations, of which 18 were in North America. Mean fecundity among these 18 populations was 0.26 female cubs/female/year, ranging from 0.18 for the Black Lake brown bear population on the Alaskan Peninsula to 0.36 for the Northern Continental Divide population in Montana (Table 13.1). Fecundity increases rapidly once females reach sexual maturity around age 4–5, peaks around age 8–9 and remains high until approximately 25 years, after which it declines sharply due to reproductive senescence (Schwartz et al. 2003a).

Survival

Population dynamics of large vertebrates such as brown bears are strongly influenced by survival rates, particularly adult female survival (Eberhardt 2002). Estimates of adult female survival are remarkably similar among North American studies, with all studies reporting annual survival rates of 0.89 or greater (Table 13.1). The few studies with evidence of stable or declining populations generally had lower levels of adult female survival. Male survival is generally lower than that of females and much more variable in both hunted and unhunted populations (Table 13.1). This is because larger males are often targeted by hunters as trophies and regulations usually prohibit the taking of maternal females. Generally, larger male home ranges and shorter denning durations also create greater exposure to mortality sources, as does dispersal of young males from their maternal home ranges (Bunnell & Tait 1980, 1981; Miller 1990; Miller & Tutterow 1999). However, adult male survival was similar to that of females in three populations: Denali National Park (Keay et al. 2018) and McNeil River in Alaska, USA (Sellers & Aumiller 1994), and the Greater Yellowstone Ecosystem in Wyoming, Montana, and Idaho, USA (Interagency Grizzly Bear Study Team 2012; Table 13.1). All three populations were unhunted and had relatively high densities. Among younger age classes, annual survival estimates vary substantially and were lowest for cubs (0.34–0.89), followed by yearlings (0.46–1.0), and subadults (0.75–1.0).

Mortality Factors

Humans are the main cause of documented mortality among bears of independent age (≥2 years old), typically representing 70–85% of all causes. However, many mortalities are unreported and, in some areas, may equal the number of reported mortalities (Cherry et al. 2002). Factors contributing to human-caused mortality vary substantially. For populations where hunting is legal (see Status in the Wild), harvest is generally the largest source of mortality and tends to be skewed toward adult males because of hunter preference and harvest regulations prohibiting the take of maternal females (Miller 1990). For example, of all grizzly bear mortalities recorded in British Columbia during 1980–2011 (n = 11,084), 86.2% were from legal harvest and 65.6% were male (McLellan et al. 2017a). Toward the southern portions of occupied range, human-related mortality is increasingly a combination of agency management removals, self-defense kills, poaching, mistaken identification by black bear hunters, and accidental deaths (Schwartz et al. 2003b). Agency removals are usually a last option in comprehensive conflict management programs involving a wide range of circumstances, from cattle depredation to site conflicts associated with unsecured anthropogenic foods to attacks on humans. Throughout the occupied range, it is legal to kill a brown bear for defense of life and, in some jurisdictions, defense of property. Accidental human-caused mortalities include collisions with vehicles and trains, electrocution, drownings in irrigation canals, and accidental ingestion of poisons. The proportion of mortalities due to conflict management, self-defense, and accidental causes increases in landscapes with greater human influence and access, ultimately determining which landscape conditions function as a population source (i.e. positive growth) or sink (Mace et al. 1996; Schwartz et al. 2010b). Spatial gradients in population trend have been shown to be associated with density of roads open to motorized traffic and this is an important management consideration (Boulanger & Stenhouse 2014). Natural mortality includes old age, intra- and interspecific killing, and accidental deaths such as rock slides, avalanches, or den collapse. Natural mortality rates for four brown bear populations were summarized by McLellan et al. (2017a) and showed high vulnerability for dependent young (cubs: 20–24%; yearlings: 16%) but limited risk for independent females (<1.7%) and males (<2.5%). Studies have also shown that males are more likely to kill dependent offspring (intraspecific infanticide) than females, although infanticide by female brown bears has been documented (Hessing & Aumiller 1994). Infanticidal males may increase fitness by mating with a female that has lost her litter earlier than if that litter was weaned after the normal period, which has been termed sexually selected infanticide or SSI. In Scandinavia, researchers reported that hunting results in an influx of immigrant males that are more likely to commit SSI than resident males because they had not fathered litters in the area (Swenson et al. 2001a, 2001b), although later research showed that infanticide was commonly committed by resident males (Bellemain et al. 2006; see Chapter 12). In contrast, decreased survival of cubs due to male-biased mortalities from hunting or other causes has not been found in North American studies of brown bears (Miller et al. 2003; McLellan 2005). Rather, North American studies indicate that cubs are likely more vulnerable to mortality, including intraspecific killings, in populations near carrying capacity, a density-dependent effect (Miller et al. 2003; van Manen et al. 2016; Keay et al. 2018; Figure 13.12). McLellan (2005) proposed that males are able to recognize females with whom they mated and avoid killing cubs they fathered. Such a mechanism would result in benefits to females that mate with multiple males, as described previously, which may function as an evolutionary counterstrategy.

Figure 13.12 Cubs and yearlings are vulnerable to mortality from adult male bears, which may contribute to reduced survivorship of dependent young in some populations that are at or near carrying capacity. (Photo courtesy of J. Davis.)

Population Growth

Reliable estimates of population growth are difficult to obtain, often requiring longer-term studies. Population growth rates have been estimated for 20 different populations in North America. Population declines were reported for three of those populations and for a portion of the study period for two additional populations, and were generally a function of low adult female survival (Table 13.1). No annual declines greater than about 4.5% have been observed. Changes in population growth estimates over decadal time frames provide important insights into factors influencing trends. The largest changes in population growth were documented in the Kuskokwim Mountains of south-western Alaska, the Flathead Valley in British Columbia, and the Greater Yellowstone Ecosystem in the conterminous US (Table 13.1). In the Flathead Valley, for example, estimates changed from approximately 6% to 8% annual growth in the first two decades to 2–4.5% decline in the third decade attributed to a decline of berry-producing communities from forest succession (McLellan 2015). Annual growth around 7–8% has only been reported for one other population, the Greater Yellowstone Ecosystem (Schwartz et al. 2006), possibly representing an upper limit for North American populations. This range of estimates is lower than those observed for European brown bear populations (see Chapter 12), likely a reflection of differences in the age young are weaned.

Density-dependent processes are notoriously difficult to study in populations of large vertebrates, but may play an important role in the demographics of brown bear populations, particularly those with little or no legal harvest (Eberhardt 2002). A literature review by McLellan (1994) did not provide conclusive evidence of density-dependent processes. Subsequent comparative studies of unhunted (national park) and nearby hunted populations in Alaska provided evidence for apparent lower (density-dependent) cub survival in the unhunted populations at carrying capacity (Miller et al. 2003; Keay et al. 2018). Recent studies in the Greater Yellowstone Ecosystem further supported these findings, indicating survival of dependent young was lower in core areas where the population was likely near carrying capacity, compared with peripheral areas of lower bear density and recently occupied areas, where human-caused mortalities were more prevalent (van Manen et al. 2016).

Behavior

Brown bears in North America are known for their aggressiveness in defending offspring, food, and personal space from other bears, wolves, and people. They are thought to be more aggressive than sympatric American black bears as a result of their evolutionary history (Stirling & Derocher 1990). Whereas American black bears primarily inhabit forested environments, the brown bear's niche also includes more open areas (Herrero 1972). Herrero (1972) hypothesized that these open, treeless areas forced female brown bears to defend their young on the ground, which would favor aggression as a defense mechanism when threatened.

Other than mother–offspring and breeding pairs, brown bears are generally considered solitary in that they do not live in social, cooperative groups (Egbert & Stokes 1976). However, bears frequently interact during the breeding season, even where population densities are low (Craighead, Sumner & Mitchell 1995; Stenhouse et al. 2005). Reproductive behavior and male intrasexual competition has been well documented for brown bears (reviewed in Steyaert et al. 2012; see also Chapter 2). Outside of the breeding season, brown bear social behavior is largely influenced by food availability, particularly

Figure 13.13 Grizzly bears are often drawn to and congregate at abundant food resources such as large ungulate carcasses. In August of 2007, 20 grizzly bears (11 pictured) were observed near a bison carcass in Yellowstone National Park, Wyoming, USA. Adult males dominate the carcasses and females with offspring (lower left) and other subordinate bears often linger in the vicinity waiting for an opportunity to feed. (Photo courtesy of Interagency Grizzly Bear Study Team.)

when food is concentrated at salmon spawning streams, ungulate or whale carcasses, and garbage dumps (Stonorov & Stokes 1972; Egbert & Stokes 1976; Craighead, Sumner & Mitchell 1995; Figure 13.13). When food is scarce at these sites, or temporally limited, increased competition can lead to agonistic encounters (Egbert & Stokes 1976). Within the dominance hierarchy that forms, rank is usually determined by sex, size, condition, aggressiveness, and number of older (1- or 2-year-old) dependent offspring (Craighead, Sumner & Mitchell 1995). Social hierarchies within aggregations of bears appear more rigid when competition is greater and the selective advantage of dominance is likely most pronounced (Gende & Quinn 2004). Indeed, Gende and Quinn (2004) found that dominance ranking had a greater effect on salmon consumption rate than the density of food available. At concentrated food sources, females with young and subadults face a trade-off between the risk of injury and increased energy consumption (Ben-David et al. 2004). Avoiding close interactions with conspecifics can lead to sexual segregation (Wielgus et al. 1995) and suboptimal foraging with energetic consequences (Ben-David et al. 2004). This may be mitigated by temporal rather than spatial avoidance of high-density areas (Nevin & Gilbert 2005). In areas where exposure to people is frequent, bears lowest in the social hierarchy may habituate to human presence to gain access to food resources (natural or anthropogenic) avoided by more dominant, but human-wary bears (Herrero et al. 2005). When food is abundant, brown bears are generally more tolerant of each other and some even seek out interactions in the form of social play (Fagen & Fagen 1990; Clapham & Kitchin 2016). Mother–offspring and sibling play behavior is common and has been linked to increased cub survival, possibly through stress resistance and enhanced neuromuscular performance (Fagen & Fagen 2004, 2009).

In North America, brown bears share their range with other members of the Carnivora, and interactions can occur. Brown bears are generally dominant over the smaller American black bear. They displace black bears from spatially restricted high-energy food sources (Belant et al. 2010) and have been documented to occasionally prey upon them (Mattson et al. 1992). In contrast, due to their larger body size, polar bears would be expected to displace brown bears in areas of terrestrial overlap. However, Miller et al. (2015) found brown bears to be the more dominant competitor at a whale carcass aggregation on the southern Beaufort Sea coast during fall. Around half of all polar–brown bear interactions led to polar bears being permanently displaced from the carcass and Miller et al. (2015) suggested that motivation to exploit the resource during fall hyperphagia and the evolutionary history for aggression in brown bears could explain this asymmetry. Interactions between brown bears and wolves at a whale carcass in southeastern Alaska indicated little aggression and were likely a function of the large amount of food available

(Lewis & Lafferty 2014). Although brown bears have been recorded to kill and displace wolves at ungulate carcasses, wolves have also been documented displacing female brown bears and killing bear cubs (Gunther & Smith 2004; Jimenez et al. 2008). Abundant food sources appear to increase tolerance and reduce interspecific aggression (Koene et al. 2002; Lewis & Lafferty 2014; see also Chapter 3).

Parasites and Disease

Brown bears are hardy and generally robust to various maladies associated with diseases or parasites. In their reviews, Rogers and Rogers (1976), LeFranc et al. (1987), and Pasitschniak-Arts (1993) summarized commonly occurring diseases and parasites found in North American brown bear populations. Whereas individual bears may be afflicted by clinical signs, major disease or parasite occurrences affecting population demographics have not been documented. However, as Schwartz et al. (2003b) stated, documenting cause of death from disease or parasites in natural conditions is difficult.

Knowles et al. (2018) reported that a brown bear cub in Alaska died from infectious canine hepatitis (caused by canine adenovirus 1, CAdV-1). Previous studies from Alaska (Zarnke & Evans 1989; Chomel et al. 1998) reported that serum antibody prevalence to infectious canine hepatitis in brown bears ranged from 12% to 14%. Zarnke and Evans (1989) noted that antibodies to the disease were not detected in bears under 2 years of age, a possible indicator of high mortality when young bears are infected. In the Cantabrian brown bear population of Spain, García Marín et al. (2018) reported that three brown bears died from canine hepatitis and considered the disease a major concern for the conservation of that small and endangered population.

Canine distemper virus is another disease agent that has been documented in North American brown bear populations (Chomel et al. 1998; Cross et al. 2018). The disease is acute, highly transmissible, and can be a significant mortality agent in canids, especially among juveniles. Population-level impacts of canine distemper virus in brown bears are unknown but warrant further investigation. Morbidity from canine distemper of a juvenile black bear in North American has also been reported (Cottrell et al. 2013).

Historically, brown bears in the US Pacific Northwest states (California, Idaho, Oregon, Washington) commonly fed on salmon (*Oncorhynchus* spp.; Robbins et al. 2018). Unlike salmon populations from more northern latitudes, salmon in this region often carry an encysted fluke (*Nanophyetus salmincola*) containing an endosymbiont (*Neorickettsia* spp.) that can produce a deadly disease in ursids and canids called salmon poisoning disease (Robbins et al. 2018). In studies with captive grizzly bears from populations with no history of salmon poisoning disease, individuals were severely sickened by consuming limited amounts of Chinook salmon (*Oncorhynchus tshawytscha*) from river drainages in the North Cascades of Washington. With restoration of grizzly bears to the North Cascades being considered (see Status in the Wild), Robbins et al. (2018) recommended additional studies to determine if free-ranging black or brown bears in the Pacific Northwest have been infected and survived.

Among endoparasites reported for brown bears in North America, *Baylisascaris* spp. are some of the most common intestinal parasites. Early researchers noted the prevalence of gastrointestinal parasites was greater in fall than spring, and speculated this was due to dietary changes (Rush 1932; Rausch 1954) or to cessation of feeding (Frechette & Rau 1978). More recently, Gau et al. (1999) confirmed these seasonal changes in prevalence of gastrointestinal parasites, but were unable to support or refute the previously proposed hypotheses regarding potential causes.

Status in the Wild

Brown bears remain abundant in portions of North America where the human footprint is limited or where significant recovery and conservation efforts have been ongoing. In Alaska, brown bears are an important game species with estimates indicating a population of about 30,000 brown bears (Miller 1993; Miller & Schoen 1999). About 25,000 brown bears live in Canada (Banci 1991), where brown bears are federally designated as a species of special concern (Committee on the Status of Endangered Wildlife in Canada 2012). Approximately 15,000 brown bears occur in British Columbia (British Columbia Ministry of Forests, Lands, and Natural Resource Operations 2012). Most populations in British Columbia are considered healthy, but a few threatened populations exist on the southern extent of their distribution (McLellan et al. 2017b). Alberta's approximately 900 bears are considered legally threatened (Alberta Environment and Parks 2016).

All grizzly bears in the US south of Canada were listed in 1975 as "threatened" under the US Endangered Species Act. Recovery efforts in the US, involving several transboundary populations, are coordinated by the Interagency Grizzly Bear Committee consisting of federal, state, and tribal agencies. The 1993 recovery plan identified six recovery zones (US Fish and Wildlife Service 1993), four of which have extant populations, one is unoccupied except for an occasional transient individual (Bitterroot Ecosystem, central Idaho), and one has unknown status (North Cascades, north-western Washington) with no recent evidence of occupancy in the US portion of the ecosystem. Since the 1970s, population numbers have increased in the Northern Continental Divide Ecosystem (Montana) from several hundred to >1000 and in the Greater Yellowstone Ecosystem (Wyoming, Montana, and Idaho) from perhaps fewer than 250 to >700. Conservation efforts contributed to achieving recovery plan criteria for the Yellowstone population and it was delisted in 2007 and 2017, but in each instance the decision was overturned by federal courts. An effort to delist the Northern Continental Divide Ecosystem population is under consideration. Populations have become more secure and slightly larger in the Cabinet-Yaak Ecosystem (Montana, Idaho, and British Columbia) and Selkirk Ecosystem (Idaho,

Washington, and British Columbia), with approximately 72 and 90 grizzly bears, respectively (Proctor et al. 2012; McLellan et al. 2017b). A plan to reintroduce grizzly bears in the Bitterroot Ecosystem was approved in the year 2000 (US Fish and Wildlife Service 2000) but was not implemented. A draft Environmental Impact Statement for grizzly bear restoration was prepared for the North Cascades Ecosystem in 2017 (US Department of the Interior 2017). As of 2019, no populations have been re-established in either the Bitterroot or North Cascades recovery zones.

Brown bear hunting in the US is allowed only in Alaska; in some areas of the state, wildlife authorities are targeting bears and other large carnivores for reductions to increase ungulate populations (Miller et al. 2017; Ripple et al. 2019). In Canada, brown bear hunting is only allowed in those areas north of the 60th parallel (i.e. Yukon, Northwest Territories, Nunavut) where bear productivity is sufficient to sustain harvest. Currently, brown bear hunting is not allowed in Alberta or British Columbia. Although scientific data indicate the brown bear harvest in British Columbia was sustainable, hunting was discontinued in 2017 based on public opposition (McLellan et al. 2017a).

Status in Captivity

Fifty-five zoo facilities hold a total of 172 brown bears in North America. The majority are in the US, with 152 brown bears across 47 facilities, 14 are in six Canadian zoos, and three each are in a Mexican and a Guatemalan zoo (Zoo Information Management System data, compiled by M. Szymańska, Poznan Zoo, Poland; March 16, 2019). Eleven captive brown bears are housed at the Bear Research, Education and Conservation Center of Washington State University in Pullman, Washington, USA. This is the only research facility in North America dedicated to the study of bear physiology, metabolism, and energetics with applications toward ecological studies, conservation, and human medicine (e.g. diabetes and obesity). Brown bears adapt well to captivity and many survive to age 30 or more. The oldest known captive brown bear was a captive-born female that died in 2018 at age 40 at the Columbus Zoo and Aquarium in central Ohio (Palmer 2015). She had cubs in captivity. Because most zoos that want brown bears already have them, it is difficult for wildlife management agencies to find placement for cubs orphaned during human–bear conflicts or as a result of management actions.

References

Alberta Environment and Parks. (2016). *Alberta grizzly bear (*Ursus arctos*) draft recovery plan. Alberta Species at Risk Recovery Plan No. 38*. Edmonton: Alberta Environment and Parks.

Ashrafzadeh, M. R., Kaboli, M. & Naghavi, M. R. (2016). Mitochondrial DNA analysis of Iranian brown bears (*Ursus arctos*) reveals new phylogeographic lineage. *Mammalian Biology* 81: 1–9.

Aune, K. E. & Kasworm, W. (1989). *East front grizzly bear study*. Final report. Helena, MT: Montana Fish, Wildlife and Parks.

Aune, K. E., Mace, R. D. & Carney D. W. (1994). The reproductive biology of female grizzly bears in the Northern Continental Divide Ecosystem with supplemental data from the Yellowstone Ecosystem. *International Conference on Bear Research and Management* 9: 451–458.

Ballard, W. B., Spraker, T. H. & Taylor, K. P. (1981). Causes of neonatal moose calf mortality in south-central Alaska. *Journal of Wildlife Management* 45: 335–342.

Ballard, W. B., Ayres, L. A., Roney, K. E., Reed, D. J. & Fancy, S. G. (1991). *Demography of Noatak grizzly bears in relation to human exploitation and mining development*. Federal Aid in Wildlife Restoration Research Final Report for Grants W-22-5, W-22-6, W-23-1, W-23-2, W-23-3 Study 4.20. Juneau, AK: Alaska Department of Fish and Game, Division of Wildlife Conservation.

Ballard, W. B., Ayres, L. A., Reed, D. J., Fancy, S. G. & Roney, K. E. (1993). *Demography of grizzly bears in relation to hunting and mining development in Northwestern Alaska*. Scientific Monograph Series NPS/NRARO/NRSM-93/23. Denver, CO: National Park Service.

Banci, V. (1991). *The status of the grizzly bear in Canada in 1990*. Ottawa: Committee on the Status of Endangered Wildlife in Canada.

Barker, O. E., Derocher, A. E. & Edwards, M. A. (2015). Use of Arctic ground squirrels (*Urocitellus parryii*) by brown bears (*Ursus arctos*). *Polar Biology* 38: 369–379.

Barnes, V. G., Jr. (1990). The influence of salmon availability on movements and range of brown bears on southwest Kodiak Island. *International Conference on Bear Research and Management* 8: 305–313.

Barnes V. G. & Van Daele, L. J. (2006). *Productivity of female brown bears on Kodiak Island, Alaska*. Job completion report for Challenge Cost Share Project. Kodiak, AK: Kodiak National Wildlife Refuge, US Fish and Wildlife Service.

British Columbia Ministry of Forests, Lands, and Natural Resource Operations. (2012). *2012 Grizzly bear population estimate for British Columbia*. Victoria: British Columbia Ministry of Forests, Lands, and Natural Resource Operations.

Belant, J. L., Griffith, B., Zhang, Y., Follmann, E. H. & Adams, L. G. (2010). Population-level resource selection by sympatric brown and American black bears in Alaska. *Polar Biology* 33: 31–40.

Bellemain, E., Swenson, J. E. & Taberlet, P. (2006). Mating strategies in relation to sexually selected infanticide in a non-social carnivore: the brown bear. *Ethology* 112: 238–246.

Ben-David, M., Titus, K. & Beier, L. R. (2004) Consumption of salmon by Alaska brown bears: a trade-off between nutritional requirement and the risk of infanticide? *Oecologia* 138: 465–474.

Bjornlie, D. D., van Manen, F. T., Ebinger, M. R., et al. (2014). Whitebark pine, population density, and home-range size of grizzly bears in the Greater Yellowstone Ecosystem. *PloS ONE* 9: e88160.

Blanchard, B. M. (1987). Size and growth patterns of the Yellowstone grizzly bear. *International Conference for Bear Research and Management* 7: 99–107.

Blanchard, B. M. & Knight, R. R. (1980). Status of grizzly bears in the Yellowstone system. *Transactions of the North American Wildlife and Natural Resources Conference* 45: 263–267.

Blanchard, B. M. & Knight, R. R. (1991). Movements of Yellowstone grizzly bears. *Biological Conservation* 58: 41–67.

Blanchard, B. M. & Knight, R. R. (1995). Biological consequences of relocating grizzly bears in the Yellowstone Ecosystem. *Journal of Wildlife Management* 59: 560–565.

Boertje, R. D., Gasaway, W. C., Grangaard, D. V. & Kellyhouse, D. G. (1987). Predation on moose and caribou by radio-collared grizzly bears in east central Alaska. *Canadian Journal of Zoology* 66: 2492–2499.

Boulanger, J. & Stenhouse, G. B. (2014). The impact of roads on the demography of grizzly bears in Alberta. *PLoS ONE* 9: e115535.

Bunnell, F. E. & Tait, D. E. N. (1980). Bears in models and reality – implications to management. *International Conference on Bear Research and Management* 3: 15–23.

Bunnell, F. E. & Tait, D. E. N. (1981). Populations dynamics of bears–implications. In: Fowler, C. W. & Smith, T. D. (Eds.), *Dynamics of large mammal populations* (pp. 75–98). New York, NY: John Wiley & Sons.

Byers, C. R. & Bettas, G. A. (1999). *Records of North American big game*, 11th edition. Missoula, MT: Boone and Crockett Club.

Campbell, B. H. (1999). Homing of translocated brown bears (*Ursus arctos*) in coastal south-central Alaska. *Northwest Naturalist* 80: 22–25.

Case, R. L. & Buckland, L. (1998). Reproductive characteristics of grizzly bears in the Kugluktuk area, Northwest Territories, Canada. *Ursus* 10: 41–47.

Cherry, S., Haroldson, M. A., Robison-Cox, J. & Schwartz, C. C. (2002). Estimating total human-caused mortality from reported mortality using data from radio-instrumented grizzly bears. *Ursus* 13: 175–184.

Chester, C. C., Hilty, J. A. & Hamilton, L. S. (2013). Mountain gloom and mountain glory revisited: a survey of conservation, connectivity, and climate change in mountain regions. *Journal of Mountain Ecology* 9: 1–34.

Chomel, B. B., Kasten, R. W., Chappuis, G., Soulier, M. & Kikuchi, Y. (1998). Serological survey of selected canine viral pathogens and zooneses in grizzly bears (*Ursus arctos horribilis*) and bears (*Ursus americanus*) from Alaska. *Revue scientifique et technique* 17: 756–766.

Ciarniello, L. M., Boyce, M. S., Heard, D. C. & Seip, D. R. (2005). Denning behavior and den site selection of grizzly bears along the Parsnip River, British Columbia, Canada. *Ursus* 16: 47–58.

Clapham, M. & Kitchin, J. (2016). Social play in wild brown bears of varying age-sex class. *Acta Ethologica* 19: 181–188.

Coleman, T. H., Schwartz, C. C., Gunther, K. A. & Creel, S. (2013). Grizzly bear and human interaction in Yellowstone National Park: an evaluation of bear management areas. *Journal of Wildlife Management* 77: 1311–1320.

Collins, G. H., Kovach, S. D. & Hinkes, M. T. (2005). Home range and movements of female brown bears in southwestern Alaska. *Ursus* 16: 181–189.

Committee on the Status of Endangered Wildlife in Canada (COSEWIC). (2012). *Assessment and status report on the grizzly bear,* Ursus arctos, *in Canada*. Ottawa: Committee on the Status of Endangered Wildlife in Canada.

Coogan, S. C. P., Raubenheimer, D., Stenhouse, G. B. & Nielsen, S.E. (2014). Macronutrient optimization and seasonal diet mixing in a large omnivore, the grizzly bear: a geometric analysis. *PLoS ONE* 9: e105719.

Costello, C. M., Cain, S. L., Pils, S. R., et al. (2016a). Diet and macronutrient optimization in wild ursids: a comparison of grizzly bears with sympatric and allopatric black bears. *PLoS ONE*: 11: e0153702.

Costello, C. M., Mace, R. D. & L. Roberts. (2016b). *Grizzly bear demographics in the Northern Continental Divide ecosystem, Montana: Research results (2004–2014) and suggested techniques for management of mortality*. Helena, MT: Montana Department of Fish, Wildlife and Parks.

Cottrell, W. O., Keel, K. M., Brooks, J. W., Mead, D. G. & Phillips, J. E. (2013). First report of clinical disease associated with canine distemper virus infection in a wild black bear (*Ursus americana). Journal of Wildlife Diseases* 49: 1024–1027.

Craighead, F. C., Jr. & Craighead, J. J. (1972). Grizzly prehibernation and denning activities as determined by radiotracking. *Wildlife Monographs* 32: 1–35.

Craighead, J. J. & Mitchell, J. A. (1982). Grizzly bear. In: Chapman J. A. & Feldhamer, G. A. (Eds.), *Wild mammals of North America* (pp. 515–555). Baltimore, MD: Johns Hopkins University Press.

Craighead, J. J., Craighead, F. C. Jr. & Sumner, J. (1976). Reproductive cycles and rates in the grizzly bear, *Ursus arctos horribilis*, of the Yellowstone Ecosystem. *International Conference on Bear Research and Management* 3: 337–356.

Craighead, J. J., Sumner, J. S. & Mitchell, J. A. (1995). *The grizzly bears of Yellowstone: Their ecology in the Yellowstone ecosystem, 1959–1992*. Washington DC: Island Press.

Craighead, L., Paetkau, D., Reynolds, H. V., Vyse, E. R. & Strobeck, C. (1995). Microsatellite analysis of paternity and reproduction in Arctic grizzly bears. *Journal of Heredity* 86: 255–261.

Cristescu, B., Stenhouse, G. B. & Boyce, M. S. (2014). Grizzly bear ungulate consumption and the relevance of prey size to caching and meat sharing. *Animal Behaviour* 92: 133–142.

Cristescu, B., Stenhouse, G. B. & Boyce, M. S. (2015). Grizzly bear diet shifting in reclaimed mines. *Global Ecology and Conservation* 4: 207–220.

Cristescu, B., Stenhouse, G. B. & Boyce, M. S. (2016a). Large omnivore movements in response to surface mining and mine reclamation. *Scientific Reports* 6: 19177.

Cristescu, B., Stenhouse, G. B., Goski, B. & Boyce, M. S. (2016b). Grizzly bear space use, survival, and persistence in relation to human habitation and access. *Human-Wildlife Interactions* 10: 240–257.

Cross, P. C., van Manen, F. T., Viana, M., et al. (2018). Estimating distemper virus dynamics among wolves and grizzly bears using serology and Bayesian state-space models. *Ecology and Evolution* 8: 8726–8735.

Darling, L. M. (1987). Habitat use by grizzly bear family groups in interior Alaska. *International Conference on Bear Research and Management* 7: 169–178.

Davison, J., Ho, S. Y. W., Bray, S. C., et al. (2011). Late-Quaternary biogeographic scenarios for the brown bear (*Ursus arctos*), a wild mammal model species. *Quaternary Science Reviews* 30: 418–430.

Deacy, W., Leacock, W., Armstrong, J. B. & Stanford, J. A. (2016). Kodiak brown bears surf the salmon red wave: direct evidence from GPS collared individuals. *Ecology* 97: 1091–1098.

Deacy, W. W., Armstrong, J. B., Leacock, W. B., et al. (2017). Phenological synchronization disrupts trophic interactions between Kodiak brown bear and salmon. *Proceedings of the National Academy of Sciences* 114: 10432–10437.

Deacy, W. W., Leacock, W. Stanford J. A. & Armstrong J. B. (2019). Variation in spawning phenology within salmon populations influences landscape-level patterns of brown bear activity. *Ecosphere* 10: e02575.

Dittrich, L. & Kronberger, H. (1963). Biologisch-anatomische Untersuchungen über die Fortpflanzungsbiologie der Braunbären (*Ursus arctos* L.) und anderer Ursiden in Gefangenschaft. *Zeitschrift für SäugetierKunde* 28: 129–155.

Eberhardt, L. L. (1995). *Population trend estimates from reproductive and survival data*. In: Knight, L. L. & Blanchard, B. (Eds.), *Yellowstone grizzly bear investigations* (pp. 13–19). Bozeman, MT: Department of Interior, National Biological Service, Interagency Grizzly Bear Study Team.

Eberhardt, L. L. (2002). A paradigm for population analysis of long-lived vertebrates. *Ecology* 83: 2841–2854.

Ebinger, M. R., Haroldson, M. A., van Manen, F. T., et al. (2015). Detecting grizzly bear use of ungulate carcasses using global positioning system telemetry and activity data. *Oecologia* 181: 695–708.

Edwards, M. A., Nagy, J. A. & Derocher, A. E. (2009). Low site fidelity and home range drift in a wide-ranging, large Arctic omnivore. *Animal Behaviour* 77: 23–28.

Edwards, M. A., Derocher, A. E., Hobson, K. A., Branigan, M. & Nagy J. A. (2011). Fast carnivores and slow herbivores: differential foraging strategies among grizzly bears in the Canadian Artic. *Oecologia* 165: 877–889.

Edwards, M. A., Derocher, A. E. & Nagy, J. A. (2013). Home range size variation in female arctic grizzly bears relative to reproductive status and resource availability. *PLoS ONE* 8: e68130.

Egbert, A. L. & Stokes, A. W. (1976). The social behaviour of brown bears on an Alaskan salmon stream. *International Conference on Bear Research and Management* 3: 41–56.

Erickson, A. W., Moosman, H. W., Hensel, R. J. & Troyer, W. A. (1968). The breeding biology of the male brown bear (*Ursus arctos*). *Zoologica* 53: 85–105.

Erlenbach, J. A., Rode, K. D., Raubenheimer, D. & Robbins, C. T. (2014). Macronutrient optimization and energy maximization determine diets of brown bears. *Journal of Mammalogy* 95: 160–168.

Ewer, R. F. (1973). *The carnivores*. Ithaca, NY: Cornell University Press.

Fagen, R. & Fagen, J. (1990). Play behavior of brown bears (*Ursus arctos*) and human presence at Pack Creek, Admiralty Island, Alaska. *International Conference on Bear Research and Management* 8: 315–319.

Fagen, R. & Fagen, J. (2004). Juvenile survival and benefits of play behaviour in brown bears, *Ursus arctos*. *Evolutionary Ecology Research* 6: 89–102.

Fagen, R. & Fagen, J. (2009). Play behaviour and multi-year juvenile survival in free-ranging brown bears, *Ursus arctos*. *Evolutionary Ecology Research* 11: 1–15.

Felicetti, L. A., Robbins, C. T. & Shipley, L. A. (2003). Dietary protein content alters energy expenditure and composition of the mass gain in grizzly bears (*Ursus arctos horribilis*). *Physiological and Biochemical Zoology* 76: 256–261.

Felicetti, L. A., Schwartz, C. C., Rye, R. O., et al. (2004). Use of naturally occurring mercury to determine the importance of cutthroat trout to Yellowstone grizzly bears. *Canadian Journal of Zoology* 82: 493–501.

Ferguson, S. H. & McLoughlin, P. D. (2000). Effect of energy availability, seasonality, and geographic range on brown bear life history. *Ecography* 23: 193–200.

Folk, G. E., Jr., Folk, M. A. & Minor, J. J. (1972). Physiological condition of three species of bears in winter dens. *International Conference on Bear Research and Management* 2:107–124.

Ford, A. T., Rettie, K. & Clevenger, A. P. (2009). Fostering ecosystem function through an international public-private partnership: a case study of wildlife mitigation measures along the Trans-Canada Highway in Banff National Park, Alberta, Canada. *International Journal of Biodiversity Science and Management* 5: 181–189.

Fortin, J. K., Schwartz, C. C., Gunther, K. A., et al. (2013a). Dietary adjustability of grizzly bears and American black bears in Yellowstone National Park. *Journal of Wildlife Management* 77: 270–281.

Fortin, J. K., Ware, J. V., Jansen, H. T., Schwartz, C. C. & Robbins, C. T. (2013b). Temporal niche switching by grizzly bears but not American black bears in Yellowstone National Park. *Journal of Mammalogy* 94: 833–844.

Frechette, J.-L. & Rau, M. E. (1978). Seasonal changes in the prevalence of ova of *Diphyllobothrium ursi* and *Baylisascaris transfuga* in the feces of the black bear (*Ursus americanus*). *Journal of Wildlife Diseases* 14: 342–344.

French, S. P., French, M. G. & Knight, R. R. (1994). Grizzly bear use of army cutworm moths in the Yellowstone Ecosystem. *International Conference on Bear Research and Management* 9: 389–399.

Gardner, C. L., Pamperin, N. J. & Benson, J. F. (2014). Movement patterns and space use of maternal grizzly bears influence cub survival in Interior Alaska. *Ursus* 25: 121–138.

Garshelis, D. L., Gibeau, M. L. & Herrero, S. (2005). Grizzly bear demographics in and around Banff National Park and Kananaskis Country, Alberta. *Journal of Wildlife Management* 69: 277–297.

Gau, R. J. (1998). Food habitats, body condition, and habitat of the barren-ground grizzly bear. MS thesis, University of Alaska.

Gau, R. J., Kutz, S. & Elkin, B. T. (1999). Parasites in grizzly bears from the central Canadian Arctic. *Journal of Wildlife Diseases* 35: 618–621.

Gau, R. J., Case, R., Penner, D. F. & McLoughlin, P. D. (2002). Feeding patterns of barren-ground grizzly bears in the central Canadian Arctic. *Arctic* 55: 339–344.

Gau, R. J., McLoughlin, P. D., Case, R., et al. (2004). Movements of subadult male grizzly bears, *Ursus arctos*, in the Central Canadian Arctic. *Canadian Field Naturalist* 118: 239–242.

Gende, S. M. & Quinn, T. P. (2004). The relative importance of prey density and social dominance in determining energy intake by bears feeding on Pacific salmon. *Canadian Journal of Zoology* 82: 75–85.

Gibeau, M. L., Clevenger, A. P., Herrero, S. & Wierzchowski, J. (2002). Grizzly bear response to human developments and activities in the Bow River Watershed, Alberta, Canada. *Biological Conservation* 103: 227–236.

Glass, B. P. (1974). *A key to the skulls of North American mammals*, 2nd edition. Stillwater, OK: Author.

Glenn, L. P. & Miller, L. H. (1980). Seasonal movements of an Alaska Peninsula brown bear population. *International Conference for Bear Research and Management* 4: 307–312.

Goldstein, M. I., Poe, A. J., Suring, L. H., Nielson, R. M. & McDonald, T. L. (2010). Brown bear den habitat and winter recreation in south-central Alaska. *Journal of Wildlife Management* 74: 35–42.

Gordon, K. R. (1977). Molar measurements as a taxonomic tool in *Ursus*. *Journal of Mammalogy* 58: 247–248.

Graham, K. & Stenhouse, G. B. (2014). Home range, movements, and denning chronology of the grizzly bear (*Ursus arctos*) in West-central Alberta. *Canadian Field-Naturalist* 128: 223–234.

Graves, T. A., Farley, S. & Servheen, C. (2006). Frequency and distribution of highway crossings by Kenai Peninsula brown bears. *Wildlife Society Bulletin* 34: 800–808.

Guilday, J. E. (1968). Grizzly bears from eastern North America. *American Midland Naturalist* 79: 247–50.

Gunther, K. A. & Smith, D. W. (2004). Interactions between wolves and female grizzly bears with cubs in Yellowstone National Park. *Ursus* 15: 232–238.

Gunther, K. A., Shoemaker, R. R., Frey, K. L., et al. (2014). Dietary breadth of grizzly bears in the Greater Yellowstone Ecosystem. *Ursus* 25: 60–72.

Hall, E. R. (1984). *Geographic variation among brown and grizzly bears (*Ursus arctos*) in North America (Special Publication 13)*. Lawrence, KS: Museum of Natural History, University of Kansas.

Hamer, D. & Herrero, S. (1987). Wildfire's influence on grizzly bear feeding ecology in Banff National Park, Alberta. *International Conference on Bear Research and Management* 7: 179–186.

Hamer, D. & Herrero, S. (1990). Courtship and use of mating areas by grizzly bears in the Front Ranges of Banff National Park, Alberta. *Canadian Journal of Zoology* 68: 2695–2697.

Harding, L. E. (1976). Den-site characteristics of arctic coastal grizzly bears (*Ursus zrctos*) on Richards Island, Northwest Territories, Canada. *Canadian Journal of Zoology* 54: 1357–1363.

Harestad, A. S. & Bunnell, F. L. (1979). Home range and body weight – a reevaluation. *Ecology* 60: 389–402.

Haroldson, M. A., Ternent, M. A., Gunther, K. A. & Schwartz, C. C. (2002). Grizzly bear denning chronology and movements in the Greater Yellowstone Ecosystem. *Ursus* 13: 29–37.

Haroldson, M. A., Schwartz, C. C., Kendall, K. C., et al. (2010). Genetic analysis of individual origins supports isolation of grizzly bears in the Greater Yellowstone Ecosystem. *Ursus* 21: 1–13.

Haroldson, M. A., Gunther, K. A., Cain, S. L., Wilmot, K. R. & Wyman, T. (2015). Grizzly cub adoptions confirmed in Yellowstone and Grand Teton National Parks. *Yellowstone Science* 23: 58–61.

Hechtel, J. L. (1985). Activity and food habits of barren-ground grizzly bears in Arctic Alaska. MS thesis, University of Montana.

Herrero, S. (1972). Aspects of evolution and adaptation in American black bears (*Ursus americanus* Pallas) and brown and grizzly bears (*U. arctos* Linné.) of North America. *International Conference for Bear Research and Management* 2: 221–231.

Herrero, S., Smith, T., Debruyn, T. D., Gunther, K. A. & Matt, C. A. (2005). From the field: brown bear habituation to people – safety, risks, and benefits. *Wildlife Society Bulletin* 33: 362–373.

Hessing, P. & Aumiller, L. (1994). Observations of conspecific predation by brown bears, *Ursus arctos*, in Alaska. *Canadian Field-Naturalist* 108: 332–336.

Hilderbrand, G. V., Jenkins, S. G., Schwartz, C. C., Hanley, T. A. & Robbins, C. T. (1999a). Effect of seasonal differences in dietary meat intake on changes in body mass and composition in wild and captive brown bears. *Canadian Journal of Zoology* 77: 1623–1630.

Hilderbrand, G. V., Schwartz, C. C., Robbins, C. T., et al. (1999b). The importance of meat, particularly salmon, to body size, population productivity, and conservation of North America brown bears. *Canadian Journal of Zoology* 77: 132–138.

Hilderbrand, G. V., Schwartz, C. C., Robbins, C. T. & Hanley, T. A. (2000). Effect of hibernation and reproductive status on body mass and condition of coastal brown bears. *Journal of Wildlife Management* 64: 178–183.

Hovey, F. W. & McLellan, B. N. (1996). Estimating population growth of grizzly bears from the Flathead River drainage using computer simulations of reproduction and survival rates. *Canadian Journal of Zoology* 74: 1409–1416.

Interagency Grizzly Bear Study Team. (2012). *Updating and evaluating approaches to estimate population size and sustainable mortality limits for grizzly bears in the Greater Yellowstone Ecosystem*. Bozeman, MT: US Geological Survey, Northern Rocky Mountain Science Center, Interagency Grizzly Bear Study Team.

Jimenez, M. D., Asher, V. J., Bergman, C., Bangs, E. E. & Woodruff, S. P. (2008). Gray wolves, *Canis lupus*, killed by cougars, *Puma concolor*, and a grizzly bear, *Ursus arctos*, in Montana, Alberta, and Wyoming. *Canadian Field Naturlist* 122: 76–78.

Judd, S. L., Knight, R. R. & Blanchard, B. M. (1986). Denning of grizzly bears in the Yellowstone National Park area. *International Conference on Bear Research and Management* 6: 111–117.

Kasworm, W. F., Proctor, M. F., Servheen, C. & Paetkau, D. (2007). Success of grizzly bear population augmentation in northwest Montana. *Journal of Wildlife Management* 71: 1261–1266.

Kasworm, W. F., Radandt, T. G., Teisberg, J. E., et al. (2018). *Cabinet-Yaak grizzly bear recovery area 2017 research and monitoring progress report*. Missoula, MT: US Fish and Wildlife Service.

Keay, J. A., Robbins, C. T. & Farley, S. D. (2018). Characteristics of a naturally regulated grizzly bear population. *Journal of Wildlife Management* 82: 789–801.

Kendall, K. C., Stetz, J. B., Roon, D. A., et al. (2008). Grizzly bear density in Glacier National Park, Montana. *Journal of Wildlife Management* 72: 1693–1705.

Kendall, K. C., Macleod, A. C., Boyd, K. L., et al. (2015). Density, distribution, and genetic structure of grizzly bears in the Cabinet-Yaak Ecosystem. *Journal of Wildlife Management* 80: 314–331.

Kingsley, M. C. S., Nagy, J. A. & Russell, R. H. (1983). Patterns of weight gain and loss for grizzly bears in northern Canada. *International Conference for Bear Research and Management* 5: 174–178.

Kite, R., Nelson, T., Stenhouse, G. B. & Darimont, C. (2016). A movement-driven approach to quantifying grizzly bear (*Ursus arctos*) near-road movement patterns in west-central Alberta, Canada. *Biological Conservation* 195: 24–32.

Klinka, D. R. & Reimchen, T. E. (2002). Nocturnal and diurnal foraging behavior of brown bears (*Ursus arctos*) on a salmon stream in coast British Columbia. *Canadian Journal of Zoology* 80: 1317–1322.

Knight, R. R., Blanchard, B. & Kendall, K. (1981). *Yellowstone grizzly bear investigations: Annual report of the Interagency Study Team, 1980*. Bozeman, MT: US Geological Survey, Northern Rocky Mountain Science Center, Interagency Grizzly Bear Study Team.

Knowles, S., Bodenstein, B. L., Hamon, T., Saxton, M. W. & Hall, J. S. (2018). Infectious canine hepatitis in a brown bear (*Ursus arctos horribilis*) from Alaska. *Journal of Wildlife Diseases* 54: 642–645.

Koch, E. (1941). Big game in Montana from early historical records. *Journal of Wildlife Management* 5: 357–370.

Koene, P., Ardesch, J., Ludriks, A., et al. (2002). Interspecific and intraspecific social interactions among brown bears and wolves in an enclosure. *Ursus* 13: 85–93.

Kovach, S. D., Collins, G. H., Hinkes, M. T. & Denton, J. W. (2006). Reproduction and survival of brown bears in southwest Alaska, USA. *Ursus* 17: 16–29.

Kumar, V., Lammers, F., Bidon, T., et al. (2017). The evolutionary history of bears is characterized by gene flow across species. *Scientific Reports* 7: 46487.

Kurtén, B. (1973). Transberingian relationships of *Ursus arctos* Linne (brown and grizzly bears). *Commentation Biology* 65: 1–10.

Ladle, A., Avgar, T., Wheatley, M., et al. (2019). Grizzly bear response to spatio-temporal variability in human recreational activity. *Journal of Applied Ecology* 56: 375–386.

Lamb, C. T., Mowat, G., Reid, A., et al. (2018). Effects of habitat quality and access management on the density of a recovering grizzly bear population. *Journal of Applied Ecology* 55: 1406–1417.

LeFranc, M. N., Jr., Moss, M. B., Patnode, K. A. & Sugg, W. C. (1987). *Grizzly bear compendium*. Washington, DC: Interagency Grizzly Bear Committee.

Lewis, T. M. & Lafferty, D. J. R. (2014). Brown bears and wolves scavenge humpback whale carcass in Alaska. *Ursus* 25: 8–13.

Linnell, J. D. C., Swenson, J. E., Anderson, R. & Barnes, B. (2000). How vulnerable are denning bears to disturbance? *Wildlife Society Bulletin* 28: 400–413.

López-Alfaro, C., Coogan, S. C. P., Robbins, C. T., Fortin, J. K. & Nielsen, S. E. (2015). Assessing nutritional parameters of brown bear diets among ecosystems gives insight into differences among populations. *PLoS ONE* 10: e0128088.

Mace, R. D. & Waller J. S. (1997). *Grizzly bear ecology in the Swan Mountains*. Final Report. Helena, MT: Montana Department of Fish, Wildlife and Parks.

Mace, R. D. & Waller, J. S. (1998). Demography and population trend of grizzly bears in the Swan Mountains, Montana. *Conservation Biology* 12: 1005–1016.

Mace, R. D., Waller, J. S., Manley, T. L., Lyon, L. J. & Zuring, H. (1996). Relationships among grizzly bears, roads, and habitat use in the Swan Mountains, Montana. *Journal of Applied Ecology* 33: 1395–1404.

Mace, R. D., Carney, D. W., Chilton-Radandt, T., et al. (2012). Grizzly bear population vital rates and trend in the Northern Continental Divide Ecosystem, Montana. *Journal of Wildlife Management* 76: 119–128.

MacHutchon, A. G. & Wellwood, D. W. (2003). Grizzly bear food habits in the northern Yukon, Canada. *Ursus* 14: 225–235.

MacHutchon, A. G., Himmer, S. & Bryden, C. A. (1993). *Khutzeymateen Valley grizzly bear study*. Wildlife Habitat Research Report No. 31. Victoria: British Columbia Ministry of Forests.

MacHutchon, A. G., Himmer, S., Davis, H. & Gallagher, M. (1998). Temporal and spatial activity patterns among coastal bear populations. *Ursus* 10: 539–546.

MacPherson, A. H. (1965). The barren-ground grizzly bear and its survival in Canada. *Canadian Audubon* 27: 2–8.

Mangipane, L. S., Belant, J. L., Hiller, T. L., et al. (2018) Influences of landscape heterogeneity on home-range sizes of brown bears. *Mammalian Biology* 88: 1–7.

García Marín, J. F., Royo, L. J., Oleaga, A., et al. (2018). Canine adenovirus type 1 (CAdV-1) in free-ranging European brown bear (*Ursus arctos arctos*): a threat for Cantabrian population? *Transboundary and Emerging Diseases* 65: 2049–2056.

Matson, G. M., Van Daele, L., Goodwin, E., et al. (1993). *A laboratory manual for cementum age determination of Alaskan brown bear first premolar teeth*. Milltown, MT: Matson's Laboratory.

Mattson, D. (1997). Use of ungulates by Yellowstone grizzly bears *Ursus arctos*. *Biological Conservation* 81: 161–177.

Mattson, D. (2004). Exploitation of pocket gophers and their food caches by grizzly bears. *Journal of Mammalogy* 85: 731–742.

Mattson, D. J. & Merrill, T. (2002). Extirpations of grizzly bears in the contiguous United States, 1850–2000. *Conservation Biology* 16: 1123–1136.

Mattson, D. J., Knight, R. R. & Blanchard, B. M. (1992). Cannibalism and predation on black bears by grizzly bears in the Yellowstone Ecosystem, 1975–1990. *Journal of Mammalogy* 73: 422–425.

Mattson, D. J., Wright, R. G., Kendall, K. C. & Martinka, C. J. (1995). Grizzly bears. In: LaRoe, E. T., Farris, G. S., Puckett, C. E., Doran, P. D. & Mac, M. J. (Eds.), *Our living resources: A report to the nation on the distribution, abundance, and health of US plants, animals, and ecosystems* (pp. 103–105). Washington, DC: US Department of the Interior, National Biological Service.

McLellan, B. N. (1989). Dynamics of a grizzly bear population during a period of industrial resource extraction I. Density and age-sex composition. *Canadian Journal of Zoology* 67: 1857–1868.

McLellan, B. N. (1994). Density-dependent population regulation of brown bears. In Taylor, M. (Ed.), *Density-dependent population regulation in black, brown, and polar bears* (pp. 15–24). International Association for Bear Research and Management Monograph Series No. 3.

McLellan, B. N. (1998). Maintaining viability of brown bears along the southern fringe of their distribution. *Ursus* 10: 607–611.

McLellan, B. N. (2005). Sexually selected infanticide in grizzly bears: the effects of hunting on cub survival. *Ursus* 16: 141–156.

McLellan, B. N. (2011). Implications of high-energy and low-protein diet on the body composition, fitness, and competitive ability of black (*Ursus americanus*) and grizzly (*Ursus arctos*) bears. *Canadian Journal of Zoology* 89: 546–558.

McLellan, B. N. (2015). Some mechanisms underlying variation in vital rates of grizzly bear on a multiple-use landscape. *Journal of Wildlife Management* 79: 749–765.

McLellan, B. N. & Banci, V. (1999). Status and management of the brown bear in Canada. In: Servheen, C., Herrero, S. & Peyton, B. (Eds.), *Bears: Status survey and conservation action plan* (pp. 46–50). Gland: IUCN/SSC Bear and Polar Bear

Specialist Groups, International Union for the Conservation of Nature.

McLellan, B. N. & Hovey, F. W. (1995). The diet of grizzly bears in the Flathead River drainage of southeast British Columbia. *Canadian Zoology of Journal* 73: 704–712.

McLellan, B. N. & Hovey, F. W. (2001). Habitats selected by grizzly bears in a multiple use landscape. *Journal of Wildlife Management* 65: 92–99.

McLellan, B. N. & Reiner, D. C. (1994). A review of bear evolution. *Ursus* 9: 85–96.

McLellan, B. N., Hovey, F. W., Mace, R. D., et al. (1999). Rates and causes of grizzly bear mortality in the interior mountains of British Columbia, Alberta, Montana, Washington, and Idaho. *Journal of Wildlife Management* 63: 911–920.

McLellan, B. N., Mowat, G., Hamilton, T. & Hatter, I. (2017a). Sustainability of the grizzly bear hunt in British Columbia, Canada. *Journal of Wildlife Management* 81: 218–229.

McLellan, B. N., Proctor, M. F., Huber, Đ. & Michel, S. (2017b). *Ursus arctos*. IUCN Red List of Threatened Species, www.iucnredlist.org/details/41688/0 (accessed April 10, 2019).

McLellan, M. L. & McLellan, B. N. (2015). Effect of season and high ambient temperature on activity levels and patterns of grizzly bears (*Ursus arctos*). *PLoS ONE* 10: e0117734.

McLoughlin, P. D. (2000). The spatial organization and habitat selection patterns of barren-ground grizzly bears in the central Arctic. PhD dissertation, University of Saskatchewan.

McLoughlin, P. D., Case, R. L., Gau, R. J., Ferguson, S. H. & Messier, F. (1999). Annual and seasonal movement patterns of barren-ground grizzly bears in the central Northwest Territories. *Ursus* 11: 79–86.

McLoughlin, P. D., Ferguson, S. H. & Messier, F. (2000). Intraspecific variation in home range overlap with habitat quality: a comparison among brown bear populations. *Evolutionary Ecology* 14: 39–60.

McLoughlin, P. D., Case, R. L., Gau, R. J., et al. (2002a). Hierarchical habitat selection by barren-ground grizzly bears in the central Canadian arctic. *Oecologia* 132: 102–108.

McLoughlin, P. D., Cluff, H. D. & Messier, F. (2002b). Denning ecology of barren-ground grizzly bears in the central Arctic. *Journal of Mammalogy* 83: 188–198.

McLoughlin, P. D., Cluff, H. D., Gau, R. J., et al. (2003a). Effect of spatial differences in habitat on home ranges of grizzly bears. *Ecoscience* 10: 11–16.

McLoughlin, P. D., Taylor, M. K., Cluff, H. D., et al. (2003b). Demography of barren-ground grizzly bears. *Canadian Journal of Zoology* 81: 294–301.

McNab, B. K. (1963). Bioenergetics and the determination of home range size. *American Naturalist* 97: 133–140.

Merriam, C. H. (1918). Review of the grizzly and big brown bears of North America (genus *Ursus*) with the description of a new genus, *Vetularctos. North American Fauna* 41: 1–136.

Milakovic, B. & Parker, K. L. (2012). Quantifying carnivory by grizzly bears in a multi-ungulate system. *Journal of Wildlife Management* 77: 39–47.

Miller, C. R., Waits, L. P. & Joyce, P. (2006). Phylogeography and mitochondrial diversity of extirpated brown bear (*Ursus arctos*) populations in the contiguous United States and Mexico. *Molecular Ecology* 15: 4477–4485.

Miller, S. D. (1990). Denning ecology of brown bears in south-central Alaska and comparisons with sympatric black bear population. *International Conference for Bear Research and Management* 8: 279–287.

Miller, S. D. (1993). *Brown bears in Alaska: A statewide management overview.* Wildlife Technical Bulletin 11. Juneau, AK: Alaska Department of Fish and Game.

Miller, S. D. & Ballard, W. B. (1982). Homing of transplanted Alaskan brown bears. *Journal of Wildlife Management* 46: 869–876.

Miller, S. D. & Schoen, J. (1999). Status and management of the brown bear in Alaska. In: Servheen, C., Herrero, S. & Peyton, B. (Eds.), *Bears: Status survey and conservation action plan* (pp. 40–46). Gland: IUCN/SSC Bear and Polar Bear Specialist Groups, International Union for the Conservation of Nature.

Miller, S. D. & Sellers, R. A. (1992). *Brown bear density on the Alaska Peninsula at Black Lake, Alaska.* Final report on completion of density estimation objectives of cooperative interagency brown bear studies on the Alaska Peninsula. Juneau, AK: Alaska Department of Fish and Game.

Miller, S. D. & Tutterow, V. L. (1999). Characteristics of nonsport mortalities to brown and black bears and human injuries from bears in Alaska. *Ursus* 11: 239–250.

Miller, S. D., Sellers, R. A. & Keay, J. A. (2003). Effects of hunting on brown bear cub survival and litter size in Alaska. *Ursus* 14: 130–152.

Miller, S. D., Schoen, J. W., Faro, J. & Klein, D. R. (2011). Trends in intensive management of Alaska's grizzly bears, 1980–2010. *Journal of Wildlife Management* 75: 1243–1252.

Miller, S. D., McLellan, B. N. & Derocher, A. E. (2013). Conservation and management of large carnivores in North America. *International Journal of Environmental Studies* 70: 383–398.

Miller, S. D., Wilder, J. & Wilson, R. R. (2015). Polar bear-grizzly bear interactions during the autumn open-water period in Alaska. *Journal of Mammalogy* 96: 1317–1325.

Miller, S. D., Schoen, J. W. & Schwartz, C. C. (2017). Tends in brown bear reduction efforts in Alaska, 1980–2017. *Ursus* 28:135–149.

Milligan, S., Brown, L., Hobson, D., Frame, P. & Stenhouse, G. (2018). Factors affecting the success of grizzly bear translocations. *Journal of Wildlife Management* 82: 519–530

Modafferi, R. D. (1984). *Review of Alaska Peninsula brown bear investigations.* Final Report Federal Aid in Wildlife Restoration Projects W-17-10, W-17-11, W-21-1, W-21-2, W-22-1. Juneau, AK: Alaska Department of Fish and Game.

Morehouse, A. T., Graves, T. A., Mikle, N. & Boyce, M. S. (2016). Nature vs. nurture: evidence for social learning of conflict behaviour in grizzly bears. *PLoS ONE* 11: e0165425.

Mowat, G. & Heard, D. C. (2006). Major components of grizzly bear diet across North America. *Canadian Journal of Zoology* 84: 473–489.

Mowat, G., Heard, D. C. & Schwarz, C. J. (2013). Predicting grizzly bear density in western North America. *PLoS ONE* 8: e82757.

Munro, R. H. M., Nielsen, S. E., Price, M. H., Stenhouse, G. B. & Boyce, M. S. (2006). Seasonal and diel patterns of grizzly bear diet and activity in west-central Alberta. *Journal of Mammalogy* 87: 1112–1121.

Nagy, J. A. & Haroldson, M. A. (1989). Comparison of some home range and population parameters among four grizzly bear populations in Canada. *International Conference for Bear Research and Management* 8: 227–235.

Nagy, J. A. & Russell, R. H. (1978). *Ecological studies of the boreal forest grizzly bear* Ursus arctos*: annual report for 1977.* Edmonton: Canadian Wildlife Service.

Nagy, J. A., Russell, R. H., Pearson, A. M., Kingsley, M. C. & Goski, B. C. (1983). *Ecological studies of the grizzly bear in arctic mountains, northern Yukon Territories, 1972 to 1975.* Edmonton: Canadian Wildlife Service.

Nelson, R. A. (1980). Protein and fat metabolism in hibernating bears. *Federation Proceedings* 39: 2955–2958.

Nelson, R. A., Folk, G. E., Jr., Pfeiffer, W. W., et al. (1983). Behavior, biochemistry, and hibernation in black, grizzly, and polar bears. *Proceedings Eastern Workshop on Black Bear Research and Management* 7: 48–53.

Nevin, O. T. & Gilbert, B. K. (2005). Measuring the cost of risk avoidance in brown bears: further evidence of positive impacts of ecotourism. *Biological Conservation* 121: 611–622.

Nielsen, S. E., Boyce, M. S., Stenhouse, G. B. & Munro, R. H. M. (2003). Development and testing of phenologically driven grizzly bear habitat models. *Ecoscience* 10: 1–10.

Nielsen, S. E., Herrero, S., Boyce, M. S., et al. (2004). Modeling the spatial distribution of human-caused grizzly bear mortalities in the Central Rockies Ecosystem of Canada. *Biological Conservation* 120: 101–113.

Nielsen, S. E., Larsen, T. A., Stenhouse, G. B. & Coogan, S. C .P. (2017). Complementary food resources of carnivory and frugivory affect local abundance of an omnivorous carnivore. *Oikos* 126: 369–380.

Northrup, J. M., Pitt, J., Muhly, T. B., et al. (2012a). Vehicle traffic shapes grizzly bear behaviour on a multiple-use landscape. *Journal of Applied Ecology* 49: 1159–1167.

Northrup, J. M., Stenhouse, G. B. & Boyce, M. S. (2012b). Agricultural lands as ecological traps for grizzly bears. *Animal Conservation* 15: 369–377.

Oftedal, O. T. & Gittleman, J. L. (1989). Patterns of energy output during reproduction in carnivores. In: Gittleman J. L. (Ed.), *Carnivore behavior, ecology, and evolution* (pp. 355–378). Boston, MA: Springer.

Paetkau, D., Shields, G. F. & Strobeck, C. (1998a). Gene flow between insular, coastal and interior populations of brown bears in Alaska. *Molecular Ecology* 7: 1283–1292.

Paetkau, D., Waits, L. P., Clarkson, P. L., et al. (1998b). Variation in genetic diversity across the range of North American brown bears. *Conservation Biology* 12: 418–429.

Palmer, K. (2015). Oldest brown bear in captivity in North America dies at Ohio zoo. Thomson Reuters. www.reuters.com/article/us-usa-ohio-bear/oldest-brown-bear-in-captivity-in-north-america-dies-at-ohio-zoo-idUSKCN0RB1UG20150911 (accessed May 29, 2019).

Pasitschniak-Arts, M. (1993). *Ursus arctos. Mammalian Species* 439: 1–10.

Pearson, A. M. (1975). *The northern interior grizzly bear* Ursus arctos *L.* Report Series No. 34. Ottawa: Canadian Wildlife Service.

Peck, C. P., van Manen, F. T., Costello, C. M., et al. (2017). Potential paths for male-mediated gene flow to and from an isolated grizzly bear population. *Ecosphere* 8: e01969.

Peterson, R. L. (1965). A well-preserved grizzly bear skull recovered from a late glacial deposit near Lake Simcoe, Ontario. *Nature* 208: 1233–1234.

Pigeon, K. E., Nielsen, S. E., Stenhouse, G. B. & Coté, S. D. (2014). Den selection by grizzly bears on a managed landscape. *Journal of Mammalogy* 95: 559–571.

Pigeon, K. E., Coté, S. D. & Stenhouse, G. B. (2016a). Assessing den selection and den characteristics of grizzly bears. *Journal of Wildlife Management* 80: 884–893.

Pigeon, K. E., Stenhouse, G. B. & Coté, S.D. (2016b). Drivers of hibernation: linking food and weather to denning behaviour of grizzly bears. *Behavioral Ecology and Sociobiology* 70: 1745–1754.

Podruzny, S. R., Cherry, S., Schwartz, C. C. & Landenburger, L. A. (2002). Grizzly bear denning and potential conflict areas in the Greater Yellowstone Ecosystem. *Ursus* 13: 19–28.

Proctor, M. F. & MacHutchon, G. (2019). *Grizzly bear management and habitat use in the agricultural Creston Valley of southeast British Columbia and a vision for long-term management. Trans-border Grizzly Bear Project.* Kaslo: Trans-border Grizzly Bear Project.

Proctor, M. F., McLellan, B. N., Strobeck, C. & Barclay, R. M. (2004). Gender-specific dispersal distance of grizzly bears estimated by genetic analysis. *Canadian Journal of Zoology* 82: 1108–1118.

Proctor, M. F., Paetkau, D., McLellan, B. N., et al. (2012). Population fragmentation and inter-ecosystem movements of grizzly bears in Western Canada and the Northern United States. *Wildlife Monographs* 180: 1–46.

Proctor, M. F., Nielsen, S. E., Kasworm, W. F., et al. (2015). Grizzly bear connectivity mapping in the Canada–US trans-border region. *Journal of Wildlife Management* 79: 544–558.

Proctor, M. F., Lamb C. T. & MacHutchon, A. G. (2017). *The grizzly dance of berries and bullets: The relationship between bottom up food resources, huckleberries, and top down mortality risk on grizzly bear population processes in southeast British Columbia.* Kaslo: Trans-border Grizzly Bear Project.

Proctor, M. F., Kasworm, W. F., Annis, K. M., et al. (2018a). Conservation of threatened Canada-USA trans-border grizzly bears linked to comprehensive conflict reduction. *Human–Wildlife Interactions* 12: 348–372.

Proctor, M. F., McLellan, B. N., Stenhouse, G. B., et al. (2018b). *Resource roads and grizzly bears in British Columbia and Alberta. Canadian Grizzly Bear Management Series, Resource Road Management.* Kaslo: Trans-border Grizzly Bear Project.

Quimby, R. & Snarski, D. J. (1974). A study of fur-bearing mammals associated with gas pipeline routes in Alaska. In: Jakimchuk, R. D. (Ed.), *Distribution of moose, sheep, muskox and fur-bearing mammals in northeastern Alaska* (pp. 11–36). Arctic Gas Biological Report Series, Volume 6. Canadian Arctic Gas Study Ltd. and Alaskan Arctic Gas Study Company.

Quinn, T. P., Gende, S. M., Ruggerone, G. T. & Rogers, D. E. (2003). Density-dependent predation by brown bears (*Ursus arctos*) on sockeye salmon (*Oncorhynchus nerka*). *Canadian Journal of Fisheries and Aquatic Sciences* 60: 553–562.

Rausch, R. L. (1953). On the status of some arctic mammals. *Arctic* 6: 91–148.

Rausch, R. L. (1954). Studies on the helminth fauna of Alaska. XXI. Taxonomy, morphological variation, and ecology of *Diphyllobothrium ursi* n. sp. provis. on Kodiak Island. *Journal of Parasitology* 40: 540–563.

Rausch, R. L. (1963). Geographic variation in size of North American brown bears, *Ursus arctos* L., as indicated by condylobasal length. *Canadian Journal of Zoology* 41: 33–45.

Rausch, R. L. (1969). Morphogenesis and age-related structure of permanent canine teeth in the brown bear, *Ursus arctos* L., in Arctic Alaska. *Zeitschrift für Morphologie der Tiere* 66: 167–188.

Reynolds, H. V. (1976). *North Slope grizzly bear studies.* Federal Aid in Wildlife Restoration Final Report for Projects W-17-6 and W-17-7, Jobs 4.8R, 4.9R, 4.10R, and 4.11R. Juneau, AK: Alaska Department of Fish and Game.

Reynolds, H. V., Curatolo, J.A. & Quimby, R. (1976). Denning ecology of grizzly bears in Northeastern Alaska. *International Conference on Bear Research and Management* 3: 403–409.

Reynolds, P. E., Reynolds, H. V. & Shideler, R. T. (2002). Predation and multiple kills of muskoxen by grizzly bears. *Ursus* 13: 79–84.

Riley, S. J., Aune, K., Mace, R. D. & Madel, M. J. (2004). Translocation of nuisance grizzly bears in northwestern Montana. *International Conference on Bear Research and Management* 9: 567–573.

Ripple, W. J., Miller, S. D., Schoen, J. W. & Rabinowitch, S. P. (2019). Large carnivores under assault in Alaska. *PLoS Biology* 17: e3000090.

Robbins, C. T., Schwartz, C. C. & Felicetti, L. A. (2004). Nutritional ecology of ursids: a review of newer methods and management implications. *Ursus* 15: 161–171.

Robbins, C. T., Fortin, J. K., Rode, K. D., Shipley, L. A. & Felicetti, L. A. (2007). Optimizing protein intake as a foraging strategy to maximize mass gain in an omnivore. *Oikos* 116: 1675–1682.

Robbins, C. T., Ben-David, M., Fortin, J. K. & Nelson, O. L. (2012). Maternal condition determines birth date and growth of newborn bear cubs. *Journal of Mammalogy* 93: 540–546.

Robbins, C. T., Woodford, N. L., Clyde, G. G., et al. (2018). Salmon poisoning disease in grizzly bears with population recovery implications. *Journal of Wildlife Management* 82:1396–1402.

Rockwell, D. (1991). *Giving voice to bear.* Niwot, CO: Roberts Rinehart publishers.

Rode, K. D. & Robbins, C. T. (2000). Why bears consume mixed diets during fruit abundance. *Canadian Journal of Zoology* 78: 1640–1645.

Rode, K. D., Robbins, C. T. & Shipley, L. A. (2001). Constraints on herbivory by grizzly bears. *Oecologia* 128: 62–71.

Rode, K. D., Farley, S. D., Fortin, J. & Robbins, C. T. (2006a). Nutritional consequences of experimentally introduced tourism in brown bears. *Journal of Wildlife Management* 71: 929–939.

Rode, K. D., Farley, S. D. & Robbins, C. T. (2006b). Sexual dimorphism, reproductive strategy, and human activities determine resource use by brown bears. *Ecology* 87: 2636–2646.

Rode, K. D., Farley, S. D. & Robbins, C. T. (2006c). Behavioral responses of brown bears mediate nutritional effects of experimentally introduced tourism. *Biological Conservation* 133: 70–80.

Rogers, L. L. & Rogers, S. M. (1976). Parasites of bears: a review. *International Conference on Bear Research and Management* 3: 411–430.

Rush, W. M. (1932). *Diphyllobothrium latum* in bear. *Journal of Mammalogy* 13: 274–275.

Sawaya, M. A., Clevenger, A. P. & Kalinowski, S. T. (2013). Wildlife crossing structures in Banff National Park. *Conservation Biology* 27: 721–730.

Schoen, J. W., Beier, L. R., Lentfer, J. W. & Johnson, L. J. (1987). Denning ecology of brown bears on Admiralty and Chichagof Islands. *International Conference on Bear Research and Management* 7: 293–304.

Schoen, J. W., Flynn, R. W., Suring, L. H., Titus, K. & Beier, L. R. (1994). Habitat-capability model for brown bear in southeast Alaska. *International Conference on Bear Research and Management* 9: 317–337.

Schullery, P. (2002). *Lewis and Clark among the grizzlies, legend and legacy in the American West.* Guildford, CT: The Globe Pequot Press.

Schwartz, C. C., Keating, K. A., Reynolds, III, H. V., et al. (2003a). Reproductive maturation and senescence in the female brown bear. *Ursus* 14: 109–119.

Schwartz, C. C., Miller, S. D. & Haroldson, M. A. (2003b). Grizzly bear. In: Feldhamer, G. A., Thompson, B. C. & Chapman, J. A. (Eds.), *Wild mammals of North America: Biology, management, and conservation* (2nd edition, pp. 556–586). Baltimore, MD: Johns Hopkins University Press.

Schwartz, C. C., Haroldson, M. A., White, G. C., et al. (2006). Temporal, spatial, and environmental influences on the demographics of grizzly bears in the Greater Yellowstone Ecosystem. *Wildlife Monographs* 161: 1–68.

Schwartz, C. C., Cain, S. L., Podruzny, S. R., Cherry, S. & Frattaroli, L. (2010a). Contrasting activity patterns of sympatric and allopatric black and grizzly bears. *Journal of Wildlife Management* 74: 1628–1638.

Schwartz, C. C., Haroldson, M. A. & White, G. C. (2010b). Hazards affecting grizzly bear survival in the greater Yellowstone ecosystem. *Journal of Wildlife Management* 74: 654–667.

Schwartz, C. C., Fortin, J. K., Teisberg, J. E., et al. (2014). Body and diet composition of sympatric black and grizzly bears in the Greater Yellowstone Ecosystem. *Journal of Wildlife Management* 78: 68–78.

Sellers, R. A. & Aumiller, L. D. (1994). Brown bear population characteristics at McNeil River, Alaska. *International Conference on Bear Research and Management* 9: 283–293.

Servheen, C. (1983). Grizzly bear food habits, movements and habitat selection in the Mission Mountains, Montana. *Journal of Wildlife Management* 47: 1026–1035.

Servheen, C. (1999). Status and management of the grizzly bear in the lower 48 United States. In: Servheen, C., Herrero, S. & Peyton, B. (Eds.), *Bears: Status survey and conservation action plan* (pp. 50–54). Gland: IUCN/SSC Bear and Polar Bear Specialist Groups, International Union for the Conservation of Nature.

Servheen, C. & Klaver, R. (1983). Grizzly bear dens and denning activity in the Mission and Rattlesnake Mountains, Montana. *International Conference on Bear Research and Management* 5: 201–207.

Servheen, C., Kasworm, W. & Christensen, A. (1987). Approaches to augmenting grizzly bear populations in the Cabinet Mountains of Montana. *International Conference on Bear Research and Management* 7: 363–367.

Shields, G. F., Adams, D., Garner, G., et al. (2000). Phylogeography of mitochondrial DNA variation in brown bears and polar bears. *Molecular Phylogenetics and Evolution* 15: 319–326.

Shores, C. R., Mikle, N. & Graves, T. A. (2019). Mapping a keystone shrub species, huckleberry (*Vaccinium membranaceum*), using seasonal colour change in the Rocky Mountains. *International Journal of Remote Sensing* 40: 5695–5715.

Sibly, R. M. & Brown, J. H. (2009). Mammal reproductive strategies driven by offspring mortality–size relationships. *American Naturalist* 173: E185–E199.

Smith, T. S. & Partridge, S. T. (2004). Dynamics of intertidal foraging by coastal brown bears in Southwestern Alaska. *Journal of Wildlife Management* 68: 233–240.

Stebler, A. M. (1972). Conservation of the grizzly: ecologic and cultural considerations. *International Conference on Bear Research and Management* 2: 297–303.

Stenhouse, G. B., Boulanger, J., Lee, J., et al. (2005). Grizzly bear associations along the eastern slopes of Alberta. *Ursus* 16: 31–40.

Stewart, B. P., Nelson, T. A., Laberee, K., et al. (2013). Quantifying grizzly bear selection of natural and anthropogenic edges. *Journal of Wildlife Management* 77: 957–964.

Steyaert, S. M. J. G., Endrestøl, A., Hackländer, K., Swenson, J. E. & Zedrosser, A. (2012). The mating system of the brown bear *Ursus arctos*. *Mammal Review* 42: 12–34.

Stirling, I. & Derocher, A. E. (1990). Factors affecting the evolution and behavioral ecology of the modern bears. *International Conference on Bear Research and Management* 8: 189–204.

Stonorov, D. & Stokes, A. W. (1972). Social behavior of the Alaska brown bear. *International Conference on Bear Research and Management* 2: 232–242.

Storer, T. I. & Tevis, L. P. (1955). *California grizzly*. Berkeley, CA: University of California Press.

Stringham, S. F. (1990). Grizzly bear reproductive rate relative to body size. *International Conference for Bear Research and Management* 8: 433–443.

Swenson, J. E. & Haroldson, M. A. (2008). Recent observations of mixed-age litters in brown bears. *Ursus* 19: 73–79.

Swenson, J. E., Dahle, B. & Sandegren, F. (2001a). Intraspecific predation in Scandinavian brown bears older than cubs-of-the-year. *Ursus* 12: 81–92.

Swenson, J. E., Sandegren, F., Brunberg, S. & Segerström, P. (2001b). Factors associated with loss of brown bear cubs in Sweden. *Ursus* 12: 69–80.

Talbot, S. L. & Shields, G. F. (1996). A phylogeny of the bears (Ursidae) inferred from complete sequences of three mitochondrial genes. *Molecular Phylogenetics and Evolution* 5: 567–575.

Tardiff, S. E. & Stanford, J. A. (1998). Grizzly bear digging: effects on subalpine meadow plants in relation to mineral nitrogen availability. *Ecology* 79: 2219–2228.

Troyer, W. A. & Hensel, R. J. (1969). *The brown bear of Kodiak Island*. Kodiak, AK: US Department of the Interior, Bureau of Sport Fish and Wildlife, Branch of Wildlife Refuges.

US Department of the Interior. (2017). Draft environmental impact statement/grizzly bear restoration plan, North Cascades Ecosystem, Washington. Federal Register 82 FR 4149–4768.

US Fish and Wildlife Service. (1993). *Grizzly bear recovery plan*. Missoula, MT: US Fish and Wildlife Service, Office of the Grizzly Bear Recovery Coordinator.

US Fish and Wildlife Service. (2000). *Record of decision and statement of findings for the Environmental Impact Statement on grizzly bear recovery in the Bitterroot Ecosystem and final rule on establishment of a nonessential experimental population of grizzly bears in the Bitterroot area of Idaho and Montana*. Missoula, MT: US Fish and Wildlife Service, Office of the Grizzly Bear Recovery Coordinator.

Van Daele, L. J., Barnes, V. G. & Smith, R. B. (1990). Denning characteristics of brown bears on Kodiak Island, Alaska. *International Conference on Bear Research and Management* 8: 257–267.

Van Daele, M. B., Robbins, C. T., Semmens, B. X., et al. (2013). Salmon consumption by Kodiak brown bears (*Ursus arctos middendorffi*) with ecosystem management implications. *Canadian Journal of Zoology* 91: 164–174.

van Manen, F. T. & Haroldson, M. A. (2017). Reproduction, survival, and population growth. In: White, P. J., Gunther, K. A. & van Manen, F. T. (Eds.), *Yellowstone grizzly bears: Ecology and conservation of an icon of wildness* (pp. 29–45). Yellowstone National Park, WY: Yellowstone Forever.

van Manen, F. T., Haroldson, M. A., Bjornlie, D. D., et al. (2016). Density dependence, whitebark pine decline, and changing vital rates of Yellowstone grizzly bears. *Journal of Wildlife Management* 80: 300–313.

Vroom, G. W., Herrero, S. & Ogilvie, R. T. (1980). The ecology of winter den sites of grizzly bears in Banff National Park, Alberta. *International Conference on Bear Research and Management* 4: 321–330.

Waits, L. P., Talbot, S. L., Ward, R. H. & Shields, G. F. (1998). Mitochondrial DNA phylogeography of the North American brown bears and implications for conservation. *Conservation Biology* 12: 408–417.

Waits, L. P., Paetkau, D. & Strobeck, C. (1999). Genetics of the bears of the world. In: Servheen, C., Herrero, S. & Peyton, B. (Eds.), *Bears: Status survey and conservation action plan* (pp. 25–32). Gland: IUCN/SSC Bear and Polar Bear Specialist Groups, International Union for the Conservation of Nature.

Wakkinen, W. L. & Kasworm, W. F. (2004). Demographics and population trends of grizzly bears in the Cabinet-Yaak and Selkirk Ecosystems of British Columbia, Idaho, Montana, and Washington. *Ursus* 15: 65–75.

Waller, J. S. & Mace, R. D. (1997). Grizzly bear habitat selection in the Swan Mountains, Montana. *Journal of Wildlife Management* 61: 1032–1039.

White, D. Jr., Berardinelli, J. G. & Aune, K. E. (1998a). Reproductive characteristics of the male grizzly bear in the continental United States. *Ursus* 10: 497–501.

White, D., Jr., Kendall, K. C. & Picton, H. D. (1998b). Grizzly bear feeding activity at alpine army cutworm moth aggregation sites in northwest Montana. *Canadian Journal of Zoology* 76: 221–227.

White, D., Jr., Kendall, K. C. & Picton, H. D. (1998c). Seasonal occurrence, body composition, and migration potential of army cutworm moths in northwest Montana. *Canadian Journal of Zoology* 76: 835–842.

Wielgus, R. B. & Bunnell, F. L. (1994a). Sexual segregation and female grizzly bear avoidance of males. *Journal of Wildlife Management* 58: 405–413.

Wielgus, R. B. & Bunnell, F. L. (1994b). Dynamics of a small, hunted brown bear

Ursus arctos population in southwestern Alberta, Canada. *Biological Conservation* 67: 161–166.

Wielgus, R. B., Bunnell, F. L., Wakkinen, W. L. & Zager, P. E. (1994). Population dynamics of Selkirk Mountain grizzly bears. *Journal of Wildlife Management* 58: 266–272.

Wielgus, R. B., Bunnell, F. L. & Press, A. (1995). Tests of hypotheses for sexual segregation in grizzly bears. *Journal of Wildlife Management* 59: 552–560.

Wilk, R. J., Solberg, J. W., Berns, V. D. & Sellers, R. A. (1988). Brown bear (*Ursus arctos*) with six young. *The Canadian Field-Naturalist* 102: 541–543.

Wilson, S. M., Madel, M. J., Mattson, D. J., Graham, J. M. & Merrill, T. (2006). Landscape conditions predisposing grizzly bears to conflicts on private agricultural lands in the western USA. *Biological Conservation* 130: 47–59.

Wirsing, A. J., Quinn, T. P., Cunningham, C. J., et al. (2018) Alaskan brown bears (*Ursus* arctos) aggregate and display fidelity to foraging neighborhoods while preying on Pacific salmon along small streams. *Ecology and Evolution* 8: 9048–9061.

Young, D. D. & McCabe, T. R. (1997). Grizzly bear predation rates on caribou calves in northeastern Alaska. *Journal of Wildlife Management* 61: 1056–1066.

Zager, P. & Beecham, J. (2006). The role of American black bears and brown bears as predators on ungulates in North America. *Ursus* 17: 95–109.

Zarnke, R. L. & Evans, M. B. (1989). Serologic survey for infectious canine hepatitis virus in grizzly bears (*Ursus arctos*) from Alaska, 1973 to 1987. *Journal of Wildlife Diseases* 25: 568–573.

Zavatsky, B. P. (1976). The use of the skull in age determination of the brown bear. *International Conference for Bear Research and Management* 4: 275–279.

Chapter 14

Polar Bear (*Ursus maritimus*)

Karyn D. Rode, Martyn Obbard, Stanislav E. Belikov, Andrew E. Derocher, George M. Durner, Gregory W. Thiemann, Morten Tryland, Robert J. Letcher, Randi Meyerson, Christian Sonne, Bjørn M. Jenssen, Rune Dietz, and Dag Vongraven

Names: polar bear, sea bear, ice bear

Genus: *Ursus*

Species: polar bear *Ursus maritimus* (Phipps, 1774)

Names in other languages: Norwegian and Danish: isbjørn; **Inuit:** Nanuq/Nanook/Nanoq; **Russian:** Полярный медведь (beliy medved); **French:** ours polaire, ours blanc; **Italian:** orso polare; **German:** Eisbär; **Spanish:** oso polar.

Subspecies: none.

Taxonomy

Polar bears belong to the order Carnivora, family Ursidae, genus *Ursus*. They are estimated to have diverged from brown bears (*Ursus arctos*) between 0.6 and 1.1 million years ago (Kumar et al. 2017). Successful first-generation hybrids with brown bears and a backcrossing of a hybrid with a brown bear have been documented (Pongracz et al. 2017). There have been multiple instances of introgressive hybridization of polar bear DNA into brown bears (Miller et al. 2012). The oldest known fossil is from Svalbard and was between 100,000 and 130,000 years old (Ingolfsson & Wiig 2009). A recent study concluded that the 19 identified polar bear subpopulations exhibit six groupings based on genetic structure: the Hudson Bay complex, the Western and Eastern Canadian Arctic Archipelago, the Western and Eastern Polar Basin, and Norwegian Bay (Malenfant et al. 2016).

Subspecies and Distribution

Polar bears are a marine species endemic to the circumpolar Arctic (Figure 14.1). There are no recognized subspecies. They occur across sea-ice covered and coastal areas as far south as James Bay, Labrador, Newfoundland in Canada, southern Greenland, and the northern Bering Sea between Alaska and Russia. The International Union for the Conservation of

Figure 14.1 Adult male polar bear following a pressure ridge on spring sea ice in the Chukchi Sea (photo credit US Fish and Wildlife Service).

Nature (IUCN)'s Polar Bear Specialist Group (PBSG) recognizes 19 subpopulations across their range based primarily on distribution patterns and management jurisdictions (Durner et al. 2018; see Figure 21.2). Although environmental and ecological conditions can vary, all subpopulations share a dependence on sea ice as a platform for hunting seals, mating, resting, and traveling (Stirling & Archibald 1977; Mauritzen et al. 2003; Stirling et al. 2011).

Sea ice extent varies seasonally and affects the annual distribution patterns of polar bears. In the late summer and autumn, sea ice extent declines, reaching an annual minimum in September (see Figure 21.1, Chapter 21). In March, sea ice extent is at its annual maximum. Spatiotemporal variation in sea ice distribution was the basis for categorizing subpopulations into four ecoregions (Figure 21.1; Amstrup et al. 2008): (1) divergent ice (DI) – composed predominantly of drifting annual sea ice that forms over Eurasian and North American Arctic continental shelves, and multiyear sea ice of the Arctic Basin. Ice over the continental shelf melts during the summer or is advected into the Arctic Basin, resulting in the retreat of sea ice off coastal areas of Russia and the United States during the summer; (2) convergent ice (CI) – which includes the northern coasts of the Canadian Archipelago, and northern and eastern Greenland. It is composed predominantly of thick, multiyear sea ice transported from the DI, *in situ* multiyear ice, and *in situ* annual ice. Sea ice generally remains throughout the year; (3) Canadian Archipelago (CA; "Archipelago") – which includes channels among the Canadian Arctic islands. Sea ice generally remains throughout the year. The CA is composed of a high proportion of thick, multiyear ice and thick annual ice; (4) seasonal ice (SI) – which includes waters of Hudson Bay and Foxe Basin in Canada, and Baffin Bay and Davis Strait between Canada and Greenland. This sea ice is annual as it melts completely each summer. Differences in seasonality of sea ice among these ecoregions result in different seasonal polar bear distribution patterns.

In the SI ecoregion where ice melts completely during the summer, nearly all bears within the subpopulations are forced to spend several months on land relying primarily on accumulated fat reserves (Atkinson & Ramsay 1995; Atkinson et al. 1996a), but consuming some terrestrial foods (Gormezano & Rockwell 2013). In the DI ecoregion polar bears either remain with the sea ice as it retreats northward into deep, Arctic Basin waters or they summer onshore (Schliebe et al. 2008; Rode et al. 2015a; Laidre et al. 2015a; Atwood et al. 2016). In the CA, polar bears congregate on land-fast ice during the summer months when sea ice extent is at its minimum (Ferguson et al. 2000). Bears in the CI tend to spend the majority of the year on the sea ice (Wiig et al. 2003).

Polar bear distribution on sea ice is driven primarily by the density and accessibility of their main prey species, ringed seal (*Pusa hispida*) and bearded seal (*Erignathus barbatus*; Smith & Stirling 1978; Stirling et al. 1982; Kingsley et al. 1985). Ocean productivity, and consequently seal densities, tend to be greater on sea ice over water less than 300 m deep, resulting in greater polar bear use of these habitats (Durner et al. 2009; Stirling et al. 2011) with some exceptions. As sea ice retreats in the late summer, polar bears often concentrate on ice edges (Ferguson et al. 2000; Durner et al. 2009; Wilson et al. 2014). Polar bears are also common in areas of glacier fronts, where they occur, due to high productivity that concentrates ringed seals (e.g. Svalbard and Greenland; Freitas et al. 2012).

Denning by pregnant females during the winter months occurs primarily on land throughout most of the polar bear's range (Kolenosky & Prevett 1983; Messier et al. 1994; Stishov 1997; Lunn et al. 2004; Richardson et al. 2005; Fischbach et al. 2007; Andersen et al. 2012; Rode et al. 2015a), although denning on sea ice also occurs in the Beaufort Sea (Fischbach et al. 2007). Females show strong fidelity to areas suitable for denning (Ramsay & Stirling 1990; Zeyl et al. 2010). The distribution of maternal dens is largely determined by ice conditions along coastal areas (Andersen et al. 2012; Rode et al. 2015a) and the availability of snow in which to dig their dens. Dens are usually located within 15–30 km of the coast (Ramsay & Stirling 1990; Andersen et al. 2012), although bears in Hudson Bay and Baffin Bay have been known to travel up to 100 km inland to den (Ramsay & Stirling 1986; Richardson et al. 2005; Escajeda et al. 2018). Declines in sea ice extent and changes in spatiotemporal distribution of ice have been associated with changes in the timing and spatial locations of dens (Derocher et al. 2011; Laidre et al. 2015a; Escajeda et al. 2018), including increased land-based denning where sea ice-based denning once predominated (Fischbach et al. 2007; Olson et al. 2017). On emerging from dens in February–March, family groups spend up to two weeks in the den vicinity before moving onto the sea ice (Smith et al. 2007). Timing of emergence appears to be related to the timing of pupping by ice-obligate seals (Smith & Stirling 1975).

Sea ice extent declined 2.2% per decade between 1979 and 1996 and more recently has declined by 10.1% per decade (Comiso et al. 2008; see Figure 21.1, Chapter 21), impacting polar bear distribution (Stirling et al. 2004; Laidre et al. 2015a; Ware et al. 2017). In addition to declines in extent, there has been a 15.1% per decade decline in the extent of multiyear ice (Comiso 2012) and a decline in the mean winter ice thickness from 3.6 to 1.9 m between 1980 and 2008 (Kwok & Rothrock 2009; Stern & Laidre 2015). The cumulative result of these changes is that the ice cover has become a seasonal phenomenon over a large part of the Arctic Ocean and the availability of preferred polar bear habitats has declined in many areas (Durner et al. 2009; Laidre et al. 2015b). Further, polar bears have responded to declines in sea ice habitat by increasingly using land in many parts of their range (Stirling et al. 1999; Schliebe et al. 2008; Cherry et al. 2013; Prop et al. 2015; Rode et al. 2015a; Atwood et al. 2016). A recent modeling study showed that in the Canadian Arctic Archipelago some regions will become ice-free for more than 5 months a year by the end of the twenty-first century assuming that warming continues at its current rate and there are no reductions in emissions (Hamilton et al. 2014). In other regions, a 5-month ice-free period is projected to be reached by the middle of the twenty-first century (Atwood et al. 2015). Sea ice loss is expected to be most significant in the Barents, Chukchi, East Greenland, and Beaufort seas (Durner et al. 2009; Laidre et al. 2015b) and to continue to result in changes in polar bear distribution patterns.

Descriptive Notes

Body measurements: **Head and body length**: adult males 220–240 cm; adult females 190–200 cm. **Tail length**: 4 cm (cubs) to 25 cm (adult max). **Body mass**: adult males 400–600 kg; adult females 180–250 kg. **Dental formula**: I 3/3, C 1/1, P 2–/2–4, M 2/3 (Derocher & Wiig 2002).

Polar bears are the largest bear species. They are covered with a pigment-free fur that gives them the appearance of being white (Figure 14.2). Polar bears are thought to grow a new hair coat and molt between early summer and autumn. Their skin, foot pads, and nose are black. Polar bears have small ears and a long slender neck and head (Figures 14.3 and 14.4). They have large forepaws that aid in swimming and short, sharp claws that provide traction on the ice. Their

Figure 14.2 A polar bear on the summer sea ice. Global warming emerged as the primary threat to polar bears and associated changes in sea ice conditions will change the abundance and availability of polar bear prey. (Photo courtesy of US Geological Survey.) (A black and white version of this figure will appear in some formats. For the color version, please refer to the plate section.)

Figure 14.3 A female polar bear at a den site on the Alaskan coast of the southern Beaufort Sea. Typically only pregnant female polar bears den while other sex and age classes continue searching for prey in the winter. Dens are most often made in snow. (Photo credit US Geological Survey.)

Figure 14.4 Polar bear by Francesco Rinaldi, with permission. (A black and white version of this figure will appear in some formats. For the color version, please refer to the plate section.)

molars are reduced compared to other bears, reflecting their carnivorous diet. Polar bears have the ability to accumulate and store large amounts of fat, but unlike other marine mammals, polar bears do not have blubber for insulation. Instead, they rely on their fur and black skin to thermoregulate in the harsh Arctic conditions. The lifespan of polar bears in the wild is approximately 25 years.

Habitat

In general, polar bears require Arctic marine habitats that have sea ice concentrations >50% for a substantial portion of the year (Stern & Laidre 2015). The spatiotemporal distribution of sea ice, physical and biological oceanography, and life history requirements of both polar bears and their prey influence the habitats used by polar bears. Much of what is known on habitat use has been derived from habitat models (e.g. resource selection functions, or RSFs; Manly et al. 2002) that merge sea ice imagery with polar bear telemetry locations obtained primarily from collars. It is primarily the finding of studies using RSFs which are summarized here.

RSFs suggest that habitat selection varies regionally and seasonally. In the Divergent and Convergent Ecoregions, polar bears select ice over the shallow continental shelves in contrast to ice over deep water (Durner et al. 2009; Wilson et al. 2014; Lone et al. 2018b). In regions with entirely shallow waters (i.e. <300 m), depth was not important in habitat selection (i.e. the CA and eastern Greenland in the DI; Ferguson et al. 2000; Laidre et al. 2015a) or bears selected ice over the deepest waters (i.e. western Hudson Bay; McCall et al. 2016). Polar bears select winter ice near land in the DI, CI, and SI (Durner et al. 2009; McCall et al. 2016; Laidre et al. 2015a, 2018) or ice furthest from land during freeze-up in Hudson Bay (McCall et al. 2016).

The concentration of sea ice influences habitat selected by polar bears, as they need relatively unconsolidated ice for accessing seals but continuous expanses of ice to support travel and resting. In western Hudson Bay (SI) polar bears departed the ice for land 24 days after the ice concentration fell below 50% (Stirling et al. 1999). RSFs are consistent with this observation, as sea ice selected by polar bears ranged between ~50% and 100% concentration (Ferguson et al. 2000; Durner et al. 2009; Wilson et al. 2014; Laidre et al. 2015a; McCall et al. 2016; Lone et al. 2018b).

Polar bears tend to select ice habitats where transitions between high and low ice concentration or open water occur. Field observations indicate that polar bears select ice where ice edges, leads, glaciers, fjord ice, and polynyas (a stretch of open water surrounded by ice) are prevalent (Stirling et al. 1993; Stirling 1997; Freitas et al. 2012; Pilfold et al. 2014). RSFs similarly support that polar bears select high concentration sea ice near ice as low as 15% concentration in eastern Greenland during winter, spring, and summer (CI; Laidre et al. 2015a), the Barents Sea during winter, spring, and autumn (DI; Lone et al. 2018b), the DI and CI region in spring through autumn (Durner et al. 2009), and western Hudson Bay in late winter and break-up (McCall et al. 2016). However, polar bears in western Hudson Bay selected ice far from low ice thresholds during early winter and especially in freeze-up (McCall et al. 2016).

In western Hudson Bay summering bears, including pregnant ones, used lake shores, stream and river banks, and beach ridges, whose height was >1.8 m above the surrounding terrain, while onshore during the summer months (Clark et al. 1997; Richardson et al. 2005). Polar bears dug shallow pits or deep-earth dens, presumably to escape insect harassment and warm air temperatures (Clark et al. 1997), or as a maternity den in the subsequent winter (Clark et al. 1997; Richardson et al. 2005). Polar bears in other regions have not been reported to excavate soil dens. The landscape features chosen by pregnant bears in other regions, however, had similarly steep and abrupt topography necessary for accumulating wind-borne snow to a depth sufficient for snow dens (Durner et al. 2003; Aars 2013). In Alaska, these landscape features were >1.4 m in height. Presumably, similar habitat is used by parturient polar bears in other portions of their range for maternal denning.

Movements and Home Range

Polar bear movement patterns coincide with the seasonal availability of sea ice as described previously in this chapter (see also Figure 21.1, Chapter 21). Total distance moved in regions where there is an extended ice-free period is >2000 km per year. Maximum movement rates in most regions where they have been measured are ~10 km/h (Parks et al. 2006).

Because the sea ice habitat of polar bears is dynamic, the location of important resources (seals, receptive females) is often unpredictable; therefore, polar bears do not defend territories. Rather, they occupy vast areas of the sea ice ranging from less than 20,000 km^2 in western Hudson Bay

(Parks et al. 2006) and coastal areas of the Barents Sea (Tartu et al. 2018) up to >185,000 km^2 in areas where ice conditions may be less predictable such as the Canadian Archipelago (Ferguson et al. 1999) and offshore habitats in the Barents Sea (Mauritzen et al. 2002; Tartu et al. 2018) and East Greenland (Wiig et al. 2003).

Activity Patterns

Polar bears remain active throughout most of the year. Pregnant females become inactive while denning, but all other sex, age, and reproductive classes remain on the sea ice throughout the winter. Activity levels for non-denning bears are influenced by sea ice and prey availability (Ware et al. 2017).

Polar bears that come onshore during the summer months in the Canadian Arctic and DI ecoregion exhibit low activity levels (Knudsen 1978; Latour 1981; Ware et al. 2017). In the Chukchi and Beaufort Seas, polar bears that remain on sea ice as it retreats off shallow continental shelf regions and into deep Arctic Basin waters similarly showed reduced activity (Whiteman et al. 2015; Ware et al. 2017). More northerly subpopulations exhibit low activity rates during October to March when heavy ice conditions occur, indicative of a behavioral response to a poor foraging environment (Messier et al. 1992). Further, use of shelter dens has been documented in response to inclement weather (Messier et al. 1994) and low seal densities during the summer and winter (Ferguson et al. 2000).

Feeding Ecology

Polar bears are primarily sit-and-wait predators (Stirling 1974; Pagano et al. 2019), which allows them to minimize energy expenditure while hunting prey from the sea ice. They are the most carnivorous of all bear species with diets composed primarily of ice-associated seals. Polar bears often strip the fat off their prey and consume less of the meat, resulting in one of the highest dietary lipid contents of any mammal. This diet allows polar bears to store large amounts of fat and fast on those reserves when alternate food is not available. Although they specialize on marine mammal prey, under various ecological circumstances polar bears can display diverse feeding habits that reflect their close evolutionary relationship to brown bears. Polar bears feed on a range of marine, terrestrial, and even freshwater foods. Yet despite the potential breadth of their diets, the evolutionary history and modern-day existence of polar bears as a species are coupled to the availability of ringed seals.

Modern scientific studies of polar bear feeding include dedicated, systematic observations, but given the remote habitat and low population density of polar bears, opportunistic reports have been an important source of information. Scat analyses (e.g. Russell 1975) have provided important insights into the feeding behavior of polar bears, but collection of fecal samples is often limited to the ice-free season, when bears are losing body mass. The most detailed information on the feeding habits of individual bears has come from biochemical analyses that track the dietary transfer of nutritional components (e.g. fatty acids, stable isotopes).

Ringed seals are the primary prey of polar bears across their range (Thiemann et al. 2008). Polar bears prey on ringed seals of all ages, but prey heavily on pups shortly after weaning (Stirling & Archibald 1977; Pilfold et al. 2012). Bearded seals are the second-most important prey of polar bears across their range. Because of their large body size, they may account for a high proportion of dietary biomass (Derocher et al. 2002) and are most vulnerable to predation by large, adult male bears (Thiemann et al. 2007).

All other species of Arctic pinnipeds and small cetaceans are also potential polar bear prey (Thiemann et al. 2008). Narwhal (*Monodon monoceros*) and beluga whales (*Delphinapterus leucas*) feed on ice-associated organisms and may be vulnerable to polar bear predation when they become entrapped in openings in the ice. Like ringed and bearded seals, harp seals (*Pagophilus groenlandicus*) and hooded seals (*Cystophora cristata*) give birth to pups on the sea ice, hauling out in large numbers in predictable whelping patches. Dense aggregations of harp seal pups in the north-west Atlantic attract bears from long distances (>1000 km). Although harbor seals (*Phoca vitulina*) typically pup on land, sea ice changes may bring them in closer proximity to polar bears, especially in southern areas. Walruses (*Odobenus rosmarus*) appear to be a rare item in the diets of polar bears, and their large size and gregarious behavior may reduce their risk of predation.

Bears of different ages and sexes have different foraging requirements and abilities. Male polar bears may have a more varied diet, thanks to their ability to subdue larger prey, whereas adult females and subadults are more limited to ringed seals or scavenging the remains of kills made by large males. Cubs learn to hunt during the 2.5 years spent with their mother, gradually increasing the frequency and success of their own hunting attempts (Stirling & Latour 1978).

Direct observation and scat samples collected from bears on land (where they congregate during the ice-free season) reflect consumption of a range of terrestrial foods, including grasses, berries, small mammals, and human garbage (Gormezano & Rockwell 2013). Caribou (*Rangifer tarandus*) and ground-nesting birds (e.g. snow geese, *Chen caerulescens*) and their eggs have been suggested as nutritionally important terrestrial foods. However, evidence to date suggests that terrestrial foods contribute relatively little to the nutritional ecology of polar bears and that bears would not be able to compensate for lost seal hunting opportunities with increased terrestrial foraging (e.g. Rode et al. 2015b).

Scavenging represents an important food source because large prey (e.g. bearded seals) cannot be consumed in a single meal, leaving remnants for other bears. Scavenging prey remains is especially important for bears with less experience or ability to hunt, and bears with higher protein requirements (e.g. immature bears, lactating females). Polar bears also scavenge on stranded marine mammals or those killed by humans and killer whales (*Orcinus* orca; Bentzen et al. 2007; Galicia et al. 2016; Laidre et al. 2018).

Polar bears have evolved as specialized predators of Arctic seals and have a limited ability to switch to other food sources. Their foraging success is tightly coupled to the population dynamics of ringed seals and the predictable availability of the sea ice platform upon which both species depend. Therefore, changes in sea ice conditions, driven by global warming, will change the abundance and availability of polar bear prey (Ferguson et al. 2005, 2017).

Reproduction and Growth

Female polar bears are sexually mature at 3–6 years of age with first parturition a year later (Ramsay & Stirling 1988). Males are able to produce viable sperm as early as 2–4 years of age (Rosing-Asvid et al. 2002). Females enter estrus in the spring, and similar to other ursids are believed to be induced ovulators (Wimsatt 1974). The mating season extends from March to June with a peak in April–May (Rosing-Asvid et al. 2002; Stirling et al. 2016). Males compete for access to females, which results in a skewed operational sex ratio, with 2–3 males available for every estrous female (Ramsay & Stirling 1986). The mating system has been variably described as female defense polygyny or serial monogamy (Derocher et al. 2010), and polyandry (Ramsay & Stirling 1986) (see also Chapter 2).

Following breeding, impregnated females deposit significant fat stores during the spring and move to den areas anytime from July to December (Derocher & Stirling 1990a; Fischbach et al. 2007; Derocher et al. 2011). Females have delayed implantation where the zygote(s) stop development at the blastocyst stage until implantation (Wimsatt 1974). Implantation occurs in the autumn and active gestation lasts about 60 days with cubs born in December to January (Derocher et al. 1992). Birth date is affected by maternal condition, with fat mothers giving birth earlier (Robbins et al. 2012). The mean litter size in the spring ranges from 1.3 to 2.3 cubs/litter with some variation between subpopulations (Derocher 1999). Litters range from one to three cubs, although triplets are uncommon and the number of cubs is related to maternal condition and age (Derocher & Stirling 1994). Younger and older females have slightly smaller litters with peak litter sizes in mid-teen females (Derocher & Stirling 1994). Reproductive senescence is reflected in smaller litters in females over 20 years old, but reproductive capacity is retained throughout life (Derocher & Stirling 1994). Multiple paternity, monozygotic twins, adoptions, and incestuous inbreeding are uncommon but documented (Zeyl et al. 2009; Malenfant et al. 2015).

At birth, cubs are altricial, covered in short hair, blind, and weigh approximately 600 g (DeMaster & Stirling 1981). The small size of cubs at birth relative to the size of their mothers is thought to be related to physiological constraints associated with maternal fasting during active gestation (Ramsay & Dunbrack 1986). At den emergence cubs have grown to 10–15 kg (DeMaster & Stirling 1981). Post-parturition growth of cubs in dens is solely from maternal stores transferred by fat-rich milk (Derocher et al. 1993). The reproductive interval is dependent on offspring survival, but varies from one year with early cub mortality that occurs before or during the breeding season to three or more years if offspring survive to weaning (Ramsay & Stirling 1988).

The growth patterns of polar bears follow a sigmoid pattern with variation between years and subpopulations that reflect variation in available prey and ecosystem productivity (Kingsley 1979; Derocher & Stirling 1998). About 97% of asymptotic size in body length is attained by 4.1–5.5 years of age by females, but is about 2 years later for males (Derocher & Stirling 1998; Derocher & Wiig 2002). Body mass growth takes about 7 years in females and 13 years in males to achieve 97% of asymptotic size (Derocher & Wiig 2002). Much of the variation in mass between individuals of the same age is related to the size of fat depots, of which subcutaneous adipocytes are 65–82% of the total in the body (Pond et al. 1992). Adult males are approximately 2.1 times the mass of females and sexual dimorphism in mated pairs is notable, with males up to three times larger than females by mass (Derocher & Wiig 2002; Derocher et al. 2010). The growth of females is related to their size at weaning, but that of males, likely due to their longer period of growth and the role of environmental variation during growth, is not related to their weaning size (Atkinson et al. 1996b). Studies suggest that polar bear growth patterns may be affected by environmental pollutants, population density, and climate change (Derocher & Wiig 2002; Rode et al. 2010; Sonne et al. 2013).

Behavior

Similar to other bears, polar bears are primarily solitary except for mating pairs, family groups, and aggregations of bears during the ice-free season. Aggregations occur at concentrated food sources, such as carcasses of bowhead whales (*Balaena mysticetus*) left on land after being harvested, where they interact with other polar bears but also with grizzly bears (Miller et al. 2015). During the ice-free season in Hudson Bay and other regions where ice melts completely in summer, polar bears segregate by age and sex, with adult males often aggregating in groups on points at the coast or on nearshore islands (Derocher & Stirling 1990b), where they show little aggression but may develop a dominance hierarchy.

Males locate estrous females by following their tracks (Owen et al. 2015) and males often isolate females in remote areas, such as fjords, bays, and hills to minimize disturbance and competition from other males (Wiig et al. 1992; Stirling et al. 2016). Competition for access to females is intense and partly due to the prolonged mother–offspring bond (typically 2.5 years) that results in females only breeding every 3 years (Ramsay & Stirling 1986). Signs of competition between males for mates, such as wounds, scars, and canine breakage, begin to occur about 6 years of age and peak in the mid-teens (Derocher et al. 2010). Mated pairs remain together for 1–4 weeks with a mean of 17.5 days (Molnár et al. 2008). The first copulations in a pair may be delayed up to 7 days and males may remain mounted on a female for up to 150 minutes with only momentary separations (Stirling et al. 2016).

Females give birth to cubs in winter dens, emerging from late February to early May (earlier at lower latitudes). Females nurse young until 2.5 years of age, although the contribution of milk to offspring energy balance declines over time (Arnould & Ramsay 1994). In western Hudson Bay, weaning of offspring at 1.5 years was common in past decades although it is now uncommon and rarely seen in other areas (Ramsay & Stirling 1988). Termination of the mother–offspring bond can result from males courting the mother or the mother abandoning her young (Stirling et al. 2016).

Polar bears are excellent swimmers and have been documented swimming for up to 9 days and covering >600 km (Durner et al. 2011), although such long-distance swims can result in increased mortality (Monnett & Gleason 2006), and swimming is energetically expensive (Griffen 2018). Some polar bears employ aquatic stalking to approach basking seals by swimming below ice floes. Bears may use multiple short dives to approach seals; the longest recorded dive by a polar bear approaching a basking seal is 3 min 10 s (Stirling & van Meurs 2015). Polar bears often dive to depths of 3–4 m apparently to feed on seaweed (Lone et al. 2018a); maximum recorded dive depth was 13.9 m and most bears dove to >6 m depth.

Parasites, Infections, and Diseases

Most disease-related investigations on wild polar bears are serological surveys documenting the occurrence of antibodies against potential pathogens. Some of these infections are associated with disease and mortality, some have an unknown impact, and some may be transmissible to humans. Concurrent exposure to two or more pathogens at the same time, as well as to persistent organic pollutants (POPs) and other compounds that may act as immunomodulators (Dietz et al. 2018a), may impact the outcome of infections. Most reports of disease in polar bears have been documented in captive animals, often with infectious agents that are less relevant for wild bears (Fagre et al. 2015). However, shrinking ice coverage of the Arctic Ocean has opened new trans-Arctic sea routes for tourism and cargo, also increasing the potential for transmission of new diseases to the region.

Parasites

A range of parasites, sometimes causing disease, have been documented in captive polar bears (Fagre et al. 2015). Whether wild polar bears are exposed to these parasites is largely unknown, because the parasitic fauna hosted by polar bears have been scarcely studied with the exception of *Trichinella* sp. and *Toxoplasma*.

Trichinella is a genus of nematodes that have a wide host range infecting polar bears and other Arctic carnivores. A seroprevalence of *Trichinella* sp., presumably against the Arctic variant, *T. nativa*, of typically >50% has been reported from polar bears on the coast of Alaska and the Bering Strait region and in Svalbard and the Barents Sea (Fay 1960; Dau & Barrett 1981; Åsbakk et al. 2010). Some disease outbreaks in humans have been associated with undercooked meat from polar bears and walrus, but the parasite is not often found in seals (Thorshaug & Rosted 1956; Tryland 2000; Tryland et al. 2014). A recent review reported 63 cases of human trichinosis linked to consumption of polar bear meat since World War II. However, although the prevalence of *Trichinella* sp. in polar bears is reported as high, the number of larvae in the muscular tissues is usually low, diminishing the risk for humans (Dupouy-Camet et al. 2017), and suggesting that the parasite has limited impact on the polar bear host.

Toxoplasma gondii is a zoonotic intestinal coccidian parasite of domestic and wild felines. Polar bears, and probably any mammal or bird, may serve as intermediate hosts, carrying extra-intestinal tissue cysts in organs such as muscle, liver, lung, and brain. Clinical toxoplasmosis has been diagnosed in a variety of captive and wild marine mammals (Tryland 2000; Tryland et al. 2014; Miller et al. 2018), affecting the central nervous system and causing reproductive failure (Miller et al. 2018). Serological investigations of polar bears from Alaska, Canada, Svalbard and the Barents Sea and Russia have indicated exposure, but clinical toxoplasmosis (i.e. disease) has not been verified (Miller et al. 2018). Exposure to *T. gondii* was found to be greater for bears that use land during the summer and fall in the Beaufort Sea (Atwood et al. 2017). However, marine mammals, including polar bears, may represent a source of infection to people, as toxoplasmosis and a high prevalence of antibodies against the parasite have been documented in Inuit communities (McDonald et al. 1990; Messier et al. 2009).

Bacterial Infections

Polar bears may host a wide range of potential pathogenic bacteria, but there are not many reports of clinical cases or specific investigations.

Many marine mammal species are exposed to bacteria of genus *Brucella* (Tryland et al. 2018). A seroprevalence of 5–14% has been found in polar bears from Svalbard and the Barents Sea (Tryland et al. 2001; Nymo et al. 2013), and a seroprevalence of 5–10% in polar bears in the Beaufort and Chukchi Seas (Rah et al. 2005; O'Hara et al. 2010). However, little is known with certainty about which *Brucella* species these animals have been exposed to, or if this exposure has health impacts on polar bears. In Phocidae (i.e. "true seals") no pathology has been associated with the isolation of *Brucella* bacteria, but in Otariidae (eared seals), *B. pinnipedialis* has been associated with pathology in reproductive organs (Sidor et al. 2008; Duncan et al. 2014).

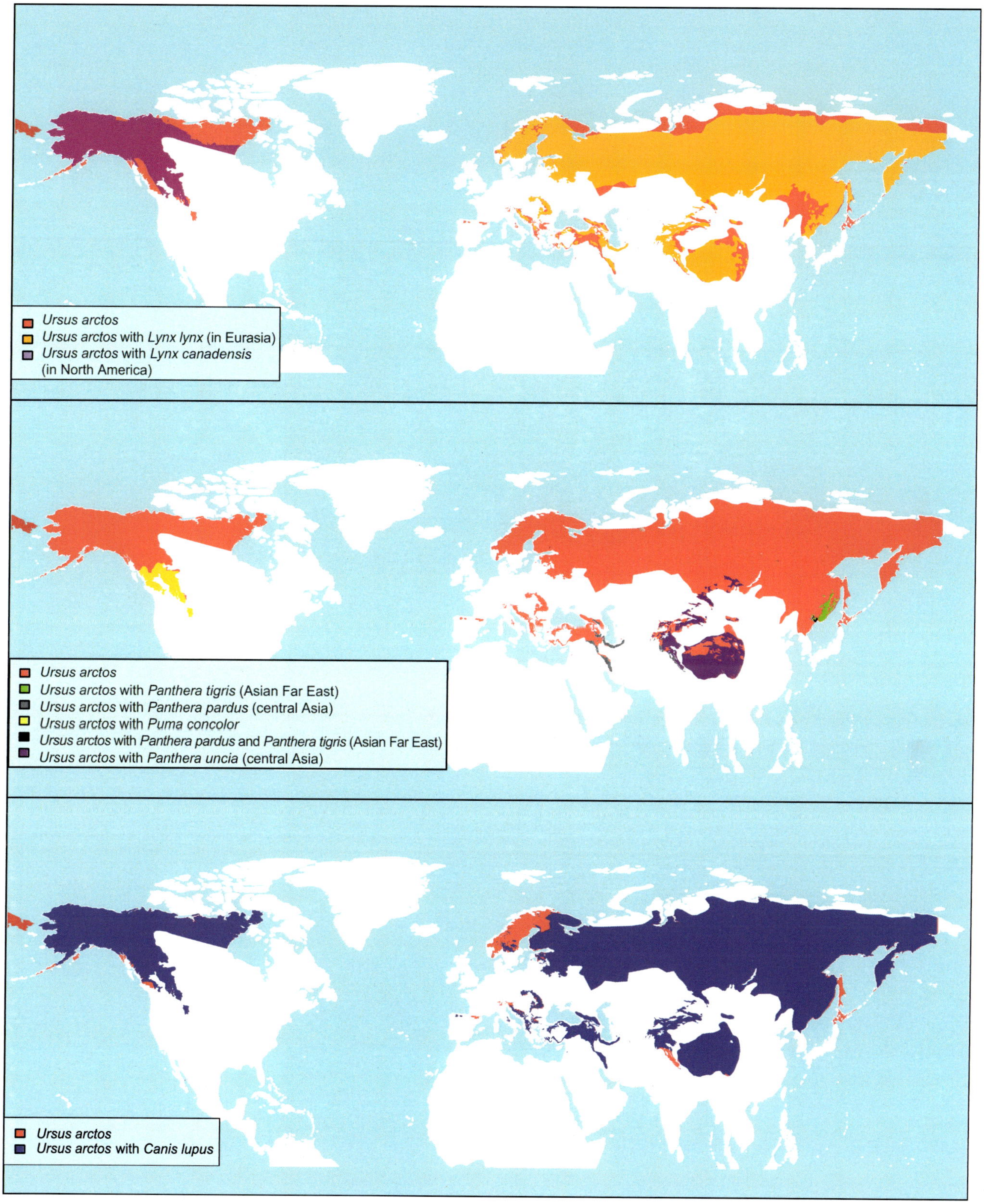

Figure 3.1 Worldwide distribution of brown bears and its overlap with the distribution of several felids (top and central panel) and wolves (bottom panel) in the Northern Hemisphere. (*Source*: IUCN 2019. The IUCN Red List of Threatened Species, Version 6.2, www.iucnredlist.org. Downloaded March 16, 2019.) (A black and white version of this figure will appear in some formats.)

Figure 5.5D Bears in Minnesota use various types of dens, here in an open nest that they construct. Individuals often switch den types year to year, and rarely reuse dens. Type of den does not affect reproduction or survival; even cubs born in open nests survive normally. Other animals (e.g. squirrels, rabbits, wolves) commonly approach dens. (Photos by D. Garshelis.) (A black and white version of this figure will appear in some formats.)

Figure 5.6 American black bears in northern Minnesota – one of the coldest places within their range – commonly spent the winter in unprotected dens that they construct. Even some females that gave birth in mid-winter, like the bears pictured here, used dens with little or no cover. Due to the insulatory properties of their thick layer of fat and dense winter coat, and reduced blood supply to the skin, snow that fell on their backs sublimated or was shaken off, rather than melting into their fur. (Photo by D. Garshelis.) (A black and white version of this figure will appear in some formats.)

Figure 6.2 Giant panda by Francesco Rinaldi, with permission. (A black and white version of this figure will appear in some formats.)

Figure 6.3 The iconic giant panda caught on a motion-activated camera. Famous for their black and white coloration, these markings have been nominated as one of the factors contributing to the "cute factor." New evidence suggests that the black and white coloration of the giant panda plays a role in crypsis against snowy backgrounds. (Photo by Jindong Zhang.) (A black and white version of this figure will appear in some formats.)

Figure 8.3 Sun bear by Francesco Rinaldi, with permission. (A black and white version of this figure will appear in some formats.)

Figure 8.7 A rescued sun bear in Cambodia. Bears rescued by Free the Bears enjoy life in large semi-natural enclosures within sanctuaries throughout South-East Asia. (Photo by Free the Bears/ Peter Yen.) (A black and white version of this figure will appear in some formats.)

Figure 9.3 Sloth bear by Francesco Rinaldi, with permission. (A black and white version of this figure will appear in some formats.)

Figure 10.2 Asiatic black bear by Francesco Rinaldi, with permission. (A black and white version of this figure will appear in some formats.)

Figure 9.4 The sloth bear is characterized by a shaggy, dusty-black coat. (Photo by Ansar Khan.) (A black and white version of this figure will appear in some formats.)

Figure 10.5 Asiatic black bear in Russia taken by a camera trap. (Photo by Ivan Seryodkin.) (A black and white version of this figure will appear in some formats.)

Figure 11.3 American black bear by Francesco Rinaldi, with permission. (A black and white version of this figure will appear in some formats.)

(A)

(B)

Figure 12.3(A,B) The Gobi bear, which lives in the mountainous desert of southern Mongolia, is the only desert-dwelling brown bear in Eurasia. (Photo by Jon Swenson.) (A black and white version of this figure will appear in some formats.)

Figure 12.4 Eurasian brown bear by Francesco Rinaldi, with permission. (A black and white version of this figure will appear in some formats.)

Figure 12.5 A brown bear climbing up a tree in northern Iran. (Photo by Hamed Tizrooyan.) (A black and white version of this figure will appear in some formats.)

Figure 12.6 Standing up on its hind legs allows a brown bear to get more information from its senses of smell, sight, and hearing. (Photo by Mario Melletti.) (A black and white version of this figure will appear in some formats.)

Figure 12.7 During autumn hyperphagia, brown bears eat large amounts of food and accumulate fat for hibernation and reproduction. (Photo by Vincenzo Penteriani.) (A black and white version of this figure will appear in some formats.)

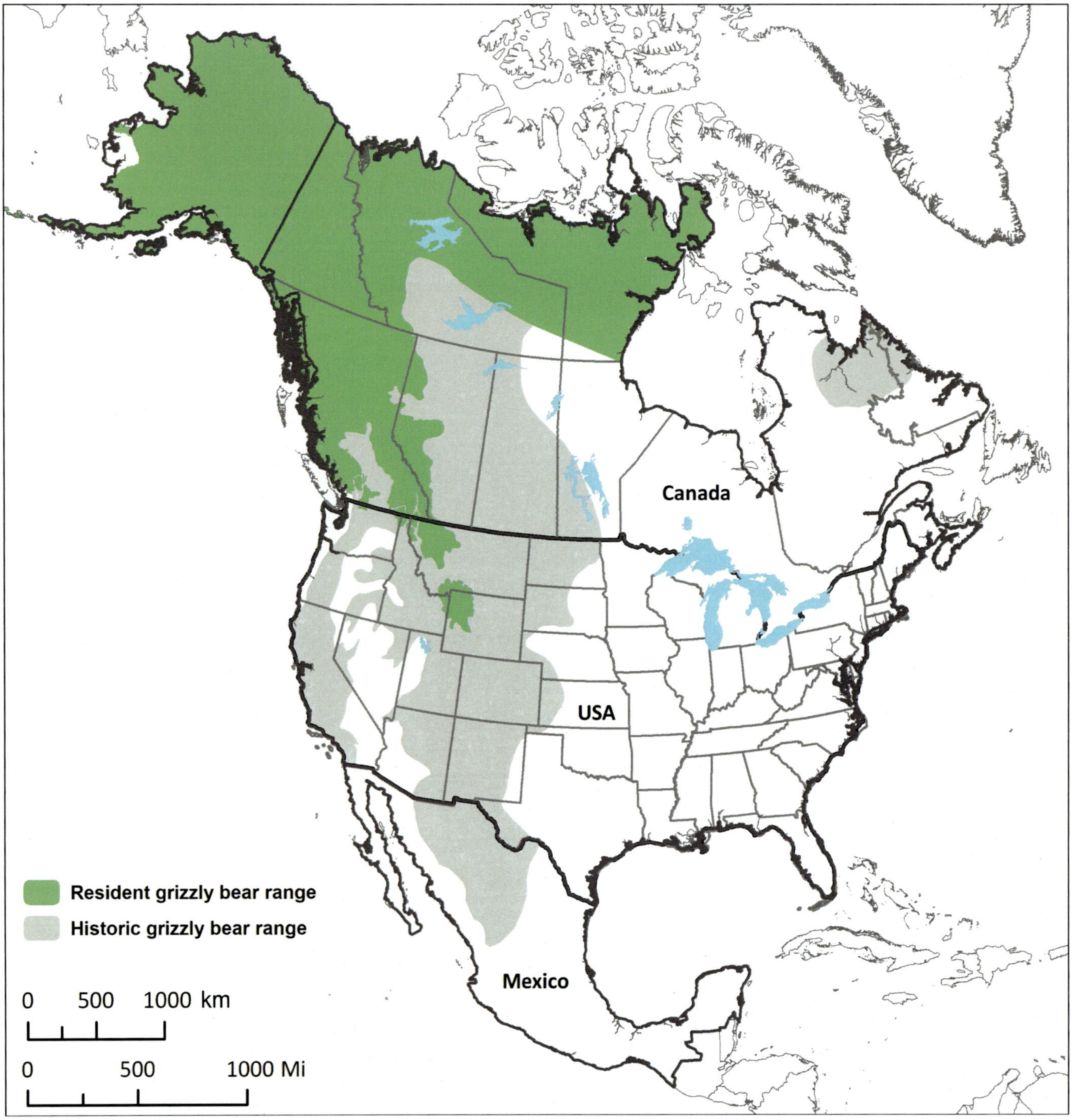

Figure 13.1 Historic and current occupied range of brown bears in North America. *Data sources*: M. Proctor, Birchdale Ecological, Ltd.; International Union for Conservation of Nature (IUCN); Interagency Grizzly Bear Study Team. Map creation: L. Landenburger, US Geological Survey and D. Baisero, Wildlife Conservation Society. (A black and white version of this figure will appear in some formats.)

Figure 13.4 A grizzly bear foraging for blueberries (*Vaccinium* spp.) on the tundra slopes within Denali National Park, Alaska, USA. In early September, blueberries are abundant and the tundra turns red and gold as fall approaches and bears target berries to build fat reserves in preparation for hibernation. (Photo courtesy of J. Schoen.) (A black and white version of this figure will appear in some formats.)

Figure 13.8 A brown bear fishing for red (sockeye) salmon (*Oncorhynchus nerka*) in the intertidal waters of the Chenik River located in the McNeil River State Game Refuge on the west side of Cook Inlet, Alaska, USA. Once the red salmon enter fresh water, they begin to turn a dark red color and start to lose significant amounts of fat and protein. The bright fish have much higher nutritional value prior to entering fresh water and brown bears often focus their fishing activities near the mouth of rivers in intertidal areas. (Photo courtesy of J. Schoen.) (A black and white version of this figure will appear in some formats.)

Figure 13.9 A North American brown bear in a coastal habitat. (Photo courtesy of M. Clapham.) (A black and white version of this figure will appear in some formats.)

Figure 14.2 A polar bear on the summer sea ice. Global warming emerged as the primary threat to polar bears and associated changes in sea ice conditions will change the abundance and availability of polar bear prey. (Photo courtesy of US Geological Survey.)
(A black and white version of this figure will appear in some formats.)

Figure 14.4 Polar bear by Francesco Rinaldi, with permission. (A black and white version of this figure will appear in some formats.)

Figure 15.4 Delay in ocean freezing due to global warming is an important extrinsic factor promoting conflicts between people and polar bears. (Photo by Marcus Elfström.) (A black and white version of this figure will appear in some formats.)

Figure 17.1 The sloth bear's aggressively defensive nature may be the result of having coevolved with large predators, such as tigers and leopards, which are known to occasionally prey upon this bear. Actually, sloth bears are strong animals that possess large canines and claws, which make them difficult prey. (Photo by Ayan Sadhu.) (A black and white version of this figure will appear in some formats.)

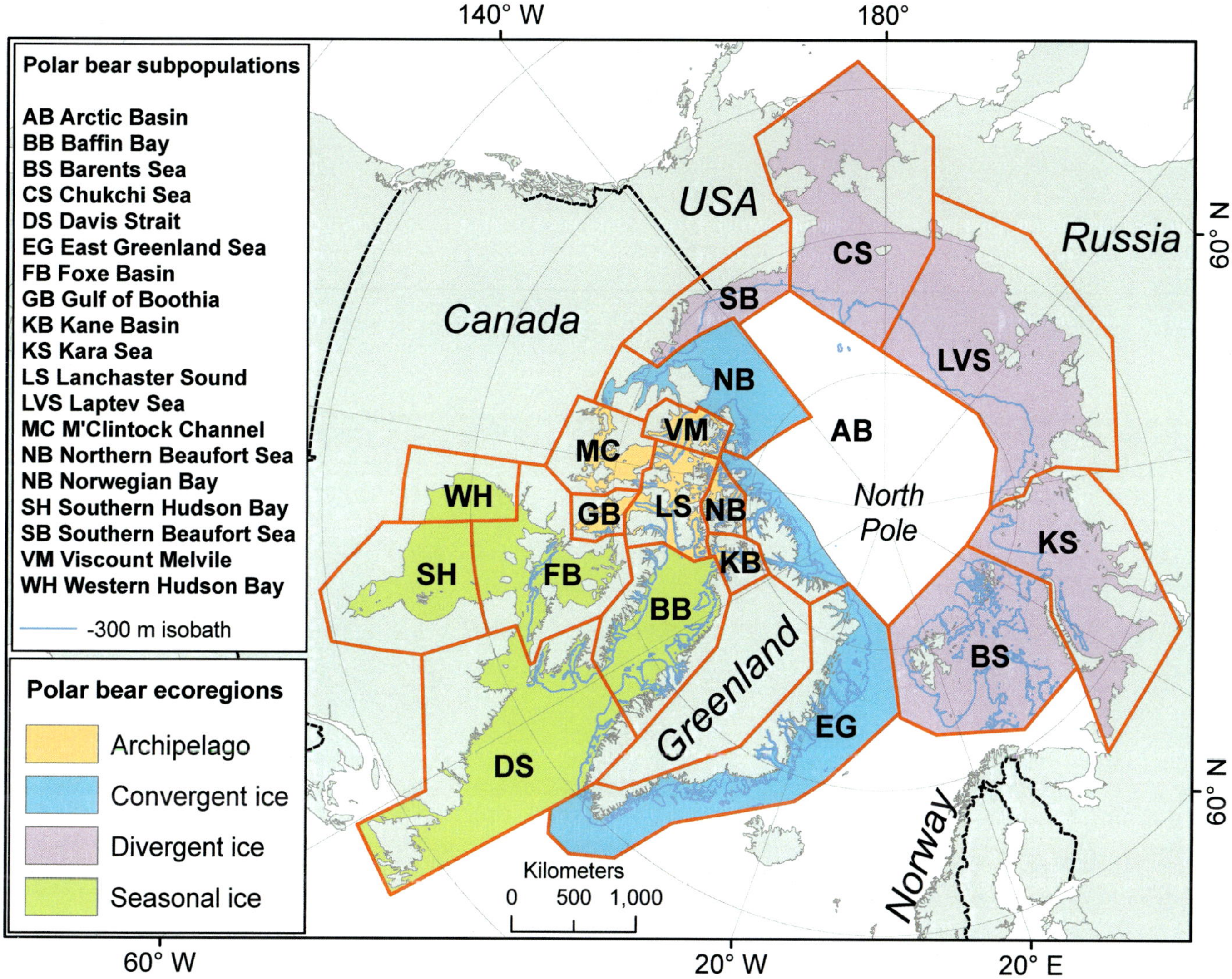

Figure 21.2 Polar bear range across the circumpolar Arctic showing the 19 subpopulations recognized by the International Union for the Conservation of Nature's Polar Bear Specialist Group and associated ecoregions as defined by Amstrup et al. (2008). The characteristics used to define ecoregions are described in Chapter 14. (A black and white version of this figure will appear in some formats.)

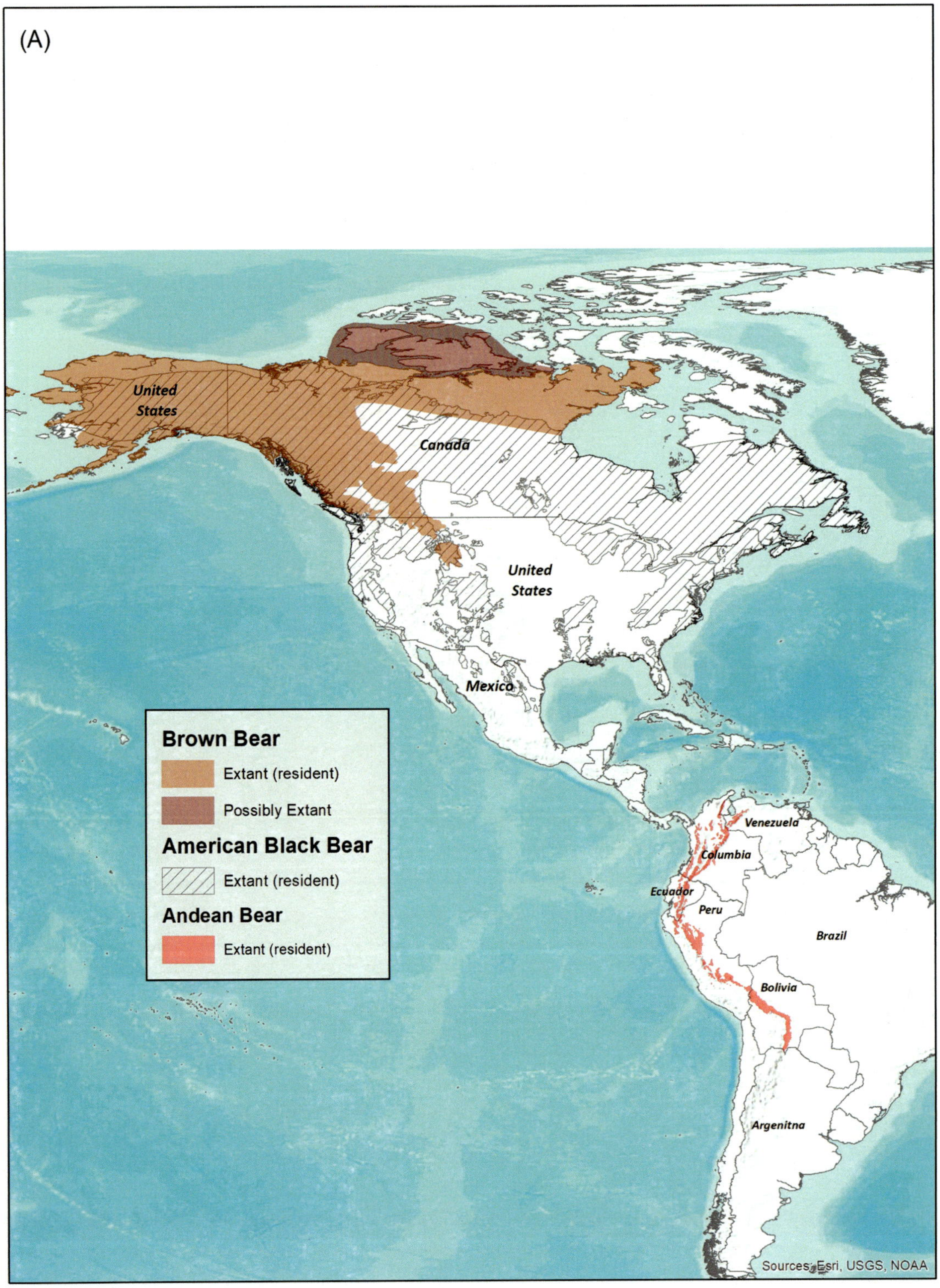

Figure 22.1 (A) Fragmented distribution of North American brown bears, American black bears, and South American Andean bears, and (B) Eurasian distribution of brown bears. (*Source*: IUCN 2017, Red List of Threatened Species.) (A black and white version of this figure will appear in some formats.)

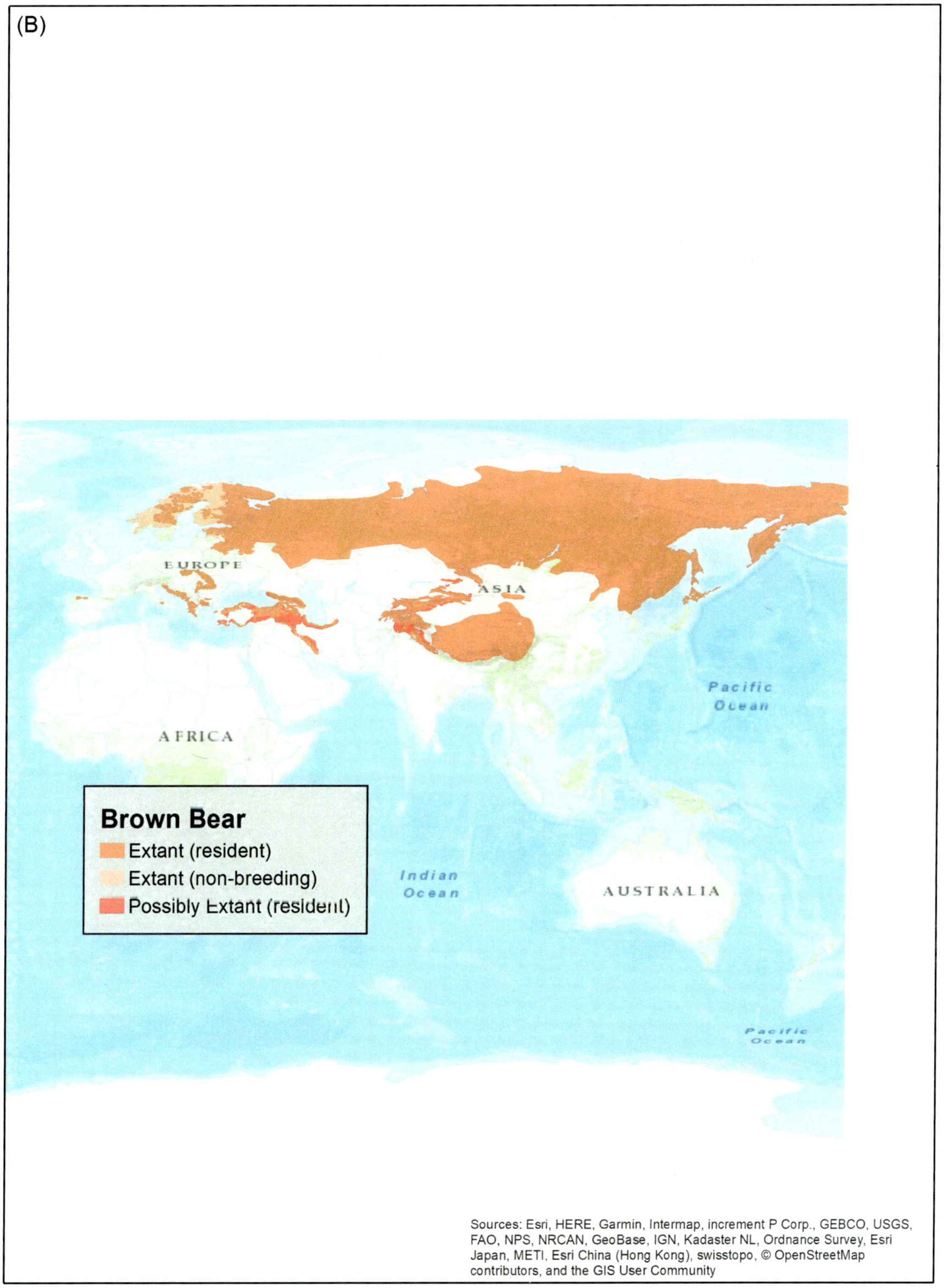

Figure 22.1 (*cont.*)

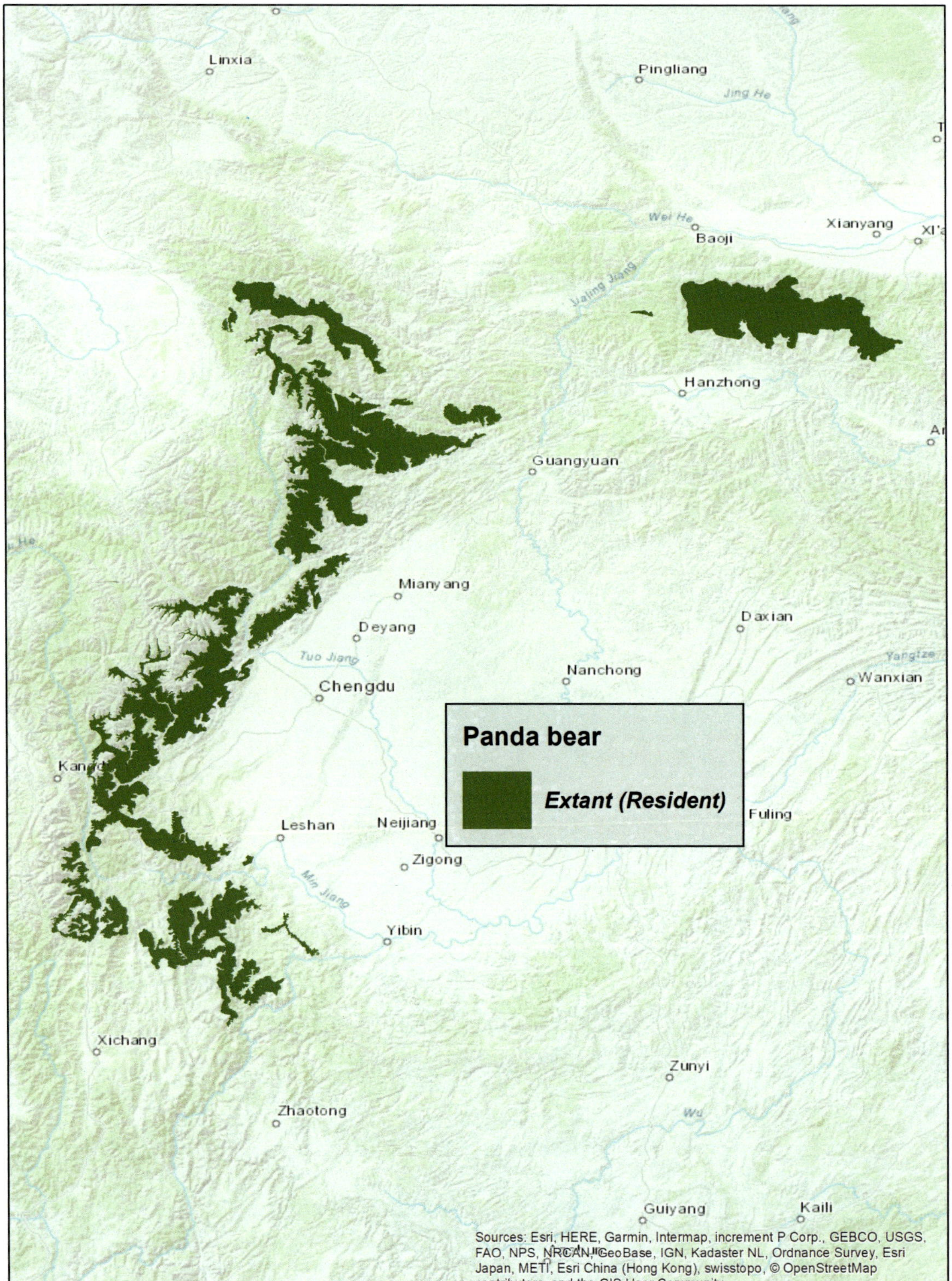

Figure 22.2 Fragmented giant panda bear distribution in western China. (*Source*: IUCN 2016, Red List of Threatened Species.) (A black and white version of this figure will appear in some formats.)

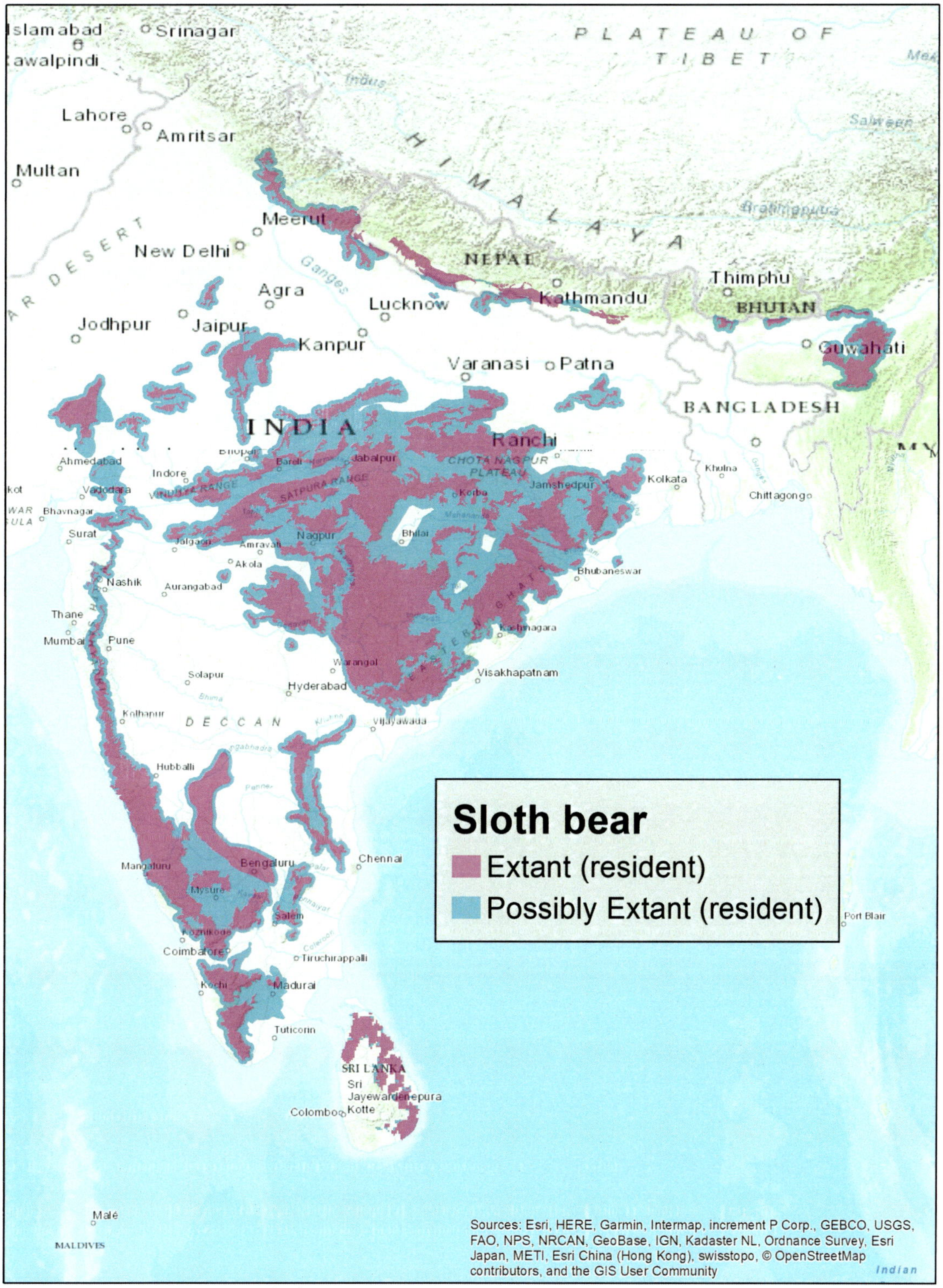

Figure 22.3 Fragmented sloth bear distribution across India and Sri Lanka. (*Source*: IUCN 2016, Red List of Threatened Species.) (A black and white version of this figure will appear in some formats.)

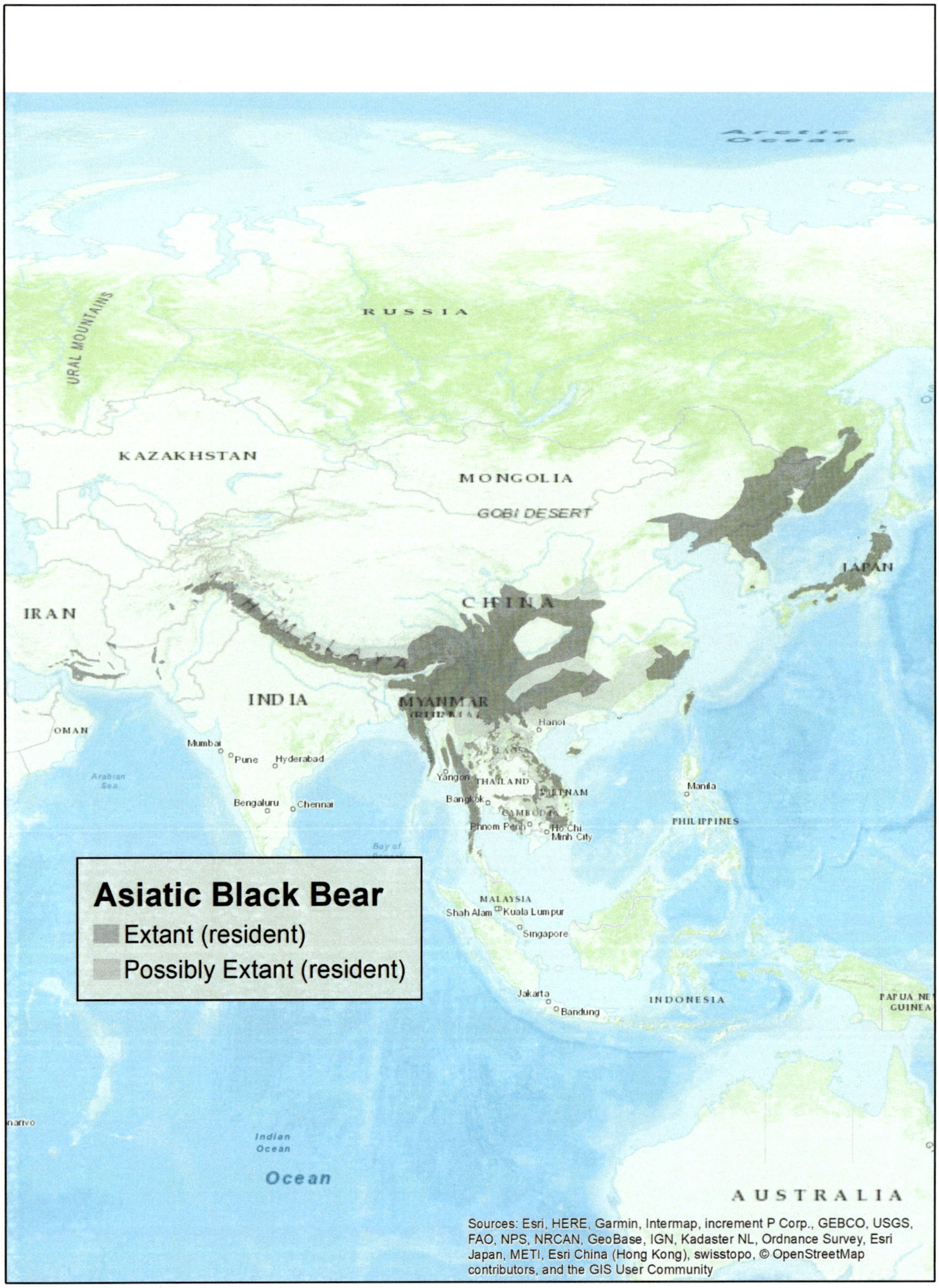

Figure 22.4 Fragmented Asiatic black bear distribution from south-eastern Russia and Japan south and west into South-East Asia and across the Himalayas into Iran. (*Source*: IUCN 2016, Red List of Threatened Species.) (A black and white version of this figure will appear in some formats.)

Figure 22.5 Fragmented sun bear distribution from South-East Asia into the islands of Sumatra and Borneo. (*Source*: IUCN 2017, Red List of Threatened Species.) (A black and white version of this figure will appear in some formats.)

Figure 23.2 The male Andean bear "Beorn" at Queens zoo is the result of the cooperation between the European Endangered Species Program and the Species Survival Plan. The father was born in the French zoo of Doué la Fontaine and transferred to Queens Zoo, USA, in 2013 to be paired with a female born at the National Zoo Park in Washington. (Photo by Megan Maher, ©WCS.) (A black and white version of this figure will appear in some formats.)

(A)

(B)

Figure 26.2 Dancing bears in India (A) and Nepal (B) with customary nose ring and ropes. (*Source*: World Animal Protection/Neil D'Cruze.) (A black and white version of this figure will appear in some formats.)

Figure 28.4 Typically secretive, sloth bears are best viewed during the day in national parks with a long history of human visitation. This adult female sloth bear was photographed at Yala National Park, Sri Lanka in 2007. (Photo by Luxshmanan Nadaraja.) (A black and white version of this figure will appear in some formats.)

Viral Infections

A wide range of viral infections have been documented and may also cause disease in marine mammals (Duignan et al. 2018), but information about such infections and possible impacts in polar bears is scarce.

Rabies virus (family *Rhabdoviridae*, genus *Lyssavirus*) is neutrotrophic, has a wide global distribution, and can cause rabies in a wide range of host species. One clinical case of rabies in a polar bear is published. The bear was found (Northwest Territories, Canada) by Inuit hunters, dragging its hind legs, and the diagnosis was based upon submission of the spinal cord and head and the finding of specific staining of inclusion bodies in neurons of the lumbar spinal cord (Loewen et al. 1990; Taylor et al. 1991). A serological screening of 266 polar bears in Svalbard and the Barents Sea revealed no seropositive animals (Tryland et al. 2005). In spite of the circumpolar distribution of the rabies virus and the Arctic fox (*Vulpes lagopus*) being its main reservoir, it is assumed polar bear exposure to rabies is low.

Morbillivirus (family *Paramyxoviridae*, genus *Morbillivirus*) has caused severe disease and epizootics in several seal and some whale species, and serological investigations have revealed exposure of many species and populations (Duignan et al. 2018). Antibodies against a variety of morbilliviruses (canine distemper, CDV; phocine distemper, PDV; dolphin morbillivirus, DMV; porpoise morbillivirus, PMV) have been detected in wild polar bears (3–50%) in Alaska, Canada, Svalbard/Barents Sea, and Russia (Fagre et al. 2015). As these viruses have similar antigenic structures and to some extent cross-react serologically, the tests have often been conducted with little evidence of which virus animals were exposed to (Fagre et al. 2015). There are indications that morbillivirus infections in polar bears may cause immunosuppression (Kirk et al. 2010), and thus may impact the course of infection with other pathogens. Antibodies against calicivirus have been reported in polar bears from Canada and from Svalbard and the Barents Sea (Philippa et al. 2004; Tryland et al. 2005), but it is not known which type of calicivirus they were exposed to or if it has caused disease. Polar bears from Canada were also exposed to canine adenovirus (CAV-2) or a similar virus (Philippa et al. 2004). In captive polar bears, it has been shown that equine herpesvirus 9 and suid herpesvirus 1 both were able to infect and cause disease, but exposure was probably through other corralled animal species or via the feed (Fagre et al. 2015), and is of little relevance for wild polar bears. A captive polar bear in Toronto, Canada was diagnosed with West Nile virus after presenting with non-ambulatory paraparesis (Dutton et al. 2009).

Industrial Chemicals

The polar bear is among the species most contaminated with mercury and persistent organic pollutants (POPs) both in the Arctic and globally due to its position at the top of the marine food web (Letcher et al. 2010; Dietz et al. 2013, 2018; Bignert et al. 2016; de Wit et al. 2017; Routti et al. 2019). Since 2001, numerous POPs have been listed, and thus globally banned, under the United Nations Environmental Programme's (UNEP's) Stockholm Convention on POPs, as well as mercury under UNEP's Minamata Convention on mercury.

Given their high trophic level, the polar bear is a useful indicator species for biomonitoring spatiotemporal trends, distribution, dynamics, fate, biomagnification, and potential effects of mercury (metals) and legacy and new POPs (Letcher et al. 2010; Dietz et al. 2013, 2018; Rigét et al. 2019; Routti et al. 2019). These POPs originate mainly from industrial and agricultural activities occurring at more southern latitudes, and are transported by air, ocean currents, and river outflows to the Arctic. Although local pollution sources exist in the Arctic, their influence is seen only in a limited geographical area and their contribution to larger-scale contamination is likely minor (Brown et al. 2018).

As has been detailed in recent reviews, the majority of contaminant-related studies have focused on contaminant "hot spots" and specifically in the Beaufort/Chukchi Sea, Hudson Bay, East Greenland, and Svalbard regions of the Arctic (Letcher et al. 2010; Bignert et al. 2016; de Wit et al. 2017; Dietz et al. 2013, 2018; Routti et al. 2019). Levels of some legacy POPs such as polychlorinated biphenyls (PCBs) in polar bears have been shown to have decreased prior and up to 2000, but have since remained relatively unchanged and in some cases are shown to be increasing due to changes in their diet and underlying food web (Dietz et al. 2013, 2018, 2019; Bignert et al. 2016; Brown et al. 2018; Letcher et al. 2018). A number of emerging environmental pollutants, such as polybrominated diphenyl ether (PBDE) flame retardants and per-/polyfluoroalkyl substances (PFASs; in particular the highly bioaccumulative perfluorooctane sulfonic acid, PFOS) among others have been reported (de Wit et al. 2017). In a recent study from Hudson Bay subpopulations, 247 POPs were screened in liver or fat samples and 210 POPs were detectable (Letcher et al. 2018). Liu et al. (2018) detected hundreds of analytes in the serum of polar bears from Hudson Bay and the Beaufort Sea, which belong to 13 classes of chemicals, including novel PCB metabolites and many fluorinated or chlorinated substances not previously detected in polar bears. These studies and numerous others (Bignert et al. 2016; de Wit et al. 2017 and references therein) exemplify that polar bears in general are exposed to an increasingly complex array of POPs via multiple routes of exposure but mainly through the diet. Although overall levels are high, concentrations of PCBs and most organochlorines show wide variation between individual polar bears, which may be related to age, sex, breeding status, changes in body fatness and energetic requirements, food web length and diet composition, and finally individual differences in biotransformation capacity (Letcher et al. 2010; de Wit et al. 2017; Routti et al. 2019).

Effects of global warming have the potential to result in interactive effects on polar bears resulting from nutritional stress, novel pathogen exposure, changing bear–human interactions (e.g. increased land use, increased shipping) and exposure to POPs, mercury, and other metals (Letcher et al. 2010; McKinney et al. 2010, 2013, 2015; Bechshøft et al. 2018; Dietz et al. 2018). Biological effects assessments provide evidence regarding the possible effects of POP exposure on physiological pathways in polar bears, including endocrine and reproductive challenges, immunological impairment, and alterations to vitamin homeostasis, among others (Letcher et al. 2010; Sonne 2010; Sonne et al. 2012; Dietz et al. 2018). Effect studies of Arctic biota that have been documented for old POPs, mercury (and a few other metals), flame retardants, and a few PFAS (mostly on PFOS) have been entirely based on correlative relationships between biomarker/effect endpoints and measurements of POPs in body tissues (liver, fat, and muscle), as well as in blood plasma (Letcher et al. 2010; Sonne 2010; Dietz et al. 2018). Thus, the suggestion of effects is based on a weight of correlative evidence rather than cause–effect relationships. Furthermore, consideration of combined or "complex mixture" effects which include targeted and as yet unidentified and new POPs (de Wit et al. 2017) will advance the understanding of the impacts of contaminants as a consequence of using these more environmentally realistic exposure scenarios for polar bears and Arctic wildlife in general.

Status in the Wild

Polar bear subpopulations vary in size from a few hundred to a few thousand bears with a total world population estimated at about 26,000 (Regehr et al. 2016). The quality of abundance estimates for individual subpopulations varies greatly from recent rigorous estimates to estimates that are dated, and others which are based on "best guesses." Because of this, the global population estimate is not used as a monitoring benchmark. Rather, it expresses a reasonable range based on a combination of the best available information and understandings of polar bear habitat.

The polar bear was listed as Vulnerable under criterion A3c (a suspected population reduction during the coming three generations based on a decline in the area of occupancy, extent, and habitat quality) by the IUCN in 2015 (Wiig et al. 2015). This designation was based on analysis that estimated 0.71 probability (range 0.20–0.95) of mean global population size decline $\geq$30% over three generations (Regehr et al. 2016). Declines in sea-ice habitat have been associated with declines in polar bear body condition, survival, and abundance in some subpopulations (Stirling et al. 1999; Regehr et al. 2007, 2010; Rode et al. 2010, 2014; Bromaghin et al. 2015; Wiig et al. 2015; Lunn et al. 2016; Obbard et al. 2016).

In Canada, where 13 of the 19 subpopulations occur and the total estimate is about 15,000 bears, the polar bear was listed as of Special Concern in 2018 (COSEWIC 2008). Three Canadian subpopulations are currently listed as "Likely Declined," two as "Likely Stable," one as "Stable," one as "Likely Increased," one as "Increased," and five are listed as "Data Deficient/Uncertain" (Environment and Climate Change Canada 2018). Polar bear harvest is legal in Canada both for subsistence and sport hunting (in some subpopulations).

In the United States, the US Fish and Wildlife Service listed the polar bear as "threatened" worldwide under the US Endangered Species Act in 2008 due to range-wide declines in sea ice (US Fish and Wildlife Service 2008). The United States shares management responsibility for the Southern Beaufort Sea subpopulation with Canada (listed by Canada as Likely Declined) and for the Chukchi Sea subpopulation with the Russian Federation. Polar bear harvesting is legal in the US by coastal Native hunters as outlined in the US Marine Mammal Protection Act. The Southern Beaufort Sea subpopulation declined by between 25% and 50% in the period 2001–2010 (Bromaghin et al. 2015). In contrast, a recent study estimated that there were about 2900 bears in the Chukchi Sea subpopulation (Regehr et al. 2018), and that the subpopulation has been productive in recent years.

Hunting of polar bears was banned in the former Soviet Union in 1956 (Larsen & Stirling 2009). Currently, the polar bear is listed in the Red Data Book of the Russian Federation (Belikov et al. 2010); thus, the polar bear is given special protection and any action that causes death or decline in the abundance of polar bears is prohibited. A US–Russia Bilateral Treaty was signed between Native and government representatives of the US and Russia in 2000 and implementation began in 2007.

The polar bear has been protected in Norway since 1973 (Vongraven et al. 2010); therefore, there is no harvest of the Barents Sea subpopulation in Norwegian territories. In 2006, the polar bear was listed as Vulnerable on the Norwegian Red List due to sea ice loss. The Barents Sea subpopulation was surveyed in 2004 and 2015, and although surveys were not directly comparable due to inadequate access to Russian territory in 2015, the subpopulation is at present considered to be stable (Aars et al. 2017), although the authors suggest that the carrying capacity of the area most likely has been significantly reduced.

The Government of Greenland is the responsible authority for the management of polar bears in Greenland, including the national legislation, national coordination and setting of hunting quotas (Jessen 2018). Greenland has sole management responsibility for the East Greenland subpopulation and shared responsibility with Canada for the Kane Basin (KB), Baffin Bay (BB), and Davis Strait (DS) subpopulations. Recent studies have produced abundance estimates for KB and BB (SWG 2016) and DS (Peacock et al. 2013), and to date no abundance estimate is available for the East Greenland subpopulation. A Memorandum of Understanding between Greenland and Canada/Nunavut on the Kane Basin and Baffin Bay subpopulations governs management of these areas, including harvest (Jessen 2018). Greenland listed polar bears as Vulnerable in 2007 (Boertmann 2008).

Table 14.1 The number, sex, and geographic locations of polar bears housed in zoos as of December 31, 2017 (Linke 2017).

	Male	Female	Unknown	Total	Institutions
Europe	66	80	5	151	59
North America	31	41	1	73	33
Asia	33	33	0	66	33
Australia	2	2	0	4	1
South America	1	1	0	2	1
Africa	0	0	0	0	0
Total	133	157	6	296	127

Status in Captivity

Some of the earliest historical notations of polar bears in human care were polar bears gifted to royalty (i.e. in 894 to King Harald of Norway, and to King Mohammed al Kamul of Egypt in 1234; Engelhard 2017). In 1252, the king of Norway gifted King Henry III of England a polar bear for the Royal Menagerie. In 1693, Fredrick I, King of Prussia used polar bears in mock "Roman arena"-like combat fights with other species. Polar bears were first exhibited in Europe at a zoo in London in 1829 and in North America in 1876 in Philadelphia (Linke 2017). Today, nearly 300 polar bears can be found in zoos on five continents (Table 14.1).

In the 1990s many US and European zoos decided to no longer house the species. Although polar bears were still popular with visitors, some zoos and other groups expressed concern about the animals' welfare and the zoo population was considered to be of low conservation value. In the time between the publication of the first North American regional polar bear studbook in 1995 and the second addition in 2006, the numbers of bears and facilities housing them decreased by almost half (Poirier & Lanthier 1995; Meyerson 2006; Linke 2017).

The Association of Zoos and Aquariums (AZA) Species Survival Program (SSP) formed in 2003 with the goal of maintaining a genetically diverse captive population while decreasing the numbers in the AZA population. Population planning for this program included breeding moratoriums and the use of contraception. Animal care and welfare professionals developed new care recommendations, which included input from field biologists and government entities. These recommendations addressed important facility design features, behavioral needs and enrichment and training programs. In 2006, the European Association of Zoos and Aquaria (EAZA) established the Polar Bear European Endangered Species Programmes (EEP). Similar population trends and welfare concerns were acknowledged by the EAZA Ursid specialists, which led to the development of *Ursid Husbandry Guidelines* (Meyerson et al. 2017). For more than a decade these two polar bear programs have worked together to enhance the health and welfare of polar bears in human care in North America and Europe.

As global warming emerged as the primary threat to polar bears, there has been renewed interest among zoos in housing polar bears as a flagship species for conservation action. With over 180 million people visiting US zoos annually, there is an opportunity to educate the public about the threats facing polar bears and the opportunities for climate change mitigation.

More recently, collaborations between zoo and field researchers have demonstrated that bears in human care provide an opportunity for research that informs the status and health of their wild counterparts. Active training programs have already supported the use of polar bears in human care for studies of sensory ecology (olfaction and hearing), energetics, emerging diseases, reproduction, and development of new methods for studying wild bears (Dutton et al. 2009; Bowen et al. 2015; Owen et al. 2015; Ware et al. 2015; Rode et al. 2016; Pagano et al. 2019). The success of these studies led to the creation of the Polar Bear Research Council and the creation of a research masterplan to facilitate research with zoo polar bears that aids in understanding wild populations. This plan was developed by members of the AZA, EAZA, government, non-government, and scientific communities.

References

Aars, J. (2013). Variation in detection probability of polar bear maternity dens. *Polar Biology* 36: 1089–1096.

Aars, J., Marques, T. A., Lone, K., et al. (2017). The number and distribution of polar bears in the western Barents Sea. *Polar Research* 36: 1374125.

Amstrup, S. C., Marcot, B. G. & Douglas D. C. (2008). A Bayesian network modelling approach to forecasting the 21st century worldwide status of polar bears. In: DeWeaver, E. T., Bitz, C. M. & Tremblay, L. B. (Eds.), *Arctic sea ice decline: Observations, projections, mechanisms, and implications* (pp. 213–268). Geophysical Monograph 180. Washington, DC: American Geophysical Union.

Andersen, M., Derocher, A. E., Wiig, Ø. & Aars, J. (2012). Polar bear (*Ursus maritimus*) maternity den distribution in Svalbard, Norway. *Polar Biology* 35: 499–508.

Arnould, J. P. Y. & Ramsay, M. A. (1994). Milk production and milk consumption in polar bears during the ice-free period in western Hudson Bay. *Canadian Journal of Zoology* 72: 1365–1370.

Åsbakk, K., Aars, J., Derocher, A. E., et al. (2010). Serosurvey for *Trichinella* in polar bears (*Ursus maritimus*) from Svalbard and the Barents Sea. *Veterinary Parasitology* 172: 256–263.

Atkinson, S. N. & Ramsay, M. A. (1995). The effects of prolonged fasting of the body composition and reproductive success of female polar bears (*Ursus maritimus*). *Functional Ecology* 9: 559–567.

Atkinson, S. N., Nelson, R. A. & Ramsay, M. A. (1996a). Changes in the body composition of fasting polar bears (*Ursus maritimus*): the effect of relative fatness on protein conservation. *Physiological Zoology* 69: 304–316.

Atkinson, S. N., Stirling, I. & Ramsay, M. A. (1996b). Growth in early life and relative body size among adult polar bears (*Ursus maritimus*). *Journal of Zoology* 239: 225–234.

Atwood, T. C., Marcot, B. G., Douglas, D. C., et al. (2015). Evaluating and ranking threats to the long-term persistence of polar bears. US Geological Survey Open-File Report 2014-1254.

Atwood, T. C., Peacock, E., McKinney, M. A., et al. (2016). Rapid environmental change drives increased land use by an Arctic marine predator. *PLoS ONE* 11: e0155932.

Atwood, T. C., Duncan, C. G., Patyk, K. A., et al. (2017). Environmental and behavioral changes may influence the exposure of an Arctic apex predator to pathogens and contaminants. *Scientific Reports* 7: 13193.

Bechshøft, T. Ø., Derocher, A. E., Viengkone, M., et al. (2018). On the integration of ecological and physiological variables in polar bear toxicology research: a systematic review. *Environmental Review* 26: 1-12.

Belikov, S. E., Boltunov, A. N., Ovsyanikov, N. G., Mordvintsev, I. & Nikiforov, V. V. (2010). Polar bear management and research in Russia, 2005–2009. In: Obbard, M. E., Thiemann, G. W., Peacock, E. & DeBruyn, T. D. (Eds.), *Polar bears: Proceedings of the 15th Working Meeting of the IUCN/SSC Polar Bear Specialist Group*, June 29–July 3, 2009, Copenhagen, Denmark. Occasional Paper of the IUCN Species Survival Commission No. 43 (pp. 165–169). Gland, Switzerland and Cambridge, UK: IUCN.

Bentzen, T. W., Follmann, E. H., Amstrup, S. C., et al. (2007). Variation in winter diet of southern Beaufort Sea polar bears inferred from stable isotope analysis. *Canadian Journal of Zoology* 85: 596–608.

Bignert, A., Hung, H., Katsoyiannis, A. A., et al. (2016). *Arctic Monitoring and Assessment Programme (AMAP) 2016: Temporal Trends in Persistent Organic Pollutants in the Arctic*. Oslo, Norway: AMAP.

Boertmann, D. (2008). *Greenland Red List – 2007*. Aarhus: Greenland Home Rule.

Bowen, L., Miles, A. K., Waters, S., et al. (2015). Gene transcription in polar bears (*Ursus maritimus*) from disparate populations. *Polar Biology* 38: 1413–1427.

Bromaghin, J. F., McDonald, T. L., Stirling, I., et al. (2015). Polar bear population dynamics in the southern Beaufort Sea during a period of sea ice decline. *Ecological Applications* 25: 634–651.

Brown, T. M., Macdonald, R., Muir, D. C. G. & Letcher, R. J. (2018). The distribution and trends of persistent organic pollutants and mercury in marine mammals from Canada's Eastern Arctic. *Science of the Total Environment* 618: 500–517.

Cherry, S. G., Derocher, A. E., Thiemann, G. W. & Lunn, N. J. (2013). Migration phenology and seasonal fidelity of an Arctic marine predator in relation to sea ice dynamics. *Journal of Animal Ecology* 82: 912–921.

Clark, D. A., Stirling, I. & Calvert, W. (1997). Distribution, characteristics, and use of earth dens and related excavations by polar bears on the western Hudson Bay Lowlands. *Arctic* 50: 158–166.

Comiso, J. C. (2012). Large decadal decline of the Arctic multiyear ice cover. *Journal of Climate* 25: 1176–1193.

Comiso, J. C., Parkinson, C. L., Gersten, R. & Stock, L. (2008). Accelerated decline in the Arctic sea ice cover. *Geophysical Research Letters* 35: L01703.

COSEWIC. (2008). COSEWIC Assessment and Update Status Report (WWW document). Available from http://sararegistry.gc.ca/virtual_sara/files/cosewic/sr_polar_bear_0808_e.pdf (accessed January 3, 2019).

Dau, J. & Barrett, R. (1981). *Trichinella*. In: Dieterich, R. A. (Ed.), *Alaskan wildlife diseases* (pp. 151–161). Fairbanks, AK: University of Alaska.

DeMaster, D. P. & Stirling, I. (1981). *Ursus maritimus. Mammalian Species* 145: 1–7.

Derocher, A. E. (1999). Latitudinal variation in litter size of polar bears: ecology or methodology? *Polar Biology* 22: 350–356.

Derocher, A. E. & Stirling, I. (1990a). Distribution of polar bears (*Ursus maritimus*) during the ice-free period in western Hudson Bay. *Canadian Journal of Zoology* 68: 1395–1403.

Derocher, A. E. & Stirling, I. (1990b). Observations of aggregating behaviour in adult male polar bears (*Ursus maritimus*). *Canadian Journal of Zoology* 68: 1390–1394.

Derocher, A. E. & Stirling, I. (1994). Age-specific reproductive performance of female polar bears. *Journal of Zoology* 234: 527–536.

Derocher, A. E. & Stirling, I. (1998). Geographic variation in growth of polar bears (*Ursus maritimus*). *Journal of Zoology* 245: 65–72.

Derocher, A. E. & Wiig, Ø. (2002). Postnatal growth in body length and mass of polar bears (*Ursus maritimus*) at Svalbard. *Journal of Zoology* 256: 343–349.

Derocher, A. E., Stirling, I. & Andriashek, D. (1992). Pregnancy rates and serum progesterone levels of polar bears in western Hudson Bay. *Canadian Journal of Zoology* 70: 561–566.

Derocher, A. E., Andriashek, D. & Arnould, J. P. Y. (1993). Aspects of milk composition and lactation in polar bears. *Canadian Journal of Zoology* 71: 561–567.

Derocher, A. E., Wiig, Ø. & Andersen, M. (2002). Diet composition of polar bears in Svalbard and the western Barents Sea. *Polar Biology* 25: 448–452.

Derocher, A. E., Andersen, M., Wiig, Ø. & Aars, J. (2010). Sexual dimorphism and the mating ecology of polar bears (*Ursus maritimus*) at Svalbard. *Behavioral Ecology and Sociobiology* 64: 939–946.

Derocher, A. E., Andersen, M., Wiig, Ø., et al. (2011). Sea ice and polar bear den ecology at Hopen Island, Svalbard. *Marine Ecology Progress Series* 441: 273–279.

de Wit, C. A., Muir, D. C. G., Vorkamp, K., et al. (2017). Arctic *Monitoring and Assessment Programme (AMAP) 2016: Chemicals of Emerging Arctic Concern*. Oslo, Norway: AMAP.

Dietz, R., Sonne, C., Basu, N., et al. (2013). What are the toxicological effects of mercury in Arctic biota? *Science of the Total Environment* 443: 775–790.

Dietz, R., Desforges, J. P., Gustavson, K., et al. (2018). Immunologic, reproductive, and carcinogenic risk assessment from POP exposure in East Greenland polar bears (*Ursus maritimus*) during

1983–2013. *Environmental International* 118: 169–178.

Dietz, R., Letcher, R. J., Desforges, J-P., et al. (2019). Current state of knowledge on biological effects from contaminants on arctic wildlife and fish. *Science of the Total Environment* 696: 133792.

Duignan, P. J., Van Bressem, M-F., Cortés-Hinojosa, G. & Kennedy-Stoskopf, S. (2018). Viruses. In: Gulland, F. M. D., Dierauf, L. A. & Whitman, K. L. (Eds.), *Handbook of marine mammal diseases*, 2nd edition (pp. 331–365). Boca Raton, FL: CRC Press.

Duncan, C. G., Tiller, R., Mathis, D., et al. (2014). *Brucella placentitis* and seroprevalence in northern fur seals (*Callorhinus ursinus*) of the Pribilof Islands, Alaska. *Journal of Veterinary Diagnostic Investigation* 26: 507–512.

Dupouy-Camet, J., Bourée, P. & Year, H. (2017). *Trichinella* and polar bears: a limited risk for humans. *Journal of Helminthology* 91: 440–446.

Durner, G. M., Amstrup, S. C. & Fischbach, A. S. (2003). Habitat characteristics of polar bear terrestrial maternal den sites in northern Alaska. *Arctic* 56: 55–62.

Durner, G. M., Douglas, D. C., Nielson, R. M., et al. (2009). Predicting 21st-century polar bear habitat distribution from global climate models. *Ecological Monographs* 79: 25–58.

Durner, G. M., Whiteman, J. P., Harlow, H. J., et al. (2011). Consequences of long-distance swimming and travel over deep-water pack ice for a female polar bear during a year of extreme sea ice retreat. *Polar Biology* 34: 975–984.

Durner, G. M., Laidre, K. L. & York, G. S. (2018). *Proceedings of the 18th working meeting of the IUCN/SSC Polar Bear Specialist Group*, June 7–11, 2016, Anchorage, AK. Gland, Switzerland and Cambridge, UK: IUCN.

Dutton, C. J., Quinnell, M., Lindsay, R., DeLay, J. & Barker, I. K. (2009). Paraparesis in a polar bear (*Ursus maritimus*) associated with West Nile virus infection. *Journal of Zoo and Wildlife Medicine* 40: 568–571.

Engelhard, M. (2017). *Ice bear*. Washington, DC: University of Washington Press.

Environment and Climate Change Canada. (2018). Canadian polar bear subpopulation and status map 2018 (WWW document). Available from www.canada.ca/en/environment-climate-change/services/biodiversity/maps-sub-populations-polar-bears-protected.html#_fig03 (accessed January 9, 2019).

Escajeda, E., Laidre, K. L., Born, E. W., et al. (2018). Identifying shifts in maternity den phenology and habitat characteristics of polar bears (*Ursus maritimus*) in Baffin Bay and Kane Basin. *Polar Biology* 41: 87–100.

Fagre, A., Patyk, K. A., Nol, P., et al. (2015). A review of infectious agents in polar bears (*Ursus maritimus*) and their long-term ecological relevance *EcoHealth* 12: 528–539.

Fay, F. H. (1960). Carnivorous walrus and some arctic zoonoses. *Arctic* 13: 111–122.

Ferguson, S. H., Taylor, M. K., Born, E. W., Rosing-Asvid, A. & Messier, F. (1999). Determinants of home range size for polar bears (*Ursus maritimus*). *Ecology Letters* 2: 311–318.

Ferguson, S. H., Taylor, M. K. & Messier, F. (2000). Influence of sea ice dynamics on habitat selection by polar bears. *Ecology* 81: 761–772.

Ferguson, S. H., Stirling, I. & McLoughlin, P. (2005). Climate change and ringed seal (*Phoca hispida*) recruitment in western Hudson Bay. *Marine Mammal Science* 21: 121–135.

Ferguson, S. H., Young, B. G., Yurkowski, D. J., et al. (2017). Demographic, ecological, and physiological responses of ringed seals to an abrupt decline in sea ice availability. *Peer Journal* 5: e2957.

Fischbach, A. S., Amstrup, S. C. & Douglas, D. C. (2007). Landward and eastward shift of Alaskan polar bear denning associated with recent sea ice changes. *Polar Biology* 30: 1395–1405.

Freitas, C., Kovacs, K. M., Andersen, M., et al. (2012). Importance of fast ice and glacier fronts for female polar bears and their cubs during spring in Svalbard, Norway. *Marine Ecology Progress Series* 447: 289–304.

Galicia, M. P., Thiemann, G. W., Dyck, M. G., Ferguson, S. H. & Higdon, J. W. (2016). Dietary habits of polar bears in Foxe Basin, Canada: possible evidence of a trophic regime shift mediated by a new top predator. *Ecology and Evolution* 6: 6005–6018.

Gormezano, L. J. & Rockwell, R. F. (2013). What to eat now? Shifts in polar bear diet during the ice-free season in western Hudson Bay. *Ecology and Evolution* 3: 3509–3523.

Griffen, B. D. (2018). Modelling the metabolic costs of swimming in polar bears (*Ursus maritimus*). *Polar Biology* 41: 491–503.

Hamilton, S. G., Castro de la Guardia, L., Derocher, A. E., et al. (2014). Projected polar bear sea ice habitat in the Canadian Arctic Archipelago. *PLoS ONE* 9: e113746.

Ingolfsson, O. &Wiig, Ø. (2009). Late Pleistocene fossil find in Svalbard: the oldest remains of a polar bear (*Ursus maritimus* Phipps, 1744) ever discovered. *Polar Research* 28: 455–462.

Jessen, A. (2018). Management on polar bears in Greenland, 2009–2016. In: Durner, G. M., Laidre, K. L. & York, G. S. (Eds.), *Polar bears: Proceedings of the 18th Working Meeting of the IUCN/SSC Polar Bear Specialist Group*, Anchorage, Alaska, June 7–11, 2016 (pp. 84–94). Gland, Switzerland and Cambridge, UK: IUCN.

Kingsley, M. C. S. (1979). Fitting the von Bertalanffy growth equation to polar bear age-weight data. *Canadian Journal of Zoology* 57: 1020–1025.

Kingsley, M. C. S., Stirling, I. & Calvert, W. (1985). The distribution and abundance of seals in the Canadian high Arctic, 1980–1982. *Canadian Journal of Fisheries and Aquatic Sciences* 42: 1189–1210.

Kirk, C. M., Amstrup, S., Swor, R., Holcomb, D. & O'Hara, T. M. (2010). Morbillivirus and *Toxoplasma* exposure and association with hematological parameters for southern Beaufort Sea polar bears: potential response to infectious agents in a sentinel species. *EcoHealth* 7: 321–331.

Knudsen, B. (1978). Time budgets of polar bears (*Ursus maritimus*) on North Twin Island, James Bay, during summer. *Canadian Journal of Zoology* 56: 1627–1628.

Kolenosky, G. B. & Prevett, J. P. (1983). Productivity and maternity denning of polar bears in Ontario. *International Conference on Bear Research and Management* 5: 238–245.

Kumar, V., Lammers, F., Bidon, T., et al. (2017). The evolutionary history of bears is characterized by gene flow across species. *Scientific Reports* 7: 46487.

Kwok, R. & Rothrock, D. A. (2009). Decline in Arctic sea ice thickness from submarine and ICESat records: 1958–2008. *Geophysical Research Letters* 36: 2009GL039035.

Laidre, K. L., Born, E. W., Heagerty, P., et al. (2015a). Shifts in habitat use by female polar bears (*Ursus maritimus*) in East Greenland. *Polar Biology* 38: 879–893.

Laidre, K. L., Stern, H., Kovacs, K.M., et al. (2015b). Arctic marine mammal population status, sea ice habitat loss, and conservation recommendations for the 21st century. *Conservation Biology* 29: 724–737.

Laidre, K. L., Stirling, I., Estes, J. A., Kochnev, A. & Roberts, J. (2018). Historical and potential future importance of large whales as a food for polar bears. *Frontiers in Ecology and the Environment* 16: 515–524.

Larsen, T. S. & Stirling I. (2009). The Agreement on the Conservation of Polar Bears – its History and Future. Norsk Polarinstittut, No. 127. 16 pp.

Latour, P. B. (1981). Spatial relationships and behavior of polar bears (*Ursus maritimus* Phipps) concentrated on land during the ice-free season of Hudson Bay. *Canadian Journal of Zoology* 56: 1763–1774.

Letcher, R. J., Bustnes, J-O., Dietz, R., et al. (2010). Exposure and effects assessment of persistent organic pollutants in Arctic wildlife and fish. *Science of the Total Environment* 408: 2995–3043.

Letcher, R. J., Morris, A. D., Dyck, M., et al. (2018). Legacy and new halogenated persistent organic pollutants in polar bears from a contamination hotspot in the Arctic, Hudson Bay Canada. *Science of the Total Environment* 610: 121–136.

Linke, K. (2017). *World Zoo & Aquarium Association International Polar Bear Studbook*. Rostock, Germany: Zoo Rostock.

Liu, Y., Richardson, E. S., Derocher, A. E., et al. (2018). Hundreds of unrecognized halogenated contaminants discovered in polar bear serum. *Angewandte Chemie* 57: 16,401–16,406.

Loewen, K., Prins, B. & Philibert, H. (1990). Northwest Territories. Rabies in a polar bear. *Canadian Veterinary Journal* 31: 457.

Lone, K., Kovacs, K. M., Lydersen, C., et al. (2018a). Aquatic behaviour of polar bears (*Ursus maritimus*) in an increasingly ice-free Arctic. *Scientific Reports* 8: 9677.

Lone, K., Merkel, B., Lydersen, C., Kovacs, K. M. & Aars, J. (2018b). Sea ice resource selection models for polar bears in the Barents Sea subpopulation. *Ecography* 41: 567–578.

Lunn, N. J., Stirling, I., Andriashek, D. & Richardson, E. (2004). Selection of maternity dens by female polar bears in western Hudson Bay, Canada and the effects of human disturbance. *Polar Biology* 27: 350–356.

Lunn, N. J., Servanty S., Regehr, E.V., et al. (2016). Demography of an apex predator at the edge of its range: impacts of changing sea ice on polar bears in Hudson Bay. *Ecological Applications* 26: 1302–1320.

Malenfant, R. M., Coltman, D. W., Richardson, E. S., et al. (2015). Evidence of adoption, monozygotic twinning, and low inbreeding rates in a large genetic pedigree of polar bears. *Polar Biology* 39: 1455–1465.

Malenfant, R. M., Davis, C. S., Cullinghman, C. I. & Coltman, D. W. (2016). Circumpolar genetic structure and recent gene flow of polar bears: a reanalysis. *PLoS ONE* 11: e0148967.

Manly, B. F. J., McDonald, L. L., Thomas, D. L., McDonald, T. L. & Erickson, W. P. (2002). *Resource selection by animals.* Dordrecht: Kluwer Academic Publishers.

Mauritzen, M., Derocher, A. E., Wiig, Ø., et al. (2002). Using satellite telemetry to define spatial population structure in polar bears in the Norwegian and western Russian Arctic. *Journal of Applied Ecology* 39: 79–90.

Mauritzen, M., Belikov, S. E., Boltunov, A. N., et al. (2003). Functional responses in polar bear habitat selection. *Oikos* 100: 112–124.

McCall, A. G., Pilfold, N. W., Derocher, A. E. & Lunn, N. J. (2016). Seasonal habitat selection by adult female polar bears in western Hudson Bay. *Population Ecology* 58: 407–419.

McDonald, J. C., Gyorkos, T. W., Alberton, B., et al. (1990) An outbreak of toxoplasmosis in pregnant women in northern Quebec. *Journal of Infectious Diseases* 161: 769–774.

McKinney, M. A., Stirling, I., Lunn, N. J., Peacock, E. & Letcher, R. J. (2010). The role of diet on long-term concentration and pattern trends of brominated and chlorinated contaminants in western Hudson Bay polar bears, 1991–2007. *Science of the Total Environment* 408: 6210–6222.

McKinney, M. A., Iverson, S., Fisk, A., et al. (2013). Global change effects on the long-term feeding ecology and contaminant exposures of East Greenland polar bears. *Global Change Biology* 19: 2360–2372.

McKinney, M. A., Pedro, S., Dietz, R., et al. (2015). Ecological impacts of global climate change on persistent organic pollutant and mercury pathways and exposures in arctic marine ecosystems: a review of initial findings. *Current Zoology* 61: 617–628.

Messier, F., Taylor, M. K. & Ramsay, M. A. (1992). Seasonal activity patterns of female polar bears (*Ursus maritimus*) in the Canadian Arctic as revealed by satellite telemetry. *Journal of Zoology* 226: 219–229.

Messier, F., Taylor, M. K. & Ramsay, M. A. (1994). Denning ecology of polar bears in the Canadian Arctic archipelago. *Journal of Mammalogy* 75: 420–430.

Messier, V., Levesque, B., Proulx, J. F., et al. (2009). Seroprevalence of *Toxoplasma gondii* among Nunavik Inuit (Canada). *Zoonoses Public Health* 56: 188–197.

Meyerson, R. (2006). *Association of Zoo and Aquariums Polar Bear Studbook*. Toledo, OH: Toledo Zoo.

Meyerson, R., Moore, D., Long, S. & Che-Castaldo, J. (2017). Welfare of captive polar bears and their value in in situ conservation efforts. In: Butterworth, A. (Ed.), *Marine mammal welfare – Human-induced change in the marine environment and its impacts on marine mammal welfare* (pp. 489–502). Bristol, UK: Springer Publishing.

Miller, M., Shapiro, K., Murray, M. J., Haulena, M. & Raverty, S. (2018). Protozoan parasites of marine mammals. In: Gulland, F. M. D., Dierauf, L. A. & Whitman, K. L. (Eds.), *Handbook of marine mammal diseases*, 3rd edition (pp. 425–469). Boca Raton, FL: CRC Press.

Miller, S., Wilder, J. & Wilson, R. R. (2015). Polar bear–grizzly bear interactions during the autumn open-water period in Alaska. *Journal of Mammalogy* 96: 1317–1325.

Miller, W., Schuster, S. C., Welch, A. J., et al. (2012). Polar and brown bear genomes reveal ancient admixture and demographic footprints of past climate change. *Proceedings of the National Academy of Sciences* 109: 2382–2390.

Molnár, P. K., Derocher, A. E., Lewis, M. A. & Taylor, M. K. (2008). Modelling the mating system of polar bears: a mechanistic approach to the Allee effect.

Proceedings of the Royal Society B – Biological Sciences 275: 217–226.

Monnett, C. & Gleason, J. S. (2006). Observations of mortality associated with extended open-water swimming by polar bears in the Alaskan Beaufort Sea. *Polar Biology* 29: 681–687.

Nymo, I., Tryland, M., Kirstine-Frie, A., et al. (2013). Age-dependent prevalence of anti-*Brucella* antibodies in hooded seals *Cystophora cristata*. *Inter-Research* 106: 187–196.

Obbard, M. E., Cattet, M. R. L., Howe, E. J., et al. (2016). Trends in body condition in polar bears (*Ursus maritimus*) from the Southern Hudson Bay subpopulation in relation to changes in sea ice. *Arctic Science* 2: 15–32.

O'Hara, T. M., Holcomb, D., Elzer, P., et al. (2010). *Brucella* species survey in polar bears (*Ursus maritimus*) of northern Alaska. *Journal of Wildlife Diseases* 46: 687–694.

Olson, J. W., Rode, K. D., Eggett, D. L., et al. (2017). Collar temperature sensor data reveal long-term patterns in southern Beaufort Sea polar bear den distribution on pack ice and land. *Marine Ecology Progress Series* 564: 211–224.

Owen, M. A., Swaisgood, R. R., Slocomb, C., et al. (2015). An experimental investigation of chemical communication in the polar bear. *Journal of Zoology* 295: 36–43.

Pagano, A., Cutting, A., Nicassio-Hiskey, N., Hash, A. & Williams, T. (2019). Energetic costs of aquatic locomotion in a sub-adult polar bear. *Marine Mammal Science* 35, 649–659.

Parks, E. K., Derocher, A. E. & Lunn, N. J. (2006). Seasonal and annual movement patterns of polar bears on the sea ice of Hudson Bay. *Canadian Journal of Zoology* 84: 1281–1294.

Peacock, E., Taylor, M. K., Laake, J. L. & Stirling, I. (2013). Population ecology of polar bears in Davis Strait, Canada and Greenland. *Journal of Wildlife Management* 77: 463–476.

Philippa, J. D., Leighton, F. A., Daoust, P. Y., *et al.* (2004). Antibodies to selected pathogens in free-ranging terrestrial carnivores and marine mammals in Canada. *Veterinary Record* 155: 135–140.

Pilfold, N. W., Derocher, A. E., Stirling, I., Richardson, E. & Andriashek, D. (2012). Age and sex composition of seals killed by polar bears in the eastern Beaufort Sea. *PLoS ONE* 7: e41429.

Pilfold, N. W., Derocher, A. E. & Richardson, E. (2014). Influence of intraspecific competition on the distribution of a wide-ranging, non-territorial carnivore. *Global Ecology and Biogeography* 23: 425–435.

Poirier, P. & Lanthier, C. (1995). *North American Regional Polar Bear Studbook*. Quebec, Canada: Jardin zoologique de Granby.

Pond, C. M., Mattacks, C. A., Colby, R. H. & Ramsay, M. A. (1992). The anatomy, chemical composition, and metabolism of adipose tissue in wild polar bears (*Ursus maritimus*). *Canadian Journal of Zoology* 70: 326–341.

Pongracz, J. D., Paetkau, D., Branigan, M. & Richardson, E. (2017). Recent hybridization between a polar bear and grizzly bears in the Canadian Arctic. *Arctic* 70: 151–160.

Prop, J., Aars, J., Bardsen, B.-J., et al. (2015). Climate change and the increasing impact of polar bears on bird populations. *Frontiers in Ecology and Evolution* 3: 33.

Rah, H., Chomel, B. B., Kasten, R. W., et al. (2005) Serosurvey of selected zoonotic agents in polar bears (*Ursus maritimus*). *Veterinary Record* 156: 7–13.

Ramsay, M. A. & Dunbrack, R. L. (1986). Physiological constraints on life-history phenomena: the example of small bear cubs at birth. *American Naturalist* 127: 735–743.

Ramsay, M. A. & Stirling, I. (1986). On the mating system of polar bears. *Canadian Journal of Zoology* 64: 2142–2151.

Ramsay, M. A. & Stirling, I. (1988). Reproductive biology and ecology of female polar bears (*Ursus maritimus*). *Journal of Zoology* 214: 601–634.

Ramsay, M. A. & Stirling, I. (1990). Fidelity of female polar bears to winter den sites. *Journal of Mammalogy* 71: 233–236.

Regehr, E. V., Lunn, N. J., Amstrup, S. C. & Stirling, I. (2007). Effects of earlier sea ice breakup on survival and population size of polar bears in western Hudson Bay. *Journal of Wildlife Management* 71: 2673–2683.

Regehr, E. V., Hunter, C. M., Caswell, H., Amstrup, S. C. & Stirling, I. (2010). Survival and breeding of polar bears in the southern Beaufort Sea in relation to sea ice. *Journal of Animal Ecology* 79: 117–127.

Regehr, E. V, Laidre, K. L., Akçakaya, H. R., et al. (2016). Conservation status of polar bears (*Ursus maritimus*) in relation to projected sea-ice declines. *Biology Letters* 12: 20160556.

Regehr, E. V., Hostetter, N. J., Wilson, R. R., et al. (2018). Integrated population modeling provides the first empirical estimates of vital rates and abundance for polar bears in the Chukchi Sea. *Scientific Reports* 8: 16780.

Richardson, E., Stirling, I. & Hik, D. S. (2005). Polar bear (*Ursus maritimus*) maternity denning habitat in western Hudson Bay: a bottom-up approach to resource selection functions. *Canadian Journal of Zoology* 83: 860–870.

Rigét, F. F., Bignert, A., Braune, B., et al. (2019). Temporal trends of persistent organic pollutants in Arctic marine and freshwater biota. *Science of the Total Environment* 649: 99–110.

Robbins, C. T., Ben-David, M., Fortin, J. K. & Nelson, O. L. (2012). Maternal condition determines birth date and growth of newborn bear cubs. *Journal of Mammalogy* 93: 540–546.

Rode, K. D., Amstrup, S. C. & Regehr, E. V. (2010). Reduced body size and cub recruitment in polar bears associated with sea ice decline. *Ecological Applications* 20: 768–782.

Rode, K. D., Regehr, E. V., Douglas, D. C., et al. (2014).Variation in the response of an Arctic top predator experiencing habitat loss: feeding and reproductive ecology of two polar bear populations. *Global Change Biology* 20: 76–88.

Rode, K. D., Wilson, R. R., Regehr, E. V., et al. (2015a). Increased land use by Chukchi Sea polar bears in relation to changing sea ice conditions. *PLoS ONE* 10: e0142213.

Rode, K. D., Robbins, C. T., Nelson, L., et al. (2015b). Can polar bears use terrestrial foods to offset lost ice-based hunting opportunities? *Frontiers in Ecology and the Environment* 13: 138–145.

Rode, K. D., Stricker, C. A., Erlenbach, J., et al. (2016). Isotopic incorporation and the effects of fasting and dietary lipid content on isotopic discrimination in large, carnivorous mammals. *Physiological and Biochemical Zoology* 89: 182–197.

Rosing-Asvid, A., Born, E. W. & Kingsley, M. C. S. (2002). Age at sexual maturity of males and timing of the mating season of polar bears (*Ursus maritimus*) in Greenland. *Polar Biology* 25: 878–883.

Routti, H., Atwood, T., Bechshoft, T. Ø., et al. (2019). State of knowledge on current exposure, fate and potential health effects of contaminants in polar bears from the circumpolar Arctic. *Science of the Total Environment* 664, 1063–1083.

Russell, R. H. (1975). The food habits of polar bears of James Bay and southwest Hudson Bay in summer and autumn. *Arctic* 28: 117–129.

Schliebe, S., Rode, K. D., Gleason, J. S., et al. (2008). Effects of sea ice extent and food availability on spatial and temporal distribution of polar bears during the fall openwater period in the Beaufort Sea. *Polar Biology* 31: 999–1010.

Sidor, I., Goldstein, T. & Whatmore, A. M. (2008). Transplacental infection with *Brucella* in California sea lion (*Zalophus californianus*) fetus. In Brucellosis International Research Conference, Chicago, USA.

Smith, T. G. & Stirling, I. (1975). The breeding habitat of the ringed seal (*Phoca hispida*). The birth lair and associated structures. *Canadian Journal of Zoology* 53: 1297–1305.

Smith, T. G. & Stirling, I. (1978). Variation in the density of ringed seal (*Phoca hispida*) birth lairs in the Amundsen Gulf, Northwest Territories. *Canadian Journal of Zoology* 56: 1066–1070.

Smith, T. S., Partridge, S. T., Amstrup, S. C. & Schliebe, S. (2007). Post-den emergence behavior of polar bears (*Ursus maritimus*) in northern Alaska. *Arctic* 60: 187–194.

Sonne, C. (2010). Health effects from long-range transported contaminants in Arctic top predators: an integrated review based on studies of polar bears and relevant model species. *Environment International* 36: 461–491.

Sonne, C., Letcher, R. J., Bechshøft, T. Ø., et al. (2012). Two decades of biomonitoring polar bear health in Greenland: a review. *Acta Veterinaria Scandinavica* 54: S15.

Sonne, C., Bechshøft, T. Ø. & Rigét, F. F. (2013). Size and density of East Greenland polar bear (*Ursus maritimus*) skulls: valuable bio-indicators of environmental changes? *Ecological Indicators* 34: 290–295.

Stern, H. L. & Laidre, K. L. (2015). Sea-ice indicators of polar bear habitat. *The Cryosphere* 10: 2027–2041.

Stirling, I. (1974). Midsummer observations on the behavior of wild polar bears (*Ursus maritimus*). *Canadian Journal of Zoology* 52: 1191–1198.

Stirling, I. (1997). The importance of polynyas, ice edges, and leads to marine mammals and birds. *Journal of Marine Systems* 10: 9–21.

Stirling, I. & Archibald, W. R. (1977). Aspects of predation on seals by polar bears. *Journal of Fisheries Research Board Canada* 34: 1126–1129.

Stirling, I. & Latour, P. B. (1978). Comparative hunting abilities of polar bear cubs of different ages. *Canadian Journal of Zoology* 56: 1768–1772.

Stirling, I. & van Meurs, R. (2015). Longest recorded underwater dive by a polar bear. *Polar Biology* 38: 1301–1304.

Stirling, I., Kingslay, M. C. S. & Calvert, W. (1982). The distribution and abundance of seals in the eastern Beaufort Sea, 1974–79. *Canadian Wildlife Service Occasional Report* 46.

Stirling, I., Andriashek, D. & Calvert, W. (1993). Habitat preferences of polar bears in the western Canadian Arctic in late winter and spring. *Polar Record* 29: 13–24.

Stirling, I., Lunn, N. J. & Iacozza, J. (1999). Long-term trends in the population ecology of polar bears in western Hudson Bay in relation to climate change. *Arctic* 52: 294–306.

Stirling, I., Lunn, N. J., Iacozza, J., Elliott, C. & Obbard, M. (2004). Polar bear distribution and abundance on the southwestern Hudson Bay coast during open water season, in relation to population trends and annual ice patterns. *Arctic* 57: 15–26.

Stirling, I., McDonald, T. L., Richardson, E. S., Regehr, E. V. & Amstrup, S. C. (2011). Polar bear population status in the northern Beaufort Sea, Canada, 1971–2006. *Ecological Applications* 21: 859–876.

Stirling, I., Spencer, C. & Andriashek, D. (2016). Behavior and activity budgets of wild breeding polar bears (*Ursus maritimus*). *Marine Mammal Science* 32: 13–37.

Stishov, M. S. (1997). Results of aerial counts of the polar bear dens on the Arctic coast of the extreme Northeast Asia. Proceedings of the working meeting of the IUCN polar bear specialist group. *International Union for the Conservation of Nature Species Survival Commission Occasional Paper* 7: 90–92.

SWG (Scientific Working Group to the Canada-Greenland Joint Commission on Polar Bear). (2016). Re-assessment of the Baffin Bay and Kane Basin polar bear subpopulations: final report to the Canada–Greenland Joint Commission on polar bear. July 31, 2016.

Tartu, S., Aars, J., Andersen, M., et al. (2018). Choose your poison – space-use strategy influences pollutant exposure in Barents Sea polar bears. *Environmental Science and Technology* 52: 3211–3221.

Taylor, M., Elkin, B., Maier, N. & Bradley, M. (1991). Observation of a polar bear with rabies. *Journal of Wildlife Diseases* 27: 337–339.

Thiemann, G., Iverson, S. J. & Stirling, I. (2007). Variability in the blubber fatty acid composition of ringed seals (*Phoca hispida*) across the Canadian Arctic. *Marine Mammal Science* 23: 241–261.

Thiemann, G. W., Iverson, S. J. & Stirling, I. (2008). Polar bear diets and arctic marine food webs: insights from fatty acid analysis. *Ecological Monographs* 78: 591–613.

Thorshaug, K. & Rosted, A. F. (1956). Researches into the prevalence of trichinosis in animals in Arctic and Antarctic waters. *Nordic Veterinary Medicine* 8: 115–129.

Tryland, M. (2000). Zoonoses of arctic mammals. *Infectious Disease Review* 2: 55–64.

Tryland, M., Derocher, A. E., Wiig, Ø. & Godfroid, J. (2001). *Brucella* sp. antibodies in polar bears from Svalbard and the Barents Sea. *Journal of Wildlife Diseases* 37: 523–531.

Tryland, M., Neuvonen, E., Huovilainen, A., et al. (2005). Serologic survey for selected virus infections in polar bears at Svalbard. *Journal of Wildlife Diseases* 41: 310–316.

Tryland, M., Nesbakken, T., Robertson, L., Grahek-Ogden, D. & Lunestad, B. T. (2014). Human pathogens in marine mammal meat – a Northern perspective. *Zoonoses and Public Health* 60: 1–18.

Tryland, M., Larsen, A. K. & Nymo, I. H. (2018). Bacterial infections and diseases. In Gulland, F. M. D., Dierauf, L. A. & Whitman, K. L. (Eds.), *Handbook of marine mammal diseases*, 2nd edition (pp. 367–388). Boca Raton, FL: CRC Press.

US Fish and Wildlife Service. (2008). Final Rule Listing the Polar Bear as a Threatened Species Under the Endangered Species Act (May 15, 2008) www.fws.gov/alaska/fisheries/mmm/polarbear/esa.htm. Accessed 3 January 2019.

Vongraven, D., Ekker, M. & Wiig, Ø. (2010). Management of polar bears in Norway, 2005–2009. In: Obbard, M. E., Thiemann, G. W., Peacock, E. & DeBruyn, T. D. (Eds.), *Polar bears: Proceedings of the 15th Working Meeting of the IUCN/SSC Polar Bear Specialist Group*, June 29–July 3, 2009, Copenhagen, Denmark. Occasional Paper of the IUCN Species Survival Commission No. 43 (pp. 149–155). Gland, Switzerland and Cambridge, UK: IUCN.

Ware, J. V., Rode, K. D., Pagano, A. M., et al. (2015). Validation of mercury tip-switch and accelerometer activity sensors for identifying resting and active behavior in bears. *Ursus* 26: 8–18.

Ware, J. V., Rode, K. D., Bromaghin, J. F., et al. (2017). Habitat degradation affects the summer activity of polar bears. *Oecologia* 184: 87–99.

Whiteman, J. P., Harlow, H. J., Durner, G. M., et al. (2015). Summer declines in activity and body temperature offer polar bears limited energy savings. *Science* 349: 295–298.

Wiig, Ø., Gjertz, I., Hansson, R. & Thomassen, J. (1992). Breeding behaviour of polar bears in Hornsund, Svalbard. *Polar Record* 28: 157–159.

Wiig, Ø., Born, E. W. & Pedersen, L. T. (2003). Movements of female polar bears (*Ursus maritimus*) in the East Greenland pack ice. *Polar Biology* 26: 509–516.

Wiig, Ø., Amstrup, S., Atwood, T., et al. (2015). *Ursus maritimus*. The IUCN Red List of Threatened Species e.T22823A1.

Wilson, R. R., Horne, J. S., Rode, K. D., Regehr, E. V. & Durner, G. M. (2014). Identifying polar bear resource selection patterns to inform offshore development in a dynamic and changing Arctic. *Ecosphere* 5: 136.

Wimsatt, W. A. (1974). Delayed implantation in the Ursidae, with particular reference to the black bear (*Ursus americanus*, Pallas). In: Enders, A. C. (Ed.), *Delayed implantation* (pp. 49–86). Chicago, IL: University of Chicago Press.

Zeyl, E., Aars, J., Ehrich, D., Bachmann, L. & Wiig, Ø. (2009). The mating system of polar bears: a genetic approach. *Canadian Journal of Zoology* 87: 1195–1209.

Zeyl, E., Ehrich, D., Aars, J., Bachmann, L. & Wiig, Ø. (2010). Denning-area fidelity and mitochondrial DNA diversity of female polar bears (*Ursus maritimus*) in the Barents Sea. *Canadian Journal of Zoology* 88: 1139–1148.

Chapter 15

Human–Bear Conflicts at the Beginning of the Twenty-First Century: Patterns, Determinants, and Mitigation Measures

Miha Krofel, Marcus Elfström, Hüseyin Ambarlı, Giulia Bombieri, Enrique González-Bernardo, Klemen Jerina, Andrés Laguna, Vincenzo Penteriani, James P. Phillips, Nuria Selva, Seth M. Wilson, Alejandra Zarzo-Arias, Claudio Groff, Djuro Huber, Alexandros A. Karamanlidis, Yorgos Mertzanis, Eloy Revilla, and Carlos Bautista

Introduction

Conflicts between humans and bears have occurred since prehistory (Zedrosser et al. 2011). Through time, the catalogue of human–bear conflicts (HBC) has been changing depending on the values and needs of human societies and their interactions with bears. Even today, conflict situations vary among the eight species of bears and geographically across these species' ranges (Can et al. 2014). It therefore comes as no surprise that there is no single global definition of HBC accepted among world practitioners and researchers. For the purposes of this book, HBCs were defined as any interaction between humans and bears that result in negative impacts on human livelihood in terms of health, economy, and sociocultural aspects.

We do, however, recognize that HBCs also importantly affect bear populations and represent a major threat to bear conservation worldwide. Conflicts may reduce human tolerance to bears, increase poaching, undermine conservation efforts, and increase demands for legal management removals. Successful conflict mitigation is therefore not only important for people's well-being, but also represents a crucial part of the modern conservation management of bears on a global scale.

Most bear species are opportunistic omnivores, they are all large-bodied and potentially dangerous to humans. They live in diverse biomes and come in contact with various human societies. This results in a broad range of interactions between bears and humans that may be considered as conflicts, including: (1) predation of domestic or semi-wild animals, including bees, hunting dogs, and pet animals; (2) damage due to foraging on cultivated berries, fruits, agricultural products, and tree bark in forest plantations; (3) economic loss due to destruction of beehives, fences, silos, houses, and other human property; (4) bear attacks on humans causing mild or fatal trauma; (5) bluff charges, bear intrusions into residential areas and direct proximity of humans or other forms of bear behavior that provoke strong unease, fear, and/or defensive reaction by humans; and (6) vehicle collisions with bears and traffic accidents indirectly provoked by bears during "bear jams" (Can et al. 2014; Bautista et al. 2017).

In this chapter we aim to outline the principal types of HBC and geographical differences in the occurrence of conflicts, show direct and indirect causes of conflicts, illuminate the human dimensions of HBC, and review methods that are used to mitigate HBC with several successful examples that resulted in improved coexistence between people and bears.

Overview of Types of Human–Bear Conflicts

Bear conflict literature, reviewed from 1970 to 2018, showed that: (1) HBCs have never been reported for giant pandas; (2) brown and black bears are the species involved in most types of conflicts; whereas (3) Andean, sun, sloth, and polar bears are involved in few, specific types of HBC (Figure 15.1). Some bear species can also represent a threat to human safety, but a specific chapter (Chapter 17) on bear attacks on humans is included in this volume and therefore they are not discussed here. In this section, we present an overview of the different types of conflicts grouped by the bear species for which conflicts have been recorded.

Andean Bear

Although Andean bear diet is mostly based on plants, this species will attack domestic animals if available in their habitat (Zukowski & Ormsby 2016; Borbón-García et al. 2017). Cattle and sheep are the most common livestock animals in the Andean bear distribution range and are also most often attacked, although also attacks on goats and horses have been reported (Goldstein 1991; Jorgenson & Sandoval 2005; Zapata-Ríos & Branch 2018). In recent years, the number of damages to livestock has increased as the best grazing lands coincide with the best bear habitats and an increasing number of locals are changing from growing crops to breeding cattle, especially for dairy (Jampel 2016; Zukowski & Ormsby 2016). Occasionally Andean bears raid crops (especially corn), mainly when fields are close to forest cover and away from human settlements (Jorgenson & Sandoval 2005; Espinosa & Jacobson 2012; Jampel 2016).

Sun Bear

Although little information is available on conflicts with sun bears, this species has been reported causing damage to crops and, more rarely, chickens and goats (Fredriksson 2005; Scotson et al. 2014; Wong et al. 2015). In Indonesian Borneo, sun bears have been reported damaging palm oil plantations and coconut trees by eating the fruits and new growth shoots (Fredriksson 2005). Sun bears are also known to feed on a variety of other fruits and horticultural products in mixed orchards such as corn, snakefruit, jackfruit, sugarcane, banana, pineapple, pumpkin, papaya, and oil palm fruits (Normua et al. 2004; Fredriksson 2005; Sethy & Chauhan 2013; Guharajan et al. 2017), as well as causing damage to farmhouses in search of household products such as rice, sugar, and palm oil (Wong et al. 2015; Figure 15.1).

Sloth Bear

Research on HBCs involving sloth bears is scarce and focuses mainly on bear attacks on humans (Can et al. 2014; Debata et al. 2017), which is the most frequently reported type of conflict (further addressed in Chapter 17). Although these bears rarely approach human areas, they can cause damage to agricultural fields when seeking crops (such as corn, sweet potatoes, potatoes, onions, or groundnuts; Rajpurohit & Krausman 2000; Bargali et al. 2005; Garcia et al. 2016).

Asiatic Black Bear

Conflicts between humans and Asiatic black bears have been reported across the whole species' range in Asia, including China (Liu et al. 2011; Huang et al. 2018), Japan (Saito et al. 2008), Pakistan (Ahmad et al. 2016; Ali et al. 2018), Iran (Ghadirian et al. 2017), India (Chauhan 2003), and Bhutan (Sangay & Vernes 2008; Jamtsho & Wangchuk 2016). The most frequently reported type of conflict is crop-raiding, especially of corn, potatoes, and various fruit trees (Figure 15.1). This kind of damage is concentrated during the summer (Charoo et al. 2011; Scotson et al. 2014), whereas attacks on livestock have been mostly documented during the autumn (Li et al. 2013; Mir et al. 2015; Jamtsho & Wangchuk 2016). Rarely, Asiatic black bears also raid beehives and fish farms (Liu et al. 2011; Jamtsho & Wangchuk 2016). Bark stripping has been reported in Japan, where bears are considered to be one of the greatest sources of conflict with forestry due to damage to tree trunks and consequent reduction of their value (Kobashikawa & Koike 2016).

American Black Bear

American black bears frequently raid crops, which represent an easy and calorically rich foraging opportunity (Mazur & Seher 2008; Merkle et al. 2013; Ditmer et al. 2015). Most frequently reported crop damage refers to cereals (Shivik et al. 2011; Ditmer et al. 2015), silage, corn, orchards, and various horticultures (Jonker et al. 1998; Baruch-Mordo et al. 2008; Treves et al. 2010; Shivik et al. 2011; Ditmer et al. 2015). Despite being less adapted to predation on large mammals compared to other large carnivores, American black bears also attack and kill livestock (Garshelis 2009). Most common victims are domestic sheep, goats, pigs, cattle, horses, and other smaller mammals and birds (Baruch-Mordo et al. 2008; Linnell et al. 2013). Apiaries are another strong attractant for black bears, which may cause considerable damage to beehives (Clark et al. 2005; McKinley et al. 2014) to obtain a high-energy food source (Beckmann & Berger 2003). American black bears can also cause considerable losses to forest landowners due to feeding or tree marking (rubbing). For example, bears strip bark from coniferous trees during spring to feed on newly forming vascular tissues for their relatively high content of free-floating sugars. This may result in significant damage (Kimball et al. 1998) and is mainly concentrated on younger stands (Stewart et al. 1999; Ziegltrum & Nolte 2001). They regularly seek out garbage, pet food, bird feeders, and other anthropogenic food and damage human property in the process, for example by breaking into houses, barns, vehicles, and other facilities (Miller & Tutterrow 1999; Breck et al. 2008, 2009; Treves et al. 2010). Other types of damage include bear–vehicle collisions and attacks on pets (Baruch-Mordo et al. 2008; Lowery et al. 2012; Wynn-Grant et al. 2018).

Brown Bear

Conflicts between people and brown bears are very diverse and are mainly related to the bears' opportunistic foraging and consumption of food. Across its range, the brown bear is the species most often reported to cause damage to livestock (Figure 15.1), which ranges from rabbits and chickens to cattle. Usually livestock damage is highest when brown bears occur in areas with sheep farming (Kaczensky 1999; Gunther et al. 2004). Besides livestock, brown bears have been reported killing pets, captive game animals, and fish (Molinari et al. 2016). When searching for insect larvae, brown bears also frequently destroy beehives (Figure 15.2), which represent the major HBC in some parts of the species' range (e.g. Groff et al. 2010; Karamanlidis et al. 2011; Naves et al. 2018). Damage in agriculture includes crop fields (especially corn), gardens, orchards, grass silage, and vineyards (Krofel & Jerina 2012). In some parts of the range, brown bears cause damage to forestry, which seems to be concentrated on mature conifer trees (Zyśk-Gorczyńska et al. 2016). Due to their large size, brown bears can also cause considerable damage to buildings, vehicles, wildlife feeders, and other human property while searching for human foods or during vehicle collisions (Krofel & Jerina 2012). Searching for anthropogenic food sources can also bring bears close to human settlements, which increases people's concern for human safety and often results in lethal removal of the intruding bear (Gunther et al. 2004).

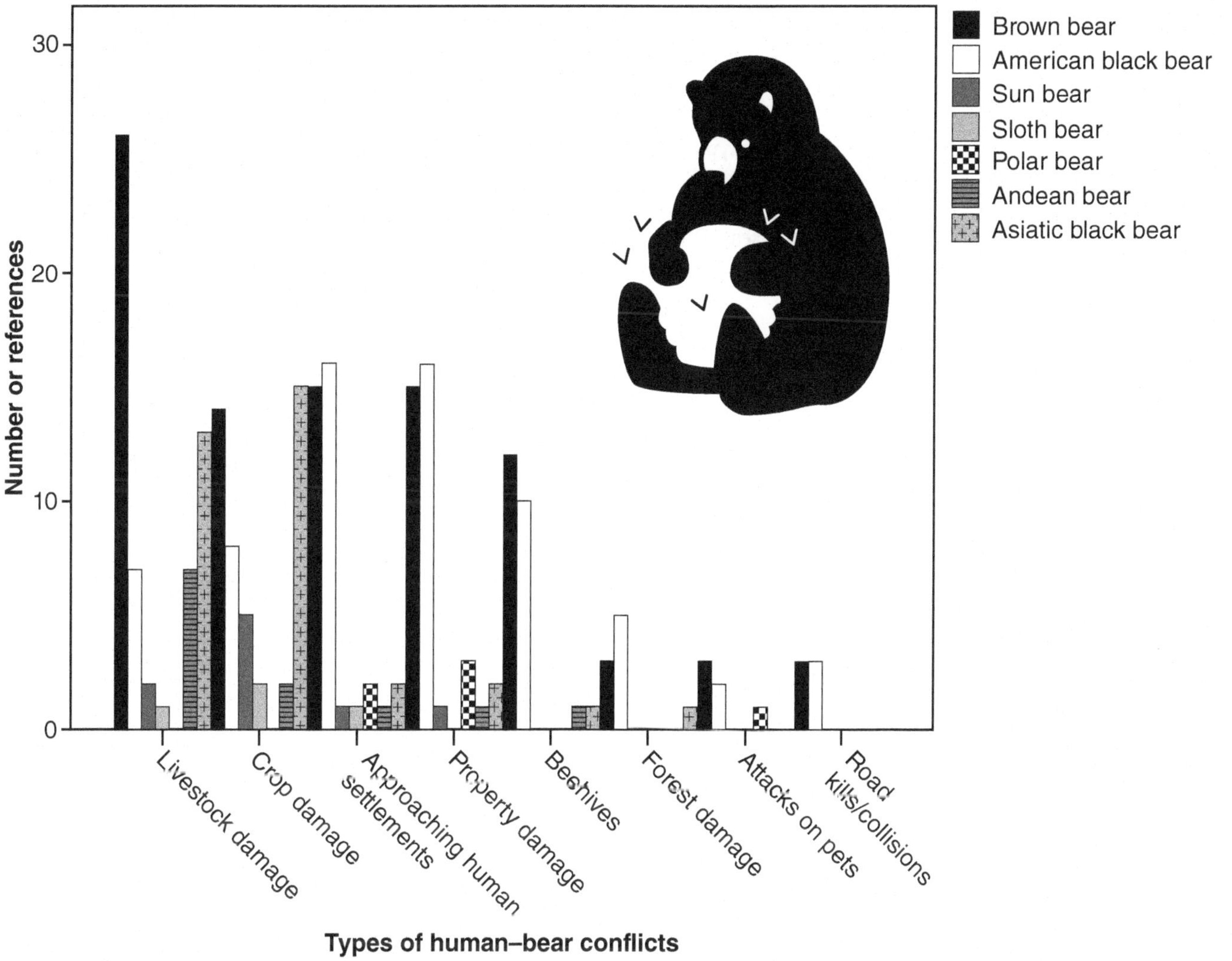

Figure 15.1 Types of human–bear conflicts reported in the scientific literature between 1970 and 2018. For each type of human–bear conflict, the number of related publications is shown. (Bear image credits: the image was downloaded from www.123rf.com, Image ID 107801431, copyright: vastard.)

Figure 15.2 Human–bear conflicts are very diverse, but mostly associated with bear foraging behavior, such as raiding beehives to obtain protein-rich bee larvae. (Photo by Miha Krofel.)

Polar Bear

Damage caused by polar bears is mainly related to human food stores, vehicles, attacks on domestic dogs, and entry into human settlements (Stenhouse et al. 1988; Dyck 2006). This damage derives from the bears' search for food, and is typically concentrated in August–November, when bears are on land due to melting sea ice (Stenhouse et al. 1988; Dyck 2006; Towns et al. 2009) and do not have access to the seals, their main prey (DeMaster & Stirling 1981).

Main Drivers and Factors That Influence Conflict Occurrence

Intrinsic Drivers

Large carnivores often avoid human activity and settlements. However, when large carnivores do occur close to people and settlements, it is often interpreted as a behavioral response, such as tolerance to human presence, and/or an association between humans and attractive foods.

The waning of a response to a stimulus (e.g. human presence) can explain the increased tolerance toward humans in animals with increasing non-threatening encounters with people, and is hereafter called *human habituation.* Food resources are unconditioned stimuli, i.e. not related with other stimuli, which usually results in the response of foraging. However, feeding can become conditioned upon an unrelated stimulus, such as human presence or settlements, after repeated association between human activity and food resources (Immelmann & Beer 1989); this process is hereafter called *food conditioning.* These learning processes can explain why some animals (habituated ones) are not wary of humans or often occur near settlements, while other conspecifics (non-habituated) still avoid them. However, human habituation and food conditioning do not predict that exposures to stimuli will vary among conspecifics (e.g. sex/age or reproductive categories). Bears can also transmit human tolerance by observational learning from mother to offspring, i.e. by social or cultural transmission (Morehouse et al. 2016). Similarly, young bears may become food-conditioned through their mother's behavior. However, regardless of the influence of cultural transmission, the development of positive associations between bears and human-derived foods requires some earlier experience with, or cues from, people, human activity, or settlements. In contrast to habituation and conditioning, if animals occur near people because they lack cumulative experience of them (Bejder et al. 2009), i.e. are naïve, this would predict a higher frequency of younger individuals near human activity and settlements. Thus, subadult bears may approach people or settlements due to their naïvety.

Human habituation and food conditioning are common responses by bears due to frequent exposure to people and human-derived foods (McCullough 1982; Herrero et al. 2005). However, human activity or settlements may also provide a refuge (sometimes termed *human shield*) for young bears and females with young against dominant conspecifics (Wielgus & Bunnell 1994; Mueller et al. 2004; Steyaert et al. 2013). Actually, predation and aggression towards conspecific young (especially cubs-of-the-year and yearlings) by older individuals, especially males, constitute a common pattern in brown bears (Swenson et al. 1997, 2001; McLellan 2005).

A common pattern across Europe, North America, and Japan is that brown bears, polar bears, and American and Asiatic black bears occurring near human activity and settlements are typically younger, and females accompanied by their offspring are more often near settlements or humans than adult males or lone adult females (Elfström et al. 2014c). The pattern of dominant bears displacing subordinate conspecifics from areas with higher habitat quality in terms of food and cover is referred to as a *despotic distribution* (Elfström et al. 2014c). In some areas, these displaced bears occurring near settlements often become human-habituated or food-conditioned (Mattson et al. 1992; Schwartz et al. 2006). In other areas, in particular among Scandinavian brown bears, food-conditioning seems to be uncommon (Elfström et al. 2014a, 2014b). Thus, the despotic distribution explains why certain bears more often occur near settlements, which may or may not result in human habituation or food conditioning. The availability of food resources and their distribution in relation to human settlements can differ considerably across the bears' range and this could explain why sometimes bears near settlements become food-conditioned, but not always.

Extrinsic Drivers

Based on a literature review we identified six main extrinsic drivers of HBC: availability of anthropogenic food, natural food failures, impacts of human activities on the landscape, climatic and meteorological events, differences in conflict management, and reduced human tolerance. Below, we give an overview of the mechanisms underlying these drivers, except for the reduced tolerance that is explained further in this chapter.

At a local scale, the availability of anthropogenic food (Figure 15.3) is a source of conflict and an ecological trap for bears in many parts of the world, for example: brown bears roaming in agriculture fields or garbage dumps in Japan, Europe, the Middle East, and North America (Narita et al. 2011; Northrup et al. 2012; Cozzi et al. 2016; Penteriani et al. 2018a); American black bears raiding mango plantations in Northern Mexico (Lira Torres et al. 2015); sun and Asiatic black bears raiding corn plantations in south-east Asia (Scotson et al. 2014); and Andean bears predating free-ranging cattle in pastures of the Ecuadorian Andes (Jampel 2016). Additionally, artificial feeding of bears, both intentionally and unintentionally, may increase the risk of HBC, which is why intentional bear feeding is generally discouraged or prohibited across North America (Garshelis et al. 2017). Artificial feeding can also shorten the hibernation period and thus

Figure 15.3 Anthropogenic food, such as garbage, is one of the most important extrinsic drivers of human–bear conflicts. (Photo by Andrej Sila, Slovenia Forest Service.)

Figure 15.4 Delay in ocean freezing due to global warming is an important extrinsic factor promoting conflicts between people and polar bears. (Photo by Marcus Elfström.) (A black and white version of this figure will appear in some formats. For the color version, please refer to the plate section.)

prolong the period when human–bear interactions occur (Krofel et al. 2017).

The frequency of bears using or searching for anthropogenic foods and consequent HBC seems to increase when natural bear food sources become scarce. For instance, human–grizzly bear conflicts increased in British Columbia, Canada in years of low availability of salmon biomass (Artelle et al. 2016) and similar patterns of increased HBC in periods of natural food failures were described for European brown bears (e.g. Molinari et al. 2016), American black bears (e.g. Baruch-Mordo et al. 2008), Asiatic black bears (e.g. Oka et al. 2004), and sun bears (Wong et al. 2015). On the other hand, years of high mast production can increase bear reproductive rates (Costello et al. 2003), which in turn can lead to increased levels of conflicts in subsequent years, due to a larger number of mothers with cubs and young individuals that look for shelter and food near humans (Elfström et al. 2014c; Obbard et al. 2014). However, an occurrence of problem bears that is unrelated with seasonal food availability has also been reported (Elfström et al. 2014b).

Human population size and the impact of human activities on the landscape also influence the occurrence of HBC. Habitat degradation due to agricultural expansion increases the rates of HBC globally (Can et al. 2014). Additionally, the use of bear habitats by humans can increase the risk of human–bear interaction. For example, direct competition for fruits led to higher rates of sloth bear attacks on humans in India (Dutta et al. 2015). Also, recreational activities, such as bear viewing and nature tourism, are increasingly demanded and without proper regulation can be a cause for more human–bear interactions which may end in HBC (Penteriani et al. 2017). On the other hand, socioeconomic changes may lead to modifications in land use and to land abandonment in rural areas, which, in turn, can facilitate the expansion of bears and thus promote conflicts. For example, in the Ecuadorian Andes shifts from crops to cattle, and from a mostly subsistence-oriented economy to a more market-oriented and capital-intensive approach, have led to higher rates of cattle predation by Andean bears (Jampel 2016). Also, decrease and aging of the local human population may enhance leaving unattended fruit trees and unharvested crops, which attract bears to villages (Kishimoto 2009; Yamazaki et al. 2009).

Various climatic and meteorological events have been associated with the increasing trend of HBC. The most outstanding example is the considerable loss of sea ice and delay in ocean freezing due to global warming, which is forcing polar bears (Figure 15.4) to increasingly use terrestrial habitats and consequently promotes encounters with people (Wilson et al. 2017) and damage caused by polar bears (Towns et al. 2009; see also Chapter 11). At lower latitudes, an increase in the frequency of crop raiding and approaches to people by sun bears in Indonesia and American black bears in New Mexico (USA) was reported during drought periods associated with the El Niño–Southern Oscillation (Zack et al. 2003; Fredriksson 2012). Climate change is predicted to increase the severity of meteorological events, such as late-spring frosts that are associated with natural food failures in temperate regions and higher use of urban areas by American and Asiatic black bears (Honda 2013; Laufenberg et al. 2018). Additionally, climate change can reduce the duration of bear hibernation and consequently prolong the period when human–bear interactions occur (Johnson et al. 2018), as well as altering the geographic range of natural food resources. As a consequence, bears might be displaced from mountainous areas toward lower, more humanized ones, where an increase in conflicts is expected (Penteriani et al. 2019).

Finally, bear-damage management greatly influences the occurrence of HBC (Bautista et al. 2017) and good husbandry practices have been the most effective and widespread technique to prevent conflict with bears and other large carnivores (van Eeden et al. 2018).

Human Dimensions of Human–Bear Conflicts

At a pragmatic level, it is vital to recognize that effectively addressing and solving HBCs requires working with people, communities, and a variety of stakeholders within a given social and ecological context (see also Chapter 20). Understanding the associated values, perceptions, social and cultural norms, and decision-making of people within a given context is important for illuminating the human dimensions of HBC. While regulatory approaches to mitigating undesirable human behaviors (e.g. poor sanitation practices, lack of use of prevention tools, or illegal poaching) are critically important to addressing HBC, collaborative approaches to decision-making offer additional means to build support and improve human behaviors and practices that address HBC.

Values Among Stakeholders are Important

Addressing HBC must have scaled levels of support, built upon effective stakeholder decision-making forums, well-designed management plans, effective public communication strategies, and institutionally backed political, legal, and financial support at national and international governmental levels. In many respects, resolving HBC rests upon *the support and acceptance of bears by communities of place and communities of interest* – where local and broad public values converge (Wilson 2016). This requires attention to be paid to the values and perspectives of people who actually live with bears and those who have strong interests in bears, often urban dwellers whose values may be different from local stakeholders.

Finding areas of convergence in values is no easy task considering that strong differences exist in terms of how groups of people or stakeholders believe bears should be managed and conserved. Moreover, in some contexts, there may be deep-rooted social issues, power inequities, historical events, or ethnic and cultural divides that cause social conflicts, resulting in a refusal by people to work collaboratively to address ways to live with bears. Failure to address these underlying sources of conflict may hinder well-intended efforts and may require facilitated processes that can expose, explore, and transform existing relationships so that meaningful HBC management can occur (Madden & McQuinn 2014). Social conflicts and the actual material damage from bears that impact individuals should therefore be understood within unique social and ecological contexts that reflect a keen understanding of political, cultural, historic, and economic conditions.

Problem Definitions Matter

Additionally, how different people orient to and define "the problem" of living with bears is an essential aspect of the human dimensions related to addressing HBC. For example, it is common for stakeholders to define the issue of HBC as one of having "too many bears" (Wilson et al. 2013). The perception that bears are too numerous can be complicated – for example, in certain European contexts, younger bears and females with cubs use human settlements as human shields from older male bears (Elfström et al. 2014c; Steyaert et al. 2016). This can make females and younger bears more regularly visible to people. Subsequently, this can cause the perception that the perceived problem is about an increase in the population of bears, when this may not be the case. If these types of perception are not addressed or corrected, then solutions to the perceived problem may be misguided – in this example, a call to reduce the bear population as a solution to HBC. This is just one example of the importance of paying attention to how local people define "the problem." In many cases, collaborative forums or decision-making processes that bring local people together with bear managers, scientists, and other stakeholders are essential for generating a shared understanding of the problem so that shared solutions are generated for effectively addressing HBC (Botetzagias & Kotilda 2018).

Community Responses to Biological Scale

The question of whether individuals and groups of people can work collaboratively to understand the problem and find solutions becomes even more imperative due to the large spatial scale of habitat that bears require. In most European landscapes, brown bear home ranges encompass human activities within a mixed mosaic of modified forest, mountain, and agricultural lands. In this context, the use of prevention tools by multiple individuals is necessary to match the biological scale at which bears live. Failure to match community-level responses to the scale of carnivore life history needs will inevitably fall short of resolving conflicts for both people and bears (Wilson et al. 2013). These challenges are further magnified in areas where large carnivore populations encompass multiple countries and where transboundary movements of bears cross multiple management jurisdictions (Bartoń et al. 2019). In these contexts, well-developed institutional partnerships and transboundary-level management agreements are central for addressing HBC across jurisdictional boundaries (Penteriani et al. 2018b; Proctor et al. 2018).

Effectiveness of Conflict Mitigation Measures

During thousands of years of coexistence with bears, humans have developed both proactive and reactive methods to prevent or mitigate HBC. While none of them is 100% effective, there are patterns of success and failure that transcend species. First and foremost, proactive methods that prevent conflict are almost always cheaper and more successful over time than reactive measures. Second, as the levels of food conditioning and human habituation increase, the effectiveness of mitigation decreases. Mast failures and water shortages also influence the effectiveness of mitigation efforts, as well as sex, age, and maternal status of individuals, and competition for resources among individuals (Majić Skrbinšek & Krofel 2015).

Preventing Access to Anthropogenic Food and Human Property

As highlighted before, anthropogenic food is a well-known primary cause of HBC; therefore, denying bears access to human food resources is perhaps the most effective and proactive way of preventing HBC. In this way, bears are not rewarded for approaching humans or developed areas, thus habituation to human presence and food conditioning are less likely to occur (Majić Skrbinšek & Krofel 2015). There are numerous approaches to effectively prevent bears from accessing anthropogenic food sources, including livestock, although reliable evidence for effectiveness of many of these methods is still lacking (van Eeden et al. 2018). Exclusion by means of bear-proof bins and electric fences prevents access to unprotected resources such as garbage and livestock, respectively (Sowka 2009). Properly installed and maintained electric fences (Figure 15.5) can have near 100% effectiveness rate and the use of guarding dogs as protection, particularly for livestock, is also a highly effective technique (Belant et al. 2011; Proctor et al. 2018; van Eeden et al. 2018). Removal of attractants such as bird feeders, pet foods, and carcasses is essential, especially in spring or during mast failures, when natural foods are not widely available. Strict legislation and its enforcement, accompanied with public education, are also crucial elements in reducing the availability of anthropogenic foods (Majić Skrbinšek & Krofel 2015).

Aversive Conditioning

Aversive conditioning is a learning process in which deterrents are continually and consistently administered to change an undesirable behavior. Negative stimuli, such as shooting with rubber bullets, are applied while an animal is engaged in undesirable behavior in order to elicit an avoidance of such behavior in the future (Gillin et al. 1994). Other methods of aversive conditioning include: taste aversion, cracker shells, propane cannons, bear dogs, shooting with bean bag rounds, and capture and release of conflictual individuals. Most of these methods have produced mixed results and depend on multiple factors, including the context in which a learning process took place, the immediacy of a consequence to a given behavioral response, and the consistency and magnitude of these consequences and rewarding of alternative behavior (Majić Skrbinšek & Krofel 2015). Pain stimuli and the use of bear dogs have proven to be the most successful in the long term. However, effectiveness is lower when an undesired behavior is already strongly established, and it is important to remember that bears may also habituate to some of the aversion techniques (Mazur 2010). A monitoring and response system that quickly detects undesired behaviors is crucial for successful application of this tool.

Lethal Removal

Lethal removal of bears has been a widespread measure used in response to bear incidents in the past (Schwartz et al. 2005). This method includes shooting, trapping, and poisoning. It can refer to management removals of specific "problem" individuals or a general culling of the population with the aim to reduce bear density. In many regions of the world, however, lethal controls are losing public approval, and for endangered populations even limited removal can have strong negative effects (Treves & Karanth 2003; Schwartz et al. 2005). Poisoning and some types of trapping are in general ill-advised due to risk of non-target kills, including endangered wildlife and domestic animals. Lethal methods are most effective when known "problem" bears are removed (Gunther et al. 2004), for which permits may be obtained from a regulatory agency. Culling as a tool for conflict mitigation is not likely an acceptable solution near settlements; however, it may be considered for rural locations with agricultural conflicts, but must also consider population and damage characteristics (Belant et al. 2011).

Non-Lethal Removal

This method includes capturing an individual and moving it to another place. Translocations of problem bears are generally more acceptable for the public than lethal removals, but agencies are increasingly reluctant to use this method (Creachbaum et al. 1998). Translocations are costly and labor-intensive with high mortality rates, as bears often return to the capture site even from several hundreds of kilometers away, present road hazards, or cause problems in the new area (Linnell et al. 1997). Noteworthy successes from translocation have considered the age and sex of the individual in addition to significant translocation distance, impassable geographical barriers, and perhaps the quality of the new habitat (Taylor & Phillips 2019). Translocating young males that are not highly food-conditioned has seen the greatest success (Belant et al. 2011; Taylor & Phillips 2019).

Diversionary Feeding of Bears

By providing food in remote areas, managers attempt to divert bears from approaching settlements and/or reduce damage to human property. Effectiveness of diversionary feeding lacks rigorous studies and existing examples from brown bears and American black bears have been met with mixed results (Steyaert et al. 2014; Garshelis et al. 2017). Diversionary feeding with carcasses to reduce livestock depredation was not effective (Kavčič et al. 2013; Morehouse & Boyce 2017). There is also an increasing list of side effects, many of which are unwanted, associated with the artificial feeding of bears (Penteriani et al. 2018a). If diversionary feeding is to be used in HBC mitigation, a comprehensive review of the literature suggests that the highest efficacy with minimal side effects is through temporary, seasonal, and as-needed placement of natural foods (Taylor & Phillips 2019). Employing this strategy encourages exploitation of natural food sources first and discourages dependence on anthropogenic supplementary foods.

Land-Use Practices

There are several potential mechanisms by which land-use practices can affect probability for occurrence of HBC. For example, limiting human access to the most crucial bear habitats in certain time periods gave positive results in American national parks (Coleman et al. 2013). Because cover is an important parameter affecting space use by bears, some authors recommended removing dense vegetation near human settlements (Elfström et al. 2014c) and crops (Sato et al. 2005). Water courses are popular travel routes for bears; therefore, removing livestock and apiaries from these areas can also be effective (Clark et al. 2005). Transition from sheep to cattle or horse breeding, from livestock breeding to other land use (e.g. agriculture, forestry), or selection of crops less attractive to bears can also reduce probability of bear damage (Gunther et al. 2004; Swenson & Andrén 2005; Wilson et al. 2006; Majić Skrbinšek & Krofel 2015). However, generating a will for such changes among stakeholders can be a considerable challenge (Linnell et al. 2013).

Figure 15.5 Electric fences and electric nets, when used properly, are a very effective way to prevent damage caused by bears to livestock, beehives, crops, or other human property. (Photo by Miha Krofel.)

Compensation

Economic compensation for claimed damage caused by bears started in the 1950s and is nowadays widespread in Europe and North America (Ravenelle & Nyhus 2017). Compensation programs differ from one country to another and can effectively reduce the economic impact of the HBC on certain stakeholders (Figure 15.6). However, they should be treated with care, as they can also decrease efforts to prevent damage and exacerbate conflicts (Bulte & Rondeau 2005). For example, in the Scandinavian brown bear population the level of conflict is considerably higher in Norway compared to Sweden, which is connected with sheep in Norway being mostly free-ranging and unprotected. In Norway, compensation is given in all cases, even if the incident is not verified and if no protection of sheep was used, whereas in Sweden compensation is conditional on the proper protection of livestock, and wildlife agencies strongly focus on subsidizing preventive measures (Swenson & Andrén 2005; Widman & Elofsson 2018). It should also be understood that compensation does not directly affect the occurrence of HBC, but only mitigates their economic impact.

Public Awareness and Education

Increasing public awareness about drivers of the HBC, demonstrating the preventive measures and their proper use, can considerably improve the effectiveness of HBC prevention programs. People also tend to follow prescribed rules more when they understand the reasons behind them (Majić Skrbinšek & Krofel 2015). Education of adults and children can also be effective to improve their attitude towards bears (Ambarlı 2016).

Figure 15.6 Economic compensation for claimed damages can effectively reduce economic impact of human–bear conflicts, as in the case of damage to fruit trees. Here is shown a brown bear scat full of domestic cherry stones. (Photo by Vincenzo Penteriani.)

Successful Examples of Conflict Prevention

North American National Parks

The first systematic measures targeting availability of human food sources were applied in the 1970s and 1980s in North American national parks following high rates of HBCs, including several human casualties (Herrero 1994). Strict garbage management, regulations on human food storage, prohibition of artificial bear feeding, and intensive public education about proper behavior in bear habitat proved very successful. After application of these measures, HBC throughout national parks decreased considerably. For example, in the Yellowstone National Park, attacks on people decreased by almost 90% and at the same time there was less need for management removals of bears (Meagher & Phillips 1983; Gunther & Hoekstra 1998). In the Denali National park, cases of bears feeding on anthropogenic food decreased by 96%, which was followed by a 77% decrease in reported HBC and 77% lower number of management removals (Schirokauer & Boyd 1998). Similarly, after the change of focus from bear management to management of people and anthropogenic food, the number of problem bears removed decreased by 94% for black and 86% for brown bears in the Jasper National Park (Ralf 1995), and by 75% for black and 70% for brown bears in the Glacier National Park (Gniadek & Kendall 1998), respectively. In the Yosemite National Park, after management was changed from reactive (lethal removals, translocations, aversive conditioning) to proactive (limiting access to anthropogenic food, education, law enforcement), the proportion of anthropogenic food and garbage in the black bear diet was considerably reduced, which was followed by a 31% decrease in the number of bear incidents and a 63% decrease in the amount of damage caused by bears (Madison 2008; Hopkins et al. 2014).

Residential Areas in North America

Limiting the availability of anthropogenic food for bears is generally more difficult in residential areas compared to national parks. However, also here considerable improvements can be achieved with public education and implementation of methods preventing access to anthropogenic food. In Blackfoot Valley (Montana, USA), a proactive, community-based project was launched to provide cost-free removal of livestock carcasses for ranchers; introduce bear-proof garbage bins, provide electric fences for beehives, cattle calving areas, and garbage dumps in parallel with intensive public education. During the three-year project the number of conflicts with brown bears decreased by 91% without the need to remove a single bear (Wilson et al. 2006; Wilson 2007). Substantial decrease in HBC and management removals of brown bears was noted also in Kennecott Valley (Alaska, USA) after local residents were provided with bear-proof garbage containers, electric fences, and targeted public education (Wilder et al. 2007). Similar successes (40–80% reduction) in reducing HBC by preventing bears feeding on human food sources were reported for the American black bears in residential areas across the USA (Tavss 2005; Leigh & Chamberlain 2008).

South America

To reduce the impact of cattle depredation and damage to corn and avocado plantations caused by Andean bears, a pilot program was developed in Ecuador to improve livestock management by providing water channels, electric fences, fertilizers, grass seeds, artificial insemination, and permanent veterinary care. This resulted in the prevention of further cattle losses and improved farmers' attitude toward bears and their conservation in the project areas (Laguna 2018). In parallel, bear-watching ecotourism is promoted, which provides additional local employment.

Middle East

Electric fencing for preventing brown bear damage to beehives and orchards in Turkey was first implemented at six locations in 2007 and 2008 (Ambarlı & Bilgin 2008). After successful prevention of damage in this pilot EU Kaçkar Mountains Conservation Project and adoption by the government authorities, electric fencing became widespread throughout the country. Today, it is locally manufactured and used at more than 5000 locations, which caused a decrease in bear-caused damage, as well as a 10-fold reduction in the price of the electric fencing equipment compared to imported fences.

Arctic

In order to prevent conflicts due to polar bears approaching human settlements, a polar bear alert program has been established in Churchill, Manitoba (Canada), since 1980. Polar bears (with the exception of family groups) that approach residential or working areas of Churchill are captured and after a minimum of 30 days released outside of the town (Hedman 2009). Recaptured individuals are translocated about 70 km north to the Hudson Bay. If translocated bears are recaptured, they may be kept in the holding facility until the ice comes in. The program relies on maintaining a high alert on bear occurrence around the urban area, as well as on educating people to reduce bear accessibility to human-derived foods and restricting human activity outside of the city to avoid disturbing bears. The ice cover is the main driver behind polar bear occurrence ashore and beyond the reach of bear management. However, the bear alert management program results in saving lives of bears and people.

Conclusions

Human–bear conflicts are complex and diverse. Consequently, there is no single one-for-all solution to effectively prevent all types of problems. Because often only few problem bears cause the majority of all bear incidents, special attention needs to be given to preventing the development of repetitive conflict behavior, especially preventing bears' access to anthropogenic food. However, when this fails and bears develop high levels of human habituation or food conditioning, conflict behavior is difficult to change. In such cases, the removal of problem bears is often the only effective solution and will therefore likely remain an important part of bear management in the foreseeable future. Nevertheless, the main focus in bear management should be prevention, as successful proactive management is considerably more acceptable to the public than reactive responses once the conflicts have already occurred. Experience from several regions suggests that this approach gives the best results when local inhabitants are actively involved and well-informed about bear biology, occurrence, and prevention of HBC. At the same time, such non-lethal management is in line with recent trends of a decrease in the public's tolerance of lethal bear removals.

The science of HBC has enabled the development of several highly successful conflict-prevention programs throughout the world and has enabled the persistence of numerous bear populations in coexistence with people. Building upon this knowledge and best-practice examples is probably the safest way to ensure the future survival of these iconic animals.

References

Ahmad, S., Hameed, S., Ali, H., et al. (2016). Carnivores' diversity and conflicts with humans in Musk Deer National Park, Azad Jammu and Kashmir, Pakistan. *European Journal of Wildlife Research* 62: 565–576.

Ali, A., Waseem, M., Teng, M., et al. (2018). Human–Asiatic black bear (*Ursus thibetanus*) interactions in the Kaghan Valley, Pakistan. *Ethology Ecology & Evolution* 30: 399–415.

Ambarlı, H. (2016). Rural and urban students' perceptions of and attitudes toward brown bears in Turkey. *Anthrozoös* 29: 489–502.

Ambarlı, H. & Bilgin, C.C. (2008). Human-brown bear conflicts in Artvin, north-eastern Turkey: encounters, damage, and attitudes. *Ursus* 19: 146–153.

Artelle, K. A., Anderson, S. C., Reynolds, J. D., et al. (2016). Ecology of conflict: marine food supply affects human–wildlife interactions on land. *Scientific Reports* 6: 25936.

Bargali, H. S., Akhtar, N. & Chauhan, N. P. S. (2005). Characteristics of sloth bear attacks and human casualties in North Bilaspur forest division, Chhattisgarh, India. *Ursus* 16: 263–267.

Bartoń, K. A., Zwijacz-Kozica, T., Zięba, F., Sergiel, A. & Selva, N. (2019). Bears without borders: long-distance movement in human-dominated landscapes. *Global Ecology and Conservation* 17: p.e00541.

Baruch-Mordo, S., Breck, S. W., Wilson, K. R. & Theobald, D. M. (2008). Spatiotemporal distribution of black bear–human conflicts in Colorado, USA. *Journal of Wildlife Management* 72: 1853–1862.

Bautista, C., Naves, J., Revilla, E., et al. (2017). Patterns and correlates of claims for brown bear damage on a continental scale. *Journal of Applied Ecology* 54: 282–292.

Beckmann, J. P. & Berger, J. (2003). Rapid ecological and behavioural changes in carnivores: the responses of black bears (*Ursus americanus*) to altered food. *Journal of Zoology* 261: 207–212.

Bejder, L., Samuels, A., Whitehead, H., Finn, H. & Allen, S. (2009). Impact assessment research: use and misuse of habituation, sensitisation and tolerance in describing wildlife responses to anthropogenic stimuli. *Marine Ecology – Progress Series* 395: 177–185.

Belant, J. L., Simek, S. L. & West, B. C. (2011). Managing human–black bear conflicts. *Human–Wildlife Conflicts Monograph* 1: 1–77.

Borbón-García, A., Reyes, A., Vives-Flórez, M. & Caballero, S. (2017). Captivity shapes the gut microbiota of Andean bears: insights into health surveillance. *Frontiers in Microbiology* 8: 1–12.

Botetzagias, I. & Kotilda, E. (2018). Bear management in Greece as a 'problem orientation' policy issue: a Q-methodology approach. *Aegean Journal of Environmental Sciences* 3: 1–36.

Breck, S. W., Williams, C. L., Beckmann, J. P., et al. (2008). Using genetic relatedness to investigate the development of conflict behavior in black bears. *Journal of Mammalogy* 89: 428–434.

Breck, S. W., Lance, N. & Seher, V. (2009). Selective foraging for anthropogenic resources by black bears: minivans in Yosemite National Park. *Journal of Mammalogy* 90: 1041–1044.

Bulte, E. H. & Rondeau, D. (2005). Why compensating wildlife damages may be bad for conservation. *Journal of Wildlife Management* 69: 14–19.

Can, Ö. E., Cruze, N. D., Garshelis, D. L., Beecham, J. J. & Macdonald, D. W. (2014). Resolving human–bear conflict: a global survey of countries, experts and key factors. *Conservation Letters* 7: 501–513.

Charoo, S. A., Sharma, L. K. & Sathyakumar, S. (2011). Asiatic black bear–human interactions around Dachigam National Park, Kashmir, India. *Ursus* 22: 106–113.

Chauhan, N. P. S. (2003). Human casualties and livestock depredation by black and brown bears in the Indian Himalaya, 1989–98. *Ursus* 14: 84–87.

Clark, J. D., Dobey, S., Masters, D. V., et al. (2005). American black bears and bee yard depredation at Okefenokee Swamp, Georgia. *Ursus* 16: 234–244.

Coleman, T. H., Schwartz, C. C., Gunther, K. A. & Creel, S. (2013). Grizzly bear and human interaction in Yellowstone National Park: an evaluation of bear management areas. *The Journal of Wildlife Management* 77: 1311–1320.

Costello, C. M., Jones, D. E., Inman, R. M., et al. (2003). Relationship of variable mast production to American black bear reproductive parameters in New Mexico. *Ursus* 14: 1–16.

Cozzi, G., Chynoweth, M., Kusak, J., et al. (2016). Anthropogenic food resources foster the coexistence of distinct life history strategies: year-round sedentary and migratory brown bears. *Journal of Zoology* 300: 142–150.

Creachbaum, M. S., Johnson, C. & Schmidt, R. H. (1998). Living on the edge: a process for redesigning campgrounds in grizzly bear habitat. *Landscape Urban Plan* 42: 269–286.

Debata, S., Swain, K. K., Sahu, H. K. & Palei, H. S. (2017). Human–sloth bear conflict in a human-dominated landscape of northern Odisha, India. *Ursus* 27: 90–98.

DeMaster, D. P. & Stirling, I. (1981). *Ursus maritimus*. *Mammalian Species* 145: 1–7.

Ditmer, M. A., Burk, T. E. & Garshelis, D. L. (2015). Do innate food preferences and learning affect crop raiding by American black bears? *Ursus* 26: 40–52.

Dutta, T., Sharma, S., Maldonado, J. E., Panwar, H. S. & Seidensticker, J. (2015). Genetic variation, structure, and gene flow in a sloth bear (*Melursus ursinus*) meta-population in the Satpura-Maikal landscape of Central India. *PLoS ONE* 10: 7–9.

Dyck, M. G. (2006). Characteristics of polar bears killed in defense of life and property in Nunavut, Canada, 1970–2000. *Ursus* 17: 52–62.

Elfström, M., Davey, M. L., Zedrosser, A., et al. (2014a). Do Scandinavian brown bears approach settlements to obtain high-quality food? *Biological Conservation* 178: 128–135.

Elfström, M., Zedrosser, A., Jerina, K., et al. (2014b). Does despotic behavior or food search explain the occurrence of problem brown bears in Europe? *Journal of Wildlife Management* 78: 881–893.

Elfström, M., Zedrosser, A., Stoen, O. G. & Swenson, J. E. (2014c). Ultimate and proximate mechanisms underlying the occurrence of bears close to human settlements: review and management implications. *Mammal Review* 44: 5–18.

Espinosa, S. & Jacobson, S. K. (2012). Human-wildlife conflict and environmental education: evaluating a community program to protect the andean bear in Ecuador. *Journal of Environmental Education* 43: 55–65.

Fredriksson, G. (2005). Human–sun bear conflicts in East Kalimantan, Indonesian Borneo. *Ursus* 16: 130–137.

Fredriksson, G. (2012). Effects of El Nino and large scale forest fires on the ecology and conservation of Malayan sun bears (*Helarctos malayanus*) in East Kalimantan, Indonesian Borneo. PhD doctoral thesis, University of Amsterdam.

Garcia, K. C., Joshi, H. M. & Dharaiya, N. (2016). Assessment of human–sloth bear conflicts in North Gujarat, India. *Ursus* 27: 5–10.

Garshelis, D. L. (2009). Family Ursidae. In: Wilson, D. E. & Mittermeier, R. A. (Eds.), *Handbook of the mammals of the world. Vol. 1. Carnivores* (pp. 448–497). Barcelona: Lynx Edicions.

Garshelis, D. L., Baruch-Mordo, S., Bryant, A., Gunther, K. A. & Jerina, K. (2017). Is diversionary feeding an effective tool for reducing human–bear conflicts? Case studies from North America and Europe. *Ursus* 28: 31–55.

Ghadirian, T., Qashqaci, A. T., Soofi, M., Abolghasemi, H. & Ghoddousi, A. (2017). Diet of Asiatic black bear in its westernmost distribution range, southern Iran. *Ursus* 28: 15–19.

Gillin, C. M., Hammond, F. M. & Peterson, C. G. (1994). Evaluation of an aversive conditioning technique used on female grizzly bears in the Yellowstone ecosystem. *International Conference on Bear Research and Management* 9: 503–512.

Gniadek, S. J. & Kendall, K. C. (1998). A summary of bear management in Glacier National Park, Montana, 1960–1994. *Ursus* 10: 155–159.

Goldstein, I. (1991). Spectacled bear predation and feeding behavior on livestock in Venezuela. *Studies on Neotropical Fauna and Environment* 26: 231–235.

Groff, C., Dalpiaz, D., Frapporti, C., Rizzoli, R. & Zanghellini, P. (2010). *2009 Bear report of the Forestry and Wildlife Department of the Autonomous Province of Trento.* Trento: Autonomous Province of Trento.

Guharajan, R., Abram, N. K., Magguna, M. A., et al. (2017). Does the vulnerable sun bear *Helarctos malayanus* damage crops and threaten people in oil palm plantations? *Oryx* 134: 635–646.

Gunther, K. A. & Hoekstra, H. E. (1998). Bear-inflicted human injuries in Yellowstone National Park, 1970–1994. *Ursus* 10: 377–384.

Gunther, K. A., Haroldson, M. A., Frey, K., et al. (2004). Grizzly bear–human conflicts in the Greater Yellowstone ecosystem, 1992–2000. *Ursus* 15: 10–22.

Hedman, D. (2009). Manitoba Conservation, Thompson, Canada. Summary of the third international bear–people conflicts workshop in Canmore, Alberta (November 15–18), and oral presentation.

Herrero, S. (1994). The Canadian National Parks and grizzly bear ecosystems: the need for interagency management. *International Conference on Bear Research and Management* 9: 7–21.

Herrero, S., Smith, T., Debruyn, T. D., Gunther, K. & Matt, C. A. (2005). From the field: brown bear habituation to people – safety, risks, and benefits. *Wildlife Society Bulletin* 33: 362–373.

Honda, T. (2013). Late spring frosts induce human–Asiatic black bear conflicts. *Mammal Study* 38: 287–292.

Hopkins, J. B. I., Koch, P. L., Ferguson, J. M. & Kalinowski, S. T. (2014). The changing anthropogenic diets of American black bears over the past century in Yosemite National Park. *Frontiers in Ecology and the Environment* 12: 107–114.

Huang, C., Li, X.-Y., Shi, L.-J. & Jiang, X.-L. (2018). Patterns of human–wildlife conflict and compensation practices around Daxueshan Nature Reserve, China. *Zoological Research* 39: 406–412.

Immelmann, K. & Beer, C. (1989). *A dictionary of ethology*. Cambridge, MA: Harvard University Press.

Jampel, C. (2016). Cattle-based livelihoods, changes in the taskscape, and human–bear conflict in the Ecuadorian Andes. *Geoforum* 69: 84–93.

Jamtsho, Y. & Wangchuk, S. (2016). Assessing patterns of human–Asiatic black bear interaction in and around Wangchuck Centennial National Park, Bhutan. *Global Ecology and Conservation* 8: 183–189.

Johnson, H. E., Lewis, D. L., Verzuh, T. L., et al. (2018). Human development and climate affect hibernation in a large carnivore with implications for human–carnivore conflicts. *Journal of Applied Ecology* 55: 663–672.

Jonker, S. A., Parehurst, J. A., Field, R. & Fuller, T. K. (1998). Black bear depredation on agricultural commodities in Massachusetts. *Wildlife Society Bulletin* 26: 318–324.

Jorgenson, J. P. & Sandoval, A. S. (2005). Andean bear management needs and interactions with humans in Colombia. *Ursus* 16: 108–116.

Kaczensky, P. (1999). Large carnivore depredation on livestock in Europe. *Ursus* 11: 59–72.

Karamanlidis, A. A., Sanopoulos, A., Georgiadis, L. & Zedrosser, A. (2011). Structural and economic aspects of human–bear conflicts in Greece. *Ursus* 22: 141–151.

Kavčič, I., Adamič, M., Kaczensky, P., Krofel, M. & Jerina, K. (2013). Supplemental feeding with carrion is not reducing brown bear depredations on sheep in Slovenia. *Ursus* 24: 111–119.

Kimball, B. A., Nolte, D. L., Engeman, R. M., Johnston, J. J. & Stermitz, F. R. (1998). Chemically mediated foraging preference of black bears (*Ursus americanus*). *Journal of Mammalogy* 79: 448–456.

Kishimoto, R. (2009). Status of the 2006 drastic rise of Asiatic black bear (*Ursus thibetanus*) intrusions into residential areas in Nagano Prefecture. In: Oi, T., Ohnishi, N., Koizumi, T. & Okochi, I. (Eds.), *Biology of bear intrusions*. Proceedings of the International Workshop "The Mechanism of the Intrusion of Bears into Residential Areas," November 21–22, 2008, Kyoto, Japan (pp. 35–39). Ibaraki: Forestry and Forest Products Research Institute.

Kobashikawa, S. & Koike, S. (2016). Spatiotemporal factors affecting bark stripping of conifer trees by Asiatic black bears (*Ursus thibetanus*) in Japan. *Forest Ecology and Management* 380: 100–106.

Krofel, M. & Jerina, K. (2012). Review of human–bear conflicts: causes and possible solutions. *Professional Journal of Forestry* 70: 235–253.

Krofel, M., Špacapan, M. & Jerina, K. (2017). Winter sleep with room service: denning behaviour of brown bears with access to anthropogenic food. *Journal of Zoology* 302: 8–14

Laguna, A. (2018). *Manual de prevención y atención de ataques por Oso Andino (*Tremarctos ornatus*) al ganado en Ecuador*. Quito: Big Mammals Conservation.

Laufenberg, J. S., Johnson, H. E., Doherty, P. F. & Breck, S. W. (2018). Compounding effects of human development and a natural food shortage on a black bear population along a human development–wildland interface. *Biological Conservation* 224: 188–198.

Leigh, J. & Chamberlain, M. J. (2008). Effects of aversive conditioning on behavior of nuisance Louisiana black bears. *Human–Wildlife Conflicts* 2: 175–182.

Li, X., Buzzard, P., Chen, Y. & Jiang, X. (2013). Patterns of livestock predation by carnivores: human–wildlife conflict in Northwest Yunnan, China. *Environmental Management* 52: 1334–1340.

Linnell, J. D. C., Aanes, R., Swenson, J. E., Odden, J. & Smith, M. E. (1997). Translocation of carnivores as a method for managing problem animals: a review. *Biodiversity and Conservation* 6: 1245–1257.

Linnell, J. D. C., Lescureux, N., Majic, A., Von Arx, M. & Salvatori, V. (2013). *Results from a stakeholder workshop on large carnivores in Brussels* (pp. 1–68). Brussels: European Commission.

Lira Torres, I., Carrera Treviño, R. & Martínez García, L.F. (2015). Primer registro de oso negro *Ursus americanus* (Carnivora: Ursidae) en un bosque tropical subperennifolio en la reserva de la biosfera "El Cielo", Tamaulipas, México. *Therya* 6: 653–660.

Liu, F., McShea, W. J., Garshelis, D. L., et al. (2011). Human–wildlife conflicts influence attitudes but not necessarily behaviors: factors driving the poaching of bears in China. *Biological Conservation* 144: 538–547.

Lowery, D. R., Morse, W. C. & Steury, T. D. (2012). Biological and social investigation of human–black bear conflicts in the panhandle of Florida. *Human Dimensions of Wildlife* 17: 193–206.

Madden, F. & McQuinn, B. (2014). Conservation's blind spot: the case for conflict transformation in wildlife conservation. *Biological Conservation* 178: 97–106.

Madison, J. S. (2008). Yosemite National Park: the continuous evolution of human–black bear conflict management. *Human–Wildlife Conflicts* 2: 160–167.

Majić Skrbinšek, A. & Krofel, M. (2015). *Defining, preventing, and reacting to problem bear behaviour in Europe* (pp. 1–56). Brussels: European Commission.

Mattson, D. J., Blanchard, B. M. & Knight, R. R. (1992). Yellowstone grizzly bear mortality, human habituation, and whitebark pine seed crops. *Journal of Wildlife Management* 56: 432–442.

Mazur, R. L. (2010). Does aversive conditioning reduce human–black bear conflict? *The Journal of Wildlife Management* 74: 48–54.

Mazur, R. & Seher, V. (2008). Socially learned foraging behaviour in wild black bears, *Ursus americanus*. *Animal Behaviour* 75: 1503–1508.

McCullough, D. R. (1982). Behavior, bears, and humans. *Wildlife Society Bulletin* 10: 27–33.

McKinley, B. K., Belant, J. L. & Etter, D. R. (2014). American black bear–apiary conflicts in Michigan. *Human–Wildlife Interactions* 8: 228–234.

McLellan, B. N. (2005). Sexually selected infanticide in grizzly bears: the effects of hunting on cub survival. *Ursus* 16: 141–156.

Meagher, M. & Phillips, J. R. (1983). Restoration of natural populations of grizzly and black bears in Yellowstone National Park. *International Association of Bear Research and Management* 5: 152–158.

Merkle, J. A., Robinson, H. S., Krausman, P. R. & Alaback, P. (2013). Food availability and foraging near human developments by black bears. *Journal of Mammalogy* 94: 378–385.

Miller, S. D. & Tutterrow, V. L. (1999). Characteristics of nonsport mortalities to brown and black bears and human injuries from bears in Alaska. *Ursus* 11: 239–252.

Mir, Z. R., Noor, A., Habib, B. & Veeraswami, G. G. (2015). Attitudes of local people toward wildlife conservation: a case study from the Kashmir Valley. *Mountain Research and Development* 35: 392–400.

Molinari, P., Krofel, M., Bragalanti, N., et al. (2016). Comparison of the occurrence of human–bear conflicts between northern Dinaric Mountains and south-eastern Alps. *Carnivore Damage Prevention News* 12: 9–17.

Morehouse, A. T. & Boyce, M. S. (2017). Evaluation of intercept feeding to reduce livestock depredation by grizzly bears. *Ursus* 28: 66–80.

Morehouse, A. T., Graves, T. A., Mikle, N. & Boyce, M. S. (2016). Nature vs. nurture: evidence for social learning of conflict behaviour in grizzly bears. *PLoS ONE* 11: e0165425.

Mueller, C., Herrero, S. & Gibeau, M. L. (2004). Distribution of subadult grizzly bears in relation to human development in the Bow River Watershed, Alberta. *Ursus* 15: 35–47.

Narita, R., Mano, T., Yokoyama, R. & Takayanagi, A. (2011). Variation in maize consumption by brown bears (*Ursus arctos*) in two coastal areas of Hokkaido, Japan. *Mammal Study* 36: 33–39.

Naves J., Ordiz A., Fernández-Gil A., et al. (2018) Patterns of brown bear damages on apiaries and management recommendations in the Cantabrian Mountains, Spain. *PLoS ONE* 13: e0206733.

Normua, F., Higashi, S., Ambu, L. & Mohamed, M. (2004). Notes on oil palm plantation use and seasonal spatial relationships of sun bears in Sabah, Malaysia. *Ursus* 15: 227–231.

Northrup, J. M., Stenhouse, G. B. & Boyce, M. S. (2012). Agricultural lands as ecological traps for grizzly bears. *Animal Conservation* 15: 369–377.

Obbard, M. E., Howe, E. J., Wall, L. L., et al. (2014). Relationships among food availability, harvest, and human–bear conflict at landscape scales in Ontario, Canada. *Ursus* 25: 98–110.

Oka, T., Miura, S., Masaki, T., et al. (2004). Relationship between changes in beechnut production and Asiatic black bears in northern Japan. *Journal of Wildlife Management* 68: 979–986.

Penteriani, V., López-Bao, J. V., Bettega, C., et al. (2017). Consequences of brown bear viewing tourism: a review. *Biological Conservation* 206: 169–180.

Penteriani, V., Delgado, M. D. M., Krofel, M., et al. (2018a). Evolutionary and ecological traps for brown bears *Ursus arctos* in human-modified landscapes. *Mammal Review* 48: 180–193.

Penteriani, V., Huber, D., Jerina, K., et al. (2018b). Trans-boundary and trans-regional management of a large carnivore: managing brown bears across national and regional borders in Europe. In: Hovardas, T. (Ed.), *Large carnivore conservation and management: Human dimensions and governance*. Oxford, UK: Routledge, Taylor & Francis Group.

Penteriani V., Zarzo-Arias A., Novo-Fernández, A., Bombieri G. & López-Sánchez, C. A. (2019). Responses of an endangered brown bear population to climate change based on predictable food resource and shelter alterations. *Global Change Biology* 25: 1133–1151.

Proctor, M. F., Kasworm, W. F., Annis, K. M., et al. (2018). Conservation of threatened Canada–USA trans-border grizzly bears linked to comprehensive conflict reduction. *Human–Wildlife Interactions* 12: 348–372.

Rajpurohit, K. S. & Krausman, P. R. (2000). Human–sloth bear conflicts in Madhya Pradesh, India. *Wildlife Society Bulletin* 28: 393–399.

Ralf, R. (1995). *History of bear–human conflict management in Jasper National Park: 1907–1995*. Jasper: Jasper National Park.

Ravenelle, J. & Nyhus, P. J. (2017). Global patterns and trends in human–wildlife conflict compensation. *Conservation Biology* 31: 1247–1256.

Saito, M., Yamauchi, K. & Aoi, T. (2008). Individual identification of Asiatic black bears using extracted DNA from damaged crops. *Ursus* 19: 162–167.

Sangay, T. & Vernes, K. (2008). Human–wildlife conflict in the Kingdom of Bhutan: patterns of livestock predation by large mammalian carnivores. *Biological Conservation* 141: 1272–1282.

Sato, Y., Mano, T. & Takatsuki, S. (2005). Stomach contents of brown bears *Ursus arctos* in Hokkaido, Japan. *Wildlife Biology* 11: 133–144.

Schirokauer, D. W. & Boyd, H. M. (1998). Bear–human conflict management in Denali National Park and Preserve, 1982–94. *Ursus* 10: 395–403.

Schwartz, C. C., Swenson, J. E. & Miller, S. D. (2005). Large carnivores, moose, and humans: a changing paradigm of predator management in the 21st century. *Alces* 39: 41–63.

Schwartz, C. C., Haroldson, M. A., White, G. C., et al. (2006). Temporal, spatial, and environmental influences on the demographics of grizzly bears in the Greater Yellowstone Ecosystem. *Wildlife Monographs* 161: 1–68.

Scotson, L., Vannachomchan, K. & Sharp, T. (2014). More valuable dead than deterred? Crop-raiding bears in Lao PDR. *Wildlife Society Bulletin* 38: 783–790.

Sethy, J. & Chauhan, N. S. (2013). Human–sun bears conflict in Mizoram, North East India: impact and conservation management. *International Journal of Conservation Science* 4: 317–328.

Shivik, J. A., Ruid, D., Willging, R. C. & Mock, K. E. (2011). Are the same bears repeatedly translocated from corn crops in Wisconsin? *Ursus* 20: 114–119.

Sowka, P. (2009). Techniques and refuse management options for residential areas, campgrounds, and group-use area. In: *Living with Black Bears, Grizzly Bears and Lions Project. 3.* Swan Valley, MT: Living with Wildlife Foundation, & Montana Fish, Wildlife and Parks.

Stenhouse, G. B., Lee, L. J. & Poole, K. G. (1988). Some characteristics of polar bear killed during conflicts with humans in the Northwest Territories, 1976–86. *Arctic* 41: 275–278.

Stewart, W. B., Witmer, G. W. & Koehler, G. M. (1999). Black bear damage to forest stands in western Washington. *Western Journal of Applied Forestry* 14: 128–131.

Steyaert, S. M. J. G., Kindberg, J., Swenson, J. E. & Zedrosser, A. (2013). Male reproductive strategy explains spatiotemporal segregation in brown bears. *Journal of Animal Ecology* 82: 836–845.

Steyaert, S. M., Kindberg, J., Jerina, K., et al. (2014). Behavioral correlates of supplementary feeding of wildlife: can general conclusions be drawn? *Basic and Applied Ecology* 15: 669–676.

Steyaert, S. M. J. G., Leclerc, M., Pelletier, F., et al. (2016). Human shields mediate sexual conflict in a top predator. *Proceedings of the Royal Society B* 283: 20160906.

Swenson, J. E. & Andrén, H. (2005). A tale of two countries: large depredation and compensation schemes in Sweden and Norway. In: Woodroffe, R., Thirgood, S. & Rabinowitz, A. (Eds.), *People and wildlife, conflict or coexistence?* (pp. 323–339). Cambridge, UK: Cambridge University Press.

Swenson, J. E., Sandegren, F., Soderberg, A., et al. (1997). Infanticide caused by hunting of male bears. *Nature* 386: 450–451.

Swenson, J. E., Dahle, B. & Sandegren, F. (2001). Intraspecific predation in Scandinavian brown bears older than cubs-of-the-year. *Ursus* 12: 81–92.

Tavss, E. A. (2005). Correlation of reduction in nuisance black bear complaints with implementation of a nonviolent program and a hunt. In; *New Jersey public hearing on the comprehensive black bear management policy*. Rutgers, NJ: State University of New Jersey.

Taylor, J. D. & Phillips, P. (2019). *Black bears.* Wildlife Damage Management Series. U.S. Department of Agriculture Animal and Plant Health Inspection Service: Wildlife Services.

Towns, L., Derocher, A. E., Stirling, I., Lunn, N. J. & Hedman, D. (2009). Spatial and temporal patterns of problem polar bears in Churchill, Manitoba. *Polar Biology* 32: 1529–1537.

Treves, A. & Karanth, K. U. (2003). Human–carnivore conflict and perspectives on carnivore management worldwide. *Conservation Biology* 17: 1491–1499.

Treves, A., Kapp, K. J. & Macfarland, D. M. (2010). American black bear nuisance complaints and hunter take. *Ursus* 21: 30–42.

van Eeden, L. M., Eklund, A., Miller, J. R., et al. (2018). Carnivore conservation needs evidence-based livestock protection. *PLoS Biology* 16: e2005577.

Widman, M. & Elofsson, K. (2018). Costs of livestock depredation by large carnivores in Sweden 2001 to 2013. *Ecological Economics* 143: 188–198.

Wielgus, R. B. & Bunnell, F. L. (1994). Sexual segregation and female grizzly bear avoidance of males. *Journal of Wildlife Management* 58: 405–413.

Wilder, J. M., DeBruyn, T. D., Smith, T. S. & Southwould, A. (2007). Systematic collection of bear–human interaction information for Alaska's national parks. *Ursus* 18: 209–216.

Wilson, R. R., Regehr, E. V., St. Martin, M., et al. (2017). Relative influences of climate change and human activity on the onshore distribution of polar bears. *Biological Conservation* 214: 288–294.

Wilson, S. M. (2007). *Community-supported conservation of grizzly bears on private agricultural lands.* Final close-out report for conservation innovation grant. Portland, OR: US Department of Agriculture – Natural Resources Conservation Service.

Wilson, S. M. (2016). *A guidebook to human–carnivore conflict: Strategies and tips for effective communication and collaboration with communities* (pp. 1–67). Ljubljana, Slovenia: Slovenia Forest Service – LIFE DINALP BEAR Project.

Wilson, S. M., Madel, M. J., Mattson, D. J., Graham, J. M. & Merrill, T. (2006). Landscape conditions predisposing grizzly bears to conflicts on private agricultural lands in the western USA. *Biological Conservation* 130: 47–59.

Wilson, S. M., Neudecker, G. A. & Jonkel, J. J. (2013). Human–grizzly bear coexistence in the Blackfoot River watershed, Montana: getting ahead of the conflict curve. In: Clark, S. & Rutherford, M. (Eds.), *Large carnivore conservation: Integrating science and policy in the North American West* (pp. 117–214). Chicago, IL: University of Chicago Press.

Wong, W. M., Leader-Williams, N. & Linkie, M. (2015). Managing human–sun bear conflict in Sumatran agroforest systems. *Human Ecology* 43: 255–266.

Wynn-Grant, R., Ginsberg, J. R., Lackey, C. W., Sterling, E. J. & Beckmann, J. P. (2018). Risky business: modeling mortality risk near the urban-wildland interface for a large carnivore. *Global Ecology and Conservation* 16: e00443.

Yamazaki, K., Koike, S., Kozakai, C., et al. (2009). Behavioral study of free ranging Japanese black bears I. Does food abundance affect the habitat use of bears? In: Oi, T., Ohnishi, N., Koizumi, T. & Okochi, I. (Eds.), *Biology of bear intrusions.* Proceedings of the International Workshop "The Mechanism of the Intrusion of Bears into Residential Areas," November 21–22, 2008, Kyoto, Japan (pp. 60–63). Ibaraki: Forestry and Forest Products Research Institute.

Zack, C. S., Milne, B. T. & Dunn, W. C. (2003). Southern oscillation index as an indicator of encounters between humans and black bears in New Mexico. *Wildlife Society Bulletin* 31: 517–520.

Zapata-Ríos, G. & Branch, L. C. (2018). Mammalian carnivore occupancy is inversely related to presence of domestic dogs in the high Andes of Ecuador. *PLoS ONE* 13: p.e0192346.

Zedrosser, A., Steyaert, S. M. J. G., Gossow, H. & Swenson, J. E. (2011). Brown bear conservation and the ghost of persecution past. *Biological Conservation* 144: 2163–2170.

Ziegltrum, G. J. & Nolte, D. L. (2001). Black bear forest damage in Washington state, USA: economic, ecological, social aspects. *Ursus* 12: 169–172.

Zukowski, B. & Ormsby, A. (2016). Andean bear livestock depredation and community perceptions in Northern Ecuador. *Human Dimensions of Wildlife* 21: 111–126.

Zyśk-Gorczyńska, E., Jakubiec, Z., Wertz, B. & Wuczyński, A. (2016). Long-term study of damage to trees by brown bears *Ursus arctos* in Poland: increasing trends with insignificant effects on forest management. *Forest Ecology and Management* 366: 53–64.

Chapter 16

Principles of Human–Bear Conflict Management in Challenging Environments

Özgün Emre Can

The problem of knowledge is that there are many more books on birds written by ornithologists than books on birds written by birds and books on ornithologists written by birds.
Nassim Nicholas Taleb

Humans and bears have coexisted since time immemorial. Ötzi the Iceman, a 5300-year-old natural mummy discovered in the Alps on the Italian–Austrian border on September 19, 1991 was found to be wearing a fur hat made from brown bear fur (O'Sullivan et al. 2016). Ötzi's hat protected his head from the elements, so in other words, brown bears made it possible for Ötzi (who died at the age of 45) and his contemporaries to survive in an extreme environment some 5300 years ago. Today, millions of people live happily alongside bears on four continents. As a result, millions of interactions occur between people and bears. Most of these interactions end without injury to a person, but there are places where problems between people and bears exist (Can et al. 2014). According to the United Nations, in just 30 years or so, 87% of the world's population, about 7.8 billion people, will be living in developing countries (United Nations 2004). Therefore, in places where bears still exist, more people will find themselves in close proximity to bears, which will likely result in increased problems between people and bears. Failure to mitigate those problems will likely reduce the tolerance of local communities and diminish conservation efforts, not only for bears but also for wider initiatives in general. Therefore, effective problem-solving, or as it is called, human–bear conflict management, will be needed more than ever within the next three decades, but what actually is human–bear conflict (Box 16.1)?

An international group of experts at a workshop in Istanbul, Turkey defined human–bear conflict as "any situation where wild bears undesirably use or damage human property, where wild bears harm people, or where people perceive bears to be a direct threat to their property or safety" (WSPA 2009). This "Istanbul definition" of conflict was later used by the Human Bear Conflict Expert Team of IUCN SSC Bear Specialist Group (IUCN SSC BSG 2019).

Box 16.1 On the term "human–bear conflict" and the need for a more accurate one

The so-called "conflict" between humans and wildlife has been ongoing since time immemorial. The word "conflict" comes from the Latin word "conflīctus" and means a fight, battle, a prolonged struggle, or clashing or variance of opposed principles, statements, arguments (OED 2015). As bears and other wildlife species are not purposefully seeking to undermine human interests, it is evident that terms such as "human–bear conflict" and "human–wildlife conflict" are more of a metaphor rather than technical terms.

Metaphors might be useful and contribute to effective communication. Militarism is a powerful mode of conceptualizing real-world challenges (Larson 2005). However, although military terminology sometimes helps us to grasp daily life concepts rapidly, it can be damaging with unintended consequences as well. For example, Elm and Diener (2007) concluded that military terminology is corrupting biomedical communication and warned physicians and researchers to avoid the cynical logic of war when writing about care of patients. Militaristic metaphors may affect the perceptions of scientific objectivity because of their resonance within contemporary political settings (Larson 2005). Using militaristic language to define the interactions between people and wildlife could weaken the current efforts for nature conservation, and cause loss of scientific credibility (Larson 2005; Redpath et al. 2015). In the case of human–wildlife conflict, the metaphor reinforces the notion that humans are apart from nature; not a part of it (Peterson et al. 2010, 2013). Although metaphors might motivate conservation action in the short term, in the long term they could be ineffective and can even cause misperceptions among conservationists themselves (Larson 2005) and the public.

Scientific research has recently shown that the language one uses shapes thinking about the fundamental dimensions of human experience such as space, time, causality, and relationships to others (Boroditsky 2011). Evidence-based science now shows us that language profoundly influences the way we see the world (Boroditsky 2010). The implication of linguistic

Box 16.1 *(cont.)*

research for wildlife management and conservation is that the way we frame the interaction between people and wildlife actually does matter. Larson (2005) warns us that linguistic wars can contribute to real wars. As history has proved with the case of people's long-lasting efforts to exterminate wolves in Europe and North America, once humans declare "war" against another species, the other species does not have much chance to survive. Therefore, those who are using the metaphor human–bear conflict have to consider how this metaphor might be affecting people's reality. This is because, as the National Academy of Sciences (Institute of Medicine 2009) puts it, "Research is based on the same ethical values that apply in everyday life, including honesty, fairness, objectivity, openness, trustworthiness, and respect for others."

There is a need for a language that is consistent with a vision of human–wildlife coexistence, not one that contributes to polarized and militaristic conceptualizations of management and conservation issues (Larson 2005). Do we have an excuse to ascribe value and attachment to the term human–wildlife conflict merely because it has been increasingly used in the literature in the last decade or so? The term "human–bear conflict" is not problem-free, it frames the interaction between people and bears as negative and we need to resist metaphoric thinking and find accurate descriptions for effective management and conservation. A widely accepted term to replace the word conflict has yet to be found. Therefore, for practical reasons, I use the term human–bear conflict in this chapter to describe the problems that arise between people and bears. I ask our readers not to allow the definitional sensitivity of the word "conflict" to distract from the focus of this chapter.

Human–bear conflicts differ geographically across the range of eight bear species (Can et al. 2014; see also Chapter 15). Although it is difficult to quantify actual trends across the range of bear species, human–bear conflict is perceived to be increasing across the globe (Hristienko & McDonald 2007). Recently, a global survey (Can et al. 2014) involving 104 experts submitting 130 completed questionnaires (as some experts provided information on more than one bear species) found that most experts (89%, n = 115) reported a high level of conflict in their area of expertise; conflicts were increasing in more than half of the areas from which survey respondents responded (56%, n = 73), and the majority (76%, n = 86) of the experts concluded that human–bear conflict has had a negative impact on conservation of bears on all four continents where bears are present (Can et al. 2014).

Later, a survey carried out among the members of the IBA also showed that conflicts are considered to be a major issue for the next decade (Zedrosser 2019). In brief, there is consensus among the world's experts that the problem of human–bear conflict is worsening across the range of bear species. However, the same problem has some quite different impacts on people and bears in Western Educated Industrialized Rich Democratic countries (or WEIRD countries, hereafter) compared to non-WEIRD countries. A whistle-stop tour around the globe indicates that: in North America conflicts do not pose a threat to the viability of American black (*Ursus americanus*) or brown bears (*U. arctos*) but cause annoyance for people; whereas in Europe, illegal killing of bears as a retaliation against real or perceived threats is the main cause of bear mortality (Can et al. 2014). On the other hand, conflicts cause hardships for people and affect the rural economy in Asia and South America (Can et al. 2014; see also Chapter 15). In return, retaliation against bears threatens the viability of Asiatic black bears and Andean bears in some of parts of their range and hinders acceptance of conservation initiatives (Can et al. 2014).

Conflict is affecting an increasing number of people in Asia and South America (Can et al. 2014). However, conservation practitioners in Asia and South America often operate in challenging circumstances, in many cases without the resources that are available in North America or Europe. By adapting the definition made by Paulus et al. (2009) for extreme environments (such as combat situations, expeditions at high altitudes, or in space), I define a challenging environment for conservation as an "environment that exposes the practitioner to demanding psychological and physical conditions and that requires optimal cognitive and behavioral performance and where institutional structure for wildlife management and conservation is limited or non-existent." Typically, such areas are in resource-limited countries, but resource-limited islands can also be found in resource-rich countries.

It is a challenge to write something useful for practitioners working on human–bear conflict in resource-limited environments in a limited space. Nevertheless, I will make an attempt to do so in this chapter by utilizing: (1) the relevant literature; (2) review of 44 management and action plans from 13 countries (Albania, Austria, Bulgaria, Canada, Croatia, Georgia, Italy, Latvia, Poland, Romania, Slovenia, USA, and Venezuela); (3) the results of the first global survey on human–bear conflicts (Can et al. 2014); and (4) the hard-earned wisdom of a series of participatory workshops that I co-organized (Turkey in 2002, 2008, 2010, and 2012; Greece in 2015; Ecuador in 2017) and was invited to (Georgia in 2010; Nepal 2015; UK in 2016 and 2017). I will first give a brief summary of how conflict studies started in international relationships and then entered the field of wildlife conservation. Then, I will provide an overall picture of human–bear conflict management and discuss how current conflict management plans can be improved by considering them as logic models. Finally, I will present 12 key factors for managing human–bear conflict in challenging environments.

How did the Conflict Resolution Studies Begin?

As Swenson in Wilson (2016) stated succinctly, managing human–carnivore conflicts "requires a knowledge of the science behind carnivore biology and human conflict management in addition to the insight that comes only from actual experience." Therefore, successful human–bear conflict management requires a melding of science (which depends on structured analytical

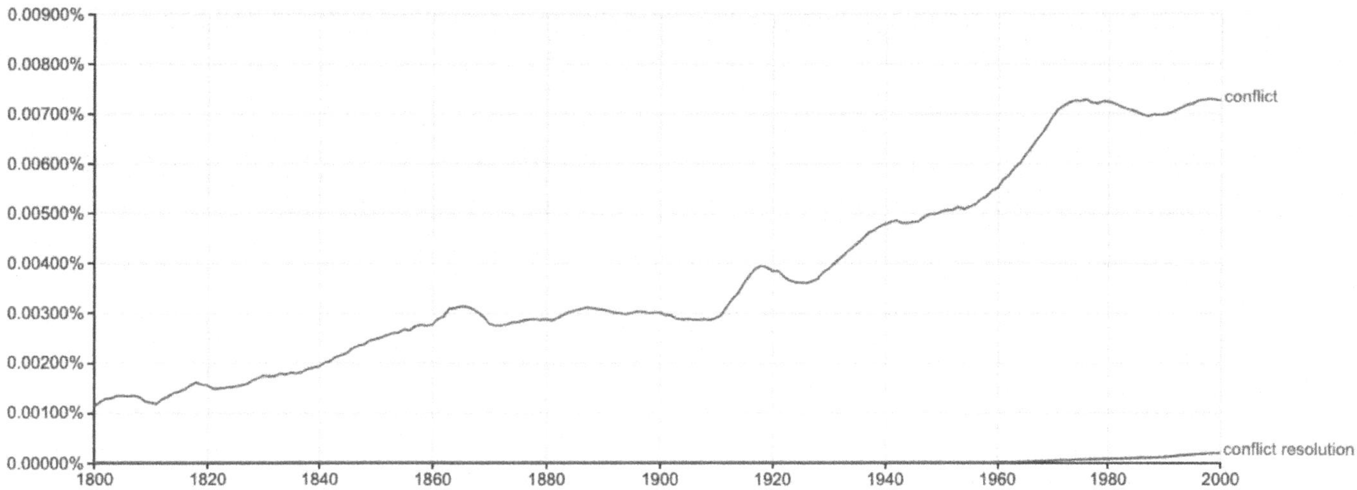

Figure 16.1 Google Books Ngram Viewer graph shows the increase in the use of the terms "conflict" and "conflict resolution" in books published since 1800 and those that are digitized by Google.

methods) and art (which depends on intuitive judgment and personal experience). The "science" and "art" of conflict management has been practiced in North America and Europe for the last seven decades.[1] In the beginning, conflict resolution had nothing to do with bears or even with wildlife.

A small group of pioneering researchers spotted the need for studying conflict itself in the 1950s after the Cold War, when a potential nuclear war was threatening the future of humanity and conflict resolution became a distinct field of study (Ramsbotham et al. 2011). The term "conflict resolution" was first used to define a new field of study with the first edition of the *Journal of Conflict* in 1957. Although it was a journal in the field of social sciences, the first editorial article of the *Journal of Conflict* (Editors 1957) was presenting the new journal surprisingly as a new species with the name *Interdisciplinaris internationalis* in Linnean terms. Conflict resolution studies progressed over time and were instrumental for making a positive difference in international relations as well as domestic politics in South Africa, the Middle East, Central America, and South-East Asia by the 1980s (Ramsbotham et al. 2011; Figure 16.1).

It is not known when conflict resolution was first used in environmental disputes, but the practice became common in the 1980s (Cohn 2002). In his landmark paper, "Tragedy of the Commons," Hardin (1968) presented how environmental conflicts arouse. Environmental and resource issues are the main reasons for global conflict today (Ramsbotham et al. 2011).

Conflict Resolution in Wildlife Conservation

Presnall (1943) was perhaps the first author who mentioned conflict in relation to wildlife conservation in a scholarly journal. According to him, understanding the conflict concepts of American Indians and Caucasians in relation to wildlife values was vital for effective game conservation programs on Indian lands. Presnall (1943) stated "the basic concepts of two races are not very different, the Indian has merely applied vocation concepts to wildlife and we think of it as an avocation." Although he was not using the term "conflict resolution," he was trying to find a common ground for effective game conservation in his article and referring to what we call today "conflict resolution" in management and conservation. Today, when the term conflict resolution is used in relation to wildlife management and conservation, what is actually implied is management of problems that arise between humans and wildlife, as resolution in the literal meaning of the word rarely happens. Even reducing impacts of conflict may not resolve conservation conflicts; in fact, perhaps no conservation conflict has ever been fully resolved by eliminating that conflict (Redpath et al. 2013; Pooley et al. 2017). Today, the phrase "human wildlife conflict" has almost replaced the term "human–wildlife relationship" and almost any animal behavior can now be described as human–wildlife conflict if that behavior can be perceived as a problem by one or more people (Peterson et al. 2013; Can et al. 2014). Moreover, human–human conflicts are now often said to be a component of human–wildlife conflict. However, a dispute between people – even if it might be related with wildlife or environmental issues – is a dispute between people. Failure to deal with disputes between people only contributes to escalations of such disputes between humans, making such disputes much harder to resolve (Peterson et al. 2013). For example, see Pooley et al. (2017) to learn how rural hunters kill wolves (*Canis lupus*) to express anger to the state in Sweden.

Human–Bear Conflict Management

Probably the first international gathering about human–bear conflict management was the international human–bear conflict

[1] See Artelle et al. (2018), who report that most management systems in North America lack indications of the basic elements of a scientific approach.

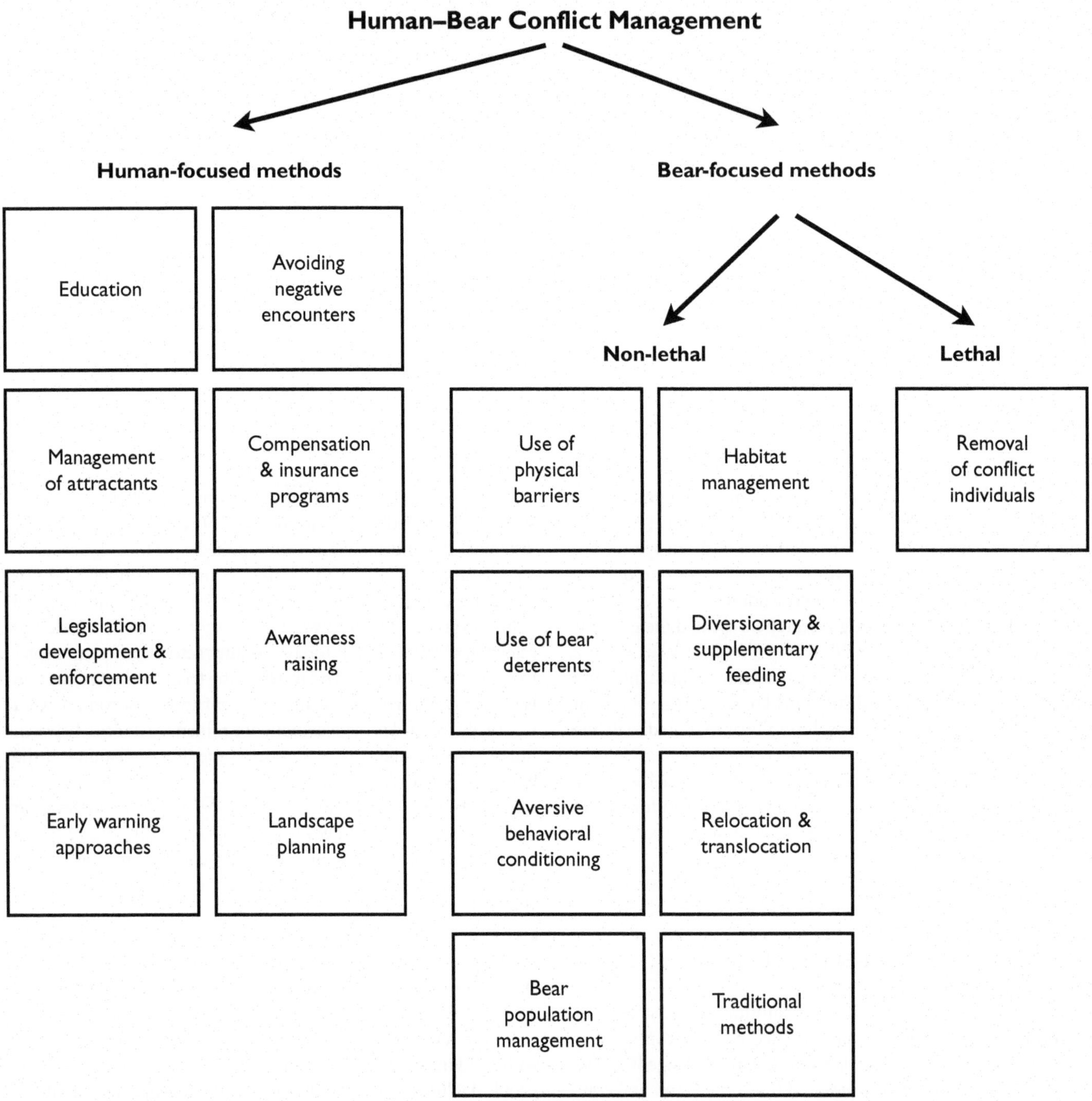

Figure 16.2 The toolbox of human–bear conflict management consists of 17 tools.

workshop in Yellowknife, Canada in 1987 where few attended (Matt 2012). Since then, human–bear conflict and its management have become a major topic for researchers and practitioners. At the heart of conflict management lies mitigating, compensating, and controlling the damage caused by bears and, in an ideal world, conflict is managed by operational plans called human–bear conflict management plans. These plans are simply a detailed road map to make the world a better place by changing the initial state of conflict to a desired state of conflict in a given area. A re-examination of data from Can et al. (2014) showed that conflict management can be said to include 17 tools that are targeted to humans and bears when 44 management plans totalling about 2375 pages are considered (Figure 16.2).

For effective conflict management, these 17 tools need to be used in a tactical way and within the framework of a carefully constructed strategy. However, there are no established standards for preparing conflict plans. Apart from 9% (n = 4), most of the reviewed plans were crowded with text and the actions listed had no specified time frame, making the documents an inventory rather than a management plan. Similarly, only 9%

(n = 4) of the plans mentioned the welfare of bears, the elephant-in-the-room aspect of conflict management. On a positive note, it is promising that 36% (n = 16) of the documents were collaborative productions of government agencies, non-governmental organizations (NGOs), and researchers, but it is less encouraging that there was little evidence of engagement with citizen stakeholders.

In most, if not in all, of the plans available, the focus is mainly on the outcomes. In fact, at the heart of a conflict management plan should be the user because conflict management plans should be operational manuals – rather than a review of the literature – that help the user to make timely decisions about the use of potential tools under various conditions (related to staff, organisational structure, equipment and gear, etc.). Moreover, as is the case in some other sectors, such as in aviation (FAA 2000), in conflict management, the greatest variability is also the human factor. Similar to the aviation sector (FAA 2000), the factors (such as stress, peer pressure, adaptive skills, physical and motivational fatigue) that affect the performance of people managing conflicts are topics that are yet to be studied by relevant academic disciplines.

Speaking about the human factor, leadership has a crucial role in conflict management. As Peter Drucker stated, "only three things happen naturally in organisations: friction, confusion, and underperformance. Everything else requires leadership" (in Sandermoen 2019). In challenging environments, what is most needed in conflict management is leadership. In countries where wildlife agencies have limited capacity to work with local communities or lack the will to do so, this leads to public frustration (Baruch-Mordo et al. 2011). Currently, managing human–bear conflicts is arguably one of the most challenging topics for wildlife managers in the USA (Spencer et al. 2007). If anything, the challenge is even bigger for those involved in managing conflicts in challenging environments, and there are two reasons for that: human–bear conflict is a wicked problem, and applicability of the 17 conflict management tools is limited in resource-limited countries.

Human–Bear Conflict is a Wicked Problem

Human–bear conflict is more of a wicked problem than a complex problem in challenging environments. Complex problems are those for which problem solvers agree on what the problem is, but for which there is no consensus on the solutions. On the other hand, for wicked problems, there is disagreement about the problem as well as the solution; planning problems are wicked problems (Rittel & Webber 1973). For wicked problems, the issue is not related to uncertainties in science or failure to consider scientific evidence, but the deep disagreement on values and world views (Balint et al. 2011). Therefore, traditional top-down approaches to conflict management where an agency produces a management plan by getting input only from scientific elites do not provide effective solutions to conflicts. As a result, the problem escalates. Identifying human–bear conflicts as wicked problems makes it clear that we need to work on the problem definition first to have a common ground regarding conflict management.

The Conflict Management Toolbox Needs Innovation

The lack of effective governance arrangements, financial mechanisms, and institutional and technical capacity make human–bear conflict management practically impossible in some resource-limited countries. Only a fraction of the conflict management tools used in North America or Europe is available to the practitioner in resource-limited environments (Table 16.1). This is sometimes the reason for the seeming unwillingness of wildlife authorities in dealing with conflicts in resource-limited countries.

Therefore, it is incumbent on the wider conservation community to develop innovative, cost-effective, thoughtful solutions (Can et al. 2014). However, nothing is easy, it seems. Although there is an increasing emphasis on the use of technology in conservation, it will take more than technology to save the world (for more information see Can & Macdonald 2017). Moreover, transferring approaches that work in resource-rich environments (WEIRD countries) to challenging environments rarely works in the long term. Gippoliti et al. (2018) thoughtfully discussed why the Western European approach of large mammal conservation is not exportable to other parts of the world and, in fact, may have negative consequences for wildlife. This applies to conflict management as well because conservation polices – as economic and foreign policies – that are mostly based on utilitarian considerations have a high chance of failure outside WEIRD countries (Can & Macdonald 2018). So is there hope for new ideas, concepts, and approaches? Deeper and contextual evidence will probably come from specific case studies done in resource-limited regions and by going beyond biological insights.

Conflict Management Plans as Logic Models

Each plan for human–bear conflict management is a proposed solution to the problem of conflict at a given site, region, or country. One way to produce better conflict management plans is to approach the plan as a logic model (for more information, see Knowlton & Philips 2009). The name may sound fancy, but a logic model is simply a detailed picture of how conflict management will be practiced on the ground. It is a tool that can be used not only for design and implementation but also for evaluation and reporting (W. K. Kellogg Foundation 2004). A logic model presents all available resources, actions that will be carried out with the available resources to solve the problem (i.e. activities), what the activities will deliver (i.e. outputs), the change that will occur as a result of the outputs (i.e. outcomes), and finally what these changes will eventually lead to (i.e. impact) (W. K. Kellogg Foundation 2004; UNDP 2009; Busjeet 2013; Figure 16.3).

For every entry in the plan, from activities to impact, the ownership, location, and time are specified; therefore, who

Table 16.1 List of 17 tools for human–bear conflict management and subjective assessment of their suitability for a resource-rich environment vs. a resource-limited environment.

Name of the tool/approach	Potential for application in a:	
	Resource-rich environment	Resource-limited environment
Aversive behavioral conditioning	Medium to high	Low to medium
Avoiding negative encounters	High	Low to medium
Awareness raising*	High	Low to high
Bear population management	Medium to high	Low
Compensation & insurance programs	Medium to high	Low
Diversionary & supplementary feeding	Medium to high	Low
Education*	High	Low to medium
Early warning approaches	Medium to high	Low
Habitat management	Medium to high	Low
Landscape planning	Medium to high	Low
Legislation development & enforcement	Medium to high	Low
Management of attractants	High	Low to medium
Relocation & translocation	Medium to high	Low
Removal of conflict individuals	Medium to high	Low to medium
Traditional methods	Low	Medium to high
Use of bear deterrents	Medium to high	Low to medium
Use of physical barriers**	Medium to high	Low to medium

* Although education and awareness are often mentioned together in the literature, they are actually two different endeavors.
** Elevated platforms, electric fences, etc.

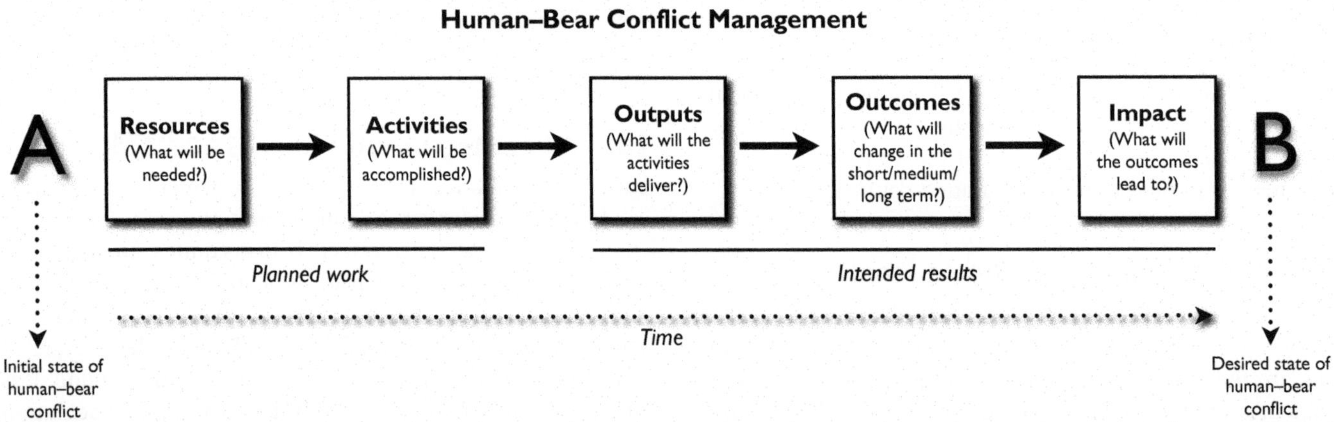

Figure 16.3 Schematic representation of human–bear conflict management as a logic model. The resources that are available to the practitioners dictate the type of activities that can be reasonably planned to deal with conflicts. Then, outputs are achieved when those activities are accomplished. Outcomes are achieved as a result of the outputs. Finally, outcomes determine the overall impact of the conflict management plan.

does what, where, and when is clearly documented. Outcomes should be specific, measurable, action-oriented, realistic, and timed (known as the SMART acronym). For an NGO, this could be signing a memorandum of understanding with the national wildlife authority, or for a wildlife agency, a change in the number of complaints received from the local community can be examples of short-term outcomes. Impact will be the ultimate state of the world desired by the people who prepared the conflict plan. Thinking of conflict management plans as logic models helps everyone (as many key contributing interest groups as necessary) involved in conflict management planning to visualize and understand how each step contributes to the ultimate intended goal of the management activities: the coexistence of people and bears. Human–bear conflict plans prepared using logic model thinking not only guide the actions on the ground, but also enable better monitoring and evaluation, and indicate where improvements are needed within the plan.

Another way to improve conflict management planning is to consider a Prevent–Respond–Recover (P2R) framework and organize the conflict management plan from a three-dimensional time perspective: before the conflict events, during the conflict events, and after the conflict events. A conflict event is when there is potential for loss of human life and/or damage to property. The Prevent phase is about how to prevent or reduce conflict before the actual conflict event occurs by using appropriate tools from the conflict management toolbox. This is when it is best to meet local people to build relationships with them, and collect data about the local community, the bears, and the habitat to fill gaps in the existing knowledge. This is also the best time to run awareness-raising and educational initiatives, to become familiar with the relevant legislation about bears, the habitat, and the legal basis of potential interventions that might be necessary in the short and medium time frame. Prioritizing the communities (i.e. specific areas on the map) by considering their recovery potential (socioeconomic conditions) from potential conflict events at this stage will also save time when large-scale conflicts occur.

Similarly, the Respond phase relates to how to best prioritize actions and respond to human–bear conflict situations while they are happening. To effectively respond to conflicts, it is vital to understand the impact of the conflict not only on the affected individuals but also on the community as a whole. Making commitments to the local community that will not be followed later on will harm the Recovery phase of conflict management. The ultimate purpose of responding to conflicts is to implement necessary measures to reduce or eliminate the risks caused by bears – in terms of human safety and damage to property – so that conflict does not escalate leading to retaliatory killings of bears. Lastly, the Recovery phase refers to the period when tools such as compensation and insurance programs can be used. The focus of this phase is managing people's expectations as much as possible to recover from the trauma.

Key Factors for Managing Human–Bear Conflicts in Challenging Environments

The following factors play a key role in managing human–bear conflicts in challenging environments.

1. **Recognize that conflict is real and is a biological phenomenon happening in the human brain.** Individuals, communities, and societies think in very diverse ways about animals; the human–animal relationship cannot be simply explained by solely measuring attitudes and behavior (Pooley et al. 2017). Sometimes social factors (such as peer group norms) but not the perceived threats from large carnivores such as bears are the reason for killing predator species (Treves & Bruskotter 2014). For the human brain, some problems (such as which bus to take to the city center) activate brain regions involved in logical problem-solving (Shenhav & Greene 2014; Eagleman 2016). Other problems, such as hearing about a bear attack on a neighbor, activate other brain networks, such as those involved in emotion.

   Many human conflicts are disputes over values but not evidence (Ginges et al. 2007). Research shows that people may not reason instrumentally when values are at stake (Tetlock 2003). This is true for human–human conflicts that are embedded in human–wildlife conflicts. Conflict is a real biological phenomenon that happens in the human brain and neuroscience has recently started to understand the neurological processes behind conflicts (Can & Macdonald 2018). Human–carnivore conflict is rarely, if ever, solely due to the damage received by large predators. When large carnivores are considered, activity in the brain networks that are involved in fear and other emotional processes increases (Can & Macdonald 2018). In brief, conflicts are about how the human brain is wired. Do not get into a fight with it, live with it.
2. **Recognize the fact that every human–bear conflict crisis contains the opportunity to make things worse.** The first principle of conflict management interventions can be perhaps thought of as the consequentialist maxim "first, do no harm" as is often stated in medicine (Marsh 2014). The planned activities for managing conflicts should not make things worse in the affected community (Chandran 2015). For example, in a community affected by damage caused by bears to apiaries, if the support (monetary or not) that the authorities provide to mitigate the negative effects of damage is not delivered to the right families (for example, due to local power dynamics), this would result in a decline in the public's trust of the authorities. This, in turn, will make effective conflict management an even bigger challenge in that particular resource-limited environment by making things worse.
3. **Recognize the negative effects of not acting when needed or acting too little or too late in significant conflicts.** Responding to conflict emergencies, e.g. where there is human fatality due to a bear attack, needs a different set of resources than slow-onset conflict situations. In conflict emergencies, there is a need for a fast flow of support to the affected family and local communities, within days. Planning conflict management across a country and ensuring multi-year resources devoted to conflict management is a dream in resource-limited countries. Nevertheless, management authorities should have short- (for example, constructing physical barriers such as elevated platforms), medium-, and long-term goals (such as habitat restoration for bears) in mind.

4. **Use the capacity to first deal with the most serious type of all conflicts: human death caused by bears.** Fatalities due to bear attacks are "conflict emergency situations" and are more common in resource-limited countries. However, to investigate deaths caused by bear attacks requires the involvement of forensic and genetic laboratories (Can et al. 2014). Except in a few countries in Europe and outside North America, wildlife agencies do not have that capacity or resources. Nevertheless, it is crucial to demonstrate to the affected family and the community that the wildlife authorities are doing all they can to prevent future tragedies. Failing to do so will cause the tolerance level toward bears to drop. Expect that this will sooner or later hinder not only conservation of bears but also other conservation initiatives in that particular area.
5. **Support families and communities that are affected by conflicts.** In poorer parts of the world, there is an absolute need to integrate human–bear conflict management with poverty alleviation strategies. Bear attacks – when not resulting in fatalities – may leave the victims disabled, leaving them unable to perform necessary activities, which in turn affects the whole family within the socioeconomic framework of rural communities (Ratnayeke et al. 2014). Assess what would be the best thing to do to support people, but be informed about cultural sensitivities and beware of unconscious biases (for more details see Brownstein 2019). Providing financial support might not be always an option due to relevant (or lack of) legislation. If that is the case, try to innovate with a solution that considers what is possible in terms of the local legislation. Providing a government service might be an option (such as providing free transportation to village children to reach school, providing free meals, providing free health service for the elderly, etc.). Prioritize and focus on areas where the conflict is a threat to the viability of bear populations.
6. **Be honest about the limits of effectiveness for each conflict management tool.** Involving and gaining the trust of the local community is a crucial step in conflict management. For effective conservation, the public needs to trust the bear managers who are acting to protect public safety (Herrero et al. 2011). Sharing responsibility in conflict management is said to be a key element for success in North America (Treves et al. 2006; Matt 2012), but in resource-limited parts of the world this is harder than gaining the trust of the local communities. There is no single magical tool that can end human–bear conflicts. As societies need a full-time police force to maintain law and order, and policing is a never-ending job, human–bear conflict management will also remain a full-time endeavor as long as bears and people coexist. If possible, practice proactive enforcement, i.e. dispensing notices to warn the public about bears, that will likely be more effective for behavioral change toward creating bear-friendly communities (Baruch-Mordo et al. 2011).
7. **Education should be tailored to the needs of the community and should have specific targets.** Education and awareness are the most mentioned tools in conflict literature (Can et al. 2014). Awareness-raising is not education and education is not just giving systematic information. The purpose of education is to achieve the desired behavioral changes in the target population. However, there is not always a direct link between knowledge and actual change in behavior (Baruch-Mordo et al. 2011). Therefore, the development of more effective educational methods is needed. Those involved in designing and running education initiatives should assess and address the values of the target population, and tailor and implement education initiatives within the framework of adaptive management (Treves et al. 2006; Gore et al. 2006, 2008; Spencer et al. 2007; Baruch-Mordo et al. 2011; Reddy et al. 2017).
8. **Pay attention to the welfare of wild bears.** Ethical dimensions of conservation science tend to be neglected in comparison to other academic fields such as medicine (Paquet & Darimont 2010). The field of conflict management is no exception. Conservation is concerned with populations and populations are emergent properties of individuals. Although it is not easy, there is a need to quantify the welfare implications of human–bear conflict. Welfare science is interested in the behavioral processes that are also of interest to conservation (Swaisgood 2007). Remember that the public more intuitively understands individual animals than abstract populations (Can et al. 2014).
9. **Search, document, and try to innovate traditional ways to alleviate bear damage.** Actively search for traditional approaches to conflict management, even if they may not be effective in reducing damage. Nevertheless, document and share them with colleagues, members of the IBA and IUCN Bear Specialist Group, and try to brainstorm ideas to innovate damage prevention techniques using locally available materials.
10. **Think big but start small.** International donor NGOs that fund the local NGOs working in wildlife conservation are in competition with each other to report success stories to their own donors and the general public (i.e. how their projects made a positive impact in the real world). Therefore, by default, most NGOs have to focus on "low-hanging fruits." Nevertheless, when like-minded people working in different institutions align, remarkable success can be achieved in challenging environments. One such example is from Turkey: the Brown Bear Research and Conservation Project run by Doğa Derneği, a local NGO, with the support of World Animal Protection (formerly known as WSPA). After four years of preliminary work, a one-year project on human–bear conflict was initiated in 2006. Successful demonstration of the use of electric fences to protect orchards, and innovative bear-proof platforms

Figure 16.4(A–D) An example of a traditional wooden beehive condominium placed on a cliff to protect beehives from the reach of bears in Turkey (A and B). Elevated bear-proof beehive platforms (C) designed by the author provided a practical and safer method to protect the beehives; electric fences protect the orchards in Turkey (D). (Photos by Ö. E. Can.)

designed by the author to protect beehives in Artvin, Rize, and Erzurum provinces, attracted nationwide interest (Can et al. 2007, 2014; Figure 16.4). As a result, the duration of the project was extended in 2008, and the author became the scientific advisor to the Ministry of Forestry, the national authority on bear research and conflict management. Within the framework of the project, the use of modern tools and techniques such as camera trapping, live trapping and handling bears, and the use of GPS-GSM collars were demonstrated for the first time in Turkey (Lise 2011). Then, the Ministry of Forestry started restoration of brown bear habitats throughout Turkey. This was possible after a paradigm change in forestry practices, as a result of the discussions held within the framework of the project. Moreover, an activity book for teachers and students prepared within the framework of the project was approved by the Ministry of Education and is used in schools. After hundreds of articles in the media reaching an estimated 10 million people (out of a country population of 73 million), the state television TRT prepared a documentary featuring the project in 2010. The documentary won several international awards and has been aired regularly since then, by now probably reaching every beekeeping family in Turkey. Later, the project team also assisted a Global Environment Facility (GEF)-funded project run by UNDP Turkey, WWF Turkey, and the Ministry of Forestry in 2011 to replicate the project in other parts of the country. The project approach was replicated by smaller NGOs too. In brief, in about six years, a small team of dedicated people from a local NGO, an international donor, and the Ministry of Forestry succeeded in changing the paradigm about human–bear conflict in a resource-limited environment. Why can't this be replicated in other challenging environments?

11. **Be positive, patient, and persistent (the mantra of conflict management in challenging environments).** Practitioners, particularly members of NGOs, increasingly find themselves under pressure to report back the success of the initiatives their donors are supporting even in the first year of projects. However, making real impact by working with local communities is harder to achieve than is reported. In the Brown Bear Research and Conservation Project example from Turkey, the use of bear-proof beehive platforms spread through the villages slowly as expected according to the diffusion of innovation theory (for more information, see Rogers 1995). It took significant time for the locals to become used to the innovative idea, but once the psychological resistance to innovation is broken, the new idea spread quickly and bear-proof beehive platforms flourished throughout the villages. The lifespan of typical NGO projects (one to two years) is too short to see the real impact of the projects. Therefore, it is necessary to negotiate with the donor organizations by explaining the diffusion of innovation theory while staying positive, patient, and persistent.
12. **Recognize that ignoring conflicts will have consequences.** Working on conflict is different from working around conflict, which is acknowledging conflict as an issue, but not actively seeking to engage in its management. If the authorities that are in charge of bear protection and conflict management do not take an active role to prevent conflict situations from escalating, then they will probably have to deal with the consequences for people and bears.

Conclusion

A recent study indicates that humans played a major role in the local extirpation and the general extinction of the European cave bear (*Ursus spelaeus*) (Gretzinger et al. 2019). However, Ötzi the Iceman and his contemporaries coexisted with brown bears some 5300 years ago. Today, humans and bears continue to coexist. However, this coexistence largely depends on changes in human behavior; therefore, like conservation, conflict management is a social and political endeavor rather than just a technical one (Peyton 1994). A wildlife manager is expected to "fix" a conflict situation in very little time. On the other hand, being a researcher provides the luxury of having the time, a relatively independent position to think about conflict, and perhaps a learned ability to think in broader terms. Therefore, it is perhaps more of a moral imperative to state this for this author: We don't know what the population status of the eight species of bears will be and the level of conflict that will be affecting bears in 50 or 100 years' time. What is more troubling is that we have no means of knowing about how people's values will change on the four continents where bears exist in that time frame, but let's hope that it will be toward a positive direction. An African proverb tells us, "Until the lion learns how to write, every story will glorify the hunter." Let us hope that future generations, when they look back, won't think that today's researchers and practitioners made this mistake.

Acknowledgments

I would like to dedicate this chapter to my dad, Professor Vedat Can, who passed away during the preparation of this chapter (I miss you a lot, Dad).

I am grateful for Professor Wilfried Buetzler (from Germany) for teaching me how to track bears 22 years ago; Dr. George Schaller (from the USA) and Victor Watkins (from the UK) for persuading me to work on human–bear conflicts; members of the International Association for Bear Research and Management (IBA) and the IUCN Bear Specialist Group for all the discussions we had over the last 20 years. I thank all researchers, game wardens, wildlife managers, decision makers, and local communities that I have worked with around the world for all the conversations we have had.

References

Artelle, K. A., Reynolds, J. D., Treves, A., et al. (2018). Hallmarks of science missing from North American wildlife management. *Science Advances* 4(3): eaao0167.

Balint, P. J., Stewart, R. E., Desai, A. & Walters, L. C. (2011). *Wicked environmental problems: Managing uncertainty and conflict.* Washington, DC: Island Press.

Baruch-Mordo, S., Breck, S. W., Wilson, K. R. & Broderick, J. (2011). the carrot or the stick? Evaluation of education and enforcement as management tools for human–wildlife conflicts. *PLoS ONE* 6(1): e15681.

Boroditsky, L. (2010, July 24). Lost in translation. *The Wall Street Journal.*

Boroditsky, L. (2011). How language shapes thought. *Scientific American*, 4.

Brownstein, M. (2019). Implicit bias. Available from https://plato.stanford.edu/entries/implicit-bias/ (accessed March 9, 2019).

Busjeet, G. (2013). *Planning, monitoring, and evaluation: Methods and tools for poverty and inequality reduction programs.* Washington, DC: World Bank. Retrieved from https://openknowledge.worldbank.org/handle/10986/16449

Can, Ö. E. & Macdonald, D. W. (2017). To protect everything, please click here: does a revolution in data collection guarantee one in conservation? *Animal Conservation*, 20(1): 1–2.

Can, Ö. E. & Macdonald, D. W. (2018). Looking under the bonnet of conservation conflicts: can neuroscience help? *Biodiversity and Conservation* 27(8): 2087–2091.

Can, Ö. E., Lise, Y. & Kandemir, İ. (2007). Bees and bears: a review of beekeeper-bear conflict in Black Sea Region, Turkey and recommendations for conflict resolution. *American Bee Journal*, July: 630–636.

Can, Ö. E., D'Cruze, N., Garshelis, D. L., Beecham, J. & Macdonald, D. W. (2014). Resolving human–bear conflict: a global survey of countries, experts, and key factors. *Conservation Letters* 7(6): 501–513.

Chandran, R. (2015, March 9). It's broke, so fix it: humanitarian response in crisis. Available from https://cpr.unu.edu/its-broke-so-fix-it-humanitarian-response-in-crisis.html (accessed April 5, 2015).

Cohn, J. P. (2002). Environmental conflict resolution. *BioScience* 52(5): 400–404.

Eagleman, D. (2016). *The brain: The story of you*. Edinburgh: Canongate.

Editors. (1957). Approaches to the study of social conflict: introduction by the editors. *Conflict Resolution* 1(2): 105–110.

Elm, E. C. & Diener, M. K. (2007). The language of war in biomedical journals. *The Lancet* 369: 1.

Federal Aviation Administration. (2000). *FAA system safety handbook*. Retrieved from www.faa.gov/regulations_policies/handbooks_manuals/aviation/risk_management/ss_handbook/

Ginges, J., Atran, S., Medin, D. & Shikaki, K. (2007). Sacred bounds on rational resolution of violent political conflict. *Proceedings of the National Academy of Sciences* 104(18): 7357–7360.

Gippoliti, S., Brito, D., Cerfolli, F., et al. (2018). Europe as a model for large carnivores conservation: is the glass half empty or half full? *Journal for Nature Conservation* 41: 73–78.

Gore, M. L., Knuth, B. A., Curtis, P. D. & Shanahan, J. E. (2006). Education programs for reducing American black bear–human conflict: indicators of success? *Ursus*, 17(1): 75–80.

Gore, M. L., Knuth, B. A., Scherer, C. W. & Curtis, P. D. (2008). Evaluating a conservation investment designed to reduce human–wildlife conflict: evaluating conservation investment. *Conservation Letters* 1(3): 136–145.

Gretzinger, J., Molak, M., Reiter, E., et al. (2019). Large-scale mitogenomic analysis of the phylogeography of the Late Pleistocene cave bear. *Scientific Reports* 9(1). doi:10.1038/s41598-019-47073-z

Hardin, G. (1968). The tragedy of the commons. *Science* 162(3859): 1243–1248.

Herrero, S., Higgins, A., Cardoza, J. E., Hajduk, L. I. & Smith, T. S. (2011). Fatal attacks by American black bear on people: 1900–2009. *The Journal of Wildlife Management* 75(3): 596–603.

Hristienko, H. & McDonald, J. E. (2007). Going into the 21st century: a perspective on trends and controversies in the management of the American black bear. *Ursus* 18(1): 72–88.

Institute of Medicine. (2009). *On being a scientist: A guide to responsible conduct in research*, 3rd edition. Washington, DC: National Academies Press.

IUCN SSC BSG. (2019). *Principles of human–bear conflict reduction*. IUCN SSC Bear Specialist Group.

Knowlton, L. W. & Phillips, C. C. (2009). *The logic model guidebook: Better strategies for great results*. Thousand Oaks, CA: SAGE.

Larson, B. M. (2005). The war of the roses: demilitarizing invasion biology. *Frontiers in Ecology and the Environment* 3(9): 495–500.

Lise, Y. (2011). Doğa Konuşmaları Yaban Hayat. *Yesil Atlas*, 8.

Marsh, H. (2014). *Do no harm: Tales of life, death and brain surgery*. London: Weidenfeld & Nicolson.

Matt, C. (2012). 4th International Human–Bear Conflicts Workshop Summary. Available from www.bearbiology.org/wp-content/uploads/2019/08/4th-human-bear-conflict-workshop-summary.pdf

OED. (2015). Conflict. *Oxford English Dictionary*. Oxford, UK: Oxford University Press. Retrieved from www.oed.com/view/Entry/38898?result=1&rskey=uilnn5&

O'Sullivan, N. J., Teasdale, M. D., Mattiangeli, V., et al. (2016). A whole mitochondria analysis of the Tyrolean Iceman's leather provides insights into the animal sources of Copper Age clothing. *Scientific Reports* 6(1). doi:10.1038/srep31279

Paquet, P. C. & Darimont, C. T. (2010). Wildlife conservation and animal welfare: two sides of the same coin. *Animal Welfare* 19(2): 177–190.

Paulus, M. P., Potterat, E. G., Taylor, M. K., et al. (2009). A neuroscience approach to optimizing brain resources for human performance in extreme environments. *Neuroscience & Biobehavioral Reviews* 33(7): 1080–1088.

Peterson, M. N., Birckhead, J. L., Leong, K., Peterson, M. J. & Peterson, T. R. (2010). Rearticulating the myth of human–wildlife conflict. *Conservation Letters* 3(2): 74–82.

Peterson, M. N., Peterson, M. J., Peterson, T. R. & Leong, K. (2013). Why transforming biodiversity conservation conflict is essential and how to begin. *Pacific Conservation Biology* 19(2): 94.

Peyton, B. (1994). Conservation in the developing world: ideas on how to proceed. *Bears: Their Biology and Management* **9**: 115.

Pooley, S., Barua, M., Beinart, W., et al. (2017). An interdisciplinary review of current and future approaches to improving human–predator relations: improving human–predator relations. *Conservation Biology* 31(3): 513–523.

Presnall, C. C. (1943). Wildlife conservation as affected by American Indian and Caucasian concepts. *Journal of Mammalogy* 24(4): 458–464.

Ramsbotham, O., Woodhouse, T. & Miall, H. (2011). *Contemporary conflict resolution*, 3rd edition. Bristol: Polity.

Ratnayeke, S., Van Manen, F. T., Pieris, R. & Pragash, V. S. J. (2014). Challenges of large carnivore conservation: sloth bear attacks in Sri Lanka. *Human Ecology* 42: 467–479. doi:10.1007/s10745-014-9643-y

Reddy, S. M. W., Montambault, J., Masuda, Y. J., et al. (2017). Advancing conservation by understanding and influencing human behavior: human behavior and nature. *Conservation Letters* 10(2): 248–256.

Redpath, S. M., Young, J., Evely, A., et al. (2013). Understanding and managing conservation conflicts. *Trends in Ecology & Evolution* 28(2): 100–109.

Redpath, S. M., Bhatia, S. & Young, J. (2015). Tilting at wildlife: reconsidering human–wildlife conflict. *Oryx* 49(2): 222–225.

Rittel, H. W. J. & Webber, M. M. (1973). Dilemmas in a general theory of planning. *Policy Sciences* 4(2): 155–169.

Rogers, E. M. (1995). *Diffusion of innovations*, 4th edition. New York, NY: Free Press.

Sandermoen, K. (2019). *Management consulting (is not) for dummies*. Alosen, Switzerland: Sandermoen Publishing.

Shenhav, A. & Greene, J. D. (2014). Integrative moral judgment: dissociating the roles of the amygdala and ventromedial prefrontal cortex. *Journal of Neuroscience* 34(13): 4741–4749.

Spencer, R. D., Beausoleil, R. A. & Martorello, D. A. (2007). How agencies respond to human–black bear conflicts: a survey of wildlife agencies in North America. *Ursus* 18(2): 217–229.

Swaisgood, R. R. (2007). Current status and future directions of applied behavioral research for animal welfare and conservation. *Applied Animal Behaviour Science* 102(3–4): 139–162.

Tetlock, P. (2003). Thinking the unthinkable: sacred values and taboo cognitions. *Trends in Cognitive Sciences* 7: 320–324.

Treves, A. & Bruskotter, J. (2014). Tolerance for predatory wildlife. *Science* 344(6183): 476–477.

Treves, A., Wallace, R. B., Naughton-Treves, L. & Morales, A. (2006). Co-managing human–wildlife conflicts: a review. *Human Dimensions of Wildlife* 11(6): 383–396.

UNDP. (2009). *Handbook on planning, monitoring and evaluating for development results.* New York, NY: United Nations Development Programme.

United Nations. (2004). *World population to 2300* (No. ST/ESA/SER.A/236). New York, NY: United Nations.

Wilson, S. M. (2016). *A guidebook to human–carnivore conflict: Strategies and tips for effective communication and collaboration with communities.* Ljubljana, Slovenia: Slovenia Forest Service – LIFE DINALP BEAR project.

W. K. Kellogg Foundation. (2004). W.K. Kellogg Foundation 2004.pdf. Michigan, USA: W.K. Kellogg Foundation.

WSPA. (2009). *Principles of human–bear conflict management.* Istanbul: World Society for the Protection of Animals.

Zedrosser, A. (2019). The major challenges facing bears. *International Bear News* 28(2): 2.

Chapter

17 Patterns of Bear Attacks on Humans, Factors Triggering Risky Scenarios, and How to Reduce Them

Vincenzo Penteriani*, Giulia Bombieri*, María del Mar Delgado, Thomas Sharp, Koji Yamazaki, Harendra Singh Bargali, Nishith Dharaiya, Ashish Kumar Jangid, Ravi Kumar Sharma, Babu Ram Lamichhane, Shyamala Ratnayeke, Ivan Seryodkin, Himanshu Shekhar Palei, Ashok Subedi, Hüseyin Ambarlı, José María Fedriani, Pedro José Garrote, Klemen Jerina, Ilpo Kojola, Miha Krofel, Prakash Mardaraj, Mario Melletti, Andrés Ordiz, Paolo Pedrini, Eloy Revilla, Luca Francesco Russo, Veronica Sahlén, Christopher Servheen, Ole-Gunnar Støen, Jon E. Swenson, and Tom Smith

Introduction

The media and scientific literature are increasingly reporting an escalation of large carnivore attacks on humans, mainly in the so-called developed countries, such as Europe and North America (Penteriani et al. 2016; Bombieri et al. 2018a). Although large carnivore populations have generally increased in developed countries (Chapron et al. 2014), increased numbers are not solely responsible for the observed rise in the number of attacks. For example, recent research has shown that people frequently engage in risk-enhancing behaviors that can increase the probability of a risky encounter and a potential attack, and perhaps even alter carnivore behavior (Penteriani et al. 2016, 2017; Garrote et al. 2017).

Of the eight bear species inhabiting the world, two (i.e. the Andean bear and the giant panda) have never, or very rarely, been reported to attack humans, whereas the other six species have: sun bears *Helarctos malayanus*, sloth bears *Melursus ursinus*, Asiatic black bears *Ursus thibetanus*, American black bears *Ursus americanus*, brown bears *Ursus arctos*, and polar bears *Ursus maritimus*. These species occur across four continents (Asia, Europe, North and South America) characterized by a huge range of social and cultural practices, e.g. from increasing leisure activities in bear areas in developed countries to daily forest works in developing countries. Such differences in the use of bear habitats by people may determine that different scenarios trigger bear attacks on humans around the world. However, even if the motivations that determine human presence in bear countries and risky encounters with bears are diverse, some triggering factors might be common in activating bears' dangerous reactions toward people, e.g. inappropriate human behaviors when sharing the landscape with bears or when encountering them at close range.

This chapter provides insights into the causes, and as a result the prevention, of bear attacks on people. Prevention and information that can encourage appropriate human behavior when sharing the landscape with bears are of paramount importance to reduce both potentially fatal human–bear encounters and their consequences to bear conservation.

Methods

We reviewed scientific/gray literature and analyzed personal databases on bear attacks on humans available from 1980 to 2018. Moreover, we also searched for PhD/MSc theses and webpages on bear attacks on humans. In addition, we collected news reports from online newspapers to complete the data set obtained from the above-mentioned sources. To this aim, for each bear species and area, we searched for news articles on Google on an annual basis using the following combination of words: "species name" + "attack" and "species name" + "attack" + "human." To prevent duplicate records in the data, we cross-checked information such as date, locality, and human characteristics. When possible, for each attack, we recorded the following information: (1) period of the attack, i.e. year, month, and time of the day; (2) location; (3) outcome, i.e. human injury or death; (4) characteristics of the person/party and bear involved in the attack, e.g. age and sex of both the person and the bear; (5) human activity at the time of the attack; and (6) the attack scenario, i.e. the factor(s) that could have triggered the attack.

An Overview of Bear Attacks in the World

Sun Bear

The sun bear, the smallest ursid in the world, is found in South-East Asia, where few ($n = 11$) attacks on humans have been reported. Although the available information is scarce and incomplete (Sethy & Chauhan 2013), attacks seem to be extremely rare and mainly the consequence of sudden encounters (i.e. with the bear being inadvertently surprised at a close distance). Such encounters mainly occur when people venture inside the forest for different purposes (Sethy & Chauhan 2016). For example, in Indonesia most attacks happen to

* These authors contributed equally to this work.

Figure 17.1 The sloth bear's aggressively defensive nature may be the result of having coevolved with large predators, such as tigers and leopards, which are known to occasionally prey upon this bear. Actually, sloth bears are strong animals that possess large canines and claws, which make them difficult prey. (Photo by Ayan Sadhu.) (A black and white version of this figure will appear in some formats. For the color version, please refer to the plate section.)

people working in the forest on a daily basis, such as rubber harvesters (55%, $n = 6$), whereas fewer cases occurred to people who work in crop fields and collect non-timber forest products (Windler 2014). In general, encounters with these bears are mainly non-fatal.

Sloth Bear

Sloth bears are known for their aggressive behavior toward humans (Burton 1856; Anderson 1957). Although the total number of people seriously injured or killed by sloth bears during a given year is not known, within one Indian state (Madhya Pradesh, central India), 48 sloth bear-inflicted fatalities and 687 maulings were documented between 1989 and 1994 (Rajpurohit & Krausman 2000). This accounts for an average of six deaths and 115 maulings per year in this Indian state. Because sloth bears occupy 19 Indian states, as well as Nepal, Sri Lanka, and possibly Bhutan, this species might be responsible for more attacks on humans than all other seven species of bears combined.

The sloth bear's motivation to attack seems to be mainly defensive. There has never been a documented predatorial attack (Sharp et al. 2017), although there have been several cases of sloth bears consuming portions of their victims (Bargali et al. 2005; Akhtar 2006). This appears to be the result of opportunistic behavior after the attack, rather than the motivation for the attack. The sloth bear's aggressively defensive nature may be the result of having coevolved with large predators, namely tigers *Panthera tigris* and leopards *Panthera pardus*, which are known to occasionally prey upon sloth bears (Littledale 1889; Fenton 1909; Kurt & Jayasuriya 1968; Laurie & Seidensticker 1977). Sloth bears are not adept climbers and are often in scrub jungle or grasslands, where climbing is not an option. They are not particularly fast runners, either. However, they are exceptionally strong animals that possess large canines and claws. Although no match for a tiger in an extended encounter, sloth bears can make themselves a particularly difficult prey (Figure 17.1). In addition to their physical attributes, sloth bears make use of intimidation tactics, such as charges coupled with vocalizations and making themselves appear larger with a bipedal stance (Bargali et al. 2005; Ratnayeke et al. 2014; Sharp et al. 2017). This aggressive–defensive nature may serve to deter predators, but impedes conservation efforts whenever they attack humans (Akhtar 2006; Dharaiya et al. 2016).

To better understand this type of human–sloth bear conflict, we compiled information on 1169 attacks that took place in the three countries with extant sloth bear populations: India, Nepal, and Sri Lanka. We also considered the reported results from individual studies conducted throughout the sloth bear's distribution. We found that the spatiotemporal patterns of these attacks varied across studies. Whereas some studies reported a spike in attacks during the monsoon season, others reported more attacks during the dry season (Rajpurohit & Krausman 2000; Bargali et al. 2005; Ratnayeke et al. 2014; Garcia et al. 2016; Dhamorikar et al. 2017; Sharp et al. 2017). These differences may be related to seasonal and daily patterns of human use of forests, or to proximity between bears and people in highly fragmented bear populations. For example, in Bilaspur (India), attacks mainly occur in villages and agricultural fields (Bargali et al. 2005), whereas in other study areas, the majority of attacks occur in forests (e.g. Ratnayeke et al. 2014; Dhamorikar et al. 2017). The high number of attacks during monsoon may be because people start agricultural practices in crop fields, collect mushrooms from the forest, and use forests for livestock grazing at the onset of this season. In monsoon, there is increased vegetation cover as well. Such a combination of different factors and the increased disturbance in and around sloth bear habitats considerably augment the probability of risky encounters with a sloth bear.

The time of day when attacks occurred also varied a great deal. Although some studies suggested that dawn and dusk were the most dangerous times to be moving through sloth bear country (Bargali et al. 2005; Mardaraj & Dutta 2011; Mardaraj 2015; Debata et al. 2017; Dhamorikar et al. 2017; Sharp et al. 2017), others reported that most attacks occurred during the middle of the day (Ratnayeke et al. 2014; Garcia et al. 2016).

There is little doubt that local factors, including human activity patterns, play a large role in the spatiotemporal patterns of sloth bear attacks. For example, although sloth bears are generally known to be more active at dusk and during the night, in many places several attacks occurred during the day. This may be partly due to resting bears being disturbed, or surprised, rather than the presence of active bears; actually, very few attacks happen during daylight hours in areas that are known to have many natural caves and hollows.

A large portion of attacks (42%) took place when humans were active in the forest (e.g. collecting forest products, walking, etc.). For example, in central India, people collect tendu leaves (*Diospyros melanoxylon*) and mahua flowers and seeds (*Madhuca indica*) for commercial use, as well as sal leaves (*Shorea robusta*), bamboo (*Dendrocalamus strictus*), chironji (*Buchanania lanzan*), and wild mushrooms, which increases the probability of sudden encounters between people and sloth bears. The second highest category (25%) involved persons who were farming or caring for orchards. These types of activities are more often performed by men than by women or children, which likely explains why adult males were found to be involved in the majority of attacks (87%).

Perhaps surprisingly, the third highest number of attacks (15%) took place when humans were defecating in the forest. This tends to occur in areas which are often frequented for this purpose. The reason for this type of attack is unclear. It is possible that fecal odours attract sloth bears to areas often used for defecation. However, this is purely speculative at this point, although it is interesting to note that the smell of human feces is known to also attract both brown and American black bears (T. Smith, unpublished data).

Sudden encounters triggering defensive–aggressive attacks accounted for 47% of the attacks. Often the victim was unaware of the sloth bear's presence until they saw the bear charging from just meters away (Figure 17.2). Like most mammals, sloth bears are protective of their young and, of the 1169 attacks compiled for this chapter, 22% involved females with cubs, although several studies reported this number to be higher (Ratnayeke et al. 2014; Sharp et al. 2017).

Most studies on sloth bear attacks report that human group size plays a role in the likelihood of attack (Ratnayeke et al. 2014; Sharp et al. 2017). We found that nearly half of all attacks involved a single person (46%). However, reported group size can be misleading, because if people are spread out, the bear may perceive each individual as a single person, rather than as part of a group. Ratnayeke et al. (2014) found that a human companion <50 m away significantly reduced the likelihood of severe injury during a sloth bear attack. Human groups are likely to be noisier, thus less likely to surprise a bear, and more intimidating to an attacking bear than a solitary person. Additionally, in some cases the person attacked by a sloth bear has been saved by other nearby people or even by animals accompanying the victims (Bargali et al. 2005; Dhamorikar et al. 2017; Silwal et al. 2017), although dogs may sometimes precipitate an attack. Indeed, Ratnayeke et al. (2014) reported three attacks that resulted from dogs running back to their owners with a bear in pursuit, putting the human directly in the path of an angry bear.

Figure 17.2 The main scenario of sloth bear attacks on humans is sudden encounters in dense forests, where the victim is often unaware of the bear's presence. (Photo by Luxshmanan Nadaraja.)

Of the 1169 sloth bear attacks we reviewed, only 5% (n = 58) resulted in the death of the victim. Based on other studies (Sharp et al. 2017), moderate to severe injuries make up another 35–50% of the attacks.

Asiatic Black Bear

Although not scientifically documented, the Asiatic black bear seems to be more nervous than its American relative, the American black bear. This might be due to the relatively small body size compared to humans or to long-term persecution by humans (Japan Bear Network 2006; Yamazaki 2010). Because Asiatic black bears are generally shy and nervous, most of their attacks on humans are defensive, especially when a female bear is with cubs or in the event of sudden or unexpected encounters (Rasool et al. 2010; Japan Bear Network 2011).

Our search resulted in a total of 747 attacks by Asiatic black bears in Pakistan, India (Kashmir, northern India), Bhutan, Nepal, Russia (Far East Region), and Japan (Honshu Island, the largest and most populous island of Japan). Although we do not know whether bears hibernate throughout these areas, most attacks occurred between May and November, with the highest frequency of attacks occurring in October (18%).

Ninety-five percent of the attacks occurred between April and November and 5% from December to March. During late autumn and early winter, when bears enter their hyperphagic period, they predominantly rely on acorns (Kozakai et al. 2011) and, when the autumn acorn production is low, bears become bolder and search for alternative, high-energy food resources (Kozakai et al. 2011; Ali et al. 2018), resulting in drastically increased human–bear conflicts, including bear attacks.

Most attacks occurred during daylight (94%), compared to twilight (4%) and night (2%). Even though Asiatic black bears are diurnal and their activity peaks are during dawn and dusk (Hwang & Garshelis 2007; Yamazaki et al. 2008), they can modify their behavior to become more nocturnal when they approach human settlements (Arimoto et al. 2014).

Most people involved in attacks were males (78%). Attacks mainly occurred while people were farming or participating in various activities in the forest (45% and 32%, respectively), and it is possible that more males are generally involved in these activities than females. The age of victims was biased towards adults (>13 years old, 94%), which is likely explained by the same phenomena as the sex bias. Most of the attacks recorded were the result of sudden encounters (86%), with bears reacting aggressively when surprised at close distances. Notably, we found that at least 30% of such encounters occurred while bears were feeding on crops or other products in people's fields or orchards. Such a scenario has been found to be the main attack circumstance in several studies, with people encountering bears at close range and consequently being attacked when visiting or working in their crop fields or orchards (Nabi et al. 2009a; Tak et al. 2009; Rasool et al. 2010; Charoo et al. 2011). Other reported scenarios involve people entering dense forest to collect wood or other forest products, or to graze livestock (Tak et al. 2009; Lal Moten et al. 2017; Ali et al. 2018).

The number of bear attacks has been increasing recently in Nepal (Acharya et al. 2016), India (Kashmir; Nabi et al. 2009b), and Japan (Yamazaki 2010). This is mostly due to human population expansion, deforestation, and destruction and fragmentation of bear habitat in most Asian countries (Japan Bear Network 2006). In contrast, the reason for increasing bear attacks in Japan is quite different, because the population and range of bears have been expanding in recent decades, due to habitat recovery as a result of aging human population and depopulation (Yamazaki 2004; Yamazaki & Sato 2014). From statistics supplied by the Ministry of Environment in Japan, 851 people were attacked (including 13 deaths) by bears in the last 10 years (2008–2017). Although the frequency of attacks by Asiatic black bears is relatively high, these attacks rarely result in fatalities (8%). Similar or lower fatality rates are reported in Nepal (Acharya et al. 2016), India–Kashmir (Nabi et al. 2009b; Rasool et al. 2010; Shah et al. 2010), and Japan (Akiyama et al. 2017). Although fatalities are rare, injuries are frequently serious, as bears often stand upright and first attack the person's neck and face using their claws, causing damage such as bone fractures and deep tissue lacerations (Nabi et al. 2009b; Rasool et al. 2010; Oshima et al. 2018). As a consequence, many victims must undergo multiple and complex facial reconstructive surgeries (Rasool et al. 2010; Shah et al. 2010).

In the northern part of Honshu Island, Akita Prefecture, Japan (Yamazaki 2017; Oshima et al. 2018), four local residents who were gathering bamboo shoots were killed and partially eaten by bears in 2016, and another local resident was attacked and partially eaten by a bear(s) in 2018. As a result, local attitudes toward bears have drastically changed within this prefecture and have become more negative, resulting in a total of 1676 bears being killed as nuisance animals between 2016 and 2018. These attacks have had a bad influence even in other prefectures, and bear management is facing serious difficulties in many areas.

American Black Bear

Outnumbering the other two species of North American bears 10-fold, the American black bear is the most ubiquitous ursid on the continent (Herrero 2018). Not surprisingly, black bears account for the majority of human–bear incidents in North America (Penteriani et al. 2016). Indeed, our search resulted in 373 attacks reported between 1980 and 2016, and only 11% were fatal. In a recent study of human–bear conflict in Alaska, Smith and Herrero (2018) reported that black bears, which outnumber brown bears three to one in the state, were responsible for only 14% (89 of 638) of bear attacks reported from 1880 to 2016. This suggests a tolerance for humans not shared by brown bears. Herrero (1972) suggested that this reluctance to engage with people is the result of their unique evolutionary past. Black bears evolved in the densely forested regions of North America and could often resolve conflict by either climbing a tree or disappearing into the dense understory (Herrero 1972). Therefore, when suddenly confronted by people, they frequently flee. Indeed, Smith and Herrero (2018) found in Alaska that brown bears accounted for nearly six times more attacks than black bears (508 vs. 89), resulting in 47 brown bear-inflicted fatalities compared to just five due to black bears (a 9.4-times higher rate for brown bears).

Nonetheless, on rare occasions, when black bears attacked and killed people, 90% of them were deemed predatory (Herrero et al. 2011). From 1900 to 2009, Herrero et al. (2011) documented 63 fatal black bear attacks in the United States and Canada. They also found that black bear-inflicted fatalities were highly correlated with human population growth in those countries, suggesting that the more people enter black bear habitat, the more likely it is that a conflict will arise. Although the vast majority of black bears clearly avoid conflict with humans, a few bears apparently perceive people as prey and attempt to take them.

Herrero and Higgins (2003) identified predatory black bears by a series of behaviors: searching, following and testing, attacking (capturing), killing, sometimes dragging a person, sometimes burying, and often feeding upon a person. By analyzing patterns of predatory attacks by all large carnivore species in North America, Penteriani et al. (2017) found that the black bear was the third species most frequently involved in this kind of encounter (16%), after cougars and coyotes, and, as expected, the targets of predatory attacks were the most vulnerable individuals, namely children and lone people.

Black bears are powerful predators quite capable of inflicting severe injury and death on humans. Herrero et al. (2011) speculated that black bears likely do not prey on people more often than they do because bears with those genes have been consistently culled from the gene pool. Importantly, fatal attacks by black bears are fundamentally different than those of brown bears because browns are rarely predatory, whereas black bear-inflicted fatalities are almost always the result of predation. Unlike brown bears, black bears have been often reported to cause conflicts and, more rarely, injure people in urban areas across North America (Bombieri et al. 2018b). In such environments, risky encounters with this species are mainly due to the presence of dogs or anthropogenic food (e.g. a bear is surprised by a person while feeding on pet food or trash in the yard and reacts aggressively). Probably due to the forest-obligate nature of the species, attacks in urban areas are more likely to occur where the vegetation cover is more abundant and less fragmented and far from buildings and roads. Half of such encounters have been found to occur at night, especially in areas where the artificial illumination is scarce (Bombieri et al. 2018b).

Brown Bear

There are always conflicts wherever humans and brown bears commingle. Several studies have shown that increases in the number of brown bear conflicts are roughly proportional to human population growth in North America (Herrero et al. 2011; Sharp et al. 2017; Smith & Herrero 2018). Although long absent from many localities, people still harbor an inordinate fear of brown bears; a fear that is deeply rooted in our past history, as explained by Wilson (1984):

> The brain evolved into its present form over a period of about 2 million years, from the time of *Homo habilis* to the late stone age of *Homo sapiens*, during which people existed in hunter-gatherer bands in intimate contact with the natural environment. The smell of water, the hum of a bee, the directional bend of the plant stalk mattered. The naturalist's trance was adaptive: the glimpse of one small animal hidden in the grass could make the difference between eating and going hungry in the evening. And a sweet sense of horror, the shivery fascination with monsters and creeping forms that so delights us today even in the sterile hearts of our cities, could see you through to the next morning. Organisms are the natural stuff of metaphor and ritual. Although the evidence is far from all in, the brain appears to have kept its old capacities, its channelled quickness. We stay alert and alive in the vanished forests of the world.

This innate fear is kept alive by the media, which often focuses an inordinate amount of attention on attacks by brown bears, thus fueling fear and undermining conservation efforts (Penteriani et al. 2016; Bombieri et al. 2018a). Human–brown bear conflicts occur throughout their range, and though few in number, have profound impacts on both the people and bears involved (Conover 2008; Penteriani et al. 2016). A study by Bombieri et al. (2019) investigated circumstances of brown bear attacks that occurred worldwide between 2000 and 2015 and highlighted common characteristics, as well as differences in attack patterns between regions of the world with different sociocultural backgrounds and histories of coexistence with this species. According to this study, people involved in attacks are almost exclusively adults (99%) and human fatalities are rare (15%). Globally, attacks are mainly the result of an encounter with a female bear with cubs (47%; Figure 17.3A), followed by sudden encounters with other bear classes (20%), dog presence (17%; Figure 17.3B), bear attacking after being shot or trapped (10%), and predatory attacks (5%). At the moment of the attack, half of the people were engaged in leisure activities (e.g. hiking, picking berries or mushrooms, camping, fishing), whereas 28% of the attacked people were working outside, i.e. farming, guarding livestock, or logging, or doing wildlife-related fieldwork, and 22% were hunting. Although patterns are similar in many regions across the wide distribution range of brown bears, local differences in attack scenarios also exist. For example, in Romania, where traditional semi-subsistence agriculture and livestock husbandry are common, half of the people involved in attacks were shepherds, cattle herders, and farmers (Bombieri et al. 2019), whereas most (75%) of the attacks that occurred in Scandinavia between 1977 and 2016 involved hunters (Støen et al. 2018). Hunters are often involved in bear attacks, because they are trying not to make noise while hunting and may consequently surprise bears, thus triggering an attack. Other factors connected to hunting may also explain the higher risk of an attack. In Scandinavia most (73%) of the hunter casualties had shot at the bear at short range (average 8 ± 11 m) before being injured and hunting dogs harassed the bears in 77% of the incidents involving hunters (Støen et al. 2018).

A recent study by Smith and Herrero (2018) chronicled 135 years of human–bear conflict involving nearly 700 incidents in Alaska, of which 78% involved brown bears. The findings of this study, although regional, offer useful insights regarding attacks, because brown bears react similarly to human confrontation regardless of locality, even though regional differences in bear responses to humans have been documented.

For example, Smith et al. (2008) reported that brown bears in Alaska respond somewhat differently toward humans in coastal areas (areas of abundant resources that support high-density bear populations, which are more tolerant of humans at very close ranges) compared to brown bears in interior areas (areas relatively scarce in resources supporting low-density populations that are more intolerant of humans). Additionally,

Figure 17.3 Two of the main brown bear attack scenarios: (A) defensive reaction of a female with cubs; and (B) unleashed dogs harassing bears. (Photos by Ivan Seryodkin.)

it has been posited that heavily hunted brown bear populations become more wary of people, although Swenson (1999) failed to find definitive evidence of that.

In general, Smith and Herrero (2018) found that increases in the number of brown bear attacks in Alaska over a 135-year period were strongly correlated with the growth of Alaska's human population. Similarly, when brown bear attack rates were regressed against a simple index created by combining both brown bear and human densities within a given region of the world, they were found to be positively correlated (data adapted from Bombieri et al. 2019).

When bear encounters occurred, bears disproportionately attacked single persons, so grouping together decreased the likelihood of an attack. A similar finding was reported by Penteriani et al. (2016) and Garrote et al. (2017). Group size may yield this effect because larger groups are noisier and alert bears to approaching humans (Herrero et al. 2011). Of the persons involved in bear conflicts in Alaska in this study (Smith & Herrero 2018; n = 1550), those without a deterrent (i.e. firearms, bear spray, etc.) resorted to desperate and ineffective evasive actions, such as running, tree-climbing, rock-throwing, and attempting to fight the animal bare-handed, strategies that rarely worked. Importantly, having a deterrent did not guarantee that the person would avoid injury; but those with deterrents suffered less injury, and less-severe injuries, than those without. Of particular note, those persons with bear spray largely escaped injury (98%), whereas persons bearing firearms suffered more (24% were injured; Smith et al. 2012).

Finally, Smith and Herrero (2018) found that attacking bears focused on the victim's head and neck regions, a finding which suggests that Herrero (2018) was justified in promoting a defensive posture of lying face-down, hands laced over the back of the neck and legs spread to provide the ability to thwart a bear's effort to flip the body over and face up. Coming to the aid of an attack victim almost always terminated the attack, perhaps because the bear felt outnumbered and left. Although domestic dogs could be helpful in protecting persons from an attack (47.5%), some abandoned their masters (40%). A few dogs (12.5%) reportedly initiated the attack.

Polar Bear

Human–polar bear conflicts have existed for as long as both species have shared the Arctic. Without a written language, these conflicts went unrecorded by indigenous people, but undoubtedly occurred with some regularity. The earliest written accounts of human–polar bear conflict are sporadic and were recorded by European mariners who kept journals. Among the earliest of these is an account of Dutch mariner William Barents (1594), who attempted to capture a polar bear to bring back alive to Holland (Wikipedia 2019). Unfortunately, the bear broke loose aboard ship, rampaged through the vessel, and was killed in the process. Additionally, Barents wrote a year later that two men in his expedition were attacked and killed by a polar bear, the earliest polar bear-inflicted fatalities we know of on record. More recently, Charlie Brower of Barrow (Alaska) witnessed a polar bear casually crush the skull and ribs of a Native guide in 1883 (Brower 1942). Such graphic accounts have unquestionably fuelled the widely held belief that it is fundamental to a polar bear's nature to "stalk and kill humans" (Ramachandran 2009). Even though brutal accounts of polar bears ravaging humans dot the pages of history books, one must ask if they truly deserve the name "stalker and killer of humans." Human–polar bear attack research does not support that claim.

In recent years, a number of studies have documented the nature of human–polar bear conflicts (Fleck & Herrero 1988; Clark 2003; Wilder et al. 2017). Both Clark (2003) and Fleck and Herrero (1988) reported that most human–polar bear interactions do not result in injury, and when persons were injured, rates of interactions resulting in injury were low (2% and 5%, respectively). Specifically, Clark (2003) reported that only one bear-inflicted injury occurred in 53 polar bear–human interactions. Fleck and Herrero (1988) documented 373 aggressive polar bear–human interactions, of which only 10 bear-inflicted injuries occurred. Because Wilder et al. (2017), reported only on polar bear attacks, their data are not comparable to the previous studies that included non-injurious interactions. However, over a 144-year period, Wilder et al. (2017) documented only 73 attacks on humans by polar bears across their entire range, and in those attacks 20 persons were killed and another 63 injured. By contrast, Smith and Herrero (2018) reported nearly the same number of attacks for brown bears in Alaska in a single decade (2000–2009). The record is clear that polar bears injure and kill far fewer humans that do black and grizzly bears (Fleck & Herrero 1988). Although some may point out that the interaction rates between polar bears and humans are far lower than those of black bears and grizzlies, injuries are low and deaths extremely rare where polar bears and people commingle considerably (e.g. Churchill, Manitoba; Kaktovik, Alaska). In all of these human–polar bear interactions, predation was rare, but when people were killed, predation was considered the primary motivation (Wilder et al. 2017).

The "Take in the Bear Home" Message: How to Avoid and Survive Bear Attacks and Specific Measures of Risk Reduction

Sloth Bear

Information on sloth bear attacks suggests that being in a group and making noise while moving through sloth bear habitats helps reduce the likelihood of startling a bear at close quarters, giving it the opportunity to leave the area without incident. Bear spray and guns are not available for protection to most who live in sloth bear country and many attacks occur too quickly for weapons to be used effectively. However, villagers carrying a heavy club or walking stick have been able to drive off an attacking bear, and this might prove to be especially effective when more than one person acts in concert (Ratnayeke et al. 2014; Sharp et al. 2017).

If a sloth bear is observed, but is not aware of the observer, people should slowly back away, giving the bear as much space as possible. Running has been reported to trigger a chase response, and several runners have been pursued and killed by sloth bears (Sharp & Sonone 2011; Ratnayeke et al. 2014; Sharp et al. 2017). Yelling or throwing stones at a non-aggressive sloth bear may also elicit an attack.

Although some have reported that the sloth bear retreated when fought, Sharp et al. (2017) reported that those who fought were more likely to be killed than people that played dead. The safest response to an attack might highly depend on factors such as the bear motivation to attack and the human characteristics as well as group size. A person that is attacked may decrease the risk of injury or death by using the protective position recommended by Herrero (2018) for grizzly bear attacks: lying face down on the ground with hands locked behind the neck and arms protecting the face. The effectiveness of this defensive position for sloth bear attacks is yet untested; however, most sloth bear attacks are defensive, cause injuries to the face and head, and limited data suggest that the attack does not persist when the bear perceives that the threat has faded.

Temporal patterns of sloth bear attacks across different geographic areas suggest that peaks in specific types of human activities increase the potential for human–sloth bear encounters and attacks. These activities may vary by location and managers should work with local people to seek solutions. For example, at locations where bears frequent village compounds and agricultural fields, villagers could be encouraged and financially supported to construct toilets and to use extreme caution moving around when bears are active (Jangid & Sharma 2018). Similarly, avoiding the collection of "non-timber forest products" during twilight hours may reduce the risk of encountering active sloth bears. Moving in groups and advertising one's presence via sound and loud conversation will give resting bears an opportunity to leave the area. In areas where bears reside within or on the periphery of human settlements, cultivating crops (e.g. maize, ground nuts) that attract bears will increase the likelihood of attacks, let alone crop losses. Training programmes aimed at sustainable livelihoods, including support for alternative forms of agriculture and income to reduce the dependency of local communities on forest resources could help to keep sloth bears and people well-separated. Proper garbage/waste management practices should be promoted in those areas where sloth bears are attracted and approach garbage sites to feed on remnants of fruits and edible materials (A. Jangid, unpublished data). In addition, provisioning of edible materials in remote temples should be stopped, so the encounters can be reduced near temple sites. Indeed, a few cases have been reported of bears breaking in and raiding houses and small temples located inside forests to feed on edible products such as oils left by pilgrims as offerings (A. Jangid, unpublished data; Singh et al. 2017; Jangid & Sharma 2018), alarming the villagers. Finally, for bears that have little alternative but to survive in forest fragments surrounded by agriculture and human settlements, guidelines and policies for safely trapping and relocating them may be the most feasible option (Figure 17.4).

Asiatic Black Bear

There are no easy solutions to reduce Asiatic black bear attacks on humans. However, public education about bears (i.e. bear habitat, behavior, and ecology) to local residents and people engaging in various activities in the forest can be an effective approach (Japan Bear Network 2006; Jamtsho &

Figure 17.4 Trapping and relocating conflictual sloth bears might represent an alternative to retaliatory killing. (Photo by Ashish Jangid.)

Wangchuk 2016). Because attacks often occur on farmers working on crop fields or orchards, suggesting that bears are attracted by agricultural products, preventive measures aimed at protecting such areas might prove effective to avoid both crop raiding and potentially dangerous encounters between farmers and bears (Jamtsho & Wangchuk 2016). Moreover, people should be cautioned about the possibility of encountering bears when working in their crop fields or moving in the forest and should be provided information on how to avoid being attacked. Moving in a group and making noise while moving in areas with poor visibility might help the bear notice human presence and leave the area to avoid the encounter (Rasool et al. 2010). Given the evident defensive strategy of attacking the head and face region commonly adopted by Asiatic black bears, if a bear reacts aggressively when encountered, people should not fight back, but protect these sensitive regions and adopt a passive position. It has been shown, indeed, that the attack terminated spontaneously in most of the cases, with the bear leaving immediately once the victim was overpowered (Rasool et al. 2010; Shah et al. 2010). Also, increasing the distribution/availability of bear avoidance equipment, such as pepper spray, at affordable prices could also be effective. Once a bear attack has occurred, comprehensive on-site verification and sharing of the obtained information among related organizations is also very important to prevent similar types of incidents in the future.

American Black Bear

Smith and Herrero (2018) and Herrero and Higgins (2003) identified a number of insights regarding safety in black bear country. These findings fall into four broad categories of bear safety messaging: (1) general information; (2) how to avoid bear encounters; (3) how to defuse encounters; and (4) how to survive attacks. First, food and garbage should be secured to avoid attracting bears. People should move in bear areas in groups of more than two and try to group together if a bear is encountered. Carrying bear spray is highly recommended. Additional specific precautions should be taken in urban areas and their proximities (Bombieri et al. 2018a). In such environments, increased attention to dogs and improved management of attractants (i.e. avoid leaving garbage, pet food, and bird feeders outside houses) would likely reduce the probability of risky encounters with this species. Moreover, to reduce the occurrence of predatory attacks, particular attention should be paid with children, who need to be constantly and strictly supervised by adults (Garrote et al. 2017; Penteriani et al. 2017). Finally, in case of an attack, being able to recognize the motivation behind it may be crucial in determining the attack outcome. That is, in the case of a predatory attack, one should try to aggressively deter the bear and fight back in any possible way. Instead, if the attack is defensive, one should be passive and adopt a defensive posture by lying face-down and protecting the neck with the hands.

Brown Bear

Based on findings by Smith and Herrero (2018), actions that avoided brown bear encounters include being wary during their most active time of day (midday through evening), avoiding areas with poor visibility when possible, making noise to alert bears to one's presence, and hiking in a party ($\geq$2 people). These safety measures have also been recommended for forest users in Scandinavia (Støen et al. 2018), where bears reside in rugged terrain (Nellemann et al. 2007), rest in dense vegetation during daytime to avoid people (Ordiz et al. 2011), and generally flee when encountered (Moen et al. 2012). Of particular note, brown bear attacks, including fatalities, have been documented in every month of the year, underscoring the fact that one must employ bear avoidance practices year-round. Similar to advice highlighted for black bear encounters, carrying bear spray and grouping together when encountering a bear is recommended to avoid being attacked. In Scandinavia, awareness and education efforts, especially among hunters, have also been highlighted to ensure human safety (Støen et al. 2018). Because brown bear attacks are mostly defensive, if attacked one must be passive and adopt a defensive posture. Importantly, Smith and Herrero (2018) reported that, when rescuers came to the aid of the victims, the mauling ended 91% of the time, highlighting the importance of coming to the aid of the victim. Although dogs might help terminate an attack, one should make sure they obey commands and, preferably, keep them on a leash to avoid them disturbing bears and therefore triggering an aggressive reaction toward the dog and its owner.

Polar Bear

Wilder et al. (2017) cautioned that, as sea ice continues to shrink, human–polar bear conflict can be expected to rise.

Importantly, Fleck and Herrero (1988) observed that the outcome of human–polar bear conflict was most often a dead bear and much more rarely an injured or dead human. Therefore, improved conflict investigation is needed to collect accurate and relevant data and communicate accurate bear safety messages and mitigation strategies to the public. With better information, people can take proactive measures in polar bear habitat to ensure their safety and prevent conflicts with polar bears.

To conclude, although rare, bear attacks on humans do occur within the whole range of bear species and undermine bear conservation efforts (Røskaft et al. 2007). Different bear species showed differing attack patterns, and although some bear species, such as the Andean bear, the sun bear, and the giant panda have never or rarely been reported to be involved in such incidents, the other five species of bears may locally represent a more serious threat to human safety. Therefore, it is of the utmost importance for bear conservation worldwide to reduce such conflicts by developing and implementing effective strategies based on both species-specific characteristics and the local socioeconomic context. In developed countries, where most attacks occur on people involved in recreational activities in bear areas, conflicts can be decreased through education and outreach (i.e. providing accurate bear safety messaging to the public). For instance, recent efforts in Yosemite National Park have demonstrated that with effective education, outreach, and appropriate penalties for the uncompliant, bear conflicts can be dramatically decreased (down by 99% since 1998), benefitting both people and bears (National Park Service 2019). A different approach should be employed in developing countries, where people involved in attacks are mostly local people who must enter bear areas for their daily work- and subsistence-related activities. Here, education on how to avoid bear encounters and attacks might not be sufficient alone and must be combined with improved facilities and financial support for local communities living in bear areas.

Acknowledgments

V.P. and A.O. were financially supported by the Excellence Project CGL2017–82782-P financed by the Spanish Ministry of Science, Innovation and Universities, the Agencia Estatal de Investigación (AEI), and the Fondo Europeo de Desarrollo Regional (FEDER, EU). G.B. was financially supported by a collaboration contract with MUSE – the Museo delle Scienze of Trento (Italy). M.M.D. was financially supported by the Spanish Ramon y Cajal grant RYC-2014-16263. J.M.F. and P.J.G. were funded by the Portuguese Foundation for Science and Technology (IF/00728/2013 and SFRH/BD/130527/2017).

References

Acharya, K. P., Paudel, P. K., Neupane, P. R. & Köhl, M. (2016). Human–wildlife conflicts in Nepal: patterns of human fatalities and injuries caused by large mammals. *PLoS ONE* 11: 1–18.

Akhtar, N. (2006). Human–sloth bear conflict: a threat to sloth bear conservation. *International Bear News* 15: 15–17.

Akiyama, G., Kuwahara, H., Asahi, R., Tosa, R. & Yokota, H. (2017). Prompt procedures have a great impact on the consequences of Asiatic black bear mauling. *Journal of Nippon Medical School* **84**: 294–300.

Ali, A., Waseem, M., Teng, M., *et al.* (2018). Human–Asiatic black bear (*Ursus thibetanus*) interactions in the Kaghan Valley, Pakistan. *Ethology Ecology and Evolution* 30: 399–415.

Anderson, K. (1957). *Man-eaters and jungle killers*. London: George Allen and Unwin Ltd.

Arimoto, I., Okamura, H., Koike, S., Yamazaki, K. & Kaji, K. (2014). Behavior and habitat of Asiatic black bear (*Ursus thibetanus*) inhabiting near settlements. *Honyuruikagaku* 54: 19–31.

Bargali, H. S., Akhtar, N. & Chauhan, N. P. S. (2005). Characteristics of sloth bear attacks and human casualties in North Bilaspur Forest Division, Chhattisgarh, India. *Ursus* 16: 263–267.

Bombieri, G., Delgado, M. del M., Russo, L. F., et al. (2018a). Patterns of wild carnivore attacks on humans in urban areas. *Scientific Reports* 8: 17728.

Bombieri, G., Nanni, V., Delgado, M. del M., et al. (2018b). Content analysis of media reports on predator attacks on humans: toward an understanding of human risk perception and predator acceptance. *BioScience* 68: 577–584.

Bombieri, G., Naves, J., Penteriani, V., et al. (2019). Brown bear attacks on humans: a worldwide perspective. *Scientific Reports* 9: 8573. https://doi.org/10.1038/s41598-019-44341-w

Brower, C. (1942). *Fifty years below zero: A lifetime of adventure in the Far North.* New York, NY: Dodd, Mead and Company.

Burton, R. G. (1856). *A book of man eaters* (first edition 1931). Delhi: Mittal Publications.

Chapron, G., Kaczensky, P., Linnell, J.D.C., et al. (2014). Recovery of large carnivores in Europe's modern human-dominated landscapes. *Science* 346: 1517–1519.

Charoo, S. A., Sharma, L. K. & Sathyakumar, S. (2011). Asiatic black bear–human interactions around Dachigam National Park, Kashmir, India. *Ursus* 22: 106–113.

Clark, D. (2003). Polar bear–human interactions in Canadian national parks, 1986–2000. *Ursus* 14: 65–71.

Conover, M.R. (2008). Why are so many people attacked by predators? *Human–Wildlife Interactions* 2: 139–140.

Debata, S., Swain, K. K., Sahu, H. K. & Palei, H. S. (2017). Human–sloth bear conflict in a human-dominated landscape of northern Odisha, India. *Ursus* 27: 90–98.

Dhamorikar, A. H., Mehta, P., Bargali, H. & Gore, K. (2017). Characteristics of human–sloth bear (*Melursus ursinus*) encounters and the resulting human casualties in the Kanha-Pench corridor, Madhya Pradesh, India. *PLoS ONE* 12: 1–18.

Dharaiya, N., Bargali, H. S. & Sharp, T. (2016). *Melursus ursinus*. The IUCN Red List of Threatened Species 2016: e.T13143A45033815 (WWW document). http://dx.doi.org/10.2305/IUCN.UK.2016-3.RLTS.T13143A45033815.en

Fenton, L. L. (1909). Tiger attacking a bear. *Bombay Natural History Society* 979.

Fleck, S. & Herrero, S. (1988). Polar bear–human conflicts. Contract report for Parks Canada and GNWT, contract 502/85/23.

Garcia, K. C., Joshi, H. M. & Dharaiya, N. (2016). Assessment of human–sloth bear conflicts in North Gujarat, India. *Ursus* 27: 5–10.

Garrote, P. J., Delgado, M. del M., López-Bao, J. V., et al. (2017). Individual attributes and party affect large carnivore attacks on humans. *European Journal of Wildlife Research* 63: 80.

Herrero, S. (1972). Aspects of evolution and adaptation in American black bears (*Ursus americanus* Pallas) and brown and grizzly bears (*U. arctos* Linné.) of North America. *Bears: Their Biology and Management* 2: 221.

Herrero, S. (2018). *Bear attacks: Their causes and avoidance*, 3rd edition. New York, NY: Lyons & Burford.

Herrero, S. & Higgins, A. (2003). Human injuries inflicted by bears in Alberta: 1960–98. *Ursus (Knoxville)* 14: 44–54.

Herrero, S., Higgins, A., Cardoza, J. E., Hajduk, L. I. & Smith, T. S. (2011). Fatal attacks by American black bear on people: 1900–2009. *Journal of Wildlife Management* 75: 596–603.

Hwang, M. H. & Garshelis, DL . (2007). Activity patterns of Asiatic black bears (*Ursus thibetanus*) in the Central Mountains of Taiwan. *Journal of Zoology* 271: 203–209.

Jamtsho, Y. & Wangchuk, S. (2016). Assessing patterns of human–Asiatic black bear interaction in and around Wangchuck Centennial National Park, Bhutan. *Global Ecology and Conservation* 8: 183–189.

Jangid, K. A. & Sharma, K. R. (2018). How locals characterize the causes of sloth bear attacks in Jawai, Rajasthan. *International Bear News* 27: 11–12.

Japan Bear Network. (2006). *Understanding Asian bears to secure their future*. Ibaraki: Japan Bear Network.

Japan Bear Network. (2011). *Report on statistics of the bear caused human injuries* (in Japanese). Ibaraki: Japan Bear Network.

Kozakai, C., Yamazaki, K., Nemoto, Y., et al. (2011). Effect of mast production on home range use of Japanese black bears. *Journal of Wildlife Management* 75: 867–875.

Kurt, F. & Jayasuriya, A. (1968). Notes on a dead bear. *Loris* 11: 182–183.

Lal Moten, T., Bhat, T. A., Gulzar, A., Mir, A. & Mir, F. (2017). Causalities of human wildlife conflict in Kashmir valley, India; a neglected form of trauma: our 10 year study. *International Journal of Research in Medical Sciences* 55: 1898–1902.

Laurie, A. & Seidensticker, J. (1977). Behavioural ecology of the sloth bear (*Melursus ursinus*). *Journal of Zoology* 182: 187–204.

Littledale, H. (1889). Bears being eaten by tigers. *Journal of Bombay Natural History Society* 4: 316.

Mardaraj, P. C. (2015). *Identifying key issues for the conservation of sloth bear (*Melursus ursinus*) in Rajnilgiri, Odisha, Eastern India*. The Rufford Foundation, United Kingdom.

Mardaraj, P. & Dutta, S.K. (2011). *Human–sloth bear conflict in Balasore Forest Division, Eastern India*. Saarbrücken, Germany: LAP Lambert Academic Publishing GmbH & Co. KG.

Moen, G. K., Støen, O. G., Sahlén, V. & Swenson, J. E. (2012). Behaviour of solitary adult Scandinavian brown bears (*Ursus arctos*) when approached by humans on foot. *PLoS ONE* 7: e31699.

Nabi, D. G., Tak, S. R., Kangoo, K. A. & Halwai, M. A. (2009a). Comparison of injury pattern in victims of bear (*Ursus thibetanus*) and leopard (*Panthera pardus*) attacks. A study from a tertiary care center in Kashmir. *European Journal of Trauma and Emergency Surgery* 35: 153–158.

Nabi, D. G., Tak, S. R., Kangoo, K. A. & Halwai, M. A. (2009b). Increasing incidence of injuries and fatalities inflicted by wild animals in Kashmir. *Injury* 40: 87–89.

National Park Service. (2019). Bear facts (WWW document). Available from www.nps.gov/yose/planyourvisit/bearfacts.htm

Nellemann, C., Støen, O. G., Kindberg, J., et al. (2007). Terrain use by an expanding brown bear population in relation to age, recreational resorts and human settlements. *Biological Conservation* 138: 157–165.

Ordiz, A., Støen, O. G., Delibes, M. & Swenson, J. E. (2011). Predators or prey? Spatio-temporal discrimination of human-derived risk by brown bears. *Oecologia* 166: 59–67.

Oshima, T., Ohtani, M. & Mimasaka, S. (2018). Injury patterns of fatal bear attacks in Japan: a description of seven cases. *Forensic Science International* 286: e14–e19.

Penteriani, V., Delgado, M. del M., Pinchera, F., et al. (2016). Human behaviour can trigger large carnivore attacks in developed countries. *Scientific Reports* 6: 20552.

Penteriani, V., Bombieri, G., Fedriani, J. M., et al. (2017). Humans as prey: coping with large carnivore attacks using a predator–prey interaction perspective. *Human–Wildlife Interactions* 11: 192–207.

Rajpurohit, K. S. & Krausman, P. R. (2000). Human–sloth bear conflicts in Madhya Pradesh, India. *Wildlife Society Bulletin* 28: 393–399.

Ramachandran, A. (2009, November 11). Frozen with fear: stranded teen v polar bears. *The Sydney Morning Herald* (WWW document). Available from www.smh.com.au/world/frozen-with-fear-stranded-teen-v-polar-bears-20091111-i9fs.html

Rasool, A., Wani, A. H., Darzi, M. A., et al. (2010). Incidence and pattern of bear maul injuries in Kashmir. *Injury* 41: 116–119.

Ratnayeke, S., Van Manen, F. T., Pieris, R. & Pragash, V. S. J. (2014). Challenges of large carnivore conservation: sloth bear attacks in Sri Lanka. *Human Ecology* 42: 467–479.

Røskaft, E., Händel, B., Bjerke, T. & Kaltenborn, B. P. (2007). Human attitudes towards large carnivores in Norway. *Wildlife Biology* 13: 172–185.

Sethy, J. & Chauhan, N.S. (2013). Human–sun bears conflict in Mizoram, North East India: impact and conservation management. *International Journal of Conservation Science* 4: 317–328.

Sethy, J. & Chauhan, N. S. (2016). Status and distribution of Malayan sun bear in Namdapha Tiger Reserve, Arunachal Pradesh, India. *International Journal of Conservation Science* 7: 533–552.

Shah, A., Mir, B., Ahmad, I., et al. (2010). Pattern of bear maul maxillofacial injuries in Kashmir. *National Journal of Maxillofacial Surgery* 1: 96.

Sharp, T. & Sonone, S. D. (2011). Sloth bear attacks: causes and consequences. *International Bear Newsletter* 20: 14–17.

Sharp, T. R., Swaminathan, S., Arun, A. S., et al. (2017). Sloth bear attack behavior and a behavioral approach to safety. Final report to International Association for Bear Research and Management.

Silwal, T., Kolejka, J., Bhatta, B. P., et al. (2017). When, where and whom: assessing wildlife attacks on people in Chitwan National Park, Nepal. *Oryx* 51: 370–377.

Singh, N., Sonone, S., Rot, J. & Dharaiya, N. A. (2017). An unusual attractant spurs sloth bear break-ins in Maharashtra, India 26: 20–21.

Smith, T. S. & Herrero, S. (2018). Human–bear conflict in Alaska: 1880–2015. *Wildlife Society Bulletin* 42: 254–263.

Smith, T. S., Herrero, S., Debruyn, T. D. & Wilder, J. M. (2008). Efficacy of bear deterrent spray in Alaska. *Journal of Wildlife Management* 72: 640–645.

Smith, T. S., Herrero, S., Layton, C. S., Larsen, R. T. & Johnson, K. R. (2012). Efficacy of firearms for bear deterrence in Alaska. *Journal of Wildlife Management* 76: 1021–1027.

Støen, O. G., Ordiz, A., Sahlén, V., et al. (2018). Brown bear (*Ursus arctos*) attacks resulting in human casualties in Scandinavia 1977–2016; management implications and recommendations. *PLoS ONE* 13: e0196876.

Swenson, J. E. (1999). Does hunting affect the behavior of brown bears in Eurasia? *Ursus* 11: 157–162.

Tak, S. R., Nabi, D. G., Halwai, M. A. & Mir, B. A. (2009). Injuries from bear (*Ursus thibetanus*) attacks in Kashmir. *Turkish Journal of Trauma & Emergency Surgery* 15: 130–134.

Wikipedia. (2019). Willem Barentsz (WWW socument). Available from https://en.wikipedia.org/wiki/Willem_Barentsz

Wilder, J. M., Vongraven, D., Atwood, T., et al. (2017). Polar bear attacks on humans: implications of a changing climate. *Wildlife Society Bulletin* 41: 537–547.

Wilson, E. O. (1984). *Biophilia*. Cambridge, MA: Harvard University Press.

Windler, A. (2014). *Views towards the sun bear and frames on the human–sun bear conflict of local people in West Sumatra, Indonesia*. Wageningen: Forest and Nature Conservation, Wageningen University & Research.

Yamazaki, K. (2004). Recent bear–human conflicts in Japan. *International Bear News* 13: 16–17.

Yamazaki, K. (2010). *Ursus thibetanus*. In: Ohdachi, S. D., Ishibashi, Y., Iwasa, M. A., Fukui, D. & Saitoh, T. (Eds.), *The wild mammals of Japan* (pp. 243–245). Kyoto: Shoukado Book Seller.

Yamazaki, K. (2017). Consecutive fatal attacks by Asiatic black bear on humans in Northern Japan. *International Bear News* 26: 16–17.

Yamazaki, K. & Sato, Y. (2014). Country-wide range mapping of Asiatic black bears reveals increasing range in Japan. *International Bear News* 23: 18–19.

Yamazaki, K., Furubayashi, K., Kasai, S., et al. (2008). A preliminary evaluation of activity-sensing GPS collars for estimating daily activity patterns of Japanese black bears. *Ursus* 19: 154–161.

Chapter 18

Effects of Human Disturbance on Brown Bear Behavior

Ole-Gunnar Støen, Andrés Ordiz, Marcus Elfström, Anne G. Hertel, Veronica Sahlén, Jonas Kindberg, and Jon E. Swenson

Introduction

Human persecution is the main cause of mortality for large carnivores in human-dominated landscapes, often due to conflicts with people, because they prey on livestock or wild game or harm people (Kaczensky 1997; Woodroffe & Ginsberg 1998; Røskaft et al. 2003; Nielsen et al. 2005; see also Chapters 15 and 17). This mortality affects population dynamics of large carnivores, leading to a strong association between high human density and the loss of regional carnivore populations (Woodroffe 2000). Human disturbance may be regarded as a predation risk, forcing large carnivores to behave as prey and consequently avoid humans to increase their survival (Lima 1998; Beale & Monaghan 2004; Ordiz et al. 2011). In the "landscape of fear" concept, this predation risk forces individual carnivores to alter their use of the landscape and invest more time in antipredator behavior (vigilance) and less time in profitable activities (foraging), which may affect reproduction (Altendorf et al. 2001; Laundré et al. 2001). Large carnivores in humanized landscapes may alter their behavior and avoid humans by being nocturnal and elusive (Basille et al. 2009; Knoppf et al. 2014; Ordiz et al. 2014). In some areas, favorable management policies allowed large carnivore populations to increase and even expand into more human-dominated landscapes (Chapron et al. 2014). Given the current scenario of human population growth, human expansion into wilderness areas, and increasing habitat loss (Hoekstra et al. 2005), the future survival of large carnivores may depend on their ability to adapt to live in human-dominated landscapes (Linnell et al. 2001).

Bears are disturbed in many ways, either directly when they encounter humans or indirectly by changing their behavior and way of life to avoid humans, human activity, and infrastructure. Here we summarize research on how brown bears normally react when encountering humans, what a human encounter may entail for a bear, and whether bears become habituated or change their behavior toward humans with increased exposure. Based on this, we also discuss: (a) how our knowledge of brown bear behavior may help people to deal with their fear of bears, and not limit their use of outdoor areas with bears; (b) how human presence, activity, and infrastructure have an indirect effect on bears, that is how bears change their movement pattern, use of terrain and vegetation, and daily activity pattern to avoid humans; (c) how human disturbance influences foraging and denning, which is crucial for brown bear growth and reproduction; and (d) apparent differences between continents in brown bear behavior toward humans and whether this may have an evolutionary cause.

Brown Bear Behavior When Encountering Humans

Brown bears can avoid humans in time and space by being mostly nocturnal and residing far from people (Nellemann et al. 2007; Martin et al. 2010). However, bears cannot always avoid encounters with humans, especially in areas close to human settlements, areas with much human activity, or even in back-country areas, where hikers may unintentionally encounter bears. Some of these encounters may result in dangerous situations, where both humans and bears can be harmed (see Chapter 17). However, the majority of such encounters are uneventful, with the bears' presence going completely unnoticed (Moen et al. 2012).

Several studies have aimed to understand how bears normally react during encounters with humans (e.g. Swenson et al. 1999). The purpose of such studies is also to be able to inform the public and management authorities about normal bear behavior, how to behave when encountering bears, and how to prevent dangerous situations from occurring. The first studies in Scandinavia, which relied on direct observations and analyzed the outcome of encounters between bears and people, concluded that bears were not aggressive toward humans and seldom attacked, and that in the cases the bear had charged at the human, these were bluff charges to scare the human away (Swenson et al. 1999).

These earlier studies were by design limited to studying how bears reacted during encounters when they were seen by the observers, and only for the time that they were being observed. Therefore, they were unable to study the bears' behavior during encounters when they were out of sight, e.g. when in dense vegetation. Further, as bears are elusive animals, these studies can only be expected to cover a small fraction of bear–human encounters, as most encounters are likely to go unnoticed by humans. The use of geographical positioning systems (GPS) technology in wildlife research has enabled researchers to study animal behavior remotely and in greater detail than in previous studies. This technology provides the opportunity to study how bears behave during encounters where they are not observed, as well as information on how often bear–human encounters occur without being detected by humans. Such studies have been carried out in the Scandinavian brown bear population, using experimental approaches that simulated the behavior of hikers in order to study how single bears (Moen et al. 2012) and females with cubs (Sahlén et al. 2015a) respond to encounters when they are aware that humans are approaching. The studies revealed that the absolute majority of encounters go unnoticed by humans (Moen et al. 2012; Ordiz et al. 2013). When bears are aware that humans are approaching, they either leave the area or hide until the humans have passed. As a result, observers only saw the bears in 15% (single bears) or 26% (females with cubs) of approaches, despite knowing exactly where the bear was located. The difference in detectability between single bears and females with cubs appears to be due to females with cubs preferring more open habitat, probably as a strategy to avoid adult infanticidal males (Sahlén et al. 2015b).

Most bears, whether active or resting in a day bed, left their location when humans approached. Whether they left depended on how close the humans passed the bears (Moen et al. 2012). The distance from the observer at which bears decided to leave, the flight initiation distance (FID), was longer for active single bears (average 115 m) that were already on their feet than for passive single bears (average 69 m) lying in their resting site. Younger single bears left at longer distances than older single bears, probably because they were less experienced with encountering people. The FID was also longer when the bears were in more open habitat, with less cover to hide in. FID did not differ between single bears and females with cubs, but a greater percentage of females with cubs left as a response to an encounter with humans (Sahlén et al. 2015a). Following the encounter, bears that left as a response to the disturbance moved on average 1.2 km before settling into a new resting site. Females with cubs displayed stronger immediate reactions, by moving farther before settling into a new resting site (Sahlén et al. 2015a).

The reactions by bears to encounters with humans reflected on their daily activity patterns, and this effect went beyond the same day of the approach. Bears were more active, i.e. they moved larger distances during nocturnal hours, and less during the day, for several days following the encounter (Ordiz et al. 2013 and Figure 18.1). The bears reduced their movement by 10% and 11% during daytime the first two days following an experimental encounter, respectively, but showed a similar increase in activity during the darkest hours of the night when they normally rest, probably to compensate for disrupted feeding during the day (Ordiz et al. 2013). The closer the observers had passed and the denser the vegetation at the encounter location, the stronger the disturbance effect. This indicates that bears react more strongly when disturbed at short distances in highly concealed sites, where bears hide from people during their day rest (Ordiz et al. 2011).

Brown bear reactions to encountering humans seem to be consistent between populations and varying human density. In Fennoscandia, a comparative study showed that the proportion of bears that fled when encountered by humans and the distance from which they initiated their flight were similar between bears in Sweden and Finland, where bears lived in areas with a higher human density (Moen et al. 2019). Similarly, in Sweden, the density of humans and roads within the home range of the bears, used as a proxy for probable encounter rates with humans, did not influence the flight responses of the bears. This consistency may be explained by a similar history in Fennoscandia, with recent population recovery after near extirpation of the brown bear population (Pulliainen 1983; Swenson et al. 1995, 2017; Kopatz et al. 2014), where bears are managed as a game species, and have experienced long-term persecution, which may make bears elusive in Fennoscandia (Ordiz et al. 2011; Zedrosser et al. 2011). However, Fennoscandia has among the lowest human population densities across the distribution range of brown bears, at least in Europe (Swenson et al. 2000), so similar studies might be needed on bears in areas with higher human density to fully understand the normal behavior of bears when encountered by humans (Moen et al. 2019).

Brown Bear Habituation and Sensitization

Habituation is a process that causes decreasing responsiveness to a repeated stimulus (Thompson 2009), whereas sensitization is an increased responsiveness to a repeated stimulus (Blumstein 2016). Although brown bears generally avoid humans, some individuals can habituate to people in close proximity. In North America, brown bear habituation to hikers and anglers, for instance, has been documented in different ecosystems and national parks (e.g. McLellan & Shackleton 1989; Herrero et al. 2005). Habituated bears face higher mortality risk, which in turn can have a negative effect on bear population dynamics. For example, habituated bears near roads or railways are more likely to be killed by vehicles or trains and to be removed legally and illegally (Herrero et al. 2005).

In Yellowstone National Park, USA, bear habituation to people and sensitization to human-derived food (i.e. food conditioning) resulted in human fatalities a few decades ago, which caused the removal of many bears and a population decline (e.g. Penteriani et al. 2018). In Europe, bears are also generally wary of humans, but this can change and requires that human activities that can alter bear behavior be actively managed to prevent habituation (Penteriani et al. 2018). Subadult bears may approach human habitation in order to avoid adult conspecifics and/or because of a lack of experience, i.e. naivety, which, after exposure to people, may result in human habituation or food conditioning (Elfström et al. 2014b). Therefore, subadults may be more vulnerable to habituation and food conditioning than older, experienced bears that generally avoid people.

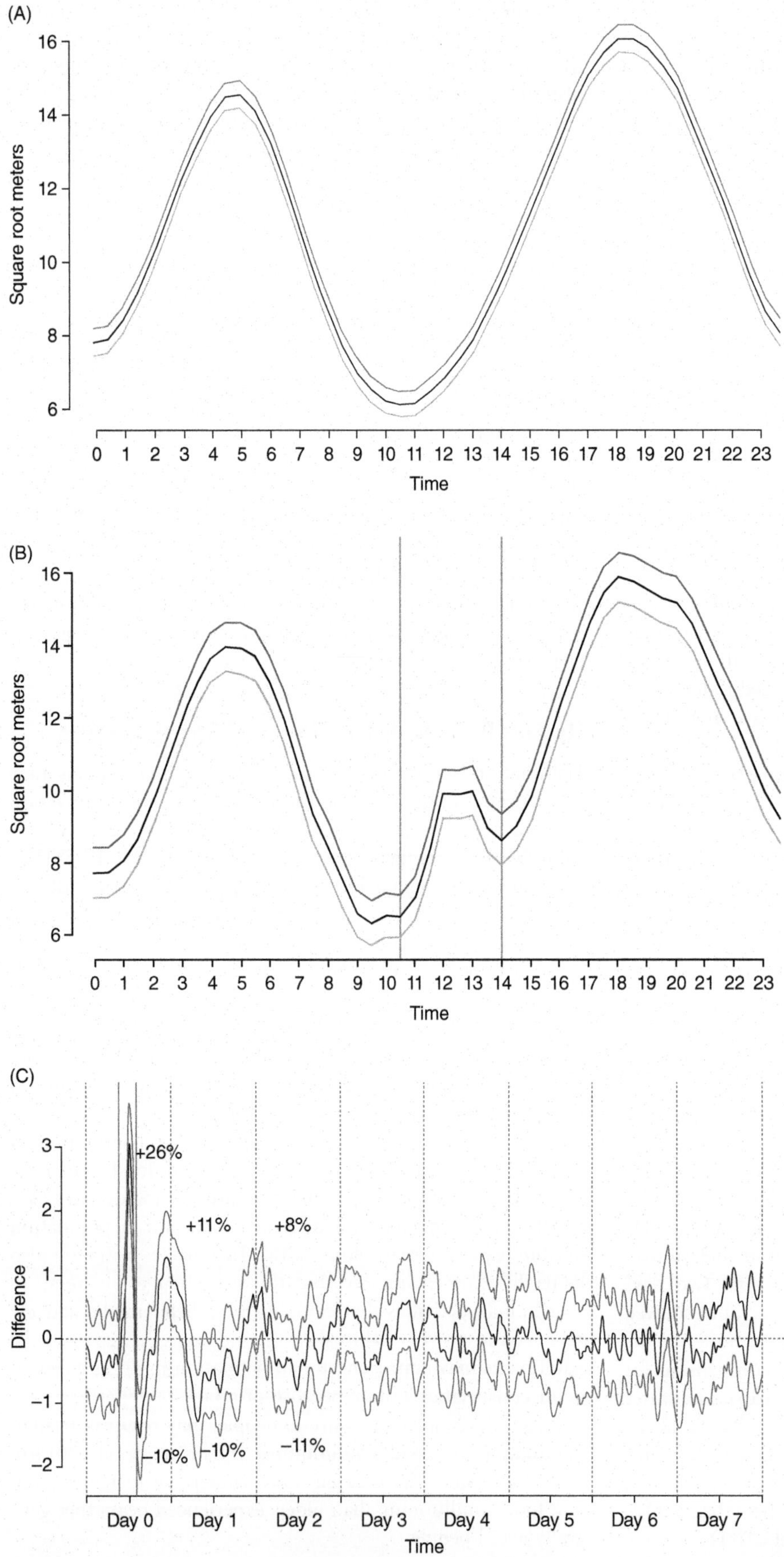

Figure 18.1 Estimated time effect, every 30 min during the 24 h day, on the daily activity pattern of brown bears in central Sweden during the week before the experimental approach. Bears typically had a main resting period during midday and a second resting period around midnight (A). The daily activity pattern of bears on the day of the approach, showing the initial escape after the encounter, followed by a reduction in movement. Vertical lines illustrate when most approaches were conducted (B). Bears became more nocturnal in the following days after the approach, compared to the daily pattern in the week before the approach. The vertical lines illustrate when most approaches were conducted (C). In all panels, curves represent the mean of the distance travelled and the 95% credible intervals. Figures from Ordiz et al. (2013). (© 2013 The Authors. *Journal of Applied Ecology* © 2013 British Ecological Society.)

A recent study in Scandinavia (Ordiz et al. 2019) performed repeated approaches to GPS-collared brown bears to document the potential habituation or sensitization to repeated, consistently similar stimuli. Bears were approached up to eight consecutive times in a few weeks, with no clear change in the bears' general response, i.e. avoiding people, in terms of flight initiation distance and proportion of bears that stayed or moved from their position before the approaches started (Ordiz et al. 2019). This suggests that, at the level of experimentally induced disturbance, bears did not show habituation or sensitization to the repeated stimuli. Interestingly, there is a remarkable scarcity of studies on wildlife habituation or sensitization that have analyzed the response of radio-collared animals, which is essential to determine if specific individuals habituate or not and, consequently, to inform managers (Blumstein 2016).

Besides the observation of clearly habituated individuals that tolerate human presence, subtle, physiological and non-visible changes that incur energetic costs can occur in response to humans (Herrero et al. 2005; Støen et al. 2015). It has been suggested that genetic differences between habituated and non-habituated individuals may exist within a bear population (Cotovelea et al. 2015). Nevertheless, bear habituation may reduce other negative effects of human activity on large carnivores, such as the total avoidance of some areas; therefore, further studies are needed to fully understand the potential population-level consequences of habituation (Wheat & Wilmers 2016). Population-level effects may be different depending on the bear population size and conservation status, degree of human encroachment, and exposure of individual bears to human activities.

Figure 18.2 A typical resting site of a brown bear in Sweden. The bear rested among the vegetation-covered rocks with good cover from surrounding trees and larger rocks. Hairs of the bears are often found in the resting bed. (Photo by O.-G. Støen.)

Human Effects on Brown Bear Activity Rhythms, Movement Patterns, and Habitat Use

Human persecution has eradicated brown bears from many places and, where bears still exist, human activities have reduced bear habitat quality and the size of many bear populations (e.g. Suring et al. 1998; Proctor et al. 2010; Ordiz et al. 2017). Therefore, human activities have effects on bear population trends and, in human-dominated landscapes, bears change or adapt their behavior to deal with humans. Bear adaptations include changes in movement patterns, terrain use, and daily activity patterns, which have been documented at different spatial and temporal scales (e.g. Ordiz et al. 2017).

Across their widespread range, brown bears select for rugged terrain, where human access is limited (Apps et al. 2004; Nellemann et al. 2007; Martin et al. 2010). At finer scales, bears select for dense vegetation cover and avoid areas near human settlements and paths when choosing resting sites, especially when human outdoor activity is greatest (Ordiz et al. 2011; Cristescu et al. 2013; Figures 18.2 and Figure 18.3). Human activity also alters the optimal daily activity patterns of brown bears. Disturbance can exclude bears completely from an area or reduce its time of use (Suring et al. 1998). Generally, bears are more active during daytime in areas with less human activity, and tend to be nocturnal in more humanized landscapes (e.g. Machutchon et al. 1998; Olson et al. 1998; Kaczensky et al. 2006; Martin et al. 2010). Reduced daytime activity is particularly clear after bears encounter people (McLellan & Shackleton 1989; Moen et al. 2012). Where bears are hunted, they also become more nocturnal when hunting seasons start (Ordiz et al. 2012).

Bears avoid human infrastructure, e.g. roads (Mace et al. 1996; Elfström et al. 2008) or are more nocturnal in areas with higher road density (Ordiz et al. 2014). Thus, limiting access or closing roads that no longer are necessary, e.g. after logging, is advisable to reduce disturbance (Ordiz et al. 2014; Sorensen et al. 2015). Adult bears, especially males, avoid human settlements, cabins, and ski resorts, whereas areas closer to people are used by younger individuals (Mueller et al. 2004; Nellemann et al. 2007; Elfström et al. 2014a, 2014b; Sorensen et al.

Figure 18.3 Across their widespread range, brown bears select for rugged terrain, where human access is limited. (Photo by V. Penteriani.)

2015). In turn, bears face higher mortality risk near people and human settlements (Steyaert et al. 2016).

In summary, bears react to all types of human activities and bear reactions are visible at all spatial and temporal scales (Ordiz et al. 2017). In bear populations living in human-dominated landscapes under long-term protection, bears continue to avoid people, yet differences in bear behavior in relation to human infrastructure may not be as clear as in hunted populations, showing a general pattern of human avoidance that does not seem to change in relation with distance to human infrastructure (Zarzo-Arias et al. 2018).

Human Effects on Brown Bear Foraging

In most animals, effective and sufficient foraging is the premise for successful reproduction. Bears give birth in the winter den and are therefore particularly dependent on finding enough food during summer prior to hibernation. Humans affect food availability and bear foraging efficiency in multiple ways. On one hand humans may intentionally or unintentionally provision bears with food. Although bears near human habitation primarily seek shelter, this may or may not result in foraging on anthropogenic foods (Elfström et al. 2014a, 2014b, 2014c). Human infrastructure provides easy access to anthropogenic food resources, like trash, and bears can readily use these resources (Bojarska & Selva 2012; Johnson et al. 2015). Ungulate hunting facilitates highly nutritious food when hunters leave carrion or gut piles of game that bears scavenge on. Further, in six of nine brown bear populations in Europe, co-occurring ungulates are given supplementary feed during winter and, despite not being the target species, bears regularly use these feeding sites (Selva et al. 2017). Elsewhere, intentional feeding of bears by humans is common. Several European countries, including Slovenia, Croatia, Serbia, and Romania, use diversionary feeding at bait sites to lure bears away from human settlements (Kavcic et al. 2013). Bait sites may likewise be used to attract bears either for hunting or for bear-viewing tourism (Bischof et al. 2008; Penteriani et al. 2017). Contrary to provisioning bears, human activity, presence, and infrastructure have also been shown to constrain bear foraging activities. Brown bears in Scandinavia have been shown to use less-productive foraging areas when avoiding humans temporally (Hertel et al. 2016; Figure 18.4) or spatially (Lodberg-Holm et al. 2019). Salmon streams in Alaska attract bears in high numbers and hence are popular for bear-viewing tourism. The presence of tourists, however, can displace bears spatiotemporally from salmon streams and therefore hamper food intake (Rode et al. 2006). Lastly, probably the most influential way humans alter food availability for bears is the transformation of land-use by mining, deforestation, or fire management (Nielsen et al. 2004; Cristescu et al. 2016). Tree or topsoil removal changes nutrient and light availability and hence fundamentally alters predominant plant communities on which bears forage. The pathways through which humans affect bear foraging depend largely on local wildlife and habitat management practices and are highly ecosystem-specific.

Cascading effects can be trait-mediated indirect effects, where an antipredation strategy by a prey species influences species in other trophic levels (Abrams 1995). One example is the brown bear in Scandinavia, which practise antipredation strategies in relation to humans (Ordiz et al. 2011) and at the same time is a major predator on red wood ants (*Formica rufa*). Red wood ants are a keystone species in the boreal forest, influencing other species in the ecosystem (Haemig 1992, 1994; Kilpelainen *et al.* 2009; Stockan & Robinson 2016). Through predation and disturbance of anthills, brown bears reduce the abundance of ants (Moen 2018). Because brown

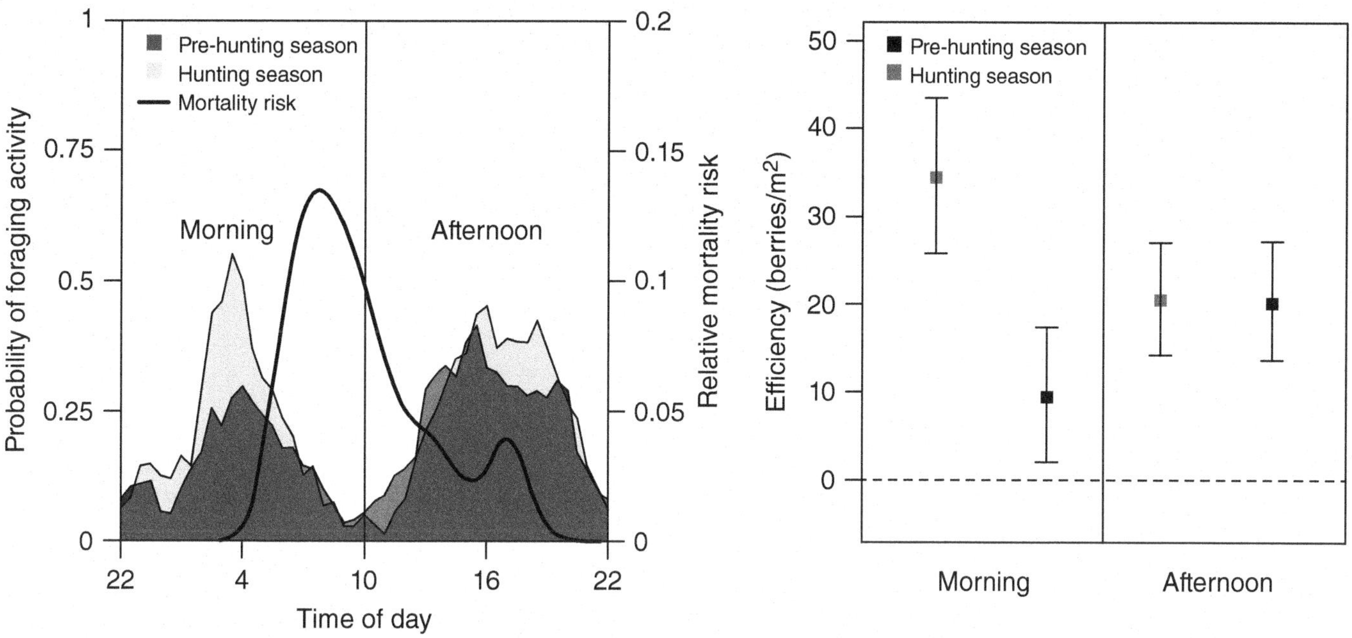

Figure 18.4 Effect of mortality risk on foraging activity (left panel) and foraging efficiency (right panel) during the first two weeks of the hunting season, as compared to last two weeks before the hunting season in south-central Sweden. Left: Foraging activity was measured as slow, meandering movements. Relative mortality risk was measured as the number of bears killed at each time of day in the study area during the hunting seasons 2006–2014 (total number of shot bears = 680). Mortality risk peaked in the morning hours after the main foraging activity peak but was low when foraging activity was high in the afternoon. Bears reduced foraging activity in the morning but not in the afternoon during the hunting season. Right: Foraging efficiency was defined as the richness of the foraging location and was measured by how many more berries were found at a bear foraging location relative to how many berries would be expected in the same habitat and time of year. Bears selected the best foraging location during their morning foraging bout prior to the hunting season but decreased after onset of the hunting season. Bears probably use denser vegetation with more vertical cover but less light availability for berry-producing shrubs. Foraging efficiency in the afternoon when mortality risk was low was the same for the pre-hunting and hunting seasons. Modified from Hertel et al. (2016).

bears avoid human activity, humans may modify the effect bears have on red wood ants, and mediate cascading effects in the boreal forest (Moen 2018).

Human Effects on Brown Bear Denning

During the winter denning period, brown bears avoid human disturbance by usually choosing den sites in steep terrain with few roads which are therefore less accessible for people (Petram et al. 2004; Elfström et al. 2008; Sahlén et al. 2011). Adult males often choose den sites in more remote areas, suggesting that the more predation-vulnerable females and subadult males also avoid dominant adult males during denning (Elfström & Swenson 2009; Elfström et al. 2014b). The level of human activity elicits increased avoidance behavior when bears choose den sites. Settlements and roads with high traffic that also provide direct human access are avoided when bears choose den sites (Elfström et al. 2008). Brown bears that den closer to plowed roads are at a higher risk of abandoning their dens compared to bears denning in areas with lower road accessibility (Elfström & Swenson 2009). As a result, bears that den closer to human activity also choose more concealed den sites, i.e. with increased horizontal cover and terrain ruggedness (Sahlén et al. 2011). Bears seem to be more sensitive to disturbance when they are about to enter their den sites, and abandonment rates are higher during the onset of the denning period (Sahlén et al. 2015a). Female brown bears that abandon their dens can experience lower reproductive success (Swenson et al. 1997). Therefore, females may tolerate greater levels of disturbance without abandoning their dens after giving birth. This suggests that roads and human activity have a negative impact and represent a disturbance for bears. After bears have entered their dens, the risk of human injuries during bear–human encounters increases due to a combination of factors, including reduced bear physiological activity that prevents them from escaping from approaching people (Sahlén et al. 2015b; Støen et al. 2018).

Human Effects on Brown Bears in an Evolutionary Perspective

We have discussed above how brown bears show plasticity in their reactions to variations in human disturbance. It is a common perception that the grizzly bear of North America is more aggressive (Herrero 1985; Swenson et al. 1999; but see Bombieri et al. 2019) and more diurnally active (Roth 1983; Ordiz et al. 2014) than the European brown bear, although they are the same species. Many factors, such as human density, bear density, habitat productivity and openness, and persecution and hunting, might be enough to explain geographic differences in brown bear behavior and life-history traits. Recent research has highlighted the potential for long-term evolutionary effects of directed selection during heavy human harvesting (Coltman et al. 2003; Festa-Bianchet 2003; Proaktor et al. 2007; Allendorf & Hard 2009; Darimot et al. 2009; Mysterud 2011). Brown bears were persecuted systematically for many centuries in Europe, but for a relatively short time (<150 years) by humans with modern weapons in North America and Asia (Zedrosser et al. 2011). Thus, it is possible that behavioral and life-history traits have been subject to greater human-induced selection in Europe, which may have contributed to some of the observed continental differences.

This hypothesis is difficult to test scientifically, but there are indications of human-induced selection in brown bears. For example, Zedrosser et al. (2011) compared brown bear populations worldwide and found that a long persecution history was associated with greater mean litter size at the population level, which is consistent with evolutionary theory and modeling results (Proaktor et al. 2007). Human-induced selection can occur rapidly, as Van de Walle et al. (2018) documented that a hunting regulation that protected female brown bears with cubs indirectly promoted slower life histories by modulating the fitness benefit of maternal care tactics during a 20-year period of relatively high hunting harvest. Therefore, it is possible that intense, long-term persecution of brown bears in Europe has selected for behaviors that increase individual survival, including increased shyness in relation to humans, more nocturnal activity patterns, and lower aggressiveness. These attributes, in addition to the higher productivity noted above, should make it easier to promote population growth in brown bears in Europe. In fact, the highest population growth ever documented in a brown bear population was in Europe (Sæther et al. 1998) and many European populations are presently increasing under enlightened management (Chapron et al. 2014).

Brown Bear Behavior and Human Fear

Brown bear populations are increasing in many parts of the world and bear populations are re-establishing in some areas where the species has been absent for several decades (Chapron et al. 2014; Penteriani et al. 2016). This increases the chance that people encounter bears during outdoor activities in areas where they previously did not encounter bears. For some people this brings feelings of interest and joy (Jacobs et al. 2014), but for others it elicits feelings of disgust and fear (Manfredo 2008; Johansson & Karlsson 2011; Jacobs et al. 2014). Even if the risk of being harmed by bears is exceedingly low (Penteriani et al. 2016; Støen et al. 2018), the fear of bears limits some people in their use of the outdoors and thereby reduces their life quality.

According to psychological theory, there are three antecedents for people's fear of large carnivores.The first is perceived danger, i.e. how much harm the animal may do to the person; the second is perceived unpredictability, i.e. the person does not know what the animal will do or how it will act during an encounter; and the third is perceived uncontrollability, i.e. the person does not know how to react or what to do when encountering the animal (Armfield 2006, 2007). Generally, interventions addressing fear of large carnivores can be categorized in four groups: information and education, exposure to animals and habitat, collaboration and participation, and financial incentives (Johansson et al. 2016; Bombieri et al. 2018). Despite many suggestions for interventions to mitigate fear of large carnivores, few of the interventions have been tested for their effect and the recommendations in the literature are contradictory (Johansson et al. 2016).

A recent study showed that people who are motivated to do something about their fear of brown bears were less afraid and less limited in their use of the outdoors after being exposed to bears in the form of forest walks in bear habitat with guides with good knowledge of bears, how bears react in encounters with humans, and what to do to be safe during an encounter (Johansson et al. 2019). People expressed that their own experience of bears, their habitat use, behavior, and dangers, and with the guide as a model of conduct, provided them with the mental tools that allowed them to handle their fear of bears (Johansson et al. 2019). This can be a management intervention that is possible to implement in all areas with bears for people who express fear of bears and are motivated to handle their fear.

References

Abrams, P. A. (1995). Implications of dynamically variable traits for identifying, classifying, and measuring direct and indirect effects in ecological communities. *The American Naturalist* 146: 112–134.

Allendorf, F. W. & Hard, J. J. (2009). Human-induced evolution caused by unnatural selection through harvest of wild animals. *Proceedings of the National Academy of Sciences USA* 106: 9987–9994.

Altendorf, K. B., Laundré, J. W., González, C. A. L. & Brown, J. S. (2001). Assessing effects of predation risk on foraging behavior of mule deer. *Journal of Mammalogy* 82: 430–439.

Apps, C. D., Bruce, N., John, G. & Michael, F. (2004). Estimating grizzly bear distribution and abundance relative to habitat and human influence. *Journal of Wildlife Management* 68: 138–152.

Armfield, J. M. (2006). Cognitive vulnerability: a model of the ethology of fear. *Clinical Psychology Review* 26: 746–768.

Armfield, J. M. (2007). Understanding animal fears: a comparison of the cognitive vulnerability and harm-looming models. *BMC Psychiatry* 7: 68. doi:10.1186/1471-244X-7-68.

Basille, M., Herfindal, I., Santin-Janin, H., et al. (2009). What shapes Eurasian lynx distribution in human dominated landscapes: selecting prey or avoiding people? *Ecography* 32: 683–691.

Beale, C. M. & Monaghan, P. (2004). Human disturbance: people as predation-free predators? *Journal of Applied Ecology* 41: 335–343.

Bischof, R., Fujita, R., Zedrosser, A., Söderberg, A. & Swenson, J. E. (2008). Hunting patterns, ban on baiting, and harvest demographics of brown bears in

Sweden. *The Journal of Wildlife Management* 72: 79–88.

Blumstein, D. T. (2016). Habituation and sensitization: new thoughts about old ideas. *Animal Behaviour* 120: 255–262.

Bojarska, K. & Selva, N.(2012). Spatial patterns in brown bear *Ursus arctos* diet: the role of geographical and environmental factors. *Mammal Review* 42: 120–143.

Bombieri, G., Nanni, V., Delgado, M. M., et al. (2018). Content analysis of media reports on predator attacks on humans: toward an understanding of human risk perception and predator acceptance. *BioScience* 68: 577–584.

Bombieri, G., Naves, J., Penteriani, V., et al. (2019). Brown bear attacks on humans: a worldwide perspective. *Scientific Reports* 9: 8573. doi.org/10.1038/s41598-019-44341-w

Chapron, G., Kaczensky, P., Linnell, J. D. C., et al. (2014). Successful recovery of large carnivores in Europe's human-dominated landscapes. *Science* 346: 1517–1519.

Coltman, D. W., O'Donoghue, P., Jorgenson, J. T., et al. (2003). Undesirable evolutionary consequences of trophy hunting. *Nature* 426: 655–658.

Cotovelea, A., Ionescu, O., Şofletea, N., et al. (2015). Testing the influence of habituation on genetic structure of brown bear (*Ursus arctos*). *Annals of Forest Research* 58: 81–90.

Cristescu, B., Stenhouse, G. B. & Boyce, M. S. (2013). Perception of human-derived risk influences choice at top of the food chain. *PLoS ONE* 8: e82738.

Cristescu, B., Stenhouse, G. B. & Boyce, M. S. (2016). Large omnivore movements in response to surface mining and mine reclamation. *Scientific Reports* 6: 19177.

Darimot, C. T., Carlson, S. M., Kinnison, M. T., et al. (2009). Human predators outpace other agents of trait changes in the wild. *Proceedings of the National Academy of Sciences USA* 106: 952–954.

Elfström, M. & Swenson, J. E. (2009). Effects of sex and age on den site use by Scandinavian brown bears. *Ursus* 20: 85–93. 10.2192/09gr005.1

Elfström, M., Swenson, J. E. & Ball, J. P. (2008). Selection of denning habitats by Scandinavian brown bears *Ursus arctos*. *Wildlife Biology* 14(2): 176–187.

Elfström, M., Davey, M. L., Zedrosser, A., et al. (2014a). Do Scandinavian brown bears approach settlements to obtain high-quality food? *Biological Conservation* 178: 128–135.

Elfström, M., Zedrosser, A., Støen, O.-G. & Swenson, J. E. (2014b). Ultimate and proximate mechanisms underlying the occurrence of bears close to human settlements: review and management implications. *Mammal Review* 44: 5–18.

Elfström, M., Zedrosser, A., Jerina, K., et al. (2014c). Does despotic behavior or food search explain the occurrence of problem brown bears in Europe? *Journal of Wildlife Management* 78: 881–893.

Festa-Bianchet, M. (2003). Exploitative wildlife management as a selective pressure for life-history evolution of large mammals. In: Apollonio, M. & Festa-Bianchet, M. (Eds.), *Animal behavior and wildlife conservation* (pp. 191–207). Washington, DC: Island Press.

Haemig, P. D. (1992). Competition between ants and birds in a Swedish forest. *Oikos* 65: 479–483.

Haemig, P. D. (1994). Effect of ants on the foraging of birds in spruce trees. *Oecologia* 97: 35–40.

Herrero, S. (1985). *Bear attacks. Their causes and avoidance*. New York, NY: Lyons Press.

Herrero, S., Smith, T., DeBruyn, T. D., Gunther, K. & Matt, C. A. (2005). Brown bear habituation to people – safety, risks, and benefits. *Wildlife Society Bulletin* 33: 362–373.

Hertel, A. G., Zedrosser, A., Mysterud, A., et al. (2016). Temporal effects of hunting on foraging behavior of an apex predator: do bears forgo foraging when risk is high? *Oecologia* 182: 1019–1029.

Hoekstra, J. M., Boucher, T. M., Ricketts, T. H. & Roberts, C. (2005). Confronting a biome crisis: global disparities of habitat loss and protection. *Ecology Letters* 8: 23–29.

Jacobs, M., Vaske, J. J., Dubois, S. & Fehres, P. (2014). More than fear: role of emotions in acceptability of lethal control of wolves. *European Journal of Wildlife Research* 60: 589–598.

Johansson, M. & Karlsson, J. (2011). Subjective experience of fear and the cognitive interpretation of large carnivores. *Human Dimensions of Wildlife* 16: 15–29.

Johansson, M., Ferreira, I., Støen, O.-G., Frank, J. & Flykt, A. (2016). Targeting human fear of large carnivores – many ideas but few known effects. *Biological Conservation* 20: 261–269.

Johansson, M., Flykt, A., Frank, J. & Støen, O.-G. (2019). Controlled exposure reduces fear of brown bears. *Human Dimensions of Wildlife* 24: 363–379. DOI:10.1080/10871209.2019.1616238.

Johnson, H. E., Breck, S. W., Baruch-Mordo, S., et al. (2015). Shifting perceptions of risk and reward: dynamic selection for human development by black bears in the western United States. *Biological Conservation* 187: 164–172.

Kaczensky, P. (1997) Large carnivore depredation on livestock in Europe. *Ursus* 11: 59–72.

Kaczensky, P., Huber, D., Knauer, F., et al. (2006). Activity patterns of brown bears (*Ursus arctos*) in Slovenia and Croatia. *Journal of Zoology* 269: 474–485.

Kavcic, I., Adamic, M., Kaczensky, P., Krofel, M. & Jerina, K. (2013). Supplemental feeding with carrion is not reducing brown bear depredations on sheep in Slovenia. *Ursus* 24: 111–119.

Kilpelainen, J., Finér, L., Neuvonen, S., et al. (2009). Does the mutualism between wood ants (*Formica rufa* group) and Cinara aphids affect Norway spruce growth? *Forest Ecology and Management* 257: 238–243.

Knopff, A. A., Knopff, K. H., Boyce, M. S. & St Clair, C. C. (2014). Flexible habitat selection by cougars in response to anthropogenic development. *Biological Conservation* 178: 136–145.

Kopatz A, Eiken, H. G., Aspi, J., et al. (2014). Admixture and gene flow from Russia in the recovering northern European brown bear (*Ursus arctos*). *PLoS ONE* 9: e97558.

Laundré, J. W., Hernández, L. & Altendorf, K. B. (2001). Wolves, elk, and bison: reestablishing the "landscape of fear" in Yellowstone National Park, U.S.A. *Canadian Journal of Zoology* 79: 1401–1409.

Lima, S. L. (1998). Nonlethal effects in the ecology of predator-prey interactions. *BioScience* 48: 25–34.

Linnell, J. D. C., Swenson, J. & Andersen, R. (2001). Predators and people: conservation of large carnivores is possible at high human densities if management policy is favourable. *Animal Conservation* 4: 345–350.

Lodberg-Holm, H. K., Wathne Gelink, H., Hertel, A. G., et al. (2019). Hunting creates a landscape of fear that influences foraging tactics of brown bears. *Basic and Applied Ecology* 35: 18–27.

Mace, R. D., Waller, J. S., Manley, T. L., Lyon, L. J. & Zuuring, H. (1996). Relationships among grizzly bears, roads and habitat in the Swan Mountains, Montana. *The Journal of Applied Ecology* 33: 1395.

Machutchon, A. G., Himmer, S., Davis, H. & Gallagher, M. (1998). Temporal and spatial activity patterns among coastal bear populations. *Ursus* 10: 539–546.

Manfredo, M. J. (2008). *Who cares about wildlife? Social science concepts for exploring human wildlife relationships and conservation issues.* New York, NY: Springer.

Martin, J., Basille, M., Van Moorter, B., et al. (2010). Coping with human disturbance: spatial and temporal tactics of the brown bear (*Ursus arctos*). *Canadian Journal of Zoology* 88: 875–883.

McLellan, B. N. & Shackleton, D. M. (1989). Immediate reactions of grizzly bears to human activities. *Wildlife Society Bulletin* 17: 269–274.

Moen, G. K. (2018). Human-mediated effects on brown bear behavior and potential cascading effects. PhD thesis, Faculty of Environmental Sciences and Natural Resource Management, Norwegian University of Life Sciences, Ås, Norway.

Moen, G. K., Støen, O.-G., Sahlén, V. & Swenson, J. E. (2012). Behaviour of solitary adult Scandinavian brown bears (*Ursus arctos*) when approached by humans on foot. *PLoS ONE* 7: e31699.

Moen, G., Ordiz, A., Kindberg, J., et al. (2019). Behavioral reactions of brown bears to approaching humans in Fennoscandia. *Écoscience* 26: 23–33. DOI: 10.1080/11956860.2018.1513387

Mueller, C., Herrero, S. & Gibeau, M. L. (2004). Distribution of subadult grizzly bears in relation to human development in the Bow River Watershed, Alberta. *Ursus* 15: 35–47.

Mysterud, A. (2011). Selective harvesting of large mammals: how often does it result in directional selection? *Journal of Applied Ecology* 48: 827–834.

Nellemann, C., Støen, O.-G., Kindberg, J., et al. (2007). Terrain use by an expanding brown bear population in relation to age, recreational resorts and human settlements. *Biological Conservation* 138: 157–165.

Nielsen, S. E., Munro, R. H. M., Bainbridge, E. L., Stenhouse, G. B. & Boyce, M. S. (2004). Grizzly bears and forestry: II. Distribution of grizzly bear foods in clearcuts of west-central Alberta, Canada. *Forest Ecology and Management* 199: 67–82.

Olson, T. L., Squibb, R. C. & Gilbert, B. K. (1998). Brown bear diurnal activity and human use: a comparison of two salmon streams. *Ursus* 10: 547–555.

Ordiz, A., Støen, O.-G., Delibes, M. & Swenson, J. E. (2011). Predators or prey? Spatio-temporal discrimination of human-derived risk by brown bears. *Oecologia* 166: 59–67.

Ordiz, A., Støen, O.-G., Sæbø, S., et al. (2012). Do bears know they are being hunted? *Biological Conservation* 152: 21–28.

Ordiz, A., Støen, O.-G., Sæbø, S., et al. (2013). Lasting behavioural responses of brown bears to experimental encounters with humans. *Journal of Applied Ecology* 50: 306–314.

Ordiz, A., Kindberg, J., Sæbø, S., Swenson, J. E. & Støen, O.-G. (2014). Brown bear circadian behavior reveals human environmental encroachment. *Biological Conservation* 173: 1–9.

Ordiz, A., Støen, O.-G., Delibes, M. & Swenson, J. E. (2017). Staying cool or staying safe in a human-dominated landscape: which is more relevant for brown bears? *Oecologia* 185: 191–194.

Ordiz, A., Moen, G. K., Sæbo, S., et al. (2019). Habituation, sensitization, or consistent behavioral responses? Brown bear responses after repeated approaches by humans on foot. *Biological Conservation* 232: 228–237.

Penteriani, V., Delgado, M. del M., Pinchera, F., et al. (2016). Human behaviour can trigger large carnivore attacks in developed countries. *Scientific Reports* 6: 20552.

Penteriani, V., López-Bao, J. V., Bettega, C., et al. (2017). Consequences of brown bear viewing tourism: a review. *Biological Conservation* 206: 169–180.

Penteriani, V., Delgado, M. del M., Krofel, M., et al. (2018). Evolutionary and ecological traps for brown bears *Ursus arctos* in human-modified landscapes. *Mammal Review* 48: 180–193.

Petram, W., Knauer, F. & Kaczensky, P. (2004). Human influence on the choice of winter dens by European brown bears in Slovenia. *Biological Conservation* 119: 129–136.

Proaktor, G., Coulson, T. & Milner-Gulland, E. J. (2007). Evolutionary responses to harvesting in ungulates. *Journal of Animal Ecology* 76: 669–678.

Proctor, M., McLellan, B., Boulanger, J., et al. (2010). Ecological investigations of grizzly bears in Canada using DNA from hair, 1995–2005: a review of methods and progress. *Ursus* 21: 169–188.

Pulliainen, E. (1983). Behaviour of an expanding population of the brown bear (*Ursus arctos*) in northern Europe. *Zeitschrift für Säugetierkunde* 48: 290–297.

Rode, K. D., Farley, S. D. & Robbins, C. T. (2006). Behavioral responses of brown bears mediate nutritional effects of experimentally introduced tourism. *Biological Conservation* 133: 70–80.

Roth, H. U. (1983). Diel activity of a remnant population of European brown bears. *International Conference on Bear Research and Management* 5: 223–229.

Røskaft, E., Bjerke, T., Kaltenborn, B., Linnell, J. D. C. & Andersen, R. (2003). Patterns of self-reported fear towards large carnivores among the Norwegian public. *Evolution and Human Behavior* 24: 184–198.

Sahlén, E., Støen, O.-G. & Swenson, J. E. (2011). Brown bear den site concealment in relation to human activity in Sweden. *Ursus* 22(2): 152–158.

Sahlén, V., Ordiz, A., Swenson, J. E. & Støen, O.-G. (2015a). Behavioural differences between single Scandinavian brown bears (*Ursus arctos*) and females with dependent young when experimentally approached by humans. *PLoS ONE* 10: e0121576.

Sahlén V., Friebe, A., Sæbø, S., Swenson, J. E. & Støen, O.-G. (2015b). Den entry behavior in Scandinavian brown bears: implications for preventing human injuries. *Journal of Wildlife Management* 79(2): 274–287.

Selva, N., Teitelbaum, C. S., Sergiel, A., et al. (2017). Supplementary ungulate feeding affects movement behavior of brown bears. *Basic and Applied Ecology* 24: 68–76.

Sorensen, A. A., Stenhouse, G. B., Bourbonnais, M. L. & Nelson, T. A. (2015). Effects of habitat quality and anthropogenic disturbance on grizzly bear (*Ursus arctos horribilis*) home-range fidelity. *Canadian Journal of Zoology* 93: 857–865.

Steyaert, S. M. J. G., Zedrosser, A., Elfström, M., et al. (2016). Ecological implications from spatial patterns in human-caused brown bear mortality. *Wildlife Biology* 22: 144–152.

Stockan, J. & Robinson, E. J. H. (2016). *Wood ant ecology and conservation.* Cambridge, UK: Cambridge University Press.

Støen, O.-G., Ordiz, A., Evans, A. L., et al. (2015). Physiological evidence for a human-induced landscape of fear in brown bears (*Ursus arctos*). *Physiology and Behavior* 152: 244–248.

Støen O.-G., Ordiz, A., Sahlén, V., et al. (2018). Brown bear (*Ursus arctos*) attacks resulting in human casualties in Scandinavia 1977–2016; management implications and recommendations. *PLoS ONE* 13(5): e0196876. https://doi.org/10.1371/journal.pone.0196876.

Suring, L., Barber, K., Schwartz, C., et al. (1998). Analysis of cumulative effects on brown bears on the Kenai Peninsula, southcentral Alaska. *Ursus* 10: 107–117.

Swenson, J. E., Wabakken, P., Sandegren, F., et al. (1995). The near extinction and recovery of brown bears in Scandinavia in relation to the bear management policies of Norway and Sweden. *Wildlife Biology* 1: 11–25.

Swenson, J. E., Sandegren, F., Brunberg, S. & Wabakken, P. (1997). Winter den abandonment by brown bears *Ursus arctos*: causes and consequences. *Wildlife Biology* 3: 35–38.

Swenson, J. E., Sandegren, F., Söderberg, A., et al. (1999). Interactions between brown bears and humans in Scandinavia. *Biosphere Conservation* 2: 1–9.

Swenson, J. E., Gerstl, N., Dahle, B. & Zedrosser, A. (2000). *Action plan for the conservation of the brown bear in Europe (*Ursus arctos*).* Strasbourg Cedex: Council of Europe Nature and Environment.

Swenson, J. E., Schneider, M., Zedrosser, A., et al. (2017). Challenges of managing a European brown bear population; lessons from Sweden, 1943–2013. *Wildlife Biology* 2017(4): wlb.00251.

Sæther, B. E., Engen, S., Swenson, J. E., Bakke, Ø. & Sandegren, F. (1998). Assessing the viability of Scandinavian brown bear, *Ursus arctos*, populations: the effects of uncertain parameter estimates. *Oikos* 83: 403–416.

Thompson, R. F. (2009). Neurobiology of learning and memory habituation: a history. *Neurobiology of Learning and Memory* 92: 127–134.

Van de Walle, J., Pigeon, G., Zedrosser, A., Swenson, J. E. & Pelletier, F. (2018). Hunting regulation favors slow life histories in a large carnivore. *Nature Communications* 9: 1100.

Wheat, R. E. & Wilmers, C. C. (2016). Habituation reverses fear-based ecological effects in brown bears (*Ursus arctos*). *Ecosphere* 7: 1–11.

Woodroffe, R. (2000). Predators and people: using human densities to interpret declines of large carnivores. *Animal Conservation* 3: 165–173.

Woodroffe, R. & Ginsberg, J. R. (1998). Edge effects and the extinction of populations inside protected areas. *Science* 280: 2126–2128.

Zarzo-Arias, A., Delgado, M., Ordiz, A., et al. (2018). Brown bear behaviour in human-modified landscapes: the case of the endangered Cantabrian population, NW Spain. *Global Ecology and Conservation* 16: e00499.

Zedrosser, A., Steyaert, S. M. J. G., Gossow, H. & Swenson, J. E. (2011). Brown bear conservation and the ghost of persecution past. *Biological Conservation* 144: 2163–2170.

Chapter 19

Bears in Human-Modified Landscapes: The Case Studies of the Cantabrian, Apennine, and Pindos Mountains

Vincenzo Penteriani, Alexandros A. Karamanlidis, Andrés Ordiz, Paolo Ciucci, Luigi Boitani, Giorgio Bertorelle, Alejandra Zarzo-Arias, Giulia Bombieri, Enrique González-Bernardo, Paola Morini, Francesco Pinchera, Néstor Fernández, María C. Mateo-Sánchez, Eloy Revilla, Miguel de Gabriel Hernando, Yorgos Mertzanis, and Mario Melletti

Introduction

Brown bears *Ursus arctos* were historically persecuted and almost eradicated from Southern Europe in the twentieth century as a result of hunting and direct persecution (Zedrosser et al. 2011; Martínez Cano et al. 2016). The effects of human-induced mortality were exacerbated by other threats, such as habitat loss and fragmentation, due to the expansion of human populations (Swenson et al. 2000). As a result, nowadays there are only small, fragmented populations of bears in Southern Europe, where brown bears frequently inhabit human-modified landscapes. These areas are characterized by the widespread presence of people and infrastructures, which potentially have ecological impacts on bears. The close coexistence of brown bears and humans generates multiple human-driven disturbances (Ordiz et al. 2017) and causes bear mortality (Bischof et al. 2009), affecting the distribution, demography, behavior, and viability of bear populations (Penteriani et al. 2018a; Zarzo-Arias et al. 2018).

Brown bears in the Cantabrian (north-western Spain), Apennine (central Italy), and Pindos (north-western Greece) mountains (Figure 19.1) represent three examples of small and threatened bear populations in human-modified landscapes (Figure 19.2). Most of their range is characterized by high human densities, widespread agricultural activities, livestock raising, and urban development, connected by dense networks of transport infrastructures (Penteriani et al. 2020; Mateo-Sánchez et al. 2016). This has resulted in a reduction of continuous habitat suitable for the species (Martínez Cano et al. 2016). Here, we summarize the past and present histories and fates of these three populations as examples of how the coexistence of bears and people in human-modified landscapes can take different turns depending on human attitudes.

Cantabrian Bears, the Partial and Ongoing Recovery of a Population

Brown bears in the Cantabrian Mountains represent the westernmost population of the species' range in Europe. Of the four isolated brown bear populations in Western Europe – i.e. Italian Central Apennines and Eastern Alps, Pyrenees (southern France and north-eastern Spain), and Cantabrian – the latter represents the only population exhibiting a natural recovery, with no reintroduction of bears from other populations. The decline of this population, which was already pronounced in the 1960s and 1970s, continued even after its protection in 1973, when population assessments estimated the presence of <100 bears separated in a western and an eastern subpopulation. Indeed, these two subpopulations were already disconnected at the beginning of the twentieth century by a 50–100 km wide strip of land, with bears in the two subpopulations declining in number until the mid-1990s (Palomero et al. 2007; Gonzalez et al. 2016). In the period 1982–1995, the western population was estimated at 50–60 bears and showed an annual decrease of 4–5% (Wiegand et al. 1998), while the eastern subpopulation was estimated in 1990 at only 20–25 bears (Palomero et al. 1993). Although concerns over an imminent extinction were expressed in the mid-1990s (Wiegand et al. 1998; Martínez Cano et al. 2016), and despite negative prospects (Wiegand et al. 1998; Naves et al. 2003a), the Cantabrian brown bear population has increased, mainly in the western part of the Cantabrian Mountains (Palomero et al. 2007; Pérez et al. 2010). It has been estimated that during the first decade of this century the western subpopulation has increased annually by 7.7% (Palomero et al. 2007).

Today, brown bears in the Cantabrian Mountains continue to show a positive trend, there is sign of connection between the two subpopulations, and reproduction has been stable during the last 30 years, at least (Pérez et al. 2010; Gonzalez et al. 2016; Penteriani et al. 2018c). A capture–mark–recapture population estimate using genetic data carried out in 2006 found that the western part of the population represented approximately 90% of the entire population, with 203 bears (confidence interval CI 95% = 168–260) in the west and only 19 bears (CI 95% = 12–40) in the east (Pérez et al. 2014). Despite the positive trends in the west, the eastern subpopulation is showing a substantially smaller increase (Martínez Cano et al. 2016).

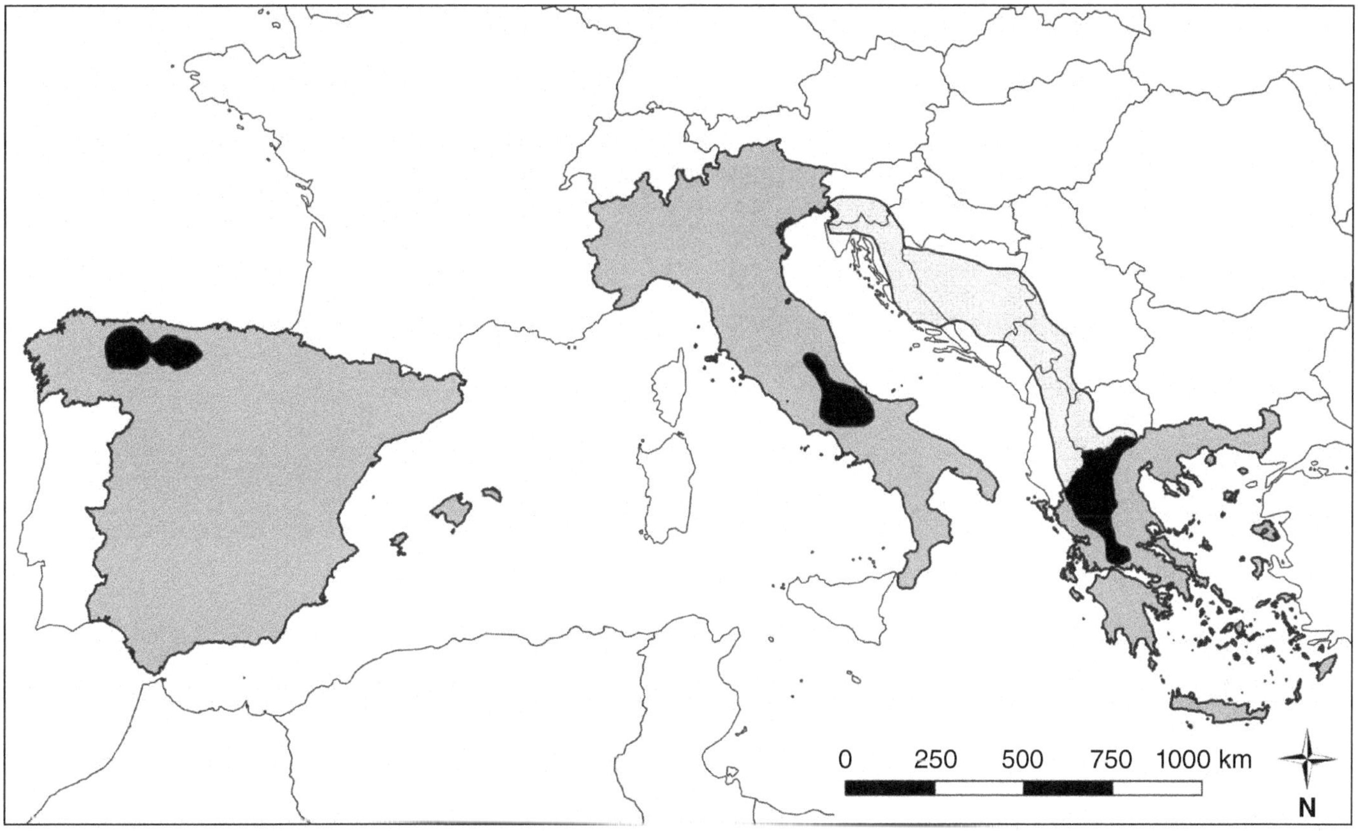

Figure 19.1 The distribution (black areas) of the Cantabrian (north-west Spain), Apennines (central Italy), and Pindos (north-west Greece) brown bear populations. The Pindos population is part of the bigger Dinaric-Pindos population (shaded in light gray) outside Greece.

The number of females with cubs, a proxy of the population trend, has also exhibited a more rapid increase in the western subpopulation than in the eastern one, e.g. 34 vs. 6 females with cubs in 2016 and an average of 25 vs. 4.3 females with cubs during the 10-year period 2007–2016 (Figure 19.3A). The Cantabrian brown bear population shows a relatively high reproductive rate (Penteriani et al. 2018c): (a) mean litter size is significantly larger in the west (1.8 ± 0.2 cubs) than in the east (1.3 ± 0.6 cubs), with litter size for the whole Cantabrian Mountains being 1.6 ± 0.3 cubs; (b) interannual variation in litter size is not significant for the western and the eastern sectors, although in the east interannual differences are more marked (Figure 19.3B); (c) apparent mean cub mortality per female per year is 0.1 ± 0.5 cubs; and (d) estimated reproductive rate for the population is 0.7 young born per year per reproductive adult female. The estimated reproductive rate for the Cantabrian population is among the highest recorded for brown bears (from 0.23 to 0.96; reviewed by Steyaert et al. 2012), which might be due to the high habitat quality of the Cantabrian Mountains (Martínez Cano et al. 2016). It is worth noting here the similarity in reproductive traits between the Cantabrian and the Apennine brown bear populations (Tosoni et al. 2017a). Indeed, both the mean litter size (1.9 cubs in the Italian population and 1.8 in the western Cantabrian Mountains) and litter size proportions for one, two, and three cubs were similar: 33%, 56%, and 11% for the Cantabrian brown bears, and 26%, 55%, and 19% for the Apennine bears, respectively. However, despite similar reproductive parameters, the trends of these two populations have been very different over the last three decades (see The Apennine bears, a demographic stagnation lasting several decades).

In the Cantabrian Mountains, brown bear mortality associated with vehicle accidents is rare and illegal poaching, the most common driver of human-induced bear mortality (Palomero et al. 2007), may have declined in recent decades, resulting in the positive trends exhibited by this population (Gonzalez et al. 2016). A series of management actions beginning in the early 1970s, which has intensified since 1989 with the initiation of several recovery plans, may partly explain the recovery of this population (Martínez Cano et al. 2016). The initiatives that have probably contributed most to the current positive trends of this bear population are: (1) increased monitoring and effective persecution of bear poaching; (2) establishment and enforcement of protected areas; and (3) establishment of compensation programs for bear damages.

However, apart from the varying reproductive rates, the contrast in population growth between the western and eastern subpopulations might suggest: (a) higher mortality rates in the eastern subpopulation; (b) lower population growth rates in the eastern subpopulation due to the smaller number of adult

Figure 19.2 Brown bears in human-modified landscapes: three examples of bear habitats with intense anthropogenic influence in the Cantabrian Mountains (A; photo by V. Penteriani), the Apennines (B; photo by P. Ciucci), and the Pindos (C; photo by A. Karamanlidis), as well as a brown bear in a mown wheat field in the Pindos (D; photo by A. Karamanlidis).

females compared to the western one; and (c) limited bear dispersal from the larger, western subpopulation. In fact, some studies have suggested a narrow connection between the two subpopulations, which seems to be maintained nowadays in spite of the development of transportation infrastructures between the two areas (García et al. 2007; Pérez et al. 2010; San Miguel et al. 2012; Mateo-Sánchez et al. 2014b). However, even if female philopatry has been highlighted as one of the potential causes of the slow recovery of the eastern sector, with little or no flow of females moving from west to east (Palomero et al. 2007), this may not solely explain the apparent lower reproductive rate or higher mortality in this subpopulation. During the last 40 years, the eastern subpopulation has always been 1/3–1/4 smaller than the western one (Palomero et al. 2007; FAPAS/FIEP 2017). If habitat use by the two brown bear subpopulations of the Cantabrian Mountains offers only limited help in explaining the evolutionary differences in population size and fecundity exhibited by the two subpopulations (Lamamy et al. 2019), other factors might be acting on the different ways in which these subpopulations have evolved over the last few decades. Indeed, the higher human presence and more fragmented forest habitat in the west (Lamamy et al. 2019) is not consistent with the higher growth rates of this subpopulation. Thus, human practices and attitudes toward bears might be undermining bear recovery more than the natural landscape, and suspicions point to direct human influence on the different trends exhibited by the two bear subpopulations. For instance, the persistence of poaching and/or bad practices during hunting might be more pronounced in the eastern sector of the Cantabrian Mountains, resulting in higher bear mortality. Less surveillance of eastern areas together with less-effective antipoaching measures might be the main causes of the low growth of the eastern subpopulation. Earlier studies have also associated the slower recovery of the eastern

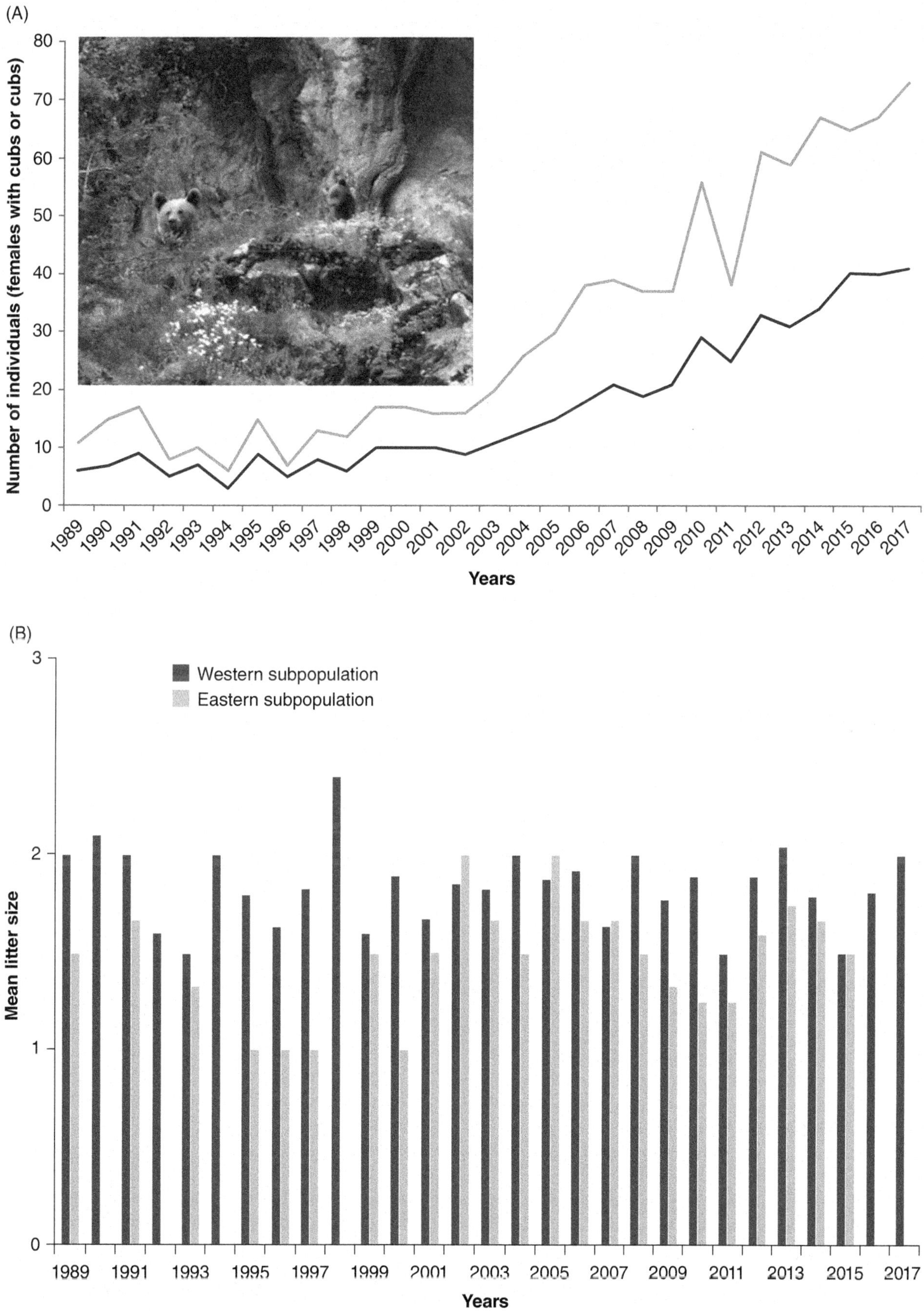

Figure 19.3 Characteristics of the brown bear population inhabiting the Cantabrian Mountains. (A) Trends of the number of females with cubs (black line) and number of cubs produced per year (gray line) from 1989 to 2017. (B) Interannual variations in brown bear mean litter size (number of cubs) for the western (1989–2017) and eastern (1989–2015) subpopulations.

subpopulation with the existence of low-quality food items in the eastern sector of the Cantabrian Mountains (Clevenger et al. 1992; Naves et al. 2003a, 2006).

Brown bear habitat suitability in the Cantabrian Mountains is determined by multiple factors at a range of scales, from local resources such as quality forest cover and food resource availability within the home range to the avoidance of human disturbance at broader spatial scales (Mateo-Sánchez et al. 2014a, 2016). To ensure the long-term viability of the species, it is critical to not only ensure the conservation of their habitat areas, but also maintain an adequate level of connectivity among them to allow the movement of individuals and genes between both subpopulations and toward new available areas of suitable habitat. Landscape features have important influences on bear movements, with areas with high forest canopy cover and shrubland cover promoting successful movements, while, on the contrary, intensively human-modified cover types present considerable resistance to species movements. Dispersing bears are more flexible in their movement behavior and less constrained by landscape conditions than when selecting areas as stable habitat, achieving long-distance movements through low-quality habitat areas even with a high human footprint (Mateo-Sánchez et al. 2015a, 2015b). This dispersal plasticity, together with the notable population growth and the implemented actions to improve landscape matrix permeability (e.g. LIFE+ programs), seem to have led to a certain level of recovery of interpopulation connectivity after a considerable period of isolation (Pérez et al. 2010, 2014). In fact, current evidence seems to show that dispersal movements between subpopulations are increasing (Pérez et al. 2010, 2014) and their frequency may lead to a sufficient amount of gene flow between populations to mitigate potential deleterious effects of isolation and inbreeding (Mills & Allendorf 1996; Vucetich & Waite 2000). Therefore, facilitating adequate levels of dispersal between the two populations should be a priority in connectivity strategies. Thus, one of the current main challenges is to foster a steady expansion of the species range into available areas that were part of the historic distribution of brown bears in the Cantabrian range and neighboring areas.

Finally, if the current positive trends continue in the Cantabrian Mountains, a large amount of habitat still seems available for range expansion of the species (Martin et al. 2012; Scharf & Fernández 2018). For example, brown bears currently occupy about 2400 km^2 in Asturias, a region at the core of the Cantabrian Mountains, while about 5000 km^2 (approximately half of the region) have been identified as suitable bear habitat (Zarzo-Arias et al. 2019). Most of the suitable areas in the western part of the region are already occupied (77% of identified areas), 41.4% of them occurring inside protected areas. Therefore, there are relatively few good areas for further bear expansion in this part of the region, although there are suitable areas in surrounding regions. However, in the eastern part of Asturias, 62% of the land is classified as suitable habitat, representing 44.3% of the total area identified as suitable bear range in the whole region. Forty-one percent of all expansion areas (suitable areas outside the current bear range) are within protected areas (Zarzo-Arias et al. 2019) in Asturias.

The Apennine Bears, a Demographic Stagnation Lasting Several Decades

Brown bears *Ursus arctos marsicanus* in the Apennines survive in a relict and isolated population of about 50 bears, including cubs (Gervasi et al. 2012; Ciucci et al. 2015), critically endangered both at the European and the national level (Ciucci et al. 2017). In fact, the risk of extinction in the next 100 years has been estimated to be between 11% and 21% (Gervasi & Ciucci 2018). Although the taxonomic status of this bear population is debated (Ciucci & Boitani 2008; Colangelo et al. 2012), their long-term isolation from other brown bear populations makes Apennine bears a unique evolutionary and conservation unit, based on genetic, morphological, and behavioral traits (Loy et al. 2008; Ciucci & Boitani 2008; Colangelo et al. 2012; Benazzo et al. 2017). Genomic data suggest that this population, originally spread throughout most of the Apennine range, lost connectivity with other brown bear populations many years ago, when forest clearing was very intense (Benazzo et al. 2017). More recently, the distribution of Apennine bears declined progressively in the seventeenth century, but most of their range contraction likely took place over the past 200 years, mainly due to human persecution and changes in land-use practices. The first effort to evaluate the census size of this population was undertaken in 1970, when 70–100 bears were estimated (possibly overestimated) within an area of about 500 km^2. In the last few decades, further estimates produced numbers between 40 and 80 individuals, a size close to the current one (Ciucci & Boitani 2008). Nowadays, the population is structured in a core and a peripheral range, the former comprising the historical Abruzzo Lazio and Molise National Park (PNALM) and surrounding areas, and the latter largely including a limited number of wandering bears, mostly males (Ciucci & Boitani 2008). Recently, the presence of bears in the peripheral areas, including females with cubs (Ciucci et al. 2017; Morini et al. 2017), has been increasingly reported (Van Gils et al. 2014; Morini et al. 2017), although it is still uncertain if this reflects an actual tendency toward range expansion or if this is the result of increased monitoring efforts (Ciucci et al. 2017). Nevertheless, these peripheral areas of bear presence, both inside and outside protected areas, are crucial for species conservation and management (Falcucci et al. 2009; Ciucci et al. 2017; Morini et al. 2017) and indicate that some connectivity with the core area exists, mainly toward the north-west, east, and north-east. Despite long-term protection and the availability of suitable habitat at the landscape scale (Posillico et al. 2004; Falcucci et al. 2009), the distribution of the Apennine population has not increased during the last several decades, nor have reproductive nuclei been stably established outside the core area (Ciucci et al. 2017).

Figures of minimum known mortality indicate that protection of Apennine bears has not been as effective as desirable for such a small bear population (reviewed by Ciucci & Boitani 2008). For example, between the establishment of the PNALM in 1923 and 1974, at least 99 bears were killed, with a higher rate of known mortality (2.4 bears/year) during the final decade of this period. Higher rates (3 bears killed/year) were reported in the following decade, when poaching and vehicle collisions accounted for 60% of known bear mortalities. However, a peak in bear mortality was recorded between 1980 and 1985, when a minimum of 32 bears were killed (5.3 bears/year), probably due to an increase in poaching. During the period 1991–2002, known bear mortality averaged 2.5 bears killed/year (with half of killed bears being female), whereas in more recent years the minimum known mortality totalled five bears (2003–2007). Three of these were killed in a single poisoning event in the core of the PNALM, most likely motivated by conflicts over livestock depredation. In addition, 16.4% of dead bears were found outside the core area (PNALM 2018), thus further hindering any attempts by young bears to expand the range of the population. Of all reported bear mortalities, 84% have been from illegal or accidental killing by humans, with poaching (mainly shooting and poisoning) being one of the primary causes of reported bear fatalities. Other causes of mortality were unintentional, but still related to the bears living in a human-modified landscape, i.e. vehicle and train accidents, mortality from snares or poison baits illegally set, and diseases (e.g. brucellosis, tuberculosis) potentially transmitted by pets and livestock. Diseases or reduced reproduction could represent an underestimated cause of mortality for Apennine bears (Ciucci & Boitani 2008); some bears have recently tested positive for canine parvovirus and canine distemper, and they can also contract *Brucella*, which can depress reproduction. In 2014 an adult female bear died of tuberculosis likely transmitted by cattle (L. Gentile, PNALM Veterinary Service, personal communication).

Currently, the bear range is estimated at 5422 km^2 across the central Apennines, mostly concentrated in a southern area (4923 km^2, 90.8% of the estimated distribution) that largely overlaps with the stronghold of the species over the last few decades (Ciucci et al. 2017). With a few exceptions (see above), reproducing female bears (1–6 family groups per year from 2005 to 2014; Tosoni et al. 2017b) occupy an area of 1460 km^2, completely encompassed within the main bear distribution. Almost 60% of the current bear range is included within protected areas. Mean litter size is 1.9 ± 0.7 (range = 1–3) cubs and the mean interval between litters is 3.7 years, with the age of first reproduction between 5 and 6 years, corresponding to a reproductive rate of 0.24 (95% CI = 0.07–0.59) (Tosoni et al. 2017a). The combination of a relatively long interbirth interval and several non-reproducing adult female bears may contribute to the relatively low reproductive rate of Apennine bears as well as partly explain the slow recovery of this population, which may also be characterized by relatively high cub mortality (Tosoni et al. 2017a; Gervasi et al. 2017). Although no clear demographic evidence of inbreeding depression has been obtained thus far, genomes are characterized by long stretches of homozygosity, and in general the genetic diversity of this population is also very low for quantitative traits (Lorenzini et al. 2004; Benazzo et al. 2017). Estimated levels of inbreeding are high, as are the number of fixed mutations predicted as deleterious that accumulated through drift during the long period of isolation of this small population (Benazzo et al. 2017). Therefore, in addition to the high levels of human-related mortality, it cannot be excluded that the relatively low reproductive performance and cub survival in Apennine bears may be due in part to genetic factors. It is interesting to note that genetic variation is retained at immune system and olfactory receptor genes, i.e. genes crucial for interacting with the environment (Benazzo et al. 2017).

Thus, Apennine bears are still highly endangered, i.e. they are a small, isolated population with limited genetic diversity and a high genetic load, with a restricted distribution of reproducing females, despite: (a) a female productivity in the core population that seems compatible with population growth (although in the lower spectrum of the reproductive performance of other non-hunted brown bear populations in Europe; Tosoni et al. 2017b; Gervasi & Ciucci 2018); (b) a relatively high density in the core range (about 40 bears/1000 km^2; 95% CI = 37–51), which, along with dietary analyses, indicate that habitat productivity is currently adequate for bears (Ciucci et al. 2014, 2015); and (c) habitat suitability and connectivity that do not seem limiting at the landscape scale (Falcucci et al. 2008, 2009; Maiorano et al. 2017). It is remarkable that, although most of the bear range falls within protected areas and productivity is high, this bear population has not grown in the last few decades, as has been the case for the Cantabrian and Pindos bears. Although bears in the Apennines have been legally protected since 1939 (well before their counterparts in Spain or Greece), they have not been efficiently supported in the last several decades by increased surveillance or protected from poachers. Furthermore, although there are several protected areas in the Central Apennines that largely overlap with suitable bear habitat, the landscape appears interspersed with ecological traps, i.e. areas where habitat suitability is high, but intimately associated with a high risk of human-caused mortality (Falcucci et al. 2009; Penteriani et al. 2018a). Therefore, as in the case for the eastern Cantabrian subpopulation, we suggest that the lack of population recovery in the Apennines is largely due to human-related mortality. In a densely populated country such as Italy, brown bear damage to human interests such as livestock or crops can be substantial and yet difficult to mitigate due to cultural and social resistance, especially outside the historical stronghold of the species, where attitudes are generally more positive and tolerance toward bears is higher. Genetic factors may play a role, especially in the near future, but their negative effects on survival and fecundity are probably still limited and require further investigation.

Even if habitat degradation is critical for the Apennine population, it is perhaps secondary to human-caused mortality (Ciucci & Boitani 2008; Falcucci et al. 2009). Antipoaching campaigns were not effective and, as a result, total mortality might have been too high to allow population recovery given the current reproductive performance. Indeed, the recent dynamics in bear range and the number of bears do not offer any clear indication of population recovery, suggesting that the Apennine brown bear may have persisted at dangerously small numbers over the last four decades. Despite protection, the current status of the Apennine brown bear population highlights how, at small bear population size, effective long-term conservation can be problematic. Across the central Apennines, including protected areas, even if human density is relatively low when compared with the rest of Italy, livestock grazing, timber harvesting, hunting (outside protected areas only), vehicle traffic, tourism, and other recreational activities are still not regulated in order to reduce their potential impact on bears (Ciucci & Boitani 2008; Falcucci et al. 2009). These additional factors have the potential to greatly influence bear occurrence, movements and feeding activity, reproduction and the survival of individuals (Ciucci et al. 2014). Consequently, the spatial scale of species conservation and management should be consistent with the scale at which human pressure and threats occur, a process which is not aided by the fragmentation of administrations in charge of bear management and conservation.

Understanding the Recovery of Brown Bears Living Next to Humans: the Case of the Pindos Mountains

In ancient times, and up until the sixteenth century, the species was widespread throughout mainland Greece, including the Peloponnese Peninsula. Following this, it appears that the brown bear range decreased dramatically from the eighteenth century onward, mainly due to the systematic persecution of the species and habitat alteration (Mertzanis 1999 and references therein). The Pindos bear population is attached to the larger Dinaric-Pindos biological population covering nine countries of the western Balkan region (Kaczensky et al. 2013).

During the period 1985–2007 the Greek brown bear population appears to have shown (especially in Pindos) small but progressive increases in population and range sizes. It is estimated that, in the mid-1990s, the total occupied range covered 13,500 km^2 and consisted of two non-communicating population nuclei in the north-western (a.k.a. Pindos population) and north-eastern (a.k.a. Rodopi population) parts of the country with a minimum population size of 130 individuals (Mertzanis 1994, 1999, 2000). Ten years later a recovery of the species was recognized, with the total population estimated to number up to 260 individuals over a total range of 21,300 km^2 (Mertzanis et al. 2009). Acknowledging the critical status of the brown bear in Greece, the species was considered "endangered" in the Red Data Book of Greece and was afforded strict protection by national and international legislation (Mertzanis et al. 2009). To alleviate human–bear conflicts, a compensation system for damages to livestock, apiaries, and crops was also established (Karamanlidis et al. 2011).

Since the beginning of this century, however, brown bears in Greece have been showing clear signs of range, genetic, and demographic recovery (Karamanlidis et al. 2015, 2018; Bonnet Lebrun et al. 2020), especially in the Pindos part of their range. These recoveries have been partly facilitated by specific adaptations by bears in their activity rhythms (De Gabriel Hernando et al. submitted a) and habitat selection patterns (De Gabriel Hernando et al. submitted a, submitted b), together with the rewilding of areas after rural abandonment by humans.

Range Recovery

A comparison between data on bear presence collected from 2004 to 2016 (within the framework of a citizen science program) and the distribution of brown bears in Pindos (Mertzanis et al. 2009) indicates that in the last ~15 years the range of the brown bear in the Pindos Mountains might have increased by as much as 100% (Bonnet Lebrun et al. 2020). This is in accordance with opportunistic, extralimital sightings that have occurred during this time (Mertzanis et al. 2005; Karamanlidis et al. 2008). The current distribution of brown bears in Greece includes recolonized former range areas and is characterized by (a) bears at lower elevations (415 m lower), (b) higher coverage by primary (>175% more) and secondary roads (100% more) as well as agricultural areas (22% more) in bear habitat, and (c) lower coverage by mature forests (25% less) in areas inhabited by bears, compared to that of the past, indicating a recolonization of primarily cultural landscapes. The ability of brown bears to recolonize cultural landscapes in Greece is most likely explained by the behavioral plasticity of the species (Ordiz et al. 2014) and the decrease in agricultural activities in less-productive areas as a consequence of the general rural abandonment, allowing a progressive naturalization of these areas (Mertzanis et al. 2009). Extensive habitat suitability analyses (Mertzanis et al. 2008, 2011) indicate that only a certain percentage of suitable bear habitats in Pindos is currently under protection within the framework of the Natura 2000 network of protected areas (Bonnet Lebrun et al. 2020) and from that, only 33% of the Natura 2000 areas implement specific management actions for the conservation of the species. However, according to very recent developments, the Hellenic Ministry of Environment and Energy has decided to affiliate all Natura 2000 sites into 37 enlarged protected areas and to conduct specific environmental studies with subsequent conservation benefits for the species. Nowadays, bear observations are often recorded in areas with highly suitable habitat currently not protected at all, thus creating a new "conservation reality" for the species in the country, which dictates a re-evaluation of the national management and conservation priorities for brown bears in Greece (Bonnet Lebrun et al. 2020).

Genetic Recovery

In 2002 a large-scale, nation-wide effort was initiated to understand the genetic processes associated with brown bear recovery in Greece (Karamanlidis et al. 2018), through the evaluation of population structure, genetic diversity, and gene flow. The results of the study indicated a clear distinction between the bear populations in the Pindos and Rodopi mountain ranges. Bears in the Rodopi Mountains, which belong to the large East Balkan population (Frosch et al. 2014), are separated from bears in Pindos by a great distance and partially unsuitable habitat, and clearly stand out as a separate population in population structure analyses. Mitochondrial DNA and microsatellite analysis indicated a further genetic distinction of the Pindos population, separating the bear population into four core areas, i.e. the Mountains of Vitsi-Varnoundas, Northern Pindos, and Central/Southern Pindos. This distinction accords with the hypothesis that isolation at the edge of a species' range favors genetic substructuring (Schaal & Leverich 1996), with this structure likely appearing when bear populations in Pindos survived in two small, isolated fragments.

The legacy of the demographic history of bears in Greece is also discernible in their genetic diversity. As is the case for neighboring populations in Bulgaria (Frosch et al. 2014), bears in Rodopi have high genetic diversity, comparable to that found in the large population in the Carpathian Mountains (Graban et al. 2013). On the other hand, the genetic diversity of bears in western Greece is the lowest recorded in the Dinaric-Pindos population (Skrbinšek et al. 2012), but higher than that recorded in the Apennine bear population (Ciucci & Boitani 2008). Finally, information on population genetics indicates an asymmetric, primarily male-mediated, gene flow among subpopulations in Pindos, which is indicative of a population recovery process through reconnection of previously isolated population fragments (Karamanlidis et al. 2018).

Demographic Recovery

The four core areas identified in the genetic study served as the main sampling areas for studying the demographic recovery of brown bears in Pindos between 2002 and 2010, using a combination of non-invasive and invasive genetic sampling (Karamanlidis et al. 2007, 2010; Tsaparis et al. 2014). Sixty-eight (Vitsi-Varnoundas; density: 54 bears/1000 km^2), 53 (Northern Pindos; 50 bears/1000 km^2), 51 (Central Pindos; 51 bears/1000 km^2), and 10 (Southern Pindos; 10 bears/1000 km^2) brown bears were estimated on average over the study period, which amounts to 182 individuals per year for all sampling areas combined and a conservative estimate of >400 bears for the entire Pindos population (Karamanlidis et al. 2015). This is indicative of a significant population recovery of the bear population in Pindos and is consistent with large carnivore population recoveries throughout Europe (Chapron et al. 2014). Bear densities obtained in the four sampling areas in Pindos were the lowest recorded for the species in the Dinara-Pindos population (Huber et al. 2008; Jerina et al. 2013) and are consistent with the assumption of a Dinara-Pindos population with a core population in the North and extending towards the South (Kaczensky et al. 2013).

Behavioral Adaptations

The range, genetic, and demographic recovery of bears in Pindos has also likely been facilitated by specific behavioral adaptations that enable bears to survive in human-modified landscapes. As it occurred during the recovery of brown bears in Scandinavia (Kindberg et al. 2011), bears in Pindos have shown high levels of behavioral plasticity as a response to the rapidly changing and ecologically challenging environment they inhabit. For example, the activity of adult bears is mainly nocturnal, enabling them to avoid human disturbance (De Gabriel Hernando et al. 2020). Similarly, bear habitat preferences revealed efforts to avoid human disturbance. Also, habitat use had a clear circadian pattern: in general, proximity to human-related habitat features (e.g. intensive crops, naturalized crops, human settlements, road networks, and unpaved roads) was higher at night, whereas proximity to natural habitat features, such as forests, shrublands, high altitudes, and areas with rough terrain was higher during the day for all sex/age classes (De Gabriel Hernando et al. in review).

Despite the population recovery exhibited by the brown bear population in Pindos, the fate of the species has not yet been totally secured. Several threats still compromise the recovery of bears despite legal and institutional protection. These threats include human-caused mortality, e.g. poaching (Mertzanis 2000; Mertzanis et al. 2009), vehicle collisions (Karamanlidis & Mertzanis 2004), retaliatory killing for damage to property (Karamanlidis et al. 2011), and habitat loss, fragmentation, and alteration at a range scale, e.g. construction of major highways (Karamanlidis et al. 2012a, 2012b).

What Can We Learn from Past and Present Histories of Small and Endangered Bear Populations in Human-Modified Landscapes?

The diverse situation, from ongoing recovery to apparent stagnation, of some brown bear populations and subpopulations in human-modified landscapes demands an improvement of our knowledge on how this species is able to coexist with humans (Carter & Linnell 2016) and, in turn, what effects human presence and activity may potentially have on them. This is particularly important in areas with recovering but still endangered brown bear populations (Treves & Karanth 2003; Ordiz et al. 2013), which inhabit areas with suitable habitat surrounded by greater human encroachment, thus creating scenarios that may become ecological traps (Naves et al. 2003b; Penteriani et al. 2018a; Scharf & Fernández 2018).

The case of the Cantabrian bear population is particularly noteworthy from a conservation perspective as human-induced mortality appears to have decreased, the potential for population recovery seems quite high, and the future of the species in the area is promising. However, this population is small and still highly vulnerable to the potential impacts of varying human activities and behaviors, as reflected by the slower demographic response of the eastern subpopulation, compared to that of the western one. A direct comparison of the genomes of Cantabrian and Apennine bears, with the latter apparently less capable of demographic growth despite similar levels of protection, could possibly clarify the role of genetic load.

On the basis of the population trajectories and current status of the three European populations presented in this study, we believe that it is necessary to include more individual-based research (e.g. telemetry in the Cantabrian Mountains) in order to extend the available information essential for effective conservation and management, such as bear rhythms of activity, dispersal, home range behavior, and high-resolution space use. Additionally, more monitoring and conservation efforts should be accorded to dispersing bears (Maiorano et al. 2017), especially in the case of females or family groups detected outside the core range. Indeed, the identification, management, and conservation of critical corridors that may allow range expansion and connectivity among core habitats should be particularly focused on the biologically most important segments of bear populations, that is, females and juveniles (Balbontín et al. 2005; Maiorano et al. 2017; Morini et al. 2017).

The future of each one of these three small populations will also depend on the cooperation and management of local administrations, irrespective of which administration has primary jurisdiction at a local scale. Specific regulations and agency responsibilities may change, but bears will require trans-regional coordination in conservation and management policies (Penteriani et al. 2018b). Knowing the areas into which these populations are likely to expand would allow authorities and conservation organizations to focus information campaigns and pre-emptive/proactive damage control actions in these areas. Such proactive approaches are important for successful large carnivore conservation and management (Ripple & Beschta 2012). In addition, as brown bears disappeared from certain areas decades ago, local communities are no longer familiar with them. Thus, local information campaigns directed at both residents and other users, e.g. hunters and tourists, of areas of potential bear expansion represent a crucial strategy to reduce human–bear conflicts and promote coexistence.

Finally, it is worthwhile to note here that, especially in the case of small, isolated, and/or endangered animal populations, the effects of climate change on trophic resources may considerably override conservation and management efforts performed at other levels, e.g. reduction of human–wildlife conflicts, threat of anthropogenic footprints and activities, poaching, and habitat fragmentation. Thus, conservation plans for species at higher trophic levels, such as brown bears, should take into account climate change vulnerability assessments of those plant communities that represent primary food resources and shelter for the target species (Penteriani et al. 2019). Thus, together with conservation actions aimed at maintaining bears in their historical and current ranges, we encourage conservation and management practices targeted at those areas potentially favorable to habitation by bears during the recovery process, taking into account the current context of climate change.

Acknowledgments

We thank the Administrations of the Gobierno del Principado de Asturias and the Junta de Castilla y León for providing the bear database. V.P. and A.O. were financially supported by the Excellence Project CGL2017-82782-P financed by the Spanish Ministry of Science, Innovation and Universities, the Agencia Estatal de Investigación (AEI), and the Fondo Europeo de Desarrollo Regional (FEDER, EU). E.R. was supported by project CGL2017-83045-R AEI/FEDER-UE from the Spanish Agencia Estatal de Investigación, Ministerio de Ciencia, Innovación y Universidades.

References

Balbontín, J., Penteriani, V. & Ferrer, M. (2005). Humans act against the natural process of breeder selection: a modern sickness for animal populations? *Biodiversity and Conservation* 14: 179–186.

Benazzo, A., Trucchi, E., Cahill, J. A., et al. (2017). Survival and divergence in a small group: the extraordinary genomic history of the endangered Apennine brown bear stragglers. *Proceedings of the National Academy of Sciences* 114: E9589–E0597.

Bischof, R., Swenson, J. E., Yoccoz, N. G., Mysterud, A. & Gimenez, O. (2009). The magnitude and selectivity of natural and multiple anthropogenic mortality causes in hunted brown bears. *Journal of Animal Ecology* 78: 656–665.

Bonnet Lebrun, A.-S., Karamanlidis, A. A., De Gabriel Hernando, M., Renner, I. & Gimenez, O. (2020). Identifying priority conservation areas for recovering large carnivores using citizen science data. *Animal Conservation* 23: 83–93.

Carter, N. H. & Linnell, J. D. C. (2016). Co-adaptation is key to coexisting with large carnivores. *Trends in Ecology and Evolution* 31: 575–578.

Chapron, G., Kaczensky, P., Linnell, J. D. C., et al. (2014). Recovery of large carnivores in Europe's modern human-dominated landscapes. *Science* 346: 1517–1519.

Ciucci, P. & Boitani, L. (2008). The Apennine brown bear: a critical review of its status and conservation problems. *Ursus* 19: 130–145.

Ciucci, P., Tosoni, E., Di Domenico, G. Quattrociocchi, F. & Boitani, L. (2014). Seasonal and annual variation in the food habits of Apennine brown bears, central Italy. *Journal of Mammalogy* 95: 572–586.

Ciucci, P., Gervasi, V., Boitani, L., et al. (2015). Estimating abundance of the remnant Apennine brown bear population using multiple noninvasive genetic data sources. *Journal of Mammalogy* 96: 206–220.

Ciucci, P., Altea, T., Antonucci, A., et al. (2017). Distribution of the brown bear (*Ursus arctos marsicanus*) in the central Apennines, Italy, 2005–2014. *Hystrix* 28: 1–6.

Clevenger, A. P., Purroy, F. J. & Pelton, M. R. (1992). Food habits of brown bears (*Ursus arctos*) in the Cantabrian Mountains, Spain. *Journal of Mammalogy* 73: 415–421.

Colangelo, P., Loy, A., Huber, D., et al. (2012). Cranial distinctiveness in the Apennine brown bear: genetic drift effect or ecophenotypic adaptation? *Biological Journal of the Linnean Society* 107: 15–26.

Falcucci, A., Maiorano, L., Ciucci, P., Garton, E. O. & Boitani, L. (2008). Land-cover change and the future of the Apennine brown bear: a perspective from the past. *Journal of Mammalogy* 89: 1502–1511.

Falcucci, A., Ciucci, P., Maiorano, L., Gentile, L. & Boitani, L. (2009). Assessing habitat quality for conservation using an integrated occurrence-mortality model. *Journal of Applied Ecology* 46: 600–609.

FAPAS/FIEP. (2017). *El Oso. Pirineos y Cordillera Cantábrica*. Gijón, Spain: Gráficas Muñiz.

Frosch, C., Dutsov, A., Zlatanova, D., et al. (2014). Noninvasive genetic assessment of brown bear population structure in Bulgarian mountain regions. *Mammalian Biology* 79: 268–276.

De Gabriel Hernando, M., Karamanlidis, A. A., Grivas, K., et al. (2020). Reducing movement in Mediterranean landscapes: a case study of brown bears in Greece. *Journal of Zoology*. https://doi .org.10.1111/jzo.12768

De Gabriel Hernando, M., Karamanlidis, A. A., Grivas, K., et al. (in review). Habitat ecology informs conservation priorities of an endangered brown bear population in a human-dominated landscape. *Journal of Nature Conservation.*

García, P., Lastra, J., Marquínez, J. & Nores, C. (2007). Detailed model of shelter areas for the Cantabrian brown bear. *Ecological Informatics* 2: 297–307.

Gervasi, V. & Ciucci, P. (2018). Demographic projections of the Apennine brown bear population *Ursus arctos marsicanus* (Mammalia: Ursidae) under alternative management scenarios. *European Zoological Journal* 85: 243–253.

Gervasi, V., Ciucci, P., Boulanger, J., et al. (2008). A preliminary estimate of the Apennine brown bear population size based on hair-snag sampling and multiple data source mark–recapture Huggins models. *Ursus* 19: 105–121.

Gervasi, V., Ciucci, P., Boulanger, J., Randi, E. & Boitani, L. (2012). A multiple data source approach to improve abundance estimates of small populations: the brown bear in the Apennines, Italy. *Biological Conservation* 152: 10–20.

Gervasi, V., Boitani, L., Paetkau, D., et al. (2017). Estimating survival in the Apennine brown bear accounting for uncertainty in age classification. *Population Ecology* 59: 119–130.

Gonzalez, E. G., Blanco, J. C., Ballesteros, F., et al. (2016). Genetic and demographic recovery of an isolated population of brown bear *Ursus arctos* L., 1758. *PeerJ* 4: e1928.

Graban, J., Kisková, J., Pepich, P. & Rigg, R. (2013). Genetic analysis for geographic isolation comparison of brown bears living in the periphery of the Western Carpathians Mountains with bears living in other areas. *Open Journal of Genetics* 3: 174–182.

Huber, Đ., Jakšić, Z., Frković, A., et al. (2008). *Brown Bear Management Plan for the Republic of Croatia.* Zagreb, Croatia: Ministry of Regional Development, Forestry and Water Management, Directorate for Hunting Ministry of Culture, Directorate for the Protection of Nature.

Jerina, K., Jonozovič, M., Krofel, M. & Skrbinšek, T. (2013). Range and local population densities of brown bear *Ursus arctos* in Slovenia. *European Journal of Wildlife Research* 59: 1–9.

Kaczensky, P., Chapron, G., von Arx, M., et al. (2013). Status, management and distribution of large carnivores – bear, lynx, wolf and wolverine – in Europe. Report to the EU Commission.

Karamanlidis, A. A. & Mertzanis, G. (2004). Fatal bear–car collision. *International Bear News* 13: 24.

Karamanlidis, A. A., Youlatos, D., Sgardelis, S. & Scouras, Z. (2007). Using sign at power poles to document presence of bears in Greece. *Ursus* 18: 54–61.

Karamanlidis, A. A., Krambokoukis, L. & Kantiros, D. (2008). Challenges and problems arising from the range expansion of brown bears in Greece. *International Bear News* 17.

Karamanlidis, A. A., Drosopoulou, E., de Gabriel Hernando, M., et al. (2010). Non-invasive genetic studies of brown bears using power poles. *European Journal of Wildlife Research* 56: 693–702.

Karamanlidis, A. A., Sanopoulos, A., Georgiadis, L. & Zedrosser, A. (2011). Structural and economic aspects of human–bear conflicts in Greece. *Ursus* 22: 141–151.

Karamanlidis, A. A., Beecham, J., Bousbouras, D., et al. (2012a). Bear-vehicle collisions: a holistic approach to evaluating and mitigating the effects of this new conservation threat to brown bears in Greece. Presented at 21st International Conference on Bear Research and Management, New Delhi, India.

Karamanlidis, A. A., Straka, M., Drosopoulou, E., et al. (2012b). Genetic diversity, structure, and size of an endangered brown bear population threatened by highway construction in the Pindos Mountains, Greece. *European Journal of Wildlife Research* 58: 511–522.

Karamanlidis, A. A., de Gabriel Hernando, M., Krambokoukis, L. & Gimenez, O. (2015). Evidence of a large carnivore population recovery: counting bears in Greece. *Journal for Nature Conservation* 27: 10–17.

Karamanlidis, A. A., Skrbinšek, T., de Gabriel Hernando, M., et al. (2018). History-driven population structure and assymetric gene flow in a recovering large carnivore at the rear-edge of its European range. *Heredity* 120: 168–182.

Kindberg, J., Swenson, J. E., Ericsson, G. & Bellemain, E. (2011). Estimating population size and trends of the Swedish brown bear *Ursus arctos* population. *Wildlife Biology* 17: 114–123.

Lamamy, C., Bombieri, G., Zarzo-Arias, A., González-Bernardo, E. & Penteriani, V. (2019). Can different habitat composition and structure help explain the different trends of Cantabrian brown bear subpopulations? *Mammal Research* 64: 559–567.

Loy, A., Genovesi, P., Galfo, M., Jacobone, M. G. & Vigna Taglianti, A. (2008). Cranial morphometrics of the Apennine brown bear (*Ursus arctos marsicanus*) and preliminary notes on the relationships with other southern European populations. *Italian Journal of Zoology* 75: 67–75.

Maiorano, L., Boitani, L., Chiaverini, L. & Ciucci, P. (2017). Uncertainties in the identification of potential dispersal corridors: the importance of behaviour, sex, and algorithm. *Basic and Applied Ecology* 21: 66–75.

Martin, J., Revilla, E., Quenette, P. Y., et al. (2012). Brown bear habitat suitability in the Pyrenees: transferability across sites and linking scales to make the most of scarce data. *Journal of Applied Ecology* 49: 621–631.

Martínez Cano, I., González Taboada, F., Naves, J., Fernández-Gil, A. & Wiegand, T. (2016). Decline and recovery of a large carnivore: environmental change and long-term trends in an endangered brown bear population. *Proceedings of the Royal Society B* 283: 20161832.

Mateo-Sánchez, M. C., Cushman, S. A. & Saura, S. (2014a). Scale dependence in habitat selection: the case of the endangered brown bear (*Ursus arctos*) in the Cantabrian Range (NW Spain). *International Journal of Geographical Information Science* 28: 1531–1546.

Mateo-Sánchez, M. C., Cushman, S. A. & Saura, S. (2014b). Connecting endangered brown bear subpopulations in the Cantabrian Range (north-western Spain). *Animal Conservation* 17: 430–440.

Mateo-Sánchez, M. C., Balkenhol, N., Cushman, S., et al. (2015a). A comparative framework to infer landscape effects on population genetic structure: are habitat suitability models effective in explaining gene flow? *Landscape Ecology* 30: 1405–1420.

Mateo-Sánchez, M. C., Balkenhol, N., Cushman, S., et al. (2015b). Estimating effective landscape distances and movement corridors: comparison of habitat and genetic data. *Ecosphere* 6(4): 1–16.

Mateo-Sánchez, M. C., Gastón, A., Ciudad, C., et al. (2016). Seasonal and temporal changes in species use of the landscape: how do they impact the inferences from multi-scale habitat modeling? *Landscape Ecology* 31: 1261–1276.

Mertzanis, G. (1994). Brown bear in Greece: distribution, present status–ecology of a northern Pindus subpopulation. *International Conference on Bear Research and Management* 9: 187–197.

Mertzanis, G. (1999). Status and management of the brown bear in Greece. In: Servheen, C., Herrero, S. & Peyton, B. (Eds.), *Bears* (pp. 72–81). Gland: IUCN – The World Conservation Union.

Mertzanis, Y. (2000). Incidence of human activities (especially hunting) on brown bear populations in Greece. In: Layna, J. F., Heredia, B., Palomero, G. & Doadrio, I. (Eds.), *La conservacion del oso pardo en Europa: un reto de cara al siglo XXI* (pp. 117–132). Seville: Fundacion Bioversidad.

Mertzanis, G., Isaak, I., Mavridis, A., et al. (2005). Movements, activity patterns and home range of the brown bear (*Ursus arctos*, L.) in Rodopi mountain range, Greece. *Belgian Journal of Zoology* 135: 217–221.

Mertzanis, G., Kallimanis, A. S., Kanellopoulos, N., et al. (2008). Brown bear (*Ursus arctos*, L.) habitat use patterns in two regions of northern Pindos. *Journal of Natural History* 42: 301–315.

Mertzanis, G., Korakis, G., Tsiokanos, K. & Aravidis, I. (2009). Expansion of brown bear range in the course of rural abandonment during the 20th century – a case study from the Pindos mountain range. In: Saratsi, E., Burgi, M., Johann, E., Kirby, K. & Moreno, D. W. (Eds.), *Woodland cultures in time and space – Tales from the past, messages for the future* (pp. 330–337). UK: Embryo Publications.

Mertzanis, G., Mazaris, A., Sgardelis, S., et al. (2011). Telemetry as a tool to study dispersal ability, habitat suitability and distribution patterns of brown bears as affected by the newly constructed Egnatia highway – N. Pindos – Greece. In: Krejcar, O. (Ed.), *Moden Telemetry* (pp. 307–329). London: InTech Open.

Mills, L. S. & Allendorf, F. W. (1996). The one-migrant-per-generation rule in conservation and management. *Conservation Biology* 10: 1509–1518.

Morini, P., Pinchera, F. P., Nucci, L. M., et al. (2017). Brown bears in Central Italy: a 15-year study on bear occurrence. *European Zoological Journal* 84: 26–33.

Naves, J., Wiegand, T., Revilla, E. & Delibes, M. (2003a). Endangered species constrained by natural and human factors: the case of brown bears in northern Spain. *Conservation Biology* 17: 1276–1289.

Naves, J., Wiegand, T., Revilla, E. & Delibes, M. (2003b). Endangered species constrained by natural and human factors: the case of brown bears in northern Spain. *Conservation Biology* 17: 1276–1289.

Naves, J., Fernández-Gil, A., Rodríguez, C. & Delibes, M. (2006). Brown bear food habits at the border of its range: a long-term study. *Journal of Mammalogy* 87: 899–908.

Ordiz, A., Bischof, R. & Swenson, J. E. (2013). Saving large carnivores, but losing the apex predator? *Biological Conservation* 168: 128–133.

Ordiz, A., Kindberg, J., Sæbø, S., Swenson, J. E. & Støen, O. G. (2014). Brown bear circadian behavior reveals human environmental encroachment. *Biological Conservation* 173: 1–9.

Ordiz, A., Støen, O. G., Delibes, M. & Swenson, J. E. (2017). Staying cool or staying safe in a human-dominated landscape: which is more relevant for brown bears? *Oecologia* 185: 191–194.

Palomero, G., Fernández, A. & Naves, J. (1993). Demografía del oso pardo en la Cordillera Cantábrica. In: Naves, J. & Palomero, G. (Eds.), *El Oso Pardo (*Ursus arctos*) en España* (pp. 55–80). Madrid, Spain.

Palomero, G., Ballesteros, F., Nores, C., et al. (2007). Trends in number and distribution of brown bear females with cubs-of-the-year in the Cantabrian Mountains, Spain. *Ursus* 18: 145–157.

Penteriani, V., Delgado, M. D. M., Krofel, M., et al. (2018a). Evolutionary and ecological traps for brown bears *Ursus arctos* in human-modified landscapes. *Mammal Review* 48: 180–193.

Penteriani, V., Huber, D., Jerina, K., et al. (2018b). Trans-boundary and trans-regional management of a large carnivore: Managing brown bears across national and regional borders in Europe. In: Hovardas, T. (Ed.), *Large carnivore conservation and management: Human dimensions and governance*. Oxford: Routledge, Taylor & Francis Group.

Penteriani, V., Zarzo-Arias, A., Bombieri, G., et al. (2018c). Density and reproductive characteristics of female brown bears in the Cantabrian Mountains, NW Spain. *European Zoological Journal* 85: 313–321.

Penteriani, V., Zarzo-Arias, A., Novo-Fernández, A., Bombieri, G. & Lopez-Sanchez, C. (2019). Responses of an endangered brown bear population to climate change based on predictable food resource and shelter alterations. *Global Change Biology* 25: 1133–1151.

Penteriani, V., Zarzo-Arias, A., Delgado, M. M., et al. (2020). Female brown bears use areas with infanticide risk in a spatially confined population. *Ursus* 2020(31e2): 1–9.

Pérez, T., Naves, J., Vázquez, J. F., et al. (2010). Evidence for improved connectivity between Cantabrian brown bear subpopulations. *Ursus* 21: 104–108.

Pérez, T., Naves, J., Vázquez, J .F., et al. (2014). Estimating the population size of the endangered Cantabrian brown bear through genetic sampling. *Wildlife Biology* 20: 300–309.

PNALM. (2018). Rapporto orso marsicano 2017 Notiziario PNALM-Natura protetta n 21 estate 2018. Available from www.parcoabruzzo.it/pdf/NaturaProtetta_RapportoOrso2017.pdf [in italian].

Posillico, M., Meriggi, A., Pagnin, E., Lovari, S. & Russo, L. (2004). A habitat model for brown bear conservation and land use planning in the central Apennines. *Biological Conservation* 118: 141–150.

Ripple, W. J. & Beschta, R. L. (2012). Trophic cascades in Yellowstone: the first 15 years after wolf reintroduction. *Biological Conservation* 145: 205–213.

San Miguel, A., Ballesteros, F., Blanco, J. C. & Palomero, G. (Eds.). (2012). *Manual de buenas prácticas para la gestión de corredores oseros en la Cordillera Cantábrica*. Madrid, Spain: Fundación Oso Pardo (FOA) and Ministerio de Agricultura, Alimentación y Medio Ambiente.

Schaal, B. A. & Leverich, W. J. (1996). Molecular variation in isolated plant populations. *Plant Species Biology* 11: 33–40.

Scharf, A. K. & Fernández, N. (2018). Up-scaling local-habitat models for large-scale conservation: assessing suitable areas for the brown bear comeback in Europe. *Diversity and Distributions* 24: 1573–1582.

Skrbinšek, T., Jelenčič, M., Waits, L. P., et al. (2012). Using a reference population yardstick to calibrate and compare genetic diversity reported in different studies: an example from the brown bear. *Heredity* 109: 299–305.

Steyaert, S. M. J. G., Endrestøl, A., Hackländer, K., Swenson, J. E. & Zedrosser, A. (2012). The mating system of the brown bear *Ursus arctos*. *Mammal Review* 42: 12–34.

Swenson, J. E., Gerstl, N., Dahle, B. & Zedrosser, A. (2000). *Action plan for the conservation of the brown bear (*Ursus arctos*) in Europe. Nature and Environment No. 114*. Strasbourg: Council of Europe Publishing.

Tosoni, E., Boitani, L., Gentile, L., et al. (2017a). Assessment of key reproductive traits in the Apennine brown bear population. *Ursus* 28: 105–116.

Tosoni, E., Boitani, L., Mastrantonio, G., Latini, R. & Ciucci, P. (2017b). Counts of unique females with cubs in the Apennine brown bear population, 2006–2014. *Ursus* 28: 1–14.

Treves, A. & Karanth, K. U. (2003). Human–carnivore conflict and perspectives on carnivore management worldwide. *Conservation Biology* 17: 1491–1499.

Tsaparis, D., Karaiskou, N., Mertzanis, Y. & Triantafyllidis, A. (2014). Non-invasive genetic study and population monitoring of the brown bear (*Ursus arctos*) (Mammalia: Ursidae) in Kastoria region – Greece. *Journal of Natural History* 49: 393–410.

Van Gils, H., Westinga, E., Carafa, M., Antonucci, A. & Ciaschetti, G. (2014). Where the bears roam in Majella National Park, Italy. *Journal for Nature Conservation* 22: 23–34.

Vucetich, J. A. & Waite, T. A. (2000). Is one migrant per generation sufficient for the genetic management of fluctuating populations? *Animal Conservation* 3: 261–266.

Wiegand, T., Naves, J., Stephan, T. & Fernandez, A. (1998). Assessing the risk of extinction for the brown bear (*Ursus arctos*) in the Cordillera Cantabrica, Spain. *Ecological Monograph* 68: 539–570.

Zarzo-Arias, A., Delgado, M., Ordiz, A., et al. (2018). Brown bear behaviour in human-modified landscapes: the case of the endangered Cantabrian population, NW Spain. *Global Change and Conservation* 16: e00499.

Zarzo-Arias, A., Penteriani, V., Delgado, M. M., et al. (2019). Identifying potential areas of expansion of the endangered brown bear population in the Cantabrian Mountains (Asturias, NW Spain). *PLoS ONE* 14(1): e0209972.

Zedrosser, A., Steyaert, S. M. J. G., Gossow, H. & Swenson, J. E. (2011). Brown bear conservation and the ghost of persecution past. *Biological Conservation* 144: 2163–2170.

Part IV **Conservation and Management**

Chapter 20

Conservation and Management of Bears

Christopher Servheen, Hüseyin Ambarlı, Harendra Singh Bargali, Stewart W. Breck, Neil D'Cruze, Claudio Groff, Gabriella M. Fredriksson, Michael L. Gibeau, Issac Goldstein Aizman, Djuro Huber, Katherine C. Kendall, Sterling D. Miller, Michael F. Proctor, Murray Rutherford, Lorraine Scotson, and Jon E. Swenson

Why Bears Need Conservation and Management: Bears as a Conservation-Reliant Species

We live in a rapidly changing world where the resiliency of the natural environment is being reduced every day by the actions of humans, whose global population is growing at a worrying rate. A recent analysis reported that land degradation seriously impacts 75% of terrestrial ecosystems and there has been an overall decline of 60% in species population sizes between 1970 and 2014 (WWF 2018). Of 8688 species on the IUCN Red List in the Near Threatened or Threatened categories, 72% are being overexploited for commerce, recreation, or subsistence. Sixty-two percent of these species are also being negatively impacted by agricultural activities including land conversion, conflicts with livestock and crops, and timber plantations (Maxwell et al. 2016). Bears are clearly among those taxa so impacted and threats to bears are increasing.

As human activities continue to increase habitat and population pressures on wild bears, we need to recognize that the only way we can continue to maintain bear populations in many areas of the world (especially those with higher human densities) is with continued conservation and management including reducing direct conflicts with humans, maintaining sufficient suitable habitat, controlling mortality to sustainable levels or reducing mortality to allow populations to increase, controlling motorized access to habitat, managing habitat and populations to allow linkage between population units for genetic and demographic rescue, and augmenting or reintroducing populations where necessary. Species requiring such ongoing and continuous management and conservation have been termed "conservation-reliant species" (Scott et al. 2005, 2010). All bear species are certainly conservation-reliant. The healthy North American black bear and most European brown bear populations are healthy throughout most of their range because of sustainable mortality limits established with updated biological data from ongoing research and monitoring. We have a long way to go if we are to establish similar bear monitoring and management systems across the range of all of the conservation-reliant bear species worldwide. The purpose of this chapter is to highlight what conservation and management can and should be implemented and how for the conservation-reliant bears across the world.

An Overview of the Basis for and Organization of Bear Conservation and Management Schemes Across the World

The management of wildlife varies worldwide in both organization and application. Understanding this variation is fundamental to recognizing the strengths and weaknesses in the management of the world's eight bear species across their range.

The relationship of human society to wildlife in general and bears in particular has evolved over time. It may be useful here to use the evolution of North American interactions with wildlife as an example of this evolution of human/wildlife interactions. In much of North America, as well as in some European countries, the management of wildlife including North American black bears, brown bears, and polar bears has evolved through three phases over time:

1. The *exploitation phase* driven by market hunting for sale of meat and hides; elimination of predators as competitors with livestock; and hunting for subsistence. This phase resulted in the functional extinction of many wildlife species and significant reductions in a wide range of wildlife, particularly predators. The three North American bears declined precipitously under this exploitation phase primarily because of aggressive predator extermination using poisons and unregulated killing. In Europe, the extermination of brown bears was encouraged by government-financed bounties or other predator control efforts (Swenson et al. 1995; Zedrosser et al. 2011).
2. The *selective conservation phase* for certain species deemed desirable by sport hunters. This was a change in focus from pure exploitation of all wildlife to the maintenance of some desirable species and the habitat they depend upon. This phase saw a change in hunter motivation from subsistence hunting to "trophy hunting" by recreational hunters. This phase of sport-hunter-driven management produced what has been called the "North American Model of Wildlife Conservation" (Geist 1995; Geist et al. 2001; Mahoney 2004). There are ongoing difficulties in the application of

this model to the conservation of wildlife not thought to be of interest to sport hunters (Organ et al. 2012), and concerns that this model depends on desires for sport hunting by a small segment of the population – mostly male sport hunters (Peterson & Nelson 2016). Some bears such as North American black bears and some brown bear populations benefited during this phase, in contrast to the intensive persecution they suffered during the exploitation phase. This was also the case for brown bears in Sweden, where the population increased rapidly after appropriate conservation measures were introduced, allowing hunting quotas to later increase accordingly (Swenson et al. 2017). Nevertheless, a major difference from the "North American Model of Wildlife Conservation" is that the owner of the hunting rights for most wildlife species in most countries in Europe is the landowner. So, even if a government agency issues the hunting quotas for bears, for example, the hunting-rights owners profit from the hunting (Knott et al. 2014). This results in bear hunting being concentrated into a more restricted group and the economic value of hunting being more important in Europe than in North America.

3. The *inclusive conservation phase* with enhanced interest in the conservation of all species, not just those of interest to sport hunters. This phase has its origins in the fact that wildlife is of value to the majority of Americans and Canadians, not just hunters. There has been a steady decline in the percentage of Americans who hunt wildlife for sport. Currently only 4.4% of adults aged 16 years and older are sport hunters (US Fish and Wildlife Service 2016). The inclusive conservation phase has been driven by public interest in the existence value of wildlife as opposed to the strictly utilitarian view common to both the exploitation phase and the selective conservation phase. The inclusive conservation phase has seen the rise of Conservation Biology (Soule 1985, 1986) and the passage of wildlife legislation such as the US Endangered Species Act to protect species not associated with hunting. Bears have benefited during the inclusive conservation phase with the recovery of small bear populations, non-lethal management of human–bear conflicts rather than removal of conflict bears, concerns about global warming and polar bears, and the reduction or elimination of sport hunting for bears in some areas. There has been erosion of public support for bear hunting unless it involves consumption of the meat of the bear. In Canada, the province of British Columbia banned grizzly bear hunting in 2017 when 91% of the public opposed trophy hunting of animals such as brown bears. The general public interest in conservation of natural resources has also been increasing in Europe, which is probably an important reason why all of the large carnivore species there are showing an increase over most of the continent (Chapron et al. 2014). Nevertheless, it appears that the poorer European countries have been most successful in conserving brown bears (Kojola et al. 2018).

Bear species in most range countries in Asia are fully protected by law, with either no, or very limited, hunting being allowed. Even with this legal protection, in some areas, Asian bear species are in decline and are listed as Threatened on the IUCN Red List. Continued conversion, degradation, and fragmentation of remaining habitat, poaching for body parts, and killing related to human conflicts are key causes of these declines. Although hard data on bear populations are lacking from all Asian range countries, forest cover maps suggest that forest range is declining at a rapid pace or is already at critically low levels, for example in Thailand. The "empty forest syndrome" is an issue in countries like Vietnam and Lao PDR where the few forests that do exist have few native large mammals.

Human attitudes toward bears have changed across the world at different rates and levels over time similar to the changes in North America. This does not mean that we have left the exploitation phase behind, as evidenced by the ongoing trade in bear parts in Asia. High prices for gall bladders and lack of or weak law enforcement make Asiatic black bears and sun bears targets for poaching, and in the case of these species it is frequently no longer a matter of tolerance when humans encounter these species outside of protected areas, but more a matter of pure survival skills under high poaching pressures.

In some areas of the world bears face continuous conflicts with people due to general public intolerance and crop raiding and livestock depredation. Even in Norway, which is a wealthy country with an advanced system of wildlife management and a generous compensation program for bear-caused depredations on sheep, the conflict level is very high (Gangaas et al. 2013). In fact, a generous compensation system can be counterproductive and actually reduce the efforts of livestock owners to prevent livestock losses.

Worldwide we are entering the inclusive conservation phase for many bear populations with increasing interest in managing the issues that cause human–bear conflicts rather than just eliminating bears. There is a big difference between documenting the problems for bear conservation and management and successfully solving these problems with effective actions. There has been progress in identifying the root causes of bear population declines for many species from loss of sea ice for polar bears to excessive exploitation for trade in bear parts, to expanding human populations that settle in bear habitat. However, much remains to do to address the causes of these problems and improve the status of bears with effective conservation and management programs in most areas of the world.

The successful conservation and management of bears is dependent on effectively addressing four factors (Figure 20.1). All too often in conservation programs, there is a tendency to focus on the collection of biological data as the primary way to conserve a species. However, without attention to developing an organized and funded management team and nurturing both public and political support, bear conservation programs are doomed to fail. In North America and in many areas of Europe, progress has been made in building all four of these

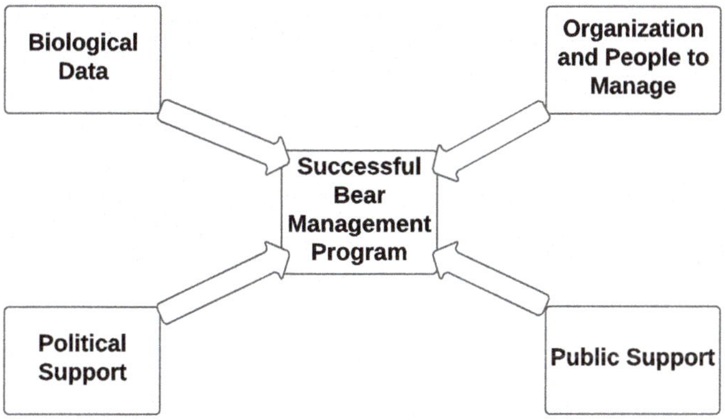

Figure 20.1 The four factors that contribute to successful bear management and conservation. Each must be addressed if any management and/or conservation program is to be successful (adapted from Servheen 1998).

Figure 20.2 Brown bear in British Columbia, Canada, being hazed with bean bags, cracker shells, and rubber bullets upon release after being relocated from a site where the bear obtained food at a human residence. Hazing is an effort to aversely condition the bear from human presence in the future. (Photo by Dave Quinn.)

factors to construct successful bear conservation and management programs. However, there continue to be setbacks in the United States and in Europe when political support wavered or changed, such as in the lack of political support to address climate change and its ultimate impact on polar bear conservation. In Europe the general situation for brown bears is good, as mentioned above; however, there are still many small and isolated populations that will require intensive management for many years to come, if they are to survive (Chapron et al. 2014). Yet, the level of conflict is high in many countries, and some have adopted a regional management approach, to allow decisions regarding bear hunting and management to be made at a more local level, within national goals, in order to reduce the level of conflict (Redpath et al. 2017; Figure 20.2).

The development of public and political support for bears is particularly a problem in much of southern Asia where effective bear conservation and management is in its infancy, with many charismatic species facing extinction (e.g. tigers, rhinoceros, and orangutans) and range country governments failing or unable to take timely and appropriate conservation action. There are few South Asian government management agencies with knowledge of or interest in bear conservation and very few skilled bear managers to address human–bear conflicts and build public support and understanding for bear conservation. Building and maintaining political and public support for bear conservation requires continuous effort because the political interests and the perceptions of people can and do change. Political and public support are closely related because political support is generally determined by public support or by pressure from outside groups.

Progress on South Asian and Andean bear (*Tremarctos ornatus*) conservation and management faces challenges in addressing these four factors. Bear management is not present in terms of American-style wildlife management in the Middle East because the brown bear is a protected species in most countries and Wildlife Departments try to protect the species against poaching. There are some efforts to offer either some mitigation measures such as implementation of electric fences or wildlife damage insurance systems. An example is that half the cost of premium insurance is subsidized by the Ministry of Agriculture and Forestry in Turkey (Tarsim 2018). If resentment among some local people arises after severe bear-caused damage, the Turkish wildlife department can adjust the hunting quota for provinces suffering this damage. However, such hunts are rarely effective in reducing conflicts because hunters are mostly foreigners who target large bears and not necessarily the individuals causing damage.

In summary, we have a long way to go to build effective conservation and management programs for bears in many areas of the world. Overcoming these challenges will require cultural sensitivity and an organized effort in both developed and developing countries if we are to secure a future for the bears of the world.

Habitat Conservation and Management

Habitat Management

Habitat quality and security from human-caused mortality are two of the most important aspects of bear conservation management in North America and likely around the world. They bring together the two driving forces thought to control all wildlife populations: bottom-up and top-down population influences. Habitat quality relates to the necessary food and energy resources important for reproduction and successful hibernation in boreal species. Delayed implantation and hyperphagia are evolutionary adaptations designed to allow bears to survive as a species through undulant interannual food supplies. There are often several natural food resources that are disproportionately important for bears' capacity to store fat resources to fuel hibernation, successful reproduction (Robbins et al. 2012; Hertel et al. 2016), and survival through low fruit productivity years in the tropics (salmon, ungulate calves, berries, acorns, figs, termites and more; Hilderbrand et al. 1999; Hashimoto et al. 2003; Schwartz et al. 2014;

Figure 20.3 Brown bear at a feeding site in Croatia. The barrel contains corn and has holes in it to minimize the amount of food obtained by the bear, thus prolonging the bear's presence at the shooting site. Hunters sit in the shooting tower and shoot bears that come to feed at the site. (Photo by Mihajlo Kovačević.)

McLellan 2015; Hertel et al. 2018a). Seasonal food resources are, of course, important, but late summer and early fall (or productive seasons in more tropical areas) hyperphagia foods are the determinants of northern bear reproductive success (Schwartz et al. 2006; McLellan 2011, 2015; Proctor et al. 2017; Hertel et al. 2018a). For bears living in more subtropical regions, where food is available year-round, fluctuations in the scale of fruiting events can play a large role in survival and life histories (Wong et al. 2005; Fredriksson et al. 2006).

Most species of bear are very susceptible to excessive human caused mortality (McLellan et al. 1999; Nielsen et al. 2004; Steyaert et al. 2016; Penteriani et al. 2018), as they generally have low reproductive rates that include late maturation, long interbirth intervals, and small litter sizes (Miller 1990; Schwartz et al. 2003; Garshelis 2009). Bears seek out easily digestible foods such as meat, fish, insects, berries, and other shrub fruits, and tree fruits (Pritchard & Robbins 1990; McLellan 2011). These needs drive bears toward human-based foods such as agricultural crops and livestock that result in human–bear negative interactions and in many cases excessive human-caused mortality (Figure 20.3).

The top-down influence on bears is usually human-caused mortality (McLellan et al. 1999; Nielsen et al. 2004; Steyaert et al. 2016). This is directly related to the probability of interaction between bears and humans: either bears seeking foods in human-dominated landscapes, or humans seeking foods or recreation in bear habitats. If it is sun bears (*Helarctos malayanus*) being drawn to oil palm plantations (Nomura et al. 2004; Fredriksson 2005), Andean bears drawn to maize (Can et al. 2014), or polar bears (*Ursus maritimus*) attracted to whale carcasses near northern seaside villages (Derocher et al. 2013; Atwood et al. 2016), the consumption of these energy-rich, easily acquired foods usually results in significant mortality risk. This mortality risk is aggravated by the fact that bear habitats have been lost across the world as humans settled and exploited lands that were previously natural habitats. Human presence in these areas converts them from natural food production to concentrated abundances of human-related energy-rich food for bears such as corn (maize), palm nuts, grains, and fruit trees. Depending on public attitudes and local norms toward human–bear coexistence, natural habitats can thus become mortality sinks for bears (Penteriani et al. 2018).

While managing front-country, human-dominated landscapes for reduced conflict mortality, it is just as important to manage habitats away from humans for reduced mortality risk, especially as connectivity of bear populations needs to be maintained over the overall landscape. This means that to be effective, mortality management must include both human-dominated landscapes and natural habitats. In South-East Asia, where wildlife law enforcement in most countries is low at best (Müller & Kaji 2016), bears are killed and poached for body parts, specifically gall bladders which reach high values at Traditional Chinese Medicine (TCM) markets (Foley et al. 2011). With this high poaching value and low law enforcement, bears are specifically targeted within protected areas, and bears are neither safe nor tolerated when venturing outside these protected areas.

Habitat Quality

Providing for habitat quality is very important but can often be challenging. Managing for habitat quality may take many forms, from the creation of a protected area network, which is more than just unproductive but beautiful landscapes (rock and ice in North America have traditionally been strong candidates for parks; Joppa & Pfaff 2009). Other resources may require management across a spectrum of habitats, such as healthy salmon stocks as bear food requires salmon management in spawning streams, but also in ocean environments and migration routes. This is the case for much of the Pacific basin from Hokkaido, Japan, to Kamchatka, Russia, north-east across the north Pacific to the famous coastal fish-bearing streams, and the rivers of the Alaska and British Columbia coasts. US west coast states of Washington, Oregon, and California currently do not support a healthy salmon–bear ecosystem due to decades of excessive mortality where bears have been extirpated regionally (Mattson & Merrill 2002). This is an example of both the habitat quality and security not being looked after. So not only do those areas need healthy salmon-bearing streams and rivers, there must also be some level of limited human access to a portion of these resources so bears may use and benefit from these resources without undue mortality risk.

Managing for other high-quality hyperphagia foods may come in the form of long-term land-use management. For example, important berry patches (*Vaccinium* spp.) in portions of western North America become more productive for bear foods after wildfire events (McLellan & Hovey 2001a), therefore managing for long-term berry supplies for bears may entail shifts in fire suppression patterns designed to save forests for timber harvests (McLellan & Hovey 2001a).

High-productivity forests in Asia are usually in the lowlands and near rivers, areas where human populations are highest, and as a result, few of these important areas remain as wildlife habitat. Managing high-quality habitats for tropical bears in Asia is now more related to establishing and managing protected areas, in remnant forests (e.g. usually in rugged mountains), reducing excessive forest harvest, halting conversion of natural forest to palm oil plantations, forest fire prevention, and retaining or restoring connectivity of landscapes between protected areas (Wong et al. 2013; Scotson et al. 2017b).

Another aspect of habitat management is managing for sea ice so polar bears can access rich seal hunting grounds. This is a global issue related to reducing worldwide CO_2 production to minimize climate change and we might be losing that battle (Derocher et al. 2013).

Security from Human-Caused Mortality

Security from human-caused mortality is well developed and documented in western North America for the grizzly bear (*Ursus arctos horribilis*). Here we discuss the principles that have been shown to result in successful habitat management and conservation successes. The southern extent of the North American brown (grizzly) bear distribution in the north-west lower 48 United States, and southern British Columbia and Alberta, Canada, has been a contentious arena for good habitat management for the past 40 years. After the conservation paradigm shift that occurred in the late 1970s, habitat management has been applied successfully in portions of this subcontinental region, with various degrees of intensity and success. A closer look reveals valuable lessons.

Grizzly bears of the North American interior have relatively large home ranges that can bring them into contact with people. Their occasional aggressive nature exacerbates their susceptibility to excessive human-caused mortality by stimulating a quicker lethal response from people (Mattson & Merrill 2002). These traits result in significantly higher mortality risk in habitats with higher back-country road densities. In North America, it has been shown across many types of habitats that higher forest road densities are associated with lower female bear survival rates (Mace et al. 1996; Schwartz et al. 2010; Boulanger & Stenhouse 2014; Steyaert et al. 2016; Proctor et al. 2017; Lamb et al. 2018). Limiting motorized access to high-quality habitats has been a cornerstone of recovery management for threatened grizzly bear populations in North America for decades and has generally been successful (Kendall et al. 2009; Schwartz et al. 2010; Mace et al. 2012). These areas have relatively high percentages of protected areas within their ecosystem borders and managers had the luxury of starting conservation efforts with hundreds of bears. A major part of the conservation program for these populations was a comprehensive program of motorized access management to close thousands of kilometers of existing timber harvest roads and improve habitat security. Motorized access management has also had a beneficial effect on the status of the isolated and much smaller (<50–100 bears) international south Selkirk Mountains population that is approaching recovery (Proctor et al. 2018a).

Managing for habitat quality is usually the long game that requires focused policy and management follow-through and is therefore challenging. Management for security from human-caused mortality can be applied over a shorter timescale with more specific management actions (such as road closures) and more immediate conservation return.

The Importance of Recognizing Geographical and Temporal Variation When It Comes to Bear Habitat Management and Security from Human-Caused Mortality

When considering habitat management and security that will promote bear conservation, it is important to consider the tremendous variation among species and among populations of the same species and that relationships can change over time. For example, Woodroffe (2000) documented that historical patterns of large carnivore extinction probability were positively related to human population density. However, this relationship was not evident using recent data from North America and Europe, suggesting that the existence of effective management structures is more important than just human densities (Linnell et al. 2001). In fact, large carnivore populations are generally increasing in Europe (Chapron et al. 2014), in spite of much higher densities of human populations, human infrastructure, forest roads, and forest exploitation than is generally found in North America.

The brown bears in Scandinavia are a good example of a population that seems to require less habitat security management than the conspecific North American grizzly bear, discussed above. There is no management of habitats, human access, roads, or human garbage that is directed to brown bear conservation in Sweden. Brown bear habitat in Sweden is heavily exploited boreal forest. In a study area in central Sweden, the forest harvest rotation time is 90–100 years. Thus, 8.4% of the forest area is clearcuts, and 42% is less than 35 years old, there is a dense network of roads (mean of 1 km/km^2, range of 0–4.6 km/km^2), and there are about 4–7 habitants/km^2 (Swenson et al. 1999; Ordiz et al. 2014). Yet, the brown bear population in central Sweden has shown the highest population growth rate ($\lambda > 1.14$) yet documented for the species (Sæther et al. 1998) and has a relatively high population density of 30 bears per 1000 km^2 (Bellemain et al. 2005). In this system, berries, especially bilberries (*Vaccinium myrtillus*), are the most important food during hyperphagia (Stenset et al. 2016). Even though the number of bilberries in a given year affects the growth and reproduction of the bears (Hertel et al. 2018a), bears do not come closer to human settlements during years of poor berry production (Hertel et al. 2018b). In fact, bears in Sweden rarely approach human settlements to obtain human-derived food (Elfström et al. 2014). This contrast in the response of brown bears to habitat management differences between North America and Sweden highlights the importance of recognizing the diversity of response that bears (even within the same species) can exhibit to human impacts across their range. It also emphasizes the need to avoid blanket application of types of habitat management as a solution to bear conservation. We must recognize that habitat management is dynamic, involving many factors including the response to and tolerance of humans toward bears, the adaptability of the local bear population, and factors such as habitat productivity that will influence how bears can survive across the world in response to human actions.

In Latin America the landscape-scale conservation approach for the conservation of the Andean bear has been implemented in Colombia, Ecuador, and Peru. In Colombia, The National Natural Parks System (PNN & WCS 2018) has implemented an Andean bear conservation strategy based on the conservation of wilderness areas and the reduction of human–bear negative interaction. The goal of the strategy is the conservation of long-term viable population by managing and conserving large landscapes composed of interconnected national parks, wilderness areas, and human dominated habitats. In Ecuador and Peru (SERNANP et al. 2014), the Andean bear conservation strategy focuses on a protected area mosaic approach where human productive activities both outside and within protected areas are managed to lower their vulnerability to bear activity and reduce human–bear negative interactions.

While there are caveats, which include regional habitat quality, cultural differences, tolerance and lethality of humans, and political differences, several general principles of habitat management are applicable across systems and species:

- Humans and bears can coexist as long as human activities are managed to limit attractants that bring bears into conflict with people resulting in excessive mortality risk and there are sufficient natural habitats protected from human settlement and agriculture. Habitats do not need to be devoid of humans to be of value to bears. Human activity must be managed in high-quality bear habitat where bears seek high-energy natural foods, usually on a seasonal basis.
- Protected areas where human landscape disturbance and human-caused mortality risk are both low are an important aspect of a successful habitat management system for most bears. Protected areas are especially important in tropical habitats where human tolerance for bears is low. In tropical bear habitats, there must be representative amounts of protected habitat containing a diversity of tropical forest habitats to serve as reserves where bears can exist with little mortality risk due to poaching and human presence.
- Understanding the variability of habitat productivity for important bear foods is very useful to inform and guide where and how to manage natural habitats. It is particularly useful to understand where the most important hyperphagia foods are located, so those areas can be managed (in some cases by seasonal road closures) to minimize risk of excessive mortality risk or habitat displacement of bears by human activity.
- Understanding non-hyperphagic seasonal habitats of importance that bring humans and bears together can be important for managing mortality risk. For example, these may be areas of spring range where bears go after exiting their dens and are concentrated in low elevation, snow-free habitats.
- Motorized access management is important to maintain habitat security to reduce negative human–bear interactions, at least in North America. One way to do this is to manage densities of roads. One well-documented example of motorized access management exists in western North America where road densities are managed to assure that 50–60% of natural habitat is more than 500 m from open motorized roads and, where roads occur, road density does not exceed 0.5–0.6 km/km^2. This may not apply in some European areas such as in Sweden where high road densities seem to have little impact on brown bear population health. Another example of motorized access management is to close timber harvest roads when timber harvest operations are complete. In much of North America, proper forest management for bears requires both (1) management of the location and type of logging to provide diverse natural foods; and (2) closure of timber harvest roads upon completion of timber harvest to maintain habitat security.

The Role of Protected Areas in Bear Conservation

Protected areas now cover between 12% and 15% of the Earth's terrestrial surface depending upon how broadly one defines "protected area." The common purpose of these protected areas is to sustain ecosystems and all their constituent species as well as to maintain ecosystem services and functions. Another way to think about the role of protected areas is to avoid, prevent, and mitigate against anthropogenic conversion of natural habitat. It is this confluence of maintaining natural habitats and avoidance of anthropogenic change that makes protected areas an important element in the conservation of the world's bears.

The conversion of natural habitats into human-dominated land use is considered to be the biggest threat to world biodiversity (Pereira et al. 2010). To counteract this phenomenon, the protected area concept has become a prominent conservation strategy in the last century. This designation of protected areas arguably constitutes "one of the most stunning conservation successes of the twentieth century" (Ervin 2003). While the concept of protecting natural habitats is a primary purpose depending upon jurisdiction, the management reality can be significantly different on the ground. Globally, relatively few protected areas are managed with biodiversity conservation as the primary objective. Of the world's 98,400 terrestrial protected areas, only 8800 (8.9%) are listed under IUCN categories I or II (Naughton-Treves et al. 2005). Indeed, many protected areas were not originally chosen for their species abundance or richness. Caughley (1994) suggested that species end up not in the habitat most favorable to them, but in the habitat least favorable to the agent of decline.

Concurrent with the rise of establishing new protected areas in the last quarter century, and especially in developing countries, the mission of protected areas has expanded from biodiversity conservation to improving human welfare (Naughton-Treves et al. 2005). This dual mandate poses challenges for the ecological value of these areas for bears, and all other wildlife for that matter.

Woodroffe and Ginsberg (1998), among others, point out that even when habitat is otherwise suitable, in isolation many protected areas simply are not large enough to maintain viable populations of species such as bears that occur at low densities.

In order to maintain populations of large mammals, protected areas need to become larger if they are to be fragmented from similar habitat in the surrounding landscape. More importantly, attempts to link protected areas need to be embedded within a broader concept of landscape conservation, in which protected areas are but one component, and in which protected areas are integrated into larger landscape planning and management frameworks (Gaston et al. 2008). In theory, a number of models address this broader view of landscape conservation including: (1) community-based conservation, where partnerships link biodiversity objectives with local development objectives; (2) conservation easements, where private landowners voluntarily restrict the types of activities on their property to enhance conservation values; and (3) co-management, where government shares power and decision making with local authorities to reach common goals.

In many areas of the world, however, such as many countries in Asia, the fate of bears rests increasingly on their persistence in protected areas. Outside of these protected areas, bears are subject to accelerating habitat loss due to conversion to plantations, degradation of habitat due to overharvesting of timber, human encroachment, and specifically human intolerance and direct persecution for their body parts. Many of these threats also occur inside protected areas due to weak laws or absence of law enforcement. The future of bears in many developing countries, particularly in much of Asia, is increasingly uncertain because effective protected areas are few, the most productive forest areas are converted to other purposes, and now forest mostly persists in mountainous regions, which may not be the optimal habitat for some of the bear species.

It is apparent now in the twenty-first century that the blueprint approach of protected areas is overly simplistic to provide the only solution to species conservation in an increasingly complex world. Likewise, the panacea of community-based conservation is probably no more effective than the panacea of exclusively state-based conservation (Berkes 2007). Conservation solutions are increasingly elusive in a globalized world. A more strategic approach (Servheen 1998) is warranted if we are to conserve the different bear species we currently have around the globe.

Population Fragmentation and its Management: The Importance of Population Connectivity to Bear Conservation

Population fragmentation is ubiquitous in modern landscapes and is a threat to biodiversity in general (Wilcove et al. 1998; Fahrig 2003; Wilson et al. 2014), and specifically to bears (McLellan et al. 2017). Extirpation risk due to the small population paradigm of isolated subpopulations (Caughley 1994) can be a mechanism of range contraction and was a likely contributor in the case for brown bears across North America in the nineteenth century (Mattson & Merrill 2002). Beyond these historic and current conservation threats, the emergence of climate change elevates the importance of connectivity conservation substantially. In a review of 22 years of climate change mitigation tools, reversing fragmentation through connectivity management was the most recommended strategy (Heller & Zaveleta 2009). We need to allow species and ecosystem components the ability to move in response to shifting habitats to survive the significant yet unpredictable changes related to climate change (Penteriani et al. 2019).

There are two primary types of fragmentation and its antithesis, connectivity: genetic and demographic. Genetic fragmentation results in a loss of genetic diversity through increased inbreeding, which occurs faster in small populations, and potentially threatens a species' ability to adapt to relatively rapid environmental changes (Keyghobadi 2007). While traditionally this had been a longer-term process (many generations), the rapid environmental changes we are experiencing, and are expected to experience, from climate change may hasten the effects of genetic fragmentation. Demographic fragmentation, the interruption of movements of each, or either, gender and the impacts that result, can be a more urgent concern. For example, an isolated small population with no immigration from either sex has no ability for demographic rescue that can result from female immigration (Proctor et al. 2005, 2012). Stochastic (chance) mortality patterns may result in excessive female mortality, leaving a population without the ability to grow fast enough to survive. Demographic rescue, in the form of female immigrants, would offer the potential to elevate reproductive rates for population increase (Proctor et al. 2005, 2012).

The example of brown bears in North America illustrates these scenarios. Because male-biased dispersal in bears often mediates gene flow (McLellan & Hovey 2001b; Proctor et al. 2004; Zedrosser et al. 2011), traditional measures of genetic discontinuity will usually miss fragmentation that primarily affects females but not males (Proctor et al. 2005). Natal dispersal in brown bears in North America is male-biased (McLellan & Hovey 2001b; Proctor et al. 2004), and therefore assessments of genetic discontinuity often cannot detect female fragmentation in systems where female movements are limited but not males'. This is the case for brown bears (Proctor et al. 2005, 2012) and black bears (Proctor et al. 2020a) in North America. However, using individual-based genetic analyses that follow the fates of individuals has allowed the detection of female fragmentation even when there is no sign of genetic discontinuity (Proctor et al. 2005, 2012, 2020a).

These situations are a result of the fact that human activities more easily restrict female brown bear movements as they naturally have shorter natal dispersal. The short natal dispersal distances of female brown bears, where one generation's new home range overlaps some portion of their mother's home range (McLellan & Hovey 2001b), means that for female (demographic) connectivity to occur, it usually takes several generations. Female bears must therefore live in the connectivity habitats linking two populations, at least, which often entails overlapping with human settlements and transportation corridors that were fragmenting the population in the first place. These are the areas where long-term patterns of human-caused mortality caused the fragmentation. Several efforts within North America have predicted brown bear linkage habitats using habitat corridor modeling from GPS telemetry to inform landscape connectivity management (Proctor et al. 2012; Peck et al. 2017). Connectivity management often takes the form of actions that reduce mortality risk of bears that are spending considerable time sharing habitats with human settlements and the related disturbance (Proctor et al. 2018b).

Brown bears in Europe are almost irreversibly fragmented into 10 separate populations. Except in the northern parts of the continent the extant populations are mostly confined to the mountain ranges. With the ever-growing highway and rail enhancements, and other infrastructure like wind turbines and ski resorts, the fragmentation of European bears is continuing. The European Union is pushing toward population-level management (Linnell et al. 2008; John et al. 2016) and efforts are underway to attempt to make new and existing infrastructure permeable to bears and other wildlife. Since 1999, in Croatia over 1000 km of new highways has been constructed in bear range. The environmental impact studies for these projects required construction of numerous over- and underpasses including 13 green bridges 100–200 m wide (Kusak et al. 2009). The good news is that in spite of all odds, brown bears and other large carnivore populations in Europe have been increasing over the past 20 years, indicating that with wildlife-friendly legislation and careful management, brown bears and other carnivores can survive and even increase in the heavily settled landscape of Europe (Chapron et al. 2014).

Another example comes from polar bears. The potential fragmentation of polar bear populations has been thoroughly assessed and the results show that they are experiencing very little human-caused fragmentation. Subpopulation boundaries have been delineated with telemetry and genetic methods (Paetkau et al. 1999; Malenfant et al. 2016). Discontinuities are most often associated with large landmasses and to a lesser extent large ice-free oceanic expanse where inter-area movements are relatively low. Climate change has been ushering in a period of declining sea ice, which has resulted in decreases in polar bear body condition (Obbard et al. 2006), and potentially changes in gene flow patterns in portions of their range (Peacock et al. 2015) but not yet extensive population declines or habitat loss. Because anthropogenic forces do not cause fragmentation of polar bears into their 19 circumpolar subpopulations (see Chapter 14), connectivity management is not thought to be part of the response to climate change as is likely the case for other bear species.

Sloth bears (*Melursus ursinus*) are endemic to the Indian subcontinent, and inhabit mostly dry forests from Nepal, through India, into Sri Lanka. Human settlements, rural farming areas, transportation corridors, and cities and towns extensively fragment their populations across India and its dense population of over 1.2 billon people (Garshelis et al. 1999a; Bargali et al. 2012). Human–bear conflict is extensive (Sharma et al. 2013; Ratnayeke et al. 2014; Dutta et al. 2015), often resulting in human injury (Bargali et al. 2005; Singh et al. 2018). There are two levels of challenges that fragmentation and connectivity conservation face in developing countries. First, is that the low level of scientific effort to address fragmentation is due to funding constraints. Radio-collaring is uncommon for sloth bear studies in India, but its use is growing (Ratneyeke et al. 2007; Bargali et al. 2012). Genetic studies that reveal patterns in fragmentation are being initiated (Dutta et al. 2015). The second challenge is dense human populations and expansion of rural farming and human settlement areas into bear habitat that fragments wildlife habitats and populations.

The extent of sloth bear fragmentation is increasing (Dutta et al. 2015). While protected areas provide some degree of security and may be the backbone of what might be termed a sloth bear metapopulation, many sloth bears live outside protected areas (Garshelis et al. 1999b). Maintaining movement opportunities for these bears has the potential to secure functional connectivity between protected areas (Dutta et al. 2015). There is some evidence of occasional sloth bear movement between larger habitat patches that are separated by up to 300 km (Dutta et al. 2015). Enhancing connectivity for sloth bears in India requires identification and management of functional corridors between protected areas to allow sloth bears to live in them. The time available to address connectivity for sloth bears is growing short because of climate change and increasing human populations and development in the Indian subcontinent.

Giant pandas are another example of the importance of fragmentation and connectivity research and management. Panda fragmentation has been studied relatively extensively compared to other bear species. They have a relatively limited diet (bamboo species) and range and are more susceptible to climate change habitat shifts (Gong et al. 2017) than any other bear species. Panda conservation effort has been significant and has greatly benefitted the species (Loucks et al. 2001; Liu et al. 2016). In 2016, their Red List rank went from Endangered to Vulnerable (Swaisgood et al. 2016), but new threats are emerging, including the expansion of livestock grazing leading to degraded habitats, infrastructure development, and increasing habitat fragmentation discussed further below (Swaisgood et al. 2018). Although fragmentation is recognized as a serious issue to panda conservation, efforts to manage for connectivity between fragmented populations are just beginning. Efforts are underway to identify core habitats, establish corridors, and prioritize areas for protection and restoration. Habitat-use data and habitat-quality models have informed the creation of 67 protected areas (Swaisgood et al. 2018). Genetic analyses have been used to estimate population size, determine population structure and fragmentation patterns, and genetic diversity, understand dispersal patterns and then predict connectivity habitat patterns (Wei et al. 2015; Figure 20.4) and the efficacy of interpopulation corridors (Wang et al. 2014; Wei et al. 2015; Swaisgood et al. 2018). Captive breeding and assisting gene flow by moving animals between fragments has been evaluated to deliver demographic rescue to isolated small populations. Despite all this conservation effort, threats to pandas continue, with expansion of livestock grazing and infrastructure development that erode panda habitat (Wang et al. 2015; Liu et al. 2016; Swaisgood et al. 2018).

Pandas might be an example where the maintenance of genetic diversity to preserve the ability to adapt to climate change may be very real and as important as demographic connectivity (Gong et al. 2017). Bamboo habitats are predicted to shift and providing suitable corridors to shifting habitats will be important. China's successful captive breeding program of pandas may also have a role to play in reintroductions into areas in need of genetic rescue and population supplementation.

In summary, our progress in managing connectivity across bear species varies tremendously. The four key drivers of anthropogenic fragmentation are human-related mortality, habitat loss, road network development, and habitat degradation (Figure 20.5). All will intensify in the future with increasing human numbers combined with climate change. These trends can only be reversed through understanding and addressing the spatially explicit causes of fragmentation. It is

Figure 20.4 A brown bear going under a barbed wire fence around a bait site. The bear passively leaves hair on the barbs and this hair can be used for analysis of DNA to identify the bear. Such hair collection fences are used for population size and trend analysis, relatedness analysis, to assess connectivity between populations, and other valuable research. (Photo by Stefan Himmer.)

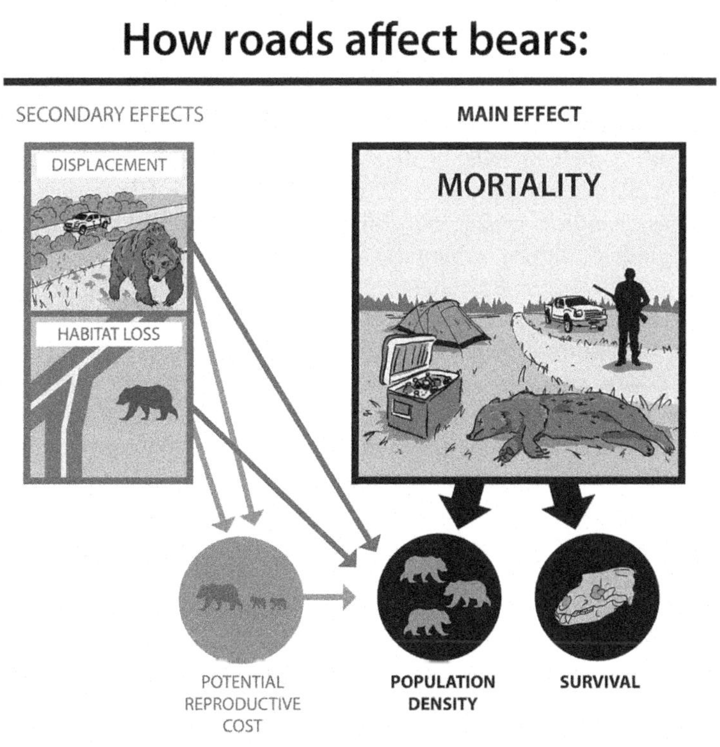

Figure 20.5 Schematic of mechanisms of bear response to roads. The main effect of mortality ultimately reduces density. Secondarily, habitat displacement and direct habitat loss potentially affect reproductive output and density. (Adapted from Proctor et al. 2020b).

important to realize that relying on protected areas alone to provide habitat and connectivity for bears, while necessary, is not sufficient in itself to produce conservation solutions in an increasingly globalized world. Connectivity management must be implemented in the matrix of protected areas and the human-impacted habitats that surround protected areas if it is to be successful.

These selected examples in this section cover the spectrum of variability of the state of fragmentation understanding and connectivity management for bears. A more thorough treatment of this topic can be found in Chapter 22, which reviews connectivity for all bear species.

Working with People to Develop Tolerance and Support for Bears and Bear Conservation

No matter how scientifically sound and well-intentioned a conservation program may be, it is unlikely to be successful if it does not have the support of the people whose behavior, property, or other interests will be affected. Support for bear conservation depends on people's values, beliefs, and attitudes about bears. It also depends on their attitudes about the agencies or other institutions behind the program, expectations about how the program will affect their broader values, and feelings about the decision-making processes through which the program is developed and implemented. Policy scientists call these factors the "social process" and the "decision process" – the sociopolitical context for a program or policy and the governance structure under which it is designed, approved, and instituted (Lasswell 1971; Clark 2002).

People's values, beliefs, and attitudes are key factors in the social process that underlie their support for or opposition to bear conservation. Research shows that people's values for wildlife are complex and multidimensional, and that these values vary with demographic factors such as age, education, and occupation (Kellert 1996). Several typologies have been developed to classify values of wildlife and nature (see, for example, Rolston 1988; Kellert 1996; Jones et al. 2016; Manfredo et al. 2017). These typologies distinguish dimensions of value such as the extent to which people believe that wildlife should be used, managed, or dominated, and the extent to which people believe that wildlife have rights and deserve respect, caring, and trust (Manfredo et al. 2017).

In addition to the values that people hold for wildlife and nature, their attitudes toward bears are also influenced by factors such as perceptions of the particular species or subspecies involved, knowledge of wildlife biology, and historical and current relationships with bears (Kellert 1994). The latter may include cultural traditions and practices, accepted patterns of utilization such as hunting, history of conflicts with people and property, and conservation status. Of course, these factors will vary with the setting in which a bear conservation program takes place. Although many people have an affinity toward bears, and research in North America and Europe shows a reasonably high degree of support for bear conservation, some people do have negative attitudes about bears and bear conservation, especially in non-urban agricultural areas where dominionistic and utilitarian values may be prevalent and there may be a history of livestock depredation or safety concerns (Kellert 1994; Dressel et al. 2015).

Specific attitudes about bears are just one of the relevant components of human perspectives in bear conservation. Conservation initiatives may also affect many of the broader values that people hold and pursue in their lives. Lasswell (1971) identifies eight main types of value at stake in human interactions: wealth, power, respect, well-being, enlightenment, rectitude, affection, and skill. The following brief hypothetical example illustrates how this wider array of values may come into play in conservation and interact with specific attitudes about bears. A decision to prohibit bear hunting in a region may be supported by conservationists because they hold moralistic and humanistic values toward bears, and they perceive the hunting prohibition as aligning with their values of rectitude (doing the right thing) and enlightenment (in their view the decision is supported by the scientific evidence). The same hunting prohibition may be opposed by guide outfitters, because their dominionistic and utilitarian values support the belief that it is not ethically or morally wrong to hunt bears (rectitude), and they predict that the prohibition will cause them to lose guiding fees (wealth) and opportunities to use their professional expertise (skill). Meanwhile, livestock producers such as ranchers may also oppose the prohibition because they believe that the absence of hunting will lead to increased property damage and depredation on livestock (wealth, respect, affection) and more safety risks for their families (well-being, affection). They may also believe that their lifestyles and livelihoods are being disrespected (respect). All of these participants will feel either indulgences or deprivations of power and respect, depending on whether they perceive that their views have been heard and they have been able to influence the decision about hunting. And all of these values and beliefs of individuals are likely to be reinforced by the social roles and groups with which they identify (for example, as livestock producers, conservationists, wildlife scientists, guide-outfitters, or hunters).

These complex value-driven dynamics of the social process and decision process are further complicated by the symbolic significance of bears. Bears are potent symbols of wilderness, power, strength, and awe. However, bears and bear conservation also often become symbols in political struggles about deeper issues and changes in society. Conservation programs may provoke anxieties and become the focus in disputes about economic loss, changing lifestyles and landscapes, recreational access, appropriate use of public lands, individual rights versus collective rights, local interests versus national interests, and other highly contested and emotional issues. People use bear conservation programs as an outlet for their frustrations with top-down governance and edicts from afar, agencies that they perceive as not listening, and decisions that they consider unfair and disempowering. If public decision-making processes (governance) do not provide opportunities for people who are affected by conservation programs to be heard and to work through such conflicts to find common ground, their disputes will likely become more bitter and their positions more polarized and entrenched. Conflict can escalate to the point that participants do not see realistic options for finding common ground, and instead focus on gaining more power to advance their own interests (Rutherford & Clark 2005). In such "politicized" settings, wildlife scientists may find that people are not really interested in improving their knowledge and understanding. Rather, people may use scientific research and recommendations only to the extent that they support their personal interests and objectives. Otherwise, they may attack not only the scientific methodology and results, but also the qualifications and integrity of the scientists (Gibeau 2012).

Given this highly complex, contested, and potentially hostile sociopolitical context in which wildlife scientists and managers must operate, how can they work with people to increase tolerance for bears, and develop conservation programs that are more broadly supported? The first step is for wildlife professionals to be self-reflective and to clarify their own standpoints – the values, cognitive biases, conceptual frames, and other factors that shape how they see the world, how they conduct their research and how they interact with others (Clark & Wallace 1999). By clarifying their own standpoints, they may be better equipped to understand the perspectives of others, and to see how and why their own views and recommendations may differ from those of the people who are affected.

A second recommendation is that wildlife professionals should map the sociopolitical context and decision-making processes that exist in the setting for a conservation program. As the discussion above makes clear, context matters, and bear conservation programs should be designed and adapted to fit the specific sociopolitical settings in which they are instituted. Local leaders, culture, institutions, traditions, decision-making practices, and other factors are all important.

Third, wildlife professionals should work with local people to address the real problems they face arising from bears and bear conservation programs. By working on tractable smaller problems, it is possible to build relationships and trust, develop capacity and begin to depoliticize situations, while implementing practical measures to protect bears and reduce conflicts (Wilson et al. 2014). Examples include installing deterrence measures such as electric fencing for beehives and calving sites, helping with changes in agricultural and husbandry practices such as cleaning up bone yards and protecting feed storage facilities, and installing bear-proof food caches or food poles in the back country and garbage facilities in the urban–wildland interface.

Fourth, people are more likely to support conservation programs if they are meaningfully engaged in the development of those programs. When wildlife professionals and their agencies are willing to share power and participate in collaborative decision-making with people who are affected or interested, there is a real possibility to develop robust and effective conservation programs (Gibeau 2012). Essentially, this type of collaborative decision-making provides a setting for dialogue, in which people are encouraged to reason about problems and discuss strategies to address them, and jointly identify the common interest (McLaughlin et al. 2005). Interest-based discussion rather than positional bargaining is encouraged.

An initial matter on which people may find it easy to agree is the standards for good decision-making. Even when people initially disagree on what should be done, they may be able to agree on how decisions should be made (Rutherford et al. 2009). As trust improves, scientific research can be reviewed, and the implications discussed in a less politicized environment (Gibeau 2012).

Such collaborative decision-making is recommended for all settings, but it is essential for settings in which indigenous people are involved. As Housty et al. (2014, p. 70) observe about the grizzly bear monitoring program they developed in the traditional territory of the Heiltsuk First Nation in Canada – a collaboration that included academics, indigenous government and community leaders, provincial and federal governments, and others – "Successful resource management by Indigenous people may require approaches like this one, in which collaborative science-based management is embedded within a socially and culturally appropriate framework for action."

People are at the center of all conservation efforts. By (1) being more self-reflective, (2) understanding the social context and decision-making processes, (3) working with local people, and (4) engaging people in meaningful dialogue, wildlife professionals can set the foundations for comprehensive problem-solving. Then, by systematically evaluating and learning from trials of such innovative approaches to conservation, they can harvest the lessons and communicate them to others, encouraging replication or adaptation in other settings.

Management of Bear Mortality

Mortality documentation is extremely important for management of the world's bear species. The numbers and ranges of all species have been significantly reduced, and excessive mortality, which is ongoing for all bear species, has been an important cause, except in American black bears, polar bears, and, in many areas, brown bears. Compared to other large mammals, high mortality rates for bears are problematic because bears generally have lower reproductive rates.

Unsustainable mortality rates in bears most commonly result from killing bears that cause damage to human interests, such as crops or livestock, of bears perceived to be threats to humans, in accidents, or for commercial sales of parts (termed "non-sport" kills in contrast to legal hunter kills). Non-sport kills may also be numerically larger than hunting kills in some areas and may also be significantly underreported. In Japan, for example, about 500 Asiatic black bears are killed annually by hunters compared with 1000–2000 kills of "nuisance" individuals (Oi & Yamazaki 2009). In Hokkaido, Japan, of 5668 brown bears recorded killed during 2010–2017 from an increasing population, 89% were non-hunting kills compared to 11% killed by hunters (unpublished data provided by T. Mano, Institute of Environmental Sciences, Sapporo, Japan). In contrast, in portions of Alaska with low human density, hunter kills predominate. "Defense of life or property" kills in Alaska were 5.2% of total reported kills of brown bears and 3.1% for American black bears (in urban areas 22.3% and 6.1%, respectively) (Miller & Tutterrow 1999). In Croatia, management plans anticipate that 80% of annual brown bear mortalities will come from hunting and 20% from accidents and nuisance bear kills (Huber et al. 2008a). Unreported bear kills occur everywhere and are thought to be high throughout the range of the world's bears, especially in areas where bear parts, such as gall bladders and paws, are illegally traded (Servheen 1990; Gong & Harris 2006).

Brown bears, American and Asiatic black bears, and polar bears are legally hunted in some areas. In these areas hunting is typically limited to sustainable levels of mortality and may focus on adult males. In terms of population growth rate, male mortalities have less impact than mortalities to adult females (e.g. Taylor et al. 1987; Harris et al. 2006). Non-sport kills, in contrast, are not as selective for larger individuals (males) (e.g. Huber et al. 2008b; Reljić et al. 2018). Hunting regulations protecting females accompanied by newborn cubs (or older offspring) also act to protect adult females. Sun bears may be hunted with special permits in Sarawak, Malaysia (Krishasamy & Shepherd 2014) but, like Andean, sloth, and panda bears, are not legally hunted elsewhere.

Throughout the world where bears are hunted (excluding Alaska) the population management objectives are to allow hunting at levels that maintain or increase existing populations. Exceptions occur in some small areas where populations are sometimes managed to reduce conflicts with humans; information campaigns on methods to reduce conflicts and other management efforts can be effective at reducing unnecessary bear mortalities (e.g. Huber et al. 2008b; Proctor et al. 2018b). In the USA, brown bears currently can be hunted only in Alaska and American black bears can be hunted in all states with significant populations. In Canada, hunting of brown bears ceased in Alberta in 2006 to allow population recovery. Hunting in British Columbia was halted in 2017, although indigenous harvest is allowed. Canadian brown bear hunting occurs in Yukon Territory, Northwest Territories, and Nunavut, and American black bear hunting is allowed in all Provinces and Territories. In Eurasia, brown bears can be legally hunted in most areas with significant populations (e.g. Sweden, the Dinaric and Carpathian Mountains, Japan, and Russia) but not in areas with small populations including Spain, Italy, or where status is unknown such as China (Huber et al. 2008b; Chapron et al. 2014). In most of Russia where brown bears are hunted, such as in Karelia and Murmansk, about 10% of the estimated population size is accounted to annual hunting rate, although in most years this quota is not achieved (Tirronen et al. 2015). In contrast to the rest of the world, in Alaska, both brown and black bears in most of the state are managed to reduce populations in the, as yet undocumented, expectation this will reduce predation on and increase hunter harvests of wild moose (*Alces alces*) and caribou (*Rangifer tarandus*) (Miller et al. 2017; Ripple et al. 2019).

Fees paid by bear hunters may promote bear conservation through protection of habitat, hiring of bear management specialists to reduce conflicts and promote coexistence, and providing a constituency for sustainable management (e.g. Miller et al. 2013). In the Karelian part of Russia a license (state fee) costs about US$50 on state land, and on private land it could reach US$1000 or more (Danilov & Tirronen 2011). In many European countries, hunters pay the trophy value of a harvested brown bear (the value is higher for larger bears and can reach as high as 10,000 EUR). Within the European Union (EU), brown bear hunting is subject to rules established by the Habitat Directive.

Most commonly, acceptable levels of harvest or mortality are established using estimates of population size combined with estimates of sustainable levels of mortality based on data on vital rates of productivity and mortality obtained from studies of radio-marked animals, hair snagging, mark–recapture, other available data, and modeling exercises (e.g. Miller 1990; Harris et al. 2006; Boulanger et al. 2008; McLellan et al. 2017; Reljić et al. 2018). Most harvest management schemes also involve efforts to document how many bears hunters in specific management areas legally kill. Sometimes collection of information on the sex and age composition of harvested individuals is also required, although this information must be used cautiously, as developed elsewhere in this chapter.

Reported impacts of hunting on survivorship of cubs vary between North America and Europe. In Europe, studies conclude that male-biased hunting reduces brown bear cub and yearling survivorship as it results in more sexually selected infanticide by male bears who benefit by having the female come into estrus earlier than if her cubs survive (e.g. Swenson et al. 1997; Swenson 2003; Gosselin et al. 2014). In contrast, North America studies of brown bear and black bears conclude that hunting does not result in changes in cub survivorship or may increase survivorship if hunting and other mortalities result in a decline in male abundance or population density (Stringham 1980; Miller et al. 2003; Schwartz et al. 2003; McLellan 2005; Czetwertynski et al. 2007).

Some jurisdictions also attempt to estimate mortalities that are not reported and incorporate these into mortality quotas. In both hunted and unhunted populations, unknown and unreported mortalities may represent a significant portion of total mortalities (Cherry et al. 2002; McLellan et al. 2018).

In North America, only aboriginal groups have rights to hunt polar bears (Prestrud & Stirling 1994; Regehr et al. 2017). In the US, the Marine Mammal Protection Act (MMPA) of

1972 eliminated polar bear hunting by the general public. This Act authorized continued hunting by subsistence users resident in coastal areas that are dependent on such hunting for subsistence. Authorized sale of authentic articles of handicrafts and clothing made from polar bears by these subsistence hunters is allowed. There are no quotas unless a population is declared "depleted" under the MMPA. In Canada, Inuit and First Nations peoples are allowed to harvest polar bears and quotas are usually allocated to villages. Hunters may utilize the meat and, in some subpopulations, Inuit hunters may sell their rights to harvest polar bears to non-indigenous people who must partake of a sport hunt led by Inuit guides using traditional means. Hides can be sold commercially and most reach the international market. Quotas are established by research on polar bear abundance and trends. There is some controversy between Inuit hunters and scientists over population trends and the harvest quotas based on them. In both the US and Canada, the ultimate threat to polar bear populations is habitat loss caused by climate change, and not overharvest (see also Chapter 14).

Population and Trend Estimation for Bears

There are no universal or widespread best ways to estimate bear abundance and trend across species and populations. Factors to be considered in selecting an approach include species and associated behaviors, habitat conditions, population density, and availability of financial and technical resources. Management needs for accuracy and precision vary widely between species and jurisdictions. In some cases available resources may be better spent in ways other than increasing the precision of population estimates, such as obtaining better estimates of the threats to populations and their habitats or controlling poaching. This is the situation for some populations of Asian and Andean bears where even the distribution of species may not be well known. In other cases, such as where American black bears are abundant, hunting quotas are set conservatively, habitat is not deteriorating, and management efforts to control unnecessary mortalities (such as to nuisance bears) are in place and successful, precise and large-scale population estimates may not be the best use of available funds.

Accurate population estimates are commonly more critical for small and/or isolated populations, where concerns over loss of genetic diversity or population extinction exist. Although precision and accuracy in population estimation techniques have improved significantly in recent decades, all techniques have actual or potential biases and the presence and direction of these must inform decisions based on estimates obtained. In some cases, these biases can be minimized with the use of multiple sampling methods (Boulanger et al. 2008). Imprecision in estimates of abundance of hunted populations generally mandates setting harvest quotas conservatively. Assessments of abundance and trends based on impressions unaccompanied by rigorous measurements should be viewed cautiously (Garshelis & Hristienko 2006).

Many techniques used to estimate abundance in bear populations employ variations of capture–mark–recapture (CMR) techniques. CMR methods establish a known number of marked individuals randomly mixed into a population, followed by a "recapture" effort, which samples both marked and unmarked individuals. The population is estimated from the ratio of marked to unmarked individuals in the recapture sample(s). There are a number of options in how animals are "marked" and "recaptured" including de-facto "marks" based on DNA in hair or fecal samples (Woods et al. 1999; Bellemain et al. 2005), physical marks like tags or telemetry transmitters, chemical marking using tetracycline-laced baits, natural markings, and group characteristics documented by direct observations or remote cameras.

Handling and fitting bears with radio transmitters and following them through time allows collection of data on age, births, deaths, movements, condition, habitat use, and other parameters that are usually extremely valuable to bear managers. Information on poor physical condition of handled animals can sometimes allow managers to anticipate declines in abundance prior to being able to detect this directly through abundance estimators (e.g. Stirling et al. 1999; Obbard et al. 2006; see also Chapter 14). In some cases, use of methods that do not involve handling bears may cost less, increase precision because of larger sample sizes, and better meet technique assumptions such as not creating trap-shy or trap-prone individuals.

CMR approaches to estimating population abundance include a variety of assumptions and characteristics (Seber 1982; Cooch & White 2019). Closed models assume equal capture probability for all individuals, and demographic and geographic closure. Open population models do not assume closure, but require at least 3 years of data to estimate abundance based on year-specific estimates of apparent survival, capture probability, and number of new individuals (Cooch & White 2019). Spatial capture–recapture (SCR) models do not assume closure and directly estimate the area to which abundance estimates apply; this permits direct estimates of density (Royle et al. 2014; Hooker et al. 2015). In contrast to non-spatial approaches, SCR models can be used to examine variation in abundance/density within study areas (Kendall et al. 2019).

Population densities are typically more valuable than just a population estimate, as they may reflect factors that vary spatially such as carrying capacity (food supply) or mortality rates. Also, estimates of density in small study areas may be extrapolated to other areas with comparable habitat and survival rates to obtain abundance estimates for larger, managerially significant geographic areas such as a park or reserve.

Techniques Requiring Handling of Bears

Trapping and Tagging

For many years the standard way to estimate abundance of American black bears was based on CMR techniques using physically captured and marked bears followed by recapture effort(s) in a grid of baited traps/snares (examples cited in Pelton 2003). Costs associated with this technique typically limit its use to small geographic areas and, corresponding, small sample sizes. As with all CMR techniques, this approach has potential assumption violations that must be addressed including presence of trap-shy or trap-prone individuals and lack of geographic closure of the population (Seber 1982).

Radio Telemetry

Most polar bear abundance estimates have involved multiyear CMR techniques (e.g. Derocher 2012). Radio-telemetry data and the Cormack–Jolly–Seber CMR models were used to estimate size and trend in various polar bear subpopulations (e.g. Amstrup et al. 2001; Stirling et al. 2011). In western Hudson Bay, Lunn et al. (2016) used live recaptures plus dead recoveries in a Bayesian implementation of multistate CMR models. Regehr et al. (2018) combined telemetry and CMR in an Integrated Population Model (IPM) for polar bears in the Chukchi Sea. In that study, multiyear capture data were used to estimate density and vital rates in a small intensively studied area with additional data added from radio telemetry. IPMs are powerful tools as they improve sampling efficiency/coverage and estimate precision by combining data from multiple sources. In Nepal, 17 radio-collared sloth bears were tracked for 2 years. Sloth bears are relatively social and sightings of nondependent unmarked bears with or near marked bears constituted the "recapture" sample (Garshelis et al. 1999a).

Known Fate Studies to Estimate Population Trend

Population growth rate (λ) can be estimated with information on natality and survivorship obtained from observations of radio-marked bears (e.g. Eberhardt 2002). This approach has been used to estimate trend in a number of bear populations in North America (e.g. Garshelis et al. 2005; Schwartz et al. 2006) and Europe (Saether et al. 1998). When growth/decline rates based on known-fate telemetry studies are made for very large areas, it is important that telemetry collars be distributed geographically in proportion to bear abundance. This was the case for a study in a 24,000 km^2 portion of the Northern Continental Divide Ecosystem (NCDE) (Mace et al. 2012). The annual minimum of 25 radio-collared bears were distributed within the NCDE based on detection results obtained from a DNA hair snaring study that applied equal sampling effort throughout the NCDE (Kendall et al. 2009, discussed below). This growth rate was applied to the initial DNA-based abundance estimate to derive annual estimates of abundance.

Using Offspring Survivorship Data to Estimate Trend

Eberhardt (2002) proposed a sequence of changes in vital rates as population density for long-lived vertebrates increases, predicting an initial decline in survival of immatures, followed by an increase in the age of primiparity, a reduction in reproductive rates of adult females, and, lastly, increased mortality rates of adults. In the Yellowstone Ecosystem, a decline in cub and yearling survival was observed in conjunction with a slowing of population growth (van Manen et al. 2016). This observation was interpreted as a possible expression of density-dependent regulation as this brown bear population neared its demographic maximal levels, or carrying capacity, in its core habitat (van Manen et al. 2016). Lower juvenile survival in unhunted National Park populations in Alaska compared to nearby hunted populations was also attributed to Park populations having reached their demographic carrying capacity while the hunted populations had not (Miller et al. 2003; Keay et al. 2018). In protected areas, adult males reach higher densities associated with higher survival, and intraspecific killing by adult males is a likely mechanism for lower survival rates of juveniles, particularly cubs. Thus, unhunted populations may approach their demographic carrying capacity at population numbers below the level that available food resources in their environment could sustain.

Unduplicated Counts of Females With Cubs to Estimate Abundance

One of the longest continuous data sets for annual estimates of bear abundance is for brown bears in an isolated population occupying 23,833 km^2 in the Greater Yellowstone Ecosystem, USA. Abundance in this area has been estimated since 1983 using unduplicated counts of females with cubs of the year (FF_{coy}) and a conservative rule set to avoid duplicate counting of groups of FF_{coy} (Schwartz et al. 2006; van Manen et al. 2018). Estimates of FF_{coy} are then extrapolated to obtain total population estimates. Similarly, FF_{coy} counts have been used since the beginning of the 1980s as an index for monitoring brown bear populations or estimating a minimum number of adult females in several small and medium-sized European populations (Ordiz et al. 2007).

Techniques Not Requiring Handling Bears

DNA Sampling

The development in the 1990s of laboratory methods that obtained useful genetic information from minute, degraded DNA samples from hair and feces (Taberlet & Bouvet 1992; Höss et al. 1992) revolutionized bear research and monitoring. Since the development of reliable techniques to identify bear species, sex, and individual identity from such samples (Taberlet et al. 1996), approaches involving sequencing of small DNA samples without the need to capture and handle individual bears have become widely used to estimate bear abundance (e.g. Woods et al. 1999; Bellemain et al. 2005; Proctor et al. 2010). Genetic detection generally samples a larger proportion of the population, resulting in more precise estimates of abundance/density than methods that involve handling. DNA-based techniques also provide information on effective population size (Pierson et al. 2018) and genetic diversity (Mikle et al. 2016). DNA from fecal samples is degraded, with lower genotyping success rates compared to hair samples (Paetkau 2003; Sorensen et al. 2017). Typically, 6–7 microsatellite loci and a sex primer are used to identify individuals. If the level of genetic diversity in a population is unknown, it is necessary to conduct a pilot study prior to setting the number of loci that need to be examined to distinguish among individuals.

Bear hair is typically obtained using hair corrals consisting of barbed wire encircling 3–6 trees positioned about 50 cm high with a scent lure at the center (Woods et al. 1999; Proctor et al. 2010; Figure 20.4). Other sampling techniques, such as hair from trees on which bears like to rub or trail hair-snares, do not require attractants and avoid potential behavioral effects on detection probability (e.g. Beier et al. 2005; Kendall et al. 2008). Scats can be obtained by systematic survey routes or grids, searches by scat-detecting dogs (Wasser et al. 2004), or volunteer collectors, such as hunters (Bellemain et al. 2005).

Sampling Hair to Estimate Abundance

One of the largest-scale studies to estimate brown and black bear population abundance using DNA was conducted in the 31,410 km^2 NCDE in western Montana including Glacier National Park (Kendall et al. 2009). Bears were sampled using two independent hair collection methods: hair corrals deployed on a grid and rub trees identified along trails, roads, and powerlines. Genotyping at seven loci identified 266 female and 182 male brown bears. The CMR model estimate of total population size was 765 (95% CI = 715–831) or 2.4/100 km^2. Obtaining hair samples from multiple methods (e.g. corrals and rubs) produces more precise and less-biased estimates than from either method alone (Boulanger et al. 2008). The same NCDE brown bear hair corrals and rub trees simultaneously collected black bear hair. This was used to estimate abundance and density in a sympatric American black bear population in Glacier National Park (4100 km^2) using closed population CMR models (Stetz et al. 2014). Based on genotyping at six loci, density was estimated at 11.4 bears/100 km^2 (95% CI = 9.9–13.0). DNA hair collection using hair corrals in small study areas resulted in density estimates for Andean bears and Asiatic black bears. For Andean bears, DNA analyses have been attempted (e.g. Viteri 2007) but require more development (Garshelis 2011). DNA techniques have also been attempted for Asiatic black bears in India (Mukesh et al. 2015).

The brown bear population in the Gobi Desert, Mongolia was sampled using hair corrals surrounding 13 feeding stations at most water sources used by the population (Tumendemberel et al. 2015). Of the 600 hair samples collected in 2009, 205 were genotyped to 21 individuals. Genetic variability was low compared to other brown bear populations, so samples were genotyped at 12 microsatellite loci to ensure reliable identification of individuals. Using CMR models, population abundance exposed to the hair corrals was estimated to be 22 bears (95% CI = 21–29).

Clustered Hair Sampling

Humm et al. (2017) estimated density of the five major American black bear subpopulations in Florida, USA, with SCR models using genetic detection data from clustered hair corrals. The efficiency of the clustered sampling design coupled with the use of several landscape covariates allowed researchers to estimate abundance (3916 bears, 95% CI = 2914–5451) in a 38,960 km^2 study area.

Sampling Feces to Estimate Abundance

Abundance estimates using DNA from feces have been obtained for brown bear populations in Eurasia. For example, in Sweden the brown bear population was estimated using CMR methods at 3298 (95% CI = 2968–3667) using six microsatellite loci that identified 1358 individuals (Bellemain et al. 2005; Kindberg et al. 2011). DNA in fecal samples was used to estimate brown bear numbers and evaluate effective population size in Slovenia and Croatia (Skrbinšek et al. 2018). Four abundance estimates in China (1974–1977, 1985–1988, 2000–2004, 2011–2014) were based on measurements of bamboo segments (assumed to represent individually specific bite sizes) found in feces. These results concluded there was a gradual national increase in giant panda abundance (Wei et al. 2018). Another approach, based on DNA found in panda feces in one area, resulted in a nationwide estimate that approximately doubled the bite size estimate (assuming the density relationship held across the panda range) (Zhan et al. 2006). The DNA-based results were considered to be biased high, because of closure bias and other considerations (Garshelis et al. 2008; see also Zhan et al. 2009).

Aerial Surveys

Aerial survey methods are most effective in open habitats where bears are readily visible. Polar bear abundance in Hudson Bay, Canada, was estimated using aerial double-observer techniques based on CMR principles (Stapleton et al. 2014, 2015). This involved independent observers in the front and rear of an aircraft; a polar bear was considered “marked” if seen by the forward observer and “recaptured” if also seen by the other observer. Transects were flown during the summer ice-free season when all polar bears are on land. A similar approach was attempted for brown bears in Alaska but had low precision (Walsh et al. 2010).

Aerial search techniques similar to those of Miller et al. (1997) but that did not require radio-collared animals were used to estimate brown bear abundance on 19,988 km^2 of treeless habitat in north-western Alaska (Schmidt et al. 2017a). Initial “marking” was based on the spatial location and characteristics (e.g. coloration, group size) of each group obtained and photographed during aerial searches. The “recapture” sample was acquired during independent aerial searches where location and these characteristics were used to determine if bears seen were “recaptures” or previously unseen bears. The cost of this method was ~10% of the cost of smaller area density estimates using radio-collared bears (Miller et al. 1997). Densities from each technique were broadly comparable, but the approach without telemetry had lower precision.

For polar bears and brown bears, double-observer aerial surveys are sometimes conducted simultaneously with distance sampling techniques (e.g. Crête et al. 1991; Stapleton et al.

2015; Obbard et al. 2015; Schmidt et al. 2017b). Distance sampling techniques are based on assumptions involving models of relative sightability of bears occurring at different distances from transect lines flown by aircraft. For polar bears, transects were perpendicular to the shore and transect spacing was based on stratification of the survey area based on pre-existing knowledge of bear density (Stapleton et al. 2015). An evaluation of approaches involving distance sampling alone and combined with CMR double-observer data concluded that distance sampling alone provided better results with less complexity (Schmidt et al. 2017b).

Combining DNA with Telemetry Data to Estimate Abundance at Edge of Range

Jurisdictions interested in estimating population size at the edge of its range need to account for part-time occupancy by transboundary bears. Norway has politically established goals for reproduction of brown bears in five management areas that occur at the edges of the bears' ranges in adjacent countries. The Norwegian goal is 13 annual reproductions in these five areas. Annual counts of individuals based on DNA gathered from scats and hair and telemetry-derived knowledge about demography, home range size and use provide annual management area-specific estimates of reproduction that can be credited to Norway (Bishof & Swenson 2012).

Using DNA to Estimate Trend

Genetic detection data were used to estimate brown bear population trend and local density based on hair samples collected at 3579–4802 natural bear rub sites along trails, roads, and fence and utility lines in western Montana's NCDE (Kendall et al. 2019). Rubs were visited 2–4 times annually during 2004 and again during 2009–2012, resulting in detection of 249–355 individual bears each year. SCR models were used to estimate an annual growth rate of 1.043 (95% CI = 1.017–1.069). During the study, the highest population densities shifted from primarily in the north to a more even distribution throughout the NCDE. Simulation studies indicated that rub tree surveys were likely to be a reliable way to detect population trends (Stetz et al. 2010). DNA-based approaches combined with SCR models promise to transform trend monitoring for bears and other difficult-to-study species across large landscapes.

Two abundance estimates conducted 10 years apart were used to estimate brown bear population trend in the Canadian Rocky Mountains (Stenhouse et al. 2015). The estimates were made with samples from hair corrals, genotyped at seven microsatellite loci, and detection data analyzed using SCR models. The population was estimated to have grown at an annual rate of λ = 1.07, but there was a large amount of uncertainty (95% CI = 0.84–1.30). Nonetheless, this study concluded that spatially explicit methods obtained robust abundance estimates with less sampling effort than traditional CMR models by enabling the use of sampling stratification, thereby reducing project costs.

Marking Bears With Ingested Tetracycline to Estimate Abundance

The number of black bears in Minnesota and parts of Michigan (USA) was estimated using a CMR estimator by hanging widely spaced baits laced with tetracycline in trees and determining how many were ingested by bears (Garshelis & Visser 1997). Tetracycline was identified by laboratory analysis in the teeth and bones of hunter-killed bears that ate these baits. This study resulted in estimates of approximately 15,600 black bears (± 1500) in Minnesota and 6600 (± 1400) in Michigan. A similar tetracycline study was also done in Michigan's Upper Peninsula (Belant et al. 2011).

Estimating Abundance and Trend Using Remote Cameras

Motion-sensitive cameras are used to detect and identify images of many wildlife species (Burton et al. 2015). Camera-based CMR studies using distinctive chest patterns to identify individuals were used to estimate density of Asiatic black bears and sympatric sun bears in two study areas in Kha Yai National Park, Thailand (Ngoprasert et al. 2012, 2015). Imprecise density estimates for Asiatic black bears were 4.3–5.9/100 km^2 and 8–29/100 km^2 for sympatric sun bears. Camera traps were used to estimate Andean bear density in a Bolivian pilot study (Rios-Uzeda & Wallace 2007). In a potentially useful advance, machine learning techniques ("deep learning") identified faces of individual brown bears in three study areas based on camera trap pictures with a claimed 93% (SE = ± 4.9%) accuracy (Clapham et al. 2018).

Lower Technology or Cost Methods

Techniques used to determine trend in bear species/populations where resources are most limited include expert opinion, surveys that enumerate bear sign, changes of habitat extent, and perceived levels of hunting pressure. It is not possible to estimate confidence intervals around trend estimates made using these methods. For example, results from several national Andean bear assessments based on an assortment of techniques, including genetics, ecological modeling, and sign surveys, were combined to give a crude range-wide estimate of 13,000–18,000 bears (5–7/100 km^2) over the species range of 260,000 km^2 (Velez-Liendo & Garcia-Rangel 2017).

Occupancy

Changes in presence of bears can be used in occupancy models to estimate increases or decreases in abundance in space or time. Presence data can come from observation (live capture, direct sightings of bears or tracks, remote camera images, or local knowledge/expert opinion) or DNA samples from hair, scats, or other sources. Occupancy by brown bears in formerly vacant habitats adjacent to growing populations in the NCDE, Greater Yellowstone Ecosystem, and Sweden was consistent with other studies, indicating growing populations expanding into former range (e.g. Schwartz et al. 2002; Kendall et al. 2019). Population trends for sun bears were measured in Sumatra using camera trap detections in an occupancy-based sampling framework (Wong et al. 2013). A trend estimate for sun bears across their global range in South-East Asia was based on the relationship between tree cover and relative density of sun bears to project population change due to habitat loss (Scotson et al. 2017a). In Borneo, sun bear presence documented with bear DNA obtained from leeches (*Haemadipsa* spp.) correlated with presence data from camera traps (Abrams et al. 2019). Because occupancy determination only requires species identification, DNA laboratory analysis is inexpensive compared to methods requiring identification of individuals.

In some portions of Asia where distribution of bears remains uncertain, useful information on occupancy as well as subjective information on trend, threats and other issues have been gathered by a combination of questionnaires and sign surveys. Typically, this involves creation of a grid of cells covering a province (or other pertinent area), asking residents of grid cells about bear occupancy, trends, and problems and then asking residents to show investigators areas of bear activity where sign can be seen. This approach has been used for Asiatic black bears in south-western China (Liu et al. 2011) and for sloth bears in Sri Lanka (Ratnayake et al. 2007). In places where Asiatic black bears and sun bears are sympatric, claw marks on trees were used to differentiate between these species (Steinmetz & Garshelis 2008).

Population estimates for Asiatic black bears and sun bears were attempted using questionnaires that asked regional experts for their subjective estimates of park populations with perceived confidence ranges (Crudge et al. 2019). For both species, the IUCN Bear Specialist Group estimated a range-wide percent decline over 30-year time windows, weighted by extent of occurrence within range. This method suggested a 30% decline of Asiatic black bears and up to 40% decline of sun bears (Garshelis & Steinmetz 2017; Scotson et al. 2017b). In India, a country-wide sloth bear population estimate of 6000–11,000 was based on questionnaire data sent to 800 wildlife researchers, managers, and naturalists, in combination with satellite-based habitat cover maps (Yoganand et al. 2006). Questionnaire surveys indicated a reduction of brown bear distribution in Pakistan of 15.7% since the 1950s (Abbas et al. 2015).

In Indonesian Borneo (Kalimantan), sign transects were used to monitor trend in sun bears during 2000–2010 in an area of regenerating burned forest. For Andean bears in Colombia, efforts are ongoing to monitor trend from presence data based on bear sign observed along non-random line transects (Acevedo et al. 2017).

Hunter Observations or Kill Data

Estimates of population trend can be obtained in jurisdictions where hunters are well organized and will report observations of bears and hours afield. In Sweden, about 200,000 hunters report the number of hours hunted and number of bears observed during the first seven days of the moose hunting season. Kindberg et al. (2009) found that the resulting bear density index documented bear distribution and showed a strong linear correlation with DNA-based population estimates at the wildlife management unit scale (*ca.* 1000–2500 km^2). Kindberg et al. (2011) combined these trend estimates with periodic DNA-based regional population estimates to obtain an estimate of brown bears in Sweden. In several European countries where bears are hunted, such as Slovenia, Croatia, and Romania, there are permanent feeding sites equipped with observation/shooting towers where hunters make visual counts of bears seen on predetermined nights. The standardized reports are summed and, for decades, were used to estimate the population size and decide the hunting quotas (Huber et al. 2008a). More recently, the same data were used only to evaluate population trend. This supplements more accurate CMR estimates based on genetic detection, which is too expensive to be done each year. A weak point of counts at feeding sites is that annual fluctuations in hard or soft mast production can change bear movement patterns (Ryan et al. 2007) and encounter frequency with feeding sites. Thus counts at feeding sites can vary in response to changes in mast production rather than changes in population numbers.

Jurisdictions where hunters are required to submit teeth from harvested bears for age determination and allow managers to inspect hides to determine gender expect that these data can be used to assess population trends (e.g. Kolenosky 1986). For bears, there is not yet a clear-cut way to interpret these data, although Population Reconstruction Analysis (Millspaugh et al. 2010; Skalski et al. 2012) may be useful in some circumstances. Harvest data are difficult to use to infer trend because of frequent and difficult-to-quantify biases in hunter-harvests associated with variable or changing vulnerabilities of sex/age categories (e.g. Garshelis 2002; Hatter et al. 2018; Newton & Obbard 2018), sometimes associated with changes in hunting regulations. Regardless, when associated with other information, these data can sometimes prove helpful in estimating trend in bear populations over large geographic scales (Fieberg et al. 2010; Allen et al. 2018) or can flag conditions where more intensive studies on abundance or trend may be needed.

The Management of Human–Bear Conflict

Human–bear conflict (HBC) is a topic that is becoming increasingly important for the conservation and management of bears throughout the world (Ambarlı & Bilgin 2008; Can et al. 2014; see also Chapter 15). A good definition of HBC is provided by the Human–Bear Conflicts Expert Team (part of the IUCN Bear Specialist Group): "HBC is any situation where wild bears undesirably use or damage human property, where wild bears harm people, or where people perceive bears to be a direct threat to their property or safety."

An important concept in the discussion of human–bear conflict is the notion of an affected stakeholder, which we define as those people living in a landscape with bears and that are impacted by HBC. Affected stakeholders are distinguished from non-affected stakeholders (e.g. people in conservation organizations) that may be invested in having bears on the landscape but are not impacted directly from their activity. Identifying and understanding impacts to affected stakeholders is important because failing to address their concerns can lead to a variety of sociopolitical impacts influencing the potential for bears and humans to coexist.

Three general strategies are used (often in conjunction) to minimize HBC: managing anthropogenic resources, managing bears, and compensating for HBC impact. Managing anthropogenic resources usually, but not always, involves efforts to reduce the availability or access to human resources. For example, reduction of anthropogenic resources formed the foundation of reducing HBC for brown bears and black bears in National Parks throughout the US and Canada by reducing the availability of garbage and other anthropogenic food in the Parks (e.g. Denali National Park 2003; Glacier National Park 2010). Applying the principle of anthropogenic food reduction to other systems (e.g. livestock spread across a landscape or agricultural crops in mixed landscapes) can be more difficult because authority to strictly manage these resources, particularly on private lands, can be limited and costs for doing so can be prohibitive. Instead of reducing anthropogenic resources, in some situations the opposite approach is taken by providing additional food sources (i.e. supplemental feeding) as a means of reducing HBC. For example, in the Black Sea forests of Turkey the decline of wild orchard trees resulted in increased conflict by brown bears. In response, the Turkish Department of Forestry planted orchard trees while implementing regeneration plantations to help reduce HBC. In the north-western US, timber industries created feeding stations to feed black bears in an effort to reduce bears feeding on the inner cambium of newly planted conifer trees (Ziegltrum 2004). The science supporting this type of effort is mixed and generally most management authorities are not in favor of supplemental feeding.

Perhaps the most effective tool for preventing conflict is electric fences, and this tool has been used effectively in a wide variety of contexts to protect agricultural crops and livestock but can be expensive to install and operate. In general, non-lethal tools become less effective as anthropogenic resources comprise a greater percentage of the landscape or are more dispersed. When non-lethal methods are ineffective it sometimes becomes necessary to lethally remove bears. In developed countries, such control actions are generally performed by management authorities, whereas in developing countries management authorities are often not available to carry out such operations and thus locals may take it upon themselves to carry out lethal control actions. In some situations, sport hunting is promoted as a method for reducing HBC. However, there is little evidence that sport hunting reduces HBC (Obbard et al. 2014) unless hunting is so excessive that it significantly reduces the bear population to very low densities.

Another strategy for dealing with HBC is paying compensation for the economic damage to affected stakeholders. For example, Turkey established an agricultural insurance pool where all damage caused by wildlife can be compensated, if the farmer pays half of the small cost of insurance (US$2 for a sheep, US$5 dollar for a cow or beehive per year) (Tarsim 2018). In Sweden, concerted effort has gone into recovering their brown bear population (Swenson et al. 1995). In areas where recovery has occurred, an important affected stakeholder group are the indigenous Sami herders that utilize landscapes in Scandinavia to maintain the traditional culture of herding semi-domestic reindeer. Brown bears prey on reindeer, resulting in negative economic impacts to the Sami herders. To minimize the impacts, national policy requires that herders are compensated for losses to bears and it is believed that this compensation is an important component of a larger management scheme that helps maintain tolerance for bears.

In the lower 48 United States, grizzly bears have been recovered in critical components of their historic range (Gunther 2015). A major part of the recovery program for the grizzly bear was assuring a clear and active response to HBC (primarily predation on livestock, impacts to beehives, and breaking into houses) (Interagency Grizzly Bear Guidelines: Mealy 1979). Agencies, in conjunction with the enhanced efforts of private groups, worked together to develop and implement a comprehensive and responsive management system, assist affected stakeholders in preventing conflicts, and increased the non-lethal management of conflict bears. Through these kinds of efforts (Proctor et al. 2018b), HBC was minimized and the number of bears that had to be lethally removed was considerably reduced, thus promoting the growth and recovery of the bear population.

Many of the management examples we present have occurred in developed countries with abundant resources available to manage conflict. These types of resources are generally not available in developing countries, which begs the question of whether the management of HBC is a conservation luxury of developed countries. For example, throughout South-East Asia, conversion of native forest for agriculture has helped cause precipitous declines for a number of bear species including the sloth bear, Asiatic black bear, and sun bear (Scotson et al. 2014; Wong et al. 2015; Debata et al. 2016). The increased presence of people and impact on the landscape has created dynamics where bears are coming into greater contact with people, which has resulted in higher rates of attacks on people and greater utilization of human food crops. Retaliatory killing of bears in an effort to reduce HBC is common and contributes to the reduction in the bear population. Compounding this problem are the economic incentives associated with poaching for sale of bear parts. In many cases, killing bears to protect crops may be a cover for the true desire to kill bears to sell their body parts. Critical to the successful management of HBC is detailed understanding of human culture where bears and humans interact and integrating affected stakeholders into the process of developing solutions to manage HBC (Redpath et al. 2017).

Managing the Trade in Bear Parts

Of the world's eight extant bear species, three species, Asiatic black bears (*Ursus thibetanus*), sun bears, and brown bears, are most threatened by the illegal wildlife trade. All are threatened by poaching and are the most commonly seized species of bears across Asia (Burgess et al. 2014). Bears are poached mainly for their gall bladders, but also for bear cubs, paws, claws, and other body parts. Bile from bear gall has been used in Chinese traditional medicine for thousands of years as a treatment for a wide range of inflammatory and degenerative ailments (Feng et al. 2009; Li et al. 2016).

Traditionally, bear bile was sourced from the gall bladders of bears hunted and killed in the wild. Since the 1980s, several Asian countries have begun keeping wild-caught bears in captivity and also farming live bears, predominantly Asiatic black bears, for their bile (Servheen & Mills 1991; Servheen et al. 1999). China is the leading consumer and producer of bear bile tapped from captive bears, with around 20,000 bears on so-called farms. Captive (often wild-caught) bears are also kept for commercial bile extraction in South Korea and Vietnam, with fewer in Lao PDR and Myanmar (Animals Asia 2011; BANCA 2017; Willcox et al. 2016; World Animal Protection 2017; Livingstone et al. 2018). Zoos and private collections also play a role in illegal bear trade. For example, Japanese "bear parks" hold more than 1000 bears of various species and some of these "parks" trade in bear bile products obtained from their captive bears (Mills & Servheen 1991; Togawa & Sakamoto 2002; Foley et al. 2011). In Lao PDR, a number of private collectors, including Chinese-owned "zoos," hold Asiatic black bears and are suspected to be involved in the illegal trade of live bears (EIA 2015; Livingstone et al. 2018).

International trade in bears and bear parts is regulated by the Convention on International Trade in Endangered Species (CITES). Asiatic black bears and sun bears are classed as CITES Appendix I, which restricts international trade in these species, unless under exceptional circumstances. Import permits can be issued if specimens are not to be used for commercial purposes, or if the import purposes are not deemed to be detrimental to the survival of the species. There may be exceptions made for specimens bred in captivity (CITES 2000). Brown bears are Appendix II, in much of their range, except for populations in Bhutan, China, Mexico (extirpated), and Mongolia, which are covered by Appendix I (www.cites.org). Differences in Appendix classifications between and within bear species create difficulties in controlling illegal trade, due to difficulties in identifying species by parts of the animals and the origin of products being traded (CITES 2000). Domestic trade in farmed bear products is permitted within China, Myanmar, and Lao PDR, providing that products come from second-generation captive-bred bears. This provision allows the promotion of bile sales for commercial purposes and creates a trade loophole due to deficiencies in monitoring the origin of captive bears and problems differentiating parts from wild versus captive-bred bears.

The impact that illegal trade in bears has on wild bear populations has not been quantified, and assessments are largely speculative (Garshelis 1997; Garshelis & Steinmetz 2017; McLellan et al. 2016; Scotson et al. 2017a), but many forest areas in South-East Asia are becoming devoid of bears and other widllife, strongly indicating that there is a poaching crisis (Harrison 2011; Gray et al. 2017). It is widely assumed by conservation scientists that bear farming has a negative impact on wild bear populations because (1) killing of wild bears for trade of their parts is unregulated and largely unknown; and (2) there is a two-tiered market for both farmed product and a more desirable wild product (Dutton et al., 2011; IUCN 2012; Livingstone & Shepherd 2014; Garshelis & Steinmetz 2017; Scotson et al. 2017a).

There is need for a clear and united strategy among organizations invested in conservation of bears to reduce consumer demand for both farmed and wild bear products. Monitoring of the trade in bear parts would benefit from a central database devoted to the bear trade, alongside enhanced communication between organizations working with bears to optimize opportunities to collate data in a meaningful way (Burgess et al. 2014). Understanding consumer motivation, promoting alternatives to bear bile in traditional medicine, and educating and changing the behavior of potential consumers of bear products could be effective tools to reduce bile trade. There have been some successes in reducing the bile trade evident in Vietnam and South Korea, where bear farming is being phased out (Willcox et al. 2016; World Animal Protection 2017).

In summary, the trade in bears and bear parts continues to be a serious problem for the conservation of Asian bears. Despite more than 25 years of effort to reduce the demand for bear bile and to decrease the number of wild bears killed for the sale of their parts, the trade continues and may be increasing in some countries.

Artificial Feeding and Bear Conservation: Advantages and Disadvantages

"Artificial feeding" of bears includes all efforts to deliberately provide food (other than garbage) in natural habitats as a bear management tool. Objectives can include baiting for hunting, diversionary feeding to decrease human–bear conflicts, concentration of bears for viewing or photography for human recreation, and/or nutritional enhancement of populations. Artificial feeding has been a tool used to promote bear conservation, particularly for threatened populations, such as the feeding program for the Gobi brown bear in Mongolia. Bear feeding can sometimes have unintended negative consequences for species and ecosystems (reviewed by Penteriani et al. 2010, 2017), such as alterations of social and trophic interactions; changed behavior, activity, and movement patterns (Krofel et al. 2017); changes in reproduction; and pathogen transmission. Therefore, bear feeding is of conservation concern and should be evaluated on a case-by-case basis if it is to continue (Skuban et al. 2016; Krofel et al. 2017; Penteriani et al. 2017).

Artificial feeding of bears is a common practice in many European countries. Increasing bear availability to hunters is the main motivation for such feeding in countries where bears are hunted, followed by increased reproduction and body size for hunted bears. Artificial feeding of bears occurs year-round in many cases and usually occurs at established feeding sites. Artificial foods for bears are also used by many non-target species of birds and mammals. Currently, more than 80% of the European countries where bears occur feed them intentionally or unintentionally. Bear feeding takes place in all European countries where brown bears occur except Albania, Andorra, Belarus, Greece, Italy, Spain, and Liechtenstein. Corn and livestock remains (whole carcasses and slaughter remains) are the most common foods used to feed bears in Europe.

More than 60% of the areas where targeted bear feeding occurs have regulations regarding the type and amount of food that can be provided and the periods when feeding can occur, although often such regulations are not fully implemented (Huber et al. 2008b). In most cases, bear feeding has not been subject to an appropriate impact assessment, even if it occurs in Natura 2000 sites (European Union 1992), Emerald Network areas, or other protected areas. In both Emerald and Natura 2000 sites, national authorities have a legal obligation to ensure that feeding is in conformity with the ecological requirements of bears and other species for which the sites are designated, and that any significant adverse effects on these species are avoided. Unintentional feeding of bears where large herbivores are fed should be minimized where possible (Selva et al. 2014).

The Large Carnivore Initiative for Europe (LCIE) completed a study on artificial feeding of bears as a management tool, which was submitted to the Council of Europe where it was adopted as a Recommendation by the Bern Convention (Convention on the Conservation of European Wildlife and Natural Habitats 2018). This report noted that there is an urgent need to reevaluate the practice of artificial feeding of bears from cultural, ecological, conservation, and legal perspectives. While recognizing that in some situations it can be an appropriate management tool in general, the LCIE recommended against artificial feeding of bears and would like to see a progressive decrease in the practice. In areas where it is currently practiced, the LCIE recommends a detailed case-by-case evaluation of the intended goals and the potential impacts on target and non-target species, as well as wider ecosystem effects. There is also a need to develop and enforce clear regulations to govern the practice that relate to periods of the year when feeding should be conducted, the location of feeding sites, and the type and amounts of food that can be used.

Critical Needs for the Conservation of Bears and Bear Habitat in South-East Asia

South-East Asia's bears would benefit most from two key management actions: (1) reducing habitat loss, particularly deforestation by timber harvest and conversion of forests to plantation agriculture; and (2) reducing illegal poaching and killing bears for the sale of their body parts. Neither action is easily achieved under the current climate of increasing and expanding human populations, minimal political support for bear conservation, and few financial resources for bear research, monitoring, and management.

Deforestation rates are highest in Myanmar, Indonesia, and Malaysia (Scotson et al. 2017b). Bear populations are negatively impacted by extensive clear-cutting for agricultural plantations (e.g. oil palm, rubber), unsustainable logging (both legal and illegal), and forest fires (Brown et al. 2005; Meijaard et al. 2005; Tumbelaka & Fredriksson 2006; Wong & Linkie 2013). Indonesia and Malaysia are both leaders in global oil palm production, which is a primary driver of bear habitat destruction (Miettinen et al. 2011; Margono et al. 2014). In Indonesia, 40% of the forest lost between 2000 and 2012 was lost in areas where logging was restricted or prohibited, including within protected forests due to deficiencies in governance (Murdiyarso et al. 2011; Margono et al. 2014). In Kalimantan alone, 56% of protected lowland forests were cleared from 1985 to 2001 (Curran et al. 2004).

A priority for bear conservation is to protect high conservation value forests from conversion to other land uses, to eliminate unsustainable logging, and prevent forest fires. Similarly, managers should prioritize restoration of degraded forest in human-modified landscapes through reforestation, connectivity management, and enhanced protected status, and enforcing buffer zones to prevent further agricultural expansion into surrounding protected areas.

Ultimately, reducing the trade in bear parts would be one of the most highly beneficial steps for the persistence and recovery of Asian bears throughout their range, especially as trade is increasingly moving toward the last strongholds in Asia (Shepherd & Shepherd 2010; Krishnasamy & Shepherd 2014). Asian bears are legally protected from hunting and trade throughout most of their range, but in many cases this is paper protection only. Deficiencies in law enforcement are recognized as the major ongoing weakness in bear conservation throughout Asia (Burgess et al. 2014; Baskin 2016). There have been a few successes, for instance in Cambodia where dedicated agencies operate with steady technical and funding support (e.g. Broadis 2011). Reduction of bear mortality by clearing forests of snares is urgently needed throughout much of the remaining bear range throughout Asia, and long-term measures are needed to prevent the problem recurring in cleared habitats (Gray et al. 2017).

Intensive Management of Brown Bears and Black Bears in Alaska

Attitudes toward hunting brown bears in North America vary from almost no hunting of a mostly healthy population in British Columbia (Canada) to hunting of a population recently delisted from Endangered Species Act protections (Wyoming and Idaho, USA – hunting put on hold in 2018 by a court decision), to widespread efforts in Alaska (USA) designed to intentionally reduce populations of both black and brown bears. The Alaskan efforts are mandated by a state "Intensive Management" law (AS 16.05.255, 1994) requiring wildlife managers to reduce the abundance of bears and wolves (*Canis lupus*) in hopes this will result in increased hunter harvests of wild ungulates such as moose, caribou, and deer (*Odocoileus hemonius sitkensis*) (Miller et al. 2017; Ripple et al. 2019). Efforts to reduce bears involve liberalized hunting regulations including all-year hunting seasons, individual annual hunting quotas of two per year for brown bears and up to five per year for black bears, allowing hunters to sell the hides and skulls of bears they shoot, elimination of requirements to purchase a tag prior to hunting, allowing hunters to bait both species of bears, allowing take of bears using snares, and relaxation of rules regarding use of aircraft for hunting. This has resulted in dramatic increases in numbers of bears harvested without corresponding efforts to document trends in bear populations (Miller et al. 2017; Ripple et al. 2019).

The Future for Bears Around the World

In this chapter we have reviewed the major conservation issues bears face and highlighted management actions that can address these conservation issues. The future of bears across the world is bright for some species but dark for others. In some areas such as North America (North American black bear and brown bear) and in parts of Europe and Asia (brown bear), bear populations have increased and stabilized because of increased management effort and increasing support for bears and their needs by the humans who share habitat with them. However, for most bear species, the future is uncertain and increasingly dire. Andean bears continue to be threatened by habitat loss and human encroachment. In much of Asia outside Japan, Asiatic black bear, sloth bear, and sun bear populations are increasingly threatened by unmanaged excessive mortality combined with habitat loss to timber harvest, plantation agriculture, and human encroachment. The long-term future for polar bears is threatened by the unmanageable threat of climate change that will irrevocably change the unique habitat on which they depend. Giant pandas are fragmented into small populations despite intense conservation efforts.

There have been advances in conservation techniques for bears, but the application of these techniques is often difficult due to lack of funding and opposition to bears by humans who are sometimes in conflict with them. The primary determinant whether bears expand and flourish or decline and disappear is not only science or the commitment of those who seek to help bears, but also the attitudes and support (or lack of support) of the people who live, work, and recreate in bear habitat and the political forces that represent these local people. Improving public and political support for bears is the most important need if we are to realize successful bear conservation and management.

References

Abbas, F., Bhatti, Z., Haider, J. & Mian, A. (2015). Bears in Pakistan: distribution, population biology and human conflicts. *Journal of Bioresource Management* 2(2): 1–13.

Abrams, J. F., Hörig, L., Brozovic, R., Axtner, J., et al. (2019). Shifting up a gear with iDNA: from mammal detection events to standardized surveys. *Journal of Applied Ecology* 56: 1637–1648.

Acevedo, C., Bernal, N., Bianchi, G., et al. (2017). Conservamos la vida: Andean bear conservation at the landscape scale. In: Molina, S., Zug, B., Velez-Liendo, X., et al. (Eds.), *25th Conference on Bear Research and Management* (p. 9). Quito, Ecuador: International Association for Bear Research and Management.

Allen, M. L., Norton, A. S., Stauffer, G., et al. (2018). A Bayesian state-space model using age-at-harvest data for estimating the population of black bears (*Ursus americanus*) in Wisconsin. *Scientific Reports* 8: 1–12.

Ambarlı, H. & Bilgin, C. C. (2008). Human–brown bear conflicts in Artvin, northeastern Turkey: encounters, damage, and attitudes. *Ursus* 19(2): 146–153.

Amstrup, S. C., McDonald, T. L. & Stirling, I.I. (2001). Polar bears in the Beaufort Sea: a 30-year mark–recapture case history. *Journal of Agricultural, Biological, and Environmental Statistics* 6: 221–234.

Animals Asia. (2011). *Bear farming industry in China*. Chengdu, China: Animals Asia Foundation.

Atwood, T. C., Peacock, E., McKinney, M. A., et al. (2016). Rapid environmental change drives increased land use by an arctic marine predator. *PLoS ONE* 11(6): e0155932. https://doi.org/10.1371/journal.pone.0155932

BANCA. (2017). The bear trade and bear farms in transboundary regions regions in Eastern Myanmar. Yangon, Myanmar: Biodiversity and Nature Conservation Association. Available from www.banca-env.org/wp-content/uploads/2018/05/the-bear-trade-and-bear-farms.pdf

Bargali, H. S., Akhtar, N. & Chauhan, M. P. S. (2005). Characteristics of sloth bear attacks and human casualties in North Balispur Forest Division, Chhattisgarh, India. *Ursus* 16: 263–267.

Bargali, H.S., Akhtar, N. & Chauhan, M. P. S. (2012). The sloth bear activity and movement in highly fragmented and disturbed habitat in central India. *World Journal of India* 7: 312–319.

Baskin, K. (2016). Hunting as sustainable wildlife management. *Mammal Study* 41: 173–180. https://doi.org/10.3106/041.0402.

Beier, L. R., Lewis, S. B., Flynn, R. W., et al. (2005). A single-catch snare to collect brown bear hair for genetic mark–recapture studies. *Wildlife Society Bulletin* 33(2): 766–773.

Belant, J. L., Etter, D. R., Mayhew, S. L., Visser, L. G. & Friedrich, P. D. (2011). Improving large-scale mark–recapture estimates for American black bear populations. *Ursus* 22(1): 9–23.

Bellemain, E., Swenson, J. E., Tallon, D., et al. (2005). Estimating population size of elusive animals with DNA from hunter-collected feces: four methods for brown bears. *Conservation Biology* 19(1): 150–161.

Berkes, F. (2007). Community-based conservation in a globalized world. *Proceedings of the National Academy of Sciences* 104(39): 15,188–15,193. http://dx.doi.org/10.1073/pnas.0702098104.

Bischof, R. & Swenson, J. E. (2012). Linking noninvasive genetic sampling and traditional monitoring to aid management of a trans-border carnivore population. *Ecological Applications* 22: 361–373.

Boulanger, J. & Stenhouse, G. B. (2014). The impact of roads on the demography of grizzly bears in Alberta. *PLoS ONE* 9(12): e115535. https://doi.org/10.1371/journal.pone.0115535

Boulanger, J., Kendall, K. C., Stetz, J. B., et al. (2008). Use of multiple data sources to improve DNA-based mark–recapture population estimates of grizzly bears. *Ecological Applications* 18(3): 577–589.

Broadis, N. (2011). Evolving threats; lessons learnt from 15 years of bear protection in Cambodia. Oral presentation, 21st International Conference for Bear Research and Management, Delhi, India.

Brown, T. H., Simangunsong, B. C. H., Sukadri, D., et al. (2005). *Restructuring and revitalization of Indonesia's wood-based industry: Synthesis of three major studies*. Jakarta: MoF-NRM-CIFOR-MFP-Forest Trends-ITTO.

Burgess, E. A., Stoner, S. S. & Foley, K. E. (2014). *Brought to bear: An analysis of seizures across Asia (2000–2011)*. Petaling Jaya, Selangor, Malaysia: TRAFFIC Southeast Asia.

Burton, A. C., Neilson, E., Moreira, D., Ladle, A., et al. (2015). Wildlife camera trapping: a review and recommendations for linking surveys to ecological processes. *Journal Applied Ecology* 52(3): 675–685.

Can, O. E., D'Cruze, N., Garshelis, D., Beecham, J. & MacDonald, D. W. (2014). Resolving human–bear conflict: a global survey of countries, experts, and key factors. *Conservation Letters* 7(6): 510–513.

Caughley, G. (1994). Directions in conservation biology. *Journal of Animal Ecology* 63(2): 215–244.

Chapron, G., Kaczensky, P., Linnell, J. D. C., et al. (2014). Successful recovery of large carnivores in Europe's human-dominated landscapes. *Science* 346(6216): 1517–1519.

Cherry, S., Haroldson, M. A., Robison-Cox, J. & Schwartz, C.C. (2002). Estimating total human-caused mortality from reported mortality using data from radio-instrumented grizzly bears. *Ursus* 13: 175–184.

CITES. (2000). Trade in bear specimens. Eleventh meeting of the Conference of the Parties Gigiri (Kenya), April 10–20, 2000. Gigiri, Kenya.

Clapham, M., Miller, E. & Nguyen, M. (2018). Developing automated face recognition technology for non-invasive monitoring of brown bears (*Ursus arctos*). In: Skrbinšek, M. A. (Ed.), *26th International Conference on Bear Research and Management*. University of Ljubljana, Abstract 237 (p. 22). Available from: https://lifewithbears.eu/book-of-abstracts/

Clark, T. W. (2002). *The policy process. A practical guide for natural resource professionals*. New Haven, CT: Yale University Press.

Clark, T. W. & Wallace, R. L. (1999). The professional in endangered species conservation: an introduction to standpoint clarification. *Endangered Species Update* 16(1): 9–13.

Convention on the Conservation of European Wildlife and Natural Habitats (2018). Recommendation No. 198 (2018) on the use of artificial feeding as a management tool of large carnivore populations and their prey, with a particular emphasis on the brown bear. 17 [document T-PVS(2018)7]. Available from https://rm.coe.int/list-of-decisions-and-adopted-textes-of-the-38th-meeting-of-the-bern-c/16808fde8d

Cooch, E. G. & White, G. C. (2019). Program MARK; a gentle introduction. Available from www.phidot.org/software/mark/docs/book/pdf/mark_book.zip

Crête, M., Vandal, D., Rivest, L. P. & Potvin, F. (1991). Double counts in aerial surveys to estimate polar bear numbers during the ice-free period. *Arctic* 44(4): 275–278.

Crudge, B., Lees, C., Hunt, M., et al. (2019). *Range-wide Conservation Action Plan for the sun bear,* Helarctos malayanus, *2018–2028*. Gland: IUCN SSC Bear Specialist Group/Free the Bears/ TRAFFIC.

Curran, L. M., Trigg, S. N., McDonald, A. K., et al. (2004). Lowland forest loss in protected areas of Indonesian Borneo. *Science* 303: 1000–1003.

Czetwertynski, S., Boyce, M. S. & Schmiegelow, F. K. (2007). Effects of hunting on demographic parameters of American black bears. *Ursus* 18: 1–17.

Danilov, P. & Tirronen, K. (2011). Large predators in the Russian North-West. In: *Beiträge zur Jagd- und Wildforschung 36 Proceedings*, April 7–10, 2011 (pp. 19–25). Camp Reinsehlen, Germany.

Debata, S., Swain, K. K., Sahu, H. K. & Palei, H. S. (2016). Human–sloth bear conflict in a human-dominated landscape of northern Odisha, India. *Ursus* 27: 90–98.

Denali National Park. (2003). Bear–human conflict management plan, Denali National Park and Preserve. Center for Resources, Science, and Learning. US National Park Service.

Derocher, A. E. (2012). *Polar bears: A complete guide to their biology and behavior*. Baltimore, MD: Johns Hopkins University Press.

Derocher, A., Aars, J. A., Amstrup, S. C., et al. (2013). Rapid ecosystem change and polar bear conservation. *Conservation Letters* 6(5): 368–375.

Dressel, S., Sandström, C. & Ericsson, G. (2015). A meta-analysis of studies on attitudes toward bears and wolves across Europe 1976–2012. *Conservation Biology* 29(2): 565–574.

Dutta, T. J., Maldanado, E., Panwar, H. S. & Seidensticker, J. (2015). Genetic variation, structure, and gene flow in a sloth bear (*Melursus ursinus*) metapopulation in the Satpura-Maikal landscape of central India. *PLoS ONE* 10(5): e0123384.

Dutton, A., Hepburn, J. C. & Macdonald, D. W. (2011). A stated preference investigation into the Chinese demand for farmed vs. wild bear bile. *PLoS ONE* 6: e21243. https://doi.org/10.1371/journal.pone.0021243

Eberhardt, L. L. (2002). A paradigm for population analysis of long-lived vertebrates. *Ecology* 83: 2841–2834.

EIA. (2015). *Sin City: Illegal wildlife trade in Laos' Golden Triangle Special Economic Zone*. London, UK: EIA.

Elfström, M., Davey, M. L., Zedrosser, A., et al. (2014). Do Scandinavian brown bears approach settlements to obtain high-quality food? *Biological Conservation* 178: 128–135.

Ervin J. (2003). Protected area assessments in perspective. *BioScience* 53(9): 819–822.

European Union. (1992). Council Directive 92/43/EEC of 21 May 1992 on the conservation of natural habitats and of wild fauna and flora. Available from https://eur-lex.europa.eu/legal-content/EN/TXT/?uri=CELEX:31992L0043

Fahrig, L. (2003). Effects of habitat fragmentation on biodiversity. *Annual Review of Ecology, Evolution, and Systematics* 34: 487–515.

Feng, Y., Siu, K., Wang, N., et al. (2009). Bear bile: dilemma of traditional medicinal use and animal protection. *Journal of Ethnobiology and Ethnomedicine* 5: 2.

Fieberg, J. R., Shertzer, K W., Conn, P. B., Noyce, K. V. & Garshelis. D. (2010). Integrated population modeling of black bears in Minnesota: implications for monitoring and management. *PLoS ONE* 5(8): e12114. https://doi.org/10.1371/jjournal.pone 0012114

Foley, K. E., Stengel, C. J. & Shepherd, C. R. (2011). *Pills, powders, vials and flakes: The bear bile trade in Asia*. Petaling Jaya, Selangor, Malaysia: TRAFFIC Southeast Asia.

Fredriksson, G. (2005). Human–sun bear conflicts in east Kalimantan, Indonesian Borneo. *Ursus* 16: 130–137.

Fredriksson, G., Wich, S. A. & Trisno. (2006). Frugivory in sun bears (*Helarctos malayanus*) is linked to El Niño-related fluctuations in fruiting phenology, East Kalimantan, Indonesia. *Biological Journal of the Linnean Society* 89(3): 489–508.

Gangaas, K. E., Kaltenborn, B. P. & Andreassen, H. P. (2013). Geo-spatial aspects of acceptance of illegal hunting of large carnivores in Scandinavia. *PLoS ONE* 8(7): e68849. https://doi.org/10.1371/journal.pone.0068849

Garshelis, D. (1997). The arrogance of ignorance – a commentary on the bear trade. *International Bear News* 6: 4–6.

Garshelis, D. (2002). Misconceptions, ironies, and uncertainties regarding trends in bear population. *Ursus* 13: 321–334.

Garshelis, D. (2009). Family Ursidae (bears). In: Wilson, D.E. & Mittermeier, R.A. (Eds.), *Handbook of the mammals of the world. Vol. 1. Carnivores* (pp. 448–497). Barcelona: Lynx Editions.

Garshelis, D. (2011). Andean bear density and abundance estimates – how reliable and useful are they? *Ursus* 22(1): 47–64.

Garshelis, D. & Hristienko, H. (2006). State and provincial estimates of American black bear numbers versus assessments of population trend. *Ursus* 17(1): 1–7.

Garshelis, D. & Steinmetz, R. (2017). *Ursus thibetanus*. IUCN Red List of Threatened Species. www.iucnredlist.org.

Garshelis, D. & Visser, L. G. (1997). Enumerating megapopulations of wild bears using an integrated biomarker. *Journal of Wildlife Management* 61(2): 466–480.

Garshelis, D., Joshi, A. R. & Smith, J. L. D. (1999a). Estimating density and relative abundance of sloth bears. *Ursus* 11: 87–98.

Garshelis, D., Joshi, A. R., Smith, J. L. D. & Rice, C. G. (1999b). Sloth bear conservation action plan. In: Servheen, C., Herrero, S. & Peyton, B. (Eds.), *Status survey and conservation action plan: Bears* (chapter 12). Gland, Switzerland and Cambridge, UK: IUCN/SSC Bear and Polar Bear Specialist Groups.

Garshelis, D., Gibeau, M. L. & Herrero, S. (2005). Grizzly bear demographics in and around Banff National Park and Kananaskis Country, Alberta. *Journal of Wildlife Management* 69(1): 277–297.

Garshelis, D., Wang, H., Wang, D., et al. (2008). Do revised giant panda population estimates aid in their conservation? *Ursus* 19: 168–176.

Gaston, K. J., Jackson, S. F., Cantu-Salazar, L. & Cruz-Pinon, G. (2008). The ecological performance of protected areas. *Annual Review of Ecology, Evolution, and Systematics* 39: 93–113.

Geist, V. (1995). North American policies of wildlife conservation. In: Geist, V. & McTaggert-Cowan, I. (Eds.), *Wildlife conservation policy* (pp. 75–129). Calgary: Detselig Enterprises Limited.

Geist, V., Mahoney, S. P. & Organ, J. F. (2001). Why hunting has defined the North American model of wildlife conservation. *Transactions of the North American Wildlife and Natural Resources Conference* 66: 175–185.

Gibeau, M. L. (2012). Of bears, chess, and checkers: moving away from pure science to solve problems. *The Wildlife Professional* 6: 62–64.

Glacier National Park. (2010). Bear management plan, Glacier National Park. US National Park Service.

Gong, J. E. & Harris, R. B. (2006). The status of bears in China: a country report. In: Japan Bear Network (Compiler), *Understanding Asian bears to secure their future* (pp. 96–101). Ibaraki, Japan: Japan Bear Network.

Gong, M., Guan, T., Hou, M., Liu, G. & Zhou, T. (2017). Hopes and challenges for giant panda conservation under climate change in the Qinling Mountains of China. *Ecology and Evolution* 7: 596–605.

Gosselin, J., Zedrosser, A., Swenson, J. E. & Peletier, F. (2014). The relative importance of direct and indirect effects of hunting mortality on the population dynamics of brown bears. *Proceedings of the Royal Society B* 282: 20141840. http://dx.doi.org/10.1098.rsph.2014.1840.

Gray, T. N. E., Lynam, A. J., Seng, T., et al. (2017). Wildlife-snaring crisis in Asian forests. *Science* 355: 255–256.

Gunther, K. (Ed.). (2015). Grizzly bear recovery in the Yellowstone Ecosystem. *Yellowstone Science* 23(2): 4–46.

Harris, R. B., Schwartz, C. C., Haroldson, M. A. & White, G. C. (2006). Trajectory of the Yellowstone grizzly bear population under alternative survival rates. In: Schwartz, C. C., Haroldson, M. A., White, G. C., et al. *Temporal, spatial and environmental influences on the demographics of grizzly bears in the Greater Yellowstone Ecosystem* (pp. 44–56). Wildlife Monographs 161.

Harrison, R. D. (2011). Emptying the forest: hunting and the extirpation of wildlife from tropical nature reserves. *BioScience* 61: 919–924.

Hashimoto, Y., Kaji, M., Sawada, H. & Takasuki, S. (2003). Five-year study on the autumn food habits of the Asiatic black bear in relation to nut production. *Ecological Research* 18(5): 485–482.

Hatter, I. W., Mowat, G. & McLellan, B. N. (2018). Statistical population reconstruction to evaluate grizzly bear trends in British Columbia, Canada. *Ursus* 29(1): 1–12.

Heller, N. E. & Zaveleta, E. S. (2009). Biodiversity management in the face of climate change: a review of 11 years of recommendations. *Biological Conservation* 142: 14–32.

Hertel, A. G., Steyaert, S. M. J. G., Zedrosser, A., et al. (2016). Bears and berries: species-species selection foraging on a patchily distributed food resource in a human-altered landscape. *Behavioral Ecology and Sociobiology* 70(6): 931–842.

Hertel, A. G., Bischof, R., Langval, O., et al. (2018a). Berry production drives bottom-up effects on body mass and reproductive success in an omnivore. *Oikos* 127: 197–207.

Hertel, A. G., Langval, O., Swenson, J. E., Kindberg, J. & Zedrosser A. (2018b). Fluctuating mast production does not drive Scandinavian brown bear behavior. *Journal of Wildlife Management* 83(4): 657–668. https://doi.org/10.1002/jwmg.21619

Hilderbrand, G. V., Schwartz, C. C., Robbins, C. T., et al. (1999). The importance of meat, particularly salmon, to body size, population productivity, and conservation of North American brown bears. *Canadian Journal of Zoology* 77(1): 132–138.

Hooker, M. J., Laufenberg, J. S., Ashley, A. K., Sylvest, J. T. & Chamberlain, M. J. (2015). Abundance and density estimation of the American black bear population in central Georgia. *Journal of Wildlife Management* 26(2): 107–115.

Höss, M., Kohn, M., Pääbo, S., Knauer, F. & Schröder W. (1992). Excrement analysis by PCR. *Nature* 359: 199.

Housty, W. G., Noson, A., Scoville, G. W., et al. (2014). Grizzly bear monitoring by the Heiltsuk people as a crucible for First Nation conservation practice. *Ecology and Society* 19(2): 70.

Huber, D., Jakšić, Z., Frković, A., et al. (2008a). *Brown bear management plan for the Republic of Croatia*. Zagreb, Croatia: Ministry of Agriculture, Forestry and Water Management, and Ministry of Culture.

Huber, D., Kusak, J., Majić-Skrbinšek, A., Majnarić, D. & Sindičić, M. (2008b). A multidimensional approach to managing the European brown bear in Croatia. *Ursus* 19(1): 22–32.

Humm, J. M., McCowen, J. W., Scheick, B. K. & Clark, J. D. (2017). Spatially explicit population estimates for black bears based on cluster sampling. *Journal of Wildlife Management* 81(7): 1187–1201. doi: 10.1002/jwmg.21294

IUCN. (2012). Resolutions and Recommendations World Conservation Congress, Jeju, Republic of Korea, September 6–15, 2012. IUCN SSC Bear Specialist Group.

John, D., Linnell, C., Trouwborst, A., et al. (2016). Border security fencing and wildlife: the end of the transboundary paradigm in Eurasia? *PLoS Biology* 14(6): e1002483.

Jones, N. A., Shaw, S., Ross, H., Witt, K. & Pinner, B. (2016). The study of human values in understanding and managing social ecological systems. *Ecology and Society* 21(1): 15.

Joppa, L. N. & Pfaff, A. (2009). High and far: biases in the location of protected areas. *PLoS ONE* 4(12): e8273. https://doi.org/10.1371/journal.pone.0008273

Keay, J. A., Robbins, C. T. & Farley, S D. (2018). Characteristics of a naturally regulated grizzly bear population. *Journal of Wildlife Management* 82(4): 789–801.

Kellert, S.R. (1994). Public attitudes toward bears and their conservation. *International Conference on Bear Research and Management* 9(1): 43–50.

Kellert, S. R. (1996). *The value of life: Biological diversity and human society*. Washington, DC: Island Press.

Kendall, K. C., Stetz, J. B., Roon, D. A., et al. (2008). Grizzly bear density in Glacier National Park, Montana. *Journal of Wildlife Management* 72(8): 1693–1705.

Kendall, K. C., Stetz, J. B., Boulanger, J., et al. (2009). Demography and genetic structure of a recovering grizzly bear population. *Journal of Wildlife Management* 73(1): 3–16.

Kendall, K. C., Graves, T. A., Royle, J. A., et al. (2019). Using bear rub data and spatial capture-recapture models to estimate trend in a brown bear population. *Scientific Reports* 9: 16804.

Keyghobadi, N. (2007). The genetic implications of habitat fragmentation for animals. *Canadian Journal of Zoology* 85: 1049–1064.

Kindberg, J., Ericsson, G. & Swenson, J. E. (2009). Monitoring rare or elusive large mammals using effort-corrected voluntary observers. *Biological Conservation* 142: 159–165.

Kindberg, J., Swenson, J. E., Ericsson, G., et al. (2011). Estimating population size and trends of the Swedish brown bear *Ursus arctos* population. *Wildlife Biology* 17(2): 114–123. doi: 10.2981/10-100

Knott, E. J., Bunnefeld, N., Huber, D., et al. (2014). The potential impacts of changes in bear hunting policy for hunting organizations in Croatia. *European Journal of Wildlife Research* 60(1): 85–97.

Kojola, I., Hallikainen, V., Helle, T. & Swenson, J. E. (2018). Can only poorer European countries afford large carnivores? *PLoS ONE* 13(4): e0194711. https://doi.org/10.1371/journal.pone.0194711

Kolenosky, G. B. (1986). The effects of hunting on an Ontario black bear population. *Bears: Their Biology and Management* 6: 45–55.

Krishnasamy, K. & Shepherd, C. (2014). A review of the sun bear trade in Sarawak, Malaysia. *TRAFFIC Bulletin* 26: 37–40.

Krofel, M., Spacapan, M. & Jerina, K. (2017). Winter sleep with room service: denning behaviour of brown bears with access to anthropogenic food. *Journal of Zoology* 302(1): 8–14.

Kusak, J., Huber, D., Gomerčić, T., Schwaderer, G. & Gužvica, G. (2009). The permeability of a highway in Gorski Kotar (Croatia) for large mammals. *European Journal of Wildlife Research* 55: 7–21.

Lamb, C. T., Mowat, G., Reid, A., et al. (2018). Effects of habitat quality and access management on the density of a recovering grizzly bear population. *Journal of Applied Ecology* 55(3): 1406–1417.

Lasswell, H. D. (1971). *A preview of policy sciences*. New York, NY: American Elsevier.

Leslie, P. H. (1945). On the use of matrices in certain population mathematics. *Biometrika* 33(3): 183–212.

Li, S., Tan, H. Y., Wang, N., et al. (2016). Substitutes for bear bile for the treatment of liver diseases: research progress and future perspective. *Evidence-based Complementary and Alternative Medicine* 2016: 4305074. https://doi.org/10.1155/2016/4305074

Linnell, J. D. C., Swenson, J. E. & Andersen, R. (2001). Predators and people: conservation of large carnivores is possible at high human densities if management policy is favourable. *Animal Conservation* 4(4): 345–349.

Linnell J., Salvatori, V. & Boitani, L. (2008). *Guidelines for population level management plans for large carnivores in Europe: A Large Carnivore Initiative for Europe report prepared for the European Commission* (contract 070501/2005/424162/MAR/B2).

Liu, F., McShea, W. J., Garshelis, D., et al. (2011). Human–wildlife conflicts influence attitudes but not necessarily behaviors: factors driving the poaching of bears in China. *Biological Conservation* 144: 538–547. doi:10.1016/j.biocon.2010.10.009.

Liu, J., Hull, V., Yang, W., et al. (2016). *Pandas and people: Coupling human and natural systems for sustainability*. Oxford: Oxford University Press.

Livingstone, E. & Shepherd, C. R. (2014). Bear farms in Lao PDR expand illegally and fail to conserve wild bears. *Oryx* 50: 1–9.

Livingstone, E., Gomez, L. & Bouhuys, J. (2018). A review of bear farming and bear trade in Lao People's Democratic Republic. *Global Ecology and Conservation* 13: e00380. https://doi.org/10.1016/j.gecco.2018.e00380

Loucks, C. J., Lu, Z., Dinerstein, E., et al. (2001). Giant pandas in a changing landscape. *Science* 294: 1465.

Lunn, N. J., Servanty, S., Regehr, E. V., et al. (2016). Demography of an apex predator at the end of its range: impacts of changing sea ice on polar bears in Hudson Bay. *Ecological Applications* 26(5): 1302–1320.

Mace, R. D., Waller, J. S., Manley, T. L., Lyon, L. J. & Zuring, H. (1996). Relationships among grizzly bears, roads, and habitat use in the Swan Mountains, Montana. *Journal of Applied Ecology* 33(6): 1395–1404.

Mace, R. D., Carney, D. W., Chilton-Radandt, T., et al. (2012). Grizzly bear population vital rates and trend in the Northern Continental Divide Ecosystem, Montana. *Journal of Wildlife Management* 76(1): 119–128.

Mahoney, S. P. (2004). The seven sisters: pillars of the North American wildlife conservation model. *Bugle* 21: 5.

Malenfant, R. M., Davis, C. S., Cullingham, C. I. & Coltman, D. W. (2016). Circumpolar genetic structure and recent gene flow of polar bears: a reanalysis. *PLoS ONE* 11(3): e0148967. doi:10.1371/journal.pone.0148967.

Manfredo, M. J., Teel, T. L., Sullivan, L. & Dietsch, A. M. (2017). Values, trust, and cultural backlash in conservation governance: the case of wildlife management in the United States. *Biological Conservation* 214: 303–311.

Margono, B. A., Potapov, P. V., Turubanova, S., Stolle, F. & Hansen, M. C. (2014). Primary forest cover loss in Indonesia over 2000–2012. *Nature Climate Change* 4: 1–6.

Mattson, D. J. & Merrill, T. (2002). Extirpations of grizzly bears in the contiguous United States, 1850–2000. *Conservation Biology* 16(4): 1123–1136.

Maxwell, S. L., Fuller, R. A., Brooks, T. M. & Watson, J. E. M. (2016). Biodiversity: the ravages of guns, nets, and bulldozers. *Nature News* 536: 143–145.

McLaughlin, G. P., Primm, S. & Rutherford, M. B. (2005). Participatory projects for coexistence: rebuilding civil society. In: Clark, T. W., Rutherford, M. B. & Casey, D. (Eds.), *Coexisting with large carnivores: Lessons from Greater Yellowstone* (pp. 177–210). Washington, DC: Island Press.

McLellan, B. N. (2005). Sexually selected infanticide in grizzly bears: the effects of hunting on cub survival. *Ursus* 16: 141–156.

McLellan, B. N. (2011). Implications of a high-energy and low-protein diet on the body composition, fitness, and competitive abilities of black (*Ursus americanus*) and grizzly (*Ursus arctos*) bears. *Canadian Journal of Zoology* 89(6): 546–558.

McLellan, B. N. (2015). Some mechanisms underlying variation on vital rates on grizzly bear in an industrial landscape. *Journal of Wildlife Management* 79(5): 749–765.

McLellan, B. N. & Hovey, F. W. (2001a). Habitats selected by grizzly bears in multiple use landscapes. *Journal of Wildlife Management* 65(1): 92–99.

McLellan, B. N. & Hovey, F. W. (2001b). Natal dispersal of grizzly bears. *Canadian Journal of Zoology* 79: 838–844.

McLellan, B. N., Hovey, F. W., Mace, R. D., et al. (1999). Rates and causes of grizzly bear mortality in the interior mountains of British Columbia, Alberta, Montana, Washington, and Idaho. *Journal of Wildlife Management* 63(3): 911–920.

McLellan, B. N., Proctor, M. F., Huber, D. & Michel, S. (2016). *Ursus arctos* Brown bear. IUCN Red List of Threatened Species. www.iucnredlist.org.

McLellan, B. N., Mowat, G., Hamilton, T. & Hatter, I. (2017). Sustainability of the grizzly bear hunt in British Columbia, Canada. *Journal of Wildlife Management* 81(2): 218–229.

McLellan, B. N., Mowat, G. & Lamb, C. T. (2018). Estimating unrecorded human-caused mortalities of grizzly bears in the Flathead Valley, British Columbia, Canada. *PeerJ* 6: e5781.

Mealy, S. (1979 [revised 1986]). *Guidelines for management involving grizzly bears in the Greater Yellowstone Area*. US Forest Service and US National Park Service.

Meijaard, E., Sheil, D., Nasi, R., et al. (2005). *Life after logging: Reconciling wildlife conservation and production forestry in Indonesian Borneo*. Bogor, Indonesia: CIFOR and UNESCO.

Miettinen, J., Shi, C. & Liew, S. C. (2011). Deforestation rates in insular Southeast Asia between 2000 and 2010. *Global Change Biology* 17: 2261–2270.

Mikle, N., Graves, T. A., Kovach, R., Kendall, K. C. & Macleod, A. (2016). Demographic mechanisms underpinning genetic assimilation of remnant groups of a large carnivore. *Proceedings of the Royal Society B* 283(1839): 20161467.

Miller, S. D. (1990). Population management of bears in North America. *International Conference on Bear Research and Management* 8: 357–373.

Miller, S. D. & Tutterrow, V. L. (1999). Characteristics of non-sport mortalities to brown and black bears and human injuries from bears in Alaska. *Ursus* 11: 239–252.

Miller, S. D., White, G. C., Sellers, R. A., et al. (1997). Brown and black bear density estimation in Alaska using radio-telemetry and replicated mark–resight techniques. *Wildlife Monographs* 133.

Miller, S. D., Sellers, R. A. & Keay, J. A. (2003). Effects of hunting on brown bear cub survival and litter size in Alaska. *Ursus* 14: 130–152.

Miller, S. D., McLellan, B. N. & Derocher, A. E. (2013). Conservation and management of large carnivores in North America. In: Mahoney, S. P. (Ed.), *Conservation and hunting in North America. The International Journal of Environmental Studies Monographs* 70(3), 383–398. http://dx.doi.org/10.1080/00207233.2013:806099.

Miller, S. D., Schoen, J. W. & Schwartz, C. C. (2017). Trends in brown bear reduction efforts in Alaska, 1980–2017. *Ursus* 28(2): 135–149.

Mills, J. & Servheen, C. (1991). *The Asian trade in bears and bear parts.* Washington, DC: World Wildlife Fund.

Millspaugh, J. J., Skalski, J. R., Townsend, R. L., et al. (2010). An evaluation of sex-age-kill (SAK) model performance. *Journal of Wildlife Management* 73(3): 442–451.

Mukesh, S., Sharma, L. K., Charoo, S. A. & Sathyakumar, S. (2015). Conflict bear translocation: investigating population genetics and fate of bear translocation in Dachigam National Park, Jammu and Kashmir, India. *PLoS ONE* 10(8): e0132005. https://doi.org/10.1371/journal.pone.0132005

Müller, A. & Kaji, K. (2018). Wildlife policy and laws in East Asia. In: Leopold, B. D., Kessler, W. B. & Cummins, J. L. (Eds.) *North American Wildlife Policy and Laws* (pp. 445–479). Missoula, MT: Boone and Crockett Club.

Murdiyarso, D., Dewi, S., Lawrence, D. & Seymour, F. (2011). Indonesia's forest moratorium: a stepping stone to better forest governance? Working Paper 76. Bogor, Indonesia: CIFOR.

Naughton-Treves L., Holland, M. B. & Brandon, K. (2005). The role of protected areas in conserving biodiversity and sustaining local livelihoods. *Annual Review of Environment and Resources* 30: 219–252.

Newton, E. & Obbard, M. E. (2018). Can population reconstruction be used to estimate black bear abundance in Ontario? Ontario Ministry of Natural Resources and Forestry, Science and Research Branch, Peterborough, ON. Science & Research Information Report IR-12. www.researchgate.net/publication/323956404

Ngoprasert, D., Reed, D. H., Steinmetz, R. & Gale, G. A. (2012). Density estimation of Asian bears using photographic capture-recapture sampling based on chest marks. *Ursus* 23(2): 117–133.

Ngoprasert, D., Gale, G. A., Vichitsoonthonkul, T. & Vaeokhaw, S. (2015). *Population demographics and genetics of Asiatic black bear and sun bear in the World Heritage Dong Phayayen-Khao Yai Forest Complex.* Bangkok, Thailand: King Mongkut's University of Technology.

Nielsen, S. E., Herrero, S., Boyce, M. S., et al. (2004). Modeling the spatial distribution of human-caused grizzly bear mortalities in the Central Rockies Ecosystem of Canada. *Biological Conservation* 120(1): 101–113.

Nomura, F., Higashi, S., Ambu, L. & Mohamed, M. (2004). Notes on oil palm plantation use and seasonal spatial relationships of sun bears in Sabah, Malaysia. *Ursus* 15(2): 227–231.

Obbard, M. E., Cattet, M. R. L., Moody, T., et al. (2006). Temporal trends in the body condition of southern Hudson Bay polar bears. *Climate Change Research Information Note #3.* Ontario: Applied Research and Development Branch.

Obbard, M. E., Howe, E. J., Wall, L. L., et al. (2014). Relationships among food availability, harvest, and human-bear conflict at landscape scales in Ontario, Canada. *Ursus* 25(2): 98–110.

Obbard, M. E., Stapleton, S., Middel, K.R., et al. (2015). Estimating the abundance of the Southern Hudson Bay polar bear subpopulation with aerial surveys. *Polar Biology* 38: 1713–1725.

Oi, T. & Yamazaki, K. (Eds.). (2009). The status of Asiatic black bears in Japan. In: *Understanding Asian bears to secure their future* (pp. 122–133). Ibaraki, Japan: Japan Bear Network.

Ordiz A., Rodríguez C., Naves C., et al. (2007). Distance-based criteria to identify minimum number of brown bear females with cubs in Europe. *Ursus* 18: 158–167.

Ordiz, A., Kindberg, J., Sæbø, S., Swenson, J. E. & Støen, O.-G. (2014). Brown bear circadian behavior reveals human environmental encroachment. *Biological Conservation* 173: 1–9.

Organ, J. F., Geist, V., Mahoney, S. P., et al. (2012). *The North American model of wildlife conservation.* The Wildlife Society Technical Review 12-04. Bethesda, MD: The Wildlife Society.

Paetkau, D. (2003). An empirical exploration of data quality in DNA-based population inventories. *Molecular Ecology* 12(6): 1375–1387.

Paetkau, D. Amstrup, S. C., Born, E. W., et al. (1999). Genetic structure of the world's polar bear populations. *Molecular Ecology* 8: 1571–1584.

Parques Nacionales Naturales de Colombia & Wildlife Conservation Society. (2018). *Estrategia para la Conservación del Oso Andino en los Parques Nacionales Naturales de Colombia (2016–2031).* Bogotá D.C.

Peacock, E., Sonsthagen, S. A., Obbard, M. L., et al. (2015). Circumpolar genetic structure and recent gene flow of polar bears: a reanalysis. *PLoS ONE* DOI:10.1371/journal.pone.011202.

Peck, C. P., Van Manen, F. T., Costello, C., et al. (2017). Potential paths for male-mediated gene flow to and from an isolated grizzly bear population. *Ecosphere* 8(10): e0196. doi.org/10.1002/ecs2.1969.

Pelton, M. R. (2003). Black bear, *Ursus americanus.* In: Feldhamer, G. A., Thompson, B. C. & Chapman, J. A. (Eds.), *Wild mammals of North America* (2nd edition, pp. 547–555). Baltimore, MD: The Johns Hopkins University Press.

Penteriani, V., Melletti, M. & Delgado, M. M. (2010). Don't feed the bears! *Oryx* 44: 169–170.

Penteriani, V., Lopez-Bao, J. V., Bettega, C., et al. (2017). Consequences of brown bear tourism: a review. *Biological Conservation* 206: 169–180.

Penteriani, V., Delgado, M. M., Krofel, M., et al. (2018). Evolutionary and ecological traps for brown bears in human-modified landscapes. *Mammal Review* 48: 180–193.

Penteriani, V., Zarzo-Arias, A., Novo-Fernández, A, Bombieri, G. & López-Sánchez, C. A. (2019). Responses of an endangered brown bear population to climate change based on predictable food resource and shelter alterations. *Global Change Biology* 25: 1133–1151.

Pereira, H. M., Leadley, P. W., Proença, V., et al. (2010). Scenarios for global biodiversity in the 21st century. *Science* 330(6010): 1496–1501. doi:10.1126/science.1196624

Peterson, N. N. & Nelson, M. P. (2016). Why the North American model for wildlife conservation is problematic for modern wildlife management. *Human Dimensions of Wildlife* 22(1): 43–54.

Pierson, J. C., Graves, T. A., Banks, S. C., Kendall, K. C. & Lindenmayer, D.B. (2018). Relationship between effective and demographic population size in continuously distributed populations. *Evolutionary Applications* 11(7): 1162–1175.

Prestrud, P. & Stirling, I. (1994). The International Polar Bear Agreement and the current status of polar bear conservation. *Aquatic Mammals* 20: 113–124.

Pritchard, G. T. & Robbins, C. T. (1990). Digestive and metabolic efficiencies of grizzly and black bears. *Canadian Journal of Zoology* 68(8): 1645–1651.

Proctor, M. F., McLellan, B. N., Strobeck, C. & Barclay, R. (2004). Gender specific dispersal distances of grizzly bears estimated by genetic analysis. *Canadian Journal of Zoology* 82: 1108–1118.

Proctor, M. F., McLellan, B. N., Strobeck, C. & Barclay, R. (2005). Genetic analysis reveals demographic fragmentation of grizzly bears yielding vulnerably small populations. *Proceedings of the Royal Society, London B* 272: 2409–2416.

Proctor, M. F., McLellan, B. N., Boulanger, J., et al. (2010). Ecological investigations of grizzly bears in Canada using DNA from hair, 1995–2005. *Ursus* 21(2): 169–188.

Proctor, M. F., Paetkau, D., McLellan, B. N., et al. (2012). Population fragmentation and inter-ecosystem movements of grizzly bears in western Canada and the northern United States. *Wildlife Monographs* 180: 1–46.

Proctor, M. F., Lamb, C. T. & MacHutchon, A. G. (2017). *The grizzly dance of berries and bullets: The relationship between bottom up food resources, huckleberries, and top down mortality risk on grizzly bear population processes in southeast British Columbia*. Kaslo, BC, Canada: Trans-border Grizzly Bear Project. Available from http://transbordergrizzlybearproject.ca/research/publications.html.

Proctor, M. F., McLellan, B. N., Stenhouse, G. B., et al. (2018a). *Resource roads and grizzly bears in British Columbia, and Alberta*. Canadian Grizzly Bear Management Series, Resource Road Management. Kaslo, BC, Canada: Trans-border Grizzly Bear Project. Available from http://transbordergrizzlybearproject.ca/research/publications.html.

Proctor, M. F., Kasworm, W. F., Annis, K. M., et al. (2018b). Conservation of threatened Canada–USA trans-border grizzly bears linked to comprehensive conflict reduction. *Human Wildlife Interactions* 12(3): 348–372.

Proctor, M. F., Kasworm, W. F., Teisberg, J. E., et al. (2020a). American black bear population fragmentation detected with pedigrees in the transborder Canada–United States region. *Ursus* 2020(31e1): 1–15.

Proctor, M. F., McLellan, B. N., Stenhouse, G. B., et al. (2020b). The effects of roads and motorized human access on grizzly bear populations in British Columbia and Alberta, Canada. *Ursus* 2019(30e2): 16–39.

Ratnayeke, S., van Manen, F. T., Pieris, R. & Praga, V. S. J. (2007). Landscape characteristics of sloth bear ranges in Sri Lanka. *Ursus* 18: 189–202.

Ratnayeke, S., van Manen, F. T., Pieris, R. & Praga, V. S. J. (2014). Challenges of large carnivore conservation: sloth bear attacks in Sri Lanka. *Human Ecology* 42: 467–479.

Redpath, S., Linnell, J. C. D., Festa-Bianchet, M., et al. (2017). Don't forget to look down – collaborative approaches to predator conservation. *Biological Reviews* 92(4): 2157–2163.

Regehr, E. V., Wilson, R. R., Rode, K. D., Runge, M. C. & Stern, H. (2017). Harvesting wildlife affected by climate change: a modelling and management approach for polar bears. *Journal Applied Ecology* 54: 1534–1543.

Regehr, E. V., Hostetter, N. J., Wilson, R. R., et al. (2018). Integrated population modeling provides the first empirical estimates of vital rates and abundance for polar bears in the Chukchi Sea. *Scientific Reports* 8: 1–12.

Reljić, S., Jerina, K., Nilsen, E. B., et al. (2018). Challenges for transboundary management of a European brown bear population. *Global Ecology and Conservation*, 16: e00488. https://doi.org/10.1016/j.gecco.2018.e00488.

Rios-Uzeda, B. & Wallace, R. B. (2007). A preliminary density estimate for Andean bear using camera-trapping methods. *Ursus* 18(1): 124–128.

Ripple, W. J., Miller, S. D., Schoen, J. W. & Rabinowitch, S. P. (2019). Large carnivores under assault in Alaska. *PLoS Biology* 17(1): e3000090. https://doi.org/10.1371/journal.pbio.3000090

Robbins, C. T., Ben-David, M., Fortin, J. K. & Nelson, O. L. (2012). Maternal condition determines birth date and growth of newborn bear cubs. *Journal of Mammalogy* 93(2): 540–546.

Rolston, H. (1988). Human values and natural systems. *Society and Natural Resources* 1(1): 271–283.

Royle, J., Chandler, A. B., Sollmann, R. & Gardner, B. (2014). *Spatial capture-Recapture*. Waltham, MA: Elsevier Academic Press.

Rutherford, M. B. & Clark, T. W. (2005). Coexisting with large carnivores: lessons from Greater Yellowstone. In: Clark, T. W., Rutherford, M. B. & Casey, D. (Eds.), *Coexisting with large carnivores: Lessons from Greater Yellowstone* (pp. 254–270). Washington, DC: Island Press.

Rutherford, M. B., Gibeau, M. L., Clark, S. G. & Chamberlain, E. C. (2009). Interdisciplinary problem solving workshops for grizzly bear conservation in Banff National Park, Canada. *Policy Sciences* 42(2): 163–187.

Ryan, C. W., Pack, J. C., Igo, W. K. & Billings, A. (2007). Influence of mast production on black bear non-hunting mortalities in West Virginia. *Ursus* 16(1): 46–53.

Sæther, B. E., Engen, S., Swenson, J. E., Bakke, Ø. & Sandegren, F. (1998). Assessing the viability of Scandinavian brown bear, *Ursus arctos*, populations: the effects of uncertain parameter estimates. *Oikos* 83(2): 403–416.

Sathyakumar, S, Kaul, R., Ashraf, N. V. K., Mookerjee, A. & Menon, V. (2012). *National bear conservation and welfare action plan*. India: Ministry of Environment and Forests, Wildlife Institute of India and Wildlife Trust of India.

Schmidt, J. H., Rattenbury, K. L., Robison, H. L., Gorn, T. S. & Shults, B. S. (2017a). Using non-invasive mark-resight and sign occupancy surveys to monitor low-density brown bear populations across large landscapes. *Biological Conservation* 207: 47–54.

Schmidt, J. H., Wilson, T. L., Thompson, W. L. & Reynolds, J. H. (2017b). Improving inference for aerial surveys of bears: the importance of assumptions and the cost of unnecessary complexity. *Ecology and Evolution* 7(13): 4812–4821. https://doi:10.1002/ece3.2912.

Schwartz, C. C., Haroldson, M. A., Gunther, K. A. & Moody, D. (2002). Distribution of grizzly bears in the greater Yellowstone Ecosystem, 1990–2000. *Ursus* 13: 203–212.

Schwartz, C. C., Miller, S. D. & Haroldson, M. A. (2003). Grizzly bear. In: Feldhamer, G. A., Thompson, B. C. & Chapman, J. A. (Eds.), *Wild mammals of North America: Biology, management, and conservation* (2nd edition, pp. 556–586). Baltimore, MD: The Johns Hopkins University Press.

Schwartz, C. C., Haroldson, M. A., White, G. C., et al. (2006). Temporal, spatial, and environmental influences on the demographics of grizzly bears in the Greater Yellowstone Ecosystem. *Wildlife Monographs* 161.

Schwartz, C. C., Haroldson, M. A. & White, G. C. (2010). Hazards affecting grizzly bear survival in the greater Yellowstone ecosystem. *Journal of Wildlife Management* 74(4): 654–667.

Schwartz, C. C., Fortin, J. K., Teisberg, J. E., et al. (2014). Body and diet composition of sympatric black and grizzly bears in the Greater Yellowstone Ecosystem. *Journal of Wildlife Management* 78(1): 68–78.

Scotson, L., Vannachomchan, K. & Sharp, T. (2014). More valuable dead then deterred? Crop-raiding bears in Lao PDR. *Wildlife Society Bulletin* 38: 783–790.

Scotson, L., Fredriksson, G., Ngoprasert, D., Wong, W. M. & Fieberg, J. (2017a). Projecting range-wide sun bear population trends using tree cover and camera-trap bycatch data. *PLoS ONE* 12(9): e0185336. https://doi.org/10.1371/journal.pone.0185336.

Scotson, L., Fredriksson, G., Augeri, D., et al. (2017b). *Helarctos malayanus* sun bear. IUCN Red List of Threatened Species. www.iucnredlist.org/species/9760/123798233.

Scott, J. M., Goble, D. D., Wiens, J. A., et al. (2005) Recovery of imperiled species under the Endangered Species Act: the need for a new approach. *Frontiers in Ecology and the Environment* 3(7): 383–389.

Scott, J. M., Goble, D. D., Haines, A. M., Weins, J. A. & Neel, M. C. (2010). Conservation reliant species and the future of conservation. *Conservation Letters* 3(2): 91–97.

Seber, G. A. F. (1982). *The estimation of animal abundance*. London: Charles Griffin.

Selva, N., Berezowska-Cnota, T. & Elguero-Claramunt, I. (2014). Unforeseen effects of supplementary feeding: ungulate baiting sites as hotspots for ground-nest predation. *PLoS ONE* 9(3): e90740. https://doi.org/10.1371/journal.pone.0090740

Servheen, C. (1990). The status and conservation of the bears of the world. International Association for Bear Research and Management, Monograph #2.

Servheen, C. (1998). Conservation of small bear populations through strategic planning. *Ursus* 10: 67–73.

Servheen, C. & Mills, J. (1991). The trade in bears and bear parts. *International Conference on Bear Research and Management* 9: 161–167.

Servheen, C., Herrero, S. & Peyton, B. (Eds). (1999). *Bears: Status survey and conservation action plan*. Gland, Switzerland and Cambridge, UK: IUCN/SSC Bear and Polar Bear Specialist Goups.

Servicio Nacional de Areas Naturales Protegidas por el Estado, Gobierno Regional del Cusco, Wildlife Conservation Society, INKATERRA. (2014). *Estrategia para la Conservación del Oso Andino en el Santuario Histórico de Machupicchu y el Área de Conservación Regional Choquequirao*. Lima, Perú.

Sharma, S., Dutta, T., Maldonado, J. F., et al. (2013). Forest corridors maintain historic gene flow in a tiger metapopulation in the highlands of central India. *Proceedings of the Royal Society* 280: 20131506.

Shepherd, C. R. & Shepherd, L. A. (2010). The poaching and trade of Malayan sun bears in peninsular Malaysia. *Traffic Bulletin* 23: 68.

Singh, N., Sonone, S. & Dharaiya, N. (2018). Sloth bear attacks on humans in central India: implications for species conservation. *Human–Wildlife Interactions* 12(3): 338–347. https://doi.org/10.26077/2mgq-fs29

Skalski, J. R., Millspaugh, J. J. & Clawson, M. V. (2012). Comparison of statistical population reconstruction using full and pooled adult age-class data. *PLoS ONE* 7 (3):e33910. https://doi.org/10.1371/journal.pone 0033910.

Skrbinšek, T., Jelenčič, M., Luštrik, R., et al. (2018). Using laboratory robotics, high-throughput sequencing and sampling with volunteers for quick and cost-effective large-scale genetic estimates of brown bear population size – transboundary case study in Slovenia. In: Skrbinšek, M. A. (Ed.), 26th International Conference on Bear Research and Management. University of Ljubljana, Abstract 211 (p. 76). Available from https://lifewithbears.eu/book-of-abstracts/

Skuban, M., Findo, S. & Kajba, M. (2016). Human impacts on bear feeding habits and habitat selection in the Poľana Mountains, Slovakia. *European Journal of Wildlife Research* 62(3): 353–364.

Sorenson, A., Milligan, S., Larsen, T., et al. (2017). Using scat DNA and citizen science to determine grizzly bear distribution and abundance in West-Central, Alberta. Report prepared for Alberta Conservation Association and Alberta Environment and Parks. Available from https://friresearch.ca/resource/using-scat-dna-and-citizen-science-determine-grizzly-bear-distribution-and-abundance-west

Soulé, M. E. (1985). What is conservation biology? A new synthetic discipline addresses the dynamics and problems of perturbed species, communities, and ecosystems. *Bioscience* 35: 727–734.

Soulé, M. E. (1986). *Conservation biology: The science of scarcity and diversity*. Sunderland, MA: Sinauer Associates.

Stapleton, S., Atkinson, S., Hedman, D. & Garshelis, D. (2014). Revisiting Western Hudson Bay using aerial surveys to update polar bear abundance in a sentinel population. *Biological Conservation* 170: 38–47.

Stapleton, S., Peacock, E. & Garshelis, D. (2015). Aerial surveys suggest long-term stability in the seasonally ice-free Foxe Basin (Nunavut) polar bear population. *Marine Mammal Science* 32(1): 181–201. https://doi.org/10.1111/mms.12251.

Steinmetz, R. & Garshelis, D. (2008). Distinguishing Asiatic black bears and sun bears by claw marks on climbed trees. *Journal of Wildlife Management* 72(3): 814–821.

Stenhouse, G. B., Boulanger, J., Efford, M., et al. (2015). Estimates of grizzly bear population size and density for the 2014 Alberta Yellowhead Population Unit (BMA 3) and south Jasper National Park Inventory Project. Report prepared for Weyerhaeuser Ltd., West Fraser Mills Ltd., Alberta Environment and Parks, and Jasper National Park. Available from https://friresearch.ca/resource/estimates-grizzly-bear-population-size-and-density-final-report.

Stenset, N. E., Lutnæs, P. N., Bjarnadóttir, V., et al. (2016). Seasonal and annual variation in the diet of brown bears (*Ursus arctos*) in the boreal forest of southcentral Sweden. *Wildlife Biology* 22(3): 107–116.

Stetz, J. B., Kendall, K. C. & Servheen, C. (2010). Evaluation of bear rub surveys to monitor grizzly bear population trends. *Journal of Wildlife Management* 74(4): 860–870.

Stetz, J. B., Kendall, K. C. & Macleod, A. C. (2014). Black bear density in Glacier

National Park, Montana. *Wildlife Society Bulletin* 38(1): 60–70.

Steyaert, S. M. J. G., Zedrosser, A., Elfstrom, M., et al. (2016). Ecological implication from spatial patterns in human-caused brown bear mortality. *Wildlife Biology* 22(4): 144–152.

Stirling, I., Lunn, N. J. & Iacozza, J. (1999). Long-term trends in the population ecology of polar bears in western Hudson Bay in relation to climate change. *Arctic* 52(3): 294–306.

Stirling, I., McDonald, T. L., Richardson, E. S., Regehr, E. V. & Amstrup, S. C. (2011). Polar bear population status in the northern Beaufort Sea, Canada 1971–2006. *Ecological Applications* 21(3): 859–876.

Stringham, S. F. (1980). Possible impacts of hunting on the grizzly/brown bear, a threatened species. *International Conference on Bear Research and Management* 5: 140–151.

Swaisgood, R., Wang, D. & Wei, F. (2016). *Ailuropoda melanoleuca*. IUCN Red List of Threatened Species. wwwiucnredlistorg/details/712/0.

Swaisgood, R., Wang, D. & Wei, F. (2018). Panda downlisted but not out of the woods. *Conservation Letters* 11: 1–9.

Swenson, J. E. (2003). Implications of sexually selected infanticide for hunting of large carnivores. In: Festa-Bianchet, M. & Apollonio, M. (Eds.), *Animal behaviour and wildlife conservation* (pp. 171–190). Covelo, CA: Island Press.

Swenson, J. E., Wabakken, P., Sandegren, F., et al. (1995). The near extinction and recovery of brown bears in Scandinavia in relation to the bear management policies of Norway and Sweden. *Wildlife Biology* 1(1): 11–25.

Swenson, J.E., Sandegren, F., Soderberg, A., et al. (1997). Infanticide caused by hunting of male bears. *Nature* 386: 450–451.

Swenson, J. E., Jansson, A., Riig, R. & Sandegren, F. (1999). Bears and ants: myrmecophagy by brown bears in central Scandinavia. *Canadian Journal of Zoology* 77(4): 551–561.

Swenson, J. E., Schneider, M., Zedrosser, A., et al. (2017). Challenges of managing a European brown bear population; lessons from Sweden, 1943–2013. *Wildlife Biology* 2017(4): 1 13. wlb.00251

Taberlet, P. & Bouvet, J. (1992). Bear conservation genetics. *Nature* 358: 197.

Taberlet, P., Griffin, S., Goossens, B., et al. (1996). Reliable genotyping of samples with very low DNA quantities using PCR. *Nucleic Acids Research* 26: 3189–3194.

Tarsim. (2018). Agricultural insurance pool for Turkey. Available from https://web.tarsim.gov.tr/havuz/homePageEng

Taylor, M., DeMaster, D., Bunnell, F. L. & Schweinsburg, R. (1987). Modeling the sustainable harvest of female polar bears. *Journal of Wildlife Management* 51: 811–820.

Tirronen, K. F., Panchenko, D. V. & Kusnecova, A. S. (2015). Brown bear (*Ursus arctos* L.) of the white sea of the Kola Peninsula. *The Herald of Game Management* 12(2): 125–136.

Togawa, K. & Sakamoto, M. (2002). *Bear markets: Japan, a summary of the findings of: "Japan's illegal trade in bear products: A threat to bears worldwide"*. Tokyo: Japan Wildlife Conservation Society.

Tumbelaka, L. & Fredriksson, G. M. (2006). *The status of sun bears in Indonesia* (pp. 73–78). Ibaraki, Japan: Japan Bear Network.

Tumendembergi, O., Proctor, M., Reynolds, H., et al. (2015). Gobi bear abundance and inter-oases movements, Gobi Desert, Mongolia. *Ursus* 26(2): 129–142.

US Fish and Wildlife Service. (2016). *National Survey of Fishing, Hunting, and Wildlife-Associated Recreation.* Washington, DC: US Fish and Wildlife Service.

van Manen, F. T., Haroldson, M. A., Bjornlie, D. D., et al. (2016). Density dependence, whitebark pine, and vital rates of grizzly bears. *Journal of Wildlife Management* 80(2): 300–313.

van Manen, F. T., Haroldson, M. A. & Karabensh B. E. (Eds.). (2018). *Yellowstone Grizzly Bear investigations: Annual report of the Interagency Grizzly Bear Study Team 2017.* US Department of the Interior, US Geological Survey.

Velez-Liendo, X. & Garcia-Rangel, S. (2017). *Tremarctos ornatus*. IUCN Red List of Threatened Species. www.iucnredlist.org.

Viteri, M. P. (2007). Conservation genetics of Andean bears (*Tremarctos ornatus*) in northeastern Ecuador: molecular tools, genetic diversity and population size. PhD dissertation, University of Idaho, Moscow, Idaho, USA.

Walsh, P., Reynolds, J., Collins, G., et al. (2010). Application of a double-observer aerial land-transect method to estimate brown bear population density in southwestern Alaska. *Journal Fish and Wildlife Management* 1(1): 47–58.

Wang, F., McShea, W. J., Wang, D., et al. (2014). Evaluating landscape options for corridor restoration between giant panda reserves. *PLoS ONE* 9: e105086. doi:10.1371/journal.pone.0105086.

Wang, F., McShea, W. J., Wang, D. & Li, S. (2015). Shared resources between giant panda and sympatric wild and domestic mammals. *Biological Conservation* 186: 319–325.

Wasser, S. K., Davenport, B., Ramage, E. R., et al. (2004). Scat detection dogs in wildlife research and management: application to grizzly and black bears in the Yellowhead Ecosystem, Alberta, Canada. *Canadian Journal of Zoology* 82(3): 475–492. doi:10.1139/Z04-020.

Wei, F., Swaisgood, R., Hu, Y., et al. (2015). Progress in the ecology and conservation of giant pandas. *Conservation Biology* 29: 1497–1507.

Wei, W., Swaisgood, R., Dai, Q., et al. (2018). Giant panda distributional and habitat-use shifts in a changing landscape. *Conservation Letters* 11: e12575. https://doi.org/10.llll/conl.12575.

Wilcove, D. S., Rothstein, D., Dubow, J., Phillips, A. & Losos, E. (1998). Quantifying threats to imperiled species in the United States. *BioScience* 48(8): 607–615.

Willcox, D., Minh, N. D. T. & Gomez, L. (2016). *An assessment of trade in bear bile and gall bladder in Viet Nam*. Selangor, Malaysia: TRAFFIC, Petaling Jaya. Available from www.traffic.org/publications/reports/bear-bile-and-gall-bladder-trade-in-viet-nam/

Wilson, S. M., Neudecker, G. A. & Jonkel, J. J. (2014). Human–grizzly bear coexistence in the Blackfoot River Watershed, Montana: getting ahead of the conflict curve. In: Clark, S.G. & Rutherford, M. B. (Eds.), *Large carnivore conservation: Integrating science and policy in the North American west* (pp. 177–213). Chicago, IL: University of Chicago Press.

Woodroffe, R. (2000). Predators and people: using human densities to interpret declines of large carnivores. *Animal Conservation* 3(2): 165–173.

Wong, S. T., Servheen, C., Ambu, L. & Norhayati, A. (2005). Impacts of fruit production cycles on Malayan sun bears and bearded pigs in lowland tropical forest of Sabah, Malaysian Borneo. *Journal of Tropical Ecology* 21(6): 627–639.

Wong, W. M. & Linkie, M. (2013). Managing sun bears in a changing tropical landscape. *Diversity and Distributions* 19: 700–709.

Wong, W. M., Leader-Williams, N. & Linke, M. (2013). Quantifying changes in sun bear distribution and their forest habitat in Sumatra. *Animal Conservation* 16(2): 216–223.

Wong, W. M., Leader-Williams, N. & Linkie, M. (2015). Managing human–sun bear conflict in Sumatran agroforest systems. *Human Ecology* 43(2): 255–266. https://doi.org/10.1007/s10745-015-9729-1

Woodroffe, R. & Ginsberg, J. R. (1998). Edge effects and the extinction of populations inside protected areas. *Science* 280(5372): 2126–2128.

Woods, J. G., Paetkau, D., Lewis, D., et al. (1999). Genetic tagging of free-ranging black and brown bears. *Wildlife Society Bulletin* 27(3): 616–627.

World Animal Protection. (2017). *Ending bear bile farming in South Korea*. London, UK: World Animal Protection.

WWF. (2018). *Living Planet Report - 2018: Aiming higher*. Grooten, M. & Almond, R. E. A. (Eds.). Gland, Switzerland: WWF.

Yoganand, K., Rice, C. G., Johnsingh, A. J. T. & Seidensticker, J. (2006). Is the sloth bear in India secure? A preliminary report on distribution, threats and conservation requirements. *Journal of the Bombay Natural History Society* 103(2–3): 172–181.

Zedrosser, A., Steyaert, S., Swenson, J. E. & Gossow, H. (2011). Brown bear conservation and the ghost of persecution past. *Biological Conservation* 144(9): 2163–2170.

Zhan, X., Ling, M., Zhang, Z., et al. (2006). Molecular censusing doubles giant panda population estimate in a key nature reserve. *Current Biology* 16(12): R451–R452

Zhan, X., Tao, Y., Li, M., et al. (2009). Accurate population size estimates are vital parameters for conserving the giant panda. *Ursus* 20(1): 56–62.

Ziegltrum, G. J. (2004). Efficacy of black bear supplemental feeding to reduce conifer damage in western Washington. *Journal of Wildlife Management*. 68(3): 470–474.

Chapter 21

How Is Climate Change Affecting Polar Bears and Giant Pandas?

Melissa Songer, Todd C. Atwood, David C. Douglas, Qiongyu Huang, Renqiang Li, Nicholas W. Pilfold, Ming Xu, and George M. Durner

Introduction

Anthropogenic greenhouse gas emissions are the primary cause of climate change (Rosenzweig et al. 2008; Kokic et al. 2014) and an estimated increase of 3.7–4.8°C is predicted by the year 2100 if emissions continue at current levels (IPCC 2014). The main predictions for the physical impacts of climate change include changes in precipitation, humidity, and sea level, with higher temperatures over land and oceans leading to decreased snow cover, glaciers, and sea ice (Rosenzweig et al. 2007; Overland et al. 2019). While climate effects are spatially heterogeneous, varying in scale, magnitude, and type (IPCC 2018), their impacts are widely recognized as disruptive to ecological processes and a major threat to biodiversity globally. Changes to phenology (Lane et al. 2012), relationships and interactions within ecological communities (Post et al. 2013), and species ranges (Barnosky et al. 2003; Parmesan & Yohe 2003; Guralnick 2006) are predicted or already underway. Shifts in climatic niches can lead to trophic mismatches and phenological desynchronization that threaten species survival (Thackeray et al. 2016). Range shifts of terrestrial species to higher latitudes and altitudes have already been observed in the Northern Hemisphere, in addition to alterations in routes and timing of bird migrations in Europe, North America, and Australia, and changes in the composition of ocean communities from cold- to warm-adapted species (IPCC 2007). The Arctic is warming two to three times faster than temperate regions (IPCC 2018), and polar ecosystems are likely to be the first to undergo rapid alterations in climatic niches, with trophic shifts and impacts to biodiversity already occurring (Rosenzweig et al. 2007; Yurkowski et al. 2018).

Polar bears (*Ursus maritimus*) and giant pandas (*Ailuropoda melanoleuca*) provide an interesting comparison study of the impact of climate change on bear species. While belonging to the same evolutionary clade, polar bears are the most recently divergent of the extant bear species, while giant pandas are the oldest (Yu et al. 2007). Both species are considered Vulnerable (Wiig et al. 2015; Swaisgood et al. 2017) and are conservation icons, polar bears in particular for climate change. Wildlife species with inflexible life histories and physiologies tightly regulated by the environment are most likely to experience declines in fitness associated with a changing climate. Examining how the environment structures life histories and physiology of polar bears and giant pandas can inform our understanding of threats posed by climate change to both species' persistence.

Polar bears have a circumpolar distribution, living in an extreme and dynamic sea ice environment. As such, their physiology and life histories have evolved to cope with a habitat that changes on multiple temporal scales, from hourly to annually. Their physiology is adapted for a feast and famine regime, including the ability to rapidly gain energy-rich fat deposits during hyperphagia (April–October; Watts & Hansen 1987), while slowing metabolism to prolong fasting capacity during hypophagia (October–May; Pilfold et al. 2016). Polar bears also synchronize reproductive activities to resource seasonality (see also Chapter 14). Mating occurs April–June, and implantation occurs August–October, provided sufficient fat stores exist (Derocher et al. 1992). Cubs are born in November–January (Ramsay & Stirling 1988), and are nursed in a den until February–March (Yee et al. 2017), when they follow their mothers onto the sea ice to hunt vulnerable ringed seal (*Pusa hispida*) pups born during the period when polar bear hyperphagia peaks (Pilfold et al. 2012). The ringed seal whelping period is vital to polar bear cub survival, and changes in availability of ringed seal pups correlate to declines in polar bear reproductive output (Stirling 2002; see also Chapter 14). These changes have been linked to region-wide environmental changes (Rode et al. 2018), indicating tight bottom-up regulation of polar bear populations.

Traits that make a species more vulnerable to disturbance include having a restricted geographic range, limited dispersal, low reproductive rates, and high dietary and habitat specialization (Peters & Darling 1985; Reid et al. 1989; McDonald & Brown 1992; IUCN 2008; Isaac et al. 2009). Giant pandas have a narrow geographic range, do not disperse over large distances, produce one cub every two to three years, and depend on bamboo for 99% of their diet (Reid et al. 1991; Zhao et al. 2013; Wei et al. 2015), suggesting vulnerability to change. Giant pandas live in temperate montane forests of central China, and although bamboo is widely available, they select it in discrete patches based on accessibility and nutrient quality (Wei et al. 2015). Giant pandas shift feeding to favor bamboo leaves from June to December and culm and shoots from February to May (Hansen et al. 2010), and their seasonal movements, in part, reflect this selection (Zhang et al. 2014). Mating occurs from February to April and birthing from July to September (Zhu et al. 2001; see Chapter 6 for details on the species biology).

Giant panda ancestors were once carnivorous and, while evolution to a bamboo diet allowed them to exploit an abundant resource, their gastrointestinal system is inefficient at digesting highly fibrous, low-calorie bamboo. Their behavior and physiology have evolved to maintain a positive energy balance while feeding on a nutrient-poor resource (Zhu et al. 2011). Despite morphological adaptations to aid in bamboo consumption (e.g. pseudo-thumb, mandible and jaw modifications, microflora in gut), giant pandas must spend much of their time feeding (Schaller et al. 1985) and their population dynamics are sensitive to stochastic changes in bamboo supply (Carter et al. 1999). Past bamboo die-offs due to flowering events have led to significant mortality (Reid et al. 1989), suggesting giant panda population dynamics are also regulated by bottom-up processes.

While polar bears and giant pandas are arguably the most distant of the bear species with regard to life histories and behavior, both are likely to be significantly impacted by the broad-scale changes to their environment that are predicted to result from climate change. Herein, we review the conservation status of both species and their habitats, and present current and predicted evidence of the impacts of a changing climate on polar bear and giant panda survival.

Polar Bears

Current Status of Polar Bears and Their Habitat

Polar bears are dependent on sea ice for nearly all aspects of their life history (see Chapter 14). Climatic warming has caused significant declines in sea ice, the primary habitat of polar bears. In 2018, Arctic sea ice extent maximum (14.48 million km^2) and minimum (4.59 million km^2) were estimated to be 7.4% and 26.2%, respectively, below the mean maximum and minimum extents for 1981–2010 (https://nsidc.org). Since 1979, annual maximum and minimum Arctic sea ice extent decreased 47,000 (3.1% decade^{-1}) and 75,000 (10.4% decade^{-1}) km^2 year^{-1}, respectively (Figure 21.1). Rates of sea ice loss have varied throughout the ranges of the estimated 20,000–25,000 wild polar bears in 19 recognized subpopulations (Figure 21.2; http://pbsg.npolar.no/), with ice-covered days declining by seven to 19 days decade^{-1} since 1979, depending on the subpopulation (Stern & Laidre 2016).

Polar bear habitat preferences are driven by the composition and distribution of sea ice relative to the annual timing of critical life history requirements. Critical habitat includes medium- to high-concentration sea ice near edges, frequently near land, and over continental shelves or shallow waters (Durner et al. 2009). In the polar basin (where six subpopulations range), optimal habitat declined annually from 1% to 14% between 1985–1995 and 1996–2006 (Durner et al. 2009), and continued to 2015 in some regions (Lone et al. 2018). Despite losses in optimal habitat, the preference by polar bears for particular sea ice types and its distribution (i.e. concentration, proximity of ice edges, shallow seas) has remained relatively unchanged (Laidre et al. 2015; Wilson et al. 2016). Habitat loss, however, has impacted polar bears in several ways, including behavioral changes (Cherry et al. 2013; Rode et al. 2015b; Atwood et al. 2016a), diminished body condition, reduced survival, and numerical declines for some subpopulations (Regehr et al. 2007; Rode et al. 2010; Bromaghin et al. 2015; Lunn et al. 2016; Obbard et al. 2016, 2018). Future declines in sea ice will increase energetic costs to polar bears using the remaining ice due to ice fragmentation, increased swimming between floes, and increased movement to

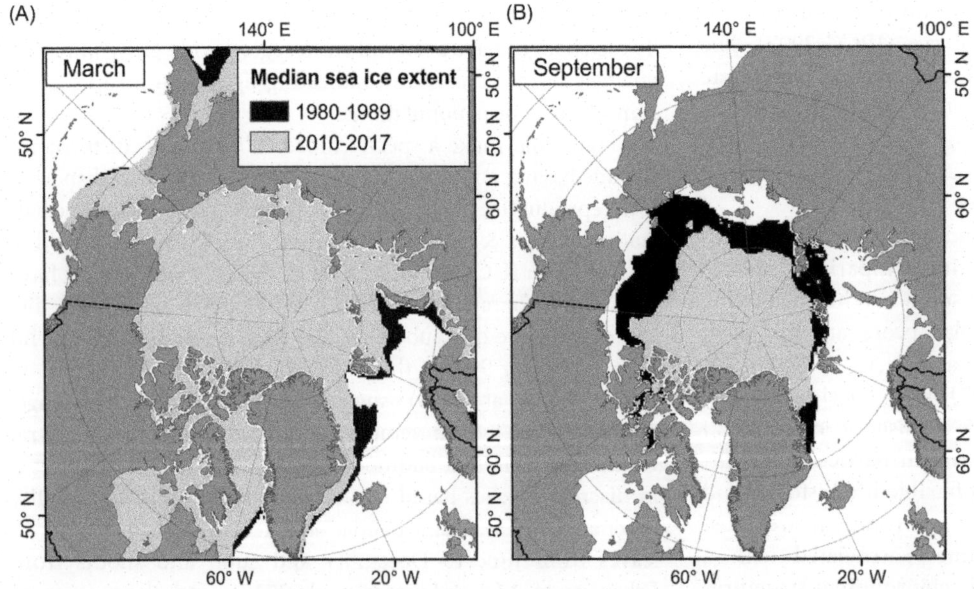

Figure 21.1 Median Arctic sea ice extent, comparing 1980–1989 to 2010–2017, for March (A) and September (B). Sea ice data source: National Snow and Ice Data Center, https://nsidc.org/.

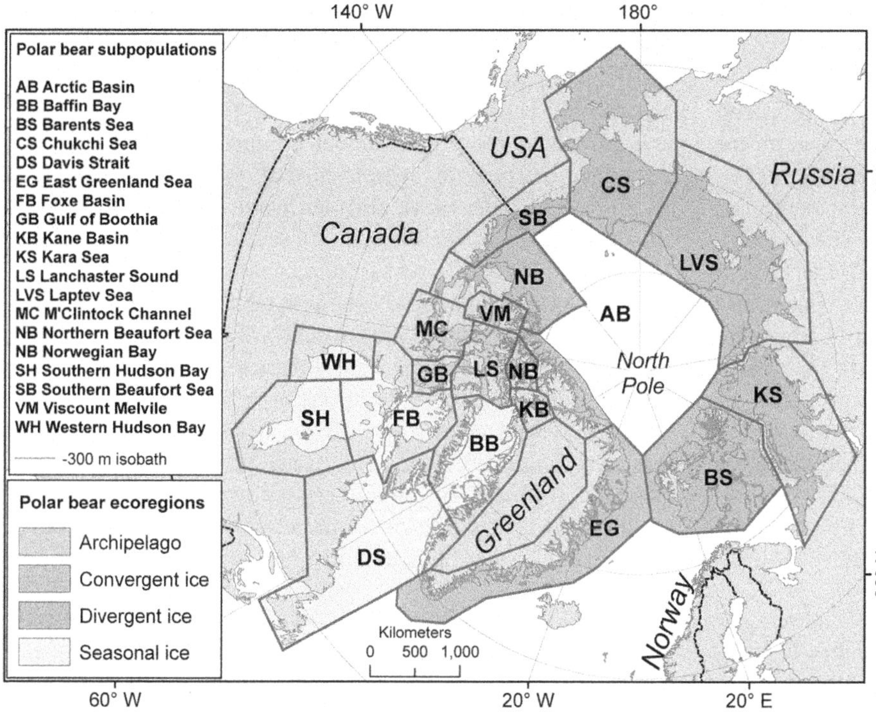

Figure 21.2 Polar bear range across the circumpolar Arctic showing the 19 subpopulations recognized by the International Union for the Conservation of Nature's Polar Bear Specialist Group and associated ecoregions as defined by Amstrup et al. (2008). The characteristics used to define ecoregions are described in Chapter 14. (A black and white version of this figure will appear in some formats. For the color version, please refer to the plate section.)

compensate for faster ice drift (Durner et al. 2017). Hence, subpopulations currently showing no negative effects may be adversely impacted should habitat loss continue (Stirling et al. 2011).

The status of some polar bear subpopulations may be a response to biological productivity rather than ice declines. The Davis Strait subpopulation increased from 1971 to 2007, despite deteriorating sea ice, due to an increasing prey base (Peacock et al. 2013). Low reproduction in Davis Strait and body condition declines in Baffin Bay bears, however, could be due to combined effects of high population density and sea ice declines (Rode et al. 2012; Peacock et al. 2013). Reproduction and body condition of Chukchi Sea polar bears remained stable between 1986–1994 and 2008–2011, despite an annual 44-day increase in ice-free days, because of high biological productivity in that region (Rode et al. 2014). This likely explains the high annual survival and abundance of Chukchi Sea polar bears (Regehr et al. 2018). While studies demonstrate that biological productivity may lessen the impacts of sea ice loss on polar bears in some regions, those mitigating effects will be moot if current rates of sea ice loss continue through the twenty-first century.

Future Impacts of Climate Change on Polar Bears and Their Habitat

The greatest twenty-first-century threat to polar bears is anthropogenic greenhouse gas (GHG)-induced loss of sea ice (Atwood et al. 2016b) as this threatens extirpation of many subpopulations by 2100 (Amstrup et al. 2008). Climate models show Arctic surface air temperatures surpassing 1986–2005 levels by 2.2°C (mitigated GHG) to 8.3°C (unabated GHG) at century's end, with sea ice reductions of 8–34% in February and 43–94% in September (Collins et al. 2013). Ninety percent of unabated GHG scenarios project a September ice-free Arctic Ocean for five consecutive years before 2100 (Collins et al. 2013). If GHG emissions continue unabated, in any year beyond 2090, the Arctic Ocean could be ice-free for 5 consecutive months (Douglas & Atwood 2017).

Ice loss will impact some polar bear subpopulations more than others. For subpopulations in the Convergent Ice (CE) and Divergent Ice (DE) ecoregions (Figure 21.2; Amstrup et al. 2008; Castro de la Guardia et al. 2013), Durner et al. (2009) predicted decreases in optimal habitat between 0.9% and 6.5% decade^{-1} by 2099 compared to 1985–1995 based on moderately abated GHG scenarios. The annual cumulative extent of optimal habitat was projected to decline by 32% (from 1.5 million to 1.0 million km^2) in the CE and DE by 2100. Castro de la Guardia et al. (2013) used a moderately abated GHG scenario to predict increased duration between annual break-up and freeze-up for sea ice habitat occupied by the Western Hudson Bay (WH) subpopulation (the period during which bears are forced ashore). Using $\geq$180 days on land as the threshold at which polar bear reproduction and survival would be compromised (Molnár et al. 2010, 2011), Castro de la Guardia et al. (2013) projected that this critical state would occur in 35 years between 2051 and 2100 and would jeopardize the persistence of WH polar bears. Unabated GHG emission scenarios had even more dire outcomes, with 41 years between 2051 and 2100 exceeding the critical threshold of 180 days.

Amstrup et al. (2008) projected that diminishing sea ice will reduce numbers of animals in polar bear subpopulations within all ecoregions by 2100. In the near term (i.e. within 25 years) subpopulations in the CE and the Canadian Arctic Archipelago Ecoregion (AE; Figure 21.2) had a relatively high probability of either remaining unchanged or increasing. Amstrup et al. (2008) suggested that the AE could provide refugium for polar bears in the late twenty-first century. However, this optimistic outlook contrasts with model projections based on aggressive emission scenarios that suggest the AE could change from a multiyear to annual ice environment, be ice-free for 2–5 months annually, and therefore unable to support polar bear populations by 2100 (Hamilton et al. 2014). The Seasonal Ice Ecoregion (SE) and DE subpopulations have a high probability of being reduced or even extirpated in the near term (Amstrup et al. 2008). Beyond 50 years, extirpation was the dominant outcome for the SE and DE subpopulations, and the dominant outcome for the CE subpopulation after 75 years.

Conservation Implications and Mitigation Strategies for Polar Bears

Projections suggest that the circumpolar abundance of polar bears is likely to decline by 30–66% by mid-century if sea ice habitat loss continues unabated (Amstrup et al. 2008; Atwood et al. 2016b; Regehr et al. 2016). However, short-term (i.e. 10–20 years) responses to environmental changes may be mediated by a variety of factors, including regional differences in ecosystem productivity, anthropogenic stressors, and transient habitat improvements (Peacock et al. 2013; Rode et al. 2014; Jenssen et al. 2015).

Because the long-term persistence of sea ice and polar bears are inextricably linked, actions that stop global warming and minimize sea ice loss must be a priority, but should not be the sole focus, of polar bear conservation efforts. Diminishing sea ice is enabling greater human access to the Arctic, facilitating expanded oil, gas, and mineral extractions (Gautier et al. 2009), establishing new shipping lanes (Smith & Stephenson 2013), and providing more recreational opportunities (Rode et al. 2018). Noise associated with industrial development and recreation could cause polar bears to avoid impacted areas and lead to changes in the distribution and productivity of prey (Kelly et al. 1988). Increasing human activities could increase exposures to pollutants and emerging pathogens (Nuijten et al. 2016; Atwood et al. 2017a), as well as increase the risk for detrimental human–polar bear interactions (Atwood et al. 2017b). While such near-term stressors are inherently more tractable to mitigate than sea ice loss, they are likely to have far less direct influence on long-term population outcomes (Atwood et al. 2016b). Nevertheless, any reduction in short-term population declines may prove crucial for maintaining viable polar bear subpopulations until such time that sea ice loss has ended.

The best long-term conservation action for minimizing reductions in the number of polar bears and the amount of sea ice habitat upon which they rely would be to hold the global mean temperature to $\leq$2°C above pre-industrial levels. With prompt, aggressive, and sustained mitigation of GHG emissions, most climate models project that Earth's average temperature would not rise more than 2°C and Arctic sea ice would persist all summer (Allen & Stocker 2014; Figure 21.3). Reductions to keep warming below 2°C would require immediate action because peak warmth from CO_2 (a prevalent and long-lived GHG) emissions occurs a decade or more after release (Ricke & Caldeira 2014), and sea ice will require even more time to stabilize (Amstrup et al. 2010). Timely

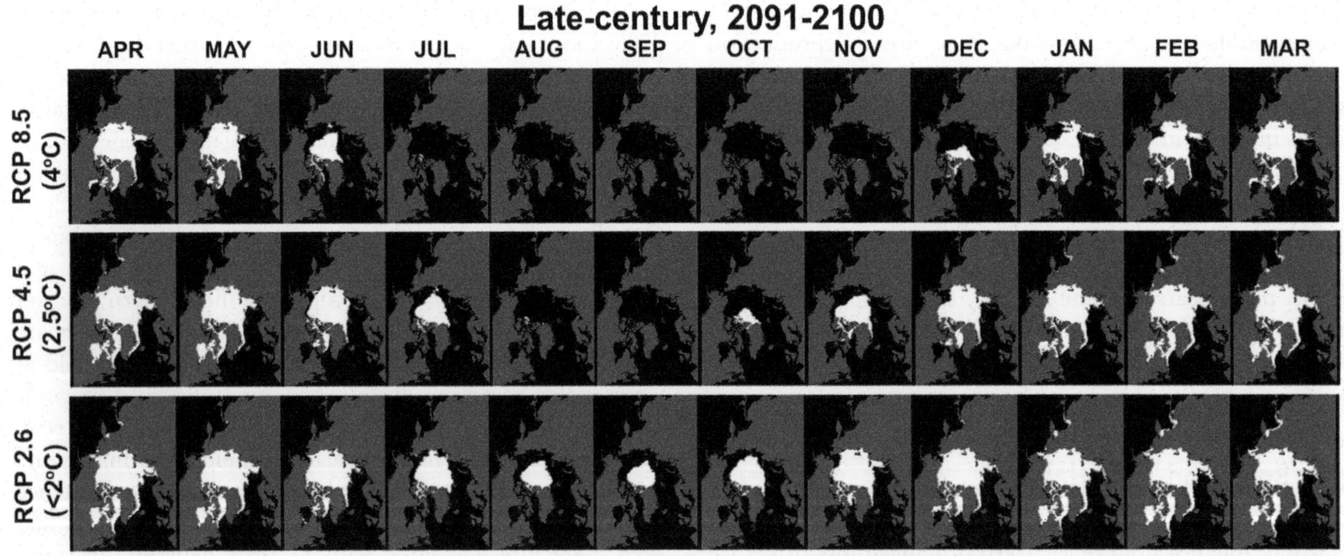

Figure 21.3 Median monthly Arctic sea ice extent projections for the last decade of the twenty-first century. Medians were derived from projections by 13 general circulation models in the Coupled Model Intercomparison Project Phase 5 (CMIP5) when forced with the "worst case" (representative concentration pathways of greenhouse-related emissions or RCP 8.5; top row), "best case" (RCP 2.6; bottom row), and "intermediate case" (RCP 4.5; middle row) of anthropogenic greenhouse gas emission scenarios. Maps view the North Pole (center) and show sea ice (white), ocean (black), and land (gray). Figure from Douglas and Atwood (2017).

intervention in the current global warming trajectory, along with focused management of near-term stressors, will likely assure the persistence of polar bears through the twenty-first century (Amstrup et al. 2010).

Giant Pandas

Current Status of Giant Pandas and Their Habitat

Giant pandas were once distributed across the Yellow, Yangtze, and Pearl river basins in southeastern China, from Zhoukoudian in Beijing in the north to neighboring Vietnam, Thailand, and Myanmar in the south. Late Pleistocene climate changes and millennia of human development have reduced their geographic distribution to <1% of their prehistorical range (Schaller et al. 1985; O'Brien et al. 1994; Loucks et al. 2001). Currently, giant pandas remain in six mountain ranges (Qinling, Minshan, Qionglai, Xiaoxangling, Daxiangling, and Liangshan) across three provinces. According to the 4th National Giant Panda Survey (NGPS), Sichuan Province comprises most of the remaining suitable habitat with ~2.03 million ha (79% of total), Shaanxi Province has ~0.36 million ha (14%), and Gansu Province has ~0.19 million ha (7%; Tang et al. 2015). The amount of suitable habitat expanded from 1.39 to 2.58 million ha between the 1980s and 2013 due primarily to expanded conservation efforts by the Chinese government and international organizations.

By the end of 2013, 1864 giant pandas survived in the wild, an increase of 17% since the 3rd NGPS in 2003, with individuals distributed across the provinces in proportions similar to remaining suitable habitat (74% in Sichuan, 19% in Shaanxi, and 7% in Gansu Province; State Forestry Administration of China 2006; Tang et al. 2015). Giant pandas are isolated into 33 subpopulations and only six of them support >100 individuals. Those larger populations are found in the northern half of the range – in the central Qinling, central Minshan, and northcentral Qionglai mountains. In the Minshan Mountains, 22 subpopulations have <30 individuals, of which 18 have <10 individuals (see also Chapter 6). There are 67 giant panda protected areas, comprising 54% of giant panda habitat and supporting 1246 individuals (68% of the total estimated population).

Over the past 3000 years, the main threats to giant panda populations came from anthropogenic activities, including broad-scale agricultural conversion, road construction, logging, mining, and hydroelectric development, along with smaller-scale threats from poaching and collection of forest products such as bamboo shoots. More recently, increased tourism brings people and impacts to an expanding number of protected areas. By 2013, there were 319 hydropower stations, 1339 km of roads, 269 km of high-voltage transmission lines, 984 residential areas, 479 mines, and 25 scenic spots found in the giant panda range (Tang et al. 2015). These developments further exacerbate habitat fragmentation and reduce gene flow between subpopulations, potentially decreasing genetic diversity and making populations less resilient to environmental changes (Schaller et al. 1985; O'Brien et al. 1994), such as those brought on by climate change (Lu et al. 2001).

Future Impacts of Climate Change on Giant Pandas and Their Habitat

Many studies have modeled impacts of climate change on geographic distributions of bamboo species and giant panda habitat, both across the geographic range of this species and within individual mountain ranges (Songer et al. 2012; Tuanmu et al. 2013; Fan et al. 2014; Li et al. 2015b; Wang et al. 2018; F. Wei et al. 2018), although results can vary widely (e.g. Figures 21.4 and 21.5; also see the Uncertainties section of this chapter). Many bamboo species consumed by giant pandas are sensitive to climate change: six of the 16 main bamboo species found within the range are projected to disappear from giant panda habitats as climate change progresses, resulting in a significant reduction in bamboo forest area (Li et al. 2015b). Tuanmu et al. (2013) predicted three dominant bamboo species would face a substantial reduction within the Qinling range during the twenty-first century. Both alpha (the number of species and their proportion in a habitat or community) and beta (the dissimilarity of species composition between two habitats or communities) diversity of bamboo forest is predicted to be severely reduced in most of the range if current climate trends continue, and the proportion of single-species bamboo forest is predicted to increase (Tuanmu et al. 2013; Li et al. 2015b), suggesting that giant pandas will have reduced availability of habitat with multiple bamboo species in the future. The reduction in bamboo forest and diversity would be most drastic in Qinling, Daxiangling, and Qionglaishan mountains; however, the amount of bamboo and bamboo diversity could increase in parts of the Minshan and Liangshan mountains (Li et al. 2015b). The projected warmer and drier climate in China could also shorten bamboo flowering cycles and increase the frequency of mass flowering events (Li et al. 2015b), which have been disastrous to the giant pandas in the past (Feng 1991).

Projections indicate that climate change could reduce the amount of suitable habitat for giant pandas throughout most of the range (Figures 21.4 and 21.5; Songer et al. 2012; Jian et al. 2014; Li et al. 2015a). An ensemble forecasting method combining both biotic (bamboo availability) and abiotic (climatic and geographic) variables projected that climate change will reduce current habitats by 50%, and 20% of the current protected areas could lose all suitable habitat in this century (Li et al. 2015a). Songer et al. (2012) integrated current giant panda habitat data with general circulation models and found that climate change could reduce habitat by nearly 60% by 2080 and predicted a shift of habitat toward higher elevation and higher latitudes, leading to an increased number of suitable habitat patches with a sharply decreased average patch size. While new areas may become suitable outside the current

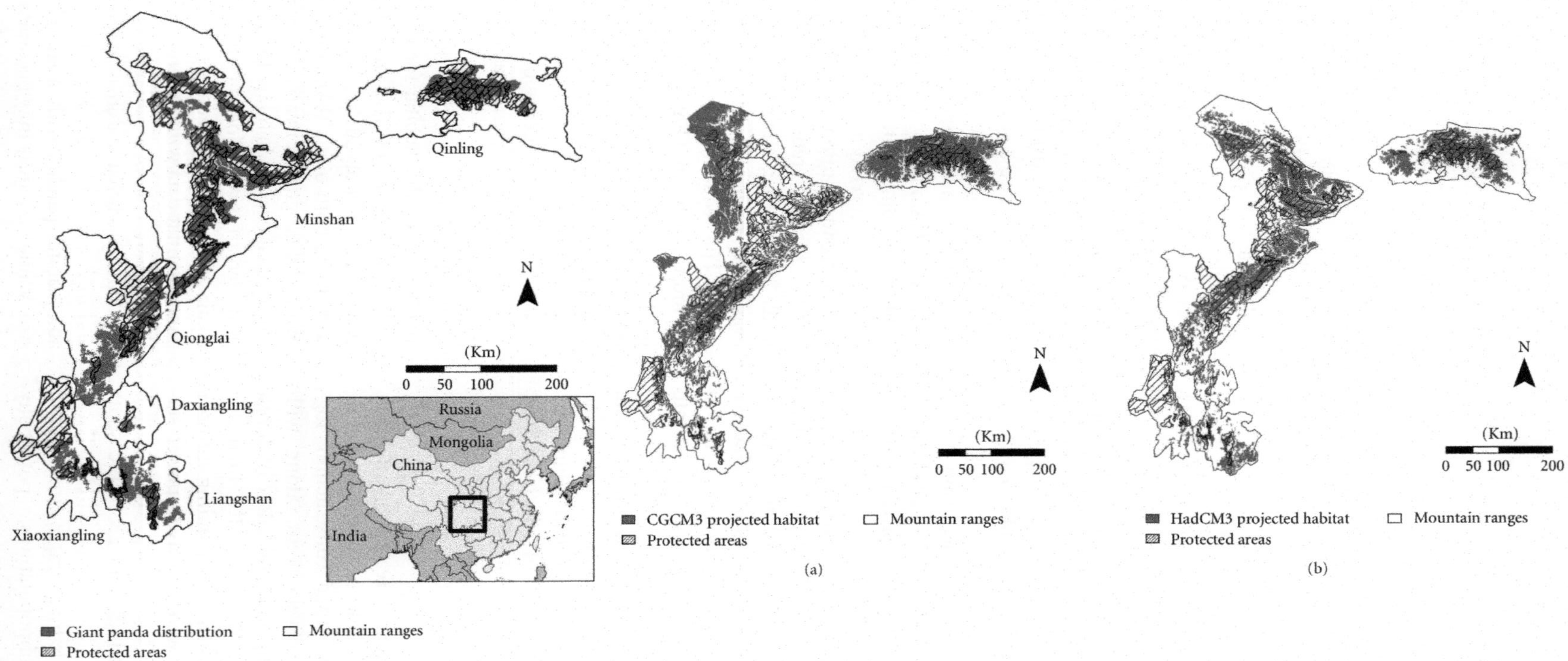

Figure 21.4 Current giant panda distribution, protected areas, and mountain ranges. Predicted suitable habitat for the year 2080 based on (a) Canadian Centre for Climate Modeling and Analysis Coupled Model, version 3 (CGCM3), (b) Hadley Center for Climate Modeling Coupled Model, version 3 (HadCM3), and protected status based on the existing protected area system. Figure from Songer et al. (2012; open access journal).

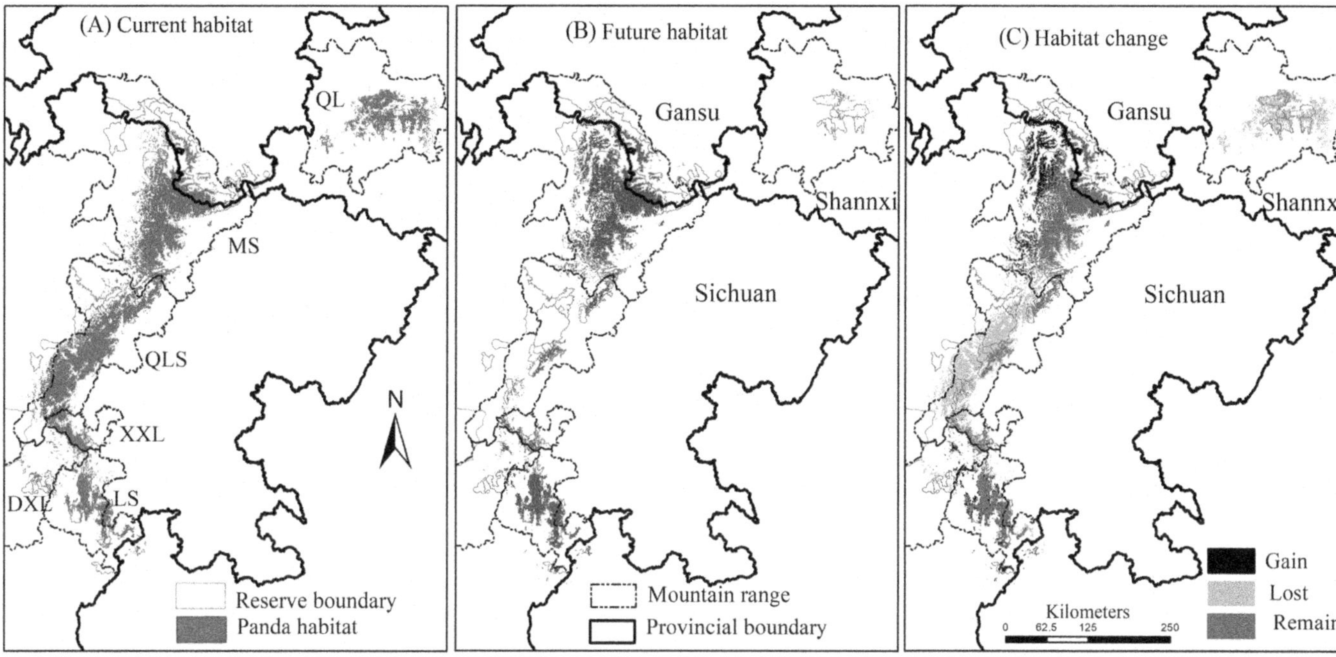

Figure 21.5 Projected geographic distribution of giant panda habitats based on consensus forecast from five global circulation models in six mountain ranges: (A) the current habitats; (B) the future distribution by 2070 for RCP 4.5; and (C) the projected changes of giant panda habitats under climate change by 2070. The projected current giant panda habitats were overlaid with future habitats to identify areas that would be lost, gained, or remain (LS, Liangshan Mountains; XXL, Xiaoxiangling Mountains; DXL, Daxiangling Mountains; QLS, Qionglaishan Mountains; MS, Minshan Mountains; QL, Qingling Mountains). Figure from Li et al. (2015a; open access journal).

geographic range, only 15% of these areas fall within the current protected area system (Songer et al. 2012). Wu et al. (2009) found similar results, predicting that the total suitable habitat of giant pandas will be reduced by 70% and become more fragmented by 2100 if current climate trends continue.

Several studies have focused on the impacts of climate change on giant panda habitats within different mountain ranges (Wang et al. 2010; Zhang et al. 2018). In the Minshan Mountains, which support the largest remaining giant panda population (43% of the total giant panda population), projected habitat gains to the west and losses near the center of the mountain range would produce an overall westward shift of giant panda habitat (Figures 21.5; Li et al. 2015a). In the same mountain range, Shen et al. (2015) predicted that climate change could lead to 67% of giant panda habitat shifting to higher altitudes and latitudes, with a 16% reduction in habitat. The Minshan giant panda habitat is also projected to become more fragmented and isolated, with the average patch size decreasing from a current average of 843 km^2 to <46–75 km^2, depending on the emission scenario (Songer et al. 2012).

In the Qionglaishan Mountains, home of the second largest giant panda population (28% of the total population), suitable giant panda habitat could decrease by 5–37% by 2070, while the northern and lower elevation limits of current suitable habitat for giant pandas could decrease (Yan et al. 2017). Climate change is predicted to bisect the future habitat range into two isolated patches, creating a major barrier to animal movement and an obstacle to gene flow in this region (Li et al. 2015a). The third largest giant panda population, found in the Qinling Mountains at the northernmost part of the range, could face a substantial contraction of habitat areas according to several studies (Fan et al. 2014; Li et al. 2015a; Gong et al. 2016). Climate change could reduce the extent of a suitable habitat for giant pandas by up to 62% and the minimum elevation of giant panda habitat could rise by 500 m (Fan et al. 2014). However, the predicted shift of habitat to northern latitudes would mean the far northern Qinling range would be less impacted than any of the other mountain ranges (Songer et al. 2012). In addition, Wang et al. (2018) found that incorporating biotic interactions reveals potential climate tolerance of giant pandas, and the biotic models predicted less habitat loss compared with climate-only models of panda distributions in this mountain range.

The three southern mountain ranges (Daxiangling, Xiaoxiangling, and Liangshan mountains) taken together support 10% of the remaining giant panda population. There are no studies targeting these mountain ranges specifically; however, range-wide studies predict that climate change will cause the most drastic reductions to both available bamboo resources and habitat in these three mountain ranges, partly due to the latitudinal northward shift in suitable habitat (Wu et al. 2009; Songer et al. 2012; Li et al. 2015a).

Conservation Implications and Mitigation Strategies for Giant Pandas

Numerous studies project that rapid climate change will adversely affect giant panda habitat, putting the population at risk and potentially reversing conservation successes of recent decades. There is a critical need to rethink the current conservation approach to incorporate planning for impacts of climate change and development of adaptive management strategies for giant pandas. First, effective conservation of giant panda habitats requires protection and restoration of bamboo forest, particularly for old-growth forests where bamboo forms the understory. Human-assisted restoration efforts to accelerate the return of bamboo and increase species diversity to previously logged areas can help ensure food security of the giant panda in a changing climate (Li et al. 2015b). Second, based on the prediction of reduced and more fragmented habitat throughout the species range under most climate change scenarios, forestry planning and management should focus on enhancing habitat connectivity by establishing habitat corridors and strengthening matrix management outside the reserves to promote gene flow and increase population viability (Wang et al. 2014). In addition, suitable habitat strongholds along with areas predicted to become suitable in key locations should be identified throughout the range and prioritized for protection, restoration, and reduction of human impact (Songer et al. 2012). Finally, an adjustment of the existing reserves may be necessary in cases where significant habitat changes within and around reserves are expected, e.g. by manipulating size, shape, and spatial orientation of reserves (Li et al. 2015a).

Giant panda conservation efforts should not only seek to ensure the recovery of the giant panda population, but also share important co-benefits to maximize return on conservation investment (Li et al. 2017; F. Wei et al. 2018a). Giant panda habitats support 293 rare and endangered species, according to the IUCN Red List, of which 133 are endemic to China and 19 are endemic to the giant panda range (Li & Pimm 2016). Giant panda habitats also play a critical role in the provision of multiple ecosystem services, with a value of US$2.6 and US$6.9 billion/year gained from giant panda services and reserves in 2010 (F. Wei et al. 2018). Approximately 90% of the giant panda reserves are more effective in increasing carbon sequestration compared to lands not declared as reserves from 1988 to 2012 (Li et al. 2017).

Although the downlisting of giant pandas from Endangered to Vulnerable is an indicator of conservation success, existing efforts should not be relaxed (Swaisgood et al. 2017). Instead, it is time to shift from traditional conservation approaches to new models, including a more expansive view of potential habitat, innovative approaches to enhancing connectivity and protection, and developing integrated plans that address multiple objectives to consider social, environmental, and resource management perspectives. The government of China is establishing a Giant Panda National Park spanning three provinces and all existing reserves. Therefore, setting regional conservation priorities should receive special attention during park planning, to foster synergies between ecosystem services and biodiversity, and to maximize the potential of species and carbon co-benefits under global change (Li et al. 2017).

Challenges

Dealing with Uncertainties of Climate Change Modeling

The modeling of future climate change is intrinsically variable. Choice of emission scenario, initial conditions such as species occurrence, model classes and parameters, and combinations of variables can all affect projections (Araújo & New 2007) and call for different conservation solutions (Heikkinen et al. 2006; Huang et al. 2018). For giant pandas, just within the Qinling range, estimates of habitat loss by 2080 ranged from 1–17% (Songer et al. 2012) to 16–42% (Wang et al. 2018) and 37–62% (Fan et al. 2014). For polar bears, habitat is predicted to decline by 2.7% decade^{-1} to 4.9% decade^{-1} for moderate and unabated emission scenarios, respectively (Durner et al. 2009).

Several methods are used to account for, and parse out, model uncertainty. Ensemble models are employed to compute multimodel means (Durner et al. 2009) or produce cumulative weighted averages based on each individual model's performance (Thuiller et al. 2009). Ensemble modeling can improve on robustness over single models, although the composition of the model suite is important (Overland et al. 2011). Studies have examined how different modeling factors contribute to variability in richness and "turnover rate" (Diniz-Filho et al. 2009; Buisson et al. 2010). New methods have also been developed to compare similarities between gridded probabilities/suitability of a single species (Huang et al. 2018). However, translating variable model predictions to practical conservation policies with a high likelihood of success remains a challenge.

Other factors can contribute to model uncertainties, including biotic interaction with conspecifics and codependent species, and adaptability to climate change (Araújo & New 2007). For example, the future distribution of polar bears may be influenced by combinations of differential accessibility of prey (ice seals), conspecific predation risk (Pilfold et al. 2014), and low nutritional value of terrestrial foods (Rode et al. 2015b), in addition to sea ice declines (Amstrup et al. 2010). Studies have revealed the giant panda's aversion to human infrastructure, which is found at lower elevations (Wang et al. 2014), although historically they were present in low-elevation regions in southeast China and part of southeast Asia (Loucks et al. 2001). It is reasonable to assume that giant panda and bamboo physiology might be resilient enough to tolerate moderate temperature increases; however, empirical studies are lacking.

While much of the existing literature focuses primarily on the impact of increased average temperatures, climate change will also increase the frequency and severity of extreme climate events which might have significant impacts on species distribution (Huang et al. 2017). Giant pandas are found in one of the most tectonically active places on earth and extreme weather events could cause further disturbances. Understanding is lacking on the impacts of extreme heat, flooding, and drought on giant panda and bamboo distributions, the impacts of summer storms and rain events on maternal dens for giant pandas, and the impacts of more fragmented and declining amounts of sea ice distribution on polar bears. Further research is urgently needed to test and validate numerous climate change predictions on polar bears and giant pandas, to understand the adaptability of both species and to better inform conservation decisions.

Data Availability and Access

The Chinese government allocates tremendous resources to the NGPS approximately every decade. The survey provides valuable information on wild populations dating back to 1974. The most recent NGPS (3rd and 4th) have similar methods and are the most comparable (Zhang et al. 2011; W. Wei et al. 2018). However, they still have somewhat different extents and transect density (Yang et al. 2017). The change in survey design creates a challenge for direct comparison and application of results to population recovery (Qiu 2015). Giant pandas are not directly counted, but abundance is inferred through estimates of spoor densities and by distinguishing individuals through bite mark sizes found on bamboo (State Forestry Administration of China 2015), and results often lack sufficient documentation on data pre-processing and error propagation methods. The lack of a variability or error estimate in population numbers was raised as a significant issue within the research community (Harris et al. 2014; Qiu 2015). In addition, NGPS raw data are not publically available, but owned by the State Forestry Administration of China and only available to scientists who sign agreements for access. Population assessments also lack estimates for age-class survival, which reduces the mechanistic inference of factors driving population changes, thus limiting use in adaptive management planning.

By contrast, data reporting and sharing from polar bear research are considerably more transparent. The International Union for the Conservation of Nature's Polar Bear Specialist Group meets on a quadrennial basis to share information and publishes their proceedings (e.g. Durner et al. 2018), providing a compendium of openly available data. Additionally, some agencies make data from publications available to the public (National Academies et al. 2018). The vast and remote circumpolar range of polar bears makes data deficiency a greater obstacle than data access. Among the 19 recognized subpopulations of polar bears, 11 currently have a data-deficient status. There remains tremendous disparity in the amount of empirical data and knowledge about their sizes and statuses due to the logistical challenges and expense of conducting fieldwork in the Arctic. This has led to an unequal amount of research across subpopulations, with the majority of empirical climate change research coming from the four subpopulations with better accessibility: western Hudson Bay, southern Beaufort Sea, Chukchi Sea, and the Barents Sea. A lack of spatial variation in measuring the response of a broadly distributed species has the potential to bias inferences about the rate or type of biological response range-wide. Of the limited studies comparing regional responses of polar bears to sea ice change, both have shown marked variation between subpopulations (Rode et al. 2014; Pilfold et al. 2017).

Conclusion

Polar bears and giant pandas are predicted to be highly susceptible to the impacts of climate change based on their life history, specialized habitat requirements, and unique geographical distributions. Habitats of both species will likely be significantly altered through reductions in area and shifts of suitable habitat. Both species are listed as Vulnerable, although giant pandas have less than a tenth of the remaining population size of polar bears. The disparity in the scale of the species' spatial distributions creates different challenges for conservation planning. Polar bears range over vast Arctic terrestrial and marine areas of 23 million km^2, while giant pandas cover less than 26,000 km^2 in mountain forests surrounded by human communities, agriculture, and roads. Giant pandas benefit from the top-down approach of the Chinese government's policies expanding protected areas, creating incentives for returning cropland to forest, and investing in research and conservation. In contrast, polar bear conservation is built on international collaboration among nations bordering the Arctic Ocean, which is more complex, but leads to synergies and resources of many nations working together towards shared goals, sometimes despite the existence of higher-level political conflicts (e.g. United States and Russia).

The most pressing question for the conservation of both species is: What can and should be done to facilitate adaptive management despite the uncertainties and variability of future climate models and emission scenarios? One of the greatest challenges posed by climate change for conservation is the shifting and dynamic changes expected for wildlife habitat. Conservation tools that depend on habitat protection may result in protected areas that do not cover key habitats in perpetuity as habitats and the species that inhabit them shift spatially due to climate change. Therefore, it is vital that adaptive management plans are based on the best available science and grounded in ongoing ecological studies that can be verified with empirical data.

Existing climate commitments complicate predictions and mitigation planning for climate change. Global mean temperature rose approximately 0.8°C between 1880 and 2012, resulting in oceanic warming and sea-level rise (Rummukainen 2015). However, there is a lag in realized climatic impacts from oceanic warming, resulting in a commitment to more climate change on top of what has already exists (Wigley 2005; Meehl et al. 2012; Rummukainen 2015; Overland et al. 2019). Non-linear changes such as the melting of ice sheets and release of permafrost carbon could have impacts for centuries to come. Our historical emissions and current socioeconomic investments such as long-lasting energy infrastructure commit us to future climate change, even in the unlikely event that our emissions are reduced to zero in the near-term. Given existing climate commitments, GHG emissions would have to go below our current levels in order to stabilize global mean temperature commitments (Wigley 2005; Meehl et al. 2012; Rummukainen 2015). These climate commitments create further uncertainty about the impacts of climate change, and a greater urgency for reducing our emissions.

However, uncertainty in future ecological scenarios and the immense challenges posed by climate change are not justification for inaction. Increasing protections, limiting development in critical areas, and enhancing connectivity among individuals within remaining habitat are priorities for both species and may buffer against the effects of change. Most importantly, limiting the continuing increase of GHG emissions is urgently needed to slow the effects of climate change already underway. This requires action at local, national, and global levels, ranging from the choices of individuals to the governance of societies. Due to their charismatic appeal, polar bears and giant pandas inspire global support of research and conservation actions. Harnessing this inspiration and raising awareness of the risks and opportunities is key for countering the predicted impacts of climate change.

Taken together, polar bears and giant pandas provide a glimpse into the wide range of challenges species are already undergoing in the face of climate change. Conservation actions and research designed to inform protection of these bears will also serve to protect their ecosystems and other vulnerable species, including other bear species that are also threatened by climate change but are not as well known. Leveraging support for polar bears and giant pandas can serve as a means to expand research on the potential effects of climate change, generate funds for research and conservation, and galvanize conservation actions to mitigate adverse impacts of climate change to wildlife populations.

References

Allen, M. R. & Stocker, T. F. (2014). Impact of delay in reducing carbon dioxide emissions. *Nature Climate Change* 4(1): 23–26.

Amstrup, S. C., Marcot, B. G. & Douglas, D. C. (2008). A Bayesian network modeling approach to forecasting the 21st century worldwide status of polar bears. In: DeWeaver, E. T., Bitz, C. M. & Tremblay, L.-B. (Eds.), *Arctic sea ice decline: Observations, projections, mechanisms, and implications. Geophysical Monograph 180* (pp. 213–268). Washington DC: American Geophysical Union.

Amstrup, S. C., DeWeaver, E. T., Douglas, D. C., et al. (2010). Greenhouse gas mitigation can reduce sea-ice loss and increase polar bear persistence. *Nature* 468(7326): 955–958.

Araújo, M. B. & New, M. (2007). Ensemble forecasting of species distributions. *Trends in Ecology & Evolution* 22(1): 42–47.

Atwood, T. C., Peacock, E., McKinney, M. A., et al. (2016a). Rapid environmental change drives increased land use by an arctic marine predator. *PLoS ONE* 11(6): e0155932.

Atwood, T. C., Marcot, B. G., Douglas, D. C., et al. (2016b). Forecasting the relative influence of environmental and anthropogenic stressors on polar bears. *Ecosphere* 7(6): e01370.

Atwood, T. C., Duncan, C., Patyk, K. A., et al. (2017a). Environmental and behavioral changes may influence the exposure of an Arctic apex predator to pathogens and contaminants. *Scientific Reports* 7(1): 13193.

Atwood, T. C., Simac, K. S., Breck, S., York, G. & Wilder, J. (2017b). Human–polar bear interactions in a changing Arctic: existing and emerging concerns. *Animal Welfare Series* 17: 397–418.

Barnosky, A. D., Hadly, E. A. & Bell, C. J. (2003). Mammalian response to global warming on varied temporal scales. *Journal of Mammalogy* 84(2): 354–368.

Bromaghin, J. F., McDonald, T. L., Stirling, I., et al. (2015). Polar bear population dynamics in the southern Beaufort Sea during a period of sea ice decline. *Ecological Applications* 25(3): 634–651.

Buisson, L., Thuiller, W., Casajus, N., Lek, S. & Grenouillet, G. (2010). Uncertainty in ensemble forecasting of species distribution. *Global Change Biology* 16(4): 1145–1157.

Carter, J., Ackleh, A. S., Leonard, B. P. & Wang, H. (1999). Giant panda (*Ailuropoda melanoleuca*) population dynamics and bamboo (subfamily Bambusoideaen) life history: a structured population approach to examining carrying capacity when the prey are semelparous. *Ecological Modelling* 123(2): 207–223.

Castro de la Guardia, L., Derocher, A. E., Myers, P. G., Terwisscha van Scheltinga, A. D. & Lunn, N. J. (2013). Future sea ice conditions in Western Hudson Bay and consequences for polar bears in the 21st century. *Global Change Biology* 19(9): 2675–2687.

Cherry, S. G., Derocher, A. E., Thiemann, G. W. & Lunn, N. J. (2013). Migration phenology and seasonal fidelity of an Arctic marine predator in relation to sea ice dynamics. *The Journal of Animal Ecology* 82(4): 912–921.

Collins, M., Knutti, R., Arblaster, J., *et al.* (2013). Long-term climate change: projections, commitments and irreversibility. In: IPCC (Ed.), *Climate Change 2013: The Physical Science Basis. Contribution of Working Group I to the Fifth Assessment Report of the Intergovernmental Panel on Climate Change* (pp. 1029–1136). Cambridge: Cambridge University Press.

Derocher, A. E., Stirling, I. & Andriashek, D. (1992). Pregnancy rates and serum progesterone levels of polar bears in western Hudson Bay. *Canadian Journal of Zoology* 70(3): 561–566.

Diniz-Filho, J. A. F., Bini, L. M., Rangel, T. F., et al. (2009). Partitioning and mapping uncertainties in ensembles of forecasts of species turnover under climate change. *Ecography* 32(6): 897–906.

Douglas, D. C. & Atwood, T. C. (2017). Uncertainties in forecasting the response of polar bears to global climate change. In: Butterworth, A. (Ed.), *Marine mammal welfare* (pp. 463–473). Cham: Springer.

Durner, G. M., Douglas, D. C., Nielson, R. M., et al. (2009). Predicting 21st-century polar bear habitat distribution from global climate models. *Ecological Monographs* 79(1): 25–58.

Durner, G. M., Douglas, D. C., Albeke, S. E., et al. (2017). Increased Arctic sea ice drift alters adult female polar bear movements and energetics. *Global Change Biology* 23(9): 3460–3473.

Durner, G. M., Laidre, K. L. & York, G. S. (2018). Polar bears: proceedings of the 18th working meeting of the IUCN/SSC Polar Bear Specialist Group, Anchorage, Alaska, June 7–11, 2016. Retrieved from https://portals.iucn.org/library/node/47667.

Fan, J., Li, J., Xia, R., et al. (2014). Assessing the impact of climate change on the habitat distribution of the giant panda in the Qinling Mountains of China. *Ecological Modelling* 274: 12–20.

Feng, W. H. (1991). The collection of research papers on Giant Panda. *Journal of Sichuan University* 3: 7–13 [in Chinese].

Gautier, D. L., Bird, K. J., Charpentier, R. R., et al. (2009). Assessment of undiscovered oil and gas in the Arctic. *Science* 324 (5931): 1175–1179.

Gong, M., Guan, T., Hou, M., Liu, G. & Zhou, T. (2016). Hopes and challenges for giant panda conservation under climate change in the Qinling Mountains of China. *Ecology and Evolution* 7(2): 596–605.

Guralnick, R. (2006). The legacy of past climate and landscape change on species' current experienced climate and elevation ranges across latitude: a multispecies study utilizing mammals in western North America. *Global Ecology and Biogeography* 15(5): 505–518.

Hamilton, S. G., de la Guardia, L. C., Derocher, A. E., et al. (2014). Projected polar bear sea ice habitat in the Canadian Arctic Archipelago. *PLoS ONE* 9(11): e113746.

Hansen, R. L., Carr, M. M., Apanavicius, C. J., et al. (2010). Seasonal shifts in giant panda feeding behavior: relationships to bamboo plant part consumption. *Zoo Biology* 29(4): 470–483.

Harris, R., Garshelis, D., McShea, W. J. & Wang, D. (2014). Introduction. In: McShea, W. J., Garshelis, D., Harris, R., et al. (Eds.), *A chance for lasting survival: Ecology and behavior of wild giant pandas* (pp. 1–25). Washington, DC: Smithsonian Institution Scholarly Press.

Heikkinen, R. K., Luoto, M., Araújo, M. B., et al. (2006). Methods and uncertainties in bioclimatic envelope modelling under climate change. *Progress in Physical Geography: Earth and Environment* 30(6): 751–777.

Huang, Q., Sauer, J. R. & Dubayah, R. O. (2017). Multidirectional abundance shifts among North American birds and the relative influence of multifaceted climate factors. *Global Change Biology* 23(9): 3610–3622.

Huang, Q., Fleming, C. H., Robb, B., Lothspeich, A. & Songer, M. (2018). How different are species distribution model predictions? – Application of a new measure of dissimilarity and level of significance to giant panda *Ailuropoda melanoleuca*. *Ecological Informatics* 46: 114–124.

IPCC. (2007). Summary for policymakers. In: Parry, M., Canziani, O., Palutikof, J., van der Linden, P. & Hanson, C. (Eds.), *Climate Change 2007: Impacts, Adaptation and vulnerability. Contribution of Working Group II to the Fourth Assessment Report of the Intergovernmental Panel on Climate Change*. Cambridge: Cambridge University Press.

IPCC. (2014). Summary for policymakers. In: Edenhofer, O., Pichs-Madruga, R., Sokona, Y. et al. (Eds.), *Climate Change 2014: Mitigation of Climate Change. Contribution of Working Group III to the Fifth Assessment Report of the Intergovernmental Panel on Climate Change*. Cambridge: Cambridge University Press.

IPCC. (2018). Summary for policymakers. *Global warming of 1.5°C* (p. 32). Geneva: World Meteorological Organization.

Isaac, J. L., Vanderwal, J., Johnson, C. N. & Williams, S. E. (2009). Resistance and resilience: quantifying relative extinction risk in a diverse assemblage of Australian tropical rainforest vertebrates. *Diversity and Distributions* 15(2), 280–288.

IUCN. (2008). *Species susceptibility to climate change impacts*. Gland, Switzerland: World Conservation Union.

Jenssen, B. M., Villanger, G. D., Gabrielsen, K. M., et al. (2015). Anthropogenic flank attack on polar bears: interacting consequences of climate warming and pollutant exposure. *Frontiers in Ecology and Evolution* 3: 16. doi:10.3389/fevo.2015.00016.

Jian, J., Jiang, H., Jiang, Z., et al. (2014). Predicting giant panda habitat with climate data and calculated habitat suitability index (HSI) map. *Meteorological Applications* 21(2): 210–217.

Kelly, B. P., Burns, J. J. & Quakenbush, L. T. (1988). Responses of ringed seals (*Phoca hispida*) to noise disturbance. In: Sackinger, W., Jeffries, M., Imm, J. & Tracey, S. (Eds.). *Port and ocean engineering under Arctic conditions. Volume II. Symposium on noise and marine mammals* (p. 13). Fairbanks, AK: Geophysical Institute, University of Alaska.

Kokic, P., Crimp, S. & Howden, M. (2014). A probabilistic analysis of human influence on recent record global mean temperature changes. *Climate Risk Management* 3: 1–12.

Laidre, K. L., Born, E. W., Heagerty, P., et al. (2015). Shifts in female polar bear (*Ursus maritimus*) habitat use in East Greenland. *Polar Biology* 38(6): 879–893.

Lane, J. E., Kruuk, L. E. B., Charmantier, A., Murie, J. O. & Dobson, F. S. (2012). Delayed phenology and reduced fitness associated with climate change in a wild hibernator. *Nature* 489(7417): 554–557.

Li, B. V. & Pimm, S. L. (2016). China's endemic vertebrates sheltering under the protective umbrella of the giant panda. *Conservation Biology* 30(2): 329–339.

Li, R., Xu, M., Wong, M. H. G., et al. (2015a). Climate change threatens giant panda protection in the 21st century. *Biological Conservation* 182: 93–101.

Li, R., Xu, M., Wong, M. H. G., et al. (2015b). Climate change-induced decline in bamboo habitats and species diversity: implications for giant panda conservation. *Diversity and Distributions* 21(4): 379–391.

Li, R., Xu, M., Powers, R., et al. (2017). Quantifying the evidence for co-benefits between species conservation and climate change mitigation in giant panda habitats. *Scientific Reports* 7(1): 12705.

Lone, K., Merkel, B., Lydersen, C., Kovacs, K. M. & Aars, J. (2018). Sea ice resource selection models for polar bears in the Barents Sea subpopulation. *Ecography* 41(4): 567–578.

Loucks, C. J., Lü, Z., Dinerstein, E., et al. (2001). Giant pandas in a changing landscape. *Science* 294(5546): 1465–1465.

Lu, Z., Johnson, W. E., Menotti-Raymond, M., et al. (2001). Patterns of genetic diversity in remaining giant panda populations. *Conservation Biology* 15(6): 1596–1607.

Lunn, N. J., Servanty, S., Regehr, E. V., et al. (2016). Demography of an apex predator at the edge of its range: impacts of changing sea ice on polar bears in Hudson Bay. *Ecological Applications: A Publication of the Ecological Society of America* 26(5): 1302–1320.

McDonald, K. A. & Brown, J. H. (1992). Using montane mammals to model extinctions due to global change. *Conservation Biology* 6(3): 409–415.

Meehl, G. A., Hu, A., Tebaldi, C., et al. (2012). Relative outcomes of climate change mitigation related to global temperature versus sea-level rise. *Nature Climate Change* 2: 576.

Molnár, P. K., Derocher, A. E., Thiemann, G. W. & Lewis, M. A. (2010). Predicting survival, reproduction and abundance of polar bears under climate change. *Biological Conservation* 143(7): 1612–1622.

Molnár, P. K., Derocher, A. E., Klanjscek, T. & Lewis, M. A. (2011). Predicting climate change impacts on polar bear litter size. *Nature Communications* 2: 186.

National Academies of Sciences, Engineering, and Medicine, Policy and Global Affairs, Board on Research Data and Information & Committee on Toward an Open Science Enterprise. (2018). Open Science by Design: Realizing a Vision for 21st Century Research. Appendix C, Office of Science and Technology Policy 2013 Memorandum: Increasing Access to the Results of Federally Funded Scientific Research. Washington, DC: National Academies Press. Available from: www.ncbi.nlm.nih.gov/books/NBK525415/

Nuijten, R. J. M., Hendriks, A. J., Jenssen, B. M. & Schipper, A. M. (2016). Circumpolar contaminant concentrations in polar bears (*Ursus maritimus*) and potential population-level effects. *Environmental Research* 151: 50–57.

Obbard, M. E., Cattet, M. R. L., Howe, E. J., et al. (2016). Trends in body condition in polar bears (*Ursus maritimus*) from the Southern Hudson Bay subpopulation in relation to changes in sea ice. *Arctic Science* 2(1): 15–32.

Obbard, M. E., Stapleton, S., Szor, G., et al. (2018). Re-assessing abundance of Southern Hudson Bay polar bears by aerial survey: effects of climate change at the southern edge of the range. *Arctic Science* 4(4): 634–655.

O'Brien, S. J., Wenshi, P. & Zhi, L. (1994). Pandas, people and policy. *Nature* 369 (6477): 179.

Overland, J. E., Wood, K. R. & Wang, M. (2011). Warm Arctic – cold continents: climate impacts of the newly open Arctic Sea. *Polar Research* 30(1): 15787.

Overland, J., Dunlea, E., Box, J. E., et al. (2019). The urgency of Arctic change. *Polar Science* 21: 6–13. doi:10.1016/j.polar.2018.11.008.

Parmesan, C. & Yohe, G. (2003). A globally coherent fingerprint of climate change impacts across natural systems. *Nature* 421(6918): 37–42.

Peacock, E. L., Taylor, M. K., Laake, J. L. & Stirling, I. (2013). Population ecology of polar bears in Davis Strait, Canada and Greenland. *Journal of Wildlife Management* 77(3): 14.

Peters, R. L. & Darling, J. D. S. (1985). The greenhouse effect and nature reserves. Global warming would diminish biological diversity by causing extinctions among reserve species. *BioScience* 35(11): 707–717.

Pilfold, N. W., Derocher, A. E., Stirling, I., Richardson, E. & Andriashek, D. (2012). Age and sex composition of seals killed by polar bears in the Eastern Beaufort Sea. *PLoS ONE* 7(7): e41429.

Pilfold, N. W., Derocher, A. E., Stirling, I. & Richardson, E. (2014). Polar bear predatory behaviour reveals seascape distribution of ringed seal lairs. *Population Ecology* 56(1): 129–138.

Pilfold, N. W., Hedman, D., Stirling, I., et al. (2016). Mass loss rates of fasting polar bears. *Physiological and Biochemical Zoology* 89(5): 377–388.

Pilfold, N. W., McCall, A., Derocher, A. E., Lunn, N. J. & Richardson, E. (2017). Migratory response of polar bears to sea ice loss: to swim or not to swim. *Ecography* 40(1): 189–199.

Post, E., Bhatt, U. S., Bitz, C. M., et al. (2013). Ecological consequences of sea-ice decline. *Science* 341(6145): 519–524.

Qiu, J. (2015). Experts question China's panda survey. *Nature News*. doi:10.1038/nature.2015.17020

Ramsay, M. A. & Stirling, I. (1988). Reproductive biology and ecology of female polar bears (*Ursus maritimus*). *Journal of Zoology* 214(4): 601–633.

Regehr, E. V., Lunn, N. J., Amstrup, S. C. & Stirling, I. (2007). Effects of earlier sea ice breakup on survival and population size of polar bears in western Hudson Bay. *Journal of Wildlife Management* 71(8): 2673–2683.

Regehr, E. V., Laidre, K. L., Akçakaya, H. R., et al. (2016). Conservation status of polar bears (*Ursus maritimus*) in relation to projected sea-ice declines. *Biology Letters* 12(12): 20160556.

Regehr, E. V., Hostetter, N. J., Wilson, R. R., et al. (2018). Integrated population modeling provides the first empirical estimates of vital rates and abundance for polar bears in the Chukchi Sea. *Scientific Reports* 8(1): 16780.

Reid, D. G., Jinchu, H., Sai, D., Wei, W. & Yan, H. (1989). Giant panda *Ailuropoda melanoleuca* behaviour and carrying capacity following a bamboo die-off. *Biological Conservation* 49(2): 85–104.

Reid, D. G., Taylor, A. H., Jinchu, H. & Zisheng, Q. (1991). Environmental influences on bamboo *Bashania fangiana* growth and implications for giant panda conservation. *Journal of Applied Ecology* 28(3): 855–868.

Ricke, K. L. & Caldeira, K. (2014). Maximum warming occurs about one decade after a carbon dioxide emission. *Environmental Research Letters* 9(12): 124002.

Rode, K. D., Amstrup, S. C. & Regehr, E. V. (2010). Reduced body size and cub recruitment in polar bears associated with sea ice decline. *Ecological Applications: A Publication of the Ecological Society of America* 20(3): 768–782.

Rode, K. D., Peacock, E. L., Taylor, M. K., et al. (2012). A tale of two polar bear populations: ice habitat, harvest, and body condition. *Population Ecology* 54(1): 318.

Rode, K. D., Regehr, E. V., Douglas, D. C., et al. (2014). Variation in the response of an Arctic top predator experiencing habitat loss: feeding and reproductive ecology of two polar bear populations. *Global Change Biology* 20(1): 76–88.

Rode, K. D., Wilson, R. R., Regehr, E. V., et al. (2015a). Increased land use by Chukchi Sea polar bears in relation to changing sea ice conditions. *PLoS ONE* 10(11): e0142213.

Rode, K. D., Robbins, C. T., Nelson, L. & Amstrup, S. C. (2015b). Can polar bears use terrestrial foods to offset lost ice-based hunting opportunities? *Frontiers in Ecology and the Environment* 13(3): 138–145.

Rode, K. D., Wilson, R. R., Douglas, D. C., et al. (2018). Spring fasting behavior in a marine apex predator provides an index of ecosystem productivity. *Global Change Biology* 24(1): 410–423.

Rosenzweig, C., Casassa, G., Karoly, D. J., et al. (2007). Assessment of observed changes and responses in natural and managed systems. In: Parry, M. L., Canziani, O. F., Palutikof, J. P., van der Linden, P. J. & Hanson, C. E. (Eds.), *Contribution of Working Group II to the Fourth Assessment Report of the Intergovernmental Panel on Climate Change* (pp. 79–131). Cambridge: Cambridge University Press.

Rosenzweig, C., Karoly, D., Vicarelli, M., et al. (2008). Attributing physical and biological impacts to anthropogenic climate change. *Nature* 453(7193): 353–357.

Rummukainen, M. (2015). Our commitment to climate change is dependent on past, present and future emissions and decisions. *Climate Research* 64(1): 7–14.

Schaller, G., Jinchu, H., Wenshi, P. & Zhu, J. (1985). The giant pandas of Wolong. *The Quarterly Review of Biology* 60(4): 524–525.

Shen, G., Pimm, S. L., Feng, C., et al. (2015). Climate change challenges the current conservation strategy for the giant panda. *Biological Conservation* 190: 43–50.

Smith, L. C. & Stephenson, S. R. (2013). New Trans-Arctic shipping routes navigable by midcentury. *Proceedings of the National Academy of Sciences* 110(13): 4871–4872.

Songer, M., Delion, M., Biggs, A. & Huang, Q. (2012). Modeling impacts of climate change on giant panda habitat. *International Journal of Ecology* 3: e108752.

State Forestry Administration of China. (2006). *The third national survey report on giant panda in China*. Beijing: Science Press.

State Forestry Administration of China. (2015). *The fourth national giant panda survey*. Beijing: Science Press.

Stern, H. L. & Laidre, K. L. (2016). Sea-ice indicators of polar bear habitat. *The Cryosphere* 10(5): 2027–2041.

Stirling, I. (2002). Polar bears and seals in the Eastern Beaufort Sea and Amundsen Gulf: a synthesis of population trends and ecological relationships over three decades. *ARCTIC* 55(5): 59–76.

Stirling, I., McDonald, T. L., Richardson, E. S., Regehr, E. V. & Amstrup, S. C. (2011). Polar bear population status in the northern Beaufort Sea, Canada, 1971–2006. *Ecological Applications* 21(3): 859–876.

Swaisgood, R. R., Wang, D. & Wei, F. (2017). *Ailuropoda melanoleuca* (errata version published in 2017). The IUCN Red List of Threatened Species. www.iucnredlist.org/en.

Tang, X., Jia, J., Wang, Z., et al. (2015). Scheme design and main result analysis of the fourth national survey on giant panda. *Forest Resource Management* 1: 11–16.

Thackeray, S. J., Henrys, P. A., Hemming, D., et al. (2016). Phenological sensitivity to climate across taxa and trophic levels. *Nature* 535(7611): 241–245.

Thuiller, W., Lafourcade, B., Engler, R. & Araújo, M. B. (2009). BIOMOD – a platform for ensemble forecasting of species distributions. *Ecography* 32(3): 369–373.

Tuanmu, M.-N., Viña, A., Winkler, J. A., et al. (2013). Climate-change impacts on understorey bamboo species and giant pandas in China's Qinling Mountains. *Nature Climate Change* 3 (3): 249–253.

Wang, F., McShea, W. J., Wang, D., et al. (2014). Evaluating landscape options for corridor restoration between giant panda reserves. *PLoS ONE* 9(8): e105086.

Wang, F., Zhao, Q., McShea, W. J., et al. (2018). Incorporating biotic interactions reveals potential climate tolerance of giant pandas. *Conservation Letters* 11(6): e12592.

Wang, R., Fan, X., Liu, Q. & Chen, W. (2010). The potential impacts of climate change on giant panda habitats in Sichuan Province. *Plateau and Mountain Meteorology Research* 30(4): 57–60.

Watts, P. & Hansen, S. (1987). Cyclic starvation as a reproductive strategy in the polar bear. *Symposium of the Zoological Society of London* 57: 305–318.

Wei, F., Costanza, R., Dai, Q., et al. (2018). The value of ecosystem services from giant panda reserves. *Current Biology* 28 (13): 2174–2180.e7.

Wei, W., Nie, Y., Zhang, Z., et al. (2015). Hunting bamboo: foraging patch selection and utilization by giant pandas and implications for conservation. *Biological Conservation* 186: 260–267.

Wei, W., Swaisgood, R. R., Dai, Q., et al. (2018). Giant panda distributional and habitat-use shifts in a changing landscape. *Conservation Letters* 11(6): e12575.

Wigley, T. M. L. (2005). The climate change commitment. *Science* 307(5716): 1766–1769.

Wiig, Ø., Amstrup, S., Atwood, T., et al. (2015). *Ursus maritimus*. The IUCN Red List of Threatened Species 2015. doi:10.2305/IUCN.UK.2015-4.RLTS.T22823A14871490.en.

Wilson, R. R., Regehr, E. V., Rode, K. D. & St Martin, M. (2016). Invariant polar bear habitat selection during a period of sea ice loss. *Proceedings of the Royal Society B: Biological Sciences* 283(1836). doi:10.1098/rspb.2016.0380.

Wu, H., Zhan, X.-J., Zhang, Z.-J., et al. (2009). Thirty-three microsatellite loci for noninvasive genetic studies of the giant panda (*Ailuropoda melanoleuca*). *Conservation Genetics* 10 (3): 649–652.

Yan, T., Ran, J., Zhao, C., Zhong, X. & Liang, M. (2017). Climate-change impacts on bamboo distribution and giant panda habitat in Qionglai Mountains. *Acta Ecologica Sinica* 37(7): 2360–2367.

Yang, H., Viña, A., Tang, Y., et al. (2017). Range-wide evaluation of wildlife habitat change: a demonstration using giant pandas. *Biological Conservation* 213: 203–209.

Yee, M., Reimer, J., Lunn, N. J., et al. (2017). Polar bear (*Ursus maritimus*) migration from maternal dens in Western Hudson Bay. *ARCTIC* 70(3): 319–327.

Yu, L., Li, Y.-W., Ryder, O. A. & Zhang, Y.-P. (2007). Analysis of complete mitochondrial genome sequences increases phylogenetic resolution of bears (Ursidae), a mammalian family that experienced rapid speciation. *BMC Evolutionary Biology* 7(1): 198.

Yurkowski, D. J., Hussey, N. E., Ferguson, S. H. & Fisk, A. T. (2018). A temporal shift in trophic diversity among a predator assemblage in a warming Arctic. *Royal Society Open Science* 5(10). doi:https://doi.org/10.1098/rsos.180259.

Zhang, Y., Mathewson, P. D., Zhang, Q., Porter, W. P. & Ran, J. (2018). An ecophysiological perspective on likely giant panda habitat responses to climate change. *Global Change Biology* 24(4): 1804–1816.

Zhang, Z., Swaisgood, R. R., Zhang, S., et al. (2011). Old-growth forest is what giant pandas really need. *Biology Letters* 7(3): 403–406.

Zhang, Z., Sheppard, J. K., Swaisgood, R. R., et al. (2014). Ecological scale and seasonal heterogeneity in the spatial behaviors of giant pandas. *Integrative Zoology* 9(1): 46–60.

Zhao, S., Zheng, P., Dong, S., et al. (2013). Whole-genome sequencing of giant pandas provides insights into demographic history and local adaptation. *Nature Genetics* 45(1): 67–71.

Zhu, L., Wu, Q., Dai, J., Zhang, S. & Wei, F. (2011). Evidence of cellulose metabolism by the giant panda gut microbiome. *Proceedings of the National Academy of Sciences* 108(43): 17,714–17,719.

Zhu, X., Lindburg, D. G., Pan, W., Forney, K. A. & Wang, D. (2001). The reproductive strategy of giant pandas (*Ailuropoda melanoleuca*): infant growth and development and mother–infant relationships. *Journal of Zoology* 253(2): 141–155.

Chapter 22

Managing for Interpopulation Connectivity of the World's Bear Species

Michael F. Proctor, Trishna Dutta, Bruce N. McLellan, Shaenandhoa Garcia Rangel, Dave Paetkau, Ronald R. Swaisgood, and Andreas Zedrosser

Introduction

The interruption of animal movement by fragmentation is a major force with far-reaching ecological and conservation consequences (Fahrig 2002). Understanding fragmentation processes underpins our ability to manage landscapes for connectivity, facilitating many ecological processes including gene flow, interpopulation dynamics, and demographic rescue. The paradigm of a slowly degrading distribution of a species into a fragmented set of isolated populations that eventually are extirpated resulting in dramatic range contraction occurs across species (van Oort et al. 2011). This is no more apparent around the world than in several bear species. For example, it has been well documented for brown bears in the USA south of Canada between 1850 and 1970 (Mattson & Merrill 2002; see also Chapter 13). In 2015 the IUCN Bear Specialist Group was reassessing the brown bear's Red List conservation status and it became apparent that while bears in their northern distribution were in large contiguous populations, much of the southern distribution was severely fragmented. This process resulted in the identification and assessment of all (or most) of the fragmented brown bear populations globally (McLellan et al. 2017). Other bear species accounts were reporting similar patterns.

Twenty-five years ago, the use of remotely sampled bear hair (and their roots) as a source of DNA to identify individuals to improve abundance estimation was developed (Woods et al. 1999). That method allowed explorations into patterns of fragmentation in grizzly bears, an emerging conservation issue (Proctor et al. 2002, 2005), and fragmentation studies were growing rapidly. The use of genetic methods to follow the fates of animals in "real time," within very recent time periods through individual-based methods (Pritchard et al. 2000; Paetkau et al. 2004), allowed migrant and fragmentation studies to reveal current conservation issues (Dixon et al. 2007) instead of detecting patterns that took centuries to develop (Steinberg & Jordan 1997).

The use of genetic sampling is not uniform across the continents occupied by bears. For some species, such as polar bears, we have very thorough genetic sampling that can be used to delineate population structure (see Chapter 14). For other species, Andean bears of South America and sun bears of Asia, there are minimal genetic sampling projects and we rely on expert opinion or turn to habitat fragmentation analyses for clues about population fragmentation.

Regardless, habitat and population-level fragmentation is prevalent across species and is influencing the conservation status of most bear species. Fragmentation's significance is also expected to grow as species, populations, and individuals attempt to adapt to our rapidly changing environment that includes the direct effects of human population growth and disturbance, but also climate change. We need to manage our natural systems and their inhabitants for movement (connectivity) so they may adapt, survive, and evolve.

Here we review the current status of fragmentation, connectivity, methods, consequences, and management of the world's eight bear species. We also consider the metapopulation paradigm, i.e. are bears being forced into some form of functioning metapopulation (McCullough 1996) or are they simply being fragmented into a series of isolated populations that, without conservation action, will likely be slowly extirpated, population by population (van Oort et al. 2011)?

Polar Bear

Polar bears (*Ursus maritimus*) make their living hunting seals from sea ice, so they can be found where sea ice persists in the Arctic (see Chapter 14). The IUCN/SSC Polar Bear Specialist Group (PBSG) recognizes 19 subpopulations, plus a central geographic catchall in the Arctic Basin that is used on a seasonal basis by bears from surrounding subpopulations (Durner et al. 2018; see also Chapter 21). Most subpopulation boundaries were defined objectively using movement data, including radiotelemetry and hunter returns of previously marked bears (e.g. Taylor et al. 2001). Despite a tremendous increase in both telemetry and genetically derived movement data, these subpopulation definitions have been remarkably stable for the past quarter century.

The majority of subpopulation boundaries are not strong barriers to movement, but rather lines across which movement is comparatively rare (Taylor et al. 2001). Even females with first-year cubs have the capacity to cross the entire polar basin (Durner & Amstrup 1995), but the norm is for individuals to remain in their home subpopulation. The geographic feature most strongly associated with subpopulation boundaries is land, but ice-free sea expanses (polynyas) coincide with some of the strongest oceanic discontinuities between conterminous subpopulations (Paetkau et al. 1999). Sea ice is far from uniform in its age, thickness and mobility, and this also influences the distribution of polar bears (Stirling et al. 1993; see also Chapter 21).

Genetic data have been used to evaluate polar bear connectivity, with a uniformity of sampling coverage that is rare for any species. The primary surprising result has been the identification of regions of reduced genetic exchange in the Canadian Arctic Archipelago, which cause the 19 functional subpopulations to be grouped into four strongly defined genetic clusters (Paetkau et al. 1999; Malenfant et al. 2016; see also Chapter 1). Three of these genetic groups span multiple subpopulations, producing estimates of >6000 individuals per genetic cluster. The fourth group contains a single subpopulation (Norwegian Bay) of roughly 200 bears at the northern extreme of the Archipelago, making it of special concern for conservation. A recent Hudson Bay study identified another small, distinct genetic group in James Bay, at the southern limit of the species range (Viengkone et al. 2016). Random genetic drift is more rapid, and divergence more extreme, for small populations; therefore, genetic observations for polar bears in Norwegian Bay and James Bay may reflect small size as much as interchange rates, but they certainly raise concern that persistence is less certain for these populations.

Current abundance estimates are available for 15 of the 19 functional subpopulations, with the sum of those estimates being 22,000 individuals (Obbard et al. 2015; Durner et al. 2018; Regehr et al. 2018). Adding expert estimates for the remaining populations, the worldwide total is approximately 26,000 (95% CI = 22,000–31,000; Durner et al. 2018) and the trend is thought to be stable. Of eight subpopulations for which PBSG provided estimates of trend in 2016, two were thought to have declined over the preceding 25 years and six to have "not reduced." In short, there is no evidence so far of a broad decline in global polar bear abundance.

While climate change has not yet manifested in dramatic decline or range loss, impacts on body condition have been documented (Obbard et al. 2016), and a significant loss of ice has occurred in all 19 subpopulations (Regehr et al. 2016). Because of the inexorable loss of sea-ice habitat, it is estimated that there is a 71% chance of the global population declining by more than 30% over the next four decades (Regehr et al. 2016); history aside, the impacts of climate change are quite probably on the cusp of producing strong negative impacts for polar bears (see also Chapter 21). While it seems unlikely that these impacts can be avoided, the intensive, internationally coordinated research effort for this species means that future changes are likely to be well documented, and as clearly understood as is possible for such a wide-ranging marine mammal.

Brown Bear

Brown bears (*Ursus* arctos) are also circumpolar and occur in the northern hemisphere across western North America, Europe, and Asia. The northern portion of their distribution is generally contiguous, except for eastern Canada. As you go south in their distribution, they become more fragmented (Figures 22.1A,B).

North America

Historically, brown bears inhabited western North America from northern Mexico to above the Arctic in Canada and Alaska (see Chapter 13). European settlement brought significant change to their southern distribution. Merriam (1922) reported the distribution of brown bears (called grizzly bears in North America) across the western portion of the lower 48 United States as severely fragmented into approximately 35 small populations. Within 50 years, most of the smaller populations were extirpated, leaving the current four populations (recovery zones; Mattson & Merrill 2002), with two being connected to Canadian populations. This example demonstrated that human settlement and behavioral patterns can severely fragment grizzly bears into small isolated populations with a high risk of extirpation. The remaining distribution of brown bears in western North America stabilized after 1970 (Mattson & Merrill 2002), but was suspected of less-obvious population fragmentation (McLellan 1998). This was eventually verified (Proctor et al. 2012), where primarily female fragmentation was documented throughout the southern peninsular distribution. Because bears have male-biased dispersal (McLellan & Hovey 2001), and males mediate gene flow, fragmentation of females became the standard for assessing the connectivity status of brown bear populations around the world (McLellan et al. 2017). Some level of population rescue required female interchange (demographic connectivity) that brought with it the ability to improve reproduction of a small fragmented and beleaguered population.

Methods in North America to detect population fragmentation were primarily genetic. DNA-based abundance survey data allowed individual microsatellite-based genotyping that informed individual-based population assignment methods, ultimately detecting migrants between geographic areas, revealing gender-specific fragmentation patterns (Proctor et al. 2005). Females were being fragmented, and males showed reduced movements. Fragmentation was associated with human settlement patterns, historic human-caused mortality, and highway traffic (Proctor et al. 2012). Similar patterns were detected in the peninsular distribution of southwest British Columbia (Apps et al. 2014).

Efforts to re-establish connectivity in North America have relied on GPS telemetry and corridor identification modeling (Proctor et al. 2015; Peck et al. 2017; McLellan 2018). Analysts developed resistance surfaces from habitat models using ecological, landscape, and human disturbance variables to identify least-cost path-derived movement corridors. Several efforts focused on female connectivity by using resource selection functions to develop habitat models (Proctor et al. 2015; McLellan 2018). Because females have short dispersal

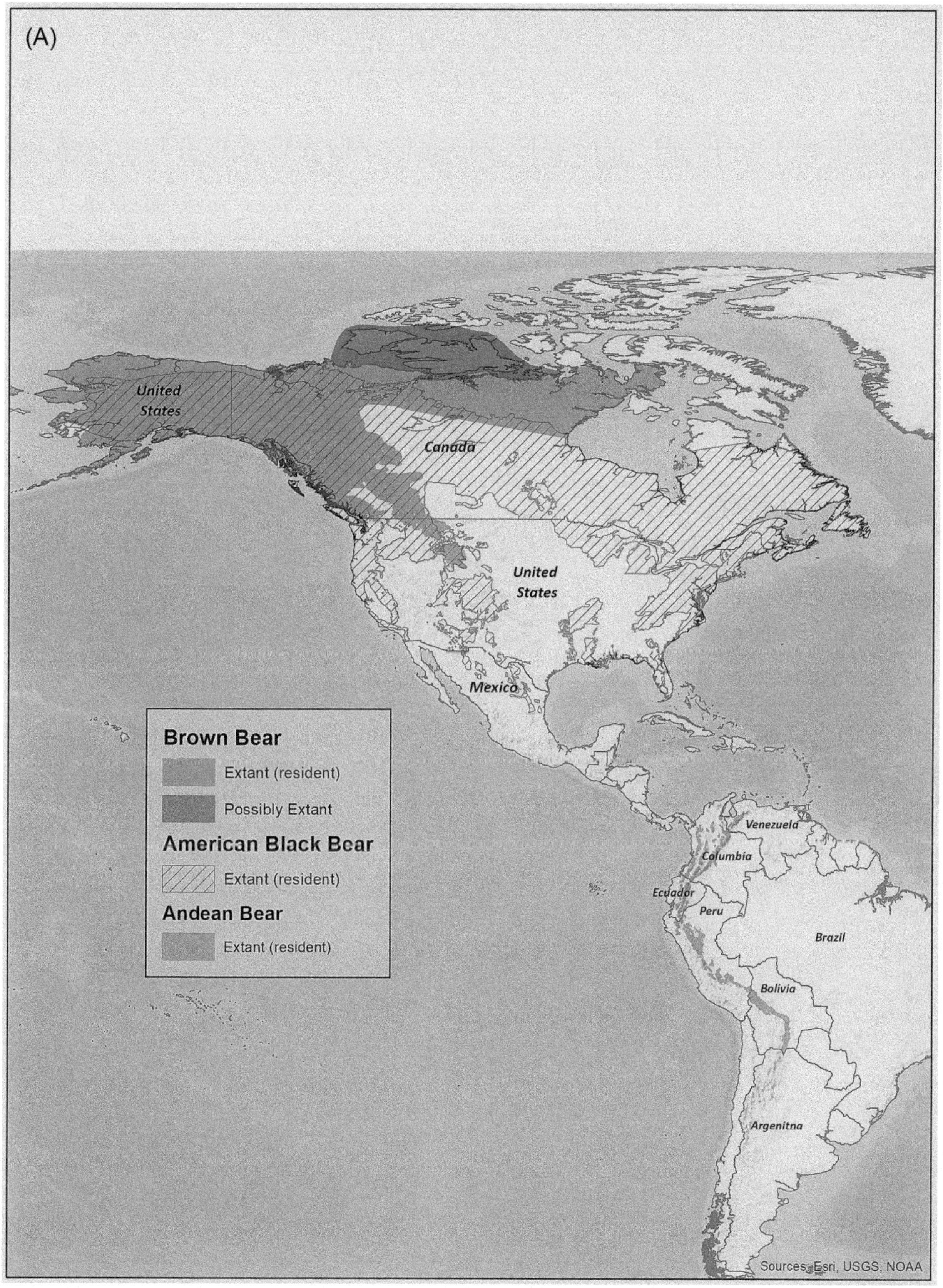

Figure 22.1 (A) Fragmented distribution of North American brown bears, American black bears, and South American Andean bears, and (B) Eurasian distribution of brown bears. (*Source*: IUCN 2017, Red List of Threatened Species.) (A black and white version of this figure will appear in some formats. For the color version, please refer to the plate section.)

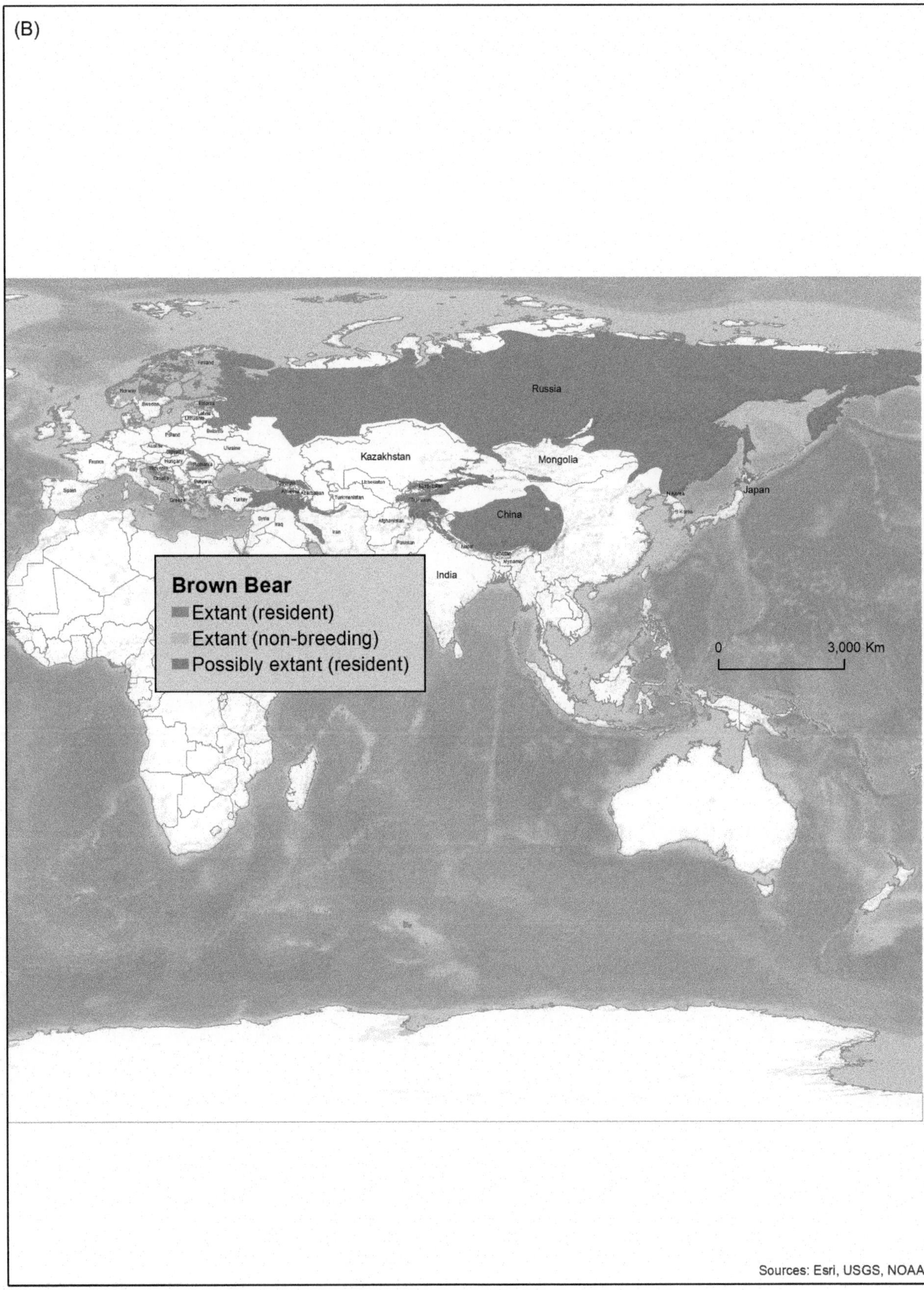

Figure 22.1 *(cont.)*

(Proctor et al. 2004), female connectivity would require females to live a portion of their lives in larger linkage areas; therefore, habitat quality would predict linkage habitat. The female-oriented approach makes sense where fragmented populations are small and require female interchange for population rescue, and fragmentation distances are within female dispersal distances (<15 km). A male-oriented approach focused on longer male dispersal with models based on movement-related habitat selection (Peck et al. 2017). The male-oriented approach was designed for larger populations >500 individuals, where female interchange is less critical, and fragmentation distances are relatively long (>100 km).

Efforts to re-establish connectivity are in various states of implementation. Using fragmentation patterns and connectivity mapping, Proctor et al. (2018) detailed over a decade of focused connectivity management including strategic land purchased by environmental NGOs, non-lethal management of potential conflict bears, extensive attractant management (e.g. cost-share electric fencing), and forestry road human-access controls that have resulted in measurable increases in direct interpopulation connectivity and genetic diversity (heterozygosity, number of alleles) in what was previously a totally isolated population for approximately six generations. Other biological metrics of connectivity were documented using population assignments, GPS telemetry movements, and breeding after inter-area movements detected with family pedigrees (Proctor et al. 2018).

The current situation in North America is less drastic than elsewhere across the southern distribution of brown bears globally. Interpopulation management of bears in much of the southern North American distribution requires efforts on populations that are only separated by 5–25 km of human-dominated mountain valleys. Human-caused mortality can be minimized with concerted effort to allow both-sex interchange (Proctor et al. 2018). Similar efforts are required across a larger area. There is some evidence that without recovery and connectivity management, populations may become extirpated (North Cascades) on the periphery of the main distribution. The more challenging, more distantly isolated (120 km) Yellowstone population is expanding and growing in number and recent trends suggest that males may establish connectivity in the near to medium future (Peck et al. 2017, C. Costello, personal communication).

Europe

The brown bear is the only bear species in Europe, and its historic distribution covered almost the entire continent (Figure 22.1B; Zedrosser et al. 2001; see also Chapter 12). Brown bears were persecuted in Europe for centuries before their gradual elimination from much of western and northern Europe; large populations remained only in eastern Europe (Swenson et al. 1994, 2000; Zedrosser et al. 2001). Today, bear populations are fragmented into 10 populations (Swenson et al. 2000; Zedrosser et al. 2001). While populations in northern (Scandinavia), eastern (Karelia, Baltic, Carpathian), and south-eastern (Dinaric–Pindos, Eastern Balkan) Europe are relatively large and partly continuous metapopulations, populations in western Europe (Spain, France, central Italy) are very small, fragmented, and isolated (Boitani et al. 2015).

Brown bears in Europe show on average longer dispersal distances compared to North America, and also substantial female dispersal can be found in some populations (Støen et al. 2006; Zedrosser et al. 2007). However, European population expansion is led by males, with spatial and temporal expansion of females lagging significantly behind (Swenson et al. 1998). Similar to North America, efforts to evaluate population fragmentation and re-establish connectivity have relied on genetic methods as well as habitat modeling based on direct observation and VHF and GPS radiotelemetry. Specific methods vary geographically, by population size, and by degree of fragmentation. Large-scale genetic studies using samples of legally harvested individuals, tissue samples from research bears, and fecal samples collected for population monitoring have evaluated connectivity of bear populations in Northern Europe, and suggest larger and continuous bear populations in Finland/Russia (Tammeleht et al. 2010; Kopatz et al. 2012; Schregel et al. 2012) and in Scandinavia (Sweden and Norway; Waits et al. 2000), as well as some genetic clusters confined to smaller geographic areas in Estonia, Southern Finland, and Leningrad oblast, Russia (Tammeleht et al. 2010).

Studies in southern Europe have revealed fragmentation causes including human activities such as road construction and human exploitation, (Karamanlidis et al. 2012). Straka et al. (2012) found that human activities have reduced and fragmented the previously large and continuous Carpathian brown bear population into larger eastern and smaller western subpopulations with limited inter-area gene flow.

Modeling approaches based on GPS locations and direct observations are commonly used to evaluate functional connectivity of bear populations in Europe. Peters et al. (2015) evaluated the connectivity of the augmented bear population in northern Italy with the large natural Dinaric–Pindos population in Slovenia, and identified two potential road crossings in north-eastern Italy, as the biggest constraint for the Italian population to expand eastward and impeding dispersal to/from Slovenia. Mateo-Sanchez et al. (2014) found connectivity between two small endangered brown bear subpopulations in the Cantabrian Mountains of Spain (see also Chapter 19) separated by only ~40 km, mediated by highly impermeable transportation infrastructure. However, Gonzalez et al. (2016) showed with genetic methods that reduction in human-caused mortality has brought about an increase in numbers, which has promoted male-mediated migration and gene flow from the western to the eastern subpopulation. Several projects exploring highways and fragmentation (Kaczensky et al. 2003; Kusak et al. 2009; Karamanlidis et al. 2012; Peters et al. 2015) have resulted in a number of wildlife highway crossing structures throughout central and south-eastern Europe to facilitate bear connectivity (e.g. Kusak et al. 2009).

Due to very poor connectivity potential in central and southern Europe, augmentation has been used as a management tool to increase population viability of small isolated populations with varying success. Successful augmentations have been carried out in the Alpine population in northeastern Italy and in the French Pyrenees (Boitani et al. 2015), but augmentation in central Austria has failed, likely due to illegal killing of bears (Kaczensky et al. 2011). The population in the French Pyrenees remains isolated with very little hope of connectivity with other bear populations.

The brown bear is a species of high conservation interest in Europe; however, the population and conservation status differ between populations and are a result of demographic history and landscape barriers. The large populations in northern and eastern Europe are relatively well-connected and not endangered at the population level. Small and/or isolated populations in western Europe (Cantabrian Mountains, Pyrenees, and Apennines) are in dire need of conservation action and there is little hope for connectivity to other populations in the future. Habitat fragmentation and loss and conflicts with humans are and remain the major threats for bears in Europe. Management and conservation of the brown bear in Europe is regulated by the Habitat Directive of the European Union as well as by a number of other international and national action and management plans (Swenson et al. 2000; Boitani et al. 2015). Boitani et al. (2015) recommend in a report to the European Commission that "Protection of bear habitat and enhancement of connectivity within and between populations" is given the highest level of urgency, and the highest level of expected benefits in terms of conservation and coexistence with locals. However, a large-scale analysis of the connectivity of brown bear populations is currently missing in Europe.

Asia

Most of the world's brown bears live in Asia (Figure 22.1B), but relatively little is known about them. Our understanding of their distribution is based largely on estimates from local biologists, and this information is often difficult to synthesize over appropriate scales. Some populations, such as those on Hokkaido, Kunashiri, Etorofu (Iturup), and Paramushir Islands are isolated due to oceanic expanses. Most populations, however, appear isolated by broad valleys that have been settled by people for many hundreds or thousands of years. McLellan et al. (2017) estimated that there were 25 populations in Asia that were likely isolated from each other at least for movement of female bears. This estimate was based on input from local biologists combined with digital imagery; there are likely some errors and more work on this topic is certainly needed. Of these, the largest population spans much of the Russian Federation from Europe to Kamchatka with likely over 200,000 bears (www.ohotcontrol.ru/resource/number/. See also Chapter 12). The population in western China (mostly Qinghai–Tibetan Plateau), is much smaller but still has perhaps 6000 bears. Other populations, such as those in the Gobi desert of Mongolia, Hindu Kush in Pakistan, Datca of Turkey, Aragat Mountains of Armenia, Mingacevir Reservoir of Georgia, and Azerbiajan have a few dozen bears or even fewer. Whether any bears remain in parts of Syria and Lebanon is uncertain, but if there are any, they are very few.

There are no clear efforts to reconnect isolated populations in Asia, although some actions to reduce human–bear conflicts that often led to the death of bears and may help with reconnection are being implemented. However, these actions are very rare, and a much greater effort will be needed to reconnect populations.

Giant Panda

Historically, giant panda (*Ailuropoda melanoleuca*) populations occupied large areas of southern and central China and parts of Myanmar, Thailand, Laos, and Viet Nam, but changing climate and centuries of expanding human populations and forest conversion have restricted the panda's range to a few isolated fragments (Loucks et al. 2001; Han et al. 2019; see also Chapters 6 and 21) in western China (Figure 22.2). Plausibly, panda populations were once characterized by a high degree of connectivity with high levels of gene flow. The current distribution of pandas is restricted to a few isolated mountain ranges in the Minshan, Qinling, and Qionglai, which constitute the largest remaining populations (SFA 2015; Figure 6.1 in Chapter 6). Smaller, more isolated populations are found further south in the Daxiangling, Liangshan, and Xiaoxiangling mountain ranges. Range-wide surveys conducted by China's State Forestry Administration indicated that panda populations were in sharp decline, but recent surveys document a stabilizing and slightly increasing population numbering an estimated 1864 adult individuals. This promising trend was sufficient to downlist the panda from Endangered to Vulnerable on the International Union for Conservation of Nature Red List of Threatened Species (Swaisgood et al. 2016).

The primary methods used to evaluate connectivity are China's circa once-a-decade national survey of panda populations and habitat and emerging genetic tools. China's State Forestry Administration (SFA 2015) has identified 33 subpopulations, with 18 of these containing fewer than 10 pandas. The largest subpopulation contains more than 400 individuals. However, some of these subpopulations are partially defined by political boundaries and are not sufficiently isolated to cause genetic divergence. A large number of genetic studies provide high spatial and temporal resolution to giant panda historical and current patterns of isolation and divergence of subpopulations, summarized in two recent reviews (Wei et al. 2012, 2015) and supplemented with whole-genome sequencing (Zhao et al. 2013). The best evidence from microsatellite analysis has identified three distinct genetic clusters. The Qinling subpopulation in the north diverged ~0.3 million years ago and is morphologically and genetically distinct from the southern populations. Qinling is separated from the rest of

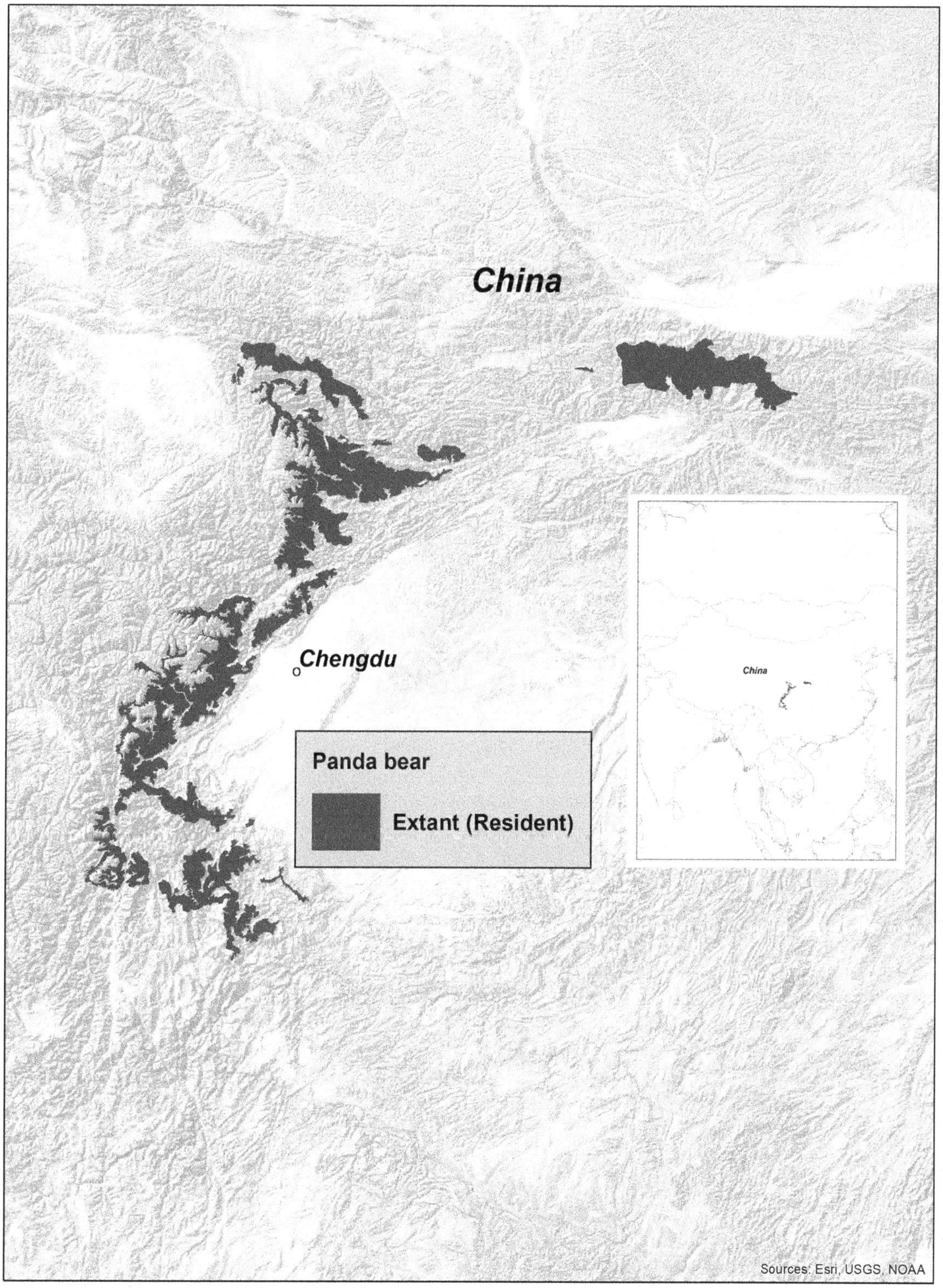

Figure 22.2 Fragmented giant panda bear distribution in western China. (*Source*: IUCN 2016, Red List of Threatened Species.) (A black and white version of this figure will appear in some formats. For the color version, please refer to the plate section.)

the panda range by distances large enough (~100 km) to preclude any connectivity. In the Minshan Mountains, the Minshan and Qionglai populations diverged ~2800 years ago. The Qionglai subpopulation complex is further subdivided into four genetic clusters, Qionglai, Daxiangling, Xiaoxiangling, and Liangshan.

The causes of population substructuring are largely anthropogenic (Wei et al. 2012, 2015; Zhao et al. 2013). However, natural barriers such as rivers have also been associated with genetic divergence; these river barriers have been made less permeable by accompanying large human settlements within the river valleys. Widespread deforestation and road construction over the past three millennia are responsible for the current distribution of pandas. These patterns of isolation may soon be exacerbated by climate change, with models forecasting increasing fragmentation of pandas' habitat (Li et al. 2015).

Despite these patterns of isolation, fragmentation, and genetic divergence, giant pandas as a species still retain high levels of genetic variability. Dispersal is female-biased and typically exceeds 20 km (Zhan et al. 2007), although the permeability of anthropogenically altered landscapes is not well understood. Because fragmentation and isolation are relatively recent phenomena, commencing some 3000 years ago and escalating in the past century, loss of genetic diversity is a concern for the near-term future. Several subpopulations are so small and isolated that genetic viability is questionable, and inbreeding depression is predicted in the near future if subpopulations are not reconnected (Hu et al. 2017). Recent trends of changing human land-use patterns, including increased mining, roads, livestock grazing, and ecotourism, threaten to further degrade and fragment panda habitat, although these trends are also partially offset by positive trends such as decreased logging and increased forest cover and age (Wei et al. 2018).

A number of efforts at genetic rescue of isolated panda populations are underway. Translocations and release of captive-bred animals into the smallest populations will help reduce loss of genetic diversity (Yang et al. 2018). More promising, the Chinese government has plans to protect much of the remaining giant panda habitat in a National Park network, and establish corridors connecting many of the isolated populations. Scientific analysis of habitat use patterns can inform policy and management for establishing corridors (Wang et al. 2014). Least-cost models indicate that pandas are most likely to move through low-elevation, flat, forested areas. Forest and bamboo restoration and automobile tunnels moving portions of roads underground are the most promising management tools for re-establishing connectivity. Other large-scale conservation initiatives also help arrest and restore the panda's forest habitat. The National Forest Conservation Program bans logging throughout the panda's range and eco-compensation programs such as the Grain to Green Program incentivize reforestation. Thus, despite some worrying trends and historical population restriction, conservation measures currently at play show promise to mitigate panda habitat fragmentation and population isolation (Swaisgood et al. 2018).

Sloth Bear

Sloth bear (*Melursus ursinus*), the only myrmecophagus species of bear (diet specialized on ants and termites) is endemic to the Indian subcontinent, with a majority of its current range in India, Sri Lanka, and the southern lowlands of Nepal (Figure 22.3; see also Chapter 9). They are rare in southern Bhutan contiguous with populations in India and Nepal (Dharaiya et al. 2015), and recently extinct in Bangladesh (Islam et al. 2013). Sloth bears across their range occupy diverse habitats from wet or dry tropical forests to savannas, shrublands, and grasslands (Garshelis et al. 1999; Yoganand et al. 2006).

While the forest areas occupied by sloth bears in India are extensive (~2000–400,000 km^2 each; Garshelis et al. 1999; Yoganand et al. 2006; Jhala et al. 2008; Sathyakumar et al. 2012), they are patchy and highly fragmented, and most patches are small (Yoganand et al. 2006). Sloth bears occupy about 11,810 km^2 or 17% of Sri Lanka, where distribution is largely contiguous except in the north and east (Ratnayeke et al. 2007). Estimation data are not available for Nepal and Bhutan.

In India, 90% of sloth bear populations are split between dry and moist deciduous forests (Yoganand et al. 2006); however, their abundance is higher in moist deciduous forests, followed by dry deciduous, scrub, and evergreen forests (Yoganand et al. 2006).

Sloth bears are sensitive to habitat fragmentation, which is rampant across their range. Fragmentation of habitat in this rapidly developing part of the world is largely due to increasing incursions of a burgeoning human population (~1.2 billion people in India) into forest habitats and rapid conversion of forestland to meet ambitious development infrastructure goals (mines, roads, and dams) in the subcontinent. A large part of the current sloth bear distribution occurs in forests outside protected areas in India (Jhala et al. 2011; Sathyakumar et al. 2012) and Sri Lanka (Ratnayeke et al. 2007). In India, some sloth bear populations have been extirpated from small (~<300 km^2) isolated (~<5 km) forest patches (Yoganand et al. 2006).

Scientific assessment of population connectivity across sloth bear range is just beginning. In southern Western Ghats, indirect sign surveys in an occupancy framework have been used to identify and map a critical corridor for sloth bears (Das et al. 2014). In central India, two studies used remotely sampled genetic approaches to assess connectivity, the number of first-generation migrants, and landscape variables associated with genetic differentiation (Dutta et al. 2015; Thatte et al., unpublished data). Both studies found moderate levels of genetic diversity and significant genetic differentiation. Dutta et al. (2015) detected five individual migrants between protected areas separated by 85–312 km. Thatte et al. (unpublished data) found that genetic connectivity between sloth bear populations was impacted by non-forested areas, high density of humans, high density of linear features (roads, railway lines, and irrigation canals), and all roads offered high resistance to sloth bear movement.

Across their range, sloth bears co-occur with wide-ranging megacharismatic species such as tigers in India (Jhala et al. 2015; Dutta et al. 2016) and elephants in Sri Lanka (De Silva 1998; Ratnayeke et al. 2007), which are often used as umbrella species for connectivity conservation or large landscape

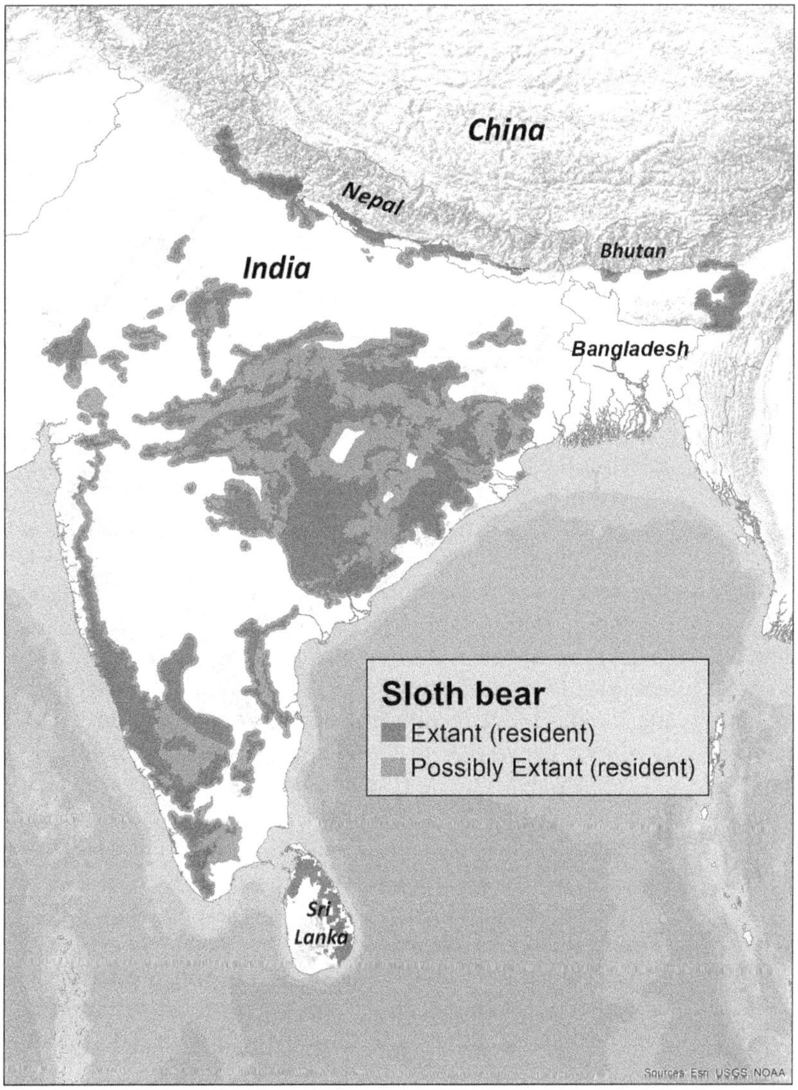

Figure 22.3 Fragmented sloth bear distribution across India and Sri Lanka. (*Source*: IUCN 2016, Red List of Threatened Species.) (A black and white version of this figure will appear in some formats. For the color version, please refer to the plate section.)

conservation planning. While this is important, it may not be an effective conservation strategy for a widely distributed species such as the sloth bear that is abundant outside protected areas, has smaller space requirements, and a completely different scale and suite of ecological requirements than the usual surrogates used for connectivity conservation (Thatte et al., unpublished data; Ratnayeke et al. 2007). Elephants and tigers need large contiguous areas for conservation, but for sloth bears, the conservation value of non-protected areas and small fragments might be high despite being embedded in a human-dominated landscape (Thatte et al., unpublished data). Therefore, sloth bears might be managed to provide some degree of connectivity between protected areas for many species with habitats needs different than those of elephants and tigers.

Despite the high rate of conflict with humans (Ratnayeke et al. 2014; Dhamorikar et al. 2017) and sensitivity to habitat fragmentation and anthropogenic influence in the landscape, sloth bears appear to be connected in large parts of their core distribution, although this may not be the case for peripheral populations. More detailed studies of connectivity across their range are a priority to identify and prioritize corridors and boundaries of functionally interconnected subpopulations. Because creating new corridors in the highly human-occupied matrix may be extremely challenging, maintaining existing connectivity and mitigating or preventing barriers in corridors appear to be the primary way to maintain viable connected sloth bear populations in the future. Carefully identified corridors and selected sites for mitigation measures that can serve connectivity needs for multiple species in the biodiversity rich areas that sloth bears inhabit will serve as a holistic conservation plan for sloth bears and other species across their range. Finally, it is imperative to anticipate and integrate mitigation measures to deal with human–bear conflict in corridors.

Andean Bear

Andean bears (*Tremarctos ornatus*) inhabit a discontinuous set of subdivided subpopulations with often unknown degrees of gene flow along the forested Andes Mountains across Venezuela, Colombia, Ecuador, Peru, Bolivia (Velez-Liendo & García-Rangel 2017; Figure 7.1, Chapter 7), and a small area in northern Argentina (Figure 22.1A; Cosse et al. 2014). They inhabit evergreen forests, which were historically distributed fairly continuously across the Andes mountain range (Garcia-Rangel 2012). Seasonally, they use high-elevation shrublands beyond the altitudinal tree line, and these habitats are naturally fragmented (García-Rangel 2011). The species is also found within pockets of semideciduous and deciduous forests, subject to natural climatic fragmentation (García-Rangel 2011).

About 30% of available habitat, in patches <400 km^2 beyond 15 km from adjacent patches, across their range is unsuitable to sustain viable populations (Velez-Liendo et al. 2014; Velez-Liendo & García-Rangel 2017). Studies flagged possible local extinctions in Venezuela, but similar analyses are required more widely (Rodríguez et al., unpublished data). Bolivia is projected to retain most of its habitat, while Venezuela will lose more core habitat patches than any other country (Velez-Liendo & García-Rangel 2017). A recent fragmentation assessment along the Venezuelan–Colombian border showed high levels of fragmentation within Colombia with more continuous habitat within Venezuela (Rodríguez et al., unpublished data). In general, a lack of detailed information about core habitat patches hinders efforts to draw plans targeting Andean bear conservation.

The need to generate information to guide Andean bear conservation has been the main driver of fragmentation and connectivity research. Structural (habitat-based) connectivity has been the main focus, but recent studies have delved into functional aspects (Velez-Liendo et al. 2013, 2014; Sánchez-Mercado et al. 2014). Initial efforts combined expert opinion and information on vegetation types (Monsalve Dam et al. 2010). Core patches were identified based on the former and data extrapolations about minimum area requirements to sustain viable populations (Monsalve Dam et al. 2010). GIS, remote sensing imagery, and ancillary spatial data helped to improve this understanding and provide more tailored assessments of habitat availability and fragmentation (Rodriguez et al. 2003; Kattan et al. 2004). These spatial variables were linked to Andean bear ecological requirements and minimum patch size thresholds were estimated (Cuesta et al. 2003; Peralvo et al. 2005). Eco-regional strategies were updated informing fragmentation and connectivity strategies (Rodriguez et al., unpublished data).

Functional connectivity was evaluated in Bolivia through resource-based modeling, female habitat patches, and cost-distance modelling (Velez-Liendo et al. 2013, 2014). A pioneering analysis of ecological traps was carried out in Venezuela to guide conservation action (Sánchez-Mercado et al. 2014). Recently, camera-traps and occupancy models provided insights into habitat use, complementing sign surveys (Monsalve Dam et al. 2010; Molina et al. 2017; Zug 2018). These tools have been used to support claims about the species' presence in protected areas, buffer zones, corridors, and other areas (Monsalve Dam et al. 2010; García-Rangel 2011; Molina et al. 2017). In the last decade, ambitious priorization analyses have been carried out in Colombia (i.e. occupancy models) and Ecuador (i.e. least-cost path analysis) to guide protected area and landscape management (Eraso Mera 2014; Acevedo et al. 2017; Utreras 2018).

Genetic analyses have been scarce (Ruiz-Garcia 2003; Ruiz-Garcia et al. 2005; Viteri & Waits 2009). Available information indicates that there are two main population blocks, and information on the spatial arrangement of the genetic structure of Andean bear populations is available at the country level (Viteri & Waits 2009).

Today's loss, degradation, and fragmentation of evergreen forests is the consequence of long-term anthropogenic impacts threatening Andean bear survival across their range (García-Rangel 2011; Sánchez-Mercado et al. 2014; Velez-Liendo & García-Rangel 2017; Rodriguez et al., unpublished data). Patch size is a concern for their long-term survival. Agricultural expansion is the main driver of this process, prompted by inadequate agrarian reforms and farming practices (García-Rangel 2011; Velez-Liendo & García-Rangel 2017). Road building, logging, infrastructure development, mining and oil extraction, land conversion to illegal crops, and social unrest also drive this fragmentation (García-Rangel 2011; Velez-Liendo & García-Rangel 2017). Access to different ecosystems across the altitudinal gradient is essential for Andean bears, but this is also hindered by fragmentation (García-Rangel 2011).

The impacts of fragmentation likely vary, but there is little documentation of genetic or demographic consequences. Human-caused mortality is expected to rise with greater access to remote areas and increased tension has increased human–bear conflicts (Sánchez-Mercado et al. 2014; Velez-Liendo & García-Rangel 2017). Poaching is associated with areas of rapid and recent forest conversion in Venezuela (Sánchez-Mercado et al. 2014). Climate change is expected to exacerbate habitat loss, and fragmentation could increase exposure, reduce resource accessibility, and limit movement of individuals (Velez-Liendo & García-Rangel 2017).

Protected areas have been widely used to counteract fragmentation of Andean bear habitat, in many cases through their design (Yerena 1998; Monsalve Dam et al. 2010; Yerena & García-Rangel 2010). Corridors have also been proposed, and recently connectivity assessments have informed landscape management plans (Yerena 1998; Yerena & García-Rangel 2010). As of 2008, 58 protected areas had been established within Andean bear habitat (García-Rangel 2011; Velez-Liendo & García-Rangel 2017), and these conservation units have halted habitat loss and degradation within their boundaries and prompted natural restoration (García-Rangel 2011). Updating the existing assessment of protected areas' status and effectiveness is now timely. Connectivity is known to be important to link areas with low poaching risk, but source–sink dynamics should be avoided when considering conservation plans (Sánchez-Mercado et al. 2014).

Asiatic Black Bear

The Asiatic black bear (*Ursus thibetanus*) distribution extends from Eastern Russia and Japan to Iran in the west and can be broadly broken up into five primary regions, each of which has varying degrees of fragmentation within them (Chapter 10, Figure 10.1; Figure 22.4). The full extent of population-level fragmentation is poorly researched and therefore unknown; however, through the expertise of regional and local researchers and managers, localized field sign surveys in some instances, and questionnaires, a picture of a discontinuous distribution has emerged.

In eastern Asia, a large occupied territory exists in extreme south-eastern Russia, into North Korea and north-eastern China. Information on the extent of internal fragmentation in this area is lacking (Garshelis & Steinmetz 2016). Aramilev (2006) estimates there may be between 5000 and 7000 Asiatic black bears in the Russian portion of the region, 40 in North Korea, and there are no estimates in the Chinese portion (Garshelis & Steinmetz 2016).

Southern China is the largest stronghold of Asiatic black bears, although no credible estimates of numbers exist (Gong & Harris 2013) and they are connected to bears living along the southern portions of the Himalaya region that extends across

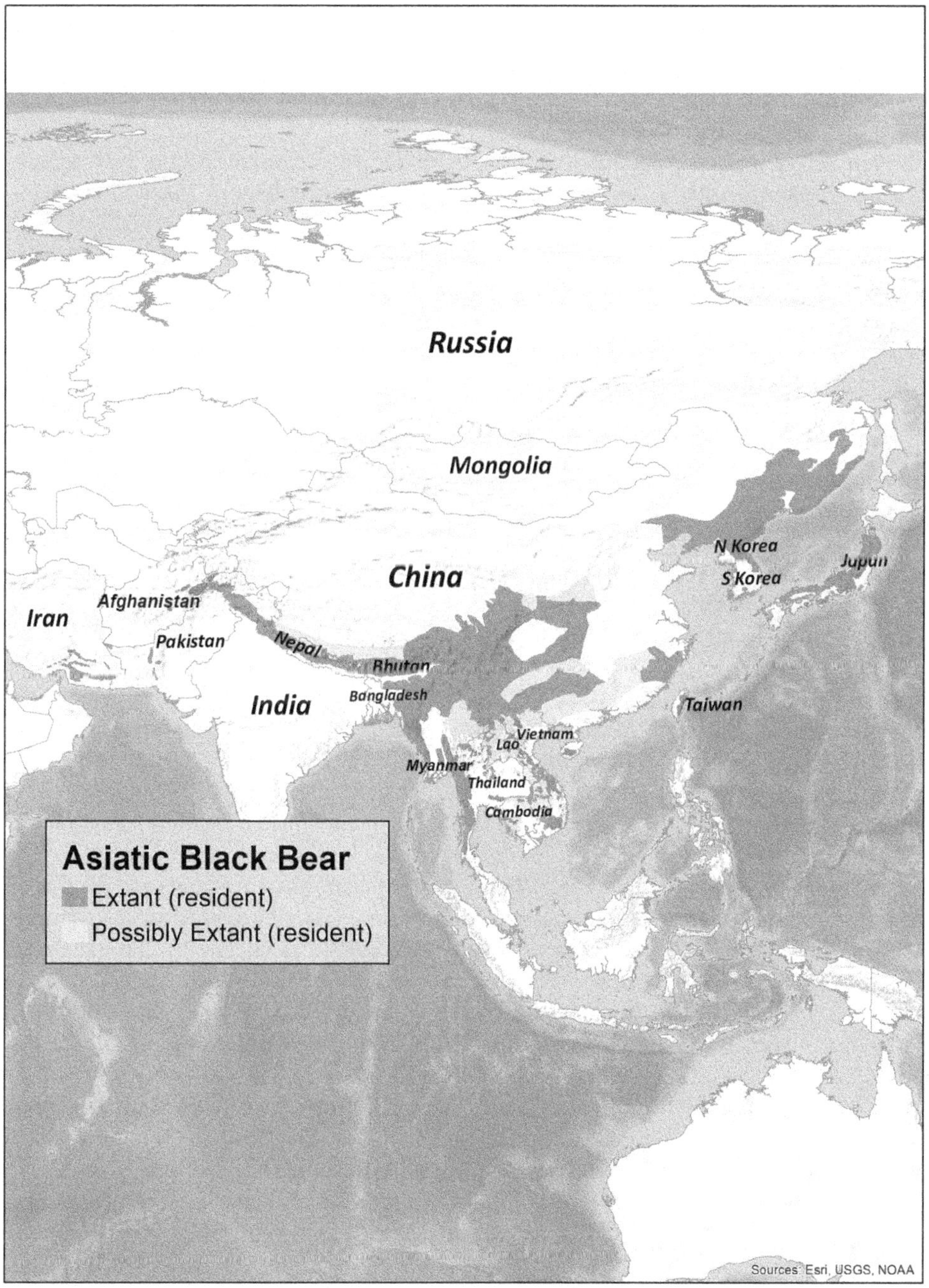

Figure 22.4 Fragmented Asiatic black bear distribution from south-eastern Russia and Japan south and west into South-East Asia and across the Himalayas into Iran. (*Source*: IUCN 2016, Red List of Threatened Species.) (A black and white version of this figure will appear in some formats. For the color version, please refer to the plate section.)

Bhutan, Nepal, and India (estimated 5000–7000 bears; Sathyakumar & Choudhury 2007). Further east along the Himalayas, scattered populations exist in Afghanistan (Ostrowski et al. 2009) and Pakistan, whose bears are likely fragmented into nine isolated populations (Khan et al. 2012; Abbas et al. 2015). Isolated patches of bears exist in Bangladesh (Garshelis & Steinmetz 2016).

South-East Asia has a heavily fragmented distribution that extends across Viet Nam, Lao, Myanmar, and Thailand (Kanchanasakha et al. 2010). Few detailed studies have been done in this region, with the exception of several rigorous population estimates in Thailand (Ngoprasert et al. 2012). The distribution of these bears is reasonably correlated with the region's forests, where fruiting tree density was a reasonable predictor of bear density (Ngoprasert et al. 2011; Stienmetz et al. 2011). Thailand has a number of small populations inside protected areas, separated by expanses of agriculture (Kanchanasakha et al. 2010).

At the western extent of their distribution, several isolated populations exist in south-eastern Iran (Yusefi 2013; Almasieh et al. 2016), with one population possibly being connected to those in western Pakistan. Across their disjunct Iranian distribution, Almasieh et al. (2016) used credible sightings, observations, camera traps, and sign surveys to estimate habitat clusters (rather than biologically isolated populations) across their range to develop a habitat suitability map predicted by a suite of ecological, climate, and human variables. They identified 31 core habitats and consolidated these into four major habitat clusters and then used least-cost path modeling to identify connectivity corridors between clusters. The result is a credible map of core Asiatic black bear habitats, and a network of corridors linking them. They recommend that this effort be the basis for conservation planning in maintaining and/or re-establishing connectivity between habitat clusters. This is an excellent example of how fragmentation and connectivity can be integrated into conservation planning, even without the advantage of expensive DNA-based surveys or GPS telemetry studies. Taking the work one step further, Farashi and Erfani (2018) used 32 variables to model potential Asiatic black bear habitat shifts due to future climate change. They found that in many cases, predicted habitat shifts may be inhibited by human environments, and recommend conservation planning teams integrate their results.

Several islands, Japan, and Taiwan and Hainan in the South China Sea also have populations of Asiatic black bears. In Japan, they occur on three islands. Hinshu, the largest island, contains a "metapopulation" of approximately 10,000 bears (Horino & Miura 2000; see Chapter 10) divided among several-to-many fragmented subpopulations with varying degrees of connectivity, some being small with <100 bears (Doko et al. 2011; see Chapter 25). The western distribution on Honshu Island contains three small populations which have been shown to be genetically isolated (Saitoh et al. 2001; Ohnishi et al. 2007). Bears on the northern tip of Honshu Island, on Shimokito Peninsula, were found to be a genetically isolated population of ~100 bears (Horino & Miura 2000; Yamamoto et al. 2012). Legal hunting, deforestation, conflict-related mortality, human settlement, and transportation corridors all contribute to fragmentation across Japan. Although most researchers recommend conservation planning to integrate connectivity, we are not aware of the extent of connectivity management at this time, except for limited programs to reduce human–bear conflicts (Sakurai & Jacobson 2011).

Very few attempts have been made to assess the distribution and state of fragmentation across the entire Asiatic black bear range. One effort (Escobar et al. 2015) used multivariate ecological niche modeling and lights-at-night data to identify areas of high human disturbance. Interestingly, their resulting map is rather similar to the latest iteration of the IUCN Red List range map.

American Black Bear

American black bears (*Ursus americanus*) occur in the primarily forested regions of three countries in the western hemisphere (Figure 22.1A; see also Chapter 11, Figure 11.1): Canada (12 provinces/territories), USA (41 states), and Mexico (six states). The northern portion of American black bear North American range, across Alaska, Canada, and northern portions of the lower 48 states, is one large, continuous population containing over 800,000 bears (Garshelis et al. 2016). Three peninsular-shaped occupied areas extend southward from this main distribution: one along the west coast, one through Idaho and Montana, and another along eastern US coastal states.

A series of variable-sized anthropogenically isolated populations exist in the south-east and central states including Georgia, Florida, Alabama, Mississippi, Arkansas, Louisiana, Oklahoma, Missouri, and through the American west including Montana, Idaho, Wyoming, Colorado, Utah, California, Nevada, Texas, New Mexico, and Arizona as well as northern Mexico (Scheick & McCown 2014). One small isolated population (<400 bears) exists in Ontario's Bruce Peninsula (Pelletier et al. 2017). American black bears are rebounding from decline in the 1900s due to unregulated hunting, human settlement patterns, and habitat loss (Garshelis et al. 2016).

Naturally isolated populations occur on several islands off both coasts of North America, Newfoundland off Canada's east coast, Vancouver Island, Haida Gwaii, and others off British Columbia's west coast, and many islands within south-east Alaska. Newfoundland and Haida Gwaii are isolated populations as the distance across water is too great for dispersal (Paetkau & Strobeck 1996).

Scheick and McCown (2014) developed a distribution map that identifies fragmentation and isolation patterns across North America from information provided through extensive surveys of working biologists and managers across black bear range. They noted sightings that were "outside" primary and secondary distribution lines may represent exploratory male movements and may inform connectivity management.

There has been significant study of connectivity in some localized areas, but less regional documentation of the state of fragmentation. While it may sometimes appear that a continuous distribution is not fragmented, that may not always be the case. Proctor et al. (2020b) used genetically derived pedigrees to determine levels of fragmentation across mountain valleys with human settlement, railroads, and highways where otherwise no genetic signal of fragmentation existed, due to a lack of genetic drift in large populations that were in fact fragmented. Larkin et al. (2004) used expert opinion, human land-use patterns, and least-cost modeling to identify dispersal corridors to re-establish a very small (<20 bears) isolated population in west central Florida and recommended land-use management consistent with bears surviving dispersal, through a network of natural areas interspersed with major highways, human settlements and rural areas. Dixon et al. (2007) detailed the state of fragmentation of Florida black bear across several isolated populations and have measured the reconnection of several of these populations in response to improved management through genetic monitoring, including the use of a corridor, and the use of population assignments to identify inter-population migrants (Dixon et al. 2006).

Louisiana has several isolated populations that were designated "Threatened" by the Federal US Fish and Wildlife Service in 1992. Laufenberg and Clark (2014) found a series of small populations (sizes varied from 50 to 300) in various states of minimal connectivity to complete isolation using genetic analyses, the clustering program *Structure*, and telemetry. They used step functions and resistance cost surfaces from telemetry data to estimate corridor locations for future connectivity management (Clark et al. 2015). Gantchoff and Belant (2017) used expert opinion to develop resistance models and circuit theory to predict black bear movement corridors across five states in south-central USA with multiple isolated populations. Corridors were evaluated using citizen sightings of bears. Protected area core habitats were interconnected by forested and riparian areas. They identified pinch points created by human disturbance, and suggested areas where connectivity management might be focused by maintaining forest cover and undeveloped riparian corridors.

In Idaho, Cushman et al. (2006) used 146 genetically sampled bears over 3000 km^2 to develop a resistance cost surface to test landscape connectivity models for black bears. They concluded that connectivity was facilitated by mid-elevation forest cover, but did not detect fragmentation across a settled rural valley with a highway. Cushman et al. (2009) used those same data with resistance and least-cost path models to identify corridors over a larger region-sized area (Idaho and Montana). Their results found narrow corridor paths of connectivity from Yellowstone National Park to the Canadian border. Considering that these corridors might be too narrow, Cushman et al. (2013) used cumulative factorial least-cost path network modeling across a two-state area to yield a network of corridors and assessed how protected these corridors were and suggested highway overpasses and underpasses. Cushman et al. (2014) evaluated their previous corridor modeling efforts with telemetry data, which provide a richer data underpinning, and found that telemetry-derived models performed better than those based on a thinner DNA sample data set.

Lewis et al. (2011) used telemetry locations to identify optimal locations for highway crossing structures for black bears and other wildlife species in northern Idaho. Beckmann et al. (2015) used scat-sniffing detection dogs, DNA genotyping, and ecological modeling to identify corridors for black bears between Yellowstone National Park and the mountains to the west in Idaho. Atwood et al. (2011) used genetic analyses, occupancy modeling, and a cost surface to assess population structure and connectivity between subpopulations in Arizona to identify corridors that might be at risk due to potential land-use changes for improved protection.

Several extirpated populations in their southern distribution are being re-established either naturally through metapopulation function or through human-assisted translocation programs. Natural recolonization after historic extirpations occurred in: Kentucky (Unger et al. 2013), Big Bend Texas, having recolonized from bears in northern Mexico (Hellgren et al. 2005; Onorato et al. 2007); California, after brown bear extirpation freed up habitat through predator release (Brown et al. 2009); and Nevada, prior to 1930 (Lackey et al. 2013). Human-assisted recolonization efforts have seen successes (Clark et al. 2002) in Arkansas (Smith et al. 1991; Wear et al. 2005), California (Brown et al. 2009), Louisiana (Benson & Chamberlain 2007), and Pennsylvania (Clark et al. 2002).

Sun Bear

Sun bears occur patchily in the tropical forests of mainland and insular South-East Asia across 11 countries in 32–40% (1.1–1.4 million km^2) of their former range (Figure 22.5) and have been locally extirpated from many areas (Scotson et al. 2017a; see also Figure 8.1 in Chapter 8). The species is categorized as Vulnerable according to the IUCN Red List assessment, and populations are believed to be declining across most of their range (Scotson et al. 2017a). Sun bears were extirpated from Singapore in the late 1800s to early 1900s (Corlett 1992; Brook et al. 2003) and it is uncertain if recent discoveries of the species in the fringes of their distribution in the Chittagong Hill tracts, Bangladesh (Creative Conservation Alliance 2016), and in Yunnan Province, China (Li et al. 2017) are resident populations. Sun bears appear to exhibit a natural increasing gradient in population size from the north to south (Steinmetz et al. 2011), even historically.

Research on sun bears has been challenging and only recently have field techniques used for other bear species shown limited success with sun bears. Research is primarily based on sign surveys to assess presence and relative abundance (Steinmetz & Garshelis 2008; Steinmetz et al. 2011, 2013); camera trapping to assess density and occupancy (Linkie et al. 2007; Ngoprasert et al. 2012; Wong et al. 2013), and

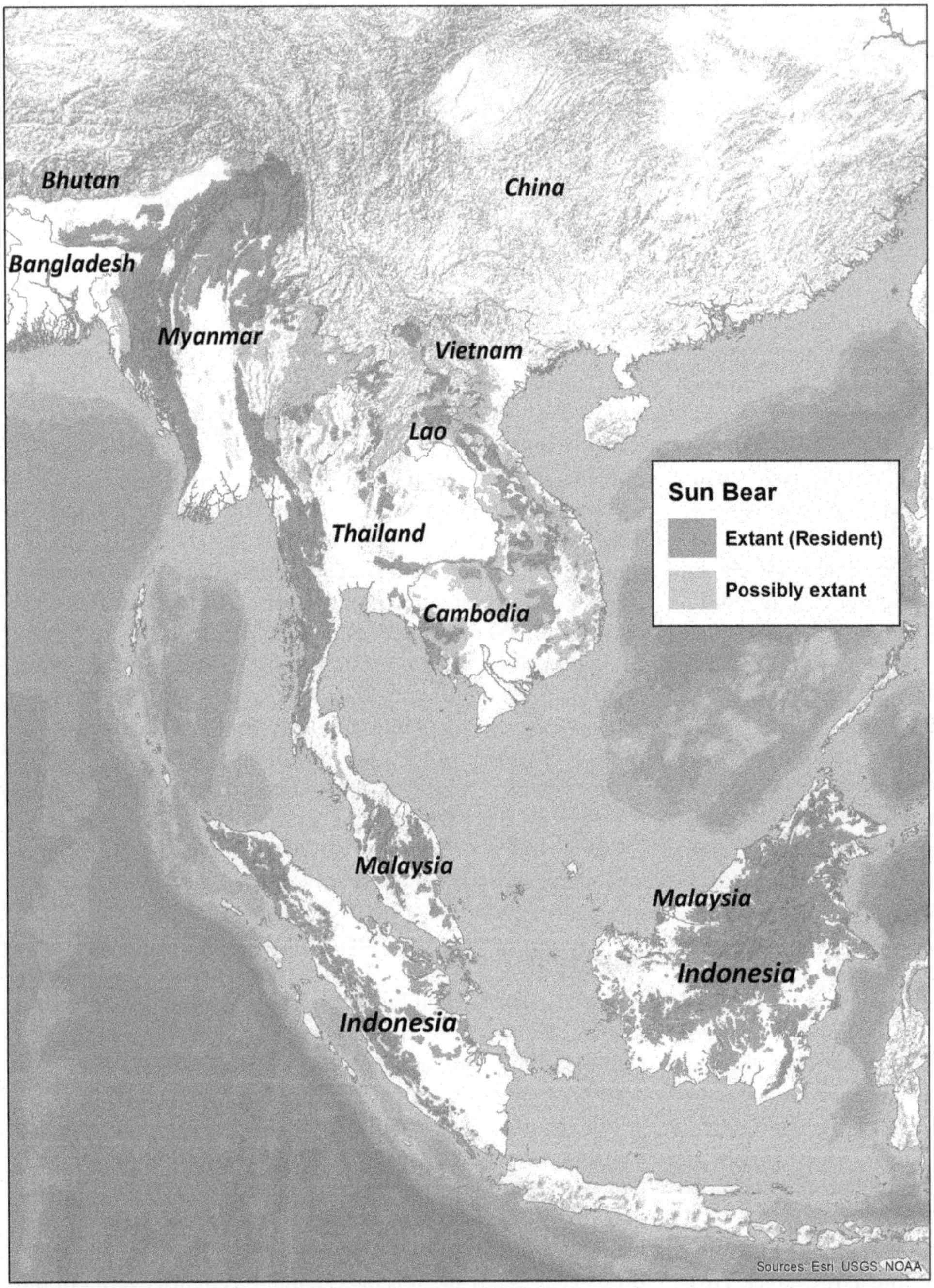

Figure 22.5 Fragmented sun bear distribution from South-East Asia into the islands of Sumatra and Borneo. (*Source*: IUCN 2017, Red List of Threatened Species.) (A black and white version of this figure will appear in some formats. For the color version, please refer to the plate section.)

multispecies landscape connectivity (Brodie et al. 2015); interviews with local people (Crudge et al. 2016; Guharajan et al. 2019); and expert opinion (Reza et al. 2013) to map distribution (Scotson et al. 2017b). Non-invasive biological sampling methods (hair snares or fecal sampling) used to genetically monitor other species of bears have not yet been successfully applied to sun bears. Recent studies have attempted to use data-driven approaches such as Maximum Entropy (MaxEnt) modeling using camera trap and known locations for peninsular Malaysia (Nazeri et al. 2012) and a compilation of camera trap by-catch data from 31 field sites across sun bear range (Scotson et al. 2017b). Scotson et al. (2017b) used a single predictor variable of tree cover to model range-wide population trends, future iterations of which could incorporate more variables that drive the species' habitat use and survivability.

Habitat conversion and forest fragmentation is among if not the biggest threat to sun bear populations. With the highest recorded rate of forest loss in the world, exceeding that of Brazil (Margono et al. 2014), South-East Asia has the lowest remaining proportion of natural tropical forest, primarily driven by logging, expansion of oil palm, and other industrial plantations (Sodhi et al. 2004; Miettinen et al. 2011; Duckworth et al. 2012; Margono et al. 2014). Cushman et al. (2017) estimated a loss of 23% of forested area of Malaysian Borneo and 15% of Kalimantan

(Indonesian Borneo) during 2000–2010. Given the diversity of fruits in the diet of sun bears in undisturbed forests (Steinmetz et al. 2011, 2013), it is expected that the extensive loss, degradation, and fragmentation of lowland forests will negatively impact this species. From the numerous camera trapping and sign survey studies that have studied habitat associations, it is believed that sun bears are a forest-dependent species and preferred habitat is often tightly linked to the availability of large fruiting trees (Normua et al. 2004; Guharajan et al. 2018). Elevation, distance from roads, settlements, and hunting intensity are other determinants of sun bear presence (Linkie et al. 2007; Steinmetz et al. 2011; Wong et al. 2013; Brodie et al. 2015). Habitat preference studies of sun bears highlight that: (1) large portions of suitable sun bear habitat exist outside protected areas (Nazeri et al. 2012; Reza et al. 2013; Scotson et al. 2017b); (2) sun bears are able to use degraded habitat patches when high-quality habitat that can sustain source populations is nearby (Normua et al. 2004; Nazeri et al. 2012; Wong et al. 2013; Guharajan et al. 2018); and (3) even small patches and narrow forest strips in highly modified landscapes are used by sun bears (Yaap et al. 2016).

Metapopulations have not been delineated, and connectivity between populations has not been studied extensively for sun bears, but it may indeed be playing a part in the persistence of the species. For example, the rediscovery site in Yunnan, China, is less than 1 km from the Kachin state of Myanmar, and the population in south-eastern Bangladesh is close to core areas in western Myanmar, making immigration from nearby source populations a strong possibility. Scotson (2019) used the sun bear range map developed from tree cover (Scotson et al. 2017b) to visualize delineation of potential subpopulations using non-habitat, habitat quality, and severity of human disturbance into discrete blocks of sun bear populations, with more subdivided areas within the blocks having a high risk from further isolation where sun bears face further barriers to movement due to habitat fragmentation, human influence, and roads. No study on sun bears examines functional connectivity, as marked or known individuals were not followed, and genetic studies on the species are not yet available.

There are only a few studies within the sun bear range that have considered landscape connectivity. Reza et al. (2013) used habitat-based connectivity between patches as a criterion to determine habitat suitability for four large mammal species in the State of Selangor in peninsular Malaysia using expert-opinion and geospatial analyses. Yaap et al. (2016) used camera traps to understand large mammal use of linear riparian corridors in an acacia (*Acacia mangium*) plantation in Sumatra, Indonesia. In addition to sun bears, they detected 16 other (several Red-Listed) mammal species in these remnant forests. Sun bears used riparian forests furthest away from the national park, regardless of whether they were surrounded by intact acacia forests or recently cleared land, and narrow remnant forests, indicating tolerance and use of highly modified landscapes. Yaap et al. (2016) recommended that linear remnant riparian forests ≤200 m in width can facilitate local (<4 km) movements of many large mammal species in Sumatra, but wider riparian remnants would likely be more effective at promoting mammal movements over longer distances. Brodie et al. (2015) used multispecies camera-trap data from within and outside protected areas, and logged forest to generate connectivity maps between eight protected areas in Sarawak. They recommended that multispecies habitat connectivity plans be tailored to groups of ecologically similar, disturbance-sensitive species to maximize their effectiveness, i.e. carnivores are good umbrella species for other carnivore species, but not so efficient for herbivore connectivity.

Several studies strongly recommend maintaining well-managed plantations containing large fruiting trees and riparian corridors. Numerous wildlife-friendly sustainable forestry policies exist, such as Sustainable Forest Management (SFM), Reduced-Impact Logging (RIL) (www.itto.int/sustainable_forest_management/logging/), and Roundtable on Sustainable Palm Oil (RSPO, https://rspo.org/), with sun bears showing a preference for sustainably managed forests in some studies (Imai et al. 2009). However, it is not yet understood if the ability of sun bears to exploit plantations is a purely ecological survival strategy, or if well planned and managed plantations can indeed help evolutionary processes such as gene flow to prevent long-term population collapse due to small, isolated populations.

Conclusion

Fragmentation is an emerging issue for bears around the world as it often results in small, isolated populations with a significantly higher extirpation risk. There is a wide variety in the extent, our understanding, conservation significance, and actual management of fragmentation and connectivity across the world's eight bear species. Resisting local and regional extirpations will require that we tackle this issue, understand it, and take steps to halt or reverse what is a global trend. In many cases, where fragmentation distances are within the dispersal distance of the species, management to enhance connectivity will be a necessity. We found a spectrum of causes – including human infrastructure, highways, and railroads, but also settlement patterns (valley bottoms in mountainous regions) and extensive rural agricultural regions, whether they be subsistence or industrial-scale agriculture. Likely the most ubiquitous pattern was that connectivity management is coupled with finding solutions to human–bear conflicts that foster coexistence and some degree of habitat overlap between people and bears.

In many cases, suites of subpopulations may not be operating as some form of metapopulation, as inter-area exchange has been cut off by extensive human activities. But often, this is not the case. For example, in densely populated India, sloth bears move between broken forest patches through human-settled agricultural areas up to several hundred kilometers (Dutta et al. 2015) and this is cause for hope and opportunity. This example argues for greater understanding of bears' connectivity habitat needs combined with spatially explicit land-use patterns: there is great possibility to maintain or re-establish interpopulation connectivity through people and land management.

This is therefore a good time to be assessing this issue, as there are still opportunities to reconnect areas through targeted management and land-use changes that allow wildlife movement. These opportunities highlight the conservation potential of population-level conservation. As we think and work at regional scales, we still have time to potentially halt and reverse population declines, shrinking habitat patches, and the intense, almost ubiquitous fragmentation that is the precursor to slow range contraction through extirpation, population by isolated population.

What we have not conquered yet is how to inspire the political will to fund more research and solutions, but also the implementation of solutions in many areas. Smaller grassroots efforts are demonstrating the success of science-informed connectivity management (Proctor et al. 2018); however, similar efforts need to be applied more broadly with the support of regional governments to be effective in the long run. For example, the current level of Protected Areas is not likely sufficient as a counter-weight to extensive fragmentation; they may need to be accompanied by a series of interconnected linkage areas of lands managed for multiple use (people and wildlife), including a wide spectrum of species movement needs (Wangchuk 2007). We need to assess levels of fragmentation, understand where it really matters, and where it is feasibly reversed. Then, identify corridors or linkage habitat and manage them in ways that allow animal movements, which also usually entails some level of human–bear overlap and sharing of landscapes in a way that is safe for humans, secure for their properties and livelihoods, yet still retains some degree of permeability allowing movement.

Genetic analysis is a powerful tool to assess regional and subcontinental fragmentation, but expert opinion and observant wildlife managers armed with camera traps can also be used to assess levels of fragmentation and connectivity to some degree. GPS radiotelemetry is a powerful tool for assessing bear movement patterns to inform corridor location and design, but reliable sign data, observations, and remote cameras can inform habitat models designed to identify wildlife corridors that inform regional conservation plans (Almasieh et al. 2016). We note a precaution to limit the spatial inference from any modeling exercise to appropriate areas where the sampling conditions are relevant. It may be more productive to use creative techniques in localized areas (Almasieh et al. 2016), rather than to extrapolate thinly informed modeling efforts over broad areas (Cushman et al. 2009, 2014). We urge connectivity modellers to use multiple relevant input variables that exert some influence over animal movements and survivability. Landscape features are certainly important (e.g. forest cover), but also human behaviors can be important (e.g. hunting patterns). Further, we were encouraged by the degree that some species make use of degraded habitats (Guharajan et al. 2018), providing opportunities for connectivity areas between more prime habitat patches.

This synthesis comes at an important time in the evolution of connectivity science and management, as it is in many respects in its infancy. It is our sincere hope that our efforts inspire a generation of graduate students, research scientists, and wildlife managers to learn from what others have done around the world and apply the assessment techniques, the science-derived solutions (corridor identification), and connectivity management to the species and regions where fragmentation is a concern.

References

Abbas, F., Bhatti, Z. I., Haider, J. & Mian, A. (2015). Bears in Pakistan: distribution, population biology and human conflicts. *Journal of Bioresource Management* 2: 1–13.

Acevedo, C., Bernal, N., Bianchi, G., et al. (2017). Conservamos la vida: Andean bear conservation at the landscape scale. 25th Conference on Bear Research and Management, Quito, Ecuador.

Akhtar, N., Bargali, H. & Chauhan, N. (2004). Sloth bear habitat use in disturbed and unprotected areas of Madhya Pradesh, India. *Ursus* 15: 203–211.

Almasieh, K., Kaboli, M. & Beier, P. (2016). Identifying habitat cores and corridors for the Iranian black bear in Iran. *Ursus* 27: 18–30.

Apps, C., Paetkau, D., Rochetta, S., et al. (2014). *Grizzly bear population abundance, distribution & connectivity across British Columbia's southern Coastal Ranges*, v. 2.2. Victoria, British Columbia: Aspen Wildlife Research and Ministry of Environment.

Aramilev, V. V. (2006). The conservation status of Asiatic black bears in the Russian Far East. In: *Understanding Asian bears to secure their future* (pp. 86–89). Ibaraki, Japan: Japan Bear Network.

Atwood, T. C., Young, J. K., Beckman, J. P., et al. (2011). Modeling connectivity of black bears in a desert sky island archipelago. *Biological Conservation* 144: 2851–2862.

Beckmann, J. P., Waites, L. P., Hurt, A., Whitelaw, A. & Bergen, S. (2015). Using detection dogs and RSFP models to assess habitat suitability for bears in greater Yellowstone. *Western North American Naturalist* 75: 396–405.

Benson, J. F. & Chamberlain, M. J. (2007). Space use, survival, movements, and reproduction of reintroduced Louisiana black bears. *Journal of Wildlife Management* 71: 2393–2403.

Boitani, L., Alvarez, F., Anders, O., et al. (2015). *Key actions for large carnivore populations in Europe.* Report to DG Environment. Brussels: European Commission.

Brodie, J. F., Giordano, A. J., Dickson, B., et al. (2015). Evaluating multispecies landscape connectivity in a threatened tropical mammal community. *Conservation Biology* 29: 122–132.

Brook, B. W., Sodhi, N. S. & Peter, K. L. N. (2003). Catastrophic extinctions follow deforestation in Singapore. *Nature* 424. Available from www.nature.com/articles/nature01795 (accessed April 23, 2019).

Brown, S. K., Hull, J. M., Updike, D. R., Fain, S. R. & Ernest, H. B. (2009). Black bear population genetics in California signatures of population structure, competitive release, and historical translocation. *Journal of Mammalogy* 90: 1066–1074.

Clark, J. D., Huber, D. & Servheen, C. (2002). Bear reintroductions: lessons and challenges. *Ursus* 13: 335–345.

Clark, J. D., Laufenberg, J. S., Davidson, M. & Murrow, J. L. (2015). Connectivity among subpopulation of a Louisiana black bears as estimated by a step function. *Journal of Wildlife Management* 79: 1347–1360.

Corlett, R. T. (1992). The ecological transformation of Singapore. *Journal of Biogeography* 19: 411–420.

Cosse, M., Del Moral Sachetti, J. F., Mannise, N. & Acosta, M.M. (2014). Genetic evidence confirms presence of Andean bears in Argentina. *Ursus* 25: 163–171.

Creative Conservation Alliance. (2016). A preliminary wildlife survey in Sangu-Matamuhuri Reserve Forest, Chittagong Hill Tracts, Bangladesh. Unpublished report submitted to Bangladesh Forest Department, Dhaka, Bangladesh.

Crudge, B. N., Wilkinson, N. M., Do, V. T., et al. (2016). Status and distribution of bears in Vietnam. Technical Report. Free the Bears/Animals Asia, Vietnam.

Cuesta, F., Peralvo, M. F. & Manen, F. T. (2003). Andean bear habitat use in the Oyacachi River Basin, Ecuador. *Ursus* 14: 198–209.

Cushman, S. A., McKelvey, K. S., Hayden, J. & Schwartz, M. K. (2006). Gene flow in complex landscapes: testing multiple hypotheses with casual modeling. *The American Naturalist* 168: 486–499.

Cushman, S. A., McKelvey, K. S. & Schwartz, M. K. (2009). Use of empirically derived source-destination models to map regional conservation corridors. *Conservation Biology* 23: 368–276.

Cushman, S. A., Lewis, J. S. & Landguth, E. L. (2013). Evaluating the intersection of a regional wildlife connectivity network with highways. *Movement Ecology* 1: 1–12.

Cushman, S. A., Lewis, J. S. & Landguth, E. L. (2014). Why did the bear cross the road? Comparing the performance of multiple resistance surfaces and connectivity modeling methods. *Diversity* 6: 844–854.

Cushman, S. A., Macdonald, E. A., Landguth, E. L., Malhi, Y. & Macdonald, D. W. (2017). Multiple-scale prediction of forest loss risk across Borneo. *Landscape Ecology* 32: 1581–1598.

Das, S., Dutta, S., Sen, S., et al. (2014). Identifying regions for conservation of sloth bears through occupancy modelling in north-eastern Karnataka, India. *Ursus* 25: 111–120.

De Silva, M. (1998). Status and conservation of the elephant and the alleviation of man–elephant conflict in Sri Lanka. *Gajah* 19: 1–24.

Dhamorikar, A. H., Mehta, P., Bargali, H. & Gore, K. (2017). Characteristics of human–sloth bear (*Melursus ursinus*) encounters and the resulting human casualties in the Kanha–Pench corridor, Madhya Pradesh, India. *PLoS ONE* 12: e0176612.

Dharaiya, N, Garshelis, D. L., Sharp, T., et al. (2015). Do sloth bears live in Bhutan? *International Bear News* 24: 10–13.

Dixon, J. D., Oli, M. K., Wooten, M. C., et al. (2006). Effectiveness of a regional corridor in connecting two Florida black bear populations. *Conservation Biology* 20: 155–162.

Dixon, J. D., Oli, M. K., Wooten, M. C., et al. (2007). Genetic consequences of habitat fragmentation and loss: the case of the Florida black bear (*Ursus americanus floridanus*). *Conservation Genetics* 8: 455–464.

Doko, T., Fukui, H., Kooiman, A., et al. (2011). Identifying habitat patches and potential ecological corridors for remnant Asiatic black bear (*Ursus thibetanus japonicus*) populations in Japan. *Ecological Modelling* 222: 748–761.

Duckworth, J.W., Batters, G., Belant, J. L., et al. (2012). Why South-East Asia should be the world's priority for averting imminent species extinctions, and a call to join a developing cross-institutional programme to tackle this urgent issue. *Surveys and Perspectives Integrating Environment and Society* 5. Available from www.researchgate.net/publication/235248971

Durner, G. M. & Amstrup, S. C. (1995). Movements of a polar bear from northern Alaska to northern Greenland. *ARCTIC* 48: 338–341.

Durner, G. M., Laidre, K. L. & York, G. S. (Eds.). (2018). *Polar bears: Proceedings of the 18th Working Meeting of the IUCN/SSC Polar Bear Specialist Group, 7–11 June 2016, Anchorage, Alaska*. Gland, Switzerland and Cambridge, UK: IUCN.

Dutta, T., Sharma, S., Maldonado, J. E., Panwar, H. S., Seidensticker, J. (2015). Genetic variation, structure, and gene flow in a sloth bear (*Melursus ursinus*) meta-population in the Satpura-Maikal landscape of Central India. *PLoS ONE* 10: e0123384.

Dutta, T., Sharma, S., McRae, B. H., Roy, P. S. & DeFries, R. (2016). Connecting the dots: mapping habitat connectivity for tigers in central India. *Regional Environmental Change* 16: 53–67.

Eraso Mera, E. (2014). Landscape connectivity between protected areas in Ecuador, identifying potential corridors for large mammals. Final Report EV5914-EV5915. JMES Cook University. http://repositorio.educacionsuperior.gob.ec/bitstream/28000/2680/1/T-SENESCYT-00863.pdf

Escobar, L. E., Awan M. N. & Qiao, H. (2015). Anthropogenic disturbance and habitat loss for the red-listed Asiatic black bears (*Ursus thibetanus*): using ecological niche modeling and nighttime light satellite imagery. *Biological Conservation* 191: 400–407.

Fahrig, L. (2002). Effects of habitat fragmentation on the extinction threshold: a synthesis. *Ecological* Applications 12: 246–353.

Farashi, A. & Erfani, M. (2018). Modelling of habitat suitability on Asiatic black bears (*Ursus thibetanus gedrosianus*) in Iran in future. *Acta Ecologica Sinica* 38: 9–14.

Gantchoff, M. G. & Belant, J. L. (2017). Regional connectivity for recolonizing American black bears (*Ursus americanus*) in southcentral USA. *Biological Conservation* 214: 66–75.

García-Rangel, S. (2011). Ecology and conservation of the Andean bear in Venezuela. PhD thesis, University of Cambridge.

García-Rangel, S. (2012) Andean bear *Tremarctos ornatus* natural history and conservation. *Mammal Review* 42(2): 85–119.

Garshelis, D. & Steinmetz, R. (2016). *Ursus thibetanus*. The IUCN Red List of Threatened Species 2016: e.T22824A114252336. http://dx.doi.org/10.2305/IUCN.UK.20163.RLTS.T22824A45034242.en

Garshelis, D. L., Joshi, A. R. & Smith, J. L. D. (1999). Estimating density and relative abundance of sloth bears. *Ursus* 11: 87–98.

Garshelis, D. L., Scheick, B. K., Doan-Crider, D. L., Beecham, J. J. & Obbard, M. E. (2016). *Ursus americanus*. The IUCN Red

List of Threatened Species 2016: e. T41687A114251609. http://dx.doi.org/10.2305/IUCN.UK.2016-3.RLTS.T41687A45034604.en

Gong, J. & Harris, R. B. (2013). The status of bears in China. In: *Understanding Asian bears to secure their future* (pp. 96–101). Ibaraki: Japan Bear Network.

Gonzalez, E. G., Blanco, J. C., Ballersteros, F., et al. (2016). Genetic and demographic recovery of an isolated population of brown bear *Ursus arctos* L., 1758. *PeerJ* 4: e1928. DOI 10.7717/peerj.1928.

Guharajan, R., Arnold, T. W., Bolongon, G., et al. (2018). Survival strategies of a frugivore, the sun bear, in a forest–oil palm landscape. *Biodiversity and Conservation* 27: 3657–3677.

Guharajan, R., Abram, N. K., Magguna, M. A., et al. (2019). Does the Vulnerable sun bear *Helarctos malayanus* damage crops and threaten people in oil palm plantations? *Oryx* 53: 611–619. https://doi.org/10.1017/S0030605317001089.

Han, H., Wei, W., Hu, Y., et al. (2019). Diet evolution and habitat contraction of giant pandas via stable isotope analysis. *Current Biology* 29: 664–669.

Hellgren, E. C., Onorato, D. P. & Skiles, J. R. (2005). Dynamics of a black bear population within a desert metapopulation. *Biological Conservation* 122: 131–140.

Horino, S. & Miura. S. 2000. Population viability analysis of a Japanese black bear population. *Population Ecology* 42: 37–44.

Hu, Y., Nie, Y., Wei, W., et al. (2017). Inbreeding and inbreeding avoidance in wild giant pandas. *Molecular Ecology* 26: 5793–5806.

Imai, N., Samejima, H., Langner, A., et al. (2009). Co-benefits of sustainable forest management in biodiversity conservation and carbon sequestration. *PLoS ONE* 4: e8267.

Islam, M. A., Uddin, M., Aziz, M. A. & Muzaffar, S. B. (2013). Status of bears in Bangladesh: going, going, gone? *Ursus* 24: 83–90.

Jhala, Y., Gopal, R. & Qureshi, Q. (2008). *Status of tigers, co-predators, and prey in India*. Dehradun: National Tiger Conservation Authority, Ministry of Environment & Forests, and Wildlife Institute of India. Available from http://projecttiger.nic.in/WriteReadData/PublicationFile/Tiger_Status_oct_2010.pdf.

Jhala, Y. V., Qureshi, Q., Gopal, R. & Sinha, P. R. (2011). *Status of tigers, co-predators and prey in India* (p. 302). TR 2011/003. Dehradun, Dehradun and New Delhi, India: National Tiger Conservation Authority, Govt. of India and Wildlife Institute of India.

Jhala, Y. V., Qureshi, Q. & Gopal, R. (2015). *The status of tigers, co-predators & prey in India 2014*. Dehradun, Dehradun and New Delhi, India: National Tiger Conservation Authority, Govt. of India, and Wildlife Institute of India.

Kaczensky, P., Gossow, H., Knauer, F., et al. (2003). The impact of high speed, high volume traffic axes on brown bears in Slovenia. *Biological Conservation* 111: 191–204.

Kaczensky, P., Jerina K., Jonozovic, M., et al. (2011). Illegal killings may hamper brown bear recovery in the Eastern Alps. *Ursus* 22: 37–46.

Kanchanasakha, B., Tanhikorn, S., Vinitpornsawan, V., Prayoon, A. & Paengphupha, G. (2010). *Status of large mammals in Thailand*. Bangkok: Department of National Parks, Wildlife, and Plant Conservation.

Karamanlidis, A. A., Straka, M., Drosopoulou, E., et al. (2012). Genetic diversity, structure, and size of an endangered brown bear population threatened by highway construction in the Pindos Mountains, Greece. *European Journal of Wildlife Research* 58: 511–522.

Kattan, G., Hernandez, O. L., Goldstein, I., et al. (2004). Range fragmentation in the spectacled bear *Tremarctos ornatus* in the northern Andes. *Oryx* 38: 155–163.

Khan, M., Rosen, T. & Zahler, P. (2012). Status and conservation of Asiatic black bears in Diamer District, Pakistan. *International Bear News* 21: 13–14.

Kopatz, A., Eiken, H. K., Hagen, S. B., et al. (2012). Connectivity and population subdivision at the fringe of a large brown bear (*Ursus arctos*) population in North Western Europe. *Conservation Genetics* 13: 681–692.

Kusak, J., Huber, D., Gomerčić, T., Schwaderer, G. & Gužvica, G. (2009). The permeability of highway in Gorski kotar (Croatia) for large mammals. *European Journal of Wildlife Research* 55: 7–21.

Lackey, C. W., Beckmann, J. P. & Sedinger, J. (2013). Bear historical ranges revisited: documenting the increase of a once-extirpated population in Nevada. *Journal of Wildlife Management* 77: 812–820.

Larkin, J. L., Maehr, D. S., Orlando, M. A., Hoctor, T. S. & Whitney, K. (2004). Landscape linkages and conservation planning for the black bear in west-central Florida. *Animal Conservation* 7: 23–34.

Laufenberg, J. S. & Clark, J. D. (2014). Population viability and connectivity of the Louisiana black bear (*Ursus americanus luteolus*). U.S. Geological Survey Open-File Report.

Lewis, J. S., Rachlow, J. L., Horne, J. S., et al. (2011). Identifying habitat characteristics to predict highway crossing areas for black bears within a human-modified landscape. *Landscape and Urban Planning* 101: 99–107.

Li, F., Zheng X., Jiang X. L. & Chan, B. P. L. (2017). Rediscovery of the sun bear (*Helarctos malayanus*) in Yingjiang County, Yunnan Province, China. *Zoological Research* 38: 206–207.

Li, R., Xu, M., Wong, M. H. G., et al. (2015). Climate change threatens giant panda protection in the 21st century. *Biological Conservation* 182: 93–101.

Linkie, M., Dinata, Y., Nugroho, A. & Haidir, I. A. (2007). Estimating occupancy of a data deficient mammalian species living in tropical rainforests: sun bears in the Kerinci Seblat region, Sumatra. *Biological Conservation* 137: 20–27.

Loucks, C. J., Lu, Z., Dinerstein, E., et al. (2001). Giant pandas in a changing landscape. *Science* 294: 1465.

Malenfant, R. M., Davis, C. S., Cullingham, C. I. & Coltman, D. W. (2016). Circumpolar genetic structure and recent gene flow of polar bears: a reanalysis. *PLoS ONE* 11: e0148967. doi:10.1371/journal.pone.0148967.

Margono, B., Potapov, P. V., Turubanova, S. A., et al. (2014). Primary forest cover loss in Indonesia over 2000–2012. *Nature Climate Change* 4: 730–735.

Mateo-Sanchez, M. C., Cushman, S. A. & Saura, S. (2014). Connecting endangered brown bear subpopulations in the Cantabrian Range (north-western Spain). *Animal Conservation* 17: 430–440.

Mattson, D. J. & Merrill, T. (2002). Extirpations of grizzly bears in the contiguous United States, 1850–2000. *Conservation Biology* 16: 1123–1136.

McCullough, D. R. (1996). Introduction. In: *Metapopulations and wildlife conservation* (pp. 1–10). Washington, DC: Island Press.

McLellan, B. N. (1998). Maintaining viability of brown bears along the southern fringe of their distribution. *Ursus* 10: 607–611.

McLellan, B. N. & Hovey, F. W. (2001). Natal dispersal of grizzly bears. *Canadian Journal of Zoology* 79: 838–844.

McLellan, B. N., Proctor, M. F., Huber, S. & Michel, S. (2017). *Ursus arctos* (amended version of 2017 assessment). International Union for the Conservation of Nature, Gland, Switzerland. www.iucnredlist.org/details/41688.

McLellan, M. L. (2018). Habitat Connectivity Mapping between the Stein-Nahatlatch and Southern Chilcotin Grizzly Bear Populations. 28 pp. Conservation Northwest, Bellingham, USA.

Merriam, C. H. (1922). Distribution of grizzly bears in U.S. *Outdoor Life* 50: 405–406.

Miettinen, J., Shi, C. & Liew, S. C. (2011). Deforestation rates in insular Southeast Asia between 2000 and 2010. *Global Change Biology* 17: 2261–2270.

Molina, S., Fuller, A. K., Morin, D. J. & Royle, J. A. (2017). Use of spatial capture–recapture to estimate density of Andean bears in northern Ecuador. *Ursus* 28: 117–126.

Monsalve Dam, D., Sánchez-Mercado, A., Yerena, E., García-Rangel, S. & Torres, D. (2010). Efectividad de las áreas protegidas para la conservación del oso andino, *Tremarctos ornatus*, en los Andes suramericanos. In: De Oliveira-Miranda, R., Lessman, J., Rodríguez-Ferraro, A. & Rojas-Suárez, F. (Eds.), *Ciencia y conservación de especies amenazadas en Venezuela: Conservación basada en evidencias e intervenciones estratégicas* (pp. 127–136). Caracas, Venezuela: Provita.

Nazeri, M., Jusoff, K., Madani, N., et al. (2012). Predictive modeling and mapping of Malayan sun bear (*Helarctos malayanus*) distribution using maximum entropy. *PLoS ONE* 7: e48104.

Ngoprasert, D., Steinmetz, R., Reed, D. H., Savini, T. & Gale, G. A. (2011). Influence of fruit on habitat selection of Asian bears in a tropical forest. *Journal of Wildlife Management* 75: 588–595.

Ngoprasert, D., Reed, D. H., Steinmetz, R. & Gale, G. A. (2012). Density estimation of Asian bears using photographic capture–recapture sampling based on chest marks. *Ursus* 23: 117–133.

Normua, F., Higashi, S., Ambu, L. & Mohamed, M. (2004). Notes on oil palm plantation use and seasonal spatial relationships of sun bears in Sabah, Malaysia. *Ursus* 15: 227–231.

Obbard, M. E., Stapleton, S., Middel, K. R., et al. (2015). Estimating the abundance of the Southern Hudson Bay polar bear subpopulation with aerial surveys. *Polar Biology* 38: 1713–1725.

Obbard, M. E., Cattet, M. R. L., Howe, E. J., et al. (2016). Trends in body condition in polar bears (*Ursus maritimus*) from the Southern Hudson Bay subpopulation in relation to changes in sea ice. *Arctic Science* 2: 15–32.

Ohnishi, N., Saitoh, T. & Ishibashi, Y. T. (2007). Low genetic diversities in isolated population of the Asiatic black bear (*Ursus thibetanus*) in Japan, in comparison with large stable populations. *Conservation Genetics* 8: 1331–1337.

Onorato, D. P., Hellgren, E. C., Van Den Bussche, R. A., et al. (2007). Genetic structure of American black bears in the desert southwest of North America: conservation implications for recolonization. *Conservation Genetics* 8: 565–576.

Ostrowski, S., Zahler, P., Dehgan, A., et al. (2009). The Asiatic black bear still survives in Nuristan, Afghanistan. *International Bear News* 18: 14–15.

Paetkau, D. & Strobeck, C. (1996). Mitochondrial DNA and the phylogeography of Newfoundland black bears. *Canadian Journal of Zoology* 74: 192–196.

Paetkau, D., Amstrup, S. C., Born, E. W., et al. (1999). Genetic structure of the world's polar bear populations. *Molecular Ecology* 8: 1571–1584.

Paetkau, D., Slade, R., Burden, M. & Estoup, A. (2004). Genetic assignment methods for the direct, real-time estimation of migration rate: a simulation based exploration of accuracy and power. *Molecular Ecology* 13: 55–65.

Peck, C. P., van Manen, F. T., Costello, C. M., et al. (2017). Potential paths for male-mediated gene flow to and from an isolated grizzly bear population. *Ecosphere* 8(10): e01969. 10.1002/ecs2.1969.

Pelletier, A., Obbard, M. E., Harnden, M., et al. (2017). Determining causes of genetic isolation in a large carnivore (*Ursus americanus*) population to direct contemporary conservation measures. *PLoS ONE* 12(2): e0172319. doi:10.1371/journal.pone.0172319.

Peralvo, M., Cuesta, F. & van Manen, F. (2005). Delineating priority habitat areas for the conservation of Andean bears in northern Ecuador. *Ursus* 162: 222–233.

Peters, W., Hebblewhite, M., Cavedon, M., et al. (2015). Resource selection and connectivity reveal conservation challenges for reintroduced brown bears in the Italian Alps. *Biological Conservation* 186: 123–133.

Pritchard, J. K., Stephens, M. & Donnelly, P. (2000). Inference of population structure using multilocus genotype data. *Genetics* 155: 945–959.

Proctor, M. F., McLellan, B. N. & Strobeck, C. (2002). Population fragmentation of grizzly bears in southeastern British Columbia, Canada. *Ursus* 13: 153–160.

Proctor, M. F, McLellan, B. N., Strobeck, C. & Barclay, R. (2004). Gender specific dispersal distances for grizzly bears analysis revealed by genetic analysis. *Canadian Journal of Zoology* 82: 1108–1118.

Proctor, M. F., McLellan, B. N., Strobeck, C. & Barclay, R. M. R. (2005). Genetic analysis reveals demographic fragmentation of grizzly bears yielding vulnerably small populations. *Proceedings of the Royal Society, London B* 272: 2409–2416.

Proctor, M. F., Paetkau, D., Mclellan, B. N., et al. (2012). Population fragmentation and inter-ecosystem movements of grizzly bears in Western Canada and the Northern United States. *Wildlife Monographs* 180: 1–46.

Proctor, M. F., Nielsen, S. E., Kasworm, W. F., et al. (2015). Grizzly bear connectivity mapping in the Canada–US trans-border region. *Journal of Wildlife Management* 79: 544–555.

Proctor, M. F., Kasworm, W. F., Annis, K. M., et al. (2018). Conservation of threatened Canada–USA trans-border grizzly bears linked to comprehensive conflict reduction. *Human Wildlife Interactions* 12: 348–372.

Proctor, M. F., McLellan, B. N., Stenhouse, G. B., et al. (2020a). The effects of roads and motorized human access on grizzly bear populations in British Columbia and Alberta, Canada. *Ursus* 2019(30e2): 16–39.

Proctor, M. F., Kasworm, W. F., Teisberg, J. E., et al. (2020b). American black bear population fragmentation detected with pedigrees in the transborder Canada–United States region. *Ursus* 2020(31e1): 1–15.

Ratnayeke, S., van Manen, F. T., Pieris, R. & Pragash, V. S. J. (2007). Landscape characteristics of sloth bear range in Sri Lanka. *Ursus* 18: 189–202.

Ratnayeke ,S., van Manen, F. T., Pieris, R. & Pragash, V. S. J. (2014). Challenges of large carnivore conservation: sloth bear attacks in Sri Lanka. *Human Ecology* 42: 467–479.

Regehr, E., Laidre, K., Akçakaya, H. R., et al. (2016). Conservation status of polar bears (*Ursus maritimus*) in relation to projected sea-ice declines. *Biology Letters* 12: 20160556.

Regehr, E. V., Hostetter, N. J., Rode, K. D., et al. (2018). Integrated population modeling provides the first empirical estimates of vital rates and abundance for polar bears in the Chukchi Sea. *Scientific Reports* 8: 16780; doi:10.1038/s41598-018-34824-7

Reza, M. I. H., Abdullah, S. A., Nor, S. B. M. & Ismail, M. H. (2013). Integrating GIS and expert judgment in a multi-criteria analysis to map and develop a habitat suitability index: a case study of large mammals on the Malayan Peninsula. *Ecological Indicators* 34: 149–158.

Rodríguez, D., Cuesta, F., Goldstein, I., et al. (2003). *Ecoregional strategy for the conservation of the spectacled bear (*Tremarctos ornatus*) in the northern Andes.* WWF Colombia, Fundación Wii, EcoCiencia, Wildlife Conservation Society, and Red Tremarctos.

Ruiz-García, M. (2003). Molecular population genetic analysis of the spectacled bear (*Tremarctos ornatus*) in the northern Andean área. *Hereditas* 138: 81–93.

Ruiz-García, M., Orozco-ter Wengel, P., Castellanos, A. & Arias, L. (2005). Microsatellite analysis of the spectacled bear (*Tremarctos ornatus*) across its range distribution. *Genes & Genetic Systems* 80: 57–69.

Saitoh, T., Ishibashi, Y., Kanamori, H. & Kitahara, E. (2001). Genetic status of fragmented populations of the Asian black bear *Ursus thibetanus* in western Japan. *Population Ecology* 43: 221–227.

Sakurai, R. & Jacobson, S. K. (2011). Public perception of bears and management interventions in Japan. *Human Wildlife Interactions* 5: 123–134.

Sánchez-Mercado, A., Ferrer-Paris, J. R., Garcia-Rangel, S., et al. (2014). Combining threat and occurrence models to predict potential ecological traps for Andean bears in the Cordillerade Merida, Venezuela. *Animal Conservation* 17: 388–398.

Sathyakumar, S. & Choudhury, A. (2007). Distribution and status on the Asiatic back bear *Ursus thibetanus* in India. *Journal of Bombay Natural History Society* 104: 316–323.

Sathyakumar S., Kaul, R., Ashraf, N. V. K., Mookerjee, A. & Menon, V. (2012). *National bear conservation and welfare action plan 2012.* Ministry of Environment and Forest, Government of India.

Scheick, B. K. & McCown, W. (2014). Geographic distribution of American black bears in North America. *Ursus* 25: 24–33.

Schregel, J., Kopatz, A., Hagen, S. B., et al. (2012). Limited gene flow among brown bear populations in far Northern Europe? Genetic analysis of the east–west border population in the Pasvik Valley. Molecular *Ecology* 21: 3474–3488.

Scotson, L. (2019). Exploring potential range connectivity of sun bear (Carnivora: Ursidae: Ursinae). *Raffles Bulletin of Zoology* 67: 67–76.

Scotson, L., Fredriksson, G., Augeri, D., et al. (2017a). *Helarctos malayanus* (errata version published in 2018). The IUCN Red List of Threatened Species 2017: e. T9760A123798233. Available from http://dx.doi.org/10.2305/IUCN.UK.2017-3.RLTS.T9760A45033547.en (accessed April 23, 2019).

Scotson, L., Fredriksson, G., Ngoprasert, D., Wong, W-M. & Fieberg, J. (2017b). Projecting range-wide sun bear population trends using tree cover and camera-trap bycatch data. *PLoS ONE* 12: e0185336.

SFA. (2015). *Report of the Fourth National Giant Panda Survey.* Beijing: Science Publishing House.

Smith, K. G., Clark, J. D. & Gipson, P. S. (1991). History of black bears in Arkansas: over-exploitation, near elimination, and successful reintroduction. *Eastern Workshop on Black Bear Research and Management* 10: 5-14.

Sodhi, N. S., Koh, L. P., Brook, B. W. & Ng, P. K. L. (2004). Southeast Asian biodiversity: an impending disaster. *Trends in Ecology & Evolution* 19: 654–660.

Steinberg, E. K. & Jordan, C. E. (1997). Using molecular genetics to learn about the ecology of threatened species: the allure and the illusion of measuring genetic structure in natural populations. In: Fiedler, P. L. & Kareiva, P. M. (Eds.), *Conservation biology for the coming decade* (2nd edition, pp. 440–460). New York, NY: Chapman & Hall.

Steinmetz, R. & Garshelis, D. L. (2008). Distinguishing Asiatic black bears and sun bears by claw marks on climbed trees. *Journal of Wildlife Management* 72: 814–821.

Steinmetz, R., Garshelis, D. L., Chutipong, W. & Seuaturien, N. (2011). The shared preference niche of sympatric Asiatic black bears and sun bears in a tropical forest mosaic. *PLoS ONE* 6: 0014509. https://doi.org/10.1371/journal.pone.0014509.

Steinmetz, R., Garshelis, D. L., Chutipong, W. & Seuaturien, N. (2013). Foraging ecology and coexistence of Asiatic black bears and sun bears in a seasonal tropical forest in Southeast Asia. *Journal of Mammalogy* 94: 1–18.

Stirling, I., Andriashek, D. & Calvert, W. (1993). Habitat preferences of polar bears in the western Canadian Arctic in late winter and spring. *Polar Record* 29: 13–24.

Støen, O. G., Zedrosser, A., Sæbø, S. & Swenson, J. E. (2006). Inversely density-dependent natal dispersal in brown bears *Ursus arctos. Oecologia* 148: 356–364.

Straka, M., Paule, L., Ionesuc, O., Stofik, J. & Ademic, M. (2012). Microsatellite diversity and structure of Carpathian brown bears (*Ursus arctos*): consequences of human caused fragmentation. *Conservation Genetics* 13: 153–164.

Swaisgood, R. R., Wang, D. & Wei, F., (2016). *Ailuropoda melanoleuca.* IUCN Red List of Threatened Species. www.iucnredlist.org/details/712/0.

Swaisgood, R. R., Wang, D. & Wei, F. (2018). Panda downlisted but not out of the woods. *Conservation Letters* 11: 1–9.

Swenson, J. E., Sandegren, F., Bjarvall, A., et al. (1994). Size, trend, distribution and conservation of the brown bear *Ursus arctos* population in Sweden. *Biological Conservation* 70: 9–17.

Swenson, J. E., Sandegren, F. & Soderberg, A. (1998). Geographic expansion of an increasing brown bear population: evidence for presaturation dispersal. *Journal of Animal Ecology* 67: 819–826.

Swenson, J. E., Gerstl, N., Dahle, B. & Zedrosser, A. (2000). *Action plan for the conservation of the brown bear in Europe (*Ursus arctos*)*. Strasbourg: Council of Europe.

Tammeleht, E., Remm, J., Korsten, M., et al. (2010). Genetic structure in large, continuous mammal populations: the example of brown bears in northwestern Eurasia. *Molecular Ecology* 19: 5359–5370.

Taylor, M. K., Akeeagok, S., Andriashek, D., et al. (2001). Delineating Canadian and Greenland polar bear (*Ursus maritimus*) populations by cluster analysis of movements. *Canadian Journal of Zoology* 79: 690–709.

Unger, D. E., Cox, J. J., Harris, H. B., et al. (2013). History and current status of the black bear in Kentucky. *Northeastern Naturalist* 20: 289–308.

Utreras, V. (2018). Aplicación del enfoque de manejo de paisajes para la conservación del oso andino (*Tremarctos ornatus*). 25th Conference on Bear Research and Management. Quito, Ecuador.

van Oort, H., McLellan, B. N. & Serrouya, R. (2011). Fragmentation, dispersal and metapopulation function in remnant populations of endangered mountain caribou. *Animal Conservation* 14: 215–224.

Velez-Liendo, X. & García-Rangel, S. (2017). *Tremarctos ornatus* (errata version published in 2018). The IUCN Red List of Threatened Species 2017: e. T22066A123792952.

Velez-Liendo, X., Strubbe, D. & Matthysen, E. (2013). Effects of variable selection on modelling habitat and potential distribution of the Andean bear in Bolivia. *Ursus* 24: 127–138.

Velez-Liendo, X., Adriaensen, F. & Matthysen, E. (2014). Landscape assessment of habitat suitability and connectivity for Andean bears in the Bolivian Tropical Andes. *Ursus* 25: 172–187.

Viengkone, M., Derocher, A. E., Richardson, E. S., et al. (2016). Assessing polar bear (*Ursus maritimus*) population structure in the Hudson Bay region using SNPs. *Ecology and Evolution* 6: 8474–8484.

Viteri, M. P. & Waits, L.P. (2009). Identifying polymorphic microsatellite loci for Andean bear research. *Ursus* 20: 102–108.

Waits, L., Taberlet, P., Swenson, J. E., Sandegren, F. & Franzen, R. (2000). Nuclear DNA microsatellite analysis of genetic diversity and gene flow in the Scandinavian brown bear (*Ursus arctos*). *Molecular Ecology* 9: 421–431.

Wang, F., McShea, W. J., Wang, D., et al. (2014). Evaluating landscape options for corridor restoration between giant panda reserves. *PLoS ONE* 9: e105086.

Wangchuk, S. (2007). Maintaining ecological resilience by linking protected areas through biological corridors in Bhutan. *Tropical Ecology* 48: 176–187.

Wear, B. J., Eastridge, R. & Clark, J. D. (2005). Factors affecting settling, survival, and viability of black bears reintroduced to Felsenthal National Wildlife Refuge, Arkansas. *Wildlife Society Bulletin* 33: 1363–1374.

Wei, F., Hu, Y., Zhu, L., et al. (2012). Black and white and read all over: the past, present and future of giant panda genetics. *Molecular Ecology* 21: 5660–5674.

Wei, F., Hu, Y., Yan, L., et al. (2015). Giant pandas are not an evolutionary cul-de-sac: evidence from multidisciplinary research. *Molecular Biology and Evolution* 32: 4-12.

Wei, W., Swaisgood, R. R., Dai, Q., et al. (2018). Giant panda distributional and habitat-use shifts in a changing landscape. *Conservation Letters* 11: e12575.

Wong, W-M., Leader-Williams, N. & Linkie, M. (2013). Quantifying changes in sun bear distribution and their forest habitat in Sumatra: sun bear population trends and deforestation in Sumatra. *Animal Conservation* 16: 216–223.

Woods, J. G., Paetkau, D., Lewis, D., et al. (1999). Genetic tagging of free-ranging black and brown bears. *Wildlife Society Bulletin* 27: 616–627.

Yaap, B., Magrach, A., Clements, G. R., et al. (2016). Large mammal use of linear remnant forests in an industrial pulpwood plantation in Sumatra, Indonesia. *Tropical Conservation Science* 9(4).

Yamamoto, T., Oka, T., Ohnishi, N., et al. (2012). Genetic characterization of northernmost isolated population of Asian black bear (*Ursus thibetanus*) in Japan. *Mammal Study* 37: 85–91.

Yang, Z., Gu, X., Nie, Y., et al. (2018). Reintroduction of the giant panda into the wild: a good start suggests a bright future. *Biological Conservation* 217: 181–186.

Yerena, E. (1998). Protected areas for the Andean bear in South America. *Ursus* 10: 101–106.

Yerena, E. & García-Rangel, S. (2010). The implementation of an interconnected system of protected areas in the Venezuelan Andes. In: Worboys, G. L., Francis, W. & Lockwood, M. (Eds.), *Connectivity conservation management: A global guide* (pp. 233–244). London: Earthscan.

Yoganand, K., Rice, C. G., Johnsingh, A. J. T. & Seidensticker, J. (2006). Is the sloth bear in India secure? A preliminary report on distribution, threats and conservation requirements. *Journal of the Bombay Natural History Society* 103: 57–66.

Yusefi, G. H. (2013). Noninvasive genetic tracking of Asiatic black bears (*Ursus thibetanus*) at its range edge in Iran – a pilot study. MS thesis, Uppsala University, Sweden.

Zedrosser, A., Dahle, B., Swenson, J. E. & Gerstl, N. (2001). Status and managment of the brown bear in Europe. *Ursus* 12: 9–20.

Zedrosser, A., Støen, O. G., Sæbø, S. & Swenson, J. E. (2007). Should I stay or should I go? Natal dispersal in the brown bear. *Animal Behaviour* 74: 369–376.

Zedrosser, A., Steyaert, S. M. J. G., Gossow, H. & Swenson, J. E. (2011). Brown bear conservation and the ghost of persecution past. *Biological Conservation* 144: 2163–2170.

Zhan, X., Zhang, Z., Wu, H., et al. (2007). Molecular analysis of dispersal in giant pandas. *Molecular Ecology* 16: 3792–3800.

Zhao, S., Zheng, P., Dong, S., et al. (2013). Whole-genome sequencing of giant pandas provides insights into demographic history and local adaptation. *Nature Genetics* 45: 67–71.

Zug, B. (2018). Andean bear (*Tremarctos ornatus*), biodiversity, and puma (*Puma concolor*) conservation on private lands in the Ecuadorian Andes: implication for conservation in a human-dominated landscape. PhD dissertation, University of Wisconsin-Madison, Madison, WI, USA.

Chapter 23

Ex Situ Conservation of Bears: Roles, Status, and Management

Lydia Kolter, Agnieszka Sergiel, Djuro Huber, and Scott Silver

This chapter presents an overview of ursids in captive facilities. Examples of how bear populations in zoos and rescue centers (i.e. *ex situ*) can support wild bear (i.e. *in situ*) conservation now and in the future are provided, and the potential for reintroduction of captive bears is also discussed.

Keeping Bears and the Conservation Roles of Zoos

Since zoos were created, bears have been one of the most popular species to be exhibited (Seidensticker 1993). With few exceptions, bears have historically been kept in small, mostly barren enclosures, rarely larger than 100–200 m^2. Such small bear facilities were still common until the 1980s, and were problematic because, despite occasional cases of closely bonded individuals (e.g. Koene & Ipema 2014), ursids are generally not social species, and need sufficient space to maintain interindividual distances during activities such as foraging and resting.

By the 1990s, the size and design of many bear enclosures started to change. Facilities exceeding 1000 m^2 and housing small groups of 2–3 bears became more common. More enclosures provided natural vegetation and plantings, as well as water features such as pools and waterfalls. Structures for climbing, hiding, and behavioral thermoregulation were also incorporated (e.g. O' Grady et al. 1990; Fischbacher & Schmid 1999; Groot 2010; Johann 2011; Figure 23.1). As technological improvements allowed for less expensive but effective fencing systems, the construction of large, naturalistic enclosures of several hectares became feasible. These enclosures often included wooded and open space areas for both breeding and non-breeding groups of bears (e.g. Koene 1996; Ganslosser 2004). Finally, environmental enrichment designed to increase the complexity of the bears' surroundings and elicit natural behaviors were increasingly integrated into the daily husbandry routines (e.g. Ames 1993; Fischbacher & Schmid 1999; Gupta et al. 2007; Law & Reid 2010).

However, the trend to meet the needs of bears kept *ex situ* – either by construction of new facilities or by redesigning and improving old ones – is not yet universal. Only by exhibiting healthy animals in environments stimulating a diverse repertoire of behaviors, which approximate activity budgets of their wild conspecifics, are bear exhibits able to meet the three conservation roles that the World Association of Zoos and Aquaria (WAZA) expects from its (more than 300) member zoos. These expectations are: (1) the promotion of conservation education and raising public awareness of conservation threats; (2) the support of conservation-related research; and (3) the maintenance of viable populations of endangered species (Barongi et al. 2015).

Bears in Zoos and Rescue Centers

Several types of databases keep track of bears in zoos and rescue centers. The Zoological Information Management System (ZIMS) and both regional and international studbooks contain information on all registered captive bears, including their sex and birth date, the identity of their parents, and any other information related to the history of each individual bear. Currently, these databases list more than 3000 bears living in zoos. The actual number is higher, as not all zoological institutions participate in ZIMS. This is particularly the case for the Asian zoos (Chua, personal communication).

Based on these databases, the majority of bears are kept in European and Asian zoos (Table 23.1). Most of the captive individuals in Asia came to zoos as cubs orphaned during legal hunts, poaching, or accidents, but there are also wild adults

Figure 23.1 Brown bear in the Berenbos in Ouwehand Zoo, Rhenen, the Netherlands. (Photo by Ouwehand Zoo.)

Table 23.1 The different species of bears registered in zoos.

Species		Number of bears							Data sources[4] (referring to 2017/2018)
Common name	Scientific name	Europe	North America	Latin America	Asia[1]	Australia	Africa	Total	
Polar bear	*Ursus maritimus*	150	71	2	65	4	0	292	ISB, ESB
Brown bear	*Ursus arctos*	820	161	18	86	0	4	1089	ESB, ZIMS, CZA, EARAZA
American black bear	*Ursus americanus*	110	255	20	1	0	4	390	ZIMS, EARAZA
Asiatic black bear	*Ursus thibetanus*	129	25	0	292	0	0	446	ESB, AZASB, CZA,ZIMS, EARAZA
Andean bear	*Tremarctos ornatus*	63	40	141	7	0	2	253	ISB, ESB
Sloth bear	*Melursus ursinus*	25	33	0	252	0	0	310	ESB, CZA, AZASB
Sun bear[2]	*Helarctos malayanus*	39	22	1	284	8	0	354	ESB, AZASB, JAZA
Giant panda[3]	*Ailuropoda melanoleuca*	17	13	0	17	2	0	49	ZIMS
Individuals per continent		1353	620	182	1004	14	10	3183	

[1] Numbers are considerably underestimated.
[2] Updated with data from Crudge et al. (2019) and W. Chua (personal communication).
[3] Giant pandas in Chinese zoos not included. For detailed information on the total *ex situ* population see Chapter 6.
[4] International studbook (ISB); European studbook (ESB); Zoological Information Management System (ZIMS); Central Zoo Authority (India) (CZA); Eurasian Regional Association of Zoos and Aquaria (EARAZA); Association of North American Zoo and Aquaria Studbook (AZASB), Japanese Association of Zoos and Aquaria (JAZA).

originally captured after being deemed "nuisance" bears, or confiscated by authorities from inappropriate facilities or private owners. More than 1300 bears live permanently in rescue centers run by welfare organizations such as Animals Asia Foundation, Baerenhilfswerk, Free The Bears, Gewerkschaft für Tiere, Four Paws, World Animal Protection, Wildlife Friends Foundation Thailand, and Wildlife SOS. Only 20% of the world-wide population of confiscated bears are found in Europe, and most of these are brown bears (*Ursus arctos*). By contrast, Asiatic black bears (*Ursus thibetanus*) make up 47% of the confiscated bears in Asian centers and only 9% are brown bears. Most of the Asiatic black bears have been rescued from bear bile farms in China and South-East Asian countries (Field and Hunt, personal communication). Sun bears (*Helarctos malayanus*) and sloth bears (*Melursus ursinus*) make up 25% and 19% of confiscated bears in Asia, respectively. The majority of the sloth bears are kept in Indian rescue facilities (Gupta, personal communication).

Management of *Ex Situ* Bear Populations

Bears living in zoos belonging to the European Association of Zoos and Aquaria (EAZA), the Eurasian Regional Association of Zoos and Aquaria (EARAZA), the North American Association of Zoos and Aquaria (AZA), the Zoo Association of Australasia (ZAA) and of Japan (JAZA) are cooperatively managed or monitored in most cases.

In order to decide the level of population management, Taxon Advisory Groups (TAGs) consisting of experts for a given taxonomic group develop Regional Collection Plans to decide which species within that taxon are to be maintained and under what degree of management (Table 23.2). Selection criteria are based on the status (both *in situ* and *ex situ*) of a given species, reproductive outputs in captivity, and the holding space available among participating institutions.

For example, the species classified as Least Concern (brown bears) and Not Threatened (American black bears, *Ursus americanus*) by the IUCN Red List are not recommended for breeding in holding institutions. Despite this recommendation, the numbers of both brown bears and American black bears are increasing in North America (Table 23.2), mostly because of the addition of orphaned cubs and non-releasable adult bears.

Coordinated programs like the European Endangered Species Program (EEP) of EAZA and the Species Survival Plan (SSP) of AZA represent the most intense form of *ex situ* management, aimed at maintaining genetic diversity and demographic stability of managed captive populations (Leus et al. 2011). To achieve this, participating institutions agree to

Table 23.2 Recommended management levels and trends of ursid populations in the European Association of Zoos and Aquaria and North American Association of Zoos and Aquaria (last census: 2018).

Species	Europe		North America	
	Management category[1]	Current trend	Management category[2]	Current trend
Polar bear	EEP	Stable	SSP	Decreasing
Brown bear	ESB non-breeding	Increasing	Monitored non-breeding	Increasing
American black bear	Monitored[3], non-breeding	Stable	Monitored non-breeding	Stable
Asiatic black bear	ESB	Decreasing	AZASB, to be phased out	Decreasing
Andean bear	EEP	Stable	SSP	Decreasing
Sloth bear	EEP	Decreasing	SSP	Stable
Sun bear	ESB	Decreasing	AZASB, to be phased out	Decreasing
Giant panda	Globally managed, increasing			

[1] European Endangered Species Program (EEP); European Studbook (ESB).
[2] Species Survival Plan (SSP); Association of Zoos and Aquaria Studbook (AZASB).
[3] Restricted to collections with focus on North American fauna.

abide by a committee's or program coordinator's annual recommendations concerning the management of the species population as a whole. Recommendations are made concerning which individuals should or should not produce cubs, the formation of breeding pairs, and where inter-institutional transfers of bears should take place in order to accomplish these goals. The criteria for these recommendations are primarily based upon the space available in zoos for the population, the levels of genetic relatedness among individuals (i.e. each bear's number of living relatives and offspring), and the demographic characteristics of the population. Recommendations are confirmed by a species committee consisting of representatives elected by the program participants.

In both North America and Europe, the first SSP/EEP established for a bear species was for Andean bears in 1986 and 1987, respectively. Today, these programs also manage polar and sloth bears in Europe and North American zoos (Table 23.2).

A less intense level of management of captive populations involves bear species for which either regional or international studbook databases are maintained but no SSP or EEP has been formed. Annually circulated studbooks inform on the demographic and genetic status of these species, but with few exceptions the studbook keepers give recommendations on breeding and transfers only on request.

Due to limitations of space and other resources, TAGs have to make difficult decisions to achieve sustainable populations of bear species in zoos. For example, the AZA bear TAG recently decided to recommend phasing out (i.e. no recommended breeding or imports) the few Asiatic black bears and sun bears from North American Zoos in order to make more holding spaces in zoos available for Andean and sloth bears. Likewise, in Europe, the recommendations are to reduce the number of American black bears kept, to increase the capacity to hold and exhibit other bear species managed by EEPs and ESBs.

Even the small sun bear populations in Australian and Japanese zoos are closely managed within their respective zoo organizations – the Zoo and Aquaria Association (ZAA) of Australasia and the Japanese Association of Zoos and Aquaria (JAZA). The South East Asian Zoo Association (SEAZA) is currently establishing a regional studbook for the species (Chua, personal communication). In India, the Central Zoo Authority (CZA) regularly monitors all captive ursids in Indian zoos (Gupta, personal communication).

Intensely Managed *Ex Situ* Bear Populations in Europe

Many of the issues associated with intensively managed bear populations are similar regardless of where the programs are located. In most cases, the populations are small, e.g. Andean bears, sloth bears, and sun bears in European zoos (Table 23.3), and have limited space available. Actually, over the last 20 years, zoo spaces available for bears have decreased, because: (a) when old bears kept in substandard conditions passed on, their small enclosures were shut down; and (b) new and larger enclosures are rebuilt slowly and intermittently. The final result is better housing conditions, but with fewer bear spaces than in the past.

Life History Parameters

Although some individual bears can reproduce very early in captivity (e.g. between the age of 2 and 4 years), the median age of first reproduction is higher (Table 23.4). The reproductive lifespan can be long, e.g. females of polar bears (*Ursus maritimus*) might give birth until the age of 28 years. However, tropical bear species rarely reproduce beyond the age of 23 years.

Sometimes advisory groups will recommend temporarily halting reproduction to adjust the number of bears to the

Table 23.3 Current numbers of males and females and genetic parameters of ursids managed 2018 in European *ex situ* breeding programs (analyses based on data from the species-specific European studbooks).

	Sex		Genetic parameters		
	Males	**Females**	**Mean inbreeding**	**Gene diversity retained**	**Number of wild-born founders**
Polar bear	50	59	0.01	96.5	49
Andean bear	32	33	0.05	90.4	23
Sloth bear	13	10	0.11	79.0	8
Sun bear	12	26	0.06	90.2	16

Table 23.4 Reproductive parameters of ursids managed in European *ex situ* breeding programs (analyses based on data from the species-specific European studbooks, time window 1998–2018).

Species	**Earliest age at reproduction**[1]		**Latest age at reproduction**[1]		**Median age at first reproduction**[2]		**Median age at reproduction**[3]		**Growth rate (λ)**[4]		**Generation length (T)**[5]	
	M	F	M	F	M	F	M	F	M	F	M	F
Polar bear	3	4	26	28	7 years 0 month (*n* = 58)	8 years 10 months (*n* = 86)	11 years 4 months (*n* = 302)	14 years 0 months (*n* = 384)	1.01	1.02	12.3 years	11.7 years
Andean bear	2	3	29	23	8 years 2 months (*n* = 27)	5 years 0 months (*n* = 41)	10 years 1 month (*n* = 95)	8 years 11 months (*n* =156)	0.98	0.99	13.3 years	9.6 years
Sloth bear	3	2	~22	20	5 years 6 months (*n* = 11)	5 years 1 month (*n* = 17)	8 years 5 months (*n* = 57)	9 years 11 months (*n* – 69)	0.99	0.98	10.6 years	10.6 years
Sun bear	4	3	~27	23	8 years 6 months (*n* = 10)	7 years 7 months (*n* = 19)	13 years 9 months (*n* = 66)	9 years 6 months (*n* = 56)	0.9	0.92	13.7 years	12.1 years

[1] Refers to single individuals.
[2] Includes the ages of males/females at first reproduction (conception for males and parturition for females) within the given time window; *n* = sample size.
[3] Includes ages of males/females over all reproductive events within the given time window; *n* = sample size.
[4] The proportional change in population size per year based on life table calculations (expected lambda), λ > 1 for increasing populations, λ < 1 for declining populations.
[5] Average age at reproduction.

available bear spaces. All managed programs ultimately aim for a stable population growth rate (λ), where births are approximately equal to the number of animals that are removed from the population through death or transfers to other programs. In Europe, most managed bear populations have been stable between 1998 and 2018 (Table 23.4). Only the sun bear population shrank by almost 10% per year during that time span (Table 23.4). In captive bears, females generally live longer than males and, in most species, females often live beyond 25, or even 30 years – well beyond their reproductive age (Table 23.5). Because of this, many spaces available for bears are occupied by older, non-breeding individuals, and there is no additional space for cubs and newly weaned offspring.

Demographic and Genetic Parameters

Sex ratios of captive bear populations are generally balanced, with the exception of sun bears, which are characterized by a strong bias toward females, and Andean bears in North America, where the SSP program has a strong male bias. In small populations, even random events such as a preponderance of male cubs over several years can result in a demographic imbalance. This in turn can lead to challenges in maintaining genetic diversity because fewer pairs can be formed. Because of the small number of breeding individuals, the inbreeding in sloth bears is higher than in the other species (Table 23.3).

Maintaining genetic diversity (GD) in any population is important to ensure that population contains enough variation to adapt to environmental change. In small populations, such as those managed in zoos, population biologists feel that 90% of the original wild population diversity should be preserved over 100 years (Frankham et al. 2002). Critical to maintaining high levels of GD is the number of "founders" in a population. "Founders" are wild-born individuals, which have descendants in the living population. In theory, 20 individuals are enough to represent 97.5% of the genetic diversity in the wild

Table 23.5 Survival parameters of ursids managed in European *ex situ* breeding programs in years (analyses based on data from the species-specific European studbooks for the time window 1998–2018).

Species	Average life expectancy at birth		Age to which 25% of the animals survive		Oldest recorded age	
	Males	Females	Males	Females	Males	Females
Polar bear	12.1	16.4	19.5	25.9	35.9	38.5
Andean bear	16.3	19.1	26.5	27.7	32.1	33
Sloth bear	14.6	13.9	19	20.7	27	32.1
Sun bear	12.9	21.3	21.9	29.6	31.7	35.8

Figure 23.2 The male Andean bear "Beorn" at Queens zoo is the result of the cooperation between the European Endangered Species Program and the Species Survival Plan. The father was born in the French zoo of Doué la Fontaine and transferred to Queens Zoo, USA, in 2013 to be paired with a female born at the National Zoo Park in Washington. (Photo by Megan Maher, ©WCS.) (A black and white version of this figure will appear in some formats. For the color version, please refer to the plate section.)

population and to maintain it at a high level in *ex situ* populations. The polar bear population still maintains 95.5% of its original GD due to the short breeding history in captivity and the considerable number of wild born founders in the EAZA population. By contrast, the genetic diversity of the captive Andean bear population, which has existed in European zoos for almost 70 years, is already close to this 90% threshold, despite 23 founders. Variation in individual reproductive success over the generations and long gaps between arrivals and reproduction of founders (e.g. the first three arrived in 1952 and reproduced in 1959, the next arrived in 1958 but did not reproduce until 1964) contributed to this. The sloth bear population was begun with a low number of founders and, consequently, its genetic diversity is low.

It is improbable that the tropical bear species in European zoos will reach self-sustainable population sizes. The solution might be to manage them globally. Yet, without the arrival of new, unrelated individuals, genetic diversity will continue to decline and levels of inbreeding will undoubtedly increase (Leus et al. 2011). In order to avoid this, close cooperation between the coordinators of the different regional programs is of crucial importance. This level of international cooperation in bear management has already begun, as both Andean and sloth bears have recently been transferred between Europe and North America (Figure 23.2). In the case of sloth and sun bears, unrelated animals were imported from Indian zoos and a Cambodian rescue facility, respectively. In addition to providing small populations with new founders, cooperation between zoos and organizations that care for bear species in their native countries has opened opportunities for zoos everywhere to support *in situ* bear conservation projects.

Conservation Roles of *Ex Situ* Bears

Education, Raising Awareness, and Support of *In Situ* Projects

One of the primary functions of keeping bears in zoos and rehabilitation centers is connecting people to bears. Employing captive populations as ambassadors for changing human behavior has specifically been identified as a goal of biologists concerned with bear conservation (Garshelis & Steinmetz 2017; Crudge et al. 2019). By offering the public the opportunity to closely observe bears and bear behavior, zoos and rehabilitation centers allow people the chance to appreciate the beauty and intrinsic value of bears, thereby enlisting their support in ensuring bear survival. Even mundane behaviors can foster public interest in bears. For example, Ueno Zoological Garden, Japan, exhibits hibernating bears to the public. The idea behind this is that even sleeping bears, when properly interpreted for the public, may convey important messages on bear ecology and conservation (Itoh et al. 2010) that the public can appreciate.

In order to develop appropriate and effective educational approaches, specific goals for conservation education should be clearly established (Thomas 2016). These goals, public values, and public attitudes might differ between regions and have to be identified to design the most effective education programs (Davis et al. 2016). For example, in regions where human activities threaten bear species, education goals should be aimed at reducing poaching and human–bear conflicts, but in regions where traditional medicine is common, educational activities aimed at reducing the consumption of bear parts for food or traditional medicine are crucial objectives (see also Chapter 26).

Figure 23.3 "The arctic research station" to educate visitors about polar bear field research at the polar bear enclosure in CERZA, Lisieux, France. (Photo by L. Kolter.)

Fundraising for conservation projects on bear species *in situ* is also an important mission for those centers and organizations displaying bears to the public. For this purpose, bear exhibits with well-designed interpretive materials, keepers' talks, and guided tours which focus information and events on bear conservation can be used to foster an appreciation for bears and their habitats (Figure 23.3). Recently, bear-focused events that specifically target support for bear conservation have been employed in some European zoos. According to EAZA's Conservation Data Base, more than € 1.3 million have been donated to bear conservation projects between 1998 and 2017 (Zimmermann, personal communication).

Conservation-Directed *Ex Situ* Research

Zoos and other bear-holding institutions are often the ideal (or only) places where certain types of important conservation research can be accomplished. Basic research questions such as those concerning ursid reproduction were examined by studies in zoos when field studies were still rare, and remote monitoring techniques not yet available (e.g. Dathe 1962; Dittrich & Kronberger 1963). Bears kept in zoos and centers offer unique opportunities to answer research questions under controlled environmental conditions. Bear physiology, genetics, nutrition, and behavior are just a few of the areas in which bears in managed collections can help uncover questions that improve the ability to protect wild bear populations. Also, several field techniques have been tested and methods validated by using bears in zoos and sanctuaries (e.g. Kaczensky et al. 2004; Schwarzenberger et al. 2004; Steinmetz & Garshelis 2008). Gut retention times, for example, have been determined in captive brown bears to help interpret field data derived from fecal samples (Elfström et al. 2013). Similarly, the hearing sensitivity of polar bears in several North American zoos has been evaluated as a first step to assessing the impact of the increasing anthropogenic noise in the Arctic environment (Owen & Bowles 2011). In AZA's polar bear research master plan (Roth et al. 2018), several conservation-related *ex situ* publications are referenced and projects planned relating to emerging technologies, physiological and behavioral ecology, human–bear coexistence, reproductive physiology, and veterinary medicine. Coordinated *ex situ* research on sun bears is being developed within in the frame of the global conservation action plan (Crudge et al. 2019). The same is ongoing for the brown bear in Europe (Kok et al. 2018).

Close cooperation between field and zoo researchers is required in order to maximize the value of managed collections for bear conservation research. In the future more conservation action plans for ursids will be created by the IUCN SSC Bear Specialist Group with input from the members who are part of the captive bear expert team, which is a part of this specialist group (www.globalbearconservation.org).

Orphaned Cubs and Their Rehabilitation

Cubs of all bear species rely heavily on their mothers for at least one year or more. Being altricial, they need the highly nutritious milk (Huber et al. 1993) and the mother's body heat to prevent hypothermia during the denning period. After cubs emerge from their dens, the mother's protection is crucial to prevent predation. Under their mother's care, cubs have to learn what foods are edible and where they are located, as well as when and how they can be obtained (Cicnjak et al. 1987; Mazur & Seher 2008). Growing cubs must also learn how to select and prepare appropriate winter denning sites. Social skills are also important. How to behave in relation to conspecifics, other species, and humans (Huber 2010; Morehouse et al. 2016) cubs learn from their mothers. For these reasons, cubs orphaned at an early age have very little likelihood of survival (Figure 23.4). Indeed, studies have shown that cub survival rates increase with the time they can spend with their mother, e.g. cubs orphaned in their first summer show higher survival rates than cubs orphaned in spring (Rogers 1985; Swenson et al. 1998; Frković et al. 2001).

The main reasons why cubs become orphaned in the wild are: (1) poaching, e.g. intentional killing of the mother in order to obtain cubs for the pet market, entertainment, or for captive breeding in Asian bile farms; (2) death of the mother during winter den hunts (in parts of Russia) or other types of hunting, due to traffic accidents or natural causes (including intraspecific killing); and (3) cubs get separated from their mother accidentally, e.g. when she is disturbed in the den, or when sliding down slopes, crossing rivers, roads, and other physical barriers (Figure 23.5).

The use of the term "rehabilitation" is in the context of welfare releases, which is "the release of individual(s) to the wild, or to an existence approximating a true wild existence, with the individual(s)' welfare being of paramount importance

Figure 23.4 An orphaned European brown bear cub has to be bottle-fed to survive and without its mother will not obtain all crucial cognitive knowledge. (Photo by A. Sergiel.)

Figure 23.5 Roads and railroads are often the source of orphaned cubs due to the mother being killed in traffic or accidental separation in these habitats fragmented by traffic routes. (Photo by A. Sergiel.)

Figure 23.6 Orphaned brown bear cub fed by bottle. It is typically the case that people feed an abandoned cub and save its life but make it heavily habituated and unable to live in the wild. (Photo by D. Huber.)

rather than conservation priorities or the potential impacts on a wild population" (Soorae 2005). Thus, as a general recommendation, breeding bears for release is not recommended. However, in recent decades, facilities have been established with the goal to care for, rehabilitate, and return orphaned bear cubs to nature. The prime motivation behind such efforts is to prevent the loss of wild-born individuals from the wild population and/or spare them a lifetime in captivity. There are also facilities that operate as "sanctuaries," with the goal of ensuring lifetime care for individual bears who are unable to return to the wild, either as a result of physical or behavioral limitations resulting from accidents or improper care by people. Additionally, some sanctuaries are focused on securing proper environments for captive bears from poor or abusive conditions.

The main approach to raising bear cubs slated for release includes minimizing human contact while providing for the husbandry needs of the cubs while they are being cared for (Figure 23.6). Although it is difficult to avoid all sources of visual and acoustic human stimulus, eliminating olfactory cues is virtually impossible (Huber 2005a, 2010). The longer bears spend in captive situations, the more persistent they can become in exhibiting behaviors that are problematic for wild bears, so minimizing the causes of these behaviors should be an important feature of programs aiming to release individuals after prolonged keeping in captivity (Vickery & Mason 2003).

Rehabilitation protocols include recommendations on proper facilities and husbandry, quarantine procedures, minimum contact with humans, and regular evaluations of the cubs' suitability for release (Latini 2015; Castellanos et al. 2016). Rehabilitation is considered successful if the orphaned released bear is in good health, does not exhibit any problem behaviors (such as coming in close contact with humans), and ultimately reproduces (e.g. Alberta Environment and Parks 2018). All this requires long-term post-release monitoring.

Finally, selection of the cub release area also requires careful consideration. Caution should be taken to ensure not to introduce foreign genes into the native population of bears as it may have an unpredictable effect (Randi et al. 1994). Although some bear cub rehabilitation success has been reported to date (e.g. Beecham et al. 2015; Komnenou et al. 2016; Penteriani et al. 2017), the reintegration of cubs in the wild continues to be controversial.

Overview on Bear Species Release Projects

Beecham et al. (2015) reported on rehabilitation protocols for five bear species (American black bears, brown bears, Andean bears, sun bears, and Asiatic black bears). These authors also evaluated 12 captive-rearing programs concerning three species of orphaned bears: seven programs for the American black bear (USA and Canada), three programs for the brown bear (Canada, Romania, and Greece), and two programs for the Asiatic black bear (South Korea and India). It appears that between 1991 and 2012 at least 550 captive-reared yearlings were released back to the wild: 245 males and 179 females of the American black bear, 37 males and 27 females of the brown bear, and 31 males and 31 females of the Asiatic black bear. Additionally, three female brown bears (two cubs-of-the-year and one 2-year-old individual) were rescued, rehabilitated, radio collared, and released back to the Cantabrian Mountains, north-western Spain, between 2008 and 2013 (Penteriani et al. 2017). Pazhetnov and Pazhetnov (2005) reported that between 1990 and 2001, they hand-raised 81 orphaned cubs, of which 61 (27 males and 34 females) were successfully released. More recently, two sun bears were rehabilitated and released as part of a reintroduction program in the Tembat Forest Reserve in Malaysia (Abidin et al. 2018). Castellanos et al. (2016) reported on 21 successfully released Andean bears of various ages.

In Europe, some rehabilitation efforts under less-than-ideal release conditions are being justified by the critical needs of small, isolated populations. Brown bear cubs in the Abruzzo Apennines, Italy (Latini 2015) have been rehabilitated and released, despite the fact that human-determined causes of mortality were still present in this area (see Chapter 19). Castellanos et al. (2016) consider the rescue and reintroduction of wild-born Andean bears into remote and untouched areas as the only alternative to ensure their recovery.

One of the most contentious topics among bear biologists is the idea of introducing captive-born bears into the wild. The wisdom and the means of introducing captive-born individuals into wild populations are highly problematic and a matter of debate and controversy (e.g. Huber 1996, 2005b, 2010; Kolter & van Dijk 2005). If support of wild populations by restocking/reintroductions becomes necessary and is considered appropriate, it makes best sense to do so with bears coming from the same or closest population because of the similarity of genetic characteristics. Additionally, extreme care has to be taken to avoid bringing in problematic individuals that can create conflicts with humans in the area of the release (Clark et al. 2002). Logistic and financial limitations, as well as gaps in our knowledge about the areas from which many of the founders originally come from, make captive breeding for reintroduction in Europe and North America a less appealing option.

There is one documented historical attempt to release captive-born cubs into the wild. Samojlik et al. (2018) described the case of the Białowieża Forest, where the release of mother-reared brown bear cubs in the times prior to and during the Second World War ultimately failed.

Post-Release Fate of Rehabilitated Bears

Data on the success of rehabilitated bears are limited, mostly because of insufficient efforts to monitor those individuals after the release (IUCN 1998; Clark et al. 2002). The majority of bear releases lack post-release monitoring, leaving outcomes unknown (Myers &Young 2018). Here, we present a limited sample of published cases. Huber et al. (1994) reported the results of the release of six orphaned, hand-raised brown bear cubs. Three had to be returned to captivity due to habituation to humans, one was killed in a traffic accident, and the fate of another bear was unknown. Only one apparently successfully survived at least the next winter after the release. This particular cub was orphaned in mid-July and was returned to the wild after only two days in captivity. Swenson et al. (1998) also stated that potentially successful releases include cubs-of-the-year that had spent at least half a year with the mother in the wild.

According to Beecham et al. (2015), approximately 30% of all released bears were not monitored after the release and, therefore, categorized as of unknown fate, 31% died, whereas 39% survived till the time of last confirmed observation, which was highly variable in length. Fifty-four percent of known mortalities of rehabilitated American black bears occurred during the first year after release, and at least two bears lived for more than 10 years in the wild. Brown bears released in Romania showed the lowest survival rates. Beecham et al. (2015) also found no difference in survival between brown bears released in spring (April–June) and those released during summer (July–September). Similar survival rates were observed for Asiatic black bears from South Korea and India. Interestingly, there appeared no difference in the survival rates between orphaned wild-born and captive-born individuals released in South Korea. Natural mortalities (37%) represented the majority of deaths for Asiatic black bears, although removals due to human conflicts were also a major cause of death for males (29%); illegal kills accounted for 26% of total deaths. Fifty-three percent of the released bears died within the first year and five individuals were alive more than eight years after release. Post-release conflicts with humans were documented for American (6.1%) and Asiatic (9.7%) black bears, most of which occurred during the first year after release (Beecham et al. 2015).

Female brown bears released in Spain exhibited marked diurnal activity and were active throughout winter. The behavior of two individuals was similar to those reported for wild bears, whereas one cub-of-the-year was recaptured after 21 days because she showed habituation to humans (Penteriani et al. 2017).

American black bears moved 1–251 km from the release site (Beecham et al. 2015). Three of six rehabilitated orphaned black bears (two females and one male) showed strong release-site fidelity. Nevertheless, long-distance dispersal was noted for two other individuals of that group, both males (Myers & Young 2018). Penteriani et al. (2017) reported that displacements and home ranges of the released females were within the range of movements of free-ranging brown bears elsewhere in Europe. The distance of movements of Asiatic black bear cubs kept and soft-released in groups from so-called den houses and supplementary feeding sites increased with age: at the age of 5 months, they settled an area within a radius of 400 m from the den house; a month later it increased to 1400 m. Then they stayed in a group and moved over 3 km away from the sites of human presence. The intragroup connections between cubs weakened gradually, with the group breaking up after first wintering (Skripova 2009).

Data from long-term monitoring of rehabilitated bears are rare. Eight American black bears monitored to reproductive age gave birth to 28 cubs in 11 litters. Age at first breeding was 2.5 years (Beecham et al. 2015). The analyses of habitat selection of six rehabilitated black bears monitored for >1 year post-release showed selection of habitat and resources similar to wild-born bears (Myers & Young 2018).

Successes and Challenges

The increased quality of bear husbandry in the last few decades (at least in most of the developed countries), as well as the establishment of stronger population management programs, can be considered a success for *ex situ* bear conservation. Moreover, communication and cooperation between regions and between *ex situ* and *in situ* communities is growing, and increasing the contribution of *ex situ* bears to wild bear conservation.

Improvement of husbandry techniques and spatial capacity, especially where they still do not meet the needs of the species, are future challenges. Another crucial objective for the future will be the establishment and evaluation of appropriate public education programs in zoos and rescue centers.

Obviously, long-term post-release monitoring should become obligatory to evaluate the effectiveness of rehabilitation projects, and to collect data on movement patterns, activity patterns, and behavior of released bears. In fact, post-release monitoring is the only objective way to assess the efficacy of rehabilitation processes, evaluate the viability of releases, the validity of behavioral development of bears in rehabilitation facilities, and alleviate concerns over post-release behavior, welfare, fitness, and potential risk to human safety and properties.

Acknowledgments

We would like to thank W. Chua, SEAZA; B.K. Gupta, CZA; N. Karpov, EARAZA; and N. Field, formerly Animals Asia; and M. Hunt, Free the Bears for providing information on bear populations kept in their regional associations and by their organizations, respectively. The keepers of the international and regional studbooks in EAZA, JAZA, and AZA are acknowledged for their efforts to maintain the data sets of the captive bear populations that were used in this chapter. We are grateful for receiving the SPARKS databases from M. Alaze, C. Barwick, F. Sicks, and J. Szanthos for the demographic and genetic analyses of sloth, sun, Andean, and polar bear populations in Europe.

References

Abidin, M. K. Z., Mohammed, A. A. & Nor, S. M. (2018). Home-range and activity pattern of rehabilitated Malayan sun bears (*Helarctos malayanus*) in the Tembat Forest Reserve, Terengganu. AIP Conference Proceedings 1940: 020036.

Alberta Environment and Parks. (2018). Alberta orphan black bear cub rehabilitation protocol (pp. 1–7).

Ames, A. (1993). *Enriching the lives of bears in zoos*. In: Stirling, I. (Ed.), *Bears* (p. 196). London: HarperCollins Publisher.

Barongi, R., Fisken, F. A., Parker, M. & Gusset, M. (Eds.). (2015). *Committing to conservation: The World Zoo and Aquarium Conservation Strategy*. Gland: WAZA Executive Office.

Beecham, J. J., De Gabriel Hernando, M., Karamanlidis, A. A., et al. (2015). Management implications for releasing orphaned, captive-reared bears back to the wild. *Journal of Wildlife Management* 79: 1327–1336.

Castellanos, A. X., Jackson, D. A. & Arias, J. L. (2016). *Guidelines for the rescue, rehabilitation, release and post-release monitoring of Andean bears* (pp. 1–113). Quito, Ecuador: Andean Bear Foundation.

Cicnjak, L., Huber, D., Roth, H. U., Ruff, R. L. & Vinovrski, Z. (1987). Food habits of brown bears in Plitvice Lakes National Park, Yugoslavia. *International Conference on Bear Research and Management* 7: 221–226.

Clark, J. D., Huber, D. & Servheen, C. (2002). Reintroducing bears: lessons and challenges (Invited Paper). *Ursus* 13: 153–163.

Crudge, B., Lees, C., Hunt, M., et al. (Eds.). (2019) *Sun bears: Global status review & conservation action plan, 2019–2028*. Gland: IUCN SSC Bear Specialist Group/IUCN SSC Conservation Planning Specialist Group/Free the Bears/ TRAFFIC.

Dathe, H. (1962). Beitrag zur Fortpflanzungsbiologie des Malaienbären, *Helarctos m. malayanus* (Raffl.). *Zeitschrift für Säugetierkunde* 28: 155–162.

Davis, E. O., O'Connor, D., Crudge, B., *et al.* (2016). Understanding public perceptions and motivations around bear part use: a study in northern Laos of attitudes of Chinese tourists and Lao PDR nationals. *Biological Conservation* 203: 282–289.

Dittrich, L. & Kronberger, H. (1963). Biologisch-anatomische Untersuchungen über die Fortpflanzungsbiologie des Braunbären (*Ursus arctos* L.) und anderer Ursiden in Gefangenschaft. *Zeitschrift für Säugetierkunde* 28: 129–155.

Elfström, M., Støen, O-G., Zedrosser, A., Warrington, W. & Swenson, J. E. (2013). Gut retention time in captive brown bears *Ursus arctos*. *Wildlife Biology* 19: 317–324.

Fischbacher, M. & Schmid, H. (1999). Feeding enrichment and stereotypic behaviour in spectacled bears. *Zoo Biology* 18: 363–371.

Frankham, R., Ballou, J. D. & Briscoe, D. (2002). *Introduction to conservation genetics*. Cambridge: Cambridge University Press.

Frković, A., Huber, D. & Kusak, J. (2001). Brown bear litter sizes in Croatia. *Ursus* 12: 103–106.

Ganslosser, U. (Ed.). (2004). *Verhaltensgerechte Bärenhaltung*. Fürth: Filander Verlag.

Garshelis, D. & Steinmetz, R. (2017). The world's least known bear species gets its day in the sun. *International Bear News* 26(3): 5–6.

Groot, de, S. (2010) Antwerp Zoo spectacled bear exhibit. In: Gunn, A. & Filby, M. (Eds.), *Zoolex*. Available from www.zoolex.org/gallery/show/1184/

Gupta, B. K., Prakash, S. & Sinha, A. K. (2007). Strategy for enriching the environment of captive Ursids. *Current Science* 93: 147–152.

Huber, D. (1996). *Bears in Zagreb*. In: Koene, P. (Ed.), *Large bear enclosures* (pp. 33–36). Bennekom, The Netherlands: International Bear Foundation.

Huber, D. (2005a). Why not introduce "rehabilitated" brown bears to the wild? In: Kolter, L. & van Dijk, J. (Eds.), *Rehabilitation and release of bears* (pp. 28–34). Cologne: Zoologischer Garten Köln.

Huber, D. (2005b). Reintroductions of bears and lynxes: procedures, limitations and risks. In: *From the EU LIFE-Nature Projects to guidelines for the reintroduction of threatened species* (pp. 67–73). Caramanico Terme, Italy: Parco Nazionale Majella.

Huber, D. (2010). Rehabilitation and reintroduction of captive-reared bears: feasibility and methodology for European brown bears *Ursus arctos*. *International Zoo Yearbook* 44: 47–54.

Huber, D., Kulier, I., Poljak, A. & Devčić-Kuhar, B. (1993). Food intake and mass gain of hand-reared brown bear cubs. *Zoo Biology* 12: 525–533.

Huber, D., Dabanović, V., Kusak, J & Frković, A. (1994). Reintroduction of hand-reared brown bears into the wild: Experiences, problems, chances. In: Dorrestein, G. M. & Kahraman, M. M. (Eds.), *International conference on aspects of bear conservation* (pp. 179–186). Bursa, Turkey: Veterinary Faculty of the Uludag University.

Itoh, K., Ide, K., Kojima, Y. & Terada, A. (2010). Hibernation exhibit for Japanese black bear *Ursus thibetanus japonicus* at Ueno Zoological Gardens. *International Zoo Yearbook* 44: 55–64.

IUCN. (1998). *Guidelines for re-introductions*. Prepared by the IUCN/SSC Re-introduction Specialist Group, approved by the 41st Meeting of the IUCN Council, Gland, Switzerland, May 1995. Gland, Switzerland and Cambridge, UK: IUCN.

Johann, A. (2011). Die neue Bären-Anlage im NaturZoo Rheine – Planung, Konzept und Erfahrungen. *Der Zoologische Garten N.F.* 80: 1–28.

Kaczensky, P., Wagner, A. & Walzer, C. (2004). Activity monitoring of a brown bear – a model approach to test field methods. *Mammalian Biology* 69: 444–448.

Koene, P. (Ed.). (1996). *Large bear enclosures*. Bennekom, The Netherlands: International Bear Foundation.

Koene, P. & Ipema, B. (2014). Social networks and welfare in future animal management. *Animals* 4: 93–118.

Kok, J., Leus, C., Fienig, E. & Voorham, M. (2018). Long-term Management Plan for the Brown bear (*Ursus arctos*). EAZA Executive Office, 73 pp.

Kolter, L. & van Dijk, J. (Eds.). (2005). *Rehabilitation and release of bears*. Cologne: Zoologischer Garten Köln.

Komnenou, A. Th., Karamanlidis, A. A., Kazakos, G. M., et al. (2016). First successful hand-rearing and release to the wild of two orphan brown bear cubs in Greece. *Journal of the Hellenic Veterinary Medical Society* 67(3): 163–170.

Latini, R. (2015). *Guidelines for releasing Marsican brown bear cubs into nature* (pp. 1–14). Pescasseroli: National Park of Abruzzo Lazio and Molise.

Law, G. & Reid, A. (2010). Enriching the lives of bears in zoos. *International Zoo Yearbook* 44: 65–74.

Leus, K., Traylor-Holzer, K. & Lacy, R. C. (2011). Genetic and demographic population management in zoos and aquariums: recent developments, future challenges and opportunities for scientific research. *International Zoo Yearbook* 45: 213–225.

Mazur, R. & Seher, V. (2008). Socially learned foraging behaviour in wild black bears, *Ursus americanus*. *Animal Behaviour* 75: 1503–1508.

Morehouse, A. T., Graves, T. A., Mikle, N. & Boyce, M. S. (2016). Nature vs. nurture: evidence for social learning of conflict behaviour in grizzly bears. *PLoS ONE* 11(11): e0165425. https://doi.org/10.1371/journal.pone.0165425.

Myers, P. J. & Young, J. K. (2018). Post-release activity and habitat selection of rehabilitated black bears. *Human-Wildlife Interactions* 12: 322–327.

O'Grady, R. J. P., Law, G., Boyle, H., Macdonald, A. & Johnstone, J, (1990), Himalayan black bear *Selenarctos thibetanus* exhibit at Glasgow zoo. *International Zoo Yearbook* 29: 233–240.

Owen, M. A. & Bowles, A. E. (2011). Auditory psychophysics and the management of threatened large carnivore, the polar bear (*Ursus maritimus*). *International Journal of Comparative Psychology* 24: 244–254.

Pazhetnov, S. & Pazhetnov, S. V. (2005): Re-introduction of orphan brown bear cubs. In: Kolter, L. & van Dijk, J. (Eds.), *Rehabilitation and release of bears* (pp. 53–61). Cologne: Zoologischer Garten Köln.

Penteriani, V., Delgado, M., López-Bao, J. V., et al. (2017). Patterns of movement of released female brown bears in the Cantabrian Mountains, northwestern Spain. *Ursus* 28: 165–170.

Randi, E., Gentile, L., Boscagli, G., Huber, D. & Roth, H. U. (1994). Mitochondrial DNA sequence divergence among some west European brown bear (*Ursus arctos* L.) populations. Lessons for conservation. *Heredity* 73: 480–489.

Rogers, L. L. (1985). Aiding the wild survival of orphaned bear cubs. *Wildlife Rehabilitation* 4: 104–111.

Roth, T., Owen, M., Cutting, A., et al. (2018). *Association of Zoos and Aquariums polar bear species survival plan.* Polar Bear Research Masterplan.

Samojlik T., Selva N., Daszkiewicz P., et al. (2018). Lessons from Białowieża Forest on the history of protection and the world's first reintroduction of a large carnivore. *Conservation Biology* 32: 808–816.

Schwarzenberger, F., Fredriksson, G., Schaller, K. & Kolter, L. (2004). Fecal steroid analysis for monitoring reproduction in the sun bear (*Helarctos malayanus*). *Theriogenology* 62: 1677–1692.

Seidensticker, J. (1993). Bears in zoos. In: Stirling, I. (Ed.), *Bears* (pp. 206–211). London: HarperCollins Publisher.

Skripova, K. V. (2009). Trends in the spatial distribution of Asiatic black bear cubs. *Russian Journal of Ecology* 40: 61–64. DOI:10.1134/S106741360901010X

Soorae, P. S. (2005). Placement options for confiscated bears. In: Kolter, L. & van Dijk, J. (Eds.), *Rehabilitation and release of bears* (pp. 17–27). Cologne: Zoologischer Garten Köln.

Steinmetz, R. & Garshelis, D. (2008). Distinguishing Asiatic black bears and sun bears by claw marks on climbed trees. *Journal of Wildlife Management* 72: 814–812.

Swenson, J. E., Franzén, R., Segerström, P. & Sandegren, F. (1998). On the age of self-sufficiency in Scandinavian brown bears. *Acta Theriologica* 43: 213–218.

Thomas, S. (2016). Editorial: Future perspectives in Conservation Education. *International Zoo Yearbook* 50: 9–15.

Vickery, S. S. & Mason, G. J. (2003). Behavioral persistence in captive bears: implications for reintroduction. *Ursus* 14: 35–43.

Chapter 24

The Challenge of Brown Bear Management in Hokkaido, Japan

Tsutomu Mano, Masami Yamanaka, Hifumi Tsuruga, and Yoshikazu Sato

Introduction

In Japan, the brown bear (*Ursus arctos*) occurs only in Hokkaido, the northernmost main island of Japan, located about 41–45°N latitude and 140–146°E longitude. The area of Hokkaido, a transition zone from the northern temperate to the subarctic zone (Tatewaki 1958), is about 78,000 km^2 and represents 22% of Japan's land area. Fifty-five percent and 71% of the island is mountainous and forested, respectively (Hokkaido Government 2017b). The population of Hokkaido is about 5.3 million as of 2018 (Hokkaido Government 2019).

The Ainu, a native people of Hokkaido, have traditionally lived by hunting, fishing, gathering, and trading with neighboring areas. Since the end of the nineteenth century, development in Hokkaido has drastically shifted from limited coastal fishery operations to massive interior development for agriculture, logging, and mining (Tabata et al. 2000). This change was due to the goal of rapid modernization proposed by the centralized Japanese government. With this development, occurrences and rumors of bears causing grisly injuries instilled a fear of bears into the minds of people living in Hokkaido (Kimura 1983). Moreover, livestock predation by bears increased in conjunction with the growth of livestock farming. As an end consequence, the Hokkaido Development Commissioner, the previous governing organization of Hokkaido, encouraged the killing of brown bears with a bounty (Yamada 2002).

Hostile policy regarding brown bears continued during the twentieth century. The Hokkaido government implemented a spring bear hunt from the 1960s through the 1980s, hoping to reduce the bear population. During the spring season, hunters can shoot bears in their dens or on the remaining snow patches after den emergence (Ohkawa et al. 1979). The spring bear hunt also represented additional hunting pressure on the reproductive segment of the population, namely females and their cubs (Kaji 1982; Mano 1995). As a consequence, brown bears were eliminated from the lowlands and were only limited to mountainous forested regions by the 1970s (Figure 24.1; Kaji 1982). Shrinkage of the distribution range continued to the beginning of the 1990s (Figure 24.1).

In 1990, the spring bear hunt system was abolished along with a major change of the bear management policy of the Hokkaido government to improve the coexistence of humans and bears. This change resulted in a drastic decline in active bear hunting in their habitat and a long-term increase in nuisance killing in and around areas of human habitation, such as farmland and urban areas (Hokkaido Institute of Environmental Sciences 2004; Mano 2009). The brown bear population has been consistently increasing since the 1990s and now brown bears occur in almost all the forested areas, which represents an important recovery in comparison to the bear distribution range in the late 1970s (Figure 24.1; Hokkaido Institute of Environmental Sciences 2004; Japan Bear Network 2014).

With the recent increase and range expansion of the bear population, conflicts between people and bears, as well as the number of controlled bear kills, have also increased. Recently, bear intrusions into urban areas, such as Sapporo City, as well as agricultural damage to corn and fruits, have increased in various parts of Hokkaido, although there has been little livestock damage (Figure 24.2). The number of bear kills has increased from 200 to 300 per year in the 1990s to over 850 by 2010 (Figure 24.2). The purpose of >90% of recent

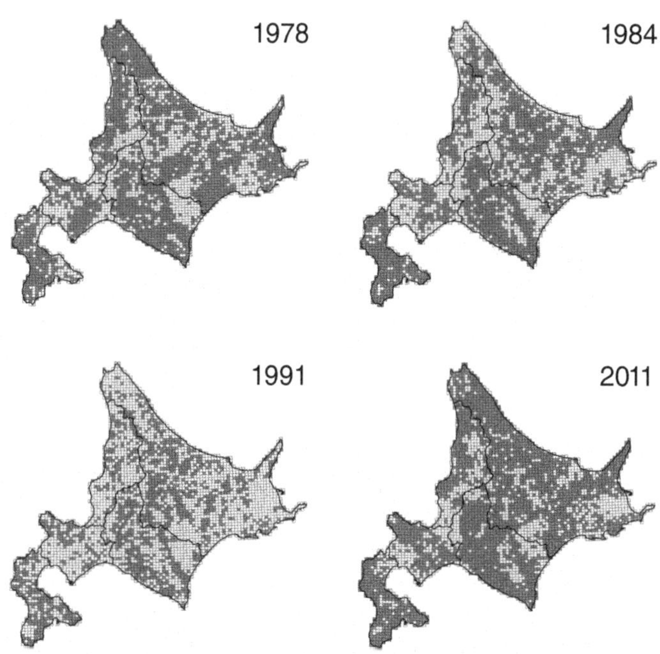

Figure 24.1 Transition of brown bear distribution ranges in Hokkaido, 1978–2011.

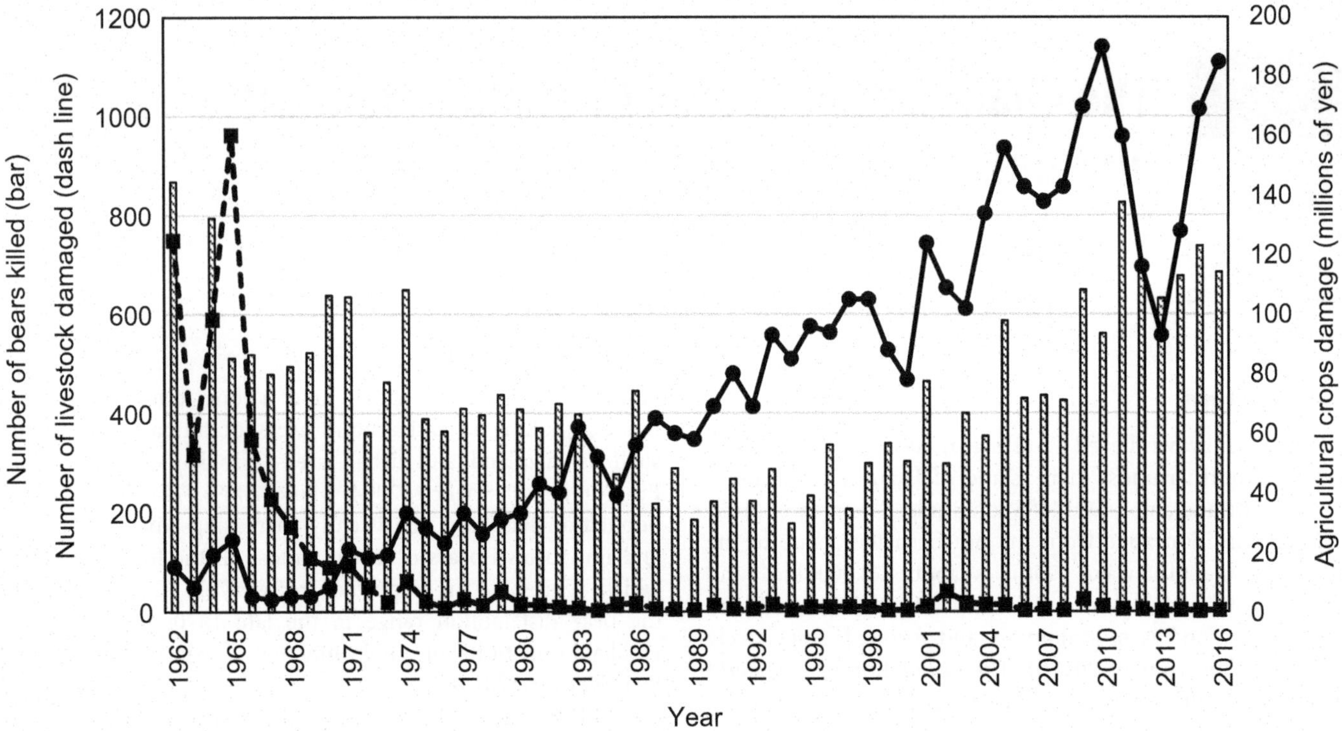

Figure 24.2 Brown bear kills, and damage to agricultural crops and livestock in Hokkaido during the period 1955–2017.

kills was damage control. The average cost of annual agricultural damage and number of bear kills between 2010 and 2016 were 13.7 million yen and 679, respectively (Hokkaido Government 2017a).

In this chapter, we present the current situation of bear management issues in Hokkaido, including the paradigm of the Brown Bear Management Plan of Hokkaido, urban bear management in Sapporo, and the human resource development and management system to develop proper brown bear management.

Brown Bear Management Plan of Hokkaido

Risk Assessment of Individual Bears and Appropriate Management Measures

The Hokkaido brown bear population is managed by the Brown Bear Management Plan of Hokkaido, which was established in 2017 (Hokkaido Government 2017a). The basic concept of the plan is that the most serious human–bear conflicts are caused by specific individuals. Thus, the control of such problematic individuals, which have become human-habituated (i.e. anthropogenic food conditioning, intrusions into houses to get food) because of wrong human behaviors, is essential for bear population management.

The nuisance status of conflictual or potentially conflictual bears is assessed following a specific flowchart (Figure 24.3) and the appropriate management measures are taken on the basis of (a) the nuisance status of the bear, and (b) the type of land use, i.e. forested, agricultural, and urban areas (Table 24.1).

Estimation of Number of Nuisance Individuals

Because a few specific nuisance individuals can be responsible for many human–bear conflicts (Tsuruga et al. 2016), the number of bear conflicts does not necessarily correspond to the number of nuisance individuals. Whereas most prefectures in Japan incorrectly employ the number of bear occurrences and/or conflict reports, crop damage, and/or damaged area as conflict indices (Kozakai et al. 2015), direct estimation of the number of nuisance individuals is employed for brown bear management in Hokkaido. The number of nuisance individuals has been estimated by taking into account the distance between the sites where damage occurred and the temporal occurrence of damage. Such standards were defined by considering the ability to travel of Hokkaido brown bears obtained by radio-collared individuals. Additional information such as foot print size and type of bears (e.g. solitary or family) was beneficial for more reliable estimations. Of course, because of the uncertainty of such estimations, both the minimum and maximum number of nuisance bears is estimated.

Control of Nuisance Individuals and Prevention of Food Conditioning

Most of the control of nuisance bears in Hokkaido is done by shooting or capturing with the help of box traps. Nevertheless, considering that the current bear population sizes are sufficiently large, we are confident that there is no immediate danger of bear extinction in the near future in Hokkaido. Actually, the

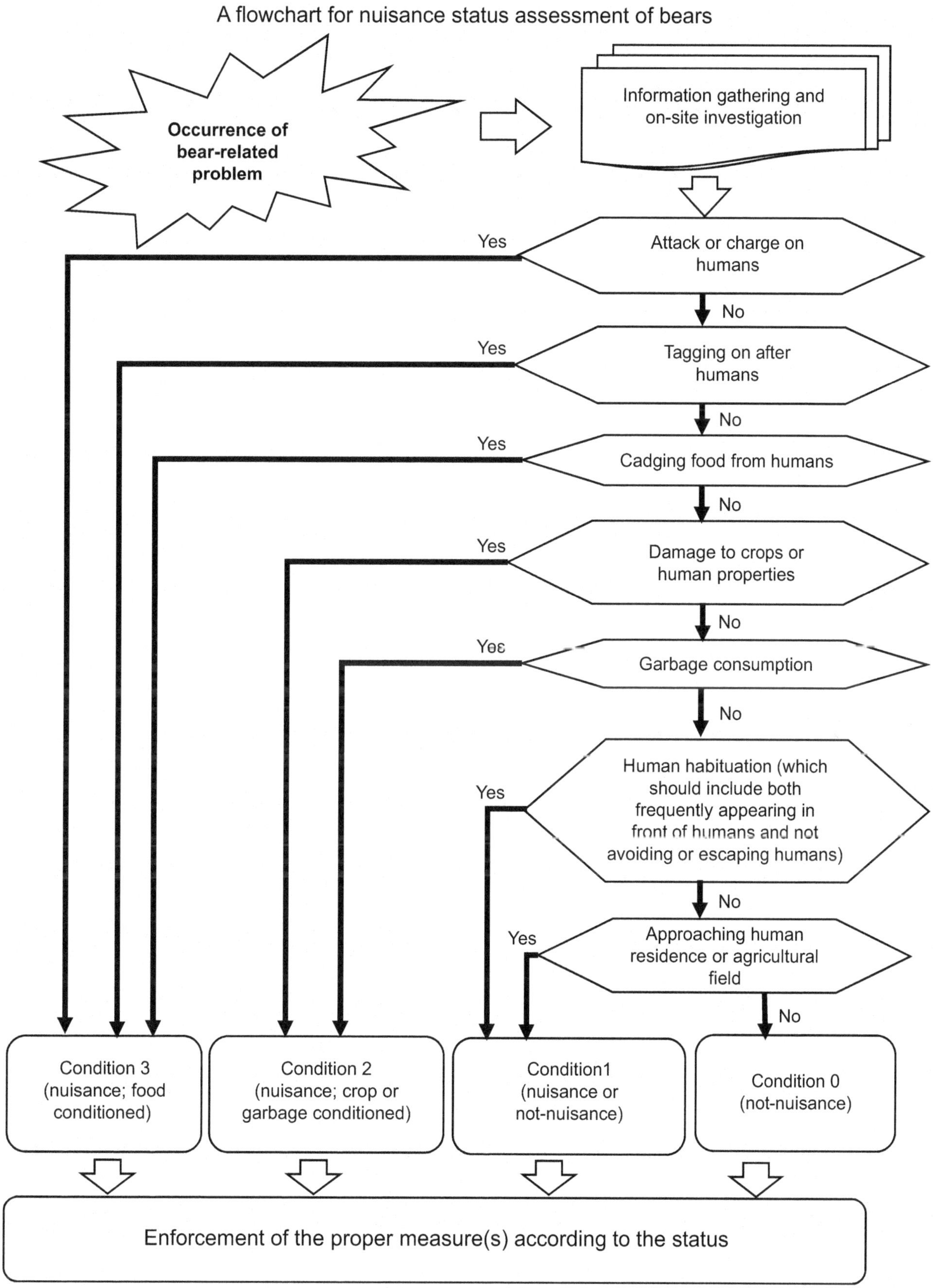

Figure 24.3 Flowchart to assess nuisance status of brown bears in Hokkaido (Hokkaido Government 2017a).

Table 24.1 Classification of brown bear nuisance status (see the text for more details) and management measures currently applied in Hokkaido, Japan (Hokkaido Government 2017a).

Bear nuisance category	Bear's attitudes toward humans or property	Nuisance status	Type of areas involved	Management measures
0	Human avoidance and absence of property damage	Non-nuisance	Urban agricultural	Public information, bear patrolling, removal of potential bear attractants (e.g. garbage, fish waste, apples)
			Forested	Public information, removal of potential bear attractants
1	Human habituation, intrusions into human settlements, but without real damage or human injury	Non-nuisance/nuisance	Urban agricultural	Public information, bear driven away from the area, removal of potential bear attractants
			Forested	Public information, restriction of forest access, bear driven away from the area, bear removal if the situation remains
2	Damage to crops or livestock, use of anthropogenic food such as livestock food, garbage, livestock carrion	Nuisance	Urban agricultural	Public information, bear patrolling, bear removal
			Forested	Public information, restriction of forest access, bear removal
3	Bluff charges on humans, attacks on humans causing injury or death	Nuisance	Urban agricultural, forested	Public information, bear patrolling, bear removal, create a task force for bear control actions

estimated population size and its 95% confidence interval in 2014 was 10,600 ± 6700 (Hokkaido Government 2017a).

Nuisance conditions were divided into conditions 2 and 3 by the two evaluation axes: condition 2 takes into account the attitude toward properties, i.e. agricultural crops and livestock, to evaluate economic damage, whereas condition 3 considers bear aggressiveness toward people (Table 24.1). Evidently, removal of individuals belonging to conditions 2 and 3 is more urgent than removal of individuals within conditions 0 and 1 that, however, may still escalate in their conflictual relationship with humans. Concurrent prevention measures to avoid bears becoming food-conditioned are also necessary to reduce nuisance individuals. Proper management of garbage and the use of electric fencing to prevent bear intrusions into agricultural field are encouraged in the management plan of conflictual individuals.

Brown Bear Management in Sapporo City

Sapporo City is the prefectural capital of Hokkaido and is Japan's fifth largest city. Forested mountainous landscapes surrounding the southern and western fringes of the urban area provide leafy scenery to the city. Satisfaction in both civil life and conservation of nature and biodiversity are important policy themes due to a high level of natural environmental civic awareness (Sapporo City Government 2013; Sato 2017).

The current urban area used to be lowland deciduous forests where brown bears lived until the late nineteenth century (Kimura 1995). As previously mentioned, the bear's distribution range had once receded to the interior mountain ranges due to intensive killing and agricultural and urban development until the beginning of the 1990s. Bears rarely visited the surrounding forested areas of Sapporo urban areas by the late twentieth century. In 1991, the brown bear population in the Western Ishikari range was officially listed in the Japanese Red Data Book (Environmental Agency 1991) and was designated as a "local population that should be protected." Brown bears have not been recognized as a threat to most citizens until the 2000s.

After the abolishment of the spring bear kill in 1990, the brown bear population has increased and their range extended into the city. Thus, bear intrusions into urban areas have repeatedly occurred since 2010. Recent camera-trap investigations in the surrounding forest revealed that there are a number of bears and some of them are regularly producing cubs (Sapporo City Government 2013).

Residents of large cities like Sapporo have little experience or practical knowledge on how to deal with bear coexistence. Consequently, whereas some people are instinctively afraid of bears, others are overly tolerant of bears. This situation is thus much different than that in rural areas of Hokkaido, where bear conflicts and control actions occur regularly. Another concern for the citizens of Sapporo is the management of outdoor activities in suburban areas during snowless seasons. There are many sightseeing and outdoor sport facilities, such as farm shops, camping and BBQ areas, and mountain biking areas, in the suburbs of Sapporo City.

The key to future management success would be improvement of the coexistence between people and bears, better information on how to behave in bear country, and clearer role allocation among related governmental organizations and stakeholders.

Shiretoko: Brown Bear Management in a National Park

Shiretoko is a 70-km long peninsula located at the north-eastern edge of Hokkaido. A primary ecosystem ranging from sea level to the mountaintop remains beyond the middle part of the peninsula (Yamanaka 2006). The area has been designated as a National Park (386 km^2) since 1964. Shiretoko National Park and surrounding areas were designated as a National Wildlife Sanctuary in 1982, and sport hunting has been forbidden since then. Kohira et al. (2009) suggested that a minimum of 200 female brown bears will inhabit the area by late twentieth century. Because bears are more visible along the coast, bear viewing from sightseeing boats is an important element of the tourism industry in the Shiretoko area.

The Shiretoko Nature Foundation, established by Shari town and Rausu town in 1988 to conserve the natural environment of the peninsula, employs personnel engaged in wildlife management and who have been working on bear monitoring and management inside and around the National Park. For the first time in Japan, Shiretoko teams have introduced non-lethal bear control measures such as cracker shells, rubber bullets, and dogs, hoping to change the behavior of nuisance bears.

Shiretoko is the first national park in Japan to introduce a control system to regulate human access to the protected area depending upon the preliminarily defined conditions. The Shiretoko Goko Lakes area, located in the central part of the peninsula, is a popular sightseeing area. Up to 4000 visitors gather on 3 km of trail each day. Because the area is densely populated by brown bears, there are many bear–human encounters and this situation has been a serious safety concern for a long time. A new system to control visitors' entry within the area was established in 2011 when an 800-m long elevated boardwalk and a gateway lecture room were completed. Visitor safety is managed in two ways: (1) a bear-defended elevated boardwalk; and (2) a trail with seasonal and bear occurrence-dependent regulations. Entry requires a bear safety speech given by park personnel before entering the area and a guide may be required depending on the season. This safety procedure for visitors is unique in Japan.

Despite bear management efforts, every year bears in the Shiretoko area have become progressively less cautious toward humans. The number of bears has increased significantly since 1995, primarily in the National Park area. The latest annual reports of bear occurrences in the whole area of Shari and Rausu towns, including outside of the National Park, is around 2000 (Figure 24.4). Of the 1088 human–bear encounters that occurred during the 4-year period of 2012–2015, people-avoiding behavior was not observed in 95.8% of cases. As a consequence of frequent harmless bear–people encounters, chiefly in the Natioanl Park area, widespread habituation of bears to people has occurred (Figure 24.5).

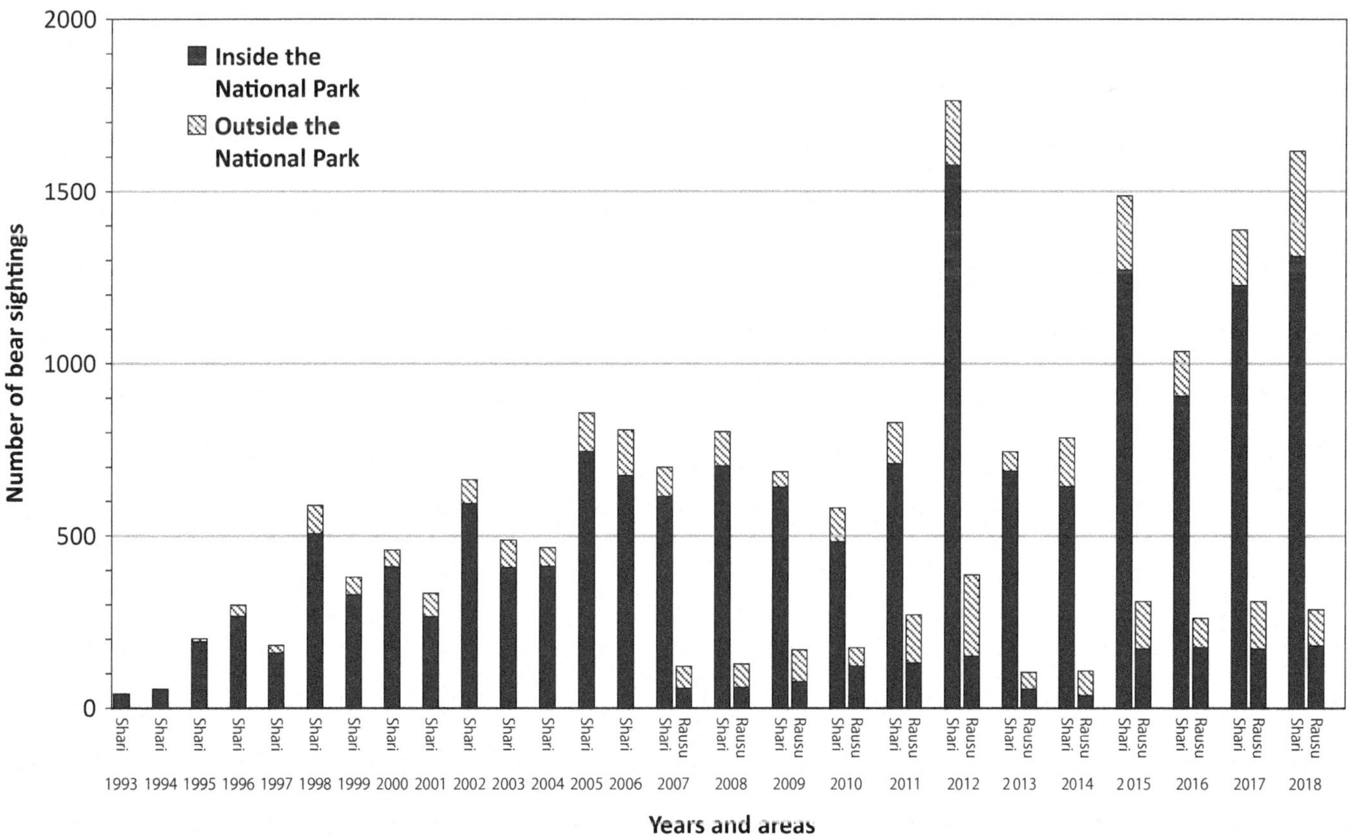

Figure 24.4 Brown bear sightings in the Shiretoko Peninsula (Shari and Rausu towns), between 1993 and 2018. Shari town is on the west and Rausu town is on the east of the Shiretoko Peninsula. No data before 2006 in Rausa town.

Figure 24.5 Tourists approaching very close to a bear in the Shiretoko National Park. This kind of behavior may be risky for people and may promote bear habituation to humans.

Issues and Challenges of Brown Bear Management in Hokkaido

First, we would like to highlight that the current estimation of the brown bear population size is unreliable; it is likely that the size is much larger than previously reported (Hokkaido Government 2017a). Therefore, an important challenge is how to assess the population trend, which represents essential information for reducing estimation biases. To improve assessment of the bear population in Hokkaido, a harvest-based estimator was thus employed (Mano et al. 2001) and correct on-site determination of the bear nuisance status is crucial to improve the accuracy of the estimation of nuisance bear numbers.

Various beneficial experiences and knowledge of bear management have already been obtained via previous monitoring and research works. The observation of drastic behavioral changes in bears in the Shiretoko National Park, for instance, is a good example for understanding how human attitudes toward bears can impact their behaviors toward us. In addition, troubles regarding habituated bear intrusions into urban areas nearby the National Park have become a serious issue in the community (Yamanaka et al. 2016).

The observations in Sapporo City provide insight into the development of Hokkaido residents' feelings of contact frequency with or distance to bears. As a consequence of the changes that occurred in the socioeconomic situation of Hokkaido in previous decades, a drift of population from rural areas to the Sapporo urban area has been continued. Nowadays, a large portion of people in Hokkaido live without worrying about brown bears. The almost complete removal of bears by the late twentieth century supported people's livelihood without consideration of bears. However, the situation has entered a new phase, where bears inhabit the neighborhood of urban people as a result of their population recovery. Thus, even for people living in a metropolitan city it is now necessary to face the reality of being neighbors with brown bears, which requires appropriate measures of bear management.

In conjunction with the drastic increase of human–bear conflicts and number of nuisances bears killed in and around human dwellings (Mano 2009), increasing bear habituation to people and damage to agricultural crops are also current local concerns. Worryingly, greater expansion of bears into human areas is expected in the future. The only bear control measure taken in most parts of Hokkaido is bear killing, and control activities have been assigned to the Japan Sports Hunter's Association, a nation-wide hunting club. However, bear removal is becoming difficult in some towns due to aging of hunters and reduction in their number.

Brown bears were once viewed as an enemy of humans in Hokkaido and were intensively killed from the beginning of the modern era until the late twentieth century. The dramatic change in bear management policy at the end of the twentieth century could be a consequence of the maturation of nature conservation ideas in Hokkaido, but the social understanding of the philosophy of coexistence with bears still needs to improve in the near future. Specifically, deployment of personnel with appropriate and specific knowledge, skills, and experience in bear management is an urgent issue in Japan.

References

Environmental Agency. (1991). *Red Data Book of Japan: Vertebrates.* Tokyo: Japan Wildlife Research Center [in Japanese].

Hokkaido Government. (2017a). *Brown bear management plan of Hokkaido.* Sapporo: Hokkaido Government [in Japanese].

Hokkaido Government. (2017b). *Forest of Hokkaido.* Sapporo: Hokkaido Government. Available from www.pref.hokkaido.lg.jp/sr/sum/kcs/indexfiles/2017hokkaido_shinrin.htm

Hokkaido Government. (2019). *Population and number of householders in Hokkaido by basic resident register in* 2018. Sapporo: Hokkaido Government [in Japanese]. Available from www.pref.hokkaido.lg.jp/ss/tuk/900brr/index2.htm0

Hokkaido Institute of Environmental Sciences. (2004). *Wildlife distribution and abundance research report; brown bears* (1999–2003). Sapporo: Hokkaido Institute of Environmental Sciences [in Japanese].

Japan Bear Network (Ed.). (2014). *Report on "Assistant Project for The Distribution Identification, Conflict Avoidance, and Restore of Endangered Local Populations of Japanese Black and Brown Bears."* Ibaraki: Japan Bear Network.

Kaji, K. (1982). Higuma no Bunpu. Brown bear distribution range in Hokkaido. *Hppoh-ringyo* 34: 108–112 [in Japanese].

Kimura, M. (1983). *Ezo Higuma Hyakka. An encyclopedia of Ezo brown bears.* Sapporo: Kyodo Bunka Sha [in Japanese].

Kimura, M. (1995). *Haru-tsuge-ju. Brown bears, who tell of the arrival of spring,* Sapporo: Kyodo Bunka Sha [in Japanese].

Kohira, M., Okada, H., Nakanishi, M. & Yamanaka, M. (2009). Modeling the effects of human-caused mortality on the brown bear population on the Shiretoko

Peninsula, Hokkaido, Japan. *Ursus* 20: 12–21.

Kozakai, C., Kondo, M., Arimoto, I., et al. (2015). II. Implementation status and issues of policy evaluation based on monitoring results. *Mammalian Science* 55: 241–263 [in Japanese].

Mano, T. (1995). Sex and age characteristics of harvested brown bears in the Oshima Peninsula, Japan. *Journal of Wildlife Management* 59: 199–204.

Mano, T. (2009). Causes and prevention of bear related conflicts in the human habitat in Hokkaido. In: Oi, T., Ohnishi, N., Koizumi, T. & Okochi, I. (Eds.), *Biology of bear intrusions* (pp. 23–27). Ibaraki: Forestry and Forest Products Research Institute.

Mano, T., Matsuda, H., Horino, S., et al. (2001). A harvest based estimator of brown bear population size and dynamics for the Oshima Peninsula, Hokkaido, Japan. Program of the 13th International Conference of Bear Research and Management, Jackson Hole, Wyoming.

Ohkawa, Y., Tajima, H. & Tanada, E. (1979). A research on wintering dens of Ezo-brown bear (*Ursus arctos yesoensis* Lydekker). *Shin Higuma Tsushin* 6: 37–51 [in Japanese].

Sapporo City Government. (2013). *A vision on biodiversity of Sapporo City*. Sapporo: Sapporo City Government [in Japanese]. Available from www.city.sapporo.jp/kankyo/biodiversity/documents/bd_vision.pdf

Sato, Y. (2017). The future of urban brown bear management in Sapporo, Hokkaido, Japan: a review. *Mammal Study* 42: 17–30.

Tabata, H., Kuwahara, M., Funazu, I. & Sekiguchi, A. (2000). *Hokkaido no Rekishi. History of Hokkaido*. Tokyo: Yamakawa Press [in Japanese].

Tatewaki, M. (1958). Forest ecology of the islands of the North Pacific Ocean. *Journal of the Faculty of Agriculture, Hokkaido University* 50: 371–486.

Tsuruga, H. & Mano, T. (2008). Brown bear management plan and monitoring programs on the Oshima Peninsula, Hokkaido. *Mammalian Science* 48: 91–100 [in Japanese with English summary].

Tsuruga, H., Kondo, M. & Tomizawa, M. (2016). Identification of problematic brown bears using bear remains gathered on conflict sites in Hokkaido, Japan. *Report of Institute of Environmental Sciences* 6: 74–78 [in Japanese].

Yamada, S. (2002). Wolves, brown bears and crows: bounty on killed "harmful birds and bears" during the Meiji Period in Hokkaido. *Memoirs of the Pioneer Memorial of Hokkaido* 41: 67–92 [in Japanese with English summary].

Yamanaka, M. (2006). Social and political problems related to wildlife and park management in Shiretoko National Park. In: McCullough, D.R., Kaji, K. & Yamanaka, M. (Eds.), *Wildlife in Shiretoko and Yellowstone National Parks: Lessons in wildlife conservation from two world heritage sites* (pp. 286–291). Shari: Shiretoko Nature Foundation.

Yamanaka, M., Masuda, Y. & Ishinazaka, T. (2016). Recent advances and challenges in the conservation and management of brown bears in the Shiretoko National Park. *Bulletin of the Shiretoko Museum Special Issue* 1: 55–78.

Chapter

25 Potential Ecological Corridors for Remnant Asiatic Black Bear Populations and its Subpopulations Linked to Management Units in Japan

Tomoko Doko, Wenbo Chen, Reina Uno, Hidetoshi B. Tamate, A.G. Toxopeus, A.K. Skidmore, and Hiromichi Fukui

Introduction

The Japanese National Biodiversity Strategy 2010 (Ministry of the Environment, Government of Japan 2010) calls for the creation of ecological networks as a biodiversity conservation policy. The development of ecological networks offers the potential to connect fragmented wildlife habitats, stem the loss of biodiversity, and promote dispersal and genetic exchange between isolated populations. An appropriate approach for preparing conservation and zoning plans requires spatially explicit information on species distributions, with more accurate and fine-resolution data at local scales. Such information would allow for suitable habitats to be identified, around which ecological networks could then be designed, if necessary. However, there is an obvious lack of information on the spatial distribution of many species and a lack of scientific methods for examining habitat requirements to establish the need for constructing these networks for target species.

The Asiatic black bear (*Ursus thibetanus*) is categorized as a Vulnerable species in the International Union for Conservation of Nature (IUCN) Red List and distributed mainly in southern and eastern Asia. This species has been threatened by overexploitation because their gall bladders sell at a high price in the traditional medicine market and their meat and fur are also highly prized. Habitat fragmentation under rapid industrialization in Asian countries could also have driven up their local extinction rate.

In Japan, a subspecies of the Asiatic black bear, the Japanese black bear (*Ursus thibetanus japonicus*), was once distributed over the three main islands: Honshu, Shikoku, and Kyushu. In Japan, there are 19 recognized local populations of Asiatic black bear (Japan Wildlife Research Center 1992). Nowadays, some local populations are listed as "critically endangered local populations" by the Ministry of the Environment, Japan. The native population in Kyushu was thought to have gone extinct in the twentieth century. In Shikoku and western Honshu, some local populations are considered extinct and the current distribution of the remaining populations is fragmented. Conversely, local populations in central and north-eastern Honshu are relatively stable and distributed over large areas.

Despite the threatened status of Japanese black bears, the incidence of bear–human conflict has increased recently, which creates serious problems for the conservation of this species. The central and north-eastern populations are often exposed to a high level of hunting pressure and nuisance controls under management plans that aim to reduce damage to agricultural crops and prevent human casualties caused by bears.

We introduce here two case studies. First, we present a quantitative method for assessing the need for ecological networks through modeling the potential geographical distributions of species based on a case study of local populations of Asiatic black bear in Fuji and Tanzawa, based on the work in Doko et al. (2011). Second, we report genetic variation of Asiatic black bear in the Tohoku region, Japan, based on the work in Uno et al. (2015). To determine how population subdivision relates to management units (MUs) proposed by the Ministry of the Environment, genetic variation in the mitochondrial DNA control region (mtDNA-CR) and seven autosomal microsatellite loci were assessed in bears captured in northern Japan. Geographical distribution of the subpopulations was assessed using landscape analyses to find the best-fit model based on maximum entropy (MaxEnt) (Phillips et al. 2006) prediction and cost of movement. Finally, we also show how human–bear conflicts, nuisance control, and traditional hunting may affect conservation and management of Asiatic black bears in the southern Tohoku area, where large suitable habitats for this species exist.

Case Study 1: Identifying Habitat Patches and Potential Ecological Corridors between Fuji and Tanzawa

Study Area and Purpose of the Work

The Fuji region is part of the Fuji–Hakone–Izu National Park, which covers an area of 121,851 ha over four prefectures and includes Mt. Fuji (3776 m; Figure 25.1). The Tanzawa region is a mountainous area that is designated as a Quasi-National Park. The Fuji and Tanzawa regions are less than 10 km apart. However, due to recent urbanization and the construction of roads, the habitat spanning the Fuji and Tanzawa regions has become fragmented. Thus, the gene flow between local populations of large mammals, particularly the Asiatic black bear, has almost ceased (Japanese Mammals Society 1997).

Case Study 1 *(cont.)*

In addition to the presence of new and upgraded roads, local black bear populations have seen their habitats fragmented by deforestation. The Tanzawa local population, in particular, has been deemed a "population which, due to its small size, should be protected" (Japanese Mammals Society 1997). Nevertheless, studies of the Tanzawa local population are rare.

Inbreeding and loss of genetic diversity are both conservation concerns, as they increase a species' risk of extinction (Frankham 2003). Loss of genetic diversity reduces a species' ability to evolve and adapt to environmental change. Inappropriate management and allocation of resources is likely to endanger animal populations if genetic factors are ignored in the management of threatened species (Frankham 2003). To maintain genetic diversity, the concept of a minimum viable population (MVP) (Gilpin & Soulé 1986) was developed, where the levels of genetic diversity necessary for adaptation and evolution are considered. MVP refers to the minimum population size needed to maintain a local population in the long term; once a population has fallen below its MVP, its local extinction becomes inevitable.

In this case study, we present a quantitative method for assessing the need for ecological networks using predictive geographic distribution models, and we apply our method to local Asiatic black bear populations in Japan's Fuji and Tanzawa regions. Here, we (1) develop a predictive distribution map of the species for identifying geographic locations of habitats, and (2) estimate the number of individuals in local and subpopulations.

Methods

Data

Presence point localities of Asiatic black bears were used from published reports (Mochizuki et al. 2005).

Following Morrison et al. (1992), we have defined bear-suitable habitats as those areas that promote occupancy by

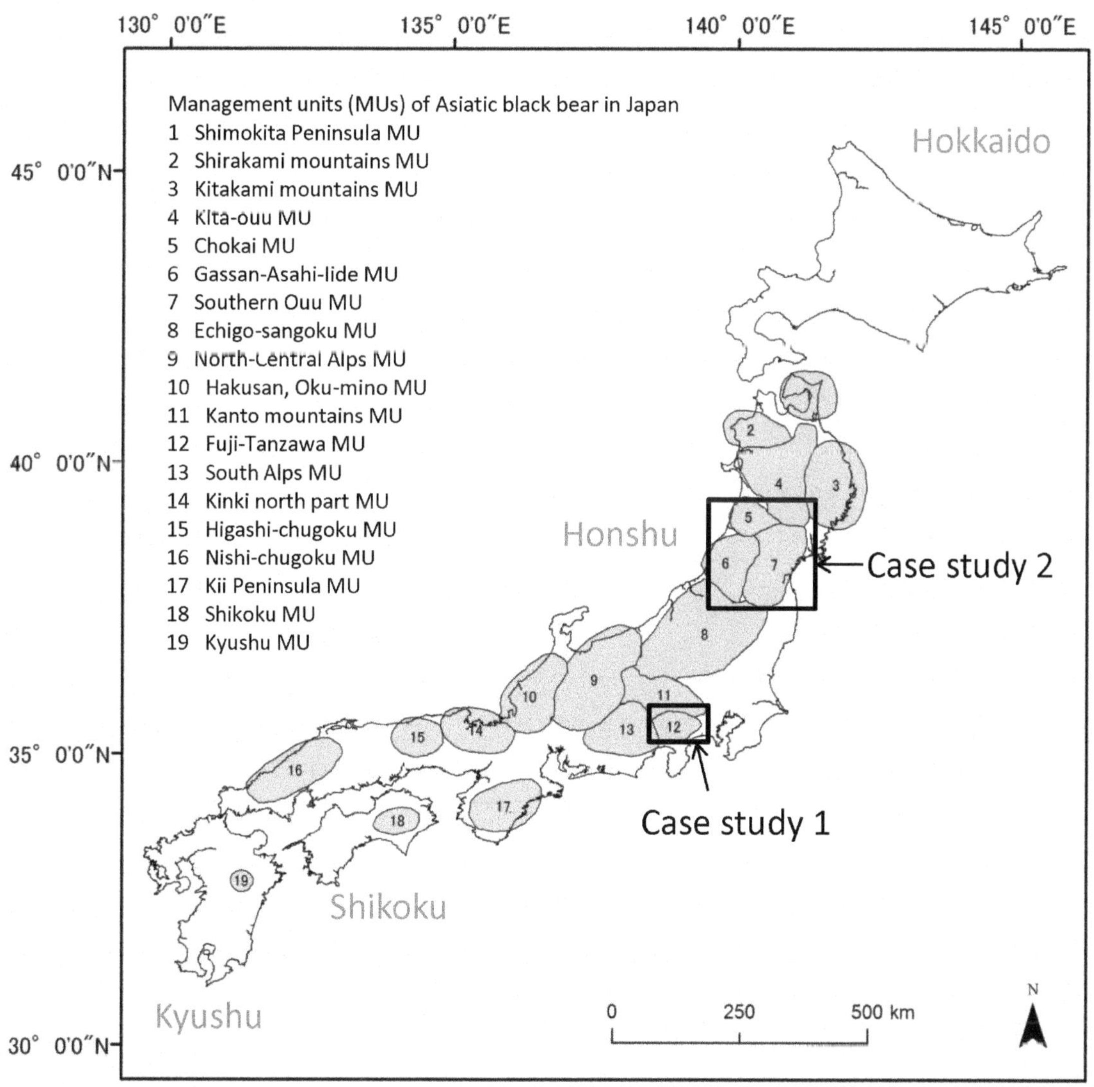

Figure 25.1 Location of the two case studies with management units (MUs) of Asiatic black bears in Japan.

Case Study 1 *(cont.)*

individuals and allow them to survive and reproduce. Based on previous studies of the ecology of Asiatic black bears, the following resources and environmental variables were selected as predictors and organized into four categories: (1) topology; (2) water resources; (3) distance to roads; and (4) vegetation. Data for the predictive variables were obtained from a variety of digital sources and stored in a GIS environment.

Modeling

Species distribution models attempt to provide detailed predictions of distributions by relating the presence or abundance of species to environmental predictors (Elith et al. 2006). Predicting species' distributions has become an important component of conservation planning in recent years, and a wide variety of modeling techniques have been developed for this purpose (Guisan & Thuiller 2005). Here, MaxEnt was used to model the occurrence probability of the Asiatic black bear and to map its spatial distribution.

The area where the species was predicted to be present was considered to represent the species' "core area." The population size of the target species was estimated based on the known population density and area. The following assessment criteria for the Asiatic black bear, as established by the Ministry of Environment of Japan (Nature Conservation Bureau in Ministry of the Environment 2000), were followed: (a) habitat patches with populations less than 100 individuals were labeled as *being in serious danger of extinction*; (b) habitat

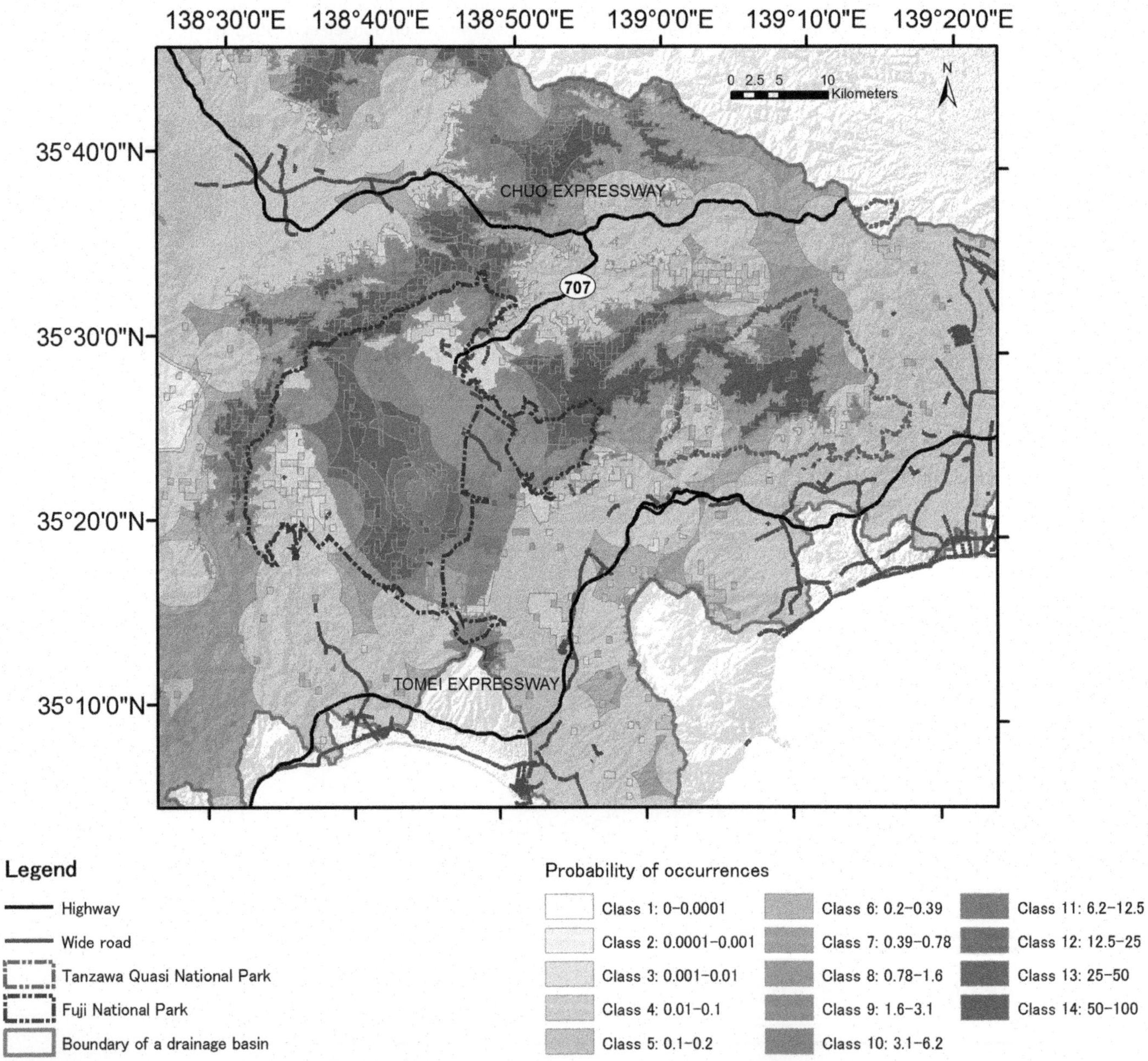

Figure 25.2 Probability of occurrence of the Asiatic black bear in the Fuji–Tanzawa region, Japan, as predicted by MaxEnt.

Case Study 1 *(cont.)*

patches with populations ranging from 100 to 400 individuals were labeled as *endangered*; and (c) habitat patches with populations >400 individuals were labeled *healthy*.

Results

The probability of Asiatic black bear occurrence was mapped for the Fuji-Tanzawa region (Figure 25.2). The grouped habitat patches were considered to represent local or subpopulations of Asiatic black bear. As pointed out in the previous section, the Fuji–Tanzawa region is considered to be isolated from other local populations, becaue the Chuo Expressway probably functions as a major barrier inhibiting the movement of bears. Thus, on this map, habitat patches for Asiatic black bear were bisected along the Chuo Expressway to provide a more accurate approximation of the distribution of continuous habitat patches.

The Fuji–Tanzawa region contained 1087 km^2 of the Asiatic black bear's core habitat area and had an estimated population size of 242 individuals. The population of the entire area was considered to be "endangered." The Fuji–Tanzawa region can be subdivided into three groups: the Fuji local population, the Tanzawa local population, and the Fuji–Tanzawa Corridor population.

Conclusions

In this case study, we have proposed a quantitative method for assessing how necessary it is for a species to be included in ecological networks through an integrated approach involving predictive distributional maps, population size estimation, and a zoning strategy. The disadvantage of this method is that it relies on existing data, such as known population densities, known MVP sizes, and point data of species' records. On the other hand, its strength is that it can complement a lack of data for a given species, and it produces objective results based on currently accepted knowledge.

The total estimated population size of Asiatic black bears in the Fuji–Tanzawa region indicates that it is endangered, and thus, even if the region was to be connected, this alone would not be sufficient to sustain a healthy population. Even the largest patch, the Fuji local population, is still endangered, and the Tanzawa local population is in serious danger of extinction. Our modeling also showed that: (1) the identified corridor, which is equivalent to the Mishoutai local population, is not protected by a wildlife protection area or a natural park; and (2) the Chuo Expressway can be considered a physical barrier to the movement between the Fuji–Tanzawa local bear population and the neighboring populations in the Chichibu–Tama region.

Case Study 2: Genetic Population Structure Within and Across Management Units in Northern Japan

Study Area and Purpose of the Work

The study area consisted of approximately 30,790 km^2 in northern Honshu (Figure 25.1). The large continuous habitat of black bears can be found from northern Honshu to middle Honshu, which is divided into seven MUs in the bear management policy proposed by the Ministry of the Environment (Yoneda 2001). There are three MUs defined by the Ministry of the Environment in the study area: Chokai, Gassan–Asahi–Iide, and southern Ouu (Figures 25.3 and 25.4).

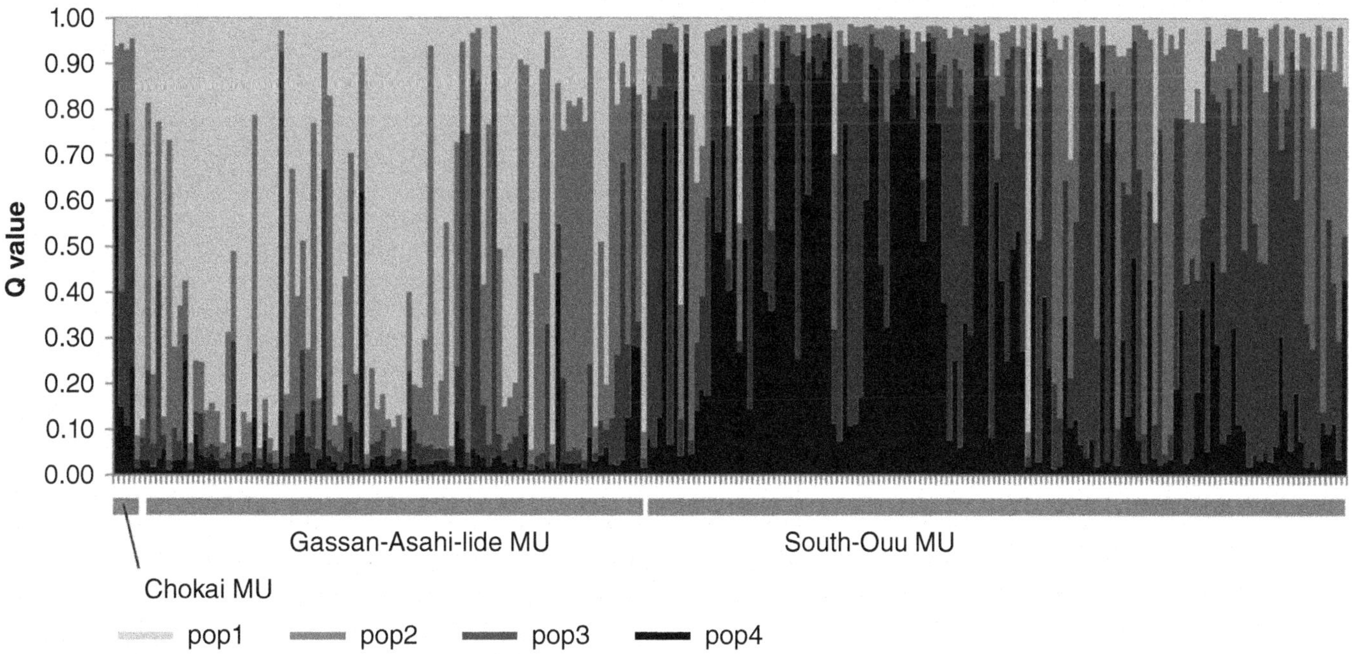

Figure 25.3 Bayesian assignment of the Asiatic black bear population structure. STRUCTURE bar plot of the estimated membership coefficient (Q value) for each individual for K (populations) = 4. Each bar represents the Q of an individual. The bars are aligned in the order of the latitude of their sampling sites and are grouped according to management unit (MU).

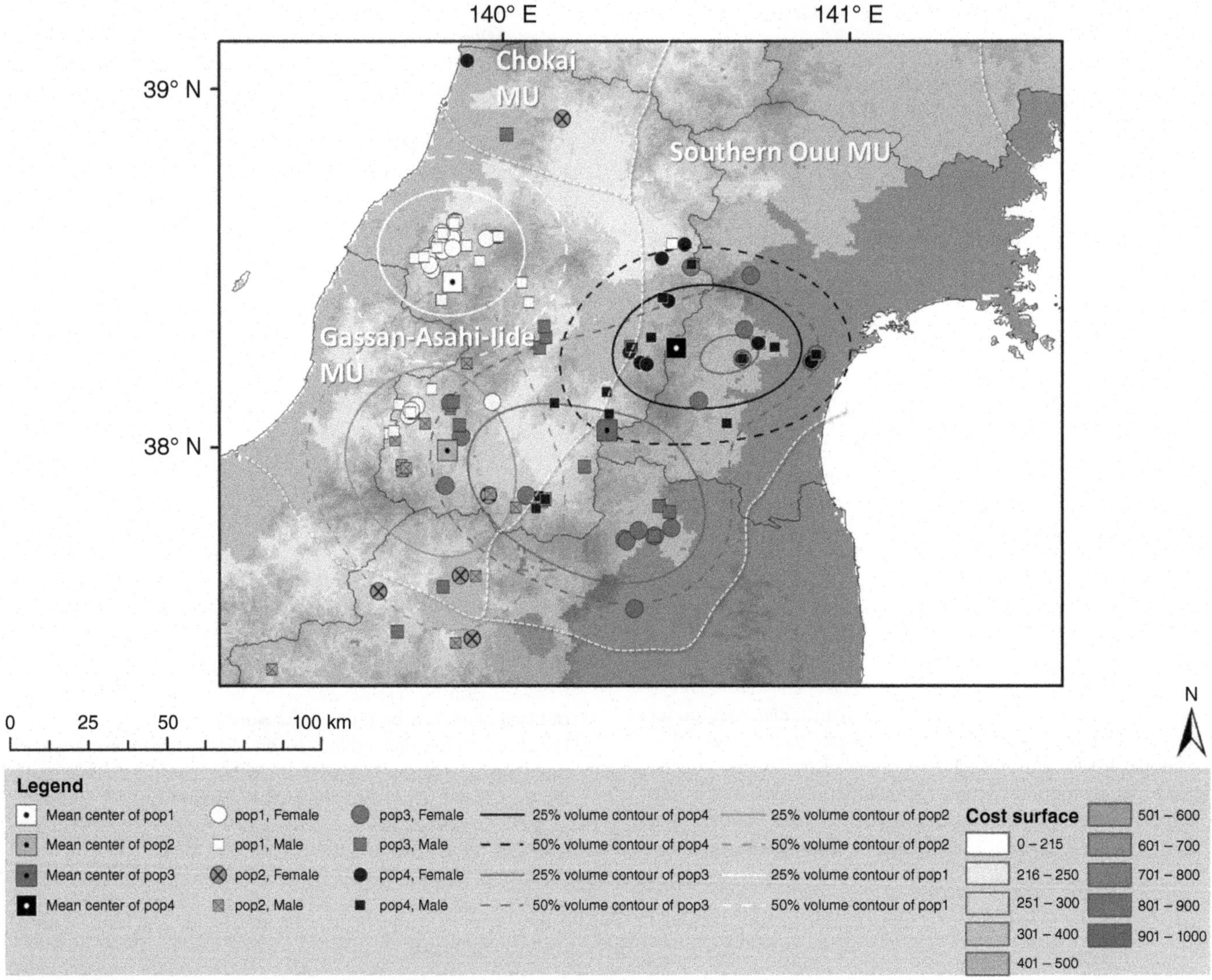

Figure 25.4 Map of the cost surface layer and individual assignment to one of the Asiatic black bear populations (K = 4) at Q > 0.6 (for more details, see the text).

Case Study 2 *(cont.)*

The first step of making a management plan is to set up a management unit: a basic unit comprising the population to be monitored and managed separately from other populations (Palsbøll et al. 2006). MUs provide a conceptual frame for the authorities to establish a bear management policy at regional and nationwide levels, although they are not easily applicable to management practice, which is usually carried out by a prefectural office. This is because the MUs are too large to be covered by a single prefecture. Each MU usually consists of more than one mountainous region so that there is environmental heterogeneity as well as subdivided local populations within an MU. Therefore, it is neither practical nor effective to apply a single management plan for the entire population within an MU, especially for a large continuous MU, such as the middle and south-east habitat areas in Japan. To overcome such difficulties, each prefectural office devises different management plans according to the different localities at a regional level. However, few empirical studies have been conducted to monitor population structures and environmental heterogeneity within an MU.

In the present case study, we focused on the black bear populations in the southern Tohoku area, which is one of the largest continuous habitats of this species. Our objectives were to detect possible population subdivisions within and across MUs, and to understand the relationship between the population structures and environmental heterogeneity within MUs.

Methods

Data

Tissue samples for DNA analyses were collected during either hunting or population control conducted in Yamagata, Miyagi, and Fukushima from 2004 to 2007.

Case Study 2 *(cont.)*

Mitochondrial DNA Analysis

A DNA fragment of approximately 1500 bp covering the mitochondrial DNA control region (mtDNA-CR), including flanking sequences, was amplified by polymerase chain reaction (PCR) using a primer pair.

Microsatellite DNA Analyses

To determine individual multilocus genotypes, seven polymorphic microsatellite loci (G1A, G10B, G10L, G10M, G10X, MSUT 1, and MSUT 6) were targeted (Paetkau et al. 1995; Kitahara et al. 2000). PCR and genotyping were carried out according to the method described previously (Uno et al. 2012).

Modeling

MaxEnt (Phillips et al. 2006) was used to predict the distribution of the bears in the study area. Three types of presence data set were used for this study, (a) national distributional map of bears, (b) observation of bears in the mountains (including direct observations and indirect signs of presence, e.g. scats), and (c) hunting maps of bears (including hunted locations and hunted numbers of bears reported by hunters in the region).

Cost Surface Layer

The cost surface represents the overall cost for the movement of an individual within the study area (Ferreras 2001; Gurrutxaga et al. 2010). A cost surface layer identifies the cost of traveling through each cell. While the habitat suitability simply corresponds to the probability of their presence at a location, the cost surface tells us how easily the bears can move in or out of the location. In order to visualize the energetic cost of movement for the Asiatic black bear, the cost surface layer was calculated and mapped. To calculate the cost surface, we assigned the resistance values to environmental variables.

The resistance values were calculated as follows. Based on the characteristics of the response curves derived from MaxEnt, environmental layers were standardized as resistance values ranging from 1 to 10, which represent criteria layers. Hence, criteria layers for seven environmental layers, i.e. distance to river, elevation, distance to road, road density, land use, snow, and snow depth, were used to calculate the weighted sum, a single cost surface layer. The coefficients for weighting criteria layers were obtained from the percent contribution (%) using MaxEnt. We used averaged values of percent contribution for individual cost surfaces for 10 MaxEnt replicates.

In order to visualize the geographical distribution of genetically distinct subpopulations inferred by STRUCTURE (Pritchard et al. 2000; Figure 25.3), we mapped sampling points of the individuals assigned to genetic clusters (pops 1–4) with an estimated membership coefficient (Q) $>$ 0.6 over the cost surface layer. Q gives a probability that an individual belongs to a subpopulation. From the spatial distribution of pops 1–4, the mean center was calculated using Spatial Statistics Tools in ArcMap. A mean center is a geographical center of capture points of individuals that are assigned to the same group, which we used as a proxy for the geographical center for each subpopulation identified by STRUCTURE. A kernel density plot for each population was generated using Hawth's Tools (Hawthorne 2012) and overlaid on the cost surface map. The kernel density contour shows a distribution range of a certain percentage of individuals that are assigned to a subpopulation.

Results

mtDNA Analysis

We found a total of nine substitutions among 708 nucleotide sites, and identified 12 haplotypes among 168 individuals.

Microsatellite Analysis

We determined genotypes at the seven microsatellite loci and identified the sex of 233 individuals.

The results demonstrated that the genetically defined subpopulations were localized to different areas: the Gassan and Asahi mountain ranges (pop1), the Iide mountain range (pop2), and the northern (pop4) and southern (pop3) parts of the Ouu mountain range (Figure 25.4).

Discussion

Genetic Diversity and Population Structure

The genetic status of the black bear populations in Japan has been studied by using mtDNA (Ishibashi & Saitoh 2004; Ohnishi et al. 2009; Yasukochi et al. 2009; Yamamoto et al. 2012) and microsatellite markers (Saitoh et al. 2001; Ohnishi et al. 2007; Yamamoto et al. 2012).

Landscape Genetic Analysis of Population Structure

Habitat suitability was mainly affected by snow and elevation; however, it was also affected to a lesser extent by factors related to human activity, namely, land use and roads. In addition, among the different types of land use, forest was identified as a highly contributing component for suitability.

A subdivided population structure in a large continuous habitat can be explained as either a consequence of restricted gene flow due to geographical and environmental barriers (Murtskhvaladze et al. 2010; Karamanlidis et al. 2012; Straka et al. 2012) or a remnant pattern of an ancestral population structure (Ohnishi et al. 2009).

The cost surface, which represents the effects of geographical and environmental barriers, accounted for some but not all of the observed population substructure. The main distribution areas of pop1 and pop2, shown by the kernel density contours, had a low-cost (cost $<$300) around the high mountains between the two populations. A high-cost zone (overall cost $>$601) was found between pop1 and pop3; however, a part of the 25% kernel density contour of pop3 overlapped with that of pop4. Similar geographical overlap of genetically identified populations was also reported for Eurasian brown bears in northwestern Eurasia (Tammeleht et al. 2010), although the size of our study area is approximately 10 times smaller than that of the brown bear study. Considering that our study area is within the known maximum dispersal range capabilities of black bears (Sato et al. 2011; Kopatz et al. 2012), this overlapping can be explained by the dispersal of individuals from their natal area.

The remnant pattern of the ancestral population structure also seems to have contributed to the observed population subdivision. Therefore, we concluded that the observed distribution of the genetically defined subpopulations has been

Case Study 2 *(cont.)*

shaped by a combination of the remnant population structure and restricted gene flow among the populations, and the genetically identified subpopulations are localized, but not clearly separated from each other.

MUs and Population Structure

Our results partially support the setting of GAI- and SO-MUs as genetically distinct units. However, the presence of subpopulations within an MU implies that each MU is not always a single demographic unit. Therefore, to establish proper management plans for Asiatic black bears in Japan, further consideration should be given to population structures that could represent separate demographic units within an MU.

Conclusions

Lessons from the history of extinct or endangered black bears in western Japan tell us the importance of maintaining a large, continuous habitat for sustainable population management. From the view of conservation genetics, it is desirable to keep these corridors in areas where gene flow prevails against the reduction in genetic diversity. In the present study, such an area was observed in the southern mountain region of Iide, where individuals assigned to the two populations coexisted. This region is likely to be the contact zone of management units, and it is worth the effort to conserve this area as a corridor to sustain the population of black bears.

References

Doko, T., Fukui, H., Kooiman, A., et al. (2011). Identifying habitat patches and potential ecological corridors for remnant Asiatic black bear (*Ursus thibetanus japonicus*) populations in Japan. *Ecological Modelling* 222: 748–761.

Elith, J. H., Graham, C. P., Anderson, R., et al. (2006). Novel methods improve prediction of species' distributions from occurrence data. *Ecography* 29: 129–151.

Ferreras, P. (2001). Landscape structure and asymmetrical inter-patch connectivity in a metapopulation of the endangered Iberian lynx. *Biological Conservation* 100: 125–136.

Frankham, R. (2003). Genetics and conservation biology. *Comptes Rendus Biologies* 326: 22–29.

Gilpin, M. E. & Soulé, M. E. (1986). Minimum viable populations: processes of species extinction. In: Soulé, M. E. (Ed.), *Conservation biology: The science of scarcity and diversity* (pp. 19–34). Sunderland, MA: Sinauer Associates.

Guisan, A. & Thuiller, W. (2005). Predicting species distribution: offering more than simple habitat models. *Ecology Letters* 8: 993–1009.

Gurrutxaga, M., Lozano, P. J. & del Barrio, G. (2010). GIS-based approach for incorporating the connectivity of ecological networks into regional planning. *Journal for Nature Conservation* 18: 318–326.

Hawthorne, B. (2012). Create minimum convex polygons tool. Hawth's analysis tools for ArcGIS. Available from www.spatialecology.com/htools/createmcp.php (accessed June 26, 2014).

Ishibashi, Y. & Saitoh, T. (2004). Phylogenetic relationships among fragmented Asian black bear (*Ursus thibetanus*) populations in western Japan. *Conservation Genetics* 5: 311–323.

Japan Wildlife Research Center. (1992). *Report from investigative commission on management of Asiatic black bear*. Tokyo: Japan Wildlife Research Center.

Japanese Mammals Society. (1997). *Red Data: Japanese mammals*. Tsukuba: Bun-ichi Co., Ltd.

Karamanlidis, A. A., Straka, M., Drosopoulou, E., et al. (2012). Genetic diversity, structure, and size of an endangered brown bear population threatened by highway construction in the Pindos Mountains, Greece. *European Journal of Wildlife Research* 58: 511–522.

Kitahara, E., Isagi, Y., Ishibashi, Y. & Saitoh, T. (2000). Polymorphic microsatellite DNA markers in the Asiatic black bear *Ursus thibetanus*. *Molecolar Ecology* 9: 1661–1662.

Kopatz, A., Eiken, H. G., Hagen, S. B., et al. (2012). Connectivity and population subdivision at the fringe of a large brown bear (*Ursus arctos*) population in North Western Europe. *Conservation Genetics* 13: 681–692.

Ministry of the Environment, Government of Japan. (2010). *The Biodiversity Strategy of Japan 2010* (p. 310). Tokyo: Ministry of the Environment Government of Japan.

Mochizuki, N., Sawai, K. & Ohba, T. (2005). *Report on investigation of the habitat of Asiatic black bear in Shizuoka Prefecture (1998–2002)*. Shizuoka Prefecture Forestry and Forest Products Research Institute.

Morrison, M. L., Marcot, B. G. & Mannan, P. W. (1992). *Wildlife–habitat relationships, concepts and applications*. Madison, WI: University of Wisconsin Press.

Murtskhvaladze, M., Gavashelishvili, A. & Tarkhnishvili, D. (2010). Geographic and genetic boundaries of brown bear (*Ursus arctos*) population in the Caucasus. *Molecolar Ecology* 19: 1829–1841.

Nature Conservation Bureau in Ministry of the Environment. (2000). *Manual of management techniques for conserving specified wildlife*. Tokyo: Nature Conservation Bureau in Ministry of the Environment.

Ohnishi, N., Saitoh, T., Ishibashi, Y. & Oi, T. (2007). Low genetic diversities in isolated populations of the Asian black bear (*Ursus thibetanus*) in Japan, in comparison with large stable populations. *Conservation Genetics* 8: 1331–1337.

Ohnishi, N., Uno, R., Ishibashi, Y., Tamate, H. & Oi, T. (2009). The influence of climatic oscillations during the Quaternary Era on the genetic structure of Asian black bears in Japan. *Heredity* 102: 579.

Paetkau, D., Calvert, W., Stirling, I. & Strobeck, C. (1995). Microsatellite analysis of population structure in Canadian polar bears. *Molecolar Ecology* 4: 347–354.

Palsbøll, P. J., Berube, M., Skaug, H. J. & Raymakers, C. (2006). DNA registers of legally obtained wildlife and derived products as means to identify illegal takes. *Conservation Biology* 20: 1284–1293.

Phillips, S. J., Anderson, R. P. & Schapire, R. E. (2006). Maximum entropy modeling of species geographic distributions. *Ecological Modelling* 190: 231–259.

Pritchard, J. K., Stephens, M. & Donnelly, P. (2000). Inference of population structure using multilocus genotype data. *Genetics* 155: 945–959.

Saitoh, T., Ishibashi, Y., Kanamori, H. & Kitahara, E. (2001). Genetic status of fragmented populations of the Asian black bear *Ursus thibetanus* in western Japan. *Population Ecology* 43: 221–227.

Sato, Y., Itoh, T., Mori, Y., Satoh, Y. & Mano, T. (2011). Dispersal of male bears into peripheral habitats inferred from mtDNA haplotypes. *Ursus* 22: 120–132.

Straka, M., Paule, L., Ionescu, O., Štofík, J. & Adamec, M. (2012). Microsatellite diversity and structure of Carpathian brown bears (*Ursus arctos*): consequences of human caused fragmentation. *Conservation Genetics* 13: 153–164.

Tammeleht, E., Remm, J., Korsten, M., et al. (2010). Genetic structure in large, continuous mammal populations: the example of brown bears in northwestern Eurasia. *Molecolar Ecology* 19: 5359–5370.

Uno, R., Kondo, M., Yuasa, T., et al. (2012). Assessment of genotyping accuracy in a non-invasive DNA-based population survey of Asiatic black bears (*Ursus thibetanus*): lessons from a large-scale pilot study in Iwate prefecture, northern Japan. *Population Ecology* 54: 509–519.

Uno, R., Doko, T., Ohnishi, N. & Tamate, H. B. (2015). Population genetic structure of the Asian black bear (*Ursus thibetanus*) within and across management units in northern Japan. *Mammal Study* 40: 231–244.

Yamamoto, T., Oka, T., Ohnishi, N., et al. (2012). Genetic characterization of northernmost isolated population of Asian black bear (*Ursus thibetanus*) in Japan. *Mammal Study* 37: 85–92.

Yasukochi, Y., Nishida, S., Han, S-H., et al. (2009). Genetic structure of the Asiatic black bear in Japan using mitochondrial DNA analysis. *Journal of Heredity* 100: 297–308.

Yoneda, M. (2001). Local population and conservation of Asian black bear; habitat preference and minimum area size. *The Japanese Institute of Landscape Architecture* 64: 314–317.

Chapter 26

Captive Bears in Asia: Implications for Animal Welfare and Conservation

Jan Schmidt-Burbach, Fakhar-i-Abbas, and Neil D'Cruze

For thousands of years, bears have been kept in captivity to generate income for people. To this day, throughout Asia, three main types of commercial use persist: bear baiting, bear dancing, and bear bile farming. Through these uses tens of thousands of bears are being subjected to conditions that are highly detrimental to their welfare through physiological and psychological impairment during capture, transport, training, and captive conditions. Furthermore, the vast majority of these practices depend on restocking their captive population from wild-caught bears or, where this is not the case, continue to drive a trade that also incentivizes poaching of bears in the wild.

While some practices have been made illegal across several range countries, the involvement of influential power players (bear-baiting landlords) or high-value industries (bear farms and bear bile manufacturers in China) pose challenges in ending these practices that are so detrimental to bears. All bear species endemic to Asia, as well as most brown bear populations in Asia, are recognized as threatened with extinction and listed under CITES Appendix I, making any international commercial trade in live animals or their parts illegal. Unfortunately, some countries maintain national legislation that allows for commercial trade or use of bears within the country. Furthermore, illegal international trade is continuing to pose challenges to law enforcement and the conservation and protection of bears worldwide. In this chapter we draw attention to the historic backgrounds, current situation, and holistic approaches that are needed to address these complex issues that are impeding bear welfare and conservation in Asia.

Bear Dancing

An animal that has been revered and persecuted, cherished and exploited, at times perhaps in almost equal measure, humans have used bears for entertainment since time immemorial (Tünaydin 2014; Figure 26.1). The earliest archaeozoological evidence suggests that bear dancing dates back to the ancient civilizations of the Near East, and bears have been used in performances since the second millennium BC (Berland & Salo 1986). Exactly why humans have found bears so compelling throughout the ages is difficult to decipher. However, it has been suggested that this fascination has been fuelled in part by virtue of their charisma, anthropomorphic posture, occasional upright position, and omnivorous diet, which have all contributed to forming legends of kinship (Tünaydin 2014).

There are abundant, albeit sporadic, accounts of bear dancing across Europe, Near and Southern Asia in the literature throughout the centuries. The practice involves training bears to "imitate human actions, dance and perform a variety of tricks, contrary to their nature" for the enjoyment and entertainment of paying audiences (Seshamani & Satyanarayan 1997; Tünaydin 2014; Kumari et al. 2016). It is a traditional occupation most commonly affiliated with a number of different nomadic peoples, including Romani communities in Europe and Kalandar communities in India (Berland & Salo 1986; Marushiakova & Popov 1998). At times, bear dancing has been equally fashionable and enjoyed by both patrons in royal courts and spectators on public streets alike (Mitrovic 1862; Ritvo 1987).

The dancing bear profession threatens the conservation of remaining wild bear populations because bear cubs are typically poached to supply this practice (D'Cruze et al. 2011; Kumari et al. 2016). In some cases, the poachers deliberately target adult bears for their claws, hair, teeth, and other body parts so that they can later be sold as amulets by dancing bear owners both during and after the performance (Seshamani & Satyanarayan 1997; D'Cruze et al. 2011). Historically, the three main bear species associated with bear dancing are the brown bear (*Ursus arctos*), the sloth bear (*Melursus ursinus*), and the Asiatic black bear (*Ursus thibetanus*), the latter two of which are listed as being Vulnerable to extinction according to the IUCN Red List of Threatened Species (Dharaiya et al. 2016).

The practice of bear dancing also raises serious animal welfare concerns. Further to the suffering endured during capture from the wild and transport along the trade chain, bear cubs are also subjected to mutilation when their muzzle is pierced with a hot iron spike and when their canines are removed (D'Cruze et al. 2011; Figure 26.2). The suffering continues thereafter when aversive conditioning is used to make them perform for public spectators (Seshamani & Satyanarayan 1997; Kumari et al. 2016). Undoubtedly, the reduced availability of wild bear cubs and negative attitudes toward nomadic communities have influenced policy decisions on bear dancing (Tünaydin 2014). However, arguably it is

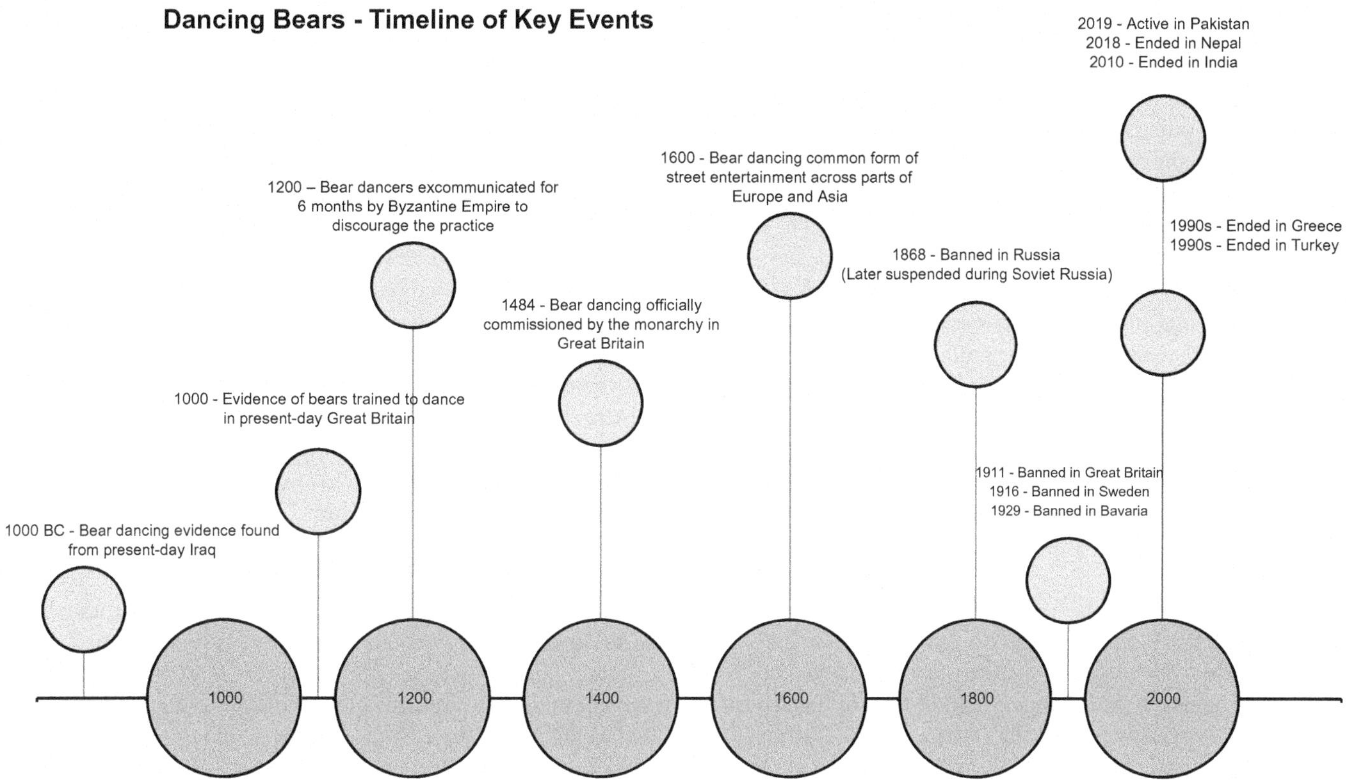

Figure 26.1 A timeline showing some of the key historical events that have taken place over the millennia in relation to the practice of bear dancing.

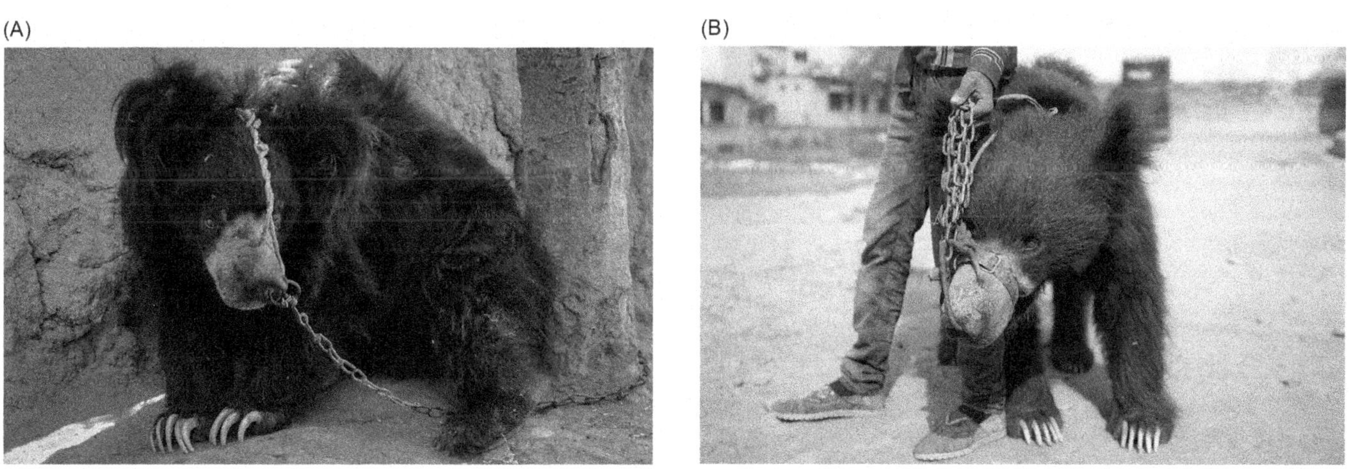

Figure 26.2 Dancing bears in India (A) and Nepal (B) with customary nose ring and ropes. (*Source*: World Animal Protection/Neil D'Cruze.) (A black and white version of this figure will appear in some formats. For the color version, please refer to the plate section.)

heightened public sensitivities to the animal cruelty involved that has been one of the main driving forces for the relatively recent global declines of this practice.

Perhaps one of the earliest expressions of the growing concern for the humane treatment of bears comes from Russia, where in 1866 Tsar Alexander issued the Imperial Edict on "Bear Comedies" that prohibited performances specifically on account of the training and mutilation methods involved. Although this prohibition was suspended indefinitely during the Soviet period, today bear dancing remains restricted to Russian circuses and is rarely practiced on the streets, if at all (Tünaydin 2014). Likewise, various governments across Europe and Asia also introduced prohibitions from the nineteenth century onward, citing animal welfare concerns but also bear conservation concerns too (*New York Times* 1929; Parliament of India 1972; Government of Nepal 1973).

Despite the almost worldwide abolition of bear dancing by the mid-1970s, in recent decades the practice remained a common and popular form of street entertainment. In response, since the late 1990s, a number of national and international wildlife organizations have focused their attention on the dancing bear issue (D'Cruze et al. 2011). Their collective efforts to improve law enforcement, provide humane lifetime care for confiscated bears, increase public awareness about the issue, and provide sustainable alternative livelihoods for bear owners resulted in a significant decrease in the number of dancing bears across Europe and Asia and have in fact ended the practice in several countries including Greece, Turkey, India, and Nepal (Kumari et al. 2016; Bale 2017). These successful models are currently being adapted to address bear dancing in remote areas of Pakistan, which remains its last remaining bastion (World Animal Protection 2018a).

Bear Baiting

The practice of bear baiting as entertainment involves the setting of dogs on a bear that has been tethered to the ground in an arena. The practice originally started in medieval Britain and rose to popular fame in the sixteenth century, where purpose-built arenas were constructed in districts of London to entertain spectators with fights between bears, dogs, chimpanzees, and bulls. It was not until the eighteenth century that bear baiting slowly fell out of favor and was banned in the UK eventually in 1835 on the basis of animal cruelty (Andrews 2017). At that time, however, the practice had already set foot on the Indian subcontinent. It is believed that when Britain was establishing its rule in the subcontinent, the colonial forces encountered resistance from warlords in Punjab. Knowing the warlords' fondness of dog fights, British army officers invited them to unique events to test their dogs' strength against tethered bears. Finding common interest, this practice paved the way to a more supportive environment for the British rulers in this area and bear baiting continued to be practiced. Toward the end of the twentieth century the cruel practice of bear baiting became internationally known again through the efforts and an investigation by World Animal Protection in 1993 identifying over 2400 bear fights held at 80 annual events throughout Pakistan (Farooq & Chaudhry 1993). Despite being illegal through federal and provincial legislation, bear baiting persisted due to the strong influence of local landlords.

Commonly, brown bears (*U. a. isabellinus*) and Himalayan black bears (*U. thibetanus laniger*) destined for this practice are poached as cubs by killing the mothers, removing most of the young bears' teeth and claws, and training them to become dancing and baiting bears. While Himalayan black bears were once widely distributed in Pakistan they were decimated heavily in the 1990s, leaving populations only in remote northern districts of Pakistan. In the traders' desperate search for more bears, they enrolled local livestock herders, who quickly learned that killing bears around their herds' remote grazing grounds was profitable, selling live cubs as well as the mother bears' skin, fat, or gall bladder to the traders. The traders, often also dealing in other wild animals, would then sell the bear cubs via elaborate trade routes to indigenous, nomadic people that specialize in using bears for entertainment. These "Kalandars" would then be hired by influential landlords throughout Pakistan's provinces of Punjab and Sindh to contribute their bear to a landlord's bear-baiting event. Until recently, such events often occurred during public fairs attracting thousands of people. Landlords hoped to increase or display their influence through hosting such public bear-baiting events. During an event the bear, unable to defend itself, would be tied down in the middle of an arena on a 5-m long rope, and two or more trained fighting dogs belonging to the landlord would be set upon the bear. After three minutes the dogs would be separated forcefully from the bear and a jury decided the winner. A bear often has to fight three times or more against different dogs during one event (Farooq & Chaudhry 1993). After an event the injured bear would receive no medical treatment and the Kalandar would lead the bear away on its leash, tied to a nose ring. Bears usually do not live longer than 10 years under these poor conditions and are prone to contracting a variety of diseases due to close proximity to people and livestock, including parasitic infections like *Trypanosoma evansi* – normally a rare occurrence in bears (Muhammad et al. 2007).

NGO-led work to end this practice has resulted in a significant decline of this practice by 2018 (Figure 26.3). Through building public opposition against the practice, encouraging religious leaders and politicians to speak out, tightening legislation, enabling alternative livelihoods for the bear owners, and providing sanctuary for surrendered or confiscated bears, the practice has essentially disappeared from the public arena (World Animal Protection 2018a). Only a handful of events continue to be carried out underground by influential landlords, but improved legislation and increasing law enforcement activities are closing in on those remaining organisers.

Nonetheless, demand from secretly organized bear-baiting events persists and the established illegal trade network that reaches far into remote bear habitats remains a threat to the endemic populations. Particularly, the targeted poaching of young bear cubs has jeopardized the future reproductive potential of wild bear populations. It is essential that Pakistan continues efforts to eradicate the last remainders of practices that exploit bears for bear baiting or dancing and maintains strong law enforcement to protect wild bears in their habitat.

Bear Bile Farming

The use of bear parts as medicine has a long history in Asia, dating back thousands of years with references in some of the earliest official pharmacopoeia in China (Huang et al. 1994; Li et al. 1995; Watkins 2002; Dutton et al. 2011). Originating in China, the medical use of bear parts was then adopted by Japan

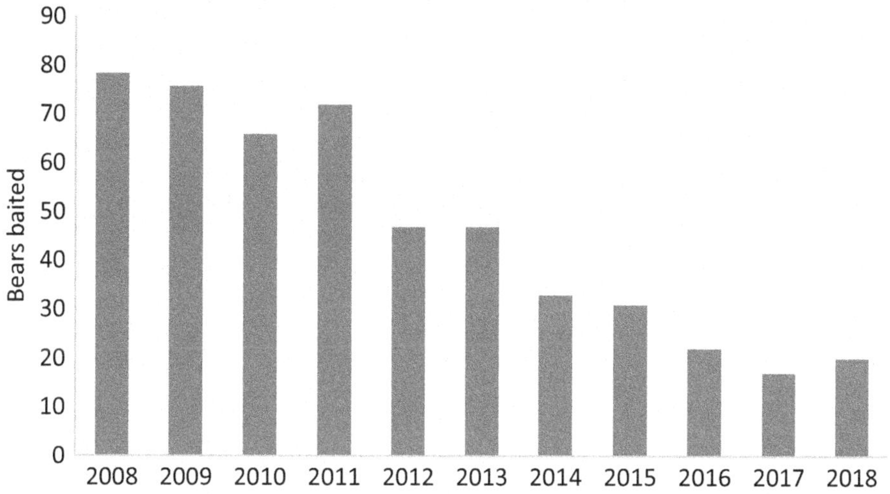

Figure 26.3 Number of bears baited per year in Pakistan. (*Source*: World Animal Protection.)

and Korea several centuries ago. Today, the consumption and trade of bear products for traditional medicinal purposes is widespread throughout Asia and Asian communities worldwide (Bhiksu et al. 2002; Feng et al. 2009). In particular, gall bladder and bile are commonly used parts in some traditional medicine products, thanks to their high concentration of ursodeoxycholic acid (UDCA) and tauroursodeoxycholic acid (TUDCA). UDCA and TUDCA are used to treat liver and gall bladder diseases, but are also claimed to be effective for some ocular or neuronal diseases (Paumgartner & Beuers 2002; Boatright et al. 2009). Herbal alternatives to bear bile exist in traditional Chinese preparations but have yet to become more widely promoted (World Animal Protection 2005).

Since the 1950s both acids can be synthesized on the basis of common cholic acid, extracted from livestock (Kanazawa et al. 1954). UDCA and TUDCA in western medicine products are entirely obtained from such synthesized sources. However, in Asia the primary source for these ingredients has remained bile and gall bladder from bears. In the 1980s North Korea developed a method for extracting bile from live bears and due to the high demand for bear bile in China, this practice quickly spread across the border, where bears were then poached in the wild to be put into cages to extract their bile (Mills & Servheen 1994; Watkins 2002). Within a few decades this practice developed into a sophisticated industry, relying on approximately 25,000 captive bears across China, South Korea, Vietnam, Laos, and Myanmar (World Animal Protection 2017). The most common bear species found in those facilities are Asiatic black bears. These bears suffer from extremely poor husbandry conditions and cruel extraction procedures, leading to the development of severe injuries, illnesses, and behavioral problems (Figure 26.4; Maas 2000; Li 2004; Loeffler et al. 2009). Several bear sanctuaries were established by NGOs in China, Vietnam, and Laos to provide up to approximately 450 confiscated bears a more adequate quality of life (Animals Asia Foundation 2018; Free the Bears 2018).

Bear farms were in the first place established to produce larger amounts of bear bile and not necessarily to protect wild bears, yet it was hoped by some conservationists that the latter would be a positive side effect by alleviating pressure on wild populations. However, this does not appear to be the case, as the increase in supply increased the demand for bear bile products at the same time; therefore, bears on farms are often restocked with wild-caught bears and there is a reported consumer preference for wild-sourced product (Dutton et al. 2011; Crudge et al. 2020). This continues to pose a threat for wild bear populations through poaching across Asia and other regions, while maintaining serious animal welfare concerns for the captive condition of bears on farms (TRAFFIC 2006; Shepherd & Nijman 2008; Shepherd & Shepherd 2010; Dutton et al. 2011; Livingstone & Shepherd 2016; Livingstone et al. 2018). Products containing bear bile were not only available within countries with legal bear farms, but also in other countries across the world. While the trade in bear bile is legal within China, Asiatic black bears and their parts are listed under CITES Appendix I and thus their international commercial trade is illegal. Despite this, investigations have evidenced illegal bear bile products in retail shops in the USA,

Figure 26.4 Asiatic black bears in a bear farm in Vietnam. (*Source*: World Animal Protection/Jan Schmidt-Burbach.)

Canada, Australia, and Japan, either illegally imported from China or Vietnam, or sourced from locally hunted wild bears (World Animal Protection 2007, 2018b; Montreal Gazette 2018). In 2012, at the World Conservation Congress in Jeju, a recommendation was issued to urge range states to either continue to phase out bear farming or at least not further expand their industry and allow for more transparency (IUCN 2012). Efforts by international NGOs have led to some milestone successes by ending bear bile farming in South Korea and reducing numbers of bears on farms in Vietnam from 4300 in 2005 to about 800 in 2018 (World Animal Protection 2017; Education for Nature – Vietnam 2018). However, in China the industry persists and maintains farms with approximately 22,000 bears, despite policy recommendations having been proposed to the State Council to phase-out the bear bile industry (Development Research Center of the State Council 2016). Progress can be seen, though, in over 1900 Chinese pharmacies pledging to not sell bear bile products and the largest manufacturer of medical bear bile products has announced its success in developing a chicken bile-based synthetic product to be available from 2022 (Animals Asia Foundation 2014; Hance 2015).

References

Andrews, E. (2017). The gruesome blood sports of Shakespearean England. History.Com. Available from www.history.com/news/the-gruesome-blood-sports-of-shakespearean-england.

Animals Asia Foundation. (2014). Thousands of Chinese pharmacies reject bear bile products. Available from www.animalsasia.org/intl/media/news/news-archive/thousands-of-chinese-pharmacies-reject-bear-bile-products.html.

Animals Asia Foundation. (2018). Bear sanctuaries. 2018. Available from www.animalsasia.org/uk/our-work/bear-sanctuaries/.

Bale, R. (2017). Nepal's last known dancing bears rescued. National Geographic, December 21, 2017, sec. Wildlife Watch. Available from https://news.nationalgeographic.com/2017/12/wildlife-watch-nepal-dancing-bears-rescued/.

Berland, J. C. & Salo, M. T. (1986). Peripatetic communities: an introduction. *Nomadic Peoples* 21/22: 1–6.

Bhiksu, W. H., Bjorkenstam, M. B., Yu-Min, C., et al. (2002). *The bear bile business: The global trade in bear products from China to Asia and beyond.* London: World Animal Protection.

Boatright, J. H., Nickerson, J.M., Moring, A. G. & Pardue, M.T. (2009). Bile acids in treatment of ocular disease. *Journal of Ocular Biology, Diseases, and Informatics* 2(3): 149–159. https://doi.org/10.1007/s12177-009-9030-x.

Crudge, B., Trang Nguyen & Tien Trung Cao. (2020). The challenges and conservation implications of bear bile farming in Viet Nam. *Oryx* 54: 252–259.

D'Cruze, N., Sarma, U. K., Mookerjee, A., et al. (2011). Dancing bears in India: a sloth bear status report. *Ursus* 22(2): 99–105.

Development Research Center of the State Council. (2016). *Research on the current situation of Chinese bear-bile industry and strategies for transition.* Beijing: Development Research Center of the State Council.

Dharaiya, N., Sharp, T. & Bargali, H. S. (2016). *Melursus ursinus.* The IUCN Red List of Threatened Species. http://dx.doi.org/10.2305/IUCN.UK.2016-3.RLTS.T13143A45033815.en.

Dutton, A. J., Hepburn, C. & Macdonald, D. W. (2011). A stated preference investigation into the Chinese demand for farmed vs. wild bear bile. *PLoS ONE* 6(7): e21243.

Education for Nature – Vietnam. (2018). Public urged to reject using bear bile and help end bear farming in Vietnam. Available from https://envietnam.org/images/News_Resources/Press_Releases/MS-Teddy-Bear-PSA.pdf.

Farooq, T. & Chaudhry, I. C. (1993). *Bear-baiting in Pakistan.* WSPA Report Series. London: World Animal Protection.

Feng, Y., Kayu, S., Ning, W., et al. (2009). Bear bile: dilemma of traditional medicinal use and animal protection. *Journal of Ethnobiology and Ethnomedicine* 5(1): 2.

Free the Bears. (2018). Laos sanctuaries. Available from https://freethebears.org/pages/laos-sanctuaries.

Gelb, I. J. (1975). *Homo ludens* in early Mesopotamia. *Studia Orientalia Electronica* 46: 43–76.

Government of Nepal. (1973). National Parks and Wildlife Conservation Act. Government of Nepal.

Hance, J. (2015). Is the end of "House of Horror" bear bile factories in sight? *The Guardian*, April 9, 2015, sec. Environment. Available from www.theguardian.com/environment/radical-conservation/2015/apr/09/bear-bile-china-synthetic-alternative.

Huang, X., Zhufan X. & Zhicen L. (1994). *Classified dictionary of traditional Chinese medicine.* Beijing: New World Press.

IUCN. (2012). *Bear farming in Asia, with particular reference to the conservation of wild populations.* Jeju, South Korea: IUCN.

Kanazawa, T., Akio, S., Tetsuo, S. & Toshio H. (1954). Syntheses of ursodesoxycholic acid and its conjugated bile acid. *Proceedings of the Japan Academy* 30(5): 391–392.

Kumari, I., Mookerjee, A., Bhagat, S., et al. (2016). *A dance to forget – The story of the eradication of sloth bear (*Melursus ursinus*) dancing from India.* Conservation Action Series 13. Delhi, India: Wildlife Trust of India.

Li, P. J. (2004). China's bear farming and long-term solutions. *Journal of Applied Animal Welfare Science* 7(1): 71–81.

Li, Ya-w., Xiu-yuan, Z., Paul Pui-Hay, B. & Hin-Wing, Y. (1995). Ethnopharmacology of bear gall bladder: I. *Journal of Ethnopharmacology* 47(1): 27–31. https://doi.org/10.1016/0378-8741(95)01249-D.

Livingstone, E. & Shepherd, C. R. (2016). Bear farms in Lao PDR expand illegally and fail to conserve wild bears. *Oryx* 50(1): 176–184.

Livingstone, E., Gomez, L. & Bouhuys, J. (2018). A review of bear farming and bear trade in Lao People's Democratic Republic. *Global Ecology and Conservation* 13: e00380.

Loeffler, I. K., Robinson, J., Cochrane, G., et al. (2009). Compromised health and welfare of bears farmed for bile in China. *Animal Welfare* 18(3): 225–235.

Maas, B. (2000). *The veterinary, behavioural and welfare implications of bear farming in Asia*. London: World Animal Protection.

Marushiakova, E. & Vesselin P. (1998). Bear-trainers in Bulgaria (tradition and contemporary situation). *Ethnologia Bulgarica* 1: 106–116.

Mills, J. & Servheen, C. (1994). The Asian trade in bears and bear parts: impacts and conservation recommendations. *International Conference on Bear Research and Management* 9: 161–167.

Mitrovic, V. V. (1862). *Adventures of Baron Wenceslas Wratislaw of Mitrowitz: What he saw in the Turkish Metropolis, Constantinople; Experienced in his captivity; And after his happy return to his country, committed to writing in the year of our Lord 1599*. Bell & Daldy.

Montreal Gazette. (2018). Black bear poaching operation busted by Quebec Wildlife Ministry. *Montreal Gazette*. Available from https://montrealgazette.com/news/local-news/black-bear-poaching-operation-busted-by-quebec-wildlife-ministry.

Muhammad, G., Saqib, M., Sajid, M. S. & Naureen, A. (2007). *Trypanosoma evansi* infections in Himalayan black bears (*Selenarctos thibetanus*). *Journal of Zoo and Wildlife Medicine* 38 1): 97–100.

New York Times. (1929). Bavaria bans dancing bears. *New York Times* November 24, 1929.

Parliament of India. (1972). The Wildlife (Protection) Act. Government of India.

Paumgartner, G. & Beuers, U. (2002). Ursodeoxycholic acid in cholestatic liver disease: mechanisms of action and therapeutic use revisited. *Hepatology (Baltimore, MD)* 36(3): 525–531. https://doi.org/10.1053/jhep.2002.36088.

Ritvo, H. (1987). *The animal estate: The English and other creatures in the Victorian age*. Cambridge, MA: Harvard University Press.

Seshamani, G. & Satyanarayan, K. (1997). *The dancing bears of India*. London: World Animal Protection (WSPA).

Shepherd, C. R. & Nijman, V. (2008). The trade in bear parts from Myanmar: an illustration of the ineffectiveness of enforcement of international wildlife trade regulations. *Biodiversity and Conservation* 17(1): 35–42.

Shepherd, C. R. & Shepherd, L. A. (2010). The poaching and trade of Malayan sun bears in peninsular Malaysia. *Traffic Bulletin* 23(1): 49–52.

TRAFFIC. (2006). *Proceedings of the Fourth International Symposium on the Trade in Bear Parts*. Nagano, Japan: TRAFFIC East Asia – Japan.

Tünaydin, P. (2014). Pawing through the history of bear dancing. *Frühneuzeit-Info* 24: 51–60.

Watkins, V. (2002). Historical perspective to the bear trade. In *The bear bile business: The global trade in bear products from China to Asia and beyond*. London: World Animal Protection.

World Animal Protection. (2005). Finding herbal alternatives to bear bile. World Animal Protection. Available from https://d31j74p4lpxrfp.cloudfront.net/sites/default/files/ca_-_en_files/findingherbalalternativestobearbilev3_tcm22-2705.pdf.

World Animal Protection. (2007). *From cage to consumer – The global trade in bear parts from China to Asia and beyond*. London: World Animal Protection.

World Animal Protection. (2017). *Ending bear bile farming in South Korea*. London: World Animal Protection. Available from www.worldanimalprotection.org/sites/default/files/int_files/bear_bile_white_paper.pdf.

World Animal Protection. (2018a). *Ending bear baiting*. London: World Animal Protection International. Available from www.worldanimalprotection.org/our-work/animals-wild/ending-bear-baiting.

World Animal Protection. (2018b). *Unbearable: The international bear bile trade*. London: World Animal Protection.

Chapter 27

Human Dimensions of Asiatic Black Bear Conflicts and Management in Japan

Ryo Sakurai

Introduction

The Asiatic black bear (*Ursus thibetanus*; hereafter black bear) is categorized as a Vulnerable Species by the IUCN Red List of Threatened Species (IUCN 2018). Black bears were widespread in three of the main Japan islands, i.e. Honshu, Shikoku, and Kyushu; however, because of a post-1950 increase in both human population and land use, black bears have probably been extirpated in Kyushu and are endangered in Shikoku (Onishi 2011). On the other hand, many rural areas in Japan have been recently experiencing depopulation and aging, leading to an increase in abandoned agricultural fields and villages. Hunter populations have also significantly decreased in Japan (Kaji et al. 2013). These trends have led to an increase in the population and habitat availability of some wildlife species, such as deer (*Cervus nippon*) and wild boars (*Sus scrofa*) in Japan, and black bear habitat availability has also increased since the 1970s (Onishi 2011).

The recent increase in human–bear conflicts has caused local residents living close to bears to often fear them and have negative attitudes toward bears in Japan (Kaneko & Kanamori 2007). Human–bear conflicts include agricultural and forestry damage, as well as human casualties (Ministry of the Environment 2008). As black bears mostly rely on mast (tree fruits), in lean years when mast production is relatively low they are known to move outside forests area to look for food (Yokoyama 2009). In those years (such as 2004, 2006, and 2010), bears frequently appear near human settlements and human–bear conflicts increase, mainly because of the increase of abandoned agricultural fields and fruit trees, as well as cover (e.g. bush) around human settlements. For example, in 2010, when a large number of bears appeared around human settlements, 3558 black bears were captured and 3074 of those were killed (Ministry of the Environment 2016a). In the same year, 142 people were injured and two people were killed by black bears (Ministry of the Environment 2016b).

Although one of the biggest challenges in sustainable and effective wildlife management is garnering and maintaining public support and understanding (Bruskotter & Shelos 2010), in Japan, residents' negative attitudes toward bears represent a major obstacle to sustainable bear management (Yokoyama 2009). This is crucial for bear conservation and management because citizen participation in implementing damage prevention activities is necessary to mitigate human–bear conflicts. Therefore, conducting human dimensions studies on bear management is important to understand public perceptions and to encourage participatory community-based wildlife management. In Japan, the field of human dimensions of wildlife management is not yet widespread, although studies in this area are increasing (Sakurai & Enari 2010; Sakurai 2019a).

This chapter describes human dimensions studies on black bear management conducted in Japan, by reporting (1) a study on the media coverage regarding black bear issues in Japan; (2) the results from interviews and surveys that were conducted on local residents of the northern part of Hyogo Prefecture regarding their attitudes and behaviors related to human–bear conflicts; (3) the effectiveness of community seminars in mitigating human–black bear conflicts; and (4) how human–bear conflicts could be used to educate university students and train them to potentially become wildlife managers.

Media Coverage on Human–Black Bear Conflicts

As people coexisting with bears tend to fear and have negative attitudes toward bears, local residents frequently call local officials to kill or capture bears when they see them (Huygens et al. 2001). Media coverage of human–large carnivore conflicts affects people's perceptions of predators (Bombieri et al. 2018). In general, the media affects people's perceptions through agenda-setting (selection of newsworthy stories that leads the audience to have "media-induced" views) and framing effects (affecting how the audience understands them) (Price et al. 1997). In Japan, human–bear conflicts are frequently covered in the media, especially in years when massive numbers of bears appeared around human settlements and many people were injured by bear attacks. Actually, Japan is said to be a "newspaper country," with some of the largest newspaper circulation in the world (WorldAtlas 2018) and, thus, mass media might negatively affect people's perceptions of bears.

In a content analysis of newspaper articles (Sakurai et al. 2013a), I chose the national newspaper *Yomiuri*, which had the largest circulation in Japan, and the local newspaper, *Shinano*

Table 27.1 Percentages of the most frequently mentioned topics and human–Asiatic black bear conflicts reported in the national (*Yomiuri*) and local (*Shinano Daily*) newspapers.[1]

		Frequencies (%)			
		National papers		Local papers	
		2005 (*n* = 99)	2006 (*n* = 100)	2005 (*n* = 31)	2006 (*n* = 118)
Most frequently mentioned topic	Problematic interactions with bears	49	57	32	69
	Government response to a bear incident	40	43	58	59
	Bear sighting	40	54	58	43
Most frequently mentioned conflicts	Fear of human injuries caused by bears	46	46	52	61
	Damage caused by bears to property	38	29	32	49
	Human injuries or deaths caused by bears	35	24	23	42

[1] Based on Sakurai et al. (2013a).

Daily, which is published in Nagano Prefecture, where the highest number of bear captures (704 bears) and human casualties occurred (16 cases) across the country in 2006. I targeted the period April 2006–March 2007, as this period experienced one of the highest records of black bear captures/killings and one of the highest numbers of human casualties. I also compared 2006 articles to 2005 articles (when relatively few bears appeared in residential areas), to understand how newspapers portrayed bears and bear issues in years with different amounts of conflict.

All article headlines that mentioned black bears were included in the analysis. The number of collected articles was 99 for *Yomiuri* in 2005 and 100 in 2006, and 31 for *Shinano Daily* in 2005 and 118 in 2006. The majority of articles (65%) focused on specific incidents of human–bear conflicts rather than on the more general thematic of management or ecological characteristics of bears, thus rarely providing information about general management actions (only 26% of the collected news). As expected by the higher number of conflicts which occurred in 2006, more articles mentioned threats or bear attacks in 2006 (61%) than in 2005 (36%). The local newspaper published more articles regarding bear attacks than did the national paper (63% for the local paper and 43% for the national paper), i.e. in years with massive bear appearances and increased number of conflicts, more negative articles on black bears were published in the local newspaper (see Table 27.1 for a list of common topics and conflicts).

These results show that newspaper articles regarding black bears were mostly about negative interactions with bears and did not provide useful information on how to solve conflicts. Thus, more objective and scientific information about bears should be reported in the media. For example, the media should also provide information on the ecological characteristics of bears, the reasons for their appearances, as well as practical management and conservation policies. This might help people to develop a better understanding of bear issues and encourage them to support sustainable bear management (Bruskotter & Wilson 2014).

Perceptions of Black Bears in Northern Hyogo Prefecture: Qualitative and Quantitative Approaches

Hyogo Prefecture (Figure 27.1) is inhabited by the "eastern Chugoku" and "northern Kinki" populations of black bears, which were categorized as endangered requiring rigorous conservation measures in the 2003 Hyogo Prefecture Red Data Book (Hyogo Prefecture 2003). Under the Black Bear Conservation and Management Plan (2003, 2007) (Hyogo Prefecture 2003, 2009) bears that cause damage are captured for behavioral modification and released in deep forest after aversive conditioning (e.g. scaring the bear with noisy firecrackers) and are only killed if this approach is ineffective or if the risk of human injury is considered high (Hyogo Prefecture 2017).

In 2007, the Wildlife Management Research Center in Hyogo was established to implement initiatives aimed to reduce the damage caused by black bears. In fact, after the black bear population increased to approximately 650 individuals in 2011 (Sakata et al. 2011), conflicts increased between local residents and bears. For example: (a) agricultural damages were calculated as a loss of approximately $180,000 in 2006 (Hyogo Prefecture 2009); (b) between 1998 and 2009 five human casualties were reported in Kami (Hyogo Prefecture 2009); and (c) in 2010, owing to low food availability in forests, 1623 bear sightings occurred around Kami and Tanto (Wildlife Management Research Center, Hyogo 2011).

The Management Plan set out by the Hyogo Prefecture involves actions to prevent bear encounters, human injury, and damage to agriculture and forestry, and to encourage local

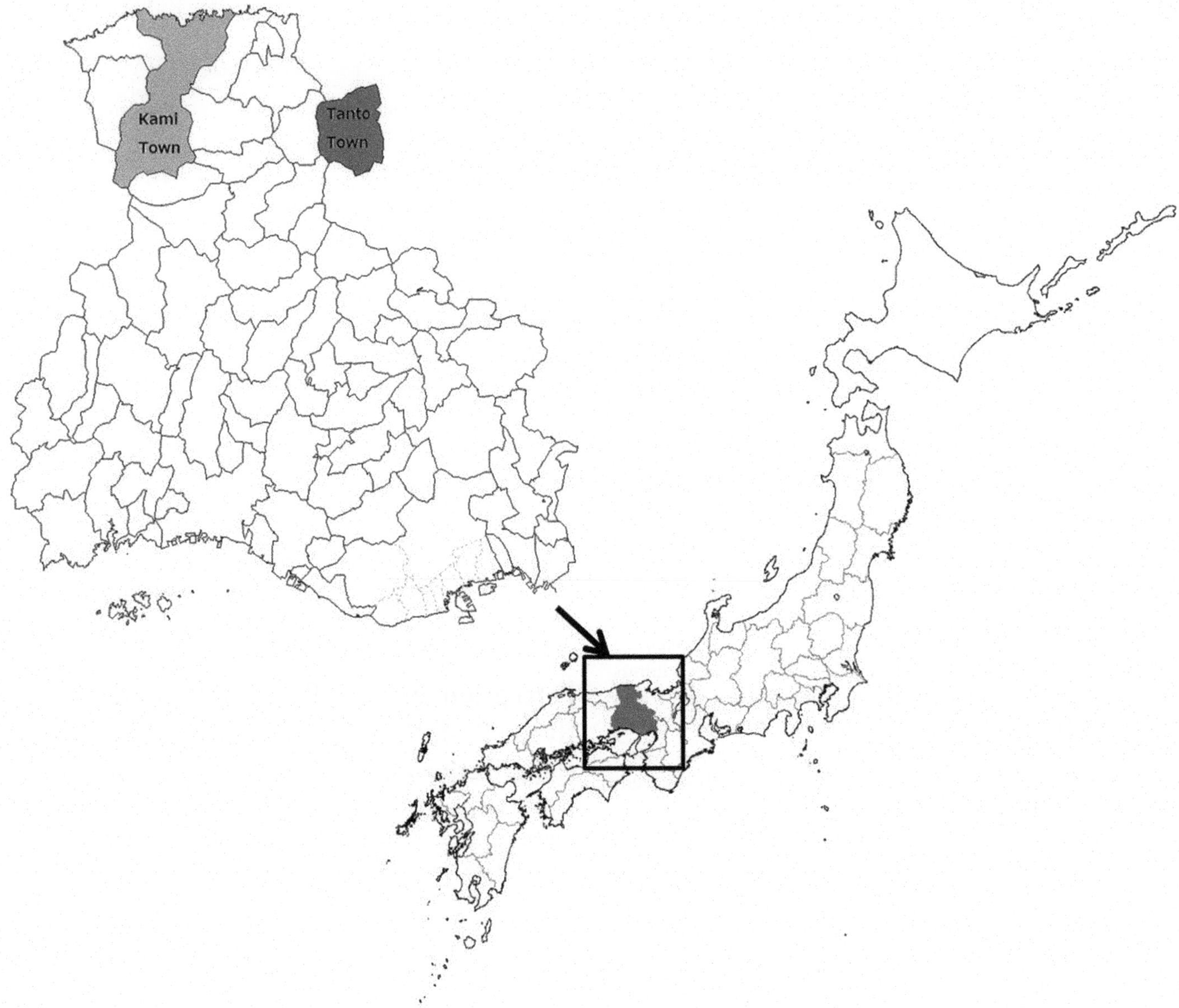

Figure 27.1 Location of Hyogo Prefecture (shaded area, right map). The inset map shows the Kami and Tanto towns in the prefecture (shaded areas, left map).

residents to report bear sightings to local government; these measures rely on the initiative of local residents. Additionally, local governments must educate citizens and encourage initiatives to reduce damage. For example, one management method is to wrap the trunks of persimmon and chestnut trees, which attract wildlife, with corrugated galvanized iron sheets that bears cannot climb. Such measures require the commitment of the entire community. However, little research has been conducted in those regions to understand how local residents were (or not) engaged in behaviors that could prevent wildlife damage and their perceptions related to these issues.

Therefore, in this study, qualitative and quantitative analyses were performed to characterize perceptions of bears and progress in their management. The first step of this research was to grasp the current situation on human–bear conflicts through understanding the relationship between local residents (communities) and bears. A qualitative study (interview) was conducted to document residents' narratives on human–bear interactions. A quantitative study (survey) aimed to reveal overall trends of residents' perceptions of bears as well as to evaluate progress in conflict resolution.

Qualitative Analysis: Interview and Group Discussion with Local Residents and Stakeholders (Based on Sakurai et al. 2011)

Highly diverse perceptions of bears make it difficult to build a consensus toward actions. In this study, semistructured interviews with residents and stakeholders were conducted to understand public perceptions. In Kami town (inhabited by the eastern Chugoku bear population) and Tanto, Toyooka City (inhabited by the northern Kinki bear population), interviews were conducted in 2010 with 22 stakeholders, including district chiefs, municipal government staff, and hunters. Each interview took approximately 90 minutes. In addition, interviews were conducted with 55 local residents in and around their homes. A group discussion was held with six stakeholders (a district chief, a president of an agricultural association, and four local farmers) for about 2 hours in one village of Tanto. In both interviews and group discussion, the in-depth interview technique was used to let people talk freely while answering questions on specific topics.

The interviews revealed that differences in perceptions mainly depended on the type and administrative charge of people who were interviewed. For example, typical statements made by the village chiefs were:

> I am not anxious about bears during the daytime, but residents may encounter bears in the morning and at night. I feel anxious when they go to work at around 6:00 a.m. I try to tell them to take safety measures.

> It is my role to properly communicate the residents' feelings to the local government.

However, local residents mainly affirmed that:

> Bears are not welcome. I am scared. Measures should be taken so that people can walk safely in bear habitats, even when alone.

> Bears come when mast is scarce. We encounter bears because the trees in the mountains have been cut down.

> Bears have been previously hunted and killed. Recently, bears are only captured and released, so the population is increasing.

Thus, residents are mostly afraid of bears and have the feeling that lifestyle changes are partially responsible for increased encounters with bears, e.g. residents recently stopped forest management, thus offering a better habitat for bears. Rapid depopulation and aging, particularly in Tanto, can partly explain these changes. Many residents also complained about bear damage to persimmon trees in both Kami and Tanto. Additionally, in a group discussion of six stakeholders including a district chief and the president of an agricultural association, negative feelings toward bears were related to the fact that bear encounters have become more frequent. Furthermore, stakeholders expressed feelings of resignation, believing that damage cannot be prevented. This leads to delayed responses by the local government, including a failure to wrap persimmon trees or report bear sightings.

As an end conclusion, local attitudes showed that residents often have negative feelings toward bears and criticize failures in management, although brief interviews are most likely to reveal negative perceptions. Yet, when taking the time for longer resident narratives, bears were considered as "neighbors" and, while scary, they were accepted as a part of daily life. Thus, if citizens feel supported by stakeholders, they will also feel the desire to take voluntary proactive actions, thereby creating a consensus toward solving community issues due to the bear presence.

Quantitative Analysis: Questionnaire Survey and the Potential Conflict Index (Based on Sakurai 2012; Sakurai et al. 2012a, 2013b)

A questionnaire survey was conducted to evaluate improvements in conflict resolution, current perceptions of bears, effects of awareness-raising campaigns by the government, and preferred government actions. The findings may guide future plans that minimize conflicts and policies that encourage engagement in damage prevention behaviors.

In total, 2316 questionnaire surveys were distributed in Tanto (where there have been no reported human casualties) and Kami (where five bear-related casualties were reported in 1998–2009). The following items were included in the survey: (1) implementation of damage prevention measures by residents; (2) damage prevention measures in each district; (3) sources of bear information and information residents would like to receive to mitigate human–bear conflicts; and (4) actions that residents want governments to implement when bears appear around human settlements. Some of the results were evaluated by a pentagon graph (Figure 27.2), which depicts the current state of each district (in terms of seriousness of human–bear conflicts) based on five evaluation items: (1) taking actions (degree of interventions implemented); (2) experience with bears (such as damage); (3) trust toward government; (4) risk perception (e.g. "human–bear conflicts are increasing"); and (5) problems in district (e.g. "reduction of manpower in the community due to aging and/or absence of young people").

The survey response rates were 48.4% (n = 868) in Tanto and 65.2% (n = 341) in Kami. Respondents were male-biased and skewed toward the elderly. More than 60% of respondents had seen at least one bear in the past 10 years and more than 10% of them reported four or more sightings. Fewer people reported black bear sightings (Kami = 59.1%, Tanto = 39.6%) and took actions to prevent damage in Tanto than in Kami (Kami = 66.4%, Tanto = 37.8%). As for the reasons for not reporting sightings, 40.1% of respondents in Kami mentioned "it is not mandatory to make a report" while in Tanto, 40.5% mentioned that "it is not surprising to see a bear." More than 80% of residents possessed persimmon and/or chestnut trees and approximately 40% suffered bear damage. Table 27.2 illustrates the information that residents would like to receive from the local administrations about bears.

The districts in Tanto and Kami with the highest rates of experience with bears did not differ in terms of experience with bears or risk perceptions or problems in the district (Figure 27.2), whereas there were considerable differences in taking action and trust toward government.

People's attitudes ranged widely from "do nothing" to "capture and kill" (Table 27.3). However, for bears on farmland that caused damage to crops or livestock, residents generally wanted bears to be captured and killed. In order to visually show the results for local residents' opinions, a Potential Conflict Index (PCI) was used (Figure 27.3). When a bear is seen in the mountains, the bubble located near "Capture and release the bear without aversive conditioning" means that, on average, local villagers have this perception about the rules that should be applied to face this situation. Moreover, the fact that the bubble is large with a high PCI score of 0.9 indicates that people's opinions were poles apart between "Capture and kill" and "Do nothing." On the other hand, when a bear appears in the district and attacks a person, the bubble is small with a PCI score of 0.2, meaning that most local residents have the same opinion ("Capture and kill" in this situation). Based on this PCI figure, we can predict which situations create polarized residential opinions, which may create difficulty in reaching consensus among residents.

When this study was conducted (2010), the government policy was to capture and release those bears that appeared on farmland and caused damage to crops or livestock. However, on average, people's perceptions were positioned between "Capture and release the bear with aversive conditioning" and "Capture and kill the bear." Therefore, the gap between people's demands and government policy becomes evident, which allows government officials to understand when residents are dissatisfied with governmental policies and interventions.

Differences in reporting and taking action between Tanto and Kami can be explained by differences in relationships with district chiefs and in human casualties. Improvements may result from projects initiated in the prefecture, such as the Wildlife Ranger Project for Damage Prevention (Sakurai 2019b), which provides volunteer support in depopulated and aging regions, and the Community Wildlife Adviser Development Program, which aims to train local experts against wildlife damage. Furthermore, the current analysis can serve as a basis for selecting districts that may benefit from future awareness-raising/educational activities.

Despite efforts worldwide to minimize conflicts between wildlife and humans, in-depth analyses of their impacts are lacking (Ferraro & Pattanayak 2006). The questionnaire survey results establish a baseline from which to measure and monitor the effectiveness of these programs (e.g. outreach activities implemented by government) in the future, as well as to set measurable goals. For example, according to the survey, less than 40% of residents report bear sightings in Tanto; accordingly, Hyogo Prefecture set a target of 80% in 2 years (unpublished information by Agriculture, Forestry and Fisheries Promotion Office of the Tajima District Administration Office, Hyogo Prefecture). Finally, it is important for researchers to clearly communicate survey results to managers and residents, which may also promote cooperative and involved residents and stakeholders.

Table 27.2 Types of information on how to mitigate human–bear conflicts[1] that local residents in Tanto and Kami towns would like to receive from local government.

	Frequencies (%)	
	Tanto town (n = 814)	Kami town (n = 319)
Regions where bears appear, frequency, and details of damage	519 (63.8)	199 (62.4)
Methods to protect crops and houses	349 (42.9)	145 (45.5)
Self-defense methods to prevent human injury	481 (59.1)	202 (63.3)
Government policies and budgets	108 (13.3)	64 (20.1)
Ecology of bears	134 (16.5)	68 (21.3)
Locations of caught and released bears	180 (22.1)	100 (31.3)
Nothing particular	61 (7.5)	12 (3.8)
Others	16 (2.0)	11 (3.4)

[1] Based on Sakurai et al. (2012a).

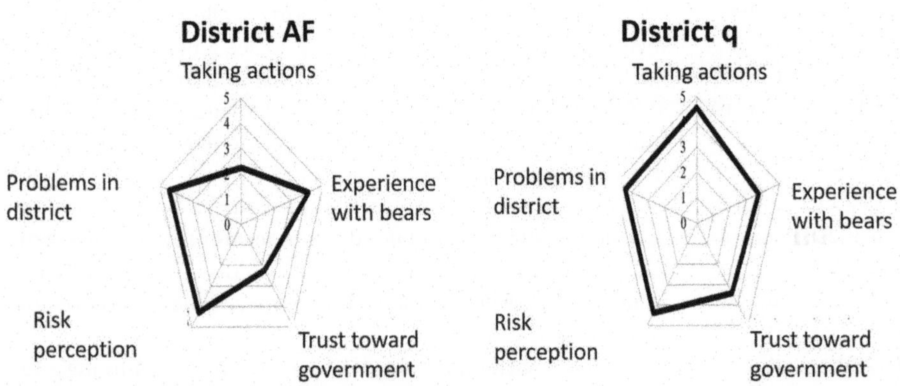

Figure 27.2 Pentagon graph of the current state of districts AF (in Tanto town) and q (in Kami town) in terms of seriousness of human–bear conflicts (based on Sakurai et al. 2012a).

Table 27.3 Actions that local residents in Tanto and Kami towns want their governments to implement when bears appear around human settlements.[1]

		Do nothing (%)	Threaten and drive the bear out (%)	Capture and release the bear without aversive conditioning (%)	Capture and release the bear with aversive conditioning (%)	Capture and kill the bear (%)
(1) When a bear is seen in the mountains	Tanto town (*n* = 773)	25.7	21.9	2.3	13.5	36.6
	Kami town (*n* = 289)	24.6	21.5	3.1	16.6	34.3
(2) When a bear appears in the district, i.e. in and around human settlements	Tanto town (*n* = 764)	4.8	23.6	2.4	18.2	51.0
	Kami town (*n* = 286)	3.5	16.8	4.5	23.8	51.4
(3) When a bear causes damage to crops or livestock	Tanto town (*n* = 770)	1.8	7.9	1.8	25.8	62.6
	Kami town (*n* = 289)	1.0	6.9	2.4	27.0	62.6
(4) When a bear attacks a person in the mountains	Tanto town (*n* = 762)	2.6	4.2	1.0	6.8	85.3
	Kami town (*n* = 289)	3.5	3.8	1.7	6.6	84.4
(5) When a bear appears in the district, i.e. in and around human settlements, and attacks a person	Tanto town (*n* = 784)	1.0	2.2	0.4	3.7	92.7
	Kami town (*n* = 297)	2.0	2.4	0.7	3.7	91.2

[1] Based on Sakurai et al. (2012a).

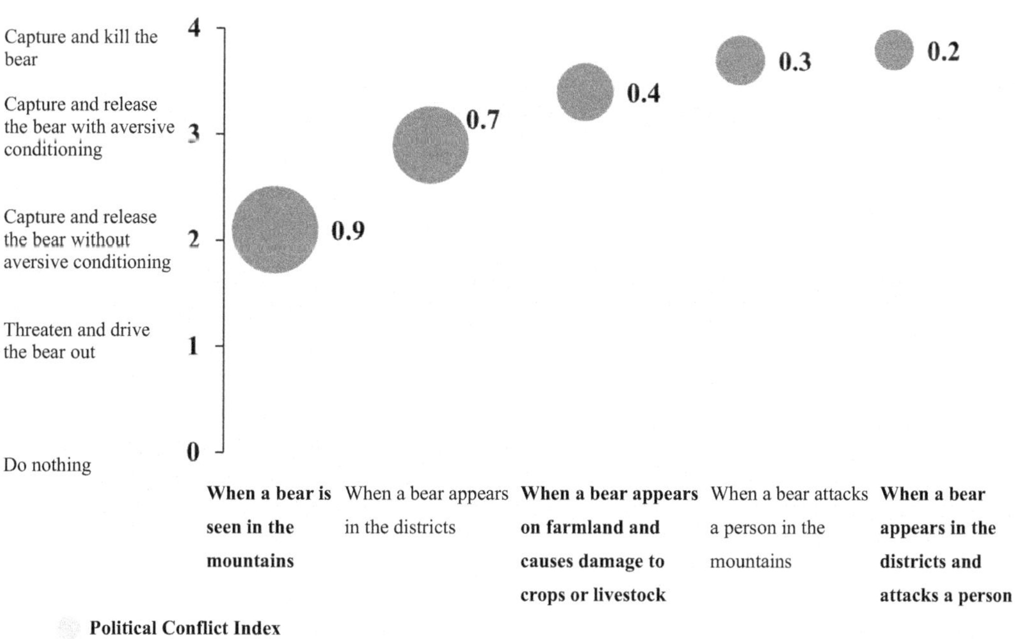

Figure 27.3 Potential for Conflict Index (PCI). The bubbles show the average scores and variances for residents' preferences in terms of implementation of governmental actions in five situations of human–bear conflicts in Tanto and Kami towns (based on Sakurai 2012).

Evaluation of the Community Education Seminars Regarding Black Bears

Based on the interview and survey results, the Agriculture, Forestry and Fisheries Promotion Office of the Tajima District Administration Office, Hyogo Prefecture, held a series of community bear education seminars for local residents and stakeholders in various regions of northern Hyogo Prefecture in 2010. The goals of these seminars were to deepen residents' understanding of bears and to encourage them to take damage prevention actions. Here, the effectiveness of these community bear education seminars is evaluated in two different studies.

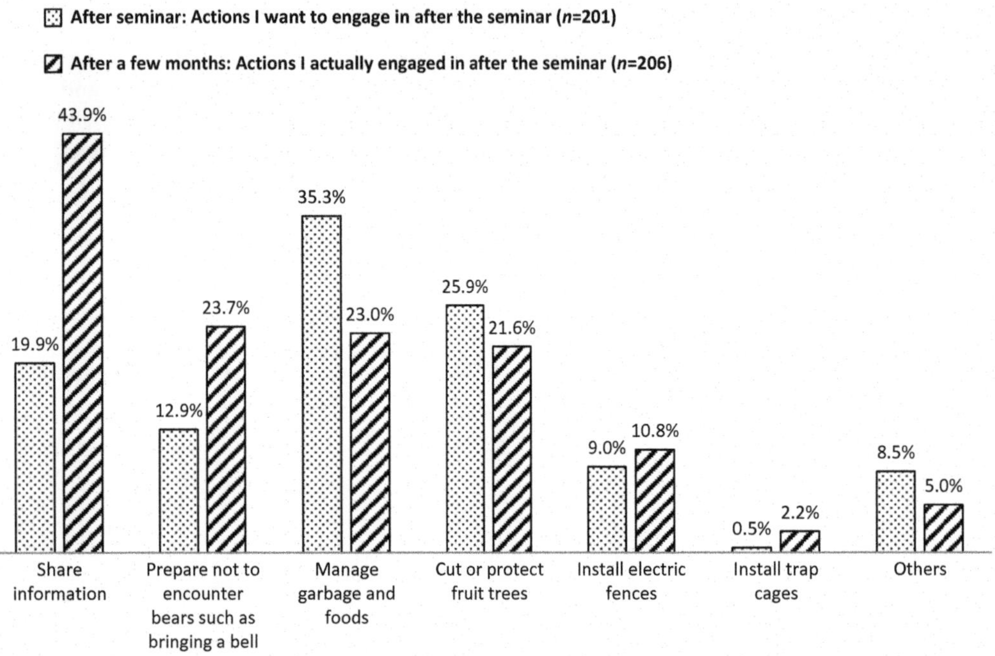

Figure 27.4 Based on Sakurai et al. (2013b), the graph shows a comparison between the actions that people intended to carry out after the seminar (based on the first survey) and the actions actually taken after the seminar (based on the second survey). Although participants originally mentioned that they would carry out actions that would have required significant manpower and time (e.g. manage trash and foods), those actions were less frequently accomplished than easier actions (e.g. information sharing).

First, a survey was distributed twice to seminar participants in two districts in Hidaka town of Toyooka City, i.e. just before the lecture started (October and November 2010) and then about 2 months after those lectures (December 2010). When the responses from the pre- and post-lecture surveys (n = 27) were compared, the confidence of surveyed people regarding their own skills to prevent bear damage, as well as their satisfaction toward the prefectural government's performance (in terms of listening to local resident opinions), significantly increased ($p < 0.05$) after the seminars (Sakurai et al. 2012b).

Then, a second study was conducted on different samples and regions. This time, the first survey (n = 335) was conducted on participants of 13 seminars held from August to October 2010 in 10 regions at the end of each seminar and a few months later (late December 2010). This study was conducted to understand if behavioral changes actually occur after the seminar as government intended and/or participants originally mentioned in the first survey. The second survey (n = 227) revealed that 61.6% of respondents took actions to prevent damage by bears, and among them, 90.6% mentioned that participating in seminars triggered their actions. Figure 27.4 shows a comparison of participants' intentions on acting against bears right after the seminar and their actual actions taken a few months after the seminar. While the highest percentages of actions that participants intended to carry out were "manage trash and foods" (35.3%) and "cut or protect fruit trees" (25.9%) right after the seminar, what were actually carried out were more likely "share information" (43.9%) and "prepare to alert bears to my presence (e.g. by using bear bells)" (23.7%). These results implicate that although participants originally mentioned that they would carry out actions that require significant manpower and time, those actions are less likely to be carried out (Sakurai et al. 2013b).

While the evaluation study identified that seminars could increase local residents' knowledge and confidence toward preventing bear damage and actually taking action, it was also found that participants were more likely to carry out easier actions (share information and carry bells) and less likely to carry out actions that require significant manpower. These findings are important because these show the potential and limitations of a one-time seminar in terms of changing people's behaviors. The findings revealed that human–bear conflicts are directly connected to the social issues that local communities face; aging and depopulation lead to a lack of manpower, which in turn affects how likely people are to carry out actions to prevent damage which require significant manpower. In such a situation, involving residents in environmental maintenance within the community could increase their confidence and encourage them to continually engage in damage-preventative behaviors (Jacobson et al. 2006). This study became one of the first steps to clarify the role of seminars among the various black bear awareness-raising activities. The results of these human dimension studies revealed local residents' and stakeholders' opinions regarding bear issues and the factors that affect those psychological aspects related to human–bear coexistence. The results also indicate how people's actions could be encouraged through outreach and education. These results also imply that human dimensions studies could potentially contribute to a participatory community-based bear management program where residents are more tolerant and knowledgeable about human–bear conflicts (for more details about a series of studies that evaluate the impacts of community bear education seminars, see Sakurai et al. 2012b, 2013b; Sakurai 2019a).

Figure 27.5 Field courses in Japan, where students participate in agricultural activities with local residents (A), make field observation of damage preventative measures (B), and interview local farmers (C).

Educating University Students to Become Professionals in the Human Dimension of Black Bear Management

It has been affirmed that 90% of wildlife management is about people management (Fazio & Gilbert 1986). This is also true when handling human–bear conflicts. For successful bear management, managers need to understand stakeholders' (e.g. local residents') attitudes, needs, and motivations, and to develop a sense of mutual trust (Sjölander-Lindqvist et al. 2015). Therefore, the human component needs to be recognized and reflected in all stages of black bear management. The significance of human dimensions studies is that it increases fairness and legitimacy of the decision-making process by incorporating stakeholders' opinions and contributes to polishing policies as well as their implementation (Decker et al. 2012; Bennett et al. 2017).

The author has been teaching a course entitled "Human Dimensions of Wildlife Conservation" at several universities in Japan, which aims to train students to become professionals in the human dimensions of wildlife management. This course covers topics that include biodiversity and endangered species, wildlife species and human–wildlife conflicts in Japan (especially focusing on black bear issues), and the academic discipline of human dimensions of wildlife management with social scientific tools to understand stakeholders' perceptions and to build stakeholder consensus. The course includes lectures, active learning, and field trips to rural mountainous villages in the northern part of Ibaraki city (Figure 27.5). In these areas, damage by wildlife is frequent and students interview local farmers and government officials to understand human–wildlife conflicts (including black bear issues). Students also engage in agricultural activities with local farmers to understand the current situation and the agricultural challenges in the region and to experience how local residents' lives are affected by wildlife. At the end of the course, students host a workshop in which they play the roles of various stakeholders in human–bear conflicts. Student roles in the workshop include facilitators, farmers (local residents who have experienced damage by black bears), researchers (e.g. ecologists, sociologists), hunters, and government officials (e.g. city, prefecture, and Ministry).

Student feedback indicates that they learned why and how human–wildlife conflicts occur. For example, by studying the methods and theories of human dimensions research and by participating in the workshop, students learned the importance of understanding the perceptions of the various stakeholders, and they also learned how difficult it is to reach a solution that incorporates divergent opinions. In reality, to mitigate human–bear conflicts, it is important that stakeholders work together and communicate about the issues and concerns because, in many cases, a bottom-up approach like community-based participatory wildlife management is more effective and sustainable than a top-down approach (Bennett et al. 2017). In this sense, managers and human-dimensions researchers should play the role of facilitators in fostering collaboration among stakeholders (Decker et al. 2012; Sjölander-Lindqvist et al. 2015). Therefore, studying social scientific methods and acquiring communication skills are some of the most important tasks for future managers and human dimensions researchers. Courses on human dimensions of wildlife management are important to train students on how to become future bear management professionals, as this academic discipline reveals not only how and why human–wildlife conflicts occur but also ways to mitigate these conflicts.

References

Bennett, N., Roth, R., Klain, S. C., et al. (2017). Conservation social science: understanding and integrating human dimensions to improve conservation. *Biological Conservation* 205: 93–108.

Bombieri, G., Nanni, V., Delgado, M. M., et al. (2018). Content analyses of media reports on predator attacks on humans: toward understanding human risk perception and predator acceptance. *Bioscience* 68: 577–584.

Bruskotter, J. T. & Shelby, L.B. (2010). Human dimensions of large carnivore conservation and management: introduction to the special issue. *Human Dimensions of Wildlife* 15: 311–314.

Bruskotter, J. T. & Wilson, R. S. (2014). Determining where the wild things will be: using psychological theory to find tolerance for large carnivores. *Conservation Letters* 7(3): 158–165.

Decker, D. J., Riley, S. J. & Siemer, W.F. (2012). *Human dimensions of wildlife management*. Baltimore, MD: The Johns Hopkins University Press.

Fazio, J. R. & Gilbert, D.L. (1986). *Public relations and communications for natural resource managers*, 2nd edn. Dubuque, IA: Kendal/Hunt.

Ferraro, P. J. & Pattanayak, S.K. (2006). Money for nothing? A call for empirical evaluation of biodiversity conservation investments. *PLoS Biology* 4(4): 482–488.

Huygens, O. C., Goto, M., Izumiyama, S., Hayashi, H. & Yoshida, T. (2001). Asiatic black bear conservation in Nagano Prefecture, central Japan: problems and solutions. *Biosphere Conservation* 3: 97–106.

Hyogo Prefecture. (2003). Hyogo Prefecture Red Data Book 2003 [in Japanese]. Available from www.pref.hyogo.jp/JPN/apr/hyogoshizen/reddata2003/index.htm (accessed December 5, 2017).

Hyogo Prefecture. (2009). Phase 2 Black Bear Conservation and Management Plan [in Japanese]. Available from http://web.pref.hyogo.jp/contents/000140140.pdf (accessed March 2, 2010).

Hyogo Prefecture. (2017). Black Bear Management Plan [in Japanese]. Available from https://web.pref.hyogo.lg.jp/nk27/documents/kuma-kanri.pdf (accessed December 27, 2018).

IUCN. (2018). Red List: Asiatic Black Bear. Available from www.iucnredlist.org/species/22824/114252336 (accessed December 19, 2018).

Jacobson, S. K., McDuff, M.D. & Monroe, M. C. (2006). *Conservation education and outreach techniques*, New York, NY: Oxford University Press.

Kaji, K., Igota, M. & Suzuki, M. (2013). *Science of hunting: For wildlife management in Japan*. Tokyo: Asakura [in Japanese].

Kaneko, A. & Kanamori, H. (2007). Residents opinion poll on Asiatic black bears (*Ursus thibetanus*) in Masuda Area. *Bulletin of the Shimane Prefecture Mountainous Region Research Center* 3: 59–69 [in Japanese].

Ministry of the Environment. (2008). A manual to respond the bear appearance: bears appearing from mountains. Ministry of the Environment, Tokyo, Japan [in Japanese]. Available from www.env.go.jp/nature/yasei/kuma_manual (accessed August 2008).

Ministry of the Environment. (2016a). Number of bears captured in 2016 [in Japanese]. Available from www.env.go.jp/nature/choju/effort/effort12/capture-qe.pdf (accessed January 12, 2017).

Ministry of the Environment. (2016b). Number of casualties caused by bears in 2016 [in Japanese]. Available from www.env.go.jp/nature/choju/effort/effort12/injury-qe.pdf (accessed January 12, 2017).

Onishi, N. (2011). Origin of population and genetic structure; comparison of east Japan and west Japan. In: Tsubota, T. & Yamazaki, K. (Eds.) *Bears in Japan: Biology of brown bears and Asiatic black bear* (pp. 189–208). Tokyo: University of Tokyo Publication [in Japanese].

Price, V., Tewksbury, D. & Powers, E. (1997). Switching trains of thought: the impact of news frames on readers' cognitive responses. *Communication Research* 24: 481–506.

Sakata, H., Kishimoto, Y. & Seki, K. (2011). Population estimate of Asiatic black bears. In: *Wildlife Monograph of Hyogo No. 3* (pp. 26–38). Hyogo, Japan: Wildlife Management Research Center [in Japanese].

Sakurai, R. (2012). Local residents' perceptions and behaviors to prevent damage: a case of Tajima region in Hyogo Prefecture. *Biological Resources* 6(3): 14–24 [in Japanese].

Sakurai, R. (2019a). *Human dimensions of wildlife management in Japan: From Asia to the world*. Singapore: Springer.

Sakurai, R. (2019b). A collaborative approach for coexistence with wildlife in rural regions of Japan. In: Frank, B. J., Glikman A. & Marchini, S. (Eds.), *Human–wildlife interactions: Turning conflict into coexistence* (chapter 9). Cambridge: Cambridge University Press.

Sakurai, R. & Enari, H. (2010). What is human dimensions? Development of social study approach in wildlife management. *Wildlife Forum* 14(3–4): 16–21 [in Japanese].

Sakurai, R., Ueda, G. & Jacobson, S. K. (2011). Bears with various meanings recounted in the narratives of local residents: interviews at Tajima region of Hyogo prefecture. *Journal of Rural Planning* 30: 399–404 [in Japanese with English abstract].

Sakurai, R., Ueda, G. & Jacobson, S. K. (2012a). Survey of residents regarding the Asiatic black bear at Tajima region of Hyogo Prefecture: designing and conducting questionnaire survey for policy application and management interventions. *Wildlife Conservation Japan* 12(2): 33–46 [in Japanese with English abstract].

Sakurai, R., Ueda, G. & Jacobson, S. K. (2012b). Effectiveness of the community bear education seminar: conducting pre and post survey at Hidaka Town of Toyooka City in Hyogo Prefecture. *Kyosei Studies* 6(1): 380–392 [in Japanese].

Sakurai, R., Jacobson, S. K. & Carlton, J. S. (2013a). Media coverage of management of the black bear *Ursus thibetanus* in Japan. *Oryx* 47(4): 519–525.

Sakurai, R., Ueda, G. & Jacobson, S. J. (2013b). Evaluation of the effectiveness of the community bear education seminar at Tajima region, Hyogo Prefecture: change of participants' awareness and behaviors after the seminar. *Wildlife and Human Society* 1(1): 29–37 [in Japanese with English abstract].

Sakurai, R., Jacobson, S. K. & Ueda, G. (2013c). Public perceptions of risk and government performance regarding bear management in Japan. *Ursus* 24(1): 70–82.

Sjölander-Lindqvist, A., Johansson, M. & Sandström, C. (2015). Individual and collective responses to large carnivore management: the roles of trust, representation, knowledge spheres, communication and leadership. *Wildlife Biology* 21: 175–185.

Wildlife Management Research Center, Hyogo. (2011). Sightings of Asiatic black bears [in Japanese]. Hyogo, Japan. Available from www.wmi-hyogo.jp/bear_info/bearinfo_h2309.pdf (accessed November 28, 2011).

WorldAtlas. (2018). Newspapers with the highest circulation in the world. Available from www.worldatlas.com/articles/newspapers-with-the-highest-circulation-in-the-world.html (accessed December 24, 2018).

Yokoyama, M. (2009). Asiatic black bear: reviving from a danger of extinction. In: Kawai, M. & Hayashi, Y. (Eds.), *Doubutsutachino Hanran [Rebellion of wildlife]* (pp. 129–158). Tokyo, Japan: PHP Research Inc. [in Japanese].

Chapter 28

Ecological and Social Dimensions of Sloth Bear Conservation in Sri Lanka

Shyamala Ratnayeke and Frank T. van Manen

Introduction

The continental island of Sri Lanka includes the southernmost limits of sloth bear range and is an important biodiversity hotspot. Topographic and climatic heterogeneity supports diverse habitats and fauna, including more than 941 species of vertebrates, of which ~40% are endemic (Gunawardene et al. 2007). Sri Lanka (65,610 km^2) lies to the south of India between 5°55′–9°50′N and 79°42′–81°52′E. Periodic land connections between southern India and Sri Lanka during the Pleistocene (2.6 MYA–11,700 years ago), and as recently as the Holocene interglacial 7000 years ago, facilitated faunal exchange between the two landmasses and influenced the present distribution of life forms in Sri Lanka (Cooray 1984; Kulatilake 2016).

The sloth bear in Sri Lanka, *Melursus ursinus inornatus* Pucheran, 1855, is a subspecies of the mainland sloth bear, *M. u. ursinus* Shaw, 1791, and is the island's only species of bear. To date, little evidence exists for differences between the subspecies, except that *M. u. inornatus* may be smaller on average than bears on the Indian peninsula, with a sparser, shorter coat (Pocock 1941). Based on field measurements of research captures at Wasgamuwa National Park, Sri Lanka, body mass was 65–85 kg for adult males and 54–67 kg for adult females (Ratnayeke et al. 2007a). Phillips (1984) reported body masses of 104 kg for an *M. u. inornatus* male and 68 kg for a female, but whether these are maximum measurements is unclear. Few data exist for *M. u. ursinus* in the southern part of its range. Considering that genetic exchange between Indian and Sri Lankan sloth bears may have been as recent as a few thousand years ago, the differences in body mass and shagginess may reflect a cline where bears are smaller and less hairy toward the southern portion of the distribution.

Balancing the needs of humans and wildlife in Sri Lanka presents enormous socioeconomic and conservation challenges. Sloth bears are legally protected, but attacks on humans generate intense fear (Ratnayeke et al. 2007, 2014), which increases the potential for human-caused bear mortality and local extirpation of bears. In this chapter, we explore what is known about the ecology and human dimensions of the sloth bear in a country with a teeming human population and a history of poverty and civil conflict. We also address whether changes in land use in the aftermath of the civil war may have impacted sloth bear distribution and discuss the conservation outlook for bears.

Natural forest cover in Sri Lanka declined by almost 40% between 1956 and 1980 (Food and Agriculture Organization Global Forest Resources Assessment 2000), eliminating much suitable sloth bear habitat. Sri Lanka's human population density, 82% of which is rural (United Nations Population Division 2018), is currently estimated at 342 people/km^2 (Department of Census and Statistics 2018). Human life expectancy (73 years) and literacy (91.6%) are high, but in 2002, >25% of the population were below the national poverty level (World Bank 2002). Almost 30 years of civil war saw the loss of nearly 70,000 lives and displaced over half a million people, causing acute humanitarian crises in the northern and eastern provinces of the country (Council on Foreign Relations [CFR] 2009). This region of civil conflict also constitutes the largest tracts of sloth bear habitat on the island.

Since 2002, economic welfare in Sri Lanka has improved, except for pockets of extreme poverty that persist in areas recovering from the civil conflict (Newhouse et al. 2016). Military presence and restrictions to forest use persist in the northern province, the former headquarters of the Tamil separatist movement. Elsewhere, subsistence living well below the poverty level forces most rural families to use forests for immediate needs or as additional sources of income. These activities, including illegal hunting for food and sale of bushmeat, contribute to forest degradation and increase the potential for human–bear conflicts (Ratnayeke et al. 2007b, 2014).

Sloth Bear Distribution at the Turn of the Twenty-First Century

Sri Lanka's climate is tropical with most rainfall received in the south-west region where most people live and where agriculture is the principal form of land use. A narrow zone of intermediate rainfall separates the wet zone from the largest region, the low country dry zone that covers the northern half of the island and the eastern and south-eastern lowlands. The dry zone climate is hot and dry for most of the year; the mean annual rainfall is ~700 mm, delivered mostly during the north-east monsoon (Domrös 1974). At the turn of the twenty-first century, forests covered 24–26% of the island (United Nations Environment Programme [UNEP] 2003), most of which

Figure 28.1 Predicted occupied range of sloth bears in Sri Lanka circa 2004 and 2015. Occupied range predictions were based on a logistic regression model and 2004 survey data and were mapped for 5 × 5-km grid cells (Ratnayeke et al. 2007b). The circa 2015 occupied range predictions were derived from the same logistic regression model with updated landscape covariate data. White circles with black inner circles represent cells with predicted range in both 2004 and 2015 (n = 808); white squares represent cells with predicted loss of occupied range from 2004 to 2015 (n = 75); solid white circles represent cells with predicted gain of occupied range (n = 3). Protected areas layer from World Database on Protected Areas (UNEP-WCMC 2019).

consisted of dry monsoon forest (Legg & Jewell 1995) in the dry zone. Much of the remaining land in the dry zone comprised sparse forests (18%) consisting of low-stature vegetation and grasslands (UNEP 2003).

Accounts from the early twentieth century indicated that sloth bears occurred almost exclusively within the dry zone (Phillips 1984). In 2004, Ratnayeke et al. (2007b) conducted a distribution survey on a 25-km^2 scale covering historic sloth bear range in Sri Lanka to verify sloth bear presence or absence by interviewing knowledgeable forest users. In 2004, sloth bears still occurred in the dry zone lowlands where there was sufficient forest cover and where human densities and road densities were substantially lower than the national average. Although sloth bears were documented at elevations ranging from 0 to 1376 m, 98% of the range was below 300 m. Ratnayeke et al. (2007b) used the 2004 distribution survey data to construct a logistic regression model to identify natural and anthropogenic landscape characteristics associated with sloth bear presence and absence and predict occupied range at a resolution of 5×5-km grid cells (Figure 28.1). That model was based on four predictive covariates (elevation, percent monsoon forest, human population density, road density) and had high correct classification rates (85–92%; Ratnayeke et al. 2007b). To update predictions of occupied range of sloth bears in Sri Lanka, we recalculated the covariates for forest cover, human population density, and road density based on contemporary data and reapplied the predictive logistic regression equation to assess if and where occupied range may have changed. We estimated percent change in forest cover based on Hansen et al. (2013), who estimated gross forest cover loss at a resolution of 30 m for the period 2000–2018. We aggregated those changes for all 25-km^2 grid cells within the historic distribution of sloth bears in Sri Lanka. We estimated percent change in human population density using 2013–2022 population growth projections at the fourth administrative level (Grama Niladhari Division) from the United Nations Office for the Coordination of Humanitarian Affairs (2018) and used those projections to backcast changes in population density for the 25-km^2 grid cells since the early 2000s. Finally, we (a) estimated proportional changes in road density at the provincial level by comparing summary statistics collected by the Road Development Authority of the Sri Lanka Ministry of Provincial Councils and Local Authorities from 2000 to 2015 (Department of National Planning 2019); and (b) applied these provincial-level estimates of change to each of the 25-km^2 grid cells. Applying the logistic regression model to the updated covariate data, we estimated approximately 20,275 km^2 of occupied range (811 grid cells), or about 31% of Sri Lanka's land mass.

This represents 1800 km^2 net loss of predicted occupied range from the early 2000s to about 2015, or 8% (Figure 28.1). Grid cells for which we predicted loss of occupied range were primarily on the edge of bear distribution, which often coincided with peripheral regions of protected areas. These findings suggest that landscape changes due to the linked effects of forest loss, road development, and population growth may be eroding the extent of occupied range. Transition or edge habitats are often the first areas where erosion of range extent becomes evident, which may be followed by encroachments into core habitat (Vogt et al. 2007). Nevertheless, these updated predictions of occupied range indicate approximately 43% of historic sloth bear range has landscape conditions conducive to occupancy and connectivity may still be high (Figure 28.1). Additionally, 69% of predicted occupied range is associated with protected areas, and the remainder is in close proximity to protected areas. The level of protection afforded by the different types of protected areas varies substantially, however, ranging from strictly protected National Parks and Strict Natural Reserves, where human access is restricted or controlled (24% of current occupied range), to forest reserves and other areas (45% of occupied range), where access is much less restricted. Indeed, predicted loss of occupied range on the periphery of some protected areas suggests that changing landscape conditions due to human activities may be affecting the persistence of bear populations. These observations point to an urgency to consider the long-term integrity of all protected areas in Sri Lanka. There are important caveats to our implementation of this update. For example, although we predicted 31% of Sri Lanka's land area is occupied range, sloth bears may not be present in some of the grid cells. Based on the field surveys conducted in 2004 (Ratnayeke et al. 2007b), bears were deemed absent in 94 of the 811 (11.6%) grid cells for which the model predicts current occupancy, suggesting this bias may be limited. Also, biases may have been introduced by downscaling some data from administrative units to 25-km^2 grid cells and backcasting human population projections. However, rather than using contemporary data that would not be comparable to the data we used to estimate occupied range in 2004, our approach of updating the original spatial data to reflect circa 2015 conditions provided a relatively unbiased assessment for trend comparisons.

Sloth bears in India and Nepal are closely tied to existing forest cover (Garshelis et al. 1999; Yoganand et al. 2006) but use a broader range of habitats than in Sri Lanka. On the Indian mainland, sloth bears are reported in montane forests of the Western Ghats (Johnsingh 2003), moist evergreen and deciduous forests of south-western India (Sreekumar & Balakrishnan 2002), and tall alluvial grasslands (Joshi et al. 1995). In contrast, there is no historical evidence in Sri Lanka of resident sloth bears in the moist, humid forests of the south-west lowlands or the hill country. A lack of preferred species of termites and fruit trees may be responsible for their exclusively low-country dry zone distribution.

Ecology of Sloth Bears in Sri Lanka

Few studies have systematically investigated the ecology and behavior of *M. u. inornatus* in Sri Lanka. To date, some of the most detailed information comes from studies at Yala (126,781 ha) and Wasgamuwa (39,385 ha) National Parks (Ratnayeke et al. 2007a; Ratnayeke & van Manen 2012). About two-thirds of Wasgamuwa National Park consists of dense primary forest, ranging from tall to medium-tall forest, but the southern region contains mostly short grasslands and scrub forest. In the southern portion of the national park, vegetation succeeded land that was previously encroached for swidden agriculture during the 1960s (Pabla et al. 1998). After the declaration of the national park in the early 1980s, the settlements were removed. The encroachments, previously covered with primary forest, are now a mosaic of grassland, scrub, and secondary forest, reflecting various stages of succession.

A capture and telemetry study of 10 adult bears in the southern region of Wasgamuwa revealed some of the smallest home ranges reported for any species of bear (Ratnayeke et al. 2007a). Annual 95% fixed kernel home ranges were 3.8 km^2 (SE = 1.01, n = 6) for males and 2.2 km^2 (SE = 0.61, n = 4) for females. Home ranges overlapped extensively within and between sexes and increased in size during the dry season as sloth bears shifted from a diet composed mostly of termites to sources of fruit. Although no data on food distribution and abundance were available, the small home ranges probably resulted from abundant and stable food sources, such as termites, which occurred in 61% of 660 sloth bear scats examined during the two-year study, and a steady supply of fruit during drier months (Ratnayeke et al. 2006). Termites composed an average of 81% of scat volume during wetter months; for the remainder of the year, fruit remains, especially those of *Cassia fistula* and *Drypetes sepiaria* (veera), composed over 70% of scat volume. Several other items appeared less frequently in scats, notably *Ficus* spp., *Diospyros ovalifolia*, and remnants of honeycombs. The sweet fruit of *Manilkara hexandra* (palu), famously known to attract bears in numbers elsewhere in Sri Lanka, rarely flowers at Wasgamuwa National Park, although large trees of this species are present. Notably, veera and palu, two important sources of food for bears, occur exclusively in the dry zone. The remains of mammals (hair and bone fragments) were found in four bear scats, suggesting that predation or scavenging by sloth bears is rare and opportunistic.

Sloth bears at Wasgamuwa were most active when temperatures were cooler, usually during 17:00–08:00 hours, but midday activity on rainy and cloudy days was common (Ratnayeke et al. 2007a). Tall, old-growth forests, which historically have experienced the least anthropogenic disturbance, were used more than expected based on availability within home ranges of radio-collared bears and compared with other vegetation types. The closed canopy restricts the growth of the understory, and older, larger trees are likely to be more productive and contain cavities for dens and daybeds. Tall forests also possess a comparatively moist ground layer, providing better conditions for decomposers and nutrient recycling. Termite abundances are presumably higher under these conditions. In contrast, sloth bears used habitats with less tree cover, particularly successional habitats dominated by short grassland, less than expected, despite this habitat comprising 30–45% of home ranges of some bears. Sloth bears limited their use of these open habitats to the scattered thickets of scrub (*Lantana* spp. and *Zizyphus* spp.) that provided dense cover and shade. Bear use of these abandoned agricultural habitats is noteworthy, indicating that swidden agriculture and similar types of human disturbance are not necessarily detrimental to bears and other wildlife in the long term if opportunities for succession and recolonization exist.

Results from a 2002–2003 remote camera survey at Yala and Wasgamuwa showed that sloth bear detection rates were considerably higher at Wasgamuwa (19 bears per 100 camera nights) than at Yala (seven bears per 100 camera nights) (Ratnayeke & van Manen 2012). Yala National Park is different from Wasgamuwa in that it includes mangrove and sand dune communities, and much greater expanses of short grasslands and thorn scrub. However, bear detection rates approach those of Wasgamuwa in the tall, old growth primary forests found at the northern end of Yala's protected area complex, further supporting the importance of old-growth primary forests in the dry zone for sloth bears. At both Yala and Wasgamuwa, the richness of other species of mammalian carnivores, notably threatened and near-threatened carnivores, was positively associated with the frequency of bear detections (Ratnayeke & van Manen 2012). Carnivores of conservation concern that co-occurred with sloth bears included the leopard (*Panthera pardus*), the endemic golden palm civet (*Paradoxurus zeylonensis*), the rusty-spotted cat (*Prionailurus rubiginosus*), and the Indian brown mongoose (*Herpestes fuscus*). Thus, sloth bears may be useful indicators of areas occupied by other carnivores and useful conservation surrogates of this ecologically valuable taxon.

Conflicts with Bears and Attitudes of Rural Men Toward Sloth Bears

Chapter 17 reviews behavioral and temporal aspects of bear attacks, including circumstances involving attacks in India and Sri Lanka. Here, we address unique aspects of human–sloth bear conflicts in Sri Lanka from interviews with survivors of sloth bear attacks, highlighting key differences between attacks reported in Sri Lanka and India. We also add perspective from data collected during independent surveys investigating human attitudes toward sloth bears in rural communities (Ratnayeke et al. 2007b). Rural villagers, predominantly men, engage in forest-related activities and are most likely to encounter and come into conflict with sloth bears (see also Chapter 17). Their perceptions and attitudes regarding bears are thus an important consideration for conservation planning and are represented in this section, but are likely not reflective of the public at large.

Between 1938 and 2004 in Sri Lanka, at least 271 people (five women, 266 men) were attacked by sloth bears, of which 71 (26%) involved serious injuries, including three human fatalities (Ratnayeke et al. 2014). Survivors of serious injuries typically sustained severe facial disfigurement and physical disabilities, including the loss of hearing, smell, and sight. Subsistence farmers and casual agricultural workers composed the majority (82%) of human victims. All but three attacks occurred away from villages when humans visited forests to hunt, gather forests products, or practice swidden agriculture. In 61 attacks, bears sustained serious injuries from guns or axes, and in 40 attacks bears were killed. Furthermore,

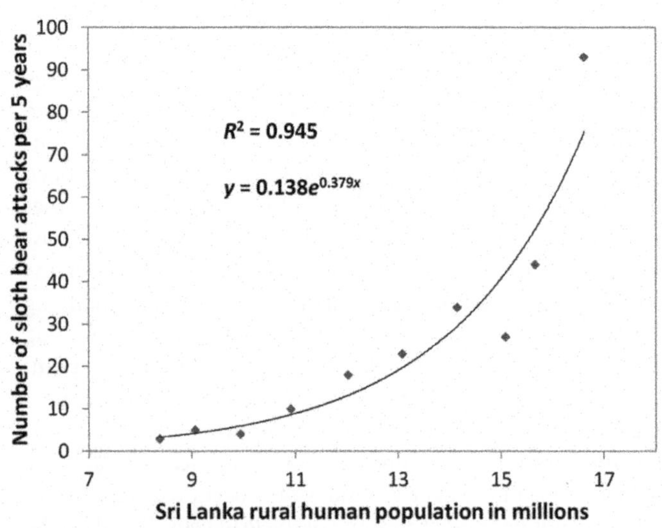

Figure 28.2 Regression of the number of sloth bear attacks in Sri Lanka per 5-year interval, 1955–2004, and rural human population in each 5-year interval. The relationship between rural population size and the number of sloth bear attacks was exponential.

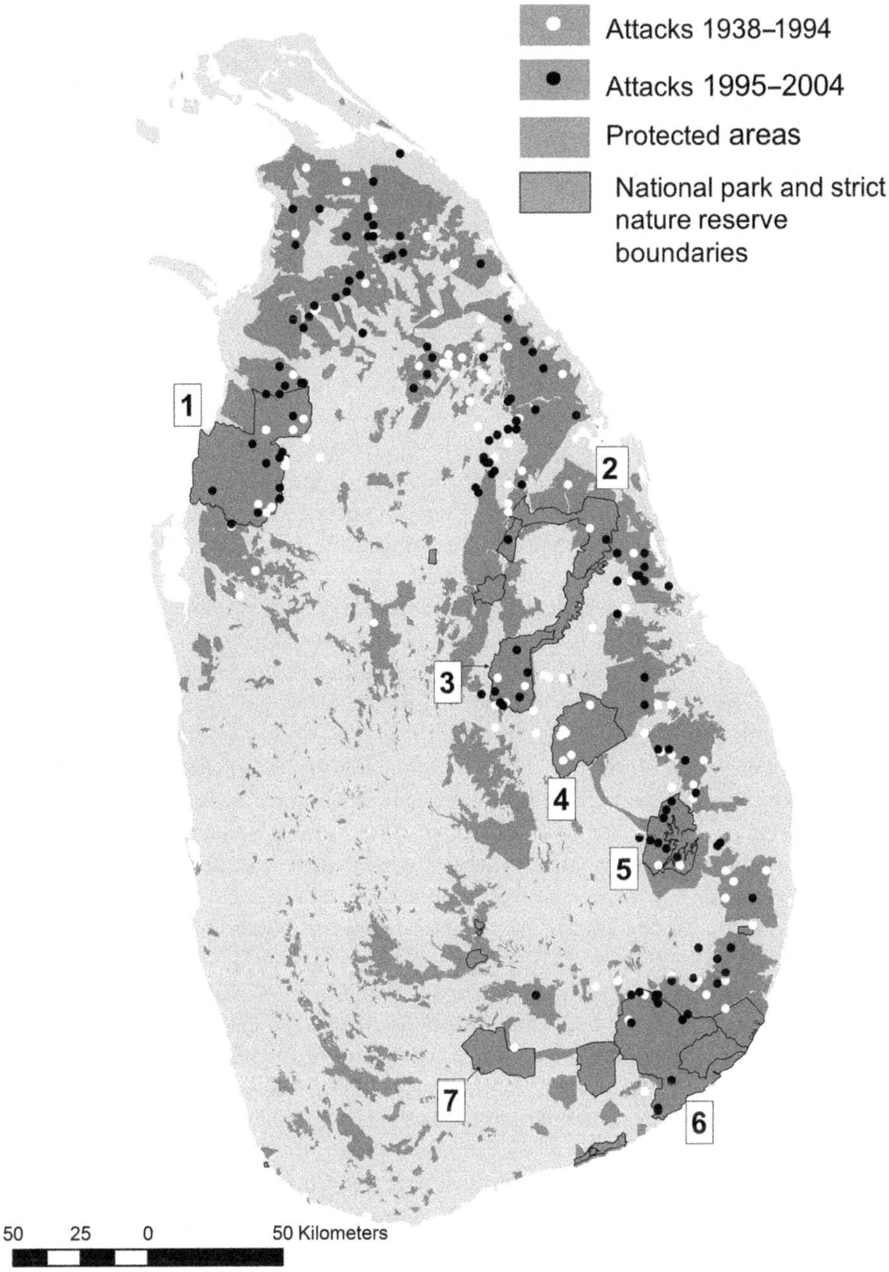

Figure 28.3 Locations of sloth bear attacks within and outside the protected area network, 1938–2004, Sri Lanka. National parks and strict nature reserves require permits for entry. (1) Wilpattu protected area complex (PC), (2) Somawathie PC, (3) Wasgamuwa National Park, (4) Maduru Oya National Park, (5) Gal Oya PC, (6) Yala PC, (7) Udawalawe National Park. Protected areas layer from World Database on Protected Areas (UNEP-WCMC 2019).

144 rural men that lived within sloth bear range confirmed that they had recently (<5 years) killed at least one sloth bear, or had knowledge of a bear that was killed locally to pre-empt a possible attack (Ratnayeke et al. 2006). Human–sloth bear encounters and bear attacks, therefore, have consequences for individual sloth bears, and could result in potential demographic effects at the population level.

Between 1990 and 2004, sloth bear attacks doubled every 5 years, showing an exponential relationship with rural human population size (Figure 28.2). The majority of sloth bear attacks occurred in forests away from human settlements (Figure 28.3) and most of these forests were under some degree of protection. Bear attacks within national parks (n = 39 during 1995–2004) may be underreported because permits are required for entry and some victims were reluctant to admit to illegally entering the park. Nevertheless, most attacks within larger national parks were close to park boundaries. Large protected areas may therefore play a crucial role in sloth bear conservation by restricting human access on foot and reducing the potential for encounters between sloth bears and people, at least in the interior.

In Sri Lanka, men in rural communities are almost exclusively the victims of sloth bear attacks. Men are more likely to make long excursions into forests, and more likely to go alone, increasing the likelihood of a severe attack (Ratnayeke et al. 2014). Attacks increase during the dry season when work in

Figure 28.4 Typically secretive, sloth bears are best viewed during the day in national parks with a long history of human visitation. This adult female sloth bear was photographed at Yala National Park, Sri Lanka in 2007. (Photo by Luxshmanan Nadaraja.) (A black and white version of this figure will appear in some formats. For the color version, please refer to the plate section.)

agricultural fields declines and humans venture into forests looking for other sources of income. In contrast, in some portions of India (e.g. Bilaspur) women and children might be attacked as well (Bargali et al. 2004) when bears enter village compounds and agricultural fields to raid crops (Rajpurohit & Krausman 2000; Akhtar et al. 2004), which is currently rare in Sri Lanka. Bargali et al. (2004) attributed crop raiding by sloth bears and the high frequency of attacks on people in Bilaspur to greatly fragmented and degraded bear habitat. In Sri Lanka, occasional visits by bears to village settlements have been reported, sometimes coinciding with severe drought conditions, and bears that made those visits were harassed or shot. Sloth bears may avoid rural settlements for these reasons (e.g. Ratnayeke et al. 2007a), and sloth bear populations may not persist near rural settlements because of the high risk of mortality.

In the early 2000s, sloth bears were seldom hunted for meat or body parts (Ratnayeke et al. 2006) and there is no tradition of consuming bear meat in Sri Lanka. Almost 83% of 266 rural males we interviewed used forests in bear habitat to supplement their livelihoods with bushmeat, honey, wood, or swidden agriculture (Ratnayeke et al. 2007b). Hunting for bushmeat on protected or unprotected government land is prohibited, yet almost a third of respondents admitted they entered forests to hunt. Many individuals that cited honey or timber gathering as the main reason to visit forests carried a firearm. Hunting was likely underreported because bystanders were present. Fear of sloth bears was a common (70%) sentiment, possibly responsible for the substantial number of respondents that admitted to shooting bears. Nevertheless, 66% supported legal protection of bears and only 35% thought that the absence of bears was preferable. This paradox may be partially explained by Buddhist and Hindu traditions in Sri Lanka favoring the protection of wild animals (Ratnayeke et al. 2007b).

Conclusions

Based on predicted occupied range circa 2015, approximately 24% of sloth bear range was within protected areas that banned hunting and regulated human access (National Parks, Strict Natural Reserves), whereas 45% was within designations (e.g. Sanctuary, Nature Reserve, Reserved Forest) that allowed free access and traditional human activities other than human settlements. Approximately 31% of occupied range is outside protected areas. Our model predicted 8% contraction of sloth bear range since the early 2000s, although it remains mostly contiguous with the largest tracts in the north and east of the island. Enforcement of protections in designated protected area networks will aid sloth bear conservation (Figure 28.4). Patterns of range loss suggest that Wasgamuwa National Park (Figure 28.3) may soon lose connectivity with adjacent bear range (Figure 28.1) unless concerted efforts are made to maintain forested corridors with protected areas at its northern and north-eastern boundaries.

Despite the extent and connectivity of remaining sloth bear range, future reductions in bear range may result from the combined effects of forest habitat loss and increasing density of humans and roads. Although forest decline in Sri Lanka has been limited in recent years (2.1% change during 2000–2016; World Bank 2019), human density increased from 232 to 342 individuals/km^2 between 1992 and 2018 (Department of Census and Statistics 2018). Although much of this increase was in areas where human density was previously high, the percentage of the population classified as rural has remained unchanged (United Nations Population Division 2018); human activity in forests has most likely increased, concomitant with the increased potential for risky encounters with sloth bears, bear mortality, and local extirpations. Sloth bear populations in Sri Lanka were not highly vulnerable to direct exploitation in the early 2000s, but vigilance for poaching networks targeting bear gall bladders and paws is needed because of the country's proximity to and strong economic links with South-East Asia and China. The most severe threats to sloth bears in Sri Lanka are habitat alteration and mortality resulting from human–bear conflicts. Enforcement of laws that regulate human access and use of protected areas is a crucial consideration for sloth bear conservation. Although sloth bears have small home ranges, large protected areas will reduce the probability of encounters between sloth bears and humans that use forests for subsistence.

Significant progress has been made in reducing extreme poverty, but people in the Northern and Eastern Provinces affected by the civil conflict account for approximately 13% of Sri Lanka's population, in a region that is also the stronghold of the sloth bear. The civil conflict caused widespread loss and destruction of private property, denial of access to agriculture, livestock, and forest assets, and landmines made previously arable land inaccessible (CFR 2009). In these provinces, 20–30% of people live on less than US$1.50 per day (Newhouse et al. 2016) and nearly 30% of children suffer from malnutrition (CFR 2009). Forests and swidden agriculture are a key resource for the rural poor, but former guerrilla control and current military presence has restricted human access to forests in the north for decades, which may account for the slight increase in predicted sloth bear range in portions of this region (i.e. north and east of Wilpattu National Park; Figures 28.1 and 28.3).

The loss of bear range in the remainder of the country is due largely to landscape changes, including expansion of road networks and human settlements along the periphery of some protected areas. Conservation efforts cannot succeed independent of human welfare in the long term. Effective social assistance programs and policies that connect rural people to productive employment opportunities may help reduce human pressure on forests and foster peace in post-civil war regions. Also, conservation programs that emphasize measures to reduce human–sloth bear encounters will aid human–bear coexistence. For example, our data on sloth bear attacks show that humans traveling in groups experience less injury from sloth bear attacks than do single travellers (Ratnayeke et al. 2014; see also Chapter 17). Although sloth bears have endured a poor reputation in Sri Lanka due to widespread fear of bears (Santiapillai & Santiapillai 1990), our data suggest that rural attitudes towards bears are positive for the most part, thus providing an important basis for enhancing conservation efforts.

References

Akhtar, N., Bargali, H. S. & Chauhan, N. P. S. (2004). Sloth bear habitat use in disturbed and unprotected areas of Madhya Pradesh, India. *Ursus* 15: 203–211.

Bargali, H. S., Akhtar, N. & Chauhan, N. P. S. (2004). Feeding ecology of sloth bears in a disturbed area in central India. *Ursus* 15: 212–217.

Cooray, P. G. (1984). Geology, with special reference to the Precambrian. In: Fernando, C. H. (Ed.), *Ecology and biogeography of Sri Lanka* (pp. 1–34). The Hague, The Netherlands: Martinus Nijhoff Publishers.

Council on Foreign Relations. (2009). The Sri Lankan conflict. Available from www.cfr.org/backgrounder/sri-lankan-conflict (accessed March 6, 2019).

Department of Census and Statistics. (2018). Chapter 2: Population. Available from www.statistics.gov.lk/Pocket%20Book/chap02.pdf (accessed March 6, 2019).

Department of National Planning. (2019). Infrastructure. Chapter 5 in Public Investment Programme 2017–2020. Available from www.mnpea.gov.lk/web/images/PIP_English/EnglishPIP-83-114.pdf (accessed March 11, 2019).

Domrös, M. (1974). *The agroclimate of Ceylon*. Wiesbaden, Germany: Franz Steiner Verlag.

Food and Agriculture Organization Global Forest Resources Assessment. (2000). Forest resources of Sri Lanka: Country report. Available from www.fao.org/forest-resources-assessment/past-assessments/fra-2000/en/ (accessed March 19, 2019).

Garshelis, D. L., Joshi, A. R., Smith, J. L. D. & Rice, C. D. (1999). Sloth bear conservation action plan (*Melursus ursinus*). In: Servheen, C., Herrero, S. & Peyton, B. (Eds.), *Bears: Status survey and conservation action plan* (pp. 225–240). Gland, Switzerland: IUCN/SSC Bear and Polar Bear Specialist Groups, IUCN.

Gunawardene, N. R., Daniels, A. E. D., Gunatilleke, I. A. U. N., et al. (2007). A brief overview of the Western Ghats–Sri Lanka biodiversity hotspot. *Current Science* 93: 1567–1572.

Hansen, M. C., Potapov, P. V., Moore, R., et al. (2013). High-resolution global maps of 21st-century forest cover change. *Science* 342: 850–853. Available from http://earthenginepartners.appspot.com/science-2013-global-forest (accessed February 28, 2019).

Johnsingh, A. J. T. (2003). Bear conservation in India. *The Journal of the Bombay Natural History Society* 100: 190–201.

Joshi, A. R., Garshelis, D. L. & Smith, J. L. D. (1995). Home ranges of sloth bears in Nepal: implications for conservation. *Journal of Wildlife Management* 59: 204–213.

Kulatilake, S. (2016). The peopling of Sri Lanka from prehistoric to historic times: biological and archaeological evidence. In: Schug. R. S. & Walimbe, S. R. (Eds.), *A companion to South Asia in the past* (pp. 426–436). Chichester: Wiley Online Library.

Legg, C. & Jewell, N. (1995). A 1:50,000 forest map of Sri Lanka: the basis for a National Forest Geographic Information System. *The Sri Lanka Forester*, 20(3–4): 35–44.

Newhouse, D. L., Suarez-Becerra, P. & Doan, D. T. T. (2016). *Sri Lanka – Poverty and welfare: Recent progress and remaining challenges* (English). Washington, DC: World Bank Group. Available from http://documents.worldbank.org/curated/en/996911467995898452/Sri-Lanka-Poverty-and-welfare-recent-progress-and-remaining-challenges (accessed March 8, 2019).

Pabla, H. S., Mathur, V. B. & Wickramasinghe, W. R. M. S. (1998). Management plan Wasgomuwa National Park and Riverine Nature Reserve: Vol 1. Project UNO/SRL/001/GEF-SRL/92/G3, Department of Wildlife Conservation, Sri Lanka.

Phillips, W. W. A. (1984). The sloth bear. In: *Manual of mammals of Sri Lanka* (pp. 290–296). Colombo, Sri Lanka: Wildlife and Nature Protection Society of Sri Lanka.

Pocock, R. I. (1941). *The fauna of British India including Ceylon and Burma: Mammalia Vol II. Family Ursidae* (pp. 163–232). London: Taylor and Francis Ltd.

Rajpurohit, K. S. & Krausman, P. R. (2000). Human–sloth bear conflicts in Madhya Pradesh, India. *Wildlife Society Bulletin* 28: 393–399.

Ratnayeke, S. & van Manen, F. T. (2012). Assessing sloth bears as surrogates for carnivore conservation in Sri Lanka. *Ursus* 23: 206–217.

Ratnayeke, S., Wijeyamohan, S. & Santiapillai, C. (2006). The status of the sloth bear in Sri Lanka. In: Oi, T., Mano T., Yamazaki, K., et al. (Eds.), *Understanding bears to secure their future* (pp. 35–40). Gifu, Japan: Japan Bear Network.

Ratnayeke, S., van Manen, F. T. & Padmalal, U. (2007a). Home ranges and habitat use of sloth bears *Melursus ursinus inornatus* in Wasgomuwa National Park, Sri Lanka. *Wildlife Biology* 13: 272–284.

Ratnayeke, S., van Manen, F. T., Pieris, R. & Pragash, V. S. J. (2007b). Landscape characteristics of sloth bear range in Sri Lanka. *Ursus* 18(2): 189–202.

Ratnayeke, S., van Manen, F. T., Pieris, R. & Pragash, V. S. J. (2014). Challenges of large carnivore conservation: sloth bear attacks in Sri Lanka. *Human Ecology* 42: 467–479.

Santiapillai, A. & Santiapillai, C. (1990). Status, distribution and conservation of the sloth bear (*Melursus ursinus*) in Sri Lanka. *Tiger Paper* 1: 13–15.

Sreekumar, P. G. & Balakrishnan, M. (2002). Seed dispersal by the sloth bear (*Melursus ursinus*) in India. *Biotropica* 34: 474–477.

United Nations Environment Program. (2003). Land cover assessment and monitoring, Sri Lanka. Environmental Assessment Technical Reports, Volume 12-A. United Nations Environment Program Regional Resource Center for Asia and the Pacific, Bangkok, Thailand.

United Nations Environment Programme World Conservation Monitoring Centre (UNEP-WCMC) and International Union for Conservation of Nature (IUCN). (2019). Protected Planet: the World Database on Protected Areas (WDPA). UNEP-WCMC and IUCN, Cambridge, UK. Available from www.protectedplanet.net (accessed March 21, 2019).

United Nations Office for the Coordination of Humanitarian Affairs-Asia and the Pacific. (2018). Sri Lanka – Population and Housing Census of 2012. Available from https://data.humdata.org/organization/ed366c60-6fc7-43b5-afca-288075bcc4d2 (accessed February 15, 2019).

United Nations Population Division. (2018). United Nations Population Division World Urbanization Prospects: 2018 revision. Available from https://data.worldbank.org/indicator/sp.rur.totl.zs (accessed March 6, 2019).

Vogt, P., Riitters, K. H., Estreguil, C., et al. (2007). Mapping spatial patterns with morphological image processing. *Landscape Ecology* 22: 171–177.

World Bank. (2002). Sri Lanka – Poverty assessment [English]. Washington, DC: World Bank. Available from http://documents.worldbank.org/curated/en/110101468777294058/Sri-Lanka-Poverty-assessment (accessed May 6, 2020).

World Bank. (2019). Forest area (% of land area). Available from https://data.worldbank.org/indicator/ag.lnd.frst.zs?end=2016&name_desc=false&start=2000 (accessed March 11, 2019).

Yoganand, K., Rice, C. G., Johnsingh, A. & Seidensticker, J. (2006). Is the sloth bear in India secure? A preliminary report on distribution, threats and conservation requirements. *Journal of the Bombay Natural History Society* 103: 172–181.

Index